A GLOSSARY OF WOOD

A GLOSSARY OF WOOD

by

THOMAS CORKHILL

M.I.Struct.E., F.B.I.C.C.

STOBART DAVIES LIMITED

Stobart House, Pontyclerc
Penybanc Road, Ammanford SA18 3HP

www.stobartdavies.com

Published April 1979

Reprinted 1981, 2004

ISBN 0 85442 010 x

STOBART DAVIES LIMITED
Stobart House
Pontyclerc, Penybanc Road
Ammanford SA18 3HP
www.stobartdavies.com

Printed in Singapore

cohesion to the structure, which is more complex in hardwoods than in softwoods. The inner cells, or springwood, are formed during the ascent of the sap. The sun perfects the sap in the leaves and the descent produces a superior and usually darker tissue, or autumn wood, to complete the ring. The cells are provided with valvular membranes which allow the sap to penetrate into the tree to nourish and thicken the cell walls, producing heartwood. Also called Growth Rings.

Annulated Column. A ring, or cluster, of small diameter columns serving as one column.

Annulet. A small moulding, usually semicircular in section, circumscribing a column.

Anobiidæ. The family of furniture beetles, *q.v.* Includes *Anobium punctatum*, etc.

Anobium. The furniture beetle.

Anogeissus. See *Axlewood* and *Yon*.

Anona. See *Corkwood*.

Anonaceæ. The Custard-apple family, which is of little timber importance except for Oxandra. The family includes Anaxagorea, Anona, Asimina, Canangium, Duguetia, Guatteria, Oxandra, Rollinia.

Antarctic Beech. *Nothofagus moorei.* Australasia. See *New Zealand Beech* and *Tasmanian Myrtle*.

Anthemion. The honeysuckle ornament.

Anthocephalus. See *Kadam*.

Antiaris. See *Oro* and *Upas Tree*.

Antifix. A vertical ornament at the eaves of a tiled roof, or as a cresting at the ridge.

Antipendium. A screen or veil for the front of an altar.

Ants. Wood-destroying insects. About ten species. *Termes fatalis*, white ant, most destructive insect to timber and tree. *Formica smaragdina*, red ant, not so destructive as white ant. The black and yellow ants are not so destructive as white and red ants. Arsenic, paraffin, and tar are protections.

Ant Tree. *Triplaris caracasana* (H.). C. America. Yellowish with reddish cast, smooth, lustrous. Light and soft, but firm ; not durable. Straight grain, medium texture. Easily wrought, somewhat like pine. Small sizes, little exported. Used for construction, joinery, etc.

Anyan. See *Apa*.

Anyaran or **Ayanran.** Nigerian satinwood. See *Ayan* and *Movingui*.

Apa. *Afzelia spp.* (*H.*). Nigeria. Also called Anyan or Ariyan. Several species, but *A. africana* is the most attractive for wrought decorative work. Brownish red, variegated streaks. Hard, heavy, strong, durable. Interlocked grain, close texture. Seasons readily, but rather difficult to work, especially *A. africana*. Finishes well, but requires careful filling. Used for structural work, wagons, dock work, flooring, carving, etc. *A. bipindensis* is straighter in grain, milder, larger in size, and suitable for structural work, framing, mouldings, etc. S.G. ·85 ; C.W. 3·5.

Apado. Ekpaghoi, *q.v.*

Apamate. See *Prima Vera*.

Apartment. A room or chamber. A suite of rooms in the same sense as " self-contained " flat.

Apaya or **Apeya.** Avodire. See *Olon*.

Apex. The vertex of a triangle. The summit or highest point.

Aphananthe. See *Grey Handlewood*.

Apitong. *Dipterocarpus grandiflorus* and other spp. (*H.*). Philippines. Reddish brown. Heavy, moderately hard, strong, not durable in contact with ground. Ribbon grain, fine texture. Requires careful seasoning. Warps and shrinks more than other lauans, *q.v.* Rather difficult to work. Used for furniture, flooring, structural work, etc. S.G. ·8 ; C.W. 4. Also see *Gurjun, Bagac, Keruing*.

Aplustre. An ornamental finish at the top of the stern post of a ship.

Apocarya. Pecan hickories. See *Hickory*.

Apocyneæ. A large family, including Alstonia, Aspidosperma, Couma, Dyera, Gonioma, Hancornia, Lacmellia, Lanugia, Lyonsia, Malouetia, Plumeria, Stemmadenia, Tabernæmontana, Thevetia, Wrightia, Zschokkea.

Apodytes. See *White Pear*.

Apophyge. The top and bottom members of the shaft of a column or pilaster. It is usually in the form of a cavetto. Also called congé, scape, or spring. See *Attic Base*.

Apopo. African walnut. See *Nigerian Walnut*.

Appayia. Avodire. See *Olon*.

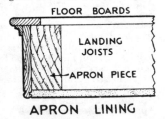

Apple. *Malus pumila, M. communis*, or *Pyrus m.* (*H.*). Britain, widespread. Brownish, hard, rather brittle, resists wear. Fine grain and texture. Not difficult to work. Resembles pear, but has darker zones and is harder. Used for saw handles, turnery, veneers. S.G. ·75 ; C.W. 3.

APPLE

Applebox. *Angophora spp.* (*H.*) Queensland. Reddish brown. Hard, heavy, strong, durable. Used locally for fencing, wheel naves, etc. Not exported. S.G. ·9.

Apple-stern. A sailing ship with rounded stern, or buttocks.

Appliqué. Planted ornamentation, *i.e.* not worked on the solid.

Apron. 1. A plank flooring on which dock gates close. 2. A projecting ornamentation on the lock rail of a door. 3. The brackets to a bracketed stair. 4. A lining tongued into the bottom edge of a stair string to hide the carriages, etc.
5. A wide, shaped rail in cabinet work.
6. A platform to assist in guiding logs in the required direction after leaving a slide. 7. A curved piece continuing the foremost end of the keel and behind the stem of a ship. It is scarphed to the fore deadwood and strengthens the stem. See *Keel* and *Stem and Head*. 8. See *Saw Bench*.

Apron Lining. A vertical lining to

FLOOR BOARDS

LANDING JOISTS

APRON PIECE

APRON   LINING

cover the *apron piece* and the joists in the well of a stair-case.

Apron Moulding. The raised ornamentation on the lock rail, or apron rail, of a door.

Aprono. Mansonia, *q.v.*

Apron Piece. **1.** A pitching piece. A timber fixed in the walls of a staircase to carry the landing joists and to support the carriage pieces. **2.** A wide, curved rail in cabinet work, when less than 3 ft. from the floor.

LOCK RAIL

APRON MOULDING

Apron Rail. See *Lock Rail* and *Apron Moulding*.

Apron Strip. A strip joining together the two halves of the fuselage of a wood aeroplane.

Apse or **Apsis.** A semicircular or polygonal end to a building, usually with a vaulted roof. The eastern end of a church.

Apuleia. See *Garapa, Ibera, Pao Ferro*.

Aqueduct. A conduit or artificial channel for water.

Aquifoliaceæ. The Holly family, which includes Ilex.

Aquilaria. See *Eaglewood*.

Araba. See *Cottonwood (Nigerian)*.

Arabesque. Ornamentation representing flowers, fruit, geometrical figures, etc., interwoven together. It is characteristic of Moham-medan architecture.

Araca. *Terminalia aff. januarensis* (*H.*). Brazil. Olive or yellowish brown with nearly black mottle or ribbon stripe. Hard, heavy, not durable. Straight, uniform grain and texture, smooth, visible rays. Similar to birch and used for similar purposes. S.G. ·77 ; C.W. 3.

Araliaceæ. The Ginseng family : Aralia, Arthrophyllum, Cheiro-dendron, Didymopanax, Fatsia, Gilibertia, Heptapleurum, Panax, Pentopanax, Schefflera, Sciadodendron.

Aranga. *Homalium luzoniense* and other spp. (*H.*). Philippines. Yellowish to pale chocolate, lustrous. Very hard, heavy, strong, and durable. Straight and interlocked grain, fine texture. Requires slow seasoning. Difficult to work due to hardness. Used for all purposes where strength and durability are important, structural work, piling, sleepers, superior joinery, etc. S.G. ·9 ; C.W. 4.

Arariba. *Centrolobium robustum, C. tomentosum,* or *Sickingia spp.* (*H.*). C. America. Also called Balustre and Canarywood. Yellowish brown to deep purplish, to deep rose colour, variegated, rather dull but smooth. Very hard and heavy. Strong but brittle, durable. Straight grain, fine uniform texture, roey, distinct rays. Rather difficult to work. Good qualities used for cabinet work, veneers, turnery, etc. Produces yellow dye. S.G. ·88 ; C.W. 4. The name Arariba is also applied to several other woods : Amarella, Putumuju, Rosa, Vermelha, etc.

Araticú. The name applied to species of *Anona* and *Cordia*.

Araucaria. See *Bunya, Chilian, Hoop,* and *Parana Pines*.

Araucariaceæ. A sub-family of Coniferæ, including *Araucaria*.

Arbalest. A strong cross-bow operated mechanically.

Arbol. A prefix to numerous S. American woods.

Arbor. 1. A temporary tapered piece on which hollow articles are fixed for turning in a lathe. 2. An axle or spindle for a wheel or pulley.

Arboreal. Connected, in any way, with trees.

Arbor Vitæ. *Thuya.* Britain. Conifer. Cupressus family. Soft variegated wood. Strong and durable. See *Thuya* and *White Cedar*.

Arbour. A bower or shady retreat ; usually of lattice work.

Arbutus. See *Madrone* and *Strawberry Tree*.

Arcade. A series of arches supported by pillars and carrying a roof to form a screened and covered walk. In a modern arcade the arches and pillars are replaced by shops.

Arch. An arrangement of wedge-shaped blocks mutually support-ing each other to span an opening and to carry the wall and load above ; or an arrangement of timbers spanning an opening for decorative purposes. An arch is named according to (*a*) the style of architecture ; (*b*) the curve of the intrados ; (*c*) its purpose in the structure ; (*d*) the number of circular arcs in the outline ; (*e*) its resemblance to some familiar object. The kinds are : camber, catenary, cycloidal, drop, elliptical, equilateral, flat, Florentine, French or Dutch, gauged, Gothic, horseshoe, hyperbolic, inverted, lancet, Moorish, ogeval, parabolic, rampant, relieving, rere, rough, Saracenic, segmental, semicircular, soldier, squinch, stilted, three (or more) centred for a false ellipse, trimmer, Tudor, Venetian.

Archangel Fir. See *Redwood*.

Archimedian Drill. Used for boring small holes in hard wood. The rotation is operated by means of a spiral stem.

Architrave. Ornamental mouldings mitred or run round a door or window opening or niche. See *Finishings*.

Architrave Cornice. An entablature without a frieze.

Archivolt. 1. A soffit with moulded edges. 2. The moulding on the face of an arch parallel to the intrados.

Arch Rib. A curved rib as in a hammer-beam roof truss. See *Angel-beam*.

Arc Shake. A cup shake. See *Shakes*.

Area. Superficial contents. The areas of regular plane figures are given alphabetically in this glossary.

Areca. A genus of tall palm trees from which catechu and betel-nut are obtained.

Arena. A place of public entertainment or contest surrounded by tiers of seats.

Arere. See *Obeche*.

Argento. *Eucalyptus regnans.* Mountain ash, *q.v.*, in veneer form.

Aria. White beam tree.

Arisauru. *Pterocarpus sp.* (*H.*). British Guiana. Yellow to reddish brown, darkens with exposure, lustrous, smooth, bitter secretion. Moderately light and soft, but fairly durable and resists insects. Difficult to work smooth and to polish. Not imported. S.G. ·75.

Ariyan. *Afzelia bipindensis.* See *Apa*.

Arjun. See *Taukkyan*.

Arkansas. See *Oilstones*.

16

Armillaria. See *Honey Fungus*.

Armoire. 1. An obsolete name for a press or wardrobe. 2. A hall fitment consisting of a box seat and wardrobe back. 3. Same as ambry, *q.v.*

Armorply. A combination of plywood and thin sheet metal. See *Plywood*.

Arms. The timber cross-pieces to telegraph poles.

Aroeira. Applied to several species of *Astronium* and to *Schinus* (H.). C. America. Cherry red, black streaks. Very hard, heavy, strong, and durable. Resists friction. Difficult to work. Used for structural work, sleepers, brake blocks, cabinet work, turnery. S.G. ·9 to 1·2 ; C.W. 5. See *Goncalo Alves* and *Pepper Tree*.

Arollo Pine. See *Pine*.

Aromilla. *Terminalia spp.* See *Nargusta*.

Arr. The bottom rail of a five-barred gate. Also called *hur*.

Arris. The salient corner where two plane surfaces meet. A sharp edge.

Arris Fillet. 1. A tilting fillet. See *Eaves*. 2. A square fillet used in place of an angle bead, *q.v.*

Arris Gutter. A V-shaped gutter.

Arris Rail. A triangular rail used for fences. See *Capping Rail* and *Spout Screed*.

Arris-wise. Material cut diagonally.

Artocarpaceæ. A family that includes *Artocarpus* of timber importance. *Artocarpus chaplasha*, Kaita-da, Sam, or Chaplash. *A. integrifolia*, Jacqueria or Jackwood. *A. lacoocha*, Lakuch. *A. lanceæfolia*, Keledang. There are other species, but without market names : *A. nobilis*, Ceylon ; *A. hirsuta*, India. Also see *Tampang*, *Tempunai*, *Terap*.

a/s. Abbreviation for *alongside*.

Asch. Abbreviation for botanist's name, Ascherson.

Ash. *Fraxinus excelsior* (H.). Europe. Whitish grey with yellow markings. Very tough and elastic. Straight grain, coarse texture. Dense summer wood, porous spring wood. Not durable when exposed to weather

ASH

and subject to insect attack. Excellent for steam bending. Polishes well. Good logs up to 20 ft. by 2 ft. diameter. Used for sports and gymnastic requisites, wheelwright's work, aircraft, etc. Mottled and burred timber used for veneers. S.G. ·65 to ·8 ; C.W. 2·8. AMERICAN ASH, *F. americana*, *F. pennsylvanica*, *F. nigra*. Not so good as *F. excelsior*. BLACK ASH, *F. nigra*. BROWN ASH, *F. nigra*. CANADIAN ASH, same as American ash. CANE ASH, American ash. EUROPEAN ASH, usually named according to the country of origin. GREEN ASH, *F. pennsylvanica lanceolata*. JAPANESE ASH, *F. mandshurica*. See *Tamo*. MANCHURIAN ASH, Tamo, *q.v.* OLIVE ASH, *F. excelsior*, Variegated brown-hearted ash. OREGON ASH, *F. oregana*. PUMPKIN ASH, *F. profunda*. WHITE ASH is the trade name for all species except *F. nigra*. Also see *Cape Ash, Maiden Ash, Mountain Ash, Silky Ash,* and *White Ash*. *Fraxinus nigra* is also called Hoop, Swamp, and Water Ash.

Ashe. The name of the botanist W. W. Ashe.

Ashlering or **Ashlaring.** Short studs, or quarterings, used to cut off the acute angle formed by a sloping roof meeting a floor, or to form a level surface with the face of the wall. The studs are lath and plastered. See *Eaves*.

Asna. Indian laurel, *q.v.*

Aspen. *Populus tremula* (*H.*). Europe. Also called Trembling Aspen. Greyish white to dull brown, odorous. Light, soft, weak, not stable or durable. Straight grain, uniform texture. Easily wrought, but some difficult to work smooth. Requires careful seasoning. Used for pulp, excelsior, plywood cores, containers, laundry appliances, matches, etc. S.G. ·34 ; C.W. 2. CANADIAN ASPEN, *P. tremuloides*. LARGE-TOOTH ASPEN, *P. grandidentata*, U.S.A. Also see *Cottonwood* and *Poplar*.

Aspidosperma. See *Amarilla, Guatambú, Paddlewood, Peroba, Piquiá, Quebracho, Yaruru*.

Assegai. A lance or spear tipped with iron.

Assegaiwood. See *Lancewood*.

Assorted. See *Brands* and *Grading*.

Asta. Unidentified, probably lancewood (*H.*). Mexico. Lemon colour. Hard, heavy, strong, durable. Fine straight grain. Used for the same purposes as hickory, and for bows, fishing-poles, etc. Also see *Lancewood*.

Astel. A plank used for overhead partitioning when tunnelling.

Astragal. A small semicircular moulding, like a raised bead. Often carved into the form of berries. Used at the shutting joint for doors in cabinet work.

Astronium. See *Aderno, Aroeira, Batingor, Ciruela, Gateado, Goncalo Alves, Urunday, Zebrawood*.

Ata or **Atagbe.** Okor, *q.v.* Also see *Okanhan*.

Ata-ata. E. Indies (*H.*). Beautiful grain and colour for cabinet work. Dark brown, mottled with black streaks.

Atako. *Strombosia pustulata* (*H.*). Nigeria. Yellowish brown, irregular streaks, reddish brown markings, smooth, lustrous. Hard, heavy, strong, fairly durable. Interlocked compact grain. Characteristics of olivewood. Rather difficult to work, polishes well. Suitable for decorative and cabinet work, flooring, etc. S.G. ·95 ; C.W. 4.

Atala. *Saccoglottis gabunense* (*H.*). Nigeria. Reddish brown to purplish red, glistening deposit, smooth, retains sharp arrises. Very hard, heavy, strong, and durable. Interlocked dense grain, close texture. Used for hard-wearing floors, rollers, bearings, canoes, etc. S.G. 1 ; C.W. 4.

Atherosperma. See *Sassafras.*

Athrotaxis selaginoides. King William pine.

Athwartship. Across from side to side.

Atlantes. Columns carved as male figures. See *Caryatides.*

Attached Column. One that is attached to a wall and only part projecting.

Attalea. A genus with numerous species of palms.

Attic. A room in the roof of a building. The ceiling is square with the walls as distinct from a garret.

Attic Base. A base to a column in the Doric or Ionic orders. It consists of torus, scotia, and lower torus, with fillets between.

Aubl. Abbreviation for the name of the botanist Aublet.

Aucoumea. See *Gaboon.*

Auction. See *Timber Exchange.*

Aucuba japonica. Japanese laurel.

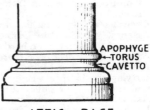

APOPHYGE
TORUS
CAVETTO

ATTIC BASE

Auditorium. That part of a public hall allotted to the audience.

Auger. A long twist bit for boring holes in timber by hand. Twist bits for the brace and machines are called auger bits.

Aumbry. Same as ambry, *q.v.*

Aunt Sally. A wood dummy at which wood balls are thrown on fair-grounds.

Australian. Applied to the many valuable timbers imported from Australia ; black bean, blackwood, blue gum, box, cedar, coolibah elm, eucalyptus, hoop pine, ironbarks, jarrah, karri, maple or silkwood, mountain ash, myrtle, sandalwood, silky oak, tallow wood, tuart, wandoo, etc.

Australian Mahogany. *Dysoxylon fraseranum* (*H.*). Also called Australian Rosewood. Reddish colour and fragrant. Characteristics of red cedar. Very durable and stable. Used for general purposes, furniture, aircraft. S.G. ·8 ; C.W. 2·75. See also *Mahogany.*

Australian Walnut. *Endiandra palmerstonii* (*H.*). See *Queensland Walnut.*

Australian White Ash. *Eucalyptus fraxinoides.* See *White Ash.*

Austrian. See *Oak.*

Austrian Pine. *Pinus nigra.* Also called Black Pine.

Autocar. A motor car.

Autoclave. An airtight cylinder in which temperature and atmospheric pressure can be controlled. It is used for steam-curing resin-bonded veneers, etc.

Autogyro. A type of aeroplane designed for vertical ascent.

Auxanometer. An instrument for measuring the growth of trees.

Auxemma. See *Páo Branco.*

Auxiliary Rafter. A second principal rafter used in large queen-post roof trusses.

Avicennia. See *Mangrove*.

av.l. Abbreviation for *average length*.

Avocado. *Persea americana* (*H.*). Laurel family. C. America, but widespread. Also called Alligator Pear and Palta. Light reddish brown, fragrant, lustrous. Fairly hard and heavy. Uniform, wavy grain, fine texture. Not difficult to work. Used for cabinet work, construction, etc. S.G. ·65 ; C.W. 3.

Avodire. *Turræanthus africana* or *T. vignei*. See *Olon*.

a.w. Abbreviation for *all widths*.

Awl. A pointed tool for boring small holes. See *Bradawl*.

Awning. A protection from the sun, usually consisting of canvas stretched on a frame.

Axemaster. Quebracho, *q.v.* Also *Krugiodendron ferreum*, C. America. Called Palo Diablo, *q.v.*, Black Ironwood, etc. A small tree, and the wood is too hard for most purposes. S.G. 1·4 ; C.W. 6.

Axim. See *Mahogany (African)*.

Axle Pulley. Used with vertically sliding sashes. The sash cords, carrying the balance weights, pass over the pulleys which are fixed near the top of the pulley stiles.

Axlewood. *Anogeissus latifolia* (*H.*). India. Similar to Yon, *q.v.*, but heavier, stronger, and more durable. S.G. ·95.

Ayan or **Ayanran.** The standard name for Nigerian Satinwood, or Movingui. See *Satinwood*. The name Ayan is also applied to species of Apa, *q.v.*

Ayinre or **Ayinreta.** *Albizzia ferruginea* and *A. zygia*. Kokko, *q.v.*

Ayo. *Holoptelea grandis* (*H.*). Nigeria. Yellow. Hard, heavy, and strong, but not durable. Mild grain and easily wrought. Suitable for joinery, turnery, etc. S.G. ·7.

Ayous. Obeche, *q.v.*

Azadirachta. See *Maranggo*.

Azobe. Ekki, *q.v.*

B

Babbar. See *Babul.*

Babia. *Cordia gerascanthoides* (*H.*). W. Indies. Greenish brown. Hard, strong, durable. Used for general purposes locally. S.G. ·75.

Babul or **Babool.** *Acacia arabica* (*H.*). India. Also called Babbar, Thorny Mimosa, and Gum Arabic tree. Light red, darkens with seasoning. Dark mottles. Hard, tough, close but coarse grain ; twisted fibres, very durable. Conspicuous medullary rays. Resists insects. Used for general purposes, sleepers, furniture, etc. S.G. ·86 ; C.W. 3·5. See *Gum Arabic.*

Baby. Applied to things of small size for their kind, as baby grand piano.

Back. 1. The top surface of an inclined or horizontal timber. 2. The convex, or upper, part of a saw tooth. 3. The surface opposite to the face side of a piece of timber. 4. The hind part of anything farthest removed from the face or important part, as the back of a drawer, cupboard, etc. 5. See *Vats.*

Back Boards. 1. Thin boards for the backs of picture frames. 2. The tailboard of a cart. 3. A board strapped to a young person with a defective back.

Backbone. The stem, keel, and stern timbers of a ship.

Back Boxing. See *Back Linings.*

Backed, Bellied, and Jointed. A stave prepared for assembling to make a barrel.

Back Fall. A piece of hardwood, up to about 3 ft. long, rotating on a centre and operated by a tracker in a pipe organ.

Back Fillet. The *return* of a projection on the face of a wall or jamb.

Back Flap. 1. The back leaves, or shutters, for folding window shutters. 2. A hinge with large wings for screwing on the face of both door and frame.

Back Foot. See *Splads.*

Backhousia. See *Lemon Ironwood* and *Carrol Ironwood.*

BACK FLAP

Backing. The shaping of a hip rafter or arch rib to the dihedral angle.

Backing Board. The last board in a converted log, to which the carriage dogs were attached.

Backing Deals. Vertical planks behind the curbs in a mine shaft.

Backings. 1. Fillets fixed on to the top edge of uneven joists to provide a plane surface for boarding. 2. Horizontal grounds fixed between vertical grounds to provide the fixings for linings to interior door jambs. See *Finishings.*

Back Iron. The cap, or cover, iron that stiffens the cutting iron for a plane. It also breaks the shavings and prevents the fibres from tearing along the grain. See *Plane Irons*.

Back Linings. 1. Thin rough boards fixed to the edges of the linings of a sash and frame window, to form an enclosed box for the balance weights and to keep the box free from mortar. See *Sash and Frame*. 2. The linings at the back of a recess for folding shutters.

Back of Stern Post. A piece bolted to the stern post of a ship when large sizes are not available. It carries the gudgeons for the rudder. See *Stern*.

Backs. 1. The principal rafters of a roof truss. 2. The waney or sappy or inferior side of a piece of timber.

Back Saw. A thin saw (tenon, dovetail, etc.) stiffened along the back edge by a strong strip of brass or steel.

Back Shore. See *Shoring*.

Back Stairs. A staircase for the servants' quarters of a large house.

Back-stay Stool. A piece fitted to secure the dead-eyes and chains for the back stays of a ship.

Bacury. *Platonia insignis* (*H*.). C. America. Also called Bacupary. Yellowish brown, variegated, smooth, fairly lustrous. Hard, heavy, tough, strong, durable, fissile. Straight grain, rather fine texture, but uneven due to laminated structure. Distinct rays. Not difficult to season or work, and polishes well. Used for construction, furniture, joinery, shipbuilding, etc. S.G. ·85 ; C.W. 3·5.

Badam. White Bombwe, *q.v.*

Badger. A large rebate plane, like a jack plane, but with the cutting iron flush with the side of the plane.

Badi. Opepe, *q.v.*

Bael. *Ægle marmelos* (*H*.). India. Yellow fading to greyish. Hard, heavy, even and fine texture, straight grain, but with interlocked fibre. Lustrous, aromatic, difficult to season and to finish, not durable. Used for tools, turnery, beetles, etc. S.G. ·9 ; C.W. 4.

Baffle. 1. A board for supporting a loud speaker. It is designed to suit the wave radiation. 2. Any arrangement to control or regulate the flow of liquids, gases, vibrations, etc.

Bagac. Apitong, *q.v.* Also see *Gurjun* and *Keruing*.

Bagasse. See *Tatajuba*.

Bagtikan. *Parashorea malaanonan* (*H*.). Philippines. One of the white lauans, *q.v.* Greyish brown. Moderately hard, heavy, and strong, not durable. Interlocked grain, fairly coarse texture. Seasons and works readily. Used for cabinet work, superior joinery, planking, pattern making, etc. S.G. ·6 ; C.W. 3.

Baguette. A bead or astragal.

Bahamas. See *Lignum vitæ*.

Bahera. *Terminalia belerica* (*H*.). India. Many local names. Pale yellow to brownish grey. Fairly hard, heavy, with straight grain, even texture, easy to work, mottled, polishes well, not durable. Used for interior work, furniture, etc. S.G. ·73 ; C.W. 2·5.

Bahia. Abura, *q.v.*

Bahiawood. Brazilwood, *q.v.* Also see *Rosewood*.

Bahut. An ornamental cabinet or chest.

Baikiæa plurijuga. See *Rhodesian " Teak."*

Bail. Abbreviation for the name of the botanist Bailey.

Bail or **Bale.** 1. A wooden bar forming a division between the stalls in a stable. 2. Thin boards used in bales of cloth. 3. A bucket or vessel for *baling out* water.

Bailey's Cypress. *Callitris baileyi* (*S.*). Queensland. See *Cypress Pine*.

Bails. Two small turned pieces laid across the tops of the stumps, or wickets, in the game of cricket.

Baing. See *Thitpok*.

Bak. Abbreviation, with distinguishing initials, for the names of the botanists Baker.

Bakau. *Rhizophora spp.* (*H.*). Malay. Several species. Brown. Very hard, heavy, and durable. Handsome, well-marked wood with distinct rays. Contains salt, from mangrove swamps. Constructional work, piling, decorative work. S.G. ·85 ; C.W. 3.

Baku. Makore, *q.v.*

Balance Beam. A long wooden lever fixed to a lock-gate, to close or open the gate against the pressure of the water.

Balanced Matched. A veneering term applied when more than two pieces are used in a single face. The pieces are of uniform size. See *Matched* and *Random Matched.*

Balanced Sashes. See *Sash and Frame.*

Balanced Shutters. Vertically sliding shutters balanced with weights like a " sash-and-frame " window.

Balanced Steps. Dancing steps, *q.v.*

Balanced Winders. Winders not radiating to a common centre, to give increased foothold at the narrow end.

BALANCED STEPS

Balance-lug. One with both boom and yard, on sailing ships.

Balance Weights. Cast-iron, or lead, weights used to balance a vertically sliding sash or shutter.

Balanocarpus. See *Chengal, Damar*, and *Penak.*

Balata. Applied to numerous woods of the Sapodilla family because of the milky juice that produces chicle, balata, or gutta-percha. See *Bulletwood, Mucuri*, and *Massaranduba.*

Balata Tree. *Mimusops balata* and *M. globosa* (*H.*). Guiana and W. Indies. Also called Bulletwood, *q.v.* Dark red, very tough, and durable. Resists insects. There are three varieties, but red is the best and used as a substitute for Greenheart. The gum, when dried, is used for the same purposes as gutta-percha. S.G. 1 ; C.W. 4.

Balau. See *Merbatu* and *Johore* " *Teak.*"

Balaustre. Arariba, *q.v.* Also see *Vermelha.*

Balconette. A small balcony.

Balcony. **1.** Tiers of seats in a theatre between circle and gallery.

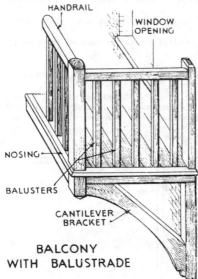

**BALCONY
WITH BALUSTRADE**

Labels on image: HANDRAIL, WINDOW OPENING, NOSING, BALUSTERS, CANTILEVER BRACKET

2. A platform projecting from a wall, supported by cantilevers or columns, and enclosed by a balustrade.

3. A projecting gallery in the stern of old wooden ships.

Baldachino or **Baldaquin.** A canopy to an altar or shrine.

Bald Cypress. *Taxodium distichum.* See *Cypress.*

Baldy. A small fishing boat, with lug-sail, used in the north of Scotland.

Balection Moulding. See *Bolection.*

Balfourodendron. See *Ivorywood* and *Pao Marfin.*

Balk or **Baulk.** A squared log. Hewn timber over 9 in. square at the middle. Squared timbers, during conversion, over 6 in. square.

Ball and Claw. A decorative feature of Queen Anne and Georgian furniture in which the feet are carved in the form of a claw clutching a ball. See *Pillar Table.*

Ball and Socket. A *movement* for a swivelled mirror to a dressing-table. A ball fixed to the mirror rotates in a tapered slot fixed to the supports. Also see *Castor.*

Ball Catch or **Bullet Catch.** A plate carrying a steel ball controlled by a spring. It is let into the edge of a swing door for holding the door in position when closed.

Ball Flower. A carved ornament resembling the petals of a flower enclosing a ball. It is characteristic of the Decorated style of architecture.

Balloon. A ball surmounting a pier or post as an ornamental terminal.

Balloon Framing. Framing for timber buildings in which the posts run through and the horizontal members are nailed to them. It is usually considered inferior work.

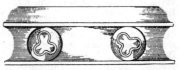

BALL   FLOWER

Balm or **Balm of Gilead Fir.** *Abies balsamea* or *A. concolor* (*S.*).

N. America. One of the species of N. American fir. The resin is an antiseptic. Consignments of spruce to England must not contain more than 20 per cent. of this species. Also called Balsam Fir. See *Cottonwood* and *Poplar*.

Baloghia. See *Ivory Birch*.

Balong. See *Mengkulang*.

Balpale. See *Banate*.

Balsa. *Ochroma lagopus* and *O. piscatoria* (*H.*). Tropical America. Yellowish white. Very light, elastic, buoyant, and soft, but difficult to split. Also called Corkwood. Used for insulating purposes, floats, fairings, hydroplanes, plywood cores, and other purposes requiring a very light wood. Probably the lightest dicotyledon. Market sizes up to 18 ft. by 18 in. by 5 in. S.G. ·15. INDIAN BALSA. See *Papita*.

Balsam. See *Balm, Colorado, Cottonwood, Western Balsam*, and *Poplar*.

Balsam Fig. *Clusia rosea* (*H.*). C. America. Reddish brown, variegated. Hard, heavy, strong, durable. Not exported.

Balsamito. Cabreuva, *q.v.*

Balsamo. *Myroxylon pereiræ* (*H.*). Called Mexican rosewood. Pink to purplish red, with variegated wavy markings, ranging to nearly black. It is similar to E. Indian rosewood, but rather lighter in colour and with the weight and texture of oak. Excellent wood for cabinet work, superior joinery, veneers, etc. The name is also applied to Copaiba, Gumbo-limbo, and Oleo Vermelho, *q.v.*

Baltic. Applied to the timbers shipped from the countries round the Baltic Sea, but chiefly to whitewood (white deal or fir) and redwood (red or yellow deal).

Baltime. A code name used in the timber trade between this country and the Baltic and White Seas. See *Charter Party Forms*.

Balusters. Small vertical pillars supporting handrails. See *Balcony*. Turned pillars round the balcony of a ship.

Balustrade. A row of balusters with base and rail, or cap, to form a protective enclosure to a balcony or stairs.

Bamboo. *Bambusa spp.* Endogenous trees or giant grasses, such as the palm, etc. Greatly valued in tropical and Eastern countries for all purposes. Used for furniture making and curtain rods, etc. They are " inward " growers and not " timber " trees, and have hollow stems. The amount of fibrous wood varies considerably in different species so that many of them provide excellent structural timbers for local uses.

Banaba. Species of *Lagerstrœmia* from the Philippines.

Banak. *Virola koschnyi* or *V. merendonis*. Also called Tapsava, *q.v.*

Banate. *Lophopetalum wightianum* (*H.*). India. Many local names. Pale yellow to brownish grey. Fairly straight grain and even texture. Not durable. Mottled. Easy to work, and polishes well. Used for general purposes, cheap furniture. S.G. ·43 ; C.W. 2.

Banawi. *Cyclostemon grandiflorus* or *Homalium spp.* (*H.*). Philippines. Many local names. Greenish brown to black, mottled. Fine texture, dense, smooth, straight grain. Hard, heavy, moderately durable. Constructional work, furniture, etc. S.G. ·75 ; C.W. 3.

Bancal or **Bangkal.** 1. *Nauclea spp.* (*H.*). There are about six species. Philippines. Canary yellow to orange, darker streaks. Moderately soft and light. Fairly straight grain and fine texture, easy to work. Feels greasy to touch. S.G. ·55 ; C.W. 2. **2.** A decorative covering for a seat.

Band. 1. Any flat member having little projection. **2.** Mouldings encircling the shaft of a column. **3.** Any continuous ornamentation along a wall or around a building.

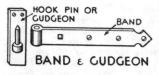

HOOK PIN OR GUDGEON   BAND

BAND ε GUDGEON

Band and Gudgeon. A long strap hinge consisting of a wrought-iron band rotating on a pin fixed to the door frame or gate pier. The pin is often called a *hook* when fixed on a door frame.

b. and b. Abbreviation for *bead and butt*

Banded Trees. Exogens, or timber trees. The wood element is formed in layers or annual rings. See *Non-banded*.

Bandelet. A narrow band, *q.v.*

b. and f.p. Abbreviation for *beaded and flush panel*.

Banding Plane. A router for cutting grooves for inlays in cabinet work.

Bandings. 1. Narrow inlays of varied patterns in cabinet work. They are usually between ¼ and 1 in. wide. **2.** Strips of wood along one or more edges of plywood to cover the core or end grain and to facilitate shaping. Also called Railing.

CHEQUERED BANDING

Band Moulding. A simple form of architrave.

Band Resaw. A band saw with roller feed for converting flitches and planks into boards.

Band Saw. An endless, or continuous, ribbon saw running around two pulley wheels, like a driving belt. Narrow saws, from ¼ in. wide and about 12 ft. long, are used for curved work. The log band saw has a blade up to 40 ft. long and 16 in. wide and is considered the most economical saw for the conversion of timber. Head band saws may be up to 20 in. wide. Some band saws are toothed on both edges.

Bandstand. A covered and partially enclosed platform on which instrumentalists give public performances in the open air.

Bangkirai. A trade name applied to Indian species of *Shorea* and *Hopea* with an air-dried specific gravity of not less than ·76.

Bania. See *Ebony*.

Banian. See *Banyan*.

Banjo. 1. A flexible adjustable rod for copying curves. It is chiefly used in ship joinery. **2.** A musical instrument of the guitar type. It has a parchment-covered body and strings played by the fingers.

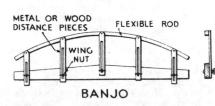

METAL OR WOOD DISTANCE PIECES   FLEXIBLE ROD

WING NUT

BANJO

Bank. A bench, or a seat, or a counter.

Banker. A platform fenced on three sides on which concrete is mixed.

Bankia. A genus of mollusk that attacks wood. See *Marine Borers*.

Banksia. 1. *Banksia marginata (H.)*. S. Australia. Light buff colour, beautiful grain. Used for veneers. See *Honeysuckle, River Banksia*, and *Wallum Oak*. 2. A Chinese climbing rose. 3. See *Jack Pine*.

Bannister. A local name for Baluster. A cabinetmaker's term for a vertical rib, or splad, in the back of a chair. See *Splad*.

Banquette. A footway on a bridge raised above the level of the roadway.

Banuyo. *Wallaceodendron celibicum (H.)*. Locust family. Philippines. Golden brown to coffee colour, beautiful grain. Moderately hard, heavy, and durable. Straight and wavy grain, fine texture, and glossy. Large sizes and easy to work. Resists insects. Used for decorative and cabinet work. S.G. ·53 ; C.W. 2·5.

Banyan. *Ficus bengalensis (H.)*. India. A variety of fig tree. Creamy to greyish brown, mottled. Coarse, light, not durable. Easy to work but difficult to finish. The timber has little economic value. The branches drop shoots to the ground which take root and form other trunks. Used for tea boxes, poles, etc. S.G. ·6.

Baobab. *Adansonia digitata (H.)*. C. Africa. The monkey-bread tree or sour gourd. A very large tree, but of little economic value as timber. Fruit valuable for cream of tartar. Another variety is the N. Australian Sour Gourd.

b.a.p. Abbreviation for *brass axle pulley*.

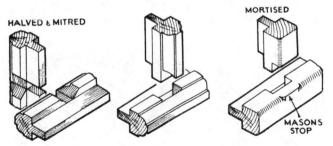

HALVED & MITRED   MORTISED   MASONS STOP

BAR JOINTS FOR SASH

Bar. 1. The intermediate members of a glazed frame, dividing the glass into smaller squares. 2. A flat piece of wrought iron sliding in *keepers* and serving as a bolt to secure a gate. 3. See *Ledged Door*. 4. See *Crowntrees*. 5. Horizontal timbers supporting the roof in a coal-mine. 6. A barrier, or rail, used as an obstruction or protection. 7. A counter at which liquid refreshments are served. 8. The members of the framing of a vehicle other than the pillars.

Barberry. *Berberis spp. (H.)*. Yellow. Hard, heavy. Firm close grain. Small sizes and seldom marketed as timber. Used for inlays and as a dye wood. *B. vulgaris*, Britain and N. America. *B. darwinii*, Himalayas. The name is also applied to *Rhamnus* and *Bumelia spp.*

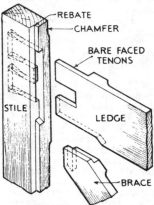

and to *Acacia gummifera*, or Morocco Gum.

Barbette. An exposed platform on board ship, usually for a gun.

Bar Clamp. A patternmaker's cramp for clamping together boards to build up a wide surface.

Bare. Slightly under size. The opposite to *full*.

Barefaced Tenon. A tenon with a shoulder on one side only, as used for the lower rails of *framed and batten* doors.

Barge. 1. A flat-bottomed freight boat used on canals and rivers. 2. A ship's light boat for conveying officers shore.

Barge Boards. The inclined timbers on the gable of a building, used to cover the ends of the purlins and ridge and to provide ornamentation. Also called Verge or Gable Boards.

BAREFACED TENON

Barge Couples. The outer couples of a roof when they project over the gable of a building.

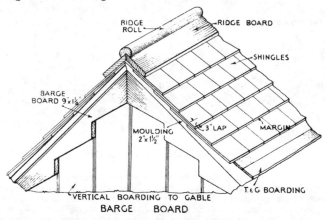

BARGE BOARD

Bari. Santa Maria, *q.v.*

Bark. 1. The outer covering of exogenous, or timber, trees. It protects the cambium layer, the removal of which destroys the tree. See *Girdling*. Tannin is extracted from bark for the leather industry;

28

and the bark of many trees has great medicinal value. The thickness varies considerably in different species. Some barks are very thin owing to periodic shedding of the outer brown bark, while in others it may be 2 ft. thick as in Sequoia and Douglas fir. See *Cork*. **2.** Any kind of boat. See *Barque*.

Bark Beetle. *Coleoptera.* An insect of the Scolytidæ family destructive to the bark of trees and logs.

Bark Blazer. See *Scratcher*.

Bark Blister. A disease specially destructive to Weymouth pine.

Barker. A French style of carriage at one time used in London.

Barking Drums. Large drums in which small softwood logs are revolved against knives to remove the bark preparatory to pulp manufacture, etc.

Barking Iron. See *Spud*.

Barking Saw. See *Rock Saw*.

Bark Peeling. A preliminary operation to felling. Also called barking, stripping, or flaying. The bark is usually removed in April or May while the sap is flowing.

Bark Pocket. A defect in timber in which the bark is wholly or partially enclosed in the wood.

Barn. A covered farm building for storing grain, hay, etc.

Barn Doors. Large double-doors to admit a loaded wagon.

Baromalli. *Catostemma fragrans (H.).* British Guiana. Pale brown, rather dull. Moderately hard and heavy, strong, shrinks greatly. Straight grain. Difficult to finish smooth. Not yet imported. S.G. ·6.

Baroque. Applied to grotesque or whimsical ornamentation.

Barouche. An obsolete horse-drawn carriage with four wheels. It had a folding hood behind, usually seated four persons, and was drawn by two horses.

Bar Posts. Posts mortised for rails and used as gate posts for a field or garden gate.

Barque. A particular type of square-rigged sailing ship with three or four masts, or with three masts without a mizzen-topsail. The name is also applied generally to small vessels.

Barquentine. A three-masted sailing ship with square-rigged foremast. See *Barque*.

Barracks. Buildings for soldiers, usually consisting of one-storey timber buildings.

Barred Doors. **1.** Doors in cabinet work in which the glass is divided into smaller squares by means of narrow bars. **2.** A local name for a ledged door, *q.v.*

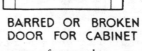

BARRED OR BROKEN DOOR FOR CABINET

Barred Gate. A field gate consisting of framing only.
Barrel. A round wooden vessel formed of curved staves bound with hoops. They are made to definite cubical capacities for most industries and named accordingly. In the beer industry a barrel is of 36 gals. capacity. See *Casks*.
Barrel Bolt. A spindle sliding in a cylindrical case for securing doors. See *Tower Bolt*.
Barrel Roof. A roof with semi-cylindrical ceiling.
Barrel Saw. A steel drum with a toothed steel band attached to the free end to form the cutting edge. Also called a Cylinder Saw.
Barrel Vaults. The simplest form of vaulting, usually semicircular. They are actually continuous arches. Also called wagon or tunnel vaults.
Barren-yard. Same as Cross-jack Yard.
Barricade. A barrier for defence purposes, to protect the enclosure. A strong wood rail across the fore part of the quarter-deck of a ship.
Barrier. A fence, railing, or gate, to form a boundary or barricade.
Barrington. See *Chee*.
Barrow. A small handcart, *q.v.* See *wheelbarrow*.
Bars. See *Bar*.
Bartizan. A small overhanging turret at the angle of a building.
Barton. A farmyard or field enclosure.
Bar Tracery. See *Tracery*.
Barwood. *Pterocarpus santalinoides*, or *P. oson*, or *P. tinctorius* (*H.*). W. Africa. Red, hard, tough, durable. Used for decorative purposes, violin bows, ramrods, etc., but principally as a dyewood. S.G. ·85. Also see *Padauk*.
Bascule. Applied to any apparatus operated on the principle of the see-saw or lever. A bridge raised or lowered on the principle of the lever, as the Tower Bridge, London.
Base. 1. The lowest member of anything. 2. A moulded skirting. 3. That part of a column between pedestal and shaft. See *Attic Base*.
Basement. 1. A storey of a building below street level. 2. The ground floor on which the columns, or order, which decorate the principal storey are placed.
Base Mouldings. The mouldings immediately above the plinth of a wall or pillar.
Base Plate. A plain skirting without mouldings.
Basil. The bevel, or grinding angle, to form the cutting edge of a tool.
Basket. A receptacle formed of plaited or interlaced rushes, twigs, canes, etc.
Basketry. The art of making baskets.
Basketweave. See *Veneer Matching*.
Basketwork. Interlaced work.
Basoolah. An adze with short haft peculiar to Eastern countries.
Basralocus. Angelique, *q.v.*
Bas Relief. Low relief. Carving projecting slightly over the face of the material of which it forms a part. See *Alto Relievo*.
Bass. 1. The inner bark, especially of the lime, or linden, tree. It is tough and fibrous bast, and is used for weaving into mats, etc. 2. See *Basswood*. 3. See *Bassine*.

Bassia latifolia. The butter tree or mahwa. Also see *Nyatoh.*
Bassine or **Bass.** *Borassus flabellifer.* A palm tree from which the leaf stalks provide a tough fibre for bass, or bast.
Bassoon. A wood-wind musical instrument for the bass register. A bass oboe.
Bass Viol. A violoncello.
Basswood. *Tilia glabra* or *T. americana, T. nigra, T. latifolia, T. canadensis, T. heterophylla* (*H.*). U.S.A. and Canada. The lime, or linden, tree. Also called American Whitewood. Large sizes. Yellowish or greyish white to pale brown. Light, soft, not strong or durable. Even grain, fine texture, soft, easy to work, shrinks and warps freely. Stains and polishes well, and used as a substitute for superior hardwoods. Used for panels, plywood, cheap cabinet work, carving, interior joinery, piano keys, containers, excelsior, etc. Inner bark very fibrous and used for rope making and matting. S.G. ·45 ; C.W. 2. See *Lime.*
Bast. See *Bass* and *Phloem.*
Bastard. Applied to woods that are not true species of the wood they substitute.
Bastard Bulletwood. *Humiria spp.* (*H.*). Tropical America. Purplish brown, fading with exposure. Very hard, heavy, strong, durable. Variable grain and texture. Used as a substitute for Bulletwood. S.G. ·9 ; C.W. 3·5.
Bastard Grain or **Sawn.** Flat sawn, or cut tangentially to the annual rings. See *Conversion.* Also see *Quarter Sawn.*
Bastard Locust. Angelique, *q.v.*
Bast Fibres. Fibres of the secondary phloem. See *Bass* and *Phloem.*
Bat. 1. A metal cramp used to fix wood frames to masonry. 2. An implement for striking the ball in the game of cricket. It consists of a willow blade and a built-up cane handle. See *Splice.*
Bataan. See *Red Lauan.*
Bate or **Bait.** Applied to the density of the annual rings in converted timber, as *wide* or *narrow* bated.
Bateau. A light river-boat.
Batete. *Kingiodendron alternifolium* (*H.*). Philippines. Many local names. Reddish brown with darker streaks due to oil secretion. Moderately hard, heavy, and strong, but not durable. Fairly fine texture and straight grain, smooth. Resists insects, stains easily. Requires careful seasoning, but works readily and polishes well. Used for constructional work, paving, furniture. S.G. ·6 ; C.W. 3.
Bathing Machine. A small hut on wheels used by sea bathers for privacy whilst dressing.
Baticulin. *Litsea obtusata* (*H.*). Philippines. Numerous species. Golden yellow, darkens with exposure. Soft to moderately hard, fragrant, fairly durable, lustrous, difficult to season but not very difficult to work. Beautiful grain and figure. Used for cabinet work and carvings. S.G. ·6 ; C.W. 3.
Batingor. *Astronium speciosa* (*H.*). S. America. Similar to *Aroeira.*
Batitinan. Bongor, *q.v.*
Batoa. West Indian boxwood, *q.v.*
Baton. 1. A glazing bead, *q.v.* 2. A staff symbolic of a particular

rank or office. **3.** A stout staff used for attack or defence by the police. **4.** A short slender stick used by the conductor of an orchestra.

Batten. 1. Applied to softwoods from 2 to 4 in. thick and from 5 to 8 in. wide. Sizes vary at different ports. **2.** Boards used to build up ledged doors, *q.v.* **3.** Wood of small section, as slate battens, etc. **4.** Floor boards. **5.** Any strip of wood used to fasten down the hatches on board ship. **6.** See *Counter Battens.* **7.** The backing of panels to a vehicle.

Batten-boards. Like laminboard, but with the core formed of strips

BATTEN BOARD

of any width up to 3 in., running at right angles to the grain of the outer veneers. **Batten Door.** A ledged, or barred, door. It consists of a number of narrow boards or battens, from 4½ to 6 in. wide, built up to the required width and held together by nailing them to three or more ledges on the back. See *Ledged Door.*

Battening. Narrow grounds fixed to a wall to carry the laths for plastering or matchboarding.

Batter. An inclination from the vertical.

Batter Post. One of the inclined timbers forming the side supports to the roof of a tunnel.

Battledore. A wood implement shaped somewhat like a tennis racket for beating clothes. A smaller type is used in the game of shuttlecock.

Battlemented. Applied to an indented parapet or capping to a screen. The indentations are called embrasures, or crenelles, and the projections are called merlons.

Batu. Borneo " teak." See *Merabau.*

Bauhinia. See *Pegunny* and *Kurâl.*

Baumier. Balsam poplar, *q.v.*

Bavins. Various local meanings : brushwood, chips, faggots, firewood, waste wood.

Bawang. See *Kulim.*

Bawk. A tie-beam, *q.v.*

Bawley. A shrimping boat used in the mouth of the Thames.

Bawn. A wooden enclosure, usually for cattle.

EMBRASURE OR CRENELLE

MERLON

OAK PINS

CORNICE

SCREEN

BATTLEMENTED CORNICE

Bay. 1. A principal compartment, or division, in the planning of a building. **2.** The floor, ceiling, or roof, between two main beams or trusses. **3.** A part of a room in the form of a recess. **4.** An opening between two pillars. **5.** A stall in a stable. **6.** See *Oak.* **7.** A reserved part of the deck of a ship for the sick or wounded.

Bay or **Bay Tree.** *Laurus nobilis* (H.). Europe. The true laurel, *q.v.* Cultivated chiefly for shrubbery ornamentation.

Bayam. See *Dedali.*

Bay Cedar. Guácima, *q.v.*

Bayott. See *Palosapis*.
Bay Poplar. See *Tupeloe*.
Bay Window. A window projecting over the face of the wall and built up from the ground. It may be square or polygonal in plan. See *Bow Window*.
Baywood. Honduras mahogany. See *Mahogany*.
b.b. Abbreviation for *bead butt*.
b/e. Abbreviation for *bill of entry*.
Beach She Oak. *Casuarina equisetifolia*. Queensland. See *She Oak*.
Bead. A round moulding, usually with small quirk, and used to remove a sharp arris or to break the joint between two boards. Also

RETURN   QUIRK

BEAD

see *Angle, Cock, Glass, Guard, Guide, Inner, Parting, Quirk, Raised, Return, Sunk* and *Staff Beads*, and *Reeds*.
Bead and Reel. An ornament used in Grecian

BEAD AND REEL

and Jacobean mouldings.
Bead Butt. Same as Bead Flush, but with the horizontal beads omitted; the beads are generally worked on the solid, and in the direction of the grain.
Bead Cuts. Applied to the preparation of the beads for a pivoted sash.
Bead Flush. Applied to thick panels that are flush with the framing on one face, and beaded all round to break the joint between frame and panel. The vertical beads may be *stuck* on the panel, but the horizontal beads, across the grain, are planted.
Bead Plane. A plane for *sticking*, or forming, beads on the edge of a piece of wood.

BEADED   PANELS

Beak. **1.** See *Bird's Mouth*. **2.** The lower edge of the *corona* forming a drip. **3.** See *Bird's Beak*.
Beak Head. A small platform at the fore part of the upper-deck of a large ship.
Beam. **1.** A strong horizontal structural timber supported at intervals by walls or columns. Beams are used to carry transverse loads and the term includes girders, cantilevers, joists, lintels, bressummers, etc. **2.** Obsolete name for tree. **3.** See *Whitebeam*. **4.** The breadth of a ship. **5.** Strong timbers running from side to side of a ship. **6.** The pole of a two-horse vehicle. **7.** A wood cylinder in a weaving loom. **8.** The part of a plough by which it is drawn.

Beam Blocking. Packing pieces nailed on joists to form sham beams in ceilings formed of *building-board*.

Beam Compass. A trammel. A wooden rod and two pointed legs, one of which carries a pencil for describing large circular arcs. The rod, or beam, may be any length and about ¾ in. × ¾ in., and the legs are adjusted on the rod to the required radius.

Beam-fillings. Short lengths of deals or logs used for filling spaces between the beams on board ship for compact stowing.

Beam Hangers. Stirrups hanging from main beams to carry secondary beams.

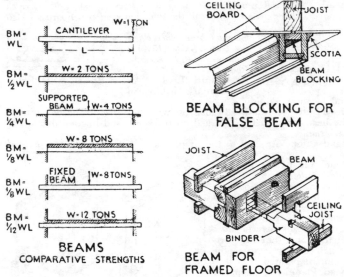

BEAM BLOCKING FOR FALSE BEAM

BEAMS
COMPARATIVE STRENGTHS

BEAM FOR FRAMED FLOOR

Beam Mould. A templet used in shipbuilding from which bevels and curves are obtained for the fittings.

Bean. See *Black Bean* and *Walnut Bean*. Also applied to several trees that produce bean-like seeds : *Carob tree, Ceratonia, Catalpa, Locust,* etc. The pods of the Queensland, or Leichhardt's, bean are sufficiently hard and wood-like to be used for small ornamental boxes.

Bearer. A general term used for small transverse structural timbers, as the joists supporting landings and winders in stairs.

Bearing. 1. The portions of a beam resting on the supports. 2. The mechanism supporting a rotating shaft.

Bearing Pile. A pile resting on a solid stratum and supporting a load. See *Friction Pile*.

Bearing Plate. A template. A horizontal member to distribute the load from another member of a structure. See *Plate*.

Beat. The ornamental grain in wood cùt tangentially to the annual rings.

Beati. See *Cassia siamea.*

Beaumontage. A stopping consisting of beeswax and resin with a few flakes of shellac. The whole is melted and well mixed, and colouring matter added as required. It can be shaped in pencil form and warmed, like sealing wax, when required.

Beaver Board. A proprietary wallboard, *q.v.*

Bebeere. Guiana greenheart.

Bechst. Abbreviation for the name of the botanist Bechstein.

Becket. A large log used with lifting tackle for loading logs.

Bed. 1. The bearing surface of anything, normal to the pressure. **2.** A piece of bedroom furniture with ornamental vertical framing at the head and foot. These are connected by two steel rails fixed to the frames, to carry a spring mattress or framed plywood. Beds are termed single, three-quarter, or double, according to the width. See *Bedsteads.* **3.** The undercarriage members to a vehicle, especially those carrying the perch bolt.

Bedara. See *Merbatu.*

Bedaru. Daru, *q.v.*

Bedd. Abbreviation for the name of the botanist Beddome.

Bedding. Placing a layer of plastic material between wood frames and the masonry or brickwork when fixing the frames. Oil putty, hair plaster, or red and white lead in oil are generally used.

Bed Moulding. 1. A moulding placed in the angle formed by a vertical surface and an overhanging horizontal surface. **2.** The cornice mouldings below the corona.

Bed Piece. A skid under a lumber pile.

Bed Posts. The legs of a bedstead, especially when continued as turned pillars above the bed to carry a canopy, as an Elizabethan "four-poster."

COVER BOARD

FRAME→

BED MOULDING

DOOR

BED MOULDING

Bedsteads. The end frames, of wood or metal, supporting a bed, *q.v*

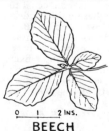

0   1   2 INS.

BEECH

Beech. *Fagus sylvatica* (*H.*). Europe. Also called Red Beech. Light reddish brown; straight, fine, close grain; even texture and lustrous. Moderately hard and heavy. Medullary rays appear as darker flecks or lines. Not difficult to work. Large sizes. Liable to insect attack and not durable if exposed. Used for furniture, tools, bent-work, turnery, airscrew packing blocks, wood-ware, beetles, butcher's blocks, rollers, charcoal, etc. S.G. ·8; C.W. 3. AMERICAN BEECH. *F. grandifolia*, similar to European. See *Blue Beech, Clinker Beech, Cudgerie, Evergreen Beech, Fagus, Myrtle, New Zealand Beech, Queensland Red Beech, White Beech, Southland Beech.*

Beef Oak. *Grevillea striata.* Queensland. See *Silky Oak.*
Beefwood. *Casuarina equisetifolia* (*H.*). India. Reddish brown.
Very hard and heavy. Straight and wavy grain, moderately fine and
even texture, difficult to season and work. Not a valuable timber
and little exported. S.G. ·9 ; C.W. 4. Other varieties are *Lophira
alata*, African beefwood, and *Roupala montana* or *C. cunninghamiana*,
Australian beefwood, and *R. brasiliensis*, C. American beefwood. They
are deep red colour and similar to Ekki, *q.v.* Also see *Balata* and
Bulletwood and *Ropala.*
Bees. Strong pieces, usually elm, bolted to the upper end of the
bowsprit of a ship.
Beetle. **1.** Many varieties of insects destructive to timber. Bark,
black carpenter, cockchafer, death-watch, elephant, furniture, long-
horn, pinhole borer, powderpost, sipalus, weevil, wharf borer, etc.
2. A heavy wooden mallet, or mall, for driving small piles, levelling
sets, etc. **3.** Pounders, used in several industries for beating and
bruising materials.
Beetle Cement. A proprietary urea-formaldehyde synthetic resin.
See *Glue* and *Plastics.*
Behra. See *Satinwood.*
Beilschmiedia. See *Blush Walnut, Canary Ash,* and *Tawa.*
Bejuco. See *Marinheiro.*
Belah She Oak. *Casuarina lepidophloia.* Queensland. See *She Oak.*
Belaying Cleat. See *Ship's Cleats.*
Belaying Pin. A wood or metal hook or pin for securing the free
end of a rope. A strong turned pin, usually of ash or iron, and about
16 in. long.
Belco. A proprietary cellulose lacquer for wood.

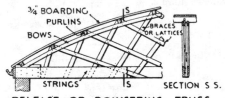

BELFAST OR BOWSTRING TRUSS

Belfast Truss. A
light roof truss for
large spans, suitable
for sheds and work-
shops. It consists of
a lower horizontal
member, or chord, a
curved top member,
or bow, connected to-
gether by thin lattices.
Also called a Bow-
string Truss. It is usually constructed by drawing the outline
of the truss on a large floor, bending a bow and securing it between
blocks fixed to the floor, nailing the lattices in position, and fixing
the top bow. The lattices are nailed between the bows and strings
and also where they cross each other.
Belfry. A bell tower, or campanile. That part of a steeple containing the timbers from which the bells are hung.
Belian. The suggested standard name for *Eusideroxylon zwageri.*
See *Billian* and *Betis.*
Bell. The body of a Corinthian or Composite capital when stripped
of its carved foliage.
Bella rosa. Palosapis, *q.v.*

Bell Gable. A bell-cote. A small bell turret on the gable of a church.
Bellied. Applied to work that is curved outwards, a convex curve. Also to panels that have buckled through dampness.
Bellows. An appliance for forcing a draught through a fire.
Bell Roof. A roof the section of which is bell-shaped, or of contrary flexure. Often used for kiosks and entertainment buildings. See *Kiosk.*
Belly. Same as Camber. The curved projecting part of a piece of work.
Belting. A strip of wood round a vessel to protect the ship when in contact with wharf or dock. Also called Fender or Rubbing-piece.
Belt Rail. A lock rail, *q.v.*
Belukar. A term applied to second growth of trees, after the destruction of a forest.
Belvedere. A projecting room above a roof, for observation purposes. An elevated summer-house.
Bema. A raised portion at the end of a church, or a transept.
Benacon. See *Charter Party Forms.*
Bench. Any rigid framing with a top on which workmen prepare the material. The form and shape vary with different trades. The usual essentials are vice, stop, and convenience for tools. **2.** A church seat or pew. See *Pew.*

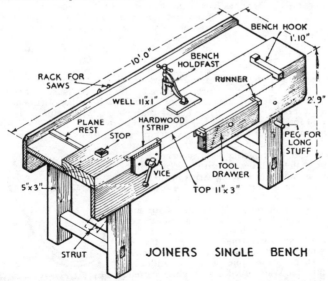

JOINERS SINGLE BENCH

Bench End. The ornamental end to a church seat or pew. See *Pew.*
Bench Holdfast. A clamp for securing the material to the bench while it is being wrought. See *Bench.*

Bench Hook. An appliance for steadying small stuff while sawing on the bench. See *Bench*.

Bench Planes. Applied to the planes that are continuously on the joiner's bench ; jack, smoothing, and try planes.

Bench Screw. The screw operating the vice to a bench.

Bench Stop. An adjustable projection to a bench against which the timber is pressed when being planed.

Bending. See *Bentwood*.

Bending Moment. A value required when designing beams, etc. At any cross-section of a beam the BM is " the algebraic sum of the moments of the external forces acting on one side of the section," and is equated against the " moment of resistance " of the beam, *i.e.* BM = MR. The four most important cases are WL, $\frac{1}{2}$WL, $\frac{1}{4}$WL, $\frac{1}{8}$WL. See *Beam*.

Benin. One of the ports for the export of W. African timbers, and which gives its name to mahogany and walnut shipped from there. See *Nigerian*.

Bent. Two or more posts or piles braced together to form a support in heavy timber construction. See *Trestle Bridge*.

Benteak. See *Nana*.

Benth. Abbreviation for the name of the botanist Bentham.

Bentwood. There has been considerable scientific research in recent

BENT CASING

years to secure the best results in the bending of solid wood for furniture and other purposes. The wood must retain a permanent set without checks or fracture developing, and the most favoured woods are ash and beech. Experiments have proved many other woods satisfactory, and these are indicated in the Glossary. Steam-heated metal forms are used, and machines may be obtained for making bends either with or across the grain. When glueing is required for laminations, rubber vacuum bags are used. See *Glue*. The older methods of bending are by steam heating or by pressure in hot sand. Other methods are kerfing on both faces and then facing with veneers when bent to the required curvature, or by kerfing on one side and glueing strips in the kerfs when bent round a drum. See *Cylinder*, *Steaming*, and *Urea*.

Bentwood Spars. Hollow spars for aircraft, sailing ships, etc. Flat sheets of silver spruce are heated and bent round suitably shaped formers. A second sheet is wrapped round the first, making a two-ply spar. More plies may be added if required. Waterproof casein glue is used. Solid blocks for bolt holes, etc., may be

BENTWOOD SPAR

left inside. This method of building up a cylinder gives the maximum resistance to buckling.

Bera. Lignum-vitæ, *q.v.*

Berangan. *Castanopsis spp.* (*H.*).   E. Indies. Malayan chestnut. Also called Kata. Yellowish grey to light brown. Moderately hard and heavy. Coarse grain with numerous pith rays. Stable but not durable and subject to insect attack. Better qualities for furniture and interior fittings. S.G. ·7 ; C.W. 3.

Berba. *Helicostylis latifolia* (*H.*).   C. America. Resembles Letterwood, *q.v.* Dark reddish brown. Hard, heavy, tough, strong, fairly durable. Irregular grain, medium texture. Difficult to work and little exported.

Berberine. *Xylopia polycarpa* (*H.*).   C. Africa. Imported for yellow dye.

Berberis vulgaris. Common Barberry. Britain.

Berline. A motor body of the limousine type with rounded glass corners at the back.

Berlinia. See *Apado* or *Ekpaghoi*.

Berombong. See *Meraga*.

Berrya. See *Hamileel* and *Trincomali*.

Bertholletia. See *Brazil Nut*.

Berth, Pullman. 1. A sleeping bunk on board ship that folds up against the bulkhead when not in use. **2.** A railway sleeping carriage.

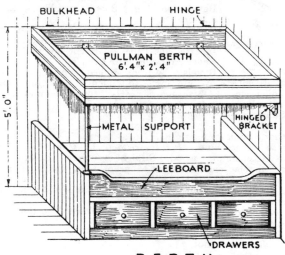

BERTH

Berus. Marketed as *Bakau*.

Besom. A broom formed of a bundle of twigs bound round the end of a long handle.

Bethel's Process. A method of preserving timber by placing it in an iron cylinder, exhausting the air, and forcing creosote into the timber under pressure. See *Preservation*.

Betis. *Payena utilis* or *Bassia betis* (*H.*). E. Indies. Reddish brown. Very hard, heavy, tough, and difficult to split but not to work. Strong and durable. Fine grain with varied coloured streaks. Used for cabinet work, tools, piling, paving, and heavy constructional work. S.G. ·85 ; C.W. 3.

Betula. Botanical name for birch, *q.v. Betula alba*, European birch. *B. lenta*, black, cherry, or sweet birch. *B. lutea*, yellow birch. *B. papyrifera*, canoe, paper, or white birch. *B. nigra*, red, or river, birch. *B. neoalashana*, Alaska or white birch. *B. pendula*, European birch. *B. populifera*, grey birch. *B. pubescens*, European birch. *B. odorata*, European birch. *B. verrucosa*, European birch.

Betulaceæ. Birch family. Includes Alnus, Betula, Carpinus, Corylus, Ostrya.

Beukenhout. See *Cape Beech*.

BEVEL

CHAMFER

BEVEL

Bevel. 1. An adjustable tool, with blade and stock like a try-square, that can be set to any angle. 2. An angle not a right angle. See *Bevels*. A sloping, or canted, surface.

Bevel Cribbing. Boards, or sidings, bevelled on both edges.

Bevel Cut. A method of preparing the wreath for a continuous handrail. The method is now superseded by the square-cut system.

Bevelled Edge. See *Chisels*.

Bevels. A term applied to timbers of which the ends are not cut at right angles to the length, as in roof timbers, shipbuilding, etc. " Obtaining the bevels " for a rafter means finding the cuts for the two ends, *i.e.* the *foot* and *head* cuts.

Bevel Siding. See *Siding*.

Bezel. See *Basil*.

b.f. Abbreviation for *bead flush*.

Bhotan or **Blue Pine.** *Pinus excelsa* (*S.*). Himalayas. Pinkish red with darker summer wood. Straight even grain. Lustrous, fairly durable. Similar characteristics to yellow deal and used for the same purposes. S.G. ·48 ; C.W. 1·5.

Bibbs. Brackets bolted to the hounds of a mast of a ship to support the trestle-trees.

Biberine. A secretion in greenheart that protects the timber from insect attack.

Bibiru. Greenheart, *q.v.*

Bibola. Nigerian walnut, *q.v.*

Bicoca. A turret or watch tower.

Bicuiba. *Virola bicuhyba* (*H.*). C. America. Pale to reddish brown with purplish tinge, striped, lustrous, smooth. Moderately hard and heavy ; stable, durable, fissile. Straight grain, medium texture, distinct rays. Easily wrought and polishes well. A substitute for birch. Used for construction, joinery, carpentry, furniture. S.G. ·65 ; C.W. 2·5.

Bier. A wood frame on which coffins are carried.
Bigleaf. Applied to *Carya sulcata* (hickory) and *Acer macrophyllum* (Pacific maple).
Bignonia. See *Live Oak*.
Bignoniaceæ. The Trumpet-creeper family : Catalpa, Chilopsis, Crescentia, Enallagma, Jacaranda, Paratecoma, Tabebuia, Tecoma.
Big Pine. See *Sugar Pine*.
Big Tree. *Sequoia gigantea.* See *Sequoia* and *Noble Fir*.
Big Shagbark. See *Hickory*.
Bihar. See *Laburnum*.
Bijasal or **Bija.** *Pterocarpus marsupium* (*H.*). India. Yellowish brown, darker streaks, darkens with exposure, lustrous. Ripple grain, interlocked fibre. Moderately hard, heavy, coarse, and durable. Difficult to work. Constructional work, furniture, veneers. S.G. ·75 ; C.W. 4.
Bijou. Anything in miniature. Small. Less than normal size.
Bilge Block. The blocks on which a vessel rests when in a dry dock for repairs.
Bilge Keel. A projecting ridge along the side of a ship, at bilge level, to check the rolling of the ship.
Bilges. 1. The widest part of a ship's bottom. The part of a vessel where the timbers take an upward turn on each side of the keel. **2.** The bulging portion of a cask.
Bilge Saw. A special type of cylinder saw for cutting small cooperage stock.
Bilinga. Opepe, *q.v.*
Bill-board. A board protecting the plank-ing of a ship when weighing anchor. Wide elm planks to receive the anchors of a ship.
Billet. 1. An ornament in Norman architecture consisting of rows of short cylinders, or prisms, usually in a hollow moulding. **2.** A tree trunk cut up the middle and the adjacent sides or edges partly dressed, or squared. **3.** A short round log. **4.** Short timbers, split, hewn, or in the round.

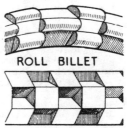

ROLL BILLET

SQUARE BILLET

Billhook. See *Bills*.
Billian. *Eusideroxylon zwageri* (*H.*). E. Indies. Also called Belian and Ironwood.
Brown, hard, heavy, very durable, and resists insects. Used for general purposes, marine constructional work, piling, etc. S.G. 1·1.
Billiard Table. A strongly framed table with slate top covered with green cloth for the game of billiards.
Bill of Lading. A formal document by which the master of a ship acknowledges receipt of timber and guarantees to deliver as stated under the terms of the bill. A " clean bill " has no marginal qualify-ing clauses. A " through bill " embraces more than one transport company.

Bills. 1. Knee timbers, *q.v.* 2. A woodcutter's tool for removing excrescences, etc. 3. The points of compasses.

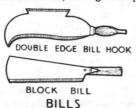

DOUBLE EDGE BILL HOOK

BLOCK BILL

BILLS

Billyboy. A Yorkshire clinker-built boat with round stern and rigged like a cutter.

Billyweb. *Sweetia panamensis.* Chichipate, *q.v.*

Bilsted. Red gum, *q.v.*

Bin. 1. A wooden box for washing sand. 2. A compartment in a wine cellar. 3. A manger. 4. A receptacle for storage purposes.

Binburra. See *White Beech.*

Bind, To, or **Binding.** Applied to any hinged frame that fits too tightly on the stop to the hanging stile, or in the rebate, and prevents easy closing. Hinge-bound.

Binder. 1. A beam carrying the common joists in a double floor. The secondary beams in a framed floor. 2. Long flexible shoots of willow, ash, etc., used for tying up bundles. 3. A springy pole used for tightening the binding chain round a load of logs.

Binding. 1. Applying size to woolly timber before glass-papering. 2. Same as To Bind.

Binding Rafter or **Timber.** A purlin. See *Roof Timbers.*

Binding Strakes. Two oak strakes near the hatch coamings of a ship's deck to strengthen the deck.

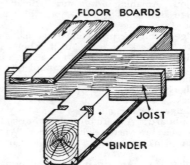

FLOOR BOARDS

JOIST

BINDER

BINDER FOR DOUBLE FLOOR

Bing. A yellowish veneer imported from France. Like bird's-eye maple with fine close grain.

Binga. *Stephegyne* or *Mitragyna diversifolia* (*H.*). India. Cream to yellow brown. Fine even texture, wavy grain. Brittle, fairly durable, not difficult to work. Should be damped before final sanding. Imports small but increasing. Used for general purposes, decorative articles, turnery, carving, light-coloured cabinet work. S.G. ·6; C.W. 2·5.

Binggas. *Terminalia sp.* Philippines. See *Kalumpit.*

Binn. Abbreviation for the name of the botanist Binnendijk.

Binnacle. The box in which the compass is kept on board ship.

Bintangor. *Calophyllum inophyllum* (*H.*). E. Indies. Evergreen. Also called Bunut, Mentangor, Pinnay, etc. About twenty-six species. Similar to Merabau. Reddish, moderately hard, strong, but not very durable. Close even grain. Used for structural work, masts, spars, cabinet work, etc. S.G. ·6; C.W. 3·5.

Biota. A species of *Thuja*, *q.v.*
Biplane. Aircraft with two main planes one above the other.
Birch. *Betula alba*,
or *B. glutinosa*, or
B. pubescens (*H.*).
N. Europe. Com-
mon silver or white
birch, whitish to light
brown. Fairly hard,
strong, tough, lus-
trous. Even and fine
grain. Not durable
if exposed. Stains
and polishes well.
Used for furniture,
fittings, turnery, carv-

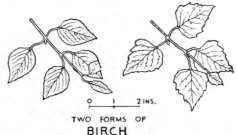

TWO FORMS OF
BIRCH

ings, plywood, etc. S.G. ·72 ; C.W. 3. Other European species
are : *B. excelsa*, *B. rubra*, *B. pendula* or *B. verucosa*, and *B. nana*
or dwarf birch. CANADIAN, QUEBEC, or YELLOW BIRCH. Also called
Black, Gold, and Hard Birch. *B. lutea*. Yellowish brown. Similar
characteristics to *B. alba*, and used for same purposes. Figured wood
valued for veneers and panels. The chief supplies of birch in this
country are Canadian. INDIAN BIRCH, *B. alnoides*, like European.
JAPANESE BIRCH, *B. alba*, *B. maximowiczei*, *B. ulmifolia*, like good
quality European. RED, or CHERRY, BIRCH, *B. lenta*. N. America.
Reddish brown. Similar characteristics to *B. lutea*. Also called
Mountain Mahogany and Black Birch. WHITE BIRCH, *B. papyrifera*.
Canada. CANOE, or PAPER, BIRCH. Inferior to *B. lutea* and only small
sizes. Chiefly used for bobbins and turnery. The terms *red* and
white often only imply heartwood and sapwood respectively. " Curly "
birch has wavy grain and is used for furniture and veneers. Figured
varieties are often marketed under special trade names : b:rnut,
flame, ice, karelian, and masur birch, *q.v.* Also see *Grey Birch, Gumbo
Limbo, New Zealand Birch, White Birch*, and *Betula*.
Birch. A cane, stick, or bundle of twigs, used for corporal punish-
ment.
Bird Cherry. *Prunus*, or *Cerasus avium*, or *C. padus*. See *Cherry*.
Bird Peck. **1.** A defect in timber, which usually increases its value,
due to birds pecking growing tree to obtain the sap. See *Bird's
Eye*. **2.** A natural inbark following a wound in deciduous trees.
Bird's Beak. A Gothic moulding resembling a bird's beak.
Bird's Eye. A decorative feature common in hard maple, and to a
lesser extent in a few other species. See *Peacock's Eye*. The figure
is due to small conical depressions in the outer annual rings, so that
the later growth follows the same contour probably for many years.
This distortion of the fibre alignment appears on rotary-cut and plain-
sawn wood as a series of small concentric circles like a bird's eye.
There are several doubtful explanations for the cause of the grain
disturbance. Embryo buds, birds in search of sap, insects, and fungi
retarding growth in small localized areas are claimed by different
authorities to be the cause. See *Maple*.

Bird's-mouth. A re-entrant angle at the end of a piece of timber to allow the end to sit astride the corner of a supporting timber.

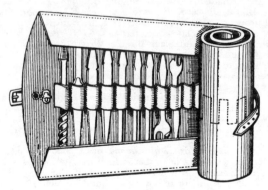

RAFTER

WALL
PLATE

BIRDSMOUTH

Birnut. A trade name for specially darkened birch for cabinet work.

Bischofia. *Bischofia javanica* (*H.*). India, Burma. Called Bishopwood, *q.v.*

Biscuit. A term used in aircraft for small plywood cover plates glued over butt joints to give a more secure fixing.

Bishop. A rammer consisting of a heavy cylindrical piece of wood with spindles at the top and side for handling.

Bishop's Mitre. A mason's mitre in which a prominent member of the moulding is produced beyond the intersection. See *Stops* and *Mitres*.

Bishopwood. *Bischofia javanica* (*H.*). India. Reddish brown, purplish tinge. Fairly hard and heavy, durable and strong. Coarse, even grain, interlocked fibres. Odorous during conversion. Easy to work when green. Resembles black walnut. Used for constructional work, piling, sleepers, carving, etc. S.G. ·74 ; C.W. 3·5.

Bit Case or **Roll.** Brace bits should be kept in a roll or box to protect

BIT CASE OR ROLL

the cutting edges. A roll of leather, moleskin, or baize is the best for small bits.

Bitch. See *Dog.*

Bit Depth Gauge. An appliance attached to the stem of a brace bit to control the depth when boring.

Bite. Applied to the cutting action of tools, especially boring tools. If the tool " bites " keenly it is in good condition.

Bits. The small interchangeable tools used in a brace, drilling machine, etc. Auger or twist, bit-stock, Broughton's, countersink, centre, Clarke's expansion, dovetailing (machine), dowel or twist, dowel rounder, expansion, flat drill, Forstner, Gedge's auger, Irwin's auger, Jenning's auger, Leadbeater's auger (for railway carriage work), nose, rimer, rose, Russel-Jenning's, sash, screwdriver, shell, slotting, snail countersink, solid wing auger, spoon, Steer's, Swiss, taper.

Bit Stop. See *Bit Depth Gauge.*

Bitt. A deck fixture for securing a line. See *Bitts.*

Bittacle. Same as binnacle, *q.v.*

Bitter Almond. See *Red Stinkwood.*

Bitternut. See *Hickory.*

Bitter Pecan. *Carya aquatica.* See *Hickory.*

Bitterwood. *Picræna officinalis (H.).* British Guiana. Called Bitter Quassia. Light yellow colour, easily wrought, resists insects, not very durable, polishes well. Very good for interior work. S.G. ·5. See *Simaruba* and *Quassia.* Also see *Macaya* and *Marupa.*

Bitterwood Family. Simarubaceæ, *q.v.*

Bitts. Two strong vertical timbers for the cables of a ship when at anchor. In addition to cable-bitts there are carrick-bitts and pawl-bitts. A fixing for the belaying of hawsers, warps, etc.

b/l. Abbreviation for *bill of lading, q.v.*

Bl. Abbreviation for the name of the botanist Blume.

Black. A prefix to certain species of the following timbers : alder, ash, bean, birch, cherry, cottonwood, cypress, gum, hickory, iron-wood, locust, maple, matai, oak, pine, poplar, spruce, stinkwood, walnut, willow. The term may be applicable to either timber or bark, and may be only used to distinguish one species from another. It often has no reference to the colour of the timber.

Black Afara. Idigbo, *q.v.*

Black Alder. Common alder. *Alnus glutinosa.*

Black Bean. *Castanospermum australe (H.).* E. Australia. Dark brown with lighter and darker streaks and mottles. Hard, fairly heavy, straight grain, close texture, strong, durable, resists insects. Not difficult to work to a lustrous finish for polishing, but its greasy nature makes gluing uncertain. Large sizes. Shrinks and warps freely. Used for high-class cabinet work, electrical appliances because of good insulating properties, veneers, panelling, and decorative joinery, carvings, turnery. S.G. ·8 ; C.W. 3·5.

Blackboard. Any dark surface for demonstrating with chalk in a schoolroom. They were originally of wood, painted black.

Black Box. *Eucalyptus bicolor (H.).* Ironbark box. One of the best timbers from Australian eucalypti. Very hard, heavy, strong, and durable. Used for all purposes from heavy engineering work to cabinet work. See *Eucalyptus* and *Ironbarks.*

Blackboy. *Xanthorrhœa preissii (H.).* W. Australia. The Australian grass tree. Provides valuable distillates, resin, and fibre, but not timber.

Black Bug. *Lecanium sp.* Larger than brown bug. Closely related to *lac* insect. Widespread. Very destructive to hardwoods as it extracts the sap of the growing tree.

Blackbutt. *Eucalyptus pilularis* and *E. patens (H.).* Australia. Pale yellow. Hard, dense, very durable and tough, difficult to split and work. Used for sleepers, paving, piling, structural work. S.G. ·9 ; C.W. 4.

Black Cabbage Bark. *Lonchocarpus castilloi* (*H.*). C. America. See *Cabbage Bark*.

Black Carpenter Ant. *Formica fuliginosa.* Very destructive to many hardwoods.

Black Carpenter Beetle. *Alaus sp.* Very destructive to timber in buildings. The beetle bores holes up to 1 in. diameter. Easily caught, and checked by paraffin and tar.

Black Check. Bark pockets containing resin. Also called Black Streak.

Black Cherry. See *Cherry*.

Black Chuglum. *Terminalia manni* (*H.*). E. Indies. Very similar to, and marketed as, white bombway, *q.v.*

Black Gum. *Nyssa sylvatica.* See *Tupelo*.

Blackheart. A defect in the heartwood of timber, especially in ash.

Black Ironbox. *Eucalyptus raveretiana.* See *Ironbox*.

Black-maire. *Olea cunninghamii.* New Zealand olive (*H.*). Light brown with beautiful figure. Characteristics of lignum vitæ. Used for veneers.

Black Mangrove. *Avicennia nitida.* See *Mangrove*.

Black Poplar. *Populus nigra.* ITALIAN BLACK POPLAR, *Populus serotina.* See *Poplar*.

Black She Oak. *Casuarina suberosa.* Queensland. See *She Oak*.

Black Strakes. Planks immediately above the wales in the side of a ship.

Blackthorn. *Prunus spinosa* (*H.*). Britain. A species of wild plum. Also called Sloe. Of little value as timber.

Black Varnish Tree. See *Rengas*.

Black Walnut. See *Juglans* and *Walnut*.

Black Wattle. *Acacia decurrens.* See *Wattle*.

Black Willow. *Salix nigra.* See *Willow*.

Blackwood. *Acacia melanoxylon* (*H.*). Australia and Tasmania. The Black Sally of N.S. Wales. Also called Hickory locally. Reddish brown to nearly black, with streaks of various colours from yellow to black. Hard, strong, lustrous, even but rather coarse texture, fiddleback, liable to warp, not difficult to work or carve. Often with interlocked fibres with ribbon and curly grain. Fiddleback very valuable and considered the most beautiful of timbers. Used for veneers and decorative work, plain timber for superior flooring, cooperage, bentwork, etc. S.G. ·8 ; C.W. 3·5. BOMBAY BLACKWOOD, *Dalbergia latifolia* (*H.*). Indian rosewood or shisham. Purplish or nut brown with markings of various colours. Hard, heavy, strong, fragrant, durable, and resists insects. Beautiful timber, especially if with ribbon grain. Even texture with smooth finish. Difficult to convert and work because of secretion and cross grain. Used for all kinds of decorative work and veneers, or where strength and durability and good appearance are required. S.G. ·85 ; C.W. 4. See *Dalbergia sissoo* and *White Bombway*. S. AFRICAN BLACKWOOD, *Royena lucida* (*H.*). Yellowish brown with darker stripes. Similar properties to Australian blackwood. The W. INDIES BLACKWOOD is a species of ebony, greenish black or brown. The E. AFRICAN BLACKWOOD, *Dalbergia* *melanoxylon*, is an excellent hardwood with similar properties to ebony, *q.v.* Also see *Logwood, Mangrove*, and *Rosewood*.

Blade. The metal part of a combined wood and metal tool, as the blade of a chisel or try-square. A thin piece of wood or metal fixed in a thicker support. The wide operating part of an appliance, as the blade of an oar, paddle, etc. See *Spokeshave*.

Blades. A term sometimes applied to principal rafters.

Blank or **Blanks.** **1.** A machine cutter before it is ground to the required shape. **2.** Wood roughly sawn to sizes for a definite purpose. Roughly sawn pieces to be turned in a lathe, or otherwise prepared as a unit.

Blank Door or **Window.** A recess in a wall having the appearance of an opening bricked up.

Blanket-leaf. *Bedfordia salicina* (*H.*). Victoria, Australia. Used for small cabinet work. Not exported.

Blanket Rate. A rate, or payment, to cover several specified services. A comprehensive quotation.

Blaze. **1.** To mark a tree so that it can be distinguished by another person, for felling or to mark a footpath or boundary. **2.** The loose wood, or core, from mortising.

Bled Timber. From trees that have been *tapped* for their gum, resin, or sap. Trees tapped for resin, etc.

Bleeding. **1.** Applied to preservatives exuding from the surface of treated timbers. **2.** Gum, resin, or sap exuding from trees.

Bleeding Heart. Maca-wood, *q.v.*

Blemish. Anything that mars the wood but is not serious enough to be classed as a defect.

Blepharocarya. See *Rose Butternut*.

Blight. Any disease that destroys plant life.

Blind. **1.** A screen to a window or door, either inside or outside, for protection or privacy. There are numerous types : Canaletti, Florentine, Helioscene, Pinoleum, Spanish, Venetian, *q.v.* **2.** Sometimes applied to defective articles, as a screw head without a slot.

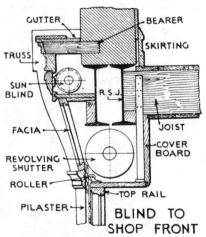

BLIND TO SHOP FRONT

Blind Bolt. A flush bolt let into the edge or back of a cabinet door.

Blind Boxes. Framed boxes into which shop window blinds roll, or fold, when not in use.

Blind Dovetail. A secret dovetail, *q.v.*

Blind Laths. Thin laths for making Venetian blinds.

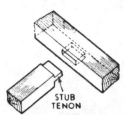

BLIND MORTISE

Blind Mortise. A mortise for a stub tenon.

Blind Rail. See *Rolling Shutter*.

Blind Storey. See *Triforium*.

Blister. 1. A small excrescence on the bark of a tree. 2. A bubble on a veneered surface. 3. A projecting observation pane in large aircraft.

Blister Figure. An uneven, or blister-like, appearance on planed timber due to unevenness in the annual rings. The figure is only obtained on flat-sawn or rotary-cut surfaces, and appears as depressed or elevated small areas of rounded contour. It is common in birch and maple.

Block. 1. A short piece of timber of large square-like section. 2. The stump of a tree. 3. See *Cutter Block*. 4. See *Mitre Block*. 5. See *Blocks*.

Blockboard. A variation of laminboard, with the core formed of square wood strips glued together, with grain alternating, and glued between two veneers. See *Laminboard*.

Block Design. See *Veneer Matching*.

Block Floor. A floor formed of wood blocks on concrete. The blocks are secured to the concrete by a bituminous mastic applied hot,

BLOCK BOARD

and they may be obtained in either softwood or hardwood. Thin pieces of hardwood are often fixed to floorboards to imitate a block floor.

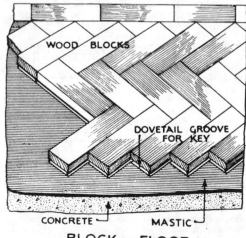

BLOCK FLOOR

48

Blockhouse. A strong building built of logs, usually for defensive purposes.

Blockings or **Blocks.** **1.** Small triangular, or square, fillets of wood glued in the angle formed by two thin pieces to secure the two pieces together. See *Angle Joints,* (*e*). **2.** The timbers supporting the roof of an underground excavation.

Block Mounts. Pieces of hardwood, usually mahogany, for mounting type in typography. See *Wood Cuts.*

Block Plane. A small metal plane useful for trimming mitres and end grain. The angle between the *single iron* and the sole is small, and the grinding angle is on top so that the iron is supported to the cutting edge. Both iron and mouthpiece are adjustable.

Blocks. Pulleys or systems of pulleys. Wood blocks consist of shell (elm), sheave and pin (metal or lignum vitæ), and strap. The various types used on board ship are : bee, bull's-eye, blunt-line, cat, cheek, clew-garnet, clew-line, dead-eye, deep-sea-line, fish, jear, jewel, leech-line, long tackle, main-sheet, nine-pin, shoulder, sister, snatch. See *Pulley Tackle, Block,* and *Blockings.*

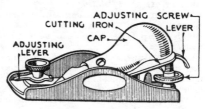

BLOCK PLANE

Block Saw.—One with a continuous grip to which the blade is screwed, instead of having a handle and back as in a dovetail saw. It is used for work on the floor near to a wall, or with a mitre block. See *Veneer Saw.*

Blolly. *Pisonia obtusata.* Also called Beefwood, Pigeonwood, and Corkwood, *q.v.*

Bloodwood. *Eucalyptus corymbosa* and *E. latifolia* (*H.*). Australia. Characteristic of other Australian eucalypti, but small trees. Very hard, heavy, and durable. Gum veins. Used for piles and heavy constructional work. Not exported. C. AMERICAN BLOODWOOD, *Pterocarpus spp.* Light brown, lustrous, variegated patches due to injury, smooth. Very variable in weight and hardness, but usually soft and light, not durable. Fairly straight grain, rather fine texture. Easily wrought. Little exported. RHODESIAN BLOODWOOD. See *Muninga* or *Kajatenhout.*

Blue. See *Ash, Mahoe,* and *Pine.*

Blue Beech. *Carpinus caroliniana* (*H.*). U.S.A. Pale brownish white. Fairly hard, heavy, and strong. Only of local use.

Blueberry Ash. See *Mountain Ash.*

Blue Fig. *Elæocarpus grandis* (*H.*). N.S. Wales. Not exported. Used for many purposes, including aircraft construction, in Australia.

Blue Goods, Lumber, or **Timber.** Coniferous timber in which the sapwood is discoloured through lack of seasoning before shipment. The blue colour is due to fungus agency (*Ceratostomella* and *Graphium spp.*). The fungus only attacks the sapwood and does not cause decay, but such timber should be treated for exterior work. A temperature of 130° F. destroys the fungi, and there will be no further infection if the wood is kept dry. See *Seasoning* and *Preservation*.

Blue Gum. *Eucalyptus globulus* (*H*.). Tasmania and S. Australia. Pinkish brown. Very hard, tough, strong, fire-resisting, and durable. Straight close grain, sometimes with interlocked fibres. Very large sizes. Used for general purposes, heavy constructional work, piling, sleepers. Ripple figure used for cabinet work. S.G. ·9 ; C.W. 4·5. See *Gum*.

Blue Mahogany. Mahoe, *q.v.*

Blue Mallee. *Eucalyptus polybractea*. Victoria. Used chiefly for oil distillation. See *Mallee*.

Blue Pine. See *Bhotan*.

Blue Print. See *Printing*.

Blue Stain. See *Blue Goods* and *Stain*.

Blush Condoo. See *Sweetbark*.

Blush Tulip Oak. *Tarrietia actinophylla*. Queensland. See *Tulip Oak*.

Blush Walnut. *Beilschmiedia obtusifolia*, or *Cryptocarya o.*, or *Nesodaphne o.* (*H*.). Queensland. Pinkish, smooth. Hard, heavy, moderately durable and strong. Fine grain and texture. Polishes well. Not difficult to work, but hard on saws in conversion. Requires careful seasoning. Used for interior joinery, flooring, turnery, etc. S.G. ·8 ; C.W. 3.

Blythe's Process. A preservative process in which the sap and moisture are extracted from the timber and replaced by carbolic acids or tar. See *Preservation*.

b.m. An abbreviation for *bird's mouth*, *bending moment*, and *board measure*.

Board. Applied to converted *softwoods* over 4 in. wide and less than 2 in. thick, and to *hardwoods* of any width and up to 1¼ in. thick. See *Market Sizes, Floor-, Match-*, and *Weather-boards*.

Board and Batten. A term applied to the boarding of timber houses when the boards are alternately thick and thin to give a recessed effect. The thin boards are usually in grooves in the thick boards like a panel.

Board Foot or **Board Measure.** The unit of measure for timber in N. America, 1 ft. square by 1 in. thick, or the equivalent of 144 cub. in. See *Measurement*.

Boarding. Any kind of thin, comparatively wide, boards for building up a wide surface.

Boarding Joists. Common joists, *q.v.*

Board Rule. A flat metal-tipped flexible rod, 3 ft. long, graduated for readily obtaining the superficial area of boards. Used more in America than in this country.

Boat. **1.** A scaffold cradle. **2.** A small open vessel propelled by oars or sails : jolly, long, racing, rowing, etc. A ship's boats may also include one or more of the following : barge, cutter, gig, launch, pinnace, lifeboats. **3.** A general name for any kind of craft, whether for use on sea, river, or lake, and no matter how propelled through the water.

Boat-house. A shed to house small boats when not in use.

Boat Skids. Long square pieces of softwood on which spare masts, boats, etc., were stowed on board wooden ships.

Boat Skin. Thin boards for boatbuilding.

Bobbin. **1.** Any type of reel, spool, or tube on which woollen, silk, or cotton, yarn or thread is wound. **2.** A boxwood ball used when bending lead pipes.

Bobbin Papering Machine. A cylinder covered with glass-paper and fixed to a revolving spindle with reciprocating motion, for *finishing off* curved work.

Bobbin Wood and Squares. Moderately hard, non-resinous, close-grained wood such as alder, ash, and birch of small sizes. The stuff is cut into " props " and then into " bobbin squares," for turning into spools, bobbins, etc.

Bobsleigh. A sledge to carry several persons down snow-covered tracks by gravity.

Bobstay. A stay used to keep the bowsprit down.

Bocote. *Cordia gerascanthus.* See *Cyp.*

Bodark. Osage orange, *q.v.*

Bodger. A turner of chair legs, etc., in woodcraft.

Body. The main part of any structure, or appliance, to which other parts are attached.

Bodying in. Filling in the grain of wood during the process of french polishing. See *Filler.*

Boehmeria. See *Sedeng.*

Bog. A small fishing boat used on the south coast.

Bogie. A four- or six-wheeled truck with a central pivot to provide free movement.

Bog Oak. Ordinary oak, turned black by chemical changes due to being buried for centuries in bogs. It is valuable for furniture and fancy articles. Other woods than oak, similarly changed, are also appreciated for various purposes.

Bohoi. Seraya, *q.v.*

Boiling. See *Preservation.*

Bois or **Boise.** A prefix to numerous S. and C. American woods. See *Amaranth, Osage Orange, Rosewood, Satiné, Simaruba, Tulipwood, Waterwood, Zebrawood.*

Bois Durci. A French process for making decorative ornaments, imitation carvings, etc., for cabinet work. Fine rosewood or ebony sawdust moistened with blood and water is pressed into moulds and baked.

Boja. See *Pyinkado.*

Bole. **1.** The stem, or trunk, of a tree when of timber size, *i.e.* over 8 in. diameter. **2.** A recess or opening in a wall.

Bolection Moulding. A rebated panel moulding, the face of which stands above the face of the framing. See *Panelled Door.*

Bolinder Cutters. A circular cutter of Swedish pattern. It has six or eight teeth backed off so that sharpening does not alter the profile of the moulding. Commonly used for tongues and grooves, beading, etc.

Bollards. **1.** Mooring posts on a wharf, pier, or jetty, or on a ship.

2. A strong wooden post at the bow of a whaleboat for the harpoon line. See *Knight Heads*.

Bolly Silkwood. *Cryptocarya oblata* (*H.*). Queensland. Resembles maple silkwood, *q.v.*, but slightly inferior. Also called Macquarie Maple and Mazlin Beech. See *Silkwood*.

Bollywood. *Litsea reticulata.* See *Queensland Sycamore*.

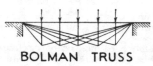

BOLMAN TRUSS

Bolman Truss. A trussed beam used for timber deck bridges. The load is carried on the top chord.

Bolo. *Ochrocarpus africanus* (*H.*). Nigeria. Also called Ologbomidu. Reddish brown. Moderately hard, heavy, strong, and durable. Interlocked fibrous grain. Mild cutting, but rather difficult to work smooth. Under investigations. S.G. ·7.

Bolongeta. *Diospyros pilosanthera.* A black cabinet wood from the Philippines. Similar to camagoon, *q.v.*

Bolster. **1.** A short horizontal timber on top of a post or pile to provide a bearing surface. See *Head Tree*. **2.** The lateral part joining the volutes of an Ionic capital. **3.** Same as baluster. **4.** See *Bunk*. **5.** A protecting timber for the cable on board ship. **6.** Same as *bed* in wheelwright's and carriage builder's work. **7.** See *Sett*.

Bolt. **1.** A segment of a short log, either sawn or split: Billets cut to the length required for some particular purpose. **2.** A cylindrical bar used to secure a door or window. **3.** Strong metal fastenings consisting of a cylindrical shank with a head, and threaded for a nut. They are named according to the shape of the head, as *square, round, hexagonal, pan,* etc., and the dimensions of the shank is the size of the bolt. **4.** A curved piece of hardwood used when bending lead pipes. See *Dead, Tower, Tie,* and *Handrail Bolts,* and *Coach Screw*.

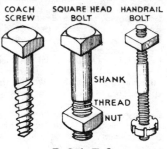

COACH SCREW SQUARE HEAD BOLT HANDRAIL BOLT

SHANK

THREAD

NUT

BOLTS

Boltel, Boultine. A quarter-round moulding.

Bolter. A machine consisting of one or more circular rip saws. It is used for cutting small squares for furniture stock, etc. See *Knee Bolter*.

Bolting Iron. A drawer-lock chisel, *q.v.*

Bolting Up. Mortising a drawer rail to receive the bolt of a lock.

Bolusanthus. See *Wistaria*.

Bombacaceæ. The Silk Cotton Tree family : Bombacopsis, Bombax, Camptostemon, Catostemma, Cavanillesia, Ceiba, Chorisia, Cœlostegia, Cumingia, Durio, Gossampinus, Hampea, Montezuma, Myrodia, Ochroma, Pachira, Quarariba.

Bombacopsis. See *Saqui*.

Bombax. *B. malabaricum* (*H.*). India. White, but darkens with exposure. Very soft, light, and not durable. S.G. ·32. See *Didu*, *Imbirussu, Pompoula, Saqui, Silk Cotton Tree.* Also see *Ceiba.*

Bombay Blackwood. Indian rosewood. See *Blackwood.*

Bombe. An arched or convex surface, especially in Louis XV. period furniture.

Bombway or **Bombwe.** See *White Bombway.*

Bonded. See *Gluing* and *Resin-bonded.*

Bonding Timbers. Long timbers built in thin walls to act as ties.

Bone Dry. Timber from which the moisture has been extracted as much as possible with normal seasoning and stoving.

Bongo. *Cavanillesia platanifolia* (*H.*). C. America. Yellow, brown streaks. Very light and soft, weak, brittle, not durable. Very porous, and coarse texture. Difficult to work smooth. Somewhat like balsa, but inferior. S.G. ·15. Also called Quipo.

Bongor. *Lagerstrœmia spp.* E. Indies. Several species. Also called Bunga or Bungor. Dark grey, brownish, or pale red. Moderately hard and heavy, coarse grain. Quality improves with age of tree. Good substitute for teak. Used for cabinet work, good-class construction, piling, sleepers. S.G. ·5 ; C.W. 3.

Bongosi. Ekki, *q.v.*

Boning. **1.** Testing a surface for twist or winding. **2.** Obtaining levels for excavations.

Boning Pegs. Small hardwood cubes used by the mason when preparing plane surfaces.

Boning Rods. Tee-shaped appliances for *sighting* levels and falls in excavations.

Bonnet. **1.** Sometimes applied to the roof of a bay window. **2.** The covering to a cage in a mine to protect the occupants from falling debris.

Booby Hatch. A portable shelter to a temporary stairs on board ship.

Bookcase. Any arrangement of encased shelves for books, and usually forming a self-contained piece of wall furniture. It is usually provided with glazed doors, either hinged or sliding, otherwise the term bookshelves is used.

Book Ends. Two decorative pieces of hardwood used to keep a number of books upright on a table. They are usually provided with sheet-metal plates bent at right angles. The outer books rest on the plates and so prevent movement.

Book Matched. A term used in veneering when adjacent sheets from a flitch are opened like opening a book. The back of one sheet is matched with the face of the next sheet. This gives a light and dark effect due to the light reflecting from the fibres which slant in opposite directions. See *Matching.*

Bookstall. A stall in the open air for the sale of books.

Bookstand. Same as bookcase, but implying a self-contained piece of furniture not necessarily encased or against a wall.

Boom. **1.** The horizontal members of a trussed girder. **2.** A floating barrier of timber across a river. **3.** A " boom of logs " is a number of logs tied together by cross timbers for floating down a river. **4.** A long pole, or spar, used to keep the bottom of a sail extended.   They are distinguished as bumkin, driver, flying-jib, jib, main, ring-tail, spanker, square-sail, and studding-sail booms. **5.** Structural members supporting the tail unit in aircraft. A beam or spar.

Boot. **1.** A box seat at the back of a chaise in which luggage can be placed. **2.** The part carrying the driving seat of a vehicle. The panel below the seat line.

Booth. A temporary shelter, either of wood or canvas.

Boot-tree. A wood last for keeping boots and shoes in shape when not in use.

Borassus flabellifer. See *Palmyra Palm*.

Border. An edging used for strength or ornament.

Bordered Pits. Thin places in the radial cell walls that allow for the conduction of moisture through the stem of a tree. See *Pits*.

Bordex. A registered fireproof, or waterproof, wallboard, *q.v.*

Borer. A hollow auger that extracts a section to show the annual rings of a standing tree. See *Boring Machine*.

Borers. Wood-boring larvæ. See *Marine, Pinhole*, and *Shothole Borers*.

Boretree. See *Elder*.

Boring. Making holes in timber with brace-bit, gimlet, bradawl, or by machine. See *Machines*.

Boring Machine or **Borer.** **1.** A machine with adjustable table and spindles carrying augers. Several holes may be bored at one operation. **2.** " Boring apparatus " applies to an adjustable table for carrying the stuff whilst boring with an auger fixed in a saw-bench spindle. See *Mortising Machine*.

Boriti. Mangrove poles from E. Africa.

Borkh. Abbreviation for the name of the botanist Borkhausen.

Borneo. Chief timber exports : billian or Borneo ironwood, camphorwood or kapor, cedar or red seriah or Borneo mahogany, ebony, kruen or kruin, sandalwood, merabau or mirabau or

BOSS

selangan batu, white cedar or white seriah or serajah or gagil. There are many other timbers, characteristic of the East Indies, but they are not exported in important quantities to this country.

Borning. See *Boning*.

Borraginaceæ. A family including Auxemma, Cordia, Echiochilon, Ehretia, Heliotrium, Lithospermum, Patagonula, Tournefortia.

Borrowed Light. An interior window obtaining light from another window.

Bosnian Maple. *Acer platanoides*. Norway maple *q.v.*

Boss. **1.** A carved block covering the intersection of two or more cross timbers in vaulting and ceilings, or at the end of a cantilever as in a balcony. **2.** A small circular projection on a pattern for moulding. **3.** A small boxwood inlay at the front of a jack or try plane for striking with the hammer to release the irons.

Bosse. *Guarea cedrata.* See *Guarea.*

Bostrichus. A small insect that attacks the alburnum of felled pines and firs. There are several varieties that propagate rapidly and are very destructive.

Bostrychidæ. A family including powderpost beetles, *q.v.*

Bosuga. *Fagara monophylla* (*H.*). C. America. Brown, lustrous. Hard, heavy, strong, tough. Similar to W. Indian satinwood. Straight grain, fine and even texture, slightly roey. Suitable for furniture, superior joinery, musical instruments, veneers. Substitute for ash or hickory. Not yet imported. S.G. ·8 ; C.W. 3.

Boswellia. See *Salai.*

Botany. The science of plants. Trees are classified according to the family, genus, and species, based on growth, structure of wood, and vessel content. The name of the authority for the classification is often given in abbreviated form. When dealing with timber the family is usually omitted, the genus is given first and then the species, *e.g.* Beech belongs to the family Fagaceæ, genus *Fagus*, species *sylvatica*, and the name of the botanist is *Linnæus*, hence the classification is **Beech,** *Fagus sylvatica, L.* A family often embraces a large number of different kinds of trees, but they all have some common peculiar characteristic. There are often several variations of a species ; and soil and climate have a great bearing on the quality of the timber. Different botanists have classified the same species differently in many cases.

Bottle Butted. See *Swell Butted.*

Bottle Turning. A detail somewhat resembling a bottle and characteristic of the William and Mary period.

Bottom Board. A board on which the bottom box is rammed over a pattern when moulding.

Bottom Rail. The lowest rail in timber framing. See *Rail.*

Bottomside. The bottom rail of railway carriage framing.

Boucherie. A preservative process in which the sap is expelled from the timber by fluid pressure and a solution of copper sulphate forced into the wood. See *Preservation.*

Bouge. The widest part of a stave for a cask.

Bough. A branch of a tree.

Boule. The French name for timber converted for superior work, cabinet making, etc.

Boule-work. Cabinet work inlaid with other materials than wood, such as ivory, tortoise-shell, mother-of-pearl, or metal.

Boultine. See *Boltel.*

Bourra courra. See *Letterwood.*

Bourtree. The elder tree.

Bow. **1.** A curved member of a frame. See *Belfast Truss.* **2.** The fore end of a ship. **3.** A bent flexible piece of wood, usually of yew, strung to propel an arrow in archery. **4.** A long slender piece of wood strung with horsehair for playing stringed instruments, as a violin.

Bowdichia. See *Sucupira.*

Bowie. A 60-gallon meal barrel.

Bowing. Applied to wood that is warped in the direction of its length only. See *Warp*.

Bowl. **1.** A basin-shaped utensil made of wood or other material. It is designed and used for a variety of purposes, usually for holding liquids or granular materials. **2.** A heavy wood ball, weighted on one side to give it a bias, to roll on grass in the game of bowls. They are usually of lignum-vitæ.

Bow Plane. A compass plane, *q.v.*

Bow Saw. A saw for cutting circular work. It consists of

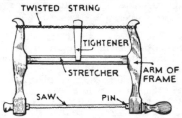

TWISTED STRING

TIGHTENER

STRETCHER — ARM OF FRAME

SAW   PIN

BOW OR TURNING SAW

two hardwood ends with a stretcher between the saw-blade and a twisted string. The string gives the required tension to the blade.

Bowser. See *Toggle*.

Bowsprit. A strong spar projecting forward over the bows or stem head of a sailing ship, or a steamship with a cut-water stem.

Bowstring Roof. See *Belfast Truss*.

Bowtell. **1.** The shaft of a clustered column. **2.** A plain round moulding.

Bow Window. A semicircular or segmental bay window.

Bow-wood. *Sageræa elliptica.* Andaman Islands (*H.*). Yellowish white. Hard, heavy, straight grain, fine texture. Bends easily. Not easy to work. Used for turnery, tools, etc. S.G. ·8. Also see *Washiba*.

Box. **1.** See *Boxwood*. **2.** See *Packing Case*. **3.** See *Sash and Frame*. **4.** See *Mitre Box*. **5.** "To box" usually means "to enclose." See *Boxed Plane* and *Boxed Heart*. **6.** The driver's seat on a carriage. **7.** See *Boxing*. **8.** A compartment in a theatre to accommodate several people.

Boxboard. Akomu. See *Nigerian Boxboard*.

Boxboards or **Caseboards.** **1.** Bundles of boards ready for assembling into boxes, cases, crates, etc. Also called Shooks. **2.** Sawn stuff from 4 to 8 ft. long and 4 in. or more in width. **3.** A classification of boards for wagon stock, governing sizes and quality, by the National Hardwood Lumber Association, U.S.A. *Boxboard* is applied to several machines used in box-making; for branding, printing, trimming, etc.

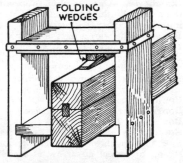

FOLDING WEDGES

BOX CLEAT

Box Cleat. A strong box-like frame inside which glued joints are wedged until the glue has set.

Box Dam. A rectangular coffer dam for underwater work.

Boxed Frames. Hollow built-up frames to receive the weights for vertically sliding sashes. Also called Cased Frames. The illustration shows a bottom corner of a boxed frame and part of the lower sash. The weights slide vertically in the box. The " feather " is used to prevent the weights from fouling each other. See *Sash and Frame*, *Inner Lining*, and *Pocket*.

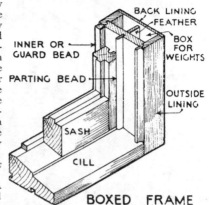

BOXED FRAME

Boxed Heart. See *Box the Heart*.

Boxed Heart Check. A heart shake in converted timber.

Boxed Hearth. One in which the concrete is supported on rough boards, on short bearers between trimmer and wall.

Boxed Mullions. Built-up mullions in which the weights slide for sliding sashes.

Boxed Plane. A wooden plane in which the mouthpiece is closed with a hardwood inlay, usually of boxwood. " To box " a plane means to insert the inlay.

Boxed Tenon. A right-angled tenon on a corner post. Also called Box Tenon, *q.v.* (See illus., p. 58.)

BOXED HEART

Box Elder. *Acer negundo.* Ash-leaved maple. It is of little commercial importance. See *Maple* and *Acer*.

Boxformers. Hollow built-up ribs of plywood to which the shell of the fuselage of an aeroplane is attached.

Box Gutter. A rectangular or parallel gutter, the sides of which are formed by a parapet wall and pole-plate. (See illus., p. 58.)

Box Horse. A box-shaped support for a barrow-run or inclined scaffold.

Boxing. 1. A recess to receive folding window shutters, or *boxing* shutters. 2. Tapping pine trees for turpentine or resin. 3. See *Box the Heart*.

Boxing Router. An improved form of coachmaker's router.

Boxing Shutters. Folding window shutters, *q.v.*

Boxing-up. Building up a large pattern from thin stuff instead of using solid wood.

Box Scarf. See *Lap Scarf*.

Box Scraper. An adjustable scraper fixed in a handle.

Box Shooks. Wood cut to special sizes for boxes.

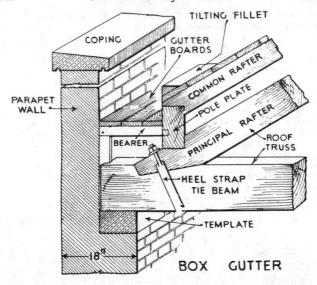

TILTING FILLET

COPING

GUTTER BOARDS

COMMON RAFTER

POLE PLATE

PRINCIPAL RAFTER

PARAPET WALL

BEARER

ROOF TRUSS

HEEL STRAP

TIE BEAM

TEMPLATE

18″

BOX GUTTER

Box Staple. The metal receiver for the bolt of a rim lock or dead lock.

Box Tenon. See *Boxed Tenon*.

Box the Heart. **1.** Converting a log by leaving a square piece at the heart. In some timbers, as eucalyptus, this is advisable, and about 4 in. square should be scrapped. **2.** Converting a log so that the centre of the heartwood is wholly contained in one piece. (See illus., p. 57.)

Boxwood. *Buxus sempervirens* (*H.*). Common box. Also called Abassian, European, Turkey, and Persian Box. S. Europe and Asia Minor. Pale yellow to orange colour. Very hard, with close even

BOX TENON

grain, and lustrous. Small sizes. Used for turnery, tools, instruments, engraver's blocks, carving, etc. S.G. 1; C.W. 4. See *Brush, Cape, Ceylon, Chinese, Indian, Kamassi, Maracaibo, Orange, Red, San Domingan, Swamp,* and *Yellow Box*. The name boxwood or box is applied to many other woods having similar characteristics : amarillo, blue gum, dogwood, erin, gardenia, ibira-nira, jacaranda, etc. Also see *Casewood*.

B.P. A registered wallboard, *q.v.*

b/p. Abbreviation for *bills payable*.

Br. Abbreviation for the names of the botanists R. and P. Brown.

b/r. Abbreviation for *bills receivable*.

Brace. **1.** A cranked mechanical contrivance for controlling boring bits. The most useful type is the ratchet brace with crocodile jaws and ball bearings. **2.** A member of a framed structure, crossing a space diagonally, and able to resist tension or compression. See *Bricklayer's Scaffold.*

Brace Bits. See *Bits.*

Brace Blocks. *Keys* used in beams that are built-up in depth. They are used to prevent sliding due to horizontal shear.

RATCHET BRACE

Brachychiton. See *Kurrajong.*

Brachylæna. See *Real Salie.*

Brachystegia. See *Aku, Okweni,* and *Zebrano.*

Brack. See *Grading.* Classifying timber into different grades.

Bracked or **Bracking.** Same as grading. Also implies a quality below the regular qualification.

Bracket. **1.** A support projecting from a vertical surface to carry a horizontal or inclined surface or a vertical load. The bracket may be solid or framed, and plain or ornamental. **2.** Rough supports nailed to the carriage to support the steps of a flight of stairs. **3.** The ornamental return of a riser for a cut string. See *Bracketed Stairs.* **4.** Pieces similar to knees fixed in the head of a ship to carry the gratings and to support the gallery. **5.** The support to the footboard of a vehicle.

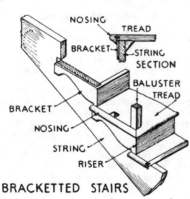

BRACKETTED STAIRS

Bracketed Cornice. A lath and plaster cornice carried by brackets fixed to wall and ceiling.

Bracketed or **Bracketted Stairs.** A flight of stairs in which the string is shaped to the outline of the steps and having ornamental brackets under the returned nosings.

Bracket Scaffold. A large bracket temporarily fixed to a wall to carry scaffold boards.

Bracking. See *Brack.*

Bracted Fir. *Abies bracteata.* See *Fir* and *Noble Fir.*

Brad. A cut nail parallel in thickness but tapering in width and with head projecting on one edge only. They are stamped from a sheet of metal and are usually **2½** in. long, for flooring. See *Nails.*

Bradawl. A hand-boring tool for small holes such as nail and screw holes.

Brail. A section of a raft of logs.

Brake. **1.** A contrivance to prevent a wheel from turning. **2.** A vehicle for pleasure parties, like a wagonette. **3.** A special type of

vehicle without body for breaking in young horses. **4.** A thicket.
5. A heavy harrow.

Brake Block. A hardwood block used to arrest motion, especially of wheels.

Brake Van. A special compartment on a railway train from which the brakes are controlled.

Branca. *Folha de Bolo* (*H.*). S. America. Red, hard, heavy, strong, and very durable. Used for general purposes, piling, sleepers, etc. S.G. ·9 ; C.W. 3. Also see *Arariba* and *Vermelha*.

Branched Knot. Two or more knots springing from a common centre.

Branched Work. Carved ornaments in the form of leaves and branches, used in panels and friezes.

Brand, Branding, Brands. See *Shipping Marks* and *Quality*.

Brandering. 1. Small fillets nailed on wide timbers before lathing for plaster. **2.** Counter-lathing.

Branding Machines. See *Machines*.

Brandreth. A rail, or fence, round the opening of a well.

Brashy. A term applied to broken or coarse fibre in timber. Short grained. Also to timber from old, or over-mature, or dead, trees.

Brass Back Saw. See *Back Saw*.

Brassie. A golf club.

Brass Plater. A merchant with an office only, and no accommodation for storage of goods.

Brattice. 1. A fence round machinery. **2.** A partition in a coal-mine.

Bratticing. Boards for forming a brattice.

Brauna do Sertao. *Buttinum ferrugineum.* Similar characteristics and properties to brauna parda, but more red in colour. Used for the same purposes.

Brauna Parda. *Melanoxylon brauna* (*H.*). C. America. Dark brown to almost black with variegated streaks, lustrous, smooth. Very hard, heavy, strong, durable, stable. Roey grain, fairly coarse texture. Rather difficult to work. Large sizes. Used for structural work, flooring, sleepers, furniture, vehicles, spokes, etc. S.G. 1 ; C.W. 3·5.

Bravaisia. See *Sancho*.

Brazil. Principal timber exports to this country : cedar, logwood, mahogany, myrtle, partridgewood, rosewood.

Brazilian Cedar. S. American or Spanish cedar, *q.v.*

Brazilian Mahogany. *Swietenia macrophylla* or *S. krukovii.* See *Mahogany.* Also see *Crabwood*.

Brazilian Pine. *Araucaria brasiliana* or *A. augustifolia*. (*S.*). S. America. Similar to Chili pine, *q.v.*

Brazilian Rosewood. *Dalbergia spp.* Also called Bahia or Rio Rosewood, Palisander, and Jacaranda, *q.v.*

Brazilian Walnut. *Phœbe porosa.* Imbuya or imbuia, *q.v.*

Brazilnut. *Bertholletia excelsa.* Somewhat like black walnut and with similar characteristics. Used for cabinet work.

Brazilwood or **Braziletto.** *Guilandina echinata* (*H.*). S. and C. America. Also called Bahia-, Para-, and Pernambuco-wood. Bright orange with darker stripes, darkens with exposure, lustrous. Very hard, heavy, and strong ; resilient and durable. Straight, interwoven, compact grain, uniform texture, visible rays. Difficult to work, polishes well. Used for decorative work, violin bows, etc. A valuable dyewood. S.G. 1·1 ; C.W. 5. There are several Brazilian woods marketed as Brazilwood, but without separate classification.

Brazing. Hard soldering. Band saws are joined into continuous bands by brazing. The two ends are filed to form a spliced joint. Spelter, moistened with borax water, is placed between and heated until it fuses with the metal.

Breadfruit Tree. *Artocarpus incisa* (*H.*). S. Sea Islands. The fruit is roasted and eaten as bread. Not imported as timber.

Breadnut. Capomo, *q.v.*

Break. A change in direction of a plane surface. A recess or projection on a wall or framing.

Breaking Down. Sawing logs into planks, boards, etc. See *Conversion*.

Breaking Joint. 1. Crossing the heading joints in built-up surfaces. Jointing the ends of matchboarding and floorboarding on different studs or joists. 2. Beading or chamfering the arrises of a joint between two boards so that shrinkage is not so obvious.

Breaking Strength. The amount of load or force that will just overcome the resistance of a structural member, as a beam, and break it. See *Safe Strength*.

Breast. 1. The projections on a wall enclosing a fireplace and sometimes containing the flues. 2. The heart side or wide side of a piece of timber.

Breast High. A recognized height in forestry of 4 ft. 3 in.

Breast-hook. A strong V-shaped timber, shaped like a knee, between the bows or stern of a vessel. See *Keel*.

Breasting. Levelling the points of saw teeth preparatory to sharpening.

Breast Lining. Interior panelled framing between a window bottom and the skirting.

Breast-rail. The upper part of the balcony on the quarter-deck of a ship.

Breeches Spout. See *Spout*.

Bressummer or **Breastsummer.** A strong wood beam serving as a lintel. It is flush with the face of the wall. See *Summerbeam*.

Briar. 1. *Erica arborea* (*H.*). S. Europe. Mottled dark brown with close grain. Roots and excrescences used for tobacco pipes and fancy articles. 2. A crosscut saw used in lumbering.

Briar Oak. *Musgravea stenostachya* (*H.*). Queensland. See *Silky Oak*.

Bricklayer's Scaffold. One built of scaffold poles, and consisting of *standards* (uprights), *ledgers* (horizontal members), and braces. *Putlogs*, usually 4½ in. by 3 in., rest on the ledgers and in the wall and carry the scaffold boards. (See illus., p. 62.)

Bricknogged. Applied to a lath and plaster partition having the lower part (to shoulder height) filled with bricks between the studs.

Bricks. Pieces building up a laminated ground, or core, for veneered cabinet work.

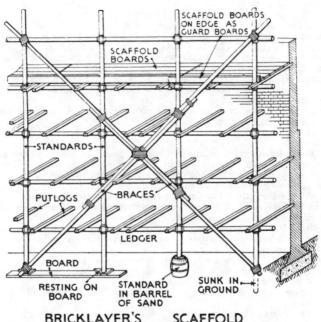

BRICKLAYER'S SCAFFOLD

Bridelia. See *Grey Birch* and *Khaja*.
Bridge. **1.** A structure for carrying traffic, supported at intervals and with open spaces below. See *Trestle Bridge*. **2.** A raised platform from which the navigator controls a ship. **3.** A piece of maple on the table of a violin to carry the strings.
Bridge Board. See *Notch Board*.
Bridge Gutter. Boards supported by bearers and covered with lead to form a gutter. See *Box Gutter*.
Bridge Shelter or **Dodger.** A shelter to the *bridge* of a ship.
Bridging or **Strutting.** Short pieces of timber fixed between joists to stiffen the joists and to prevent " canting over." See *Strutting* and *Nogging*.
Bridging Joists. See *Common Joists*.
Bridging Over. Placing timbers, such as *bridging joists*, across a space.

Bridging Piece. A short timber fixed between two joists to carry a load, as from a partition.

Bridging Run. A run, bridging a gap between two scaffolds, for the passage of workmen. It should be at least 18 in. wide if over 6 ft. from the floor, and the boards should be *tied* to prevent unequal sagging.

BRIDGING PIECE

Bridle. 1. A trimmer joist 2. A short rope with hooks and clamp for controlling the speed of logs when skidding.

BRIDLE JOINT

Bridle Joint. 1. A joint similar to the stub mortise and tenon joint, but with the positions of the mortise and tenon reversed. It is often used for angles less than right angles. 2. Applied to an open mortise and tenon joint in cabinet work.

Brig or **Brigantine.** A small merchant sailing ship with two masts, usually square-rigged.

Brigalow Spearwood. *Acacia harpophylla* (*H.*). Queensland. See *Spearwood.* Also see *Myall.*

Bright. Applied to newly sawn wood. Not discoloured.

Bright Deals. Applied to Canadian softwoods that are conveyed direct from sawmill to ship without " floating," as floated rafts usually get discoloured in transit.

Bright Dry and Flat. A market specification denoting no discoloration or warping, and reasonably seasoned.

Bright Floated. Applied to timber floated in clear water and on rafts above water level to avoid discoloration.

British Columbia. Large exports to this country of Douglas fir, western hemlock, western white pine, western red cedar, and sitka or silver spruce.

British Guiana. Exports to this country : greenheart, logwood, mahogany or red crabwood, mora, and small quantities of lancewood, letterwood, and zebrawood.

British Honduras. See *Honduras.*

British Standards Institution. An organisation for preparing specifications (B.S.S.) to standardise materials (names, sizes, qualities, etc.) used by the different industries in this country. Over 1,400 standards have been issued, including a number for wood.

British Trees. (*Indigenous*) alder, ash, aspen, beech, birch, elder, grey poplar, hawthorn, mountain ash, oak, sallow, Scots elm, Scots pine, white poplar, yew. (Introduced by Romans) apple, apricot, box, cherry, chestnut, English elm, hazel, lime, mulberry, peach, pear, plane, poplar, quince, service. (Up to 16th century) hornbeam, maple, sycamore, willow. (16th century) evergreen oak, holly, juniper, laburnum, spruce, stone pine, walnut. (17th century) acacia, buckthorn, horse chestnut, larch, silver fir. (18th century) cedar, maritime pine, pitch pine.   (19th century) Austrian pine, Chile pine, deodar, Douglas fir, Lawson's cypress.

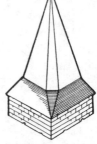

Britt. Abbreviation for the name of the botanist Britton.

Brittle. Applied to timber that breaks or snaps easily across the grain.

Broach. 1. A tapered bit, as a rimer. 2. A local term for a spire.

Broaches. The inclined corners from a square base to intersect an octagonal spire.

Broach Spire. A spire rising from a tower without a parapet.

Broad. A turner's tool for soft woods.

Broad-leaved. See *Ironbark* and *Maple*.

Broad-leaved Trees. Dicotyledons. Deciduous trees, as distinct from needle- or narrowleaved trees. They are technically classed as hardwoods. See *Hardwoods* and *Trees*. The

BROACH SPIRE

timber is usually hard and the structure is more complex than in narrow-leaved trees. They are not resinous but many species yield gums.

Broads. Canadian deals 12 in. or over in width.

Broadside. The part of a ship between the bow and the quarter, and above water level.

Brob. A spike with a right-angled head, so that it can be withdrawn from the wood by a lever.

Broke. A part of a forest or wood marked for felling.

Broken Pediment. See *Pediment*.

Broken Specification. A variation from the original specification of imported timber, by deleting certain dimensions.

Broken Stowage. See *Stowage*.

Broken Stripe. A modification of ribbon stripe. The markings taper out, due to twisted grain, so that the ribbon stripe is not continuous.

Broken-up Door. A door, in cabinetmaking, with glazing bars. See *Barred Door*.

Broker. A person who arranges contracts between buyer and seller and who is compensated by a percentage called " brokerage." He is not responsible for the fulfilment of the contract.

Broker's Sale Measure. A method adopted for measuring square logs of mahogany and furniture woods generally. An expert measurer allows for defects in addition to the customary specified allowances for waste in conversion. See *Hoppus Measure*.

Bronze Sheathed. Thin drawn bronze on a hardwood core, used extensively for shop fronts and fittings. See *Kalamein Sheathing*.

Brooming. The bruising of a pile-head with the pile-driver.

Brosimopsis. See *Leiteira*.

Brosinum. See *Guayamero, Ogechi, Satiné*, and *Snakewood*.

Brousewood. See *Brushwood*.

Brown. A prefix applied to numerous woods, usually to distinguish a species. See *Malletwood, Penda, Quandong, Tea Tree*, and *Tulip Oak*.

Brown Ash. *Fraxinus nigra.* Black ash. See *Ash.*
Brown Beech. *Pennantia cunninghamii* (*H.*). Queensland. Very similar to English beech and used for similar purposes. S.G. ·6.
Brown Bloodwood. *Eucalyptus trachyphloia* (*H.*). Queensland. See *Bloodwood.*
Brown Cugerie. *Protium australasicum* or *Bursera australasica* (*H.*). Queensland. Also called Carrotwood. Pinkish, smooth, fragrant, lustrous. Fairly light and soft, not strong or durable. Straight grain, fine close texture, but rather spongy. Easily wrought, but requires careful seasoning. Used for interior work as a substitute for softwoods and for cheap cabinet work. S.G. ·56 ; C.W. 2·5.
Brown Ebony. *Brya ebenus.* See *Cocus.* Also *Cæsalpinia granadillo.* See *Partridgewood.*
Brown Hazelwood. *Lysicarpus ternifolius.* See *Mountain Oak.*
Brown Heart. Angelin, *q.v.*
Brown Hickory. See *Hickory.*
Brown Mahogany. Gedunoha, *q.v.*
Brown Oak. Rich dark brown heartwood of English oak, valuable for cabinet work. The darkened colour may be due to fungus agency, or to an acid peculiar to certain trees. It has no effect on the strength or durability of the wood.
Brown Pine. *Podocarpus elata* (*S.*). New S. Wales. Soft, close-grained, easily wrought. Similar to hoop pine, *q.v.* S.G. ·6 ; C.W. 1·5.
Brown Rot. Decay in timber due to fungi removing cellulose compounds, which leaves a brown friable mass.
Brown or **Scaly Bug.** *Lecanium sp.* Closely allied to the *lac* insect. Very·destructive to hardwoods.
Brown Stringybark. *Eucalyptus capitellata.* Victoria. See *Stringybark.*
Browsings. Mangers, *q.v.*
Bruguiera. See *Black Mangrove, Kakra, Tumu.*
Bruinheart. Acapú, *q.v.*
Brummer. A proprietary stopping for woodwork.
Brush Box. *Tristania conferta* (*H.*). S. Australia. Brown to grey when seasoned. Strong, tough, durable. Close even grain. Also called Red or Bastard Box. Used for tools, machine parts, etc. S.G. ·9.
Brush Treatment. Applying preservatives to timber by brush only. See *Preservation.*
Brushwood. **1.** Small branches and twigs. **2.** Small round hardwood for making brush *heads* or *backs.* **3.** A thicket of shrubs and small trees.
Bruyère. Briar, *q.v.*
Bruzze. A wheelwright's V-shaped chisel.
Brya. See *Cocus.*
b.s. Abbreviation for *both sides* and for *breaking strength.*
B.S.I. Abbreviation for British Standards Institution.
B.S.P. Abbreviation for the names of the botanists Britton, Sterns, and Poggenburg.
B.S.S. Abbreviation of British Standards Specification.
b.t. Abbreviation for *berth terms.*

Bua-bua. *Guettarda sp.* (*H.*). E. Indies. Reddish yellow. Hard, heavy, durable. Suitable for furniture but not imported.

Bubinga. *Copaifera spp., Brachystegia spp.,* and *Didelotia sp.* (*H.*). W. Africa. Also called Kevazingo, Okweni, etc. Resembles padauk, *q.v.* Reddish to dark brown, wavy streaks, smooth. Hard, heavy, strong, durable, stable. Interlocked grain, close uniform texture, rays, roe, and mottle. Difficult to work. Used for furniture, interior fittings, turnery, veneers. S.G. ·85 ; C.W. 4·5. The name bubinga is usually applied to sliced veneers and kevazingo to rotary cut veneers.

Buchanania. See *Char.*

Bucida buceras. See *Olivier* and *Jucaro.*

Buck. 1. An interior door post recessed to receive the end of a breeze block partition and to provide a fixing for door linings. **2.** Sawing felled trees into logs. A crosscutter.

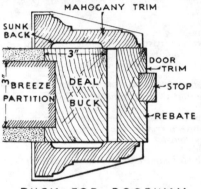

MAHOGANY TRIM

SUNK BACK

3"

DOOR TRIM

BREEZE PARTITION

DEAL BUCK

STOP

REBATE

BUCK FOR DOORWAY IN PARTITION

Buckboard. A lightly constructed four-wheeled carriage with long body.

Bucket. A receptacle for holding and carrying liquids. See *Pail* and *Rack.*

Bucket Rack. A rack on board ship for storing buckets.

Buckeye. *Æsculus spp.* (*H.*). N. America. Pale yellow, lustrous. Light, soft, not durable. The species include : yellow buckeye, *Æ. octandra* ; Californian buckeye, *Æ. californica* ; Ohio buckeye, *Æ. glabra.* The wood is often sold mixed with other woods, such as yellow poplar. It is very similar to basswood and used for similar purposes. Also see *Horse Chestnut.*

Buckl. Abbreviation for the name of the botanist Buckley.

Bucklandia. See *Pipli.*

Buckled or **Buckling. 1.** Applied to warped boards, especially when the edges are undulating. **2.** A saw permanently bent through careless handling and which must be corrected by hammering.

Bucklers. Two pieces fitted together for the hawse-holes on board ship, with just enough room for the cable. They are used to avoid taking in water in heavy seas.

Buckles. 1. Large wooden pins used for thatching. **2.** See *Tiller.*

Buckthorn. *Rhamnus cathartica* (*H.*). A British shrub. ALDER BUCKTHORN, *R. frangula,* is another British species. Small sizes, very hard. Close grain. Used for small ornamental work, walking sticks, charcoal, etc. The berries are valuable for the colouring called sap-green, and for medicine. The Canadian wild cherry is also called buckthorn.

Bucrania. A carving representing the head of an ox decorated with a wreath. It is used in the Corinthian and Ionic frieze.

Budongo. *Entandrophragma angolense.* See *Gedunohor.*

Buey. E. Indies ebony. Also called Arang, Kayu, Meribut, etc.

Buff Box. See *Orange Box.*

Buffer. Any contrivance for deadening concussion. See *Bumping Block.*

Buffet. 1. A sideboard or small cupboard for a dining-room. 2. A refreshment bar. 3. A low stool.

Buggy. A light two-wheeled vehicle drawn by one horse, and resembling a dogcart. The name is also applied to a low handcart for heavy packages.

Buhl. See *Boule.*

Builder's Scaffold or **Staging.** A strong scaffold built of square timbers.

Building Boards. Applied to the numerous types of wallboards used for lining walls and ceilings. They are made from all kinds of materials such as asbestos, bamboo, cane, gypsum, sawdust, straw, wood pulp, etc., and have various finishes to the surfaces. They are usually sold under proprietary names, and some varieties may be veneered with ornamental woods, bakelite, celluloid, or metal, and may be obtained in large sizes and various thicknesses. See *Plywood* and *Wallboards.*

Built Rib. A built-up, or laminated, member of a frame or structure. The term is usually applied to a curved member. See *Angel-beam.*

Built Up. 1. A term used by woodworkers when the width or thickness is comprised of several pieces. 2. Plywood faced with superior hardwood and manufactured in one operation, as distinct from ordinary plywood veneered.

Built-up Roof. A plywood top to a carriage.

Bulbous. Pillars, legs, feet, etc., having a bulb-like swelling characteristic of the Jacobean style of furniture.

Bulkhead. 1. A partition on a ship, or in aircraft, to form separate compartments. 2. A sloping top or ceiling to a flight of stairs leading to a basement or under a shop window. 3. Vertical partitions erected when tunnelling.

Bullace. A species of plum.

Bull Chain. A heavy chain to which shorter chains, with hooks and dogs, are attached. The appliance is used to draw logs up a gangway from the pond.

Bulldog. See *Timber Connectors.*

Bull Donkey. A donkey engine for hauling logs.

Buller Nails. Nails with short shanks and large lacquered heads, for attaching tapestry, etc., to furniture.

Bullet Tree. *Terminalia buceras.* Also called Bucaro, Black Olive, Foocadie, and Jacaro, *q.v.*

BULBOUS
LEG

Bulletwood. The name is applied to numerous species of Mimusops and the wood is known by various names : balata, beefwood, horseflesh-wood, red lancewood, S.A. mahogany, wild dilly, etc. Species of Terminalia are also called bulletwood. The woods are very hard,

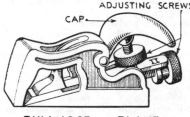

heavy, strong, and durable ; with straight and wavy grain, and fine texture, but rather harsh. Difficult to work and require careful seasoning, polish well. Used for structural work, sleepers, furniture, tools, spokes, fishing rods, etc. S.G. 1 ; C.W. 4. See *Balata, Bastard Bulletwood, Humiria, Kaya, Mohwa, Nargusta, Olivier, Towaronero,* and *Yellow Boxwood.*

BULLNOSE   PLANE

Bullnose. A small metal rebate plane.
Bullnose Step. A step at the foot of a flight of stairs with one or both ends rounded to a quarter circle.

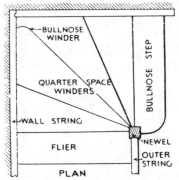

BULLNOSE   STEP & WINDER

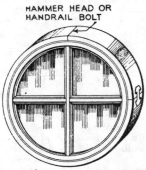

BULL'S   EYE   WINDOW

Bullnose Winder. Used to avoid weakening the joint between the strings and the difficulty of cleaning in the angle. It is better to use a kite winder if possible.
Bull Oak. See *Shea* or *Shee Oak.*
Bull Pine. See *Western Yellow Pine.*
Bull's Eye. 1. A small circular or elliptical window or opening. See *Turret.* 2. A wood thimble or ring with a groove in the circumference.
Bulnesia. See *Vera-wood* and *Maracaibo Lignum-vitæ.*
Buloke. *Casuarina luehmanni* (H.). Victoria. See *She Oak* and *Ru.*
Bulwark. 1. A sea-defence wall, often built of heavy timbers. 2. The side of a vessel above the deck.
Bumboat. A boat used to remove refuse from ships or to carry stores to the ships.

Bumelia. See *Ibira-nira* and *Jiqui*.

Bumkins. Short booms projecting from the bows of a ship to extend the lower edge of the fore-sail. Also a spar projecting over the stern for the mizzen-sheet. A mizzen-boom.

Bummer. A truck with two low wheels and long pole for hauling logs.

Bumping Blocks. Strong wooden frames firmly fixed to resist the impact of wagons running on rails, and to bring rolling stock to rest.

Bung. A tapered stopper for closing a hole in a cask. It often includes the tap for the barrel.

Bunga. See *Bongor*.

Bungalow. A domestic building of one storey.

Bunk. 1. A recess or box in a ship's cabin. A sleeping berth. 2. A dog, to the chain of a log jack, which grips the log when it is required to drag the log to the sawmill. See *Log Jack*. 3. A strong beam on which logs rest on a truck or sled.

Bunker. A large hopper-like structure for storing granular materials. It is like an inverted truncated pyramid, with an opening so that the material can be loaded into carts, etc.

Buntons. Horizontal timbers across a mine shaft.

Bunut. Bintangor, *q.v.*

Bunya Pine. *Araucaria bidwillii* (S.). Australia. A pinkish cream timber with beautiful figure. Also called Bunya Bunya. Used for veneers and general purposes. See *Queensland Pine*.

Buranhem. See *Sweetwood*.

Bureau. 1. A writing table with drawers, or a chest of drawers. 2. A room, office, or department for secretarial business or inquiries.

Burin. An engraving tool.

Burk. Abbreviation for the name of the botanist Burkill.

Burkea. See *Wilde Sering*.

Burl. Same as burr, *q.v.*, but sometimes applied to the wood immediately surrounding a knot.

Burly. Timber with irregular grain or burls.

Burma. A large variety of timbers are exported to this country from Burma, including box, eng, gurjun, haldu, koko, padauk, pyinkado, pyinma, rosewood, teak, thitka or sal or B. mahogany, white bombway.

Burma Padauk. *Pterocarpus macrocarpus*. See *Padauk*.

Burma Tulipwood. *Dalbergia oliveri*. See *Tamalan*.

Burnettizing. A preservative process in which zinc chloride is forced into the timber. See *Preservation*.

Burnishing. Forming a glazed surface on hardwoods by means of friction through rubbing with a hard, smooth substance such as bone, agate, polished steel, or close-grained boxwood. The prepared surface is suitable for wax polishing but not for french polishing.

Bur Oak. *Quercus macrocarpa*. White oak. See *Oak*.

Burr. Abnormal growth, or excrescences, common to most trees, especially to those that can reproduce by stooling. Irritation or injury forms an interwoven, contorted, or gnarly mass of dense woody tissue. In some species the burls are large and decorative and valued for veneers, especially in *Betula*, *Erica*, *Juglans*, *Quercus*, *Sequoia*, *Ulmus*. See *Sap*, *Twig*, *Walnut*, and *Root Burr*.

Burr Woods. Timbers in which the burr forms an important decorative feature, such as amboyna, ash, elm, oak, thuya, walnut, yew, etc. In some timbers burrs have no decorative value and in all cases they decrease the strength.

Bursera. See *Gumbo*, *Mulato*, *Murtenga*, and *Brown Cugerie*.

Burseraceæ. The Torchwood family : Boswellia, Bursera, Canarium, Commophora, Elaphrium, Garuga, Protium (Icaca), Santiria.

Buruta. *Chloroxylon swietenia* (*H*.). E. Indian satinwood. Also called Mutirai or Flowered Satinwood. A decorative wood used for veneers. See *Satinwood* (*Ceylon*).

Bus. Abbreviation for *omnibus*.

Bush. **1.** Small trees in which the branches are near the ground or from the roots instead of from a central stem **2.** A cylindrical lining round a rotating shaft in machinery.

Bushel. An 8-gal. measure for grain, fruit, etc.

Buss. An old type of large fishing vessel used in the North Sea.

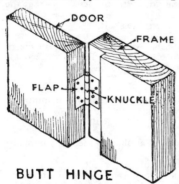

BUTT HINGE

But. See *Abut*.

Butt. **1.** The base of a tree. **2.** Hinges intended to be sunk into the edge of a door or casement. **3.** The bottom end of a shingle. **4.** The largest size of beer barrel of 108 gal. capacity. **5.** See *Butted*.

Butt and Break. A plasterer's term for lathing in which the ends of the laths are butted and grouped on different joists or studs.

Butt Cut or **Butt-log.** The log obtained from the first cut above the stump or butt.

Butted. Timbers placed end to end with no preparation of the joint beyond squaring the ends.

Butt-end. The end of a log nearest the butt.

Butternut. *Juglans cinerea* (*H*.). White walnut. N. America. Similar grain to black walnut, but not so heavy or strong. Used as a substitute for black walnut when stained. Durable. S.G. ·45 ; C.W. 3. Also see *Souari*.

Butter Pear. See *Pear*.

Butter Tree. *Bassia latifolia* or *B. longifolia* (*H*.). India, Burma. Called Mahwa, Moha, etc. Rose red to nut brown, some lighter streaks, smooth. Very hard, heavy, strong, and durable. Close interlocked, even grain, firm texture, distinct fine rays. Difficult to work. Used for general purposes, shipbuilding, furniture, turnery, etc. S.G. ·95 ; C.W. 4.

Buttery Hatch. A half-door over which provisions are served.

Butt Gauge. A tool with three cutters for gauging door linings for butt hinges.

Buttinum.　See *Brauna*.

Butt Joint.　Two pieces of wood joined together along the edges with a plain square joint.　Also see *Angle Joints* and *Butted*.

Butt Match.　See *End Match*.

Buttock.　The convex curve of a ship under the stern, between the counter and the after part of the bilge and between the quarter and the rudder.

Button.　**1.** A small piece of wood or metal secured by one screw so that it is free to revolve. It is used to secure one piece of framing to another and to allow for expansion or contraction, as a counter top to the battens and pedestals.　See *Battens*.　**2.** A cupboard turn.

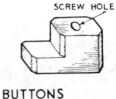

BUTTONS

Buttonwood.　*Platanus occidentalis* (*H.*).　U.S.A.　Now called American Plane and Sycamore.　Yellowish brown.　Hard, moderately heavy, tough, stiff, strong.　Often cross-grained.　Quarter-sawn wood is called Lacewood.　*P. racemosa* is a similar wood.　See *Sycamore* and *Mangrove* (*White*).

Buttress.　An extraneous growth on many tropical trees that supports the trunk.

Butt Saw.　A crosscut saw, either circular or reciprocating.

Buxus.　See *Cape Box* and *Boxwood*.

Buzz.　A large type of spokeshave or scraper used by coopers for cleaning up the outside of a cask.

By-products.　The following are the by-products obtained from timber and trees : acetic acid, bark for tanning, cellulose, charcoal for gunpowder and crayons, etc., dyes, excelsior, gutta-percha, medicines, oils, pitch, plastics, potash, pulp for paper, etc., rosin, rubber, scents, turpentine, wood meal and flour, wood spirit, etc.

Byre.　A cow-house.

Byrsonima.　See *Spoon* (*Golden*).

C

c. Abbreviation for *compression* and *constant*.

Čaa. South American holly.

Cab. A hackney carriage. It may be a hansom, brougham, or taxi.

Cababa. *Khaya anthotheca.* W. Africa. See *Mahogany*.

Cabane. See *Pylon*.

Cabbage Bark. Partridgewood. Pheasantwood, *q.v.* Also called Andira, Angelin, Wild Olive, etc. See *Angelin*.

Caber. A pole, or spar, used in Scottish sports.

Cabeti. Species of *Luehea*, *q.v.*

Cabilma. *Guarea trichiliodes.* C. America. Very similar to San Domingo mahogany.

Cabima. See *Copaiba*.

Cabin. 1. A temporary wooden dwelling, or hut, as used for the convenience of workmen during the erection of buildings. 2. A compartment on a ship. 3. A railway building from which points and signals are controlled.

Cabinet. 1. A small room or private apartment. 2. A nest of drawers or a small ornamental cover, frame, or display cupboard with glass doors.

Cabinet Cherry. *Prunus serotina.* American black cherry. See *Cherry*.

Cabinet Joints. See *Joinery Joints*.

Cabinetmaking or **Cabinet Work.** The craft of making furniture. The cabinetmaker is usually concerned only with self-contained articles. When the work is part of the fabric it is joinery.

Cabin Hook. A hook for securing an open door or casement.

Cabiuna. *Dalbergia nigra.* Brazil. See *Jacaranda*.

Cable Moulding. A cylindrical moulding carved to imitate a rope.

CABLE MOULDING

Cable Tier. The storage place on the orlop-deck of a ship.

Caboose. A cook-room on a ship's deck. A galley.

Cabraiba. See *Oity Oity*.

Cabralea. See *Cancharana*.

Cabreuva. *Myrocarpus frondosus* or *Myroxylon sp.* (*H.*). Tropical America. Also called *Oleo pardo* and *O. vermelho*, etc. Walnut brown, lustrous, dark streaks, fragrant. Hard, heavy, strong, and durable. Fine roey grain and fairly difficult to work. Used for superior joinery, cabinet work, turnery, shipbuilding, sleepers. S.G. ·9 ; C.W. 3·5.

Cabriole. Applied to uprights in Queen Anne furniture in which the upper part swells outwards and the lower part bends inwards. See *Ball and Claw*.

Cabriolet. A light two-wheeled carriage drawn by one horse.

Cabui. *Enterolobium lutescens*. S. Brazil. See *Goncalo*.

Cacao. See *Coca*.

Caco. *Jacaranda, q.v.*

Cactaceæ. Cactus family. Great variety, but few of use as timber. See *Cardon*.

Cadoga. *Eucalyptus torelliana* (*H.*). Queensland. Similar to spotted irongum and swamp messmate. Pale brown. Hard, heavy, strong, tough, flexible, but not very durable. Used for structural work, construction, decking, poles, etc. S.G. ·9.

RAIL

CABRIOLE

Cæsalpinia. See *Guayacan, Brazilwood, Sappan, Divi-divi, Ironwood, Juca, Partridgewood*.

Cæsalpinioideæ. A sub-family of Leguminosæ and including Apuleia, Cæsalpinia, Cassia, Cynometra, Copaifera, Dialium, Dicorynia, Dimorphandra, Eperua, Gleditschia, Hardwickia, Hæmatoxylon, Holocalyx, Hymenæa, Melanoxylon, Peltogyne, Peltophorum, Prioria, Pterogyne, Swartzia, Zollernia.

Cafeteria. A restaurant in which customers wait on themselves.

Cage. 1. A lift for conveying miners up and down the shaft of a mine. 2. An open-work receptacle for animals or birds.

Cairn's Ash. *Flindersia ifflaiana*. Also called Hickory Ash or Queensland Hickory. See *Hickory*.

Caisson. 1. A deeply sunk panel in a ceiling or soffit. 2. A watertight structure used when laying foundations below water level.

Cajeput. Pacific myrtle, *q.v.*

cal. Abbreviation for *caliper measure*.

Calaba. See *Santa Maria*.

Calabash. *Crescentia cujete* (*H.*). Tropical America. Yellowish brown, darker veins. Moderately hard, heavy, tough, and strong. Not durable. Coarse texture with ripple marks. Easy to work to a smooth finish. Tools, turnery, vehicle parts, pipes, etc. Utensils from gourd-like fruits. S.G. ·6 ; C.W. 2.

Calamander. *Diospyros quæsita* (*H.*). E. Indies, Macassar. Ebony family. Hazel brown, striped and mottled with black. Hard, heavy, and close grain. Similar to Arang, Buey, Marblewood, and Zebrawood. Used for cabinet work and veneers. S.G. ·9 ; C.W. 3·5. The name is applied generally to variegated ebony.

Calamansanay. *Nauclea sp.* (*H.*). Philippines. Yellowish red to rose colour, varied markings, lustrous. Hard, dense, fine texture, durable, and resists insects. Tough and difficult to work. Used for structural work, furniture, etc. S.G. ·64 ; C.W. 4·5.

Calantas. *Toona calantas* (*H.*). E. Indies. Philippine mahogany. Like Spanish cedar. Light to dark red. Soft, light, durable, odorous, resists insects. Straight grain, moderately fine texture, smooth, and

73

easy to work. Used for wardrobes, musical instruments, cigar boxes, carving. S.G. ·43 ; C.W. 2.

Caledonian Oak. *Carnarvonia araliæfolia* (*H.*). Queensland. Also called Red and Black Oak. Chocolate red, lustrous. Hard, heavy, strong, tough, fissile, not durable in ground. Straight grain, firm uniform texture, ray, and mottle. A dark decorative cabinet wood. Requires careful seasoning, not difficult to work. Used for superior joinery, cabinet work, turnery. S.G. ·7 ; C.W. 3.

Caley's Ironbark. *Eucalyptus caleyi* (*H.*). Queensland. Similar to other ironbarks, *q.v.*

Calf. The wedge-shaped part of a scarfed joint, as used for lengthening timbers.

California. Chief timber exports to this country are sequoia, sugar pine, white pine, myrtle, red fir. Also see *U.S.A.*

Californian Nutmeg. *Torreya californicum.* See *Stinking Cedar.*

Californian Redwood. Sequoia, *q.v.*

Caliper Measure. A method of measurement for softwood squared logs. The length is taken to $\frac{1}{2}$ ft. and width and thickness to the nearest inch, and the volume is calculated to $\frac{1}{2}$ cub. ft.

Calipers. A two-legged appliance for measuring the diameters of cylindrical and spherical objects. They are like compasses, but the legs are arched, or bowed, in some types and straight in others. The type used for measuring logs is the " Hull calipers," which consists of a graduated beam with one fixed, and one sliding, arm.

Callitris. See *Cypress Pine, Murray Pine,* and *Oyster Bay Pine.*

Callus. The tissue formed over a wound in a tree. An unnatural growth in the tree that has incorporated with the wood growth.

Calocarpum. See *Sapote.*

Calodendron. See *Cape Chestnut.*

Calophyllum. See *Bintangor, Canoe Tree, Galba, Nicobar, Penago, Pinnay, Poon, Santa Maria,* and *Touriga.*

Calycophyllum. See *Degame* and *Palo Blanco.*

Cam. An irregular curved projection on a revolving shaft, to give intermittent motion.

Camachile. *Pithecolobium dulce* (*H.*). Tropical America. Reddish brown. Hard, heavy, strong but brittle, durable. Interwoven grain and fine texture. Polishes well. Rather difficult to work. Chiefly used for constructional work and for yellow dye.

Camagon. *Diospyros discolor.* **Camagoon.** *D. pilosanthera* (*H.*). E. Indies. Ebony family. Nearly black, with streaks and mottle. Very hard, heavy, dense, and durable. Straight grain, fine texture, smooth. Used for same purposes as ebony. S.G. 1 ; C.W. 3·5.

Cambala. Iroko, *q.v.*

Camber. The rising at the middle of a horizontal structural member, for appearance or to counteract sagging. The rise in a flat segmental arch. The athwartship curvature of a ship's deck.

Camber Arch. A straight arch with the soffit raised a little at the middle. See *Camber.*

Camber Beam. The tie-beam of a roof truss. See *King Post Truss.*

Camber Slip. 1. A single piece of wood used as a centre for an arch when the rise is small, as in a camber arch. Also called a Turning Piece. **2.** A templet to describe a flat arc of a circle.

Cambium. The wood-forming layer, by cell division, in exogenous trees, immediately under the bark which protects the layer, between xylem and phloem. An injury to the cambium prevents growth of new cells. See *Annual Rings*.

Cambium-miner. An insect, *Agromyza carbonaria*, that burrows in the cambium and causes pith flecks in numerous woods.

Cambleted. A term applied to wood obtained from roots, especially of ash, that is curiously veined.

Camdeboo. *Celtis rhamifolia*. S. Africa. See *Stinkwood*.

Camellia. *Thea japonica*. Japan.

Campana. See *Bell*.

Campanile. A detached clock, or bell, tower.

Camp Bed. A narrow portable folding bed. A camp bedstead is a portable piece of light folding framework that can readily be prepared as a bed.

Camp Ceiling. A sloping or convex ceiling. An attic ceiling in the form of a truncated pyramid.

Camp Chair. A light portable folding chair of sticks and canvas.

Campeachy-wood. *Hæmatoxylon campechianum*. See *Logwood*.

Camphorwood. *Cinnamomum camphora (H.)*. E. Asia. Laurel family. Several species. Light yellowish olive to orange brown, with darker streaks. Moderately light and soft. Straight, wavy, and interlocked grain, with pith flecks. Fragrant, smooth, not difficult to work. Little marketed. Used for cheap furniture, constructional work, planking, tea boxes. S.G. ·64 ; C.W. 2·5. BORNEO CAMPHOR, *Dryobalanops aromatica*. Not a true camphor. Also called Keledan or Kapur. Brownish red. Fairly hard, heavy, tough, and durable. Close texture, interlocked grain. Odorous. Care required for a smooth finish. Resembles mahogany when polished. Used for constructional work, flooring, exterior work, superior fittings. S.G. ·75 ; C.W. 3·5. FORMOSAN CAMPHOR, *C. camphora* and *Machilus thumbergii*. The latter species has few of the properties of camphorwood. INDIAN CAMPHOR, *C. glanduliferum*. Very similar to *C. camphora*. KENYA CAMPHOR, *Ocotea usambarensis*, E. African Camphor or *Muzaiti*. Pale yellow, seasoning to deep nut brown. Moderately hard and strong. Even texture, interlocked grain, stable. Not very difficult to work and polishes fairly well. Obtainable in large sizes, but rather short lengths. Used for cabinet work, superior joinery, carriage construction, veneers, etc. Similar to light-coloured W. African mahogany, but fragrant, and contains a gum secretion. S.G. ·6 ; C.W. 3·5. NEPAL CAMPHOR, same as Indian. QUEENSLAND CAMPHOR, *C. oliveri*. Similar to Asiatic. Not exported.

Campnosperma. See *Terentang*.

Campshedding or **Campsheeting.** **1.** Sheet piling in loose soil or to a wharf wall or for underwater work. **2.** The cap or sill of a wharf wall.

Campstool. A folding stool of wood, or metal, and canvas.

Camwood. *Baphia nitida, B. pubescens*, or *Pterocarpus osun*. Nigeria. Also called Barwood and Osun. Rich claret red, darker streaks and zones. Fades on exposure unless treated to preserve colour. Fairly hard and heavy. Close firm texture. Easy to work. Suitable for cabinet work and superior joinery, but used chiefly as a dyewood. S.G. ·7 ; C.W. 2·5. Also see *Padauk*.

Canada. The woods exported to this country include ash, aspen, basswood, beech, birch, butternut, cedar, cottonwood, Douglas fir, elm, fir, hemlock, hickory, larch, maple, pine, spruce, tamarack, walnut, yellow poplar.

Canada Balsam. A turpentine from *Abies balsamea*. It has the same refractive index as glass and is used as a jointing compound in airtight showcases, etc., and mounting of slides for microscopic work.

Canada Measure. 1000 ft.× 12 in.× 1 in. Nearly half of a Petrograd Standard.

Cana fistula. *Cassia brasiliana* or *Peltophorum vogelianum* (*H.*). Tropical America. Also called Horse Cassia and Stinking Toes. Reddish with darker streaks. Hard, heavy, durable, lustrous finish, gum deposit. Irregular grain and warps freely. Polishes well. Used for constructional work, cabinet work, turnery, sleepers. S.G. ·8 to 1 ; C.W. 4.

Canal. A flute, *q.v.*, especially when in the soffit of a cornice.

Canalete or **Canaletta.** *Cordia sp.* See *Bocote* and *Cyp*.

Canarium. See *Indian White Mahogany* or *Dhup*, *Ekpakpogo*, and *Papo*.

Canary Ash. *Beilschmiedia bancroftii* (*H.*). Queensland. Also called Yellow Walnut. Lemon yellow, smooth, lustrous. Moderately hard, heavy, strong, and stable, not durable. Resembles walnut bean. Dulls saws and knives in conversion, due to silica content. Useful wood for superior joinery, flooring, matching, cabinet work. S.G. ·7 ; C.W. 3·5.

Canary Cheesewood. *Sacrocephalus cordatus* (*H.*). Queensland. Also called Leichardt Pine. Saffron, odorous. Light, soft, not strong or durable. Excellent for carving because of its cheese-like texture. Also used for turnery and as a substitute for softwoods. Scarce and not exported. S.G. ·48.

Canary Sassafras. *Doryphora sassafras* and *Daphnandra micrantha* (*H.*). Queensland. Yellow, lustrous, smooth. Fairly light and soft, not durable but resists insects, brittle. Fine even grain and texture. Easily wrought and seasoned. Used for flooring, joinery, brush stock, toys, turnery, carving, etc. S.G. ·55 ; C.W. 2.

Canary Whitewood. *Liriodendron tulipifera* (*H.*). U.S.A. Variable colour, from greyish white through yellow to green in the same piece. Soft, light, with uniform even grain and texture. Easy to work, not durable. Stains and polishes well. Used for engineer's patterns, carvings, joinery, cabinet work. It is marketed as American whitewood, saddle tree, and yellow poplar. S.G. ·45 ; C.W. 1·5. Also see *Cucumber*.

Canary Wood. *Centrolobium robustum, Euxylophora paraensis.* Brazil. See *Arariba* and *Vermelha*. Also see *Canary Whitewood*.

Cancharana. *Cabralea cangerana* (*H.*). C. America. Resembles cedar. Maroon or dull red, lighter and darker markings. Moderately hard, heavy, and strong but brittle. Very durable. Straight and wavy grain. Medium texture, smooth and fragrant. Easy to work. Used for furniture, carpentry, structural work, shipbuilding. Provides a red dye. S.G. ·6 ; C.W. 2·5.

c. and e.b., or **V.** Abbreviation for *centre and edge bead*, or *vee*.

Candleboard. A small shelf fitting, characteristic of Sheraton tables, for holding a candlestick.

Candlenut Siris. *Aleurites moluccana* (*H.*). Queensland. Silver white. Very light and soft, not strong but fairly tough. Straight grain, rather spongy. Used chiefly for shelving and kitchen dressers, etc. S.G. ·4.

Candlestick. An appliance for holding a candle. It may be of wood, metal, or porcelain, and it is usually square or circular turned.

Cane Ash. *Fraxinus americana.* Also called American White Ash and Ground Ash. It is very similar to European ash and used for the same purposes. See *Ash.*

Canec. A registered wallboard, *q.v.*

Canellaceæ. Wild Cinnamon family. See *Canella, Cinnamomum, Cinnamon, Pleodendron, Warburgia.*

Canello, Canela, or **Canella.** *Cumamodendron oxillare* and *Mespilodaphe opifera* (*H.*). Tropical America. Laurel family. Numerous species of *Nectandra* and *Ocotea* are known as Canella. Also see *Imbuia.* The name Canell is also used in a general sense for woods suitable for cabinet work, and includes woods of several genera and numerous species.

Canelon. *Rapanea lætevirens* or *Myrsine sp.* (*H.*). C. America. Grey, some with rosy hue and stripes. Fairly light and soft but firm, not durable. Fairly straight grain, coarse texture, distinct rays. Easily wrought. Used for carpentry, cabinet work, cases, pulp, etc. S.G. ·5 ; C.W. 2.

Caney. Applied to the soft part of a growth ring ; or to wood, especially ash, having a certain percentage of springwood.

Cangerana. See *Cancharana.*

Cangica. *Piratinera guianensis.* See *Letterwood.*

Cangu. *Shorea tumbuggi* (*H.*). India. Reddish. Very hard, heavy, tough, and durable. Similar to Sal. Used chiefly for structural work. S.G. 1·1.

Canjocas. See *Pegui.*

Canker. A fungoid disease in trees.

Cannon. See *Split Baluster.*

Canoe Cedar. See *Red Cedar.*

Canoe Tree. *Calophyllum spectabile* (*H.*). Andaman Islands. Pale light red to reddish brown, darker streaks, lustrous, smooth. Fairly hard, heavy, strong, and durable. Interlocked grain, but not difficult to work. Resists insects. Constructional work, joinery, spars. S.G. ·8 ; C.W. 3.

Canopy. 1. A projecting hood, supported on brackets to a door or window. A covering to a niche, altar, font, or bed. Also see *Sounding Board.* 2. The same as *Crown* (2). 3. The forest roof. (See illus., p. 78.)

Cant. 1. An external splayed angle. 2. To turn over at an angle not a right angle. 3. An alternative name for Wane or Waney, *q.v.* 4. See *Cants.* 5. A log slabbed on one or more sides. To *cant* a log is to remove slabs from one or more sides.

Cant Bay. A bay window with octagonal or hexagonal sides.

Cant Board. **1.** The sloping side of a gutter. Also called a Learboard, *q.v.* **2.** A coachbuilder's setting-out board.

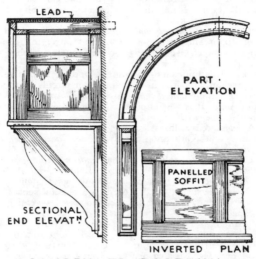

CANOPY TO DOORWAY

Cant Column. A column with polygonal sides.

Cant Dog. A handspike, about 6 ft. long, for turning or levering logs. It is provided with toe ring and lip.

Canted. Not square ; splayed ; on the bevel.

Canteen. **1.** A building for refreshments. The name is generally applied to a temporary building or to one in a works or barracks. **2.** A case containing table accessories, cutlery, etc.

Canterbury. A piano seat with " well " to hold music.

Canthium. See *Ajelera* and *Ceylon Boxwood*.

Cantilever. A beam supported at one end only. See *Beams*.

Cantilever Bracket. A bracket built into the wall to withstand the leverage of the load it has to support.

Canting Livre. A ship's console bracket.

Canting Over. Applied to vertical members that are forced over from the vertical.

Cant Moulding. One with a bevelled face.

Cant Rail. **1.** The top rail to the side framing of a railway carriage, or any type of carriage. **2.** See *Capping Rail*.

Cants. **1.** The layers building up a pattern for a propeller. **2.** Sole pieces following the inclination of a ship's deck. They are rebated for the framing such as bulkheads, etc. See *Carling Sole.*

Cant Timbers. The timbers, at the ends of a ship, placed obliquely on the keel.

Caoba. Peruvian and C. American mahogany. See *Mahogany.*

Cap. 1. A piece planted on the top of a post for ornamentation or weathering. **2.** See *Capping.* **3.** An iron cone, with a hole at the top for a chain. The cone fits over the end of a log, for snaking, to avoid catching an obstruction during dragging operations. **4.** A block to tie the top of a lower mast to the bottom of a top mast. It is fitted with eye-bolts for the pulley blocks, to hoist the top mast. **5.** Abbreviation for the name of the botanist Capanema. **6.** Same as *Bar* (5), also called a Lid. **7.** A rotating box-like structure in which the shaft was placed for carrying the sails of a windmill.

Capa. Fiddlewood, *q.v.*

Cape. 1. Applied to numerous S. African woods. There is little exported to this country, with the exception of cape box, owing to home demands. **2.** Also applied to several Tropical American species of *Clusia.*

Cape Ash. *Ekebergia capensis* (*H.*). S. Africa. Also called Dog Plum and Essenhout. Appearance and characteristics of European ash and used for the same purposes.

Cape Beech. *Myrsine melanophleos* (*H.*). S. Africa. Similar to European beech. Quarter-sawn for panels and furniture. S.G. ·8.

Cape Box. *Buxus macowanii* (*H.*). S. Africa. East London boxwood. Similar to European boxwood. Also *Gonioma kamassi*, now called Kamassi, *q.v.*

Cape Chestnut. *Calodendron capense* (*H.*). S. Africa. Whitish, soft, tough. Used for panelling, yokes, etc.

Cape, or **Natal, Mahogany.** *Trichilia emetica* (*H.*). S. Africa. Also called Stinkwood. Appearance and characteristics of mahogany except for odour. Polishes well, and used for cabinet work and interior joinery.

Cape Olive. *Olea capensis* (*H.*). S. Africa. Olive colour with dark streaks. Hard, heavy, close texture, lustrous. Difficult to season and work but good for turnery and small fancy articles. S.G. 1·1. Also see *Stinkwood.*

Cape Sandal. *Excœcaria lucida.* See *Sandaleen.*

Cape " Teak." *Strychnos atherstonei* (*H.*). S. Africa. Like teak in some respects. Chiefly used for wagon-building.

Capillarity. The phenomenon of water climbing above its own level in confined spaces, due to surface tension. Throatings are used in wood work to prevent capillarity.

Capital. The head of a pilaster or column immediately under the entablature. Each architectural

CAPITAL (CARVED)

order has its own characteristic capital, otherwise there is considerable variation in design. See *Order*.

Capoma. See *Ogechi*.

Capparidaceæ. The Caper family of trees, including Belencita Cadaba, Capparis, Cratæva, Niebuhria, Mærua, Morisonia, Forch-hammeria, Roydsia.

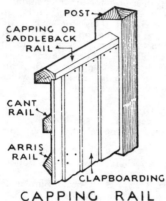

POST

CAPPING OR
SADDLEBACK
RAIL

CANT
RAIL

ARRIS
RAIL

CLAPBOARDING

CAPPING RAIL

Capping or **Capping Piece.** A long length serving as a cap, or coping, for framing, gates, etc. See *Cap*.

Capping Rail. A saddleback rail for the top of a fence. It may be rebated or grooved for the boards.

Caprifoliaceæ. A family of trees including elder, guelder rose, wayfaring tree, etc.

Capstan. A machine, based on the principle of the wheel and axle, for hauling heavy loads.

Capstan Bars. Strong wood bars for turning the drumhead of a capstan.

Capur Barros. See *Camphor* (*Borneo*).

Car. Applied to numerous types of vehicles or compartments on railway trains : motor, tram, jaunting, luncheon, side, sleeping, etc.

Caraba. See *Carapa*.

Carab Tree. See *Locust*.

Caracole. A spiral staircase.

Caracoli. See *Espavé*.

Carallia. *Carallia integerrima* (*H.*). India and E. Indies. Numerous species. Pale reddish brown. Moderately hard, heavy, and durable. Rather brittle. Straight grain with large rays. Medium texture and smooth. Resists insects. Not difficult to work when green. Should be quarter-sawn. Prominent figure. Used for construction, joinery, furniture, panelling, flooring. S.G. ·65 ; C.W. 4.

Carapa. A genus of Meliaceæ (mahogany) family, widely distributed. Colour varies from wine red to reddish brown, fairly lustrous. Moderately hard, heavy, and durable. Straight with some interlocked grain and ripple marks. Fairly difficult to work. The best known species are andiroba, Brazil mahogany, crabwood, karambola, nyireh, pussur.

Caravan. 1. A covered wagon or cart used by gypsies, a travelling circus, etc. 2. A trailer to a motor car containing bed and living conveniences.

Carbeen Bloodwood. *Eucalyptus tesselaris* (*H.*). Queensland. Dark chocolate brown. Similar to other eucalypti except for colour. Easily wrought, due to greasy nature. S.G. 1.

Carbolineum. See *Peterlineum*.

Carbonero. *Piptadenia pittieri* (*H.*). Venezuela. Greyish to yellowish brown, streaked on radial surface, lustrous.   Very hard, heavy, strong, and stable.   Roey grain.   Not yet imported.   S.G. ·85. See *Piptadenia* for similar species.

Carborundum.   Carbide of silicon, SiC.   It is made by heating coke with sand in an electric furnace, and is used for oilstones and as an abrasive.

Carbunk.   A trolley for carrying timber in seasoning kilns.

Carcase or **Carcass.**   The main parts of structures and framing before the coverings or decorative features are applied.   The frame only of a piece of cabinet work.   See *Wardrobe.*

Carcassing.   1. Preparing the framing of a structure.   2. The carpentry in a building.

Carder.   A pair of boards, with handles, and faced with fine hook-like pins.   They are used to dress fleece, etc.

Cardinal Wood.   *Brosimum paraense.*   Satiné, *q.v.*

Cardon.   *Cereus sp.* (*H.*).   Giant cactus.   Tropical America. Yellowish with ray flecks.   Characteristics and uses of sycamore.

Card Process.   A preservative pressure treatment with an oil and salt-solution mixture, usually creosote and zinc chloride.   See *Preservation.*

Card Table.   A table specially designed for card playing.   The usual type has a baize-covered top, about 2 ft. 6 in. square, and folding legs.

Cardwellia.   See *Silky Oak* and *Lacewood.*

Careya.   See *Kumbi.*

Cargo Liner.   A superior type of cargo vessel allowed to carry up to twelve passengers.

Caribbean Pine.   Cuban Pine.   See *Pine.*

Carica.   See *Papaw.*

Cariniana.   See *Jequitiba.*

Caripe.   Oity and chozo, *q.v.*

Carling Sole or **Runner.**   A grooved timber, or head, to secure the top edges of framing on ship's decks, and to form the structural part of a cornice. It is fixed to the carlings, or beams, of the deck above.

Car Load.   A load of hardwood consisting of 15,000 ft. super of 1 in. stuff, or its equivalent.   See *Load.*

Carmine.   Used for dyeing wood red colour.   See *Dyes.*

Carnarvonia.   See *Caledonian Oak.*

Carne d'Anta.   Tapirwood, *q.v.*

Carob.   *Ceratonia siliqua.*   Cyprus.   See *Locust.*

Caroba.   See *Fotui* and *Jacaranda.*

Carol.   A recess, or small closet, for quiet study.

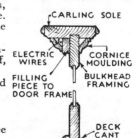

CARLING SOLE

Carolina Pine. See *Pine*.

Carpathian Beech. *Fagus sylvatica*. See *Beech*.

Carpenter's Boast. The half-dovetail joint between the collar-tie and rafter in a collar-tie roof truss, *q.v.*

Carpentry. 1. The structural part of the woodwork in a building, or heavy engineering woodwork. 2. The craft of preparing constructional and structural woodwork.

Carpentry Joints. These are constructional joints, and strength is the chief requirement, hence they should be designed against tension, compression, bearing, and shear. The resistance of the various parts should be equal to obtain the maximum strength of the timber, and the joints should be simple and accurate. Surfaces receiving pressure should be at right angles to the line of pressure ; abutting surfaces should be large enough to resist the thrust ; shrinkage must be considered. The various types are : bird's mouth, bridle, cogged, dovetail (simple), fished, halving, joggle, keyed, mitred, mortise, saddle, scarfed, step, tabled, tenon, trenched, tusk tenon, and their variations. See *Joinery Joints*.

Carpinus. See *Hornbeam* and *Blue Beech*.

Carr. Abbreviation for the name of the botanist Carrière.

Carrack. A three-masted ship.

Carriage. 1. A framed support for the conveyance of goods or men. 2. The wood framing carrying a large bell and its accessories. 3. The cost of conveying goods. 4. The under gear of a vehicle. 5. A vehicle for private use. 6. The framework to which a log is secured during conversion. The name is applied to the appliances used with the carriage : dogs, head blocks, trailer, etc.

Carriage Gates. A pair of wide ornamental gates at the opening of a drive to a house, etc.

Carriage Piece. A rough bearer under a flight of stairs to strengthen wood steps at the middle of their length.

Carriole. 1. A two-wheeled wagon. 2. A small open carriage for one person. 3. A sledge.

Carrol Ironwood. *Backhousia myrtifolia* (*H.*). Queensland. Pink to pale red, smooth. Very hard, heavy, strong. Dense, straight grain, fine even texture. Too hard for ordinary purposes. Used for turnery, handles, etc. S.G. 1·1.

Carroty. A term applied to short-grained wood.

Cart. 1. A horse-drawn vehicle for carrying loads. 2. A horse-drawn two-wheeled carriage, as a dogcart, etc.

Cart Ladder. A framework fixed to the sides of a wagon or cart to increase the carrying capacity for light bulky materials, such as hay.

Carton. A very light box or container. It is usually of cardboard, but may be of very thin wood, like a band box.

Cartouche. 1. A carved tablet in the form of a scroll, or roll, for an inscription. 2. A modillion, *q.v.*

Carvel Built. Applied to a boat in which the side planking presents a flush surface.

Carvel Work. Fitting the edges of side planking of a boat together and caulking them. See *Clincher Work*.

Carya. See *Hickory, Bitternut, Pignut*.

Caryatides. Columns in the form of female figures. See *Atlantis*.

Caryocar. See *Piquia*.

Cascara. *Rhamnus purshiana* (*H*.). U.S.A. Yellowish brown. Moderately hard and heavy, but not strong. Stable and durable. Used locally for posts, turnery, etc. Chief importance is the medicinal properties of the bark. S.G. ·54.

Casco. A proprietary waterproof cold-water glue.

Case. 1. See *Packing Case*. 2. See *Chozo*.

Casearia. *Gossypiospermum*. West Indies boxwood. See *Zapatero*.

Case Bays. The joists framed between two girders. The bays of a floor or roof, except the end, or tail, bays.

Caseboards. See *Boxboards*.

Case Compression. See *Case-hardened*.

Cased Frames. Hollow built-up frames, as for a "sash and frame" window. See *Boxed Frame*.

Case-hardened. A condition of stress, after seasoning, between adjacent layers. It is due to too rapid surface drying in the kiln,

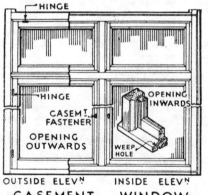

CASEMENT
FASTENER

OUTSIDE ELEV^N   INSIDE ELEV^N
C A S E M E N T   W I N D O W

and causes warping of the wood when further converted. See *Honeycombing* and *Seasoning*.

Casein Glue. See *Glues*.

Casemate. A strong building for defensive purposes.

Casement. 1. A hinged or pivoted sash. 2. A hollow moulding, characteristic of the Gothic style of architecture.

Casement Fastener. A metal fastening for a casement window.

Casement Lights. Windows with hinged or pivoted sashes. Not sliding sashes.

Casement Stay. A metal bar used to fix an open casement in any required position.

FIXED TO FRAME

FIXED TO CASEMENT

CASEMENT STAY

Caserne. A building, or barracks, for housing soldiers.

Case Tension. See *Case-hardened*.

Casewood. Small stuff, in soft white woods, called " off products." It is usually of fir or spruce. The wood must saw cleanly, be free from loose and large knots, not split with nailing, and must be cheap.

Cashaw. *Prosopis juliflora.* See *Mesquite*.

Cashew. *Anacardium occidentale.* See *Espavé*.

Casings. **1.** An alternative name for jamb linings. **2.** Linings to cover rough timbers, pipes, etc. **3.** Framing in cabinet work. **4.** The built-up boxes for the weights for sliding sashes.

Cask. A vessel formed of staves and headings, and held together by hoops. Cask is a general term and includes barrel, brandy, butt, firkin, hogshead, kilderkin, pin, pipe. See *Cooper* and *Staves*.

Casket. **1.** A small box or chest for jewels, etc. **2.** A coffin.

Caskwood. The wood for caskmaking is selected to suit the contents of the cask, as the wood may affect the characteristics of the contents.

Cassia. *Cinnamomum cassia* (*H.*). China. Important for medicine. The wood is distilled for febrifuge. There are other species, W. Indies, etc., but they are not marketed as timber. See *Cana fistola, Ingartu, Indian Laburnum,* and *Urimidi.*

Cassia siamea. (*H.*). India. Also called Beati. Various shades of brown with vary-coloured, to almost black, streaks. Used for turnery, inlays, small fancy articles, etc. S.G. ·8. Also see *Indian Laurel.*

Castanea. See *Chestnut.*

Castanets. A pair of small pieces of hardwood hollowed out to make a clapping noise. They are used to accompany the music in certain dances.

Castanopsis. See *Berangan, Golden Chinquapin, Indian Chestnut, Katus,* and *Wu-tien.*

Castanospermum. See *Black Bean.*

Castilla. See *Tunu.*

Casting. **1.** Twisting or warping. **2.** An object made by pouring the material, while in a fluid state, into a mould. The material sets hard either by cooling or by chemical action. The name is especially applied to iron and brass objects. **3.** The operation of pouring the fluid material into the mould.

Castors. Small swivelled wheels or revolving balls for the feet of furniture.

Casuarina. See *Beefwood, Chow, Oak, Ru,* and *Shea Oak.*

Catalin. A thermo-setting urea-formaldehyde plastic in numerous colours.

Catalpa. *Catalpa bignoniodes* (*H.*). U.S.A. Common catalpa. Greyish brown. Resembles butternut. Light, soft, durable, stable, but not strong. Distinct growth rings, coarse texture. Used locally for construction, joinery, cabinet work, and for exterior work. Of little commercial importance. S.G. ·4 ; C.W. 2. HARDY CATALPA, *C. speciosa*. Similar to above, and marketed together. Also see *Roble* and *Yokewood*.

Catalpa bungei. See *Chiu*.

Catamaran. 1. A raft with grapple and windlass for retrieving sunken logs. 2. A boat propelled by sails, paddles, or steam, and consisting of two hulls bridged by framework. 3. An Eastern raft.

Catblock. One with two or three sheaves and with iron strap and hook attached. It was used to draw the anchor up to the cathead.

Catboat. A small sailing boat with centre board and single mast stepped in the eyes of the boat.

Catch. The part of a fastening that secures the latch. See *Suffolk Latch*.

Catch Boom. A boom across a river to hold floating logs.

Cateban. Mempening, *q.v.*

Caterpillar. Larva. The stage in the development of an insect between egg and chrysalis. They feed on the wood tissue at this stage and are very destructive to tree and timber.

Catface. A partially healed fire scar on a tree.

Cathead. 1. A projecting cantilever used for hoisting purposes. 2. Strong timbers projecting over a ship's bow, to which the anchors are hoisted.

Catherine Wheel. A circular, or rose, window with radiating divisions or bars. See *Rose Window*.

Cathetus. The centre of the Ionic volute.

Catigua. *Trichilia spp.* See *Pimenteira*.

Cativo. See *Cynometra* and *Prioria*.

Cat Ladder. A vertical ladder fixed to a wall for access to a loft, etc.

Catmon. *Dillenia philippinensis* (*H.*). E. Indies. Light to dark reddish brown. Hard, heavy, dense, and durable. Curly and flake grain and coarse. Fairly difficult to work. Used for constructional work, cabinetmaking, veneers. S.G. ·75 ; C.W. 4.

Catostemma. See *Baromalli*.

Cat Spruce. Canadian white spruce. See *Spruce*.

Cattle Guards. Pieces of wood, 7 ft. long, fixed to the bracing wires of telegraph poles.

Caucho. Tumu, *q.v.*

Caul. A plate or shaped appliance of wood, zinc, or aluminium used in veneering. It is applied warm and keeps the veneer in position until the glue has set. See *Veneering*.

Caulicoli. Small volutes under the flowers on the abacus of the Corinthian capital.

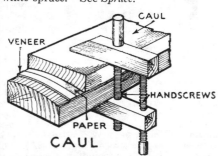

CAUL

Caulk, To. **1.** The act of cogging a joist or binder on to a wall-plate. **2.** To fill, or stop, an open joint to make it water- or steam-tight.
Cav. Abbreviation for the botanist's name, Cavanilles.
Cavanillesia. See *Bongo*.
Cavetto. A hollow moulding in the form of a quarter circle. Often called a Scotia. See *Apophyge*.
Cavils. Strong wood posts on board ship.
Cavity Lath or **Batten.** A piece of wood used to keep a cavity clean from mortar during the construction of a brick cavity wall.
Cavity Moisture. See *Moisture Content*.
c.b. Abbreviation for *centre-beaded*.
c.c. Abbreviation for *country cut*.
C.DC. Abbreviation for the botanist's name, Casimar de Candolle.
Ceanothus. See *Cogwood*.
Ceathus. See *Lilac*.
Cecropia. Several species (*H*.). Tropical America. A light brown wood, light, soft, tough, and strong for its weight. Straight grain, coarse texture, easy to work. The wood is called Ambay, Shakewood, Pumpwood, Imbauba, etc. S.G. ·44. See *Ambay*.
Cedar. A large variety of woods, including both softwoods and hardwoods. Most of them are soft, light, fragrant, very durable, resist insects, and are easy to work. C.W. 1·5 to 3. AFRICAN CEDAR, *Juniperus procera* (*S*.). E. African pencil cedar. Largest of junipers. Rich reddish brown after seasoning. Soft, light, fragrant, durable.

CEDAR

Fine, straight, even grain. Ant-proof. Easy to work and polishes well. Used for pencils, cabinet work. S.G. ·54; C.W. 2. Also see *Agba* and *Guarea*. ALASKA CEDAR. Yellow cedar, *q.v.* ATLAS CEDAR. Atlantic cedar. See *Cedrus*. AUSTRALIAN CEDAR, *Agonis juniperiana*. W. Australia. Yellowish brown. Very strong and elastic. Good substitute for ash and hickory. BASTARD CEDAR.

White cedar. See *Copa* and *Roble*. BAY CEDAR, *Suriana maritima* (*H*.). C. America. Not exported. BORNEO CEDAR. See *Seraya*. C. AMERICAN CEDAR. The standard name for *Cedrela mexicana*. Very often named after the port of shipment or source of origin. See *Honduras Cedar*. CEYLON CEDAR. Lunumidella, *q.v.* FALSE CEDAR. *Cupressus, Juniperus, Thuya*. FORMOSAN CEDAR. *Chamæcyparis*. Like Port Orford cedar. GOLD COAST CEDAR. See *Sapele*. HIMALAYAN CEDAR. See *Deodar*. HONDURAS CEDAR, *Cedrela odorata* or *C. mexicana*. W. Indian, British Guiana, Spanish, or cigar-box cedar, or Indian mahogany (*H*.). C. America. Like soft mahogany. Light red, sometimes mottled, fragrant, durable, stable, easy to work. Used for superior joinery, cabinet work, cigar boxes aircraft, planking. S.G. ·6 ; C.W. 2·5. INCENSE or WHITE CEDAR, *Libocedrus decurrens (S.).* U.S.A. Reddish to dull brown, odorous. Moderately soft, light, and strong. Stable and very durable, fissile. Straight grain, uniform texture. Easily wrought. Sometimes sold as western red cedar. Used for fence posts, pencil slats, etc. S.G. ·35. INDIAN or MOULMEIN CEDAR. See *Toon.* JAPANESE CEDAR. See *Sugi.* LEBANON or TRUE CEDAR, *Cedrus libanotica (S.).* S. Europe and Asia Minor. Yellow brown, light, soft, durable, brittle, and splits easily. Fine, close grain. Easy to work. Used for furniture interiors, etc. See *Deodar.* NIGERIAN CEDAR. See *Moboran* and *Guarea.* NORTHERN WHITE CEDAR. White cedar, *q.v.* PORT ORFORD or OREGON CEDAR, *Cupressus lawsoniana* and *Chamæcyparis l. (S.).* Western U.S.A. Lawson's cypress. Yellowish white, close grain, fragrant, very durable. Large sizes and easy to work. Used for furniture, joinery, aircraft. C.W. 2·5. RED CEDAR. Applied to several varieties. See *Juniper* and *Cypress.* S. AFRICAN CEDAR. See *Boom, Clanwilliam,* and *Mlanje.* S. AMERICAN CEDAR. The standard name for *Cedrela spp.,* or Spanish cedar, from Tropical S. America : Brazil, Guiana, Peru. See *Honduras Cedar.* SOUTHERN WHITE CEDAR, *Chamæcyparis thyoides (S.).* U.S.A. Similar to northern white cedar. See *White Cedar.* SPANISH CEDAR, Honduras cedar, *q.v.* VIRGINIAN or PENCIL CEDAR, *Juniperus virginiana (S.).* Reddish brown, soft, light, fragrant, very durable, fine even grain. Used for pencils, floats, furniture. WESTERN RED CEDAR, *Thuya plicata (S.).* N.W. America. Also called Giant Cedar, Pacific Red Cedar, Western Arbor vitæ. Pink to light reddish brown, weathering to grey. Light, soft, very durable, strong, but brittle. Close, straight, even grain. Stable and easy to work. Resists insects. Stains and polishes. Used for joinery, shingles, plywood, flooring, cheap furniture. S.G. ·45 ; C.W. 1·75. W. INDIAN CEDAR. Honduras cedar. WHITE CEDAR, *Thuya occidentalis (S.)* and *Chamæcyparis thyoides (S.).* U.S.A. Light reddish brown, fragrant. Soft, light, stable, very durable, brittle, and not strong. Straight grain, fine even texture. Numerous knots and subject to wind shakes. Used for shingles, woodware, containers, sidings, etc. S.G. ·32 ; C.W. 1·5. YELLOW CEDAR, *Cupressus nootkatensis* and *Chamæcyparis n. (S.).* Also called Alaska Cedar, Yellow Cypress, Nootka Cypress, etc. Similar to Port Orford cedar. Also see *Cedrela, Cedrus, Ceylon Cedar, Cupressus, Cypress, Juniper, Kaikawaka, Kurana, Podocarpus, Thuya, White Cedar,* and *Stinking Cedar.*

Cedarmah. A name implying cedar-mahogany. It is an unclassified Nigerian timber, and is a species of the Mahogany family with characteristics of cedar. The wood is a chance import and only obtainable as veneer.

Cedar Mangrove. *Carapa moluccensis (H.).* Queensland. A mangrove of the Mahogany family. See *Pussurwood.*

Cedrela, Cedro. Meliaceæ, or Mahogany, family. *C. australis.* See *Toon. C. brasiliensis,* similar to Honduras cedar (Tropical America). *C. fissilis* and *C. herreræ.* S. American cedar. See *Honduras Cedar. C. microcarpa,* similar to toon (India). *C. odorata.* Honduras cedar. *C. serrata,* similar to toon (India).

C. toona. See *Toon*. *C. vermelho* (Brazil). Like Honduras cedar. Also see *Calantas, Cedar, Toon, Chinese Mahogany,* and *Kurana.*

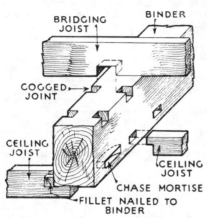

CEILING JOISTS

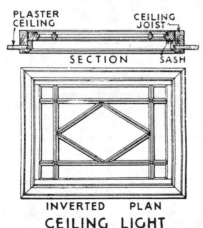

SECTION   SASH

INVERTED   PLAN

CEILING LIGHT

Cedron. Agba, *q.v.*

Cedrus. See *Deodar* and *Lebanon Cedar.*

Ceiba. *Ceiba pentandra* and *Eriodendron anfractuosum.* Also called Araba, Bombax, Kapok, Okha, Silk Cotton Tree, Honduras Cottonwood. Very light. Used for fairings. See *Silk Cotton Tree* and *Silk Cottonwood.*

Ceibo. *Erythrina cristagalli* (H.). C. America. Yellowish white, liable to blue stain. Very light and soft, tough and strong for weight, not durable. Straight and irregular grain, coarse texture, fibrous, feels harsh but smooth. Easily wrought, does not polish easily. Used for floats, canoes, wood shoes, stable floors, etc. S.G. ·25. Also see *Ceiba.*

Ceiling. 1. The inside planks, or lining, of a wooden ship. The floor of a hold. 2. The surface covering to the top of a room. See *Solid Construction.*

Ceiling Boards. Matchboarding, *q.v.,* or building boards, *q.v.*

Ceiling Joists. Light joists, below the floor joists, to carry a lath and plaster ceiling or boards. See sketch.

Ceiling Light. A borrowed light in a ceiling. A lay-light.

Ceiling, Matched. A ceiling formed of tongued and grooved matchboarding.

Ceiling Straps. Narrow strips of wood, nailed to rafters or floor joists, from which ceiling joists are suspended.

Ceiling Trimming. Framing ceiling joists round chimney breasts, or for openings for lay-lights, trap doors, etc.

Celastraceæ. The Staff-tree family : Celastrus, Euonymus, Goupia, Gyminda, Lophopetalum, Maytenus, etc.

Celastrus. See *Orangebark.*

Celcurising. A proprietary method of preservation that is effective against dry rot and insect attack.

Celery Top Pine. *Phyllocladus rhomboidalis* (*S.*). Tasmania. Strong, durable, and stable. Used for carriage building, joinery, and cabinet work chiefly. S.G. ·62 ; C.W. 2.

Cell. 1. A small enclosed cavity. A unit mass of living matter. A structural unit of plant tissue. See *Annual Rings.* **2.** A small room in a monastery or prison. **3.** The panels in a rib and panel vault.

Cellar. A basement, or underground room.

Cellarette. A cupboard, usually to a sideboard, for holding wine bottles.

Cellar Fungus. See *Coniophora* or *Dry Rot.*

Cellular Structure. Varies with different species of wood. Softwoods have a more elementary structure than hardwoods and belong to a lower botanical order. The wood consists chiefly of tracheids with thin walls and large cavities in the springwood and thicker walls in the summerwood. Hardwoods have a more complicated structure with elongated cells, known as fibres and rays, and conducting cells, called vessels.

Cellulose. The principal chemical constituent of the cell walls of wood, forming 70 per cent. of the whole. It is used for a variety of purposes : cellophane, celluloid, clothing, explosives, glass substitutes, lacquer, etc.

Cell Walls. The membrane enclosing the cell contents of wood. It consists of several layers when the cell is matured.

Celotex. A registered wallboard, *q.v.*

Celtis. See *Hackberry, Itako, Nettlewood, Ohia, Stinkwood* (*White*), and *Tala.*

Cembran Pine. *Pinus cembra.* Europe. Also called Alpine, Arolla, Siberian Yellow, Stone, and Swiss Pine. See *Pine.*

Centering or **Centring.** Temporary timbering used as a support for the component parts of an arch or vault during erection. (See illus., p. 90.)

Centering Slips. Tapered supporting strips of wood that allow for adjusting the position of a frame or square of glass, etc., supported by the tapered strips. When the frame or glass is centred correctly the strips are fixed in position.

Central America. The important timbers exported to this country are bulletwood, cocobolo, mahogany, greenheart. Many others are described in the Glossary.

Central American Cedar. The standard name for *Cedrela mexicana.* See *Cedar* (*Honduras*).

Central American Mahogany. The standard name for *Swietenia macrophylla.* See *Mahogany* (*Honduras*).

Centre. 1. A temporary frame or mould used as a support while turning, or building, an arch. **2.** The pith, or core, of a log.

Centre Board. A movable keel, on a yacht, that can be raised or lowered through a slot in the bottom of the craft.

Centre Hinge. Two plates, one with a pin and the other with a socket, used as hinges in cabinet work. They are fixed on the ends of the hanging stile. A *necked centre hinge* is cranked to bring the pin to the face of the door.

Centre Hung Sash. A pivoted sash.

Centre Matched. A term used in veneering when two pieces of equal size are matched, with the joint in the middle of the panel. See *Veneer Matching*.

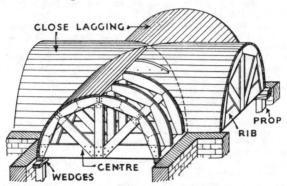

CENTRES AND CENTERING
FOR VAULTS

Centre of Gravity (C.G.). A point in or near a body about which all its parts balance each other. See *Centroid*.

Centre Plates. Wood or metal plates fixed to the ends of heavy pieces to be turned in a lathe. The plates are in pairs.

Centre Points. The conical points for centering the stuff when turning in a lathe.

Centres. Pivots, especially when for swing doors.

Centre-to-Centre. The method of spacing timbers, such as joists, spars, studs, etc., by measuring from the centres or from *in to out*.

Centroid. The " centre of gravity " of a very thin lamina or " area."

Centrolobium. See *Arariba*, *Kartang*, *Muiracoatiara*, *Putumujus*, and *Vermelho*.

Cephalotaxaceæ. A sub-family of Coniferæ, *q.v.*

Cerambycidæ. A family that includes longhorn beetles, *q.v.*

Cerasus. Cherry, *q.v.*

Ceratopetalum. See *Coachwood* and *Satin Sycamore*.

Ceratostomella. Fungi that cause blue stain, *q.v.*

Cercidiphyllum. See *Katsura*.

Cercocarpus. *Cercocarpus ledifolius* (*H.*). N. America. Called Mountain Mahogany. Small sizes and of little commercial importance.

Cereus. See *Cardon*.

Cerillo. Princewood, *q.v.*

Ceriops. See *Tengar*.

Certus Glue. A proprietary casein glue. See *Glues*.

Cesspool. **1.** A lead-lined box in a roof gutter in which the water is collected before it enters the down-pipe. See *Parapet Gutter*. **2.** A large chamber for collecting sewage.

Ceylon. The important timbers exported to this country include cedar or mahogany, ebony, halmillila, palu, satinwood, *q.v.*

Ceylon Boxwood. *Canthium didymum* (*H.*). S. India. Light brown, lustrous, dulls with age. Moderately heavy, but hard and strong. Durable and resists insects. Straight even grain, fine texture, and smooth. Not difficult to work. Turnery and carving. S.G. ·76 ; C.W. 3·5.

Ceylon Cedar or **Mahogany.** See *Lunumidella*.

c.f.m. Abbreviation for *customs fund measure*.

Cha. *Cudrinia triloba* (*H.*). China. Cultivated chiefly for silkworm industry. Wood used for small fancy articles and decorative work.

Cha-cha. East Indian walnut. See *Kokko*.

Chai or **Chooi.** *Sagræa elliptica* or *S. listeri* (*H.*). Andaman bowwood. Yellowish white, lustrous, smooth. Hard, heavy, elastic, strong. Straight grain, fine even texture. Liable to tear in working. Turnery, tool handles, ribs, bentwood. S.G. ·81.

Chain and Barrel. A fastening for a door that allows the door to open a few inches only.

Chain Dogs. Two steel hooks connected by a chain to the crane hook for hoisting heavy blocks.

Chain Figure. A succession of uniform cross markings in decorative woods, suggesting the links of a chain.

Chain Locker. A box on board ship for the anchor chain, etc., when the anchor is not in use.

Chain Mortiser. A machine for cutting mortises in wood by means of an endless revolving chain cutter.

Chain Moulding. A carved moulding imitating the links of a chain.

Chain Timbers. Lengths of wood used to bond and strengthen thin walls, or walls built of small units, especially when the wall is circular in plan.

CHAIN MOULDING

Chair. **1.** A portable single seat with back and sometimes arms. See *Chippendale*. There is a variety of types : arm, bedroom, child's, deck, dining-room, drawing-room, easy, invalid, kitchen, *prie-Dieu*, rocking, tub, Windsor, etc. They are also named according to the style or their special use, as dentist's, etc. Also see *Chaise Longue*. **2.** A small hand-drawn carriage for one person as *bath* or *sedan* chair. **3.** Applied generally to any form of raised support.

Chair Rail. A horizontal moulding fixed to a wall to protect the wall from chair backs, and for ornamentation. A dado rail.

Chaise. A small, light, low carriage with four wheels but no driver's box.

Chaise Longue. A long, low chair on which a person reclines.
Chalet. A small cottage or hut characteristic of a Swiss herdsman's cottage.
Chalk Line. A long string, coated with chalk, used for setting out long straight lines on plane surfaces. It is stretched taut and pulled outwards at the middle ; when released it strikes a straight line on the surface.
Chamæcyparis. Cypress. See *Cedar* (*Alaska, Port Orford,* and *Formosan*), *Southern White Cedar,* and *White Cedar.*
Chamber. 1. An assembly room or an apartment. 2. An enclosed space. A vault. 3. Same as *throat* in saw teeth.
Chambico. *Platymiscium spp.* An attractive hardwood from Mexico. It is like a heavy dark mahogany, with nearly black straight stripes.
Chamfer. A corner bevelled so that the wood is removed equally on each face of the material. See *Bevel.*
Chamfer Bit. A countersink, *q.v.*
Chamfer Plane. A plane specially shaped with an interior right-angled sole for forming chamfers.
Champ. 1. A flat surface. 2. See *Champac.*
Champac. *Michelia champaca* and other spp. (*H.*). India and E. Indies. Yellowish brown to olive brown. Soft, fairly light, lustrous, strong, and durable. Straight even grain and smooth. Polishes well. Difficult to season unless girdled before felling. Used for joinery, carriage building, bentwood, cabinet work. S.G. ·6 ; C.W. 2·5.
Chancel. The area round a church altar enclosed by a screen or communion rail.
Chancel Screen. An ornamental parclose screen used to separate the chancel from the rest of the church.
Chang. Chinese camphorwood. Excellent timber, easy to work, resists insects. Used for wardrobes, chests, coffins, and sleepers to resist white ant. Supplies diminishing rapidly due to distillation of wood for camphor.
Channel. 1. A horizontal timber on the outside of a ship to extend the shrouds from each other and to clear the gunwale. 2. A duct, tube, hollow, groove, etc., especially if used to convey liquids.
Chanter. The holed pipe to a bagpipe for producing the notes.
Chantlate. See *Sprocket-piece.*
Chantry. An endowed chapel for the singing of masses.
Chaparral. A thicket of dwarf evergreen oaks.
Chapel. 1. A nonconformist place of worship. 2. A church subordinate to a larger one, a chapel-of-ease. 3. Part of a church or cathedral specially dedicated with an altar, as a lady-chapel. 4. A private place of worship to a hospital, college, or private house.
Chapiter. The upper part of the capital of a column.
Chaplash. *Artocarpus chaplasha* (*H.*). India. Yellow to golden brown, lighter and darker streaks, darkens with exposure, lustrous but dulls with age. Moderately hard and durable, and resists insects. Straight with slightly interlocked grain. Even texture but coarse, and rather difficult to finish smooth. Resembles satinwood when polished. Liable to check and warp with seasoning unless girdled. Used for construction, joinery, shipbuilding, furniture, wheelwright's work, turnery. S.G. ·56 ; C.W. 3·5.

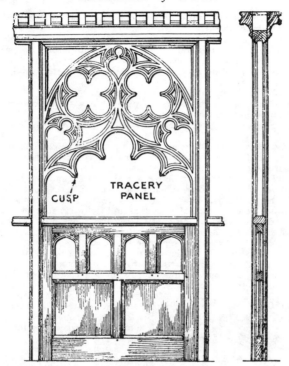

CHANCEL SCREEN

Chaplet. A small cylindrical moulding carved in the form of beads.
Chaplet Block. A wood block supporting a chaplet nail which supports a core independently of the prints in patternmaking.
Chaps. Cracks or fissures in the bark of trees due to frost or very hot sun. The cambium layer is liable to disease due to the exposure.
Chapter House. A building attached to a cathedral or abbey for the assembly of dean and canons.
Chaptrell. An impost supporting an arch.
Char. *Buchanania latifolia* (*H.*). India, Burma. Numerous vernacular names. Brownish grey. Light, soft, not durable, and subject to insect attack. A second-class wood, not imported. S.G. ·5.
Characteristics. Distinguishing features of wood that assist in identification : colour, grain, hardness, lustre, odour, resin content, structure, taste, texture, weight, etc. *Gross* features are those distinguished by observation not requiring more than a pocket lens.

The *minute* structure requires higher magnification. See *Photomicrograph* and *Properties*.

Charcoal. The carbonaceous residue after partial combustion of vegetable, animal, or mineral substances. An important by-product of wood. It is used as a gas absorber (respirators), and for gunpowder, brushes for electrical work, packing, and in steel production, etc.

Chariot. The sliding table to a cross-cut saw.

Chariot Plane. Like a small block plane, but with the mouth close to the front as in a bull-nose plane.

Charley Forest. A good stone for oilstones, quarried in Leicestershire.

Charring. Burning the bottom of wood posts. The charcoal formed on the surface is a good protection against decay. See *Preservation*.

Chartering Agent. One who charters a vessel to transport timber.

Charter Party. A contract between a shipowner and the freighter, or charterer, for the conveyance of goods. There are numerous printed Charter Party Forms covering the sections of the timber trade and between certain ports of shipment and delivery, such as *Baltime* (Baltic and White Seas), *Baltwood* (Baltic and Norway), *Benacon* (N. America, Atlantic), *Merblanc* (White Sea), *Pitwoodcon* (France to Bristol Channel), *Pixpinus* (pitch-pine trade), etc.

Chase or **Chasing.** **1.** A groove or channel cut or formed in the material. **2.** A decorative incising of metal ornamentation.

Chase Form. A wooden mould for forming a channel, or chase, in concrete.

MORTISE

CHASE

CHASE
MORTISE

Chase Mortise. A mortise for a stub tenon with a chase so that the tenon can slide into position. See *Ceiling Joists*.

Chaser. A lumberer engaged in guiding the logs to the drag.

Chasmood. A patent drawer runner. See *Tracks for Drawers*.

Chassis. The body of a coach or wagon. The framework, with engine, wheels, etc., but without coachwork, for a motor car. The undercarriage of an aeroplane.

Chataigne. Chestnut, *q.v.*

Chatiyan. *Alstonia scholaris* (*H.*). India. Shaitanwood. White, darkens to pale brown. Fairly light, soft, lustrous ; not durable. Straight grain, even texture, smooth ; easy to work. Water seasoning best. Should be converted when green. Used for cheap furniture, plywood, carvings, packing cases, tea boxes. S.G. ·47 ; C.W. 2.

Chattering. The vibration of badly fitted plane irons which causes ridges on the surface of the wood.

Che or **Chea.** *Butea superba*. Chinese oaks. Indo-China.

Chechem. *Metopium brownii* (*H.*). British Honduras. See *Honduras Walnut*.

Check. **1.** A rabbet or rebate. **2.** A longitudinal shake in wood that does not go through the whole of the cross-section, due to rapid and faulty seasoning. **3.** Oil or pneumatic floor springs for swing doors. **4.** To check means to compare with results previously obtained, to test for accuracy. **5.** A pattern formed of squares, like a chessboard.

Check Brace. A strap to check, or prevent, swaying in a vehicle.

Checked Joint. A joint to allow for the rounding of internal angles.

Checked Sarking. Sawn boards prepared for overlapping and used for sarking under slates.

Checkerboard and **Chequerlap.** Registered types of wovenboard fencing, *q.v.* Also see *Chequer*.

Chee. *Barringtonia acutangula* or *Butea superba* (*H.*). Indo-China. Marketed as Chinese oak. Like ash, with silver grain. Reddish grey. Moderately light, soft, lustrous, durable, and strong.

CHECKED JOINTS

Straight grain, medium texture, smooth, and fairly easy to work. Silver grain. Liable to warp. Used for cabinet work, panelling, boatbuilding. S.G. ·58 ; C.W. 2·5.

Cheek Blocks. Half shells bolted to the mast-head of a ship, with one bolt used as the pin for the sheave.

Cheek Boards. *Formwork* for the vertical sides of beams.

Cheek Cut. The side-bevel of an inclined member with a splayed cut, as a jack rafter.

Cheeks. **1.** The sides of a mortise or the removed sides of a tenon. **2.** The splayed sides to a fireplace. **3.** The vertical sides of dormer windows. **4.** Jambs, stiles of door frames, etc. **5.** Knee pieces fixed to the bow of a ship and to the head knee. **6.** The sides of the shell of a pulley block. **7.** The face, or projecting part, of a mast carrying the trestle tree on which the top frame and top mast rests.

Cheese Block. A wedge or block to prevent a log from rolling.

Cheesewood. *Pittosporum spp.* Little commercial value. Also see *Whitewood* and *Canary Cheesewood*.

Chelura terebrans. Wood-boring shrimp. It bores just below the surface, which flakes away. See *Gribble*.

Chemical Stains. Stains in wood caused by chemical changes in the materials present in the lumina of xylary cells. Also called Oxidation Stains. They do not affect the strength. When the stains occur during seasoning they are distinguished as *yard brown* or *kiln brown* stains.

Chemnen or **Chamfuta.** *Afzelia quanzensis.* E. Africa. Appearance and characteristics of mahogany. Fairly easy to season and work. Used for panelling, furniture, etc.

Chempedak. See *Keledang*.

Chêne-limba. Afara, *q.v. Chêne* is French for oak.

Chengal or **Chengai.** *Balanocarpus heimii* and *B. maximus* (*H.*). Malay. Also called Penak. Brownish yellow, to dark brown on exposure. Hard, heavy, very strong, and durable. Fine distinct rays, ripple marks. Subject to pinhole borers and dry rot. Not difficult to work. Used for construction, furniture, sleepers, etc. S.G. ·6 to ·8 ; C.W. 3. Also see *Giam* and *Merawan*.

Chequer. **1.** A decoration in the form of squares like a chessboard. **2.** The small cylindrical pieces used in the game of draughts.

Chequerboard. See *Checkerboard*.

Chequered Banding. See *Banding*.

Cherry. *Prunus* (*H.*). Great Britain. Pinkish or yellowish brown, deepening with exposure, resembles birch. Also called Gean, Mazzard, and European, or Wild, Cherry. Three species : *P. avium*, *P. cerasus*, and *P. padus*. Fairly hard and heavy, tough and strong. Small sizes and not durable. Fairly fine even texture. Requires careful seasoning. Curly grain, difficult to work. Used for furniture, turnery, and decorative ware. S.G. ·7 ; C.W. 3. AUSTRALIAN CHERRY, *Exocarpus cupressiformis*. Hard and durable. Used for gun-stocks, tool handles, turnery. BLACK CHERRY, *P. serotina* (*H.*). N. America. Also called Cabinet, Rum, and American Cherry. Pinkish red to dark reddish brown, beautiful wavy grain, lustrous, fine distinct rays. Not difficult to work. Used for cabinet work, airscrews, etc. S.G. ·5. CHERRY ALDER, *Eugenia luehmanni*. Queensland. Similar to English alder, but more suitable for general purposes. S.G. ·55. CHERRY BIRCH, *Betula lenta*. See *Birch*. Also *Schizomeria ovata*. Queensland. Similar to birch (*Betula alba*), but inferior. CHERRY MAHOGANY. See *Makore*. CHOKE CHERRY, *P. virginiana*. U.S.A. Light brown. Hard, heavy, but not strong. Close even grain. PERFUMED CHERRY. *P. mahaleb*. Austria. Reddish brown with greenish streaks. Hard and fragrant. Used for decorative articles, walking sticks, pipes, etc.

Cherry Pickers. Men engaged in towing timber rafts on the River Thames.

Chessel. A cheese vat. A wood tub used in cheesemaking.

Chess-trees. Two vertical pieces on each side of a ship for the clews of the main-sail.

CHESTNUT
(SWEET)

Chest. 1. A large strong box. 2. A chest of drawers is a piece of furniture consisting of a number of long drawers for storing clothes, linen, etc. 3. The air chamber to a pipe organ.

Chesterfield. A form of couch or settee with double ends.

Chestnut. *Castanea vulgaris*, *C. sativa*, or *C. vesca* (*H.*). Britain. Sweet or Spanish chestnut. Resembles oak, but without silver grain. Moderately hard, heavy, strong, and durable. Easy to work and season. Large sizes. Used for construction, joinery, furniture, fencing, casks, coffins, plywood, wheelwright's work. S.G. ·62 ; C.W. 3. AMERICAN CHESTNUT, *C. dentata*. U.S.A. Characteristics and uses as *C. sativa*. INDIAN CHESTNUT, *Castanopsis indica*. Greyish yellow to light greyish brown. Lustrous but dulls later. Moderately hard and heavy. Interlocked grain, coarse uneven texture. Easy to work but not to finish. Fine silver mottle prominent when polished. Used for cheap furniture, panel-ling, shingles. JAPANESE CHESTNUT. Valued for a beautiful "water - mark," which is due to a fungus. MALAY CHESTNUT. See *Berangan*. Also see *Horse Chestnut*.

Chestnut Oak. *Quercus montana.* See *Oak*.

Chev. Abbreviation for the name of the botanist Chevalier.

Cheval Glass. A long, swivelled mirror.

Chevron. A zigzag mould-ing characteristic of Norman architecture.

Chewing-gum Tree. Sapo-dilla, *q.v.*

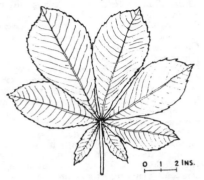

CHESTNUT (HORSE)

CHEVRON

Chewstick. See *Waika*.

chfd. Abbreviation for *chamfered*.

Chi. *Byrsonima crassifolia* (*H.*). C. America. Also called Golden Spoon, Nance, Yaca, etc. Reddish colour. Fairly hard, heavy, durable. Strong but brittle. Roey grain and coarse texture. Small sizes and difficult to work to a smooth finish. Some species esteemed for cabinet work. CHINESE CHI, *Rhus vernicifera.* Called the Varnish Tree. Important for lacquer.

Chicarron. *Comocladia sp.* (*H.*). C. America. Brownish or olive green. Hard, heavy. Irregular hard and soft bands giving strong figure. Interlocked grain. Suitable for superior joinery. S.G. 1.

Chichipate. *Sweetia panamensis* (*H.*). C. America. Variegated brown with stripes, lustrous. Very hard, heavy, strong, tough, and durable. Roey grain, fine texture. Difficult to work. Polishes well. Good substitute for hickory. S.G. 1 ; C.W. 4.

Chicozapote. Sapodilla, *q.v.*

Chicrasei or **Chickrassy.** *Chukrasia tabularis* (*H.*). India, Burma. Mahogany family. Also called Chittagong and Yinma. Light brown and lustrous. Moderately hard and heavy. Interlocked grain, roey and wavy. Even, medium texture. Stable, fairly strong, but not durable. Not difficult to work. Used for construction, furniture, carving, panelling, veneers. S.G. ·62 ; C.W. 3·5.

Chiffonier. An ornamental cupboard. A small sideboard.

Chil or **Chir.** *Pinus longifolia* (*S.*). N. India. Long-leaved pine. Light reddish brown, darker markings, mottled. Fairly lustrous and smooth. Resinous but not durable. Straight and twisted and uneven grain. Medium texture. Moderately easy to work. Used for construction, joinery, veneers, cheap furniture. **S.G.** ·6 ; C.W. 2.

Chili. Timber exports to this country are chiefly cedar, cocobolo, and mahogany.

Chili or **Chilian Pine.** *Araucaria imbricata* or *A. araucana* (*S.*). S. America. The monkey-puzzle tree. Similar wood to hoop pine, *q.v.*

Chimarrhis. See *Waterwood.*

Chimney. A vertical opening in a timber pile to help air circulation.

Chimneypiece. A shelf over a fireplace. Also see *Mantelpiece.*

China. A large variety of timbers exported, especially from Manchuria. The trees are similar to N. European in the northern belt and sub-tropical and tropical in the south. Several trees are peculiar to China.

China-wood Oil. See *Tung Oil.*

Chinchona. See *Cinchona.*

Chine. 1. The projecting rim of a cask or the bevels on the ends of the staves. 2. Part of a ship's waterway above the deck to allow for the caulking of the lower seam of the spirketting. 3. The lowest strake of planking of a barge. 4. The angle formed by the meeting of the top sides and the bottom of V-bottom boats.

Chinese Box or **Myrtle.** *Murraya exotica* (*H.*). Also called Andaman Satinwood. Greyish yellow. Lustrous and smooth, like box. Heavy, hard, not durable, difficult to season. Straight or curly, close grain. Fine even texture. Easy to saw but not to work. Used for tools, inlays, turnery. S.G. ·9 ; C.W. 4.

Chinese Mahogany. *Cedrela spp.* Called Chinensis or Hsiang Chun. Yellowish brown, red markings, fragrant. Stable, strong, and durable. Easy to work. Used for joinery, furniture, cigar boxes, etc.

Chine Stringer. A longitudinal member in a boat between keel and gunwale

Chink. Silver grain, *q.v.*

Chinquapin or **Chinquipin.** See *Golden Chinquapin.*

Chip Basket. One made of thin plaited wood or thick shavings.

Chip Breaker. The pressure bar in front of the cutters of a planing machine to prevent the tearing of the grain.

Chip Carver. A woodcarver's knife.

Chip Marks. Bruises due to chips getting under the pressure rollers in careless machining.

Chipped Grain. Torn grain due to faulty machining.

Chippendale. A term applied to a light delicate-looking type of drawing-room furniture, characteristic of, and named after, a well-known eighteenth century cabinetmaker, *Chippendale*, 1745-1780.

Chir. See *Chil.*

Chirimoya. *Poulsenia armata* (*H.*). C. America. A liana, but it provides useful wood. Greyish yellow. Fairly hard and heavy, not durable. Straight and irregular grain, rather coarse texture, distinct rays. Easily wrought and smooth. Small sizes. Also see *Pond Apple* or *Corkwood.*

Chisels. The following are the types used in woodworking : bevel-edged, coachmaker's, drawer-lock, firmer, millwright's, mortise, mortise-lock, paring, parting, round-backed, round-nosed, sash or pocket, socket, turning. Also see *Mortise Chisel* and *Gouge.*

Chitonga. *Strychnos mellodora* (*H.*). S. Africa. Yellow. Hard,

CHIPPENDALE MAHOGANY ARM-CHAIR

CHIPPENDALE

tough, flexible. Used for spokes, tool handles, etc.

Chittagong. Indian mahogany. See *Chicrasei.*

Chittamwood. *Cotinus americana* (*H.*). U.S.A. Also called Smoke Tree. Orange colour, light, soft, durable; coarse grain. Of little commercial importance. Also see *Ibira.*

Chiu. *Stillengia sebifera* (*H.*). China. The tallow tree. White, close even grain. Used for furniture. Seeds important for tallow.

CHIU TZU SHU, *Catalpa bungei* (*H.*). China. An excellent timber, very durable and resists insects. Used for decorative work, musical instruments, sleepers, poles, etc.

Chive. A cooper's tool for preparing the grooves for the heads of a cask.

Chlorophora. See *Iroko* and *Fustic.*

Chloroxylon. See *Satinwood* (*Ceylon*).

Chock. **1.** A small wood block to resist thrust or pressure or to prevent movement. **2.** A shaped hardwood support, such as anchor, boat, or rudder chock. A deck fixture to secure lines on board ship. **3.** See *Chockwood.*

Chockwood. Squared timbers built into triangular or square formation to form a stack, as a permanent support for the roof of a coal-mine.

Choir Stalls. The pews in that part of a church reserved for the trained singers.

Chokers. A lumberman's term for small logs.

Choking. A term applied to a plane in which the shavings wedge themselves between the two irons due to a badly fitted back-iron.

Chooi. See *Chai.*

Chops or **Saw Chops.** A frame for clamping the edge of a saw whilst the teeth are sharpened.

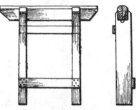

CHOPS, SAW

Chord. 1. A straight line crossing a circle and terminated at each end by the circumference. See *Circle*. 2. A long horizontal main member of a framed girder or truss.

Chores. Waste wood ; firewood.

Chorisia. See *Samohu*.

Chou Chun. Tree of heaven, *q.v.*

Chow. *Casuarina equisetifolia*. Menkabang Penang tree (*H*.). Borneo. Straw colour, hard, heavy, tough, and durable. Close straight grain. Resists insects. Used for piling, structural work, etc. S.G. 1·1.

Chozo. *Licania hypoleuca* (*H.*). Tropical America. Brownish grey to yellowish, very hard, heavy, strong, fairly durable. Roey grain, medium texture. Difficult to work and splits easily. Used for structural work chiefly. S.G. ·7 ; C.W. 4.

Christiania Standard. A measure of timber, consisting of 103½ cub. ft. Now seldom used.

Chrysophyllum. See *Satin-leaf, Osanko*, and *Saffron Tree*.

Chuck. 1. A wood or metal plate, or cup, for securing a piece to be turned in a lathe. 2. The head of a recessing spindle machine. 3. *Chucking* means to fix the stuff to the lathe chucks.

Chuglum, Black. *Terminalia manii* (*H.*). India. Olive grey aging to brownish grey, purplish streaks, fairly lustrous and smooth. Hard, heavy, durable, and resists insects. Straight grain and medium texture. Difficult to season and not easy to work owing to the picking-up of the grain. Very similar to white bombway. Used for construction, furniture, and decorative work. S.G. ·79 ; C.W. 3 5. WHITE CHUGLUM, *Terminalia bialata*. Chrome yellow to greyish yellow on exposure. Logs are selected for false heart, which is grey to olive brown with dark grey streaks and mottles and very ornamental and called *Indian Silver Greywood*. Lustrous, smooth, brittle, fairly durable, but liable to insect attack before conversion. Fairly straight grain, coarse texture, and not difficult to work. Used for furniture, panelling, superior joinery. Non-selected wood used for construction, flooring, etc. S.G. ·7 ; C.W. 3.

Chukrasia. See *Chickrassy*.

Chump Pieces. Short return side rails at the front of a carriage.

Chuncoa. See *Olivier*.

Church Fittings. The interior wood, stone, or metal furniture, such as pews, pulpit, screens, altar table, reading desk, rostrum, etc.

Churn. 1. A large milk container. 2. A butter-making machine.

Churn Butted. See *Swell Butted*.

Chute. An inclined trough down which materials are slid from a higher to a lower level. See *Shoot*.

Chytroma. See *Jarana* and *Guatequero*.

Ciborium. The cover, or canopy, to an altar.

c.i.f. Abbreviation for *cost, insurance, freight*. The contract includes cost of materials, insurance, and conveyance.

Cigar Box Cedar. Honduras cedar. See *Cedar*.

Cilery. Foliage or drapery carved on the heads of columns.

Cill. See *Sill*.

Cimatium. See *Cymatium*.

Cimbia. See *Cincture*.

Cincfoil or **Cinquefoil.** An ornament with five cuspidated divisions. See *Tracery Panel*.

Cinchona. Several species of trees. Important for bark (Peruvian bark) for quinine.

Cincture. A ring, or listel, at the top or bottom of a column.

Cinnamomum. See *Cinnamon, Camphorwood, Karawé, Malligiri*, and *Wild Cinnamon*.

Cinnamon, Wild. *Canella winterana* (*H.*). Tropical America. Olive brown. Very hard, heavy, strong, and durable. Fine grain and texture. The wood is of little commercial value except the species called taggarwood from Madagascar. INDIAN CINNAMON, *Cinnamomum tavoyanum*. Laurel family. Brownish grey to olive grey, lustrous, smooth, fragrant, dur-

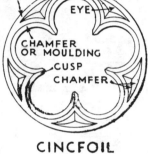

FACE OF TIMBER

EYE

CHAMFER OR MOULDING

CUSP

CHAMFER

CINCFOIL

able, and resists insects. Fairly hard, heavy, and strong. Close straight grain with medium fine and even texture. Easy to work. Used for construction, joinery, furniture. S.G. ·57; C.W. 2·5. Also see *Camphorwood* and *Medang*.

Cippus. A small pillar used as a boundary mark.

Cipre. See *Cypre*.

Circassian Walnut. *Juglans regia*. See *Walnut*.

Circis. See *Judas Tree*.

Circle. The various parts of a circle in geometry are : radius, diameter, circumference, arc, chord, normal, tangent, segment, sector. Area of circle $= \pi r^2$, and length of circumference $= 2\pi r$, where $r =$ radius and $\pi = 3\cdot142$.

Circle-on-Circle. A term applied to work of double curvature, *i.e.* when it is curved in plan and elevation.

Circular Saw. A machine-driven circular disc of steel with teeth round the periphery, for cutting wood. See *Machines, Saw Bench*, and *Saw Teeth*.

cir.f. Abbreviation for *circular face*.

Cirouaballi. *Nectandra speciosa, Licaria spp.* or *Ocotea sp.* (*H.*). Guiana. Also called Silverballi. A species of greenheart. Hard, very durable, and resists insects. Polishes. Used for exterior woodwork, cabinet-making, aircraft, turnery. S.G. ·8; C.W. 3·5.

Ciruelo. *Astronium spp.* (*H.*). Honduras. Cherry red, dark brown or black stripes. Very hard, heavy, strong, and durable. Fine uniform grain. Excellent wood, used chiefly for cabinet work.

Cistanthera. See *Danta* and *Otutu*.

Cistern. 1. A large wood trough on board ship. Used to hold sea-water to wash the decks, and placed in the well below the orlop-deck. 2. A storage tank for water.

Cither or **Cittern.** A musical stringed instrument like a lute. A zither.

Citrus. A genus of trees of the Rutaceæ family, cultivated for fruit rather than timber : citron, grapefruit, kumquat, lemon, lime, orange, tangerine. The wood is orange-yellow colour to brownish, and of fine and even texture. The density varies considerably with different species, but it is usually hard and heavy. Some species are fragrant and some oily. Also see *White Cornelwood*.

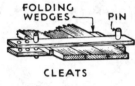

FOLDING WEDGES — PIN

CLEATS

IRON CLAMP

CLAMPS or **CLEATS**

City-wood. Applied to the best quality mahogany, shipped from San Domingo.

Cladrastis. See *Yellow-wood*.

Clamp. 1. A temporary wood or iron cramp for holding two or more pieces together until the glue has set. See *Cleat*. 2. A batten fixed across the grain of wood to prevent warping. See *Counter Cramp*. 3. S-shaped, or corrugated steel, clamps driven in the ends of logs to prevent checks before conversion. 4. Strong timbers on the inside of a ship's side, from stem to stern, to carry the beams. 5. See *Veneer Caul*.

Clamp Carrier. See *Clamping Machine* (2).

Clamping. Holding together by means of a clamp.

Clamping Machine. 1. A machine that automatically cramps up doors or sashes. 2. A machine for holding glue-jointed boards together until the glue has set. It consists of a number of revolving arms or carriers on which the glued boards are placed. Foot pressure brings the clamp into operation and swings it out of position to be replaced by the next carrier. Another type is on a travelling table, in which the carriers swing vertically.

Clanwilliam Cedar. *Widdringtonia juniperoides*. S. Africa. A light, durable, fragrant softwood. Excellent timber, but local demands equal supplies.

Clap Boards. 1. Weather boards on the sides of a house. Quarter-sawn boards tapering in thickness for overlapping. Roofing boards larger than shingles. See *Weather Boarding*. 2. Also applied to a stave for a cask.

Clara Walnut. *Juglans hindsii* or *J. californica*. Californian walnut, *q.v.*

6" x 7/8" TO 1/4"

CLAP BOARDING

Clarionet or **Clarinet.** A wood-wind musical instrument, or an organ reed stop with similar sounding notes.
Clarisia. See *Oiti.*
Clark's Bit. A patent expansion bit for use with the brace.
Clash. Silver grain, *q.v.*
Clasp Nails. Wrought or cut nails with large, nearly flat, heads to clasp the wood.
Classical. Applied to Greek and Roman Orders of architecture, or to anything that has characteristic features of these Orders.
Classification. See *Family*, *Genus*, *Species*, and *Botany*. Also see *Hardwoods*, *Softwoods*, *Grading*, and *Shipping Marks.*
Clavichord. A musical instrument similar to, and the forerunner of, the piano.
Claw. See *Ball and Claw* and *Hammer*. Also see *Timber Connectors.*
Cleading. 1. The timbering for an excavated shaft. 2. A jacket, formed of narrow strips of wood, round a hot-water cylinder or boiler, to prevent loss of heat by radiation. A layer of non-conducting material is usually packed between the boiler and the cleading. 3. The lining to a cart or wagon.
Clean Bill. See *Bill of Lading.*
Clean Cargo. One consisting of regular-sized timbers, such as deals.
Cleaning Up. The finishing of joinery and cabinet work, which includes smoothing, scraping, and sandpapering. See *Working Qualities.*
Clean Stuff. A term applied to good quality wood, free from knots and other defects.
Clearance. 1. A space between two surfaces to prevent contact, as between the edge of a door and the rebate of the frame. 2. A custom's certificate that allows for the departure of a ship from port. 3. See *Saw Teeth.*
Clear and Better. A marketing term in the grading of N. American woods such as western red cedar, etc.
Clear Length, or **Trunk.** The lower part of the stem of a tree that is free from branches.
Clear Lumber. Applied to timber free from defects such as knots, shakes, etc.
Clear Space. The actual space between two members, as between two joists, spars, etc. An alternative method of spacing to *centre-to-centre.*
Clear Span. The distance between the walls supporting a roof-truss or beam.
Cleats. 1. Small battens. 2. A piece of wood fixed to a wall to carry a bracket, hook, shelf, etc. 3. A bearing block nailed under a beam. 4. Temporary wood clamps used when jointing long timbers. See *Clamp.* 5. A block receiving the thrust from an inclined member. See *Raking Shore.* 6. Narrow pieces nailed on an inclined platform or scaffold to give foothold. 7. See *Ship's Cleats*. Also see *Straining Beam.*
Cleavability. A tendency to split. Splitting properties of wood, depending upon transverse strength and tensile strength across the grain. See *Fissile.*

Cleft. Applied to wood that is split instead of sawn, such as plasterer's laths, shingles, etc.

Clench. See *Clinch*.

Clerestory. 1. A row of windows near ceiling level in a high hall. The top tier of windows lighting the choir and nave of a church. 2. Headroom projecting above the top of a vehicle.

Clerk of Works. The person representing the owner and architect on a building during erection. His duties are to see that the work and materials are as specified.

Clevis. The forked end of a rod forming a connection.

Clew-garnet Block. One with single sheave suspended from the yards of a ship by a strap, with two eyes for the lashing. The clew-line block is similar.

Clinch. Turning over the projecting point of a nail to form a kind of rivet to prevent withdrawal.

Clincher. See *Clinker Built*.

Clinker Beech. See *New Zealand Beech*.

Clinker Built. Applied to a boat built of overlapping boards.

Clipper. A sailing ship with sharp bow for speed.

Clipping. Trimming the edges of veneers or squaring the ends of boards.

Cloakscreen. A screen in a cloakroom with hooks for hats and coats.

Clock. See *Moulding Box*.

Clock-case. The wood frame enclosing the mechanism of a clock. The name is especially applied to one standing on the floor, as a grandfather clock.

Clogs. Footwear with wood soles, usually of birch or alder.

Cloister. A covered arcade to a college or monastery. A form of corridor round a quadrangle.

Close Bark Willow. See *Willow*.

Close Boarded. Applied to a roof that is boarded before slating.

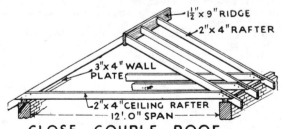

CLOSE COUPLE ROOF

Close Couple. Roof timbers consisting of pairs of rafters tied together at the feet to prevent spreading.

Close Cut Hip. See *Mitred Hip*.

Close-fights. Protecting bulkheads in old ships, fore and aft, in the event of the enemy boarding the ship.

Close Grain. Applied to wood with narrow growth rings. See *Grain*.

Close String. A string straight on the top edge and into which the treads and risers are housed. See *Cut Strings* and *Strings*.

Cloth Boards. See *Lapping Boards*.

Clothes Horse. Two skeleton frames hinged together by webbing for airing clothes.

Clothes Press. A cupboard for storing clothes.

Clothes Rail. Slender rails framed together and suspended from a ceiling by means of pulleys and cord, and used for airing clothes.

Clothes Tree. A stand with branched arms for hats and coats.

Clout Nail. A short nail with large flat head for fixing felt and cords. See *Nails*.

clr. Abbreviation for *clear*.

Club. A stick with head, or knob, used as a weapon ; or specially shaped for such games as hockey and golf.

SINGLE PULLEY
CEILING
DOUBLE PULLEYS
WALL
CLEAT AND HOOK
CLOTHES RAIL

Club Foot. A bulbous foot to a cabriole leg, *q.v.* It is a characteristic feature of early Chippendale and Queen Anne furniture.

Clusia. *Clusia rosea* (H.). Tropical America. Reddish brown with streaks. Hard, heavy, strong, durable. Fine, even grain. Also called Aralee, Awasakuli, Balsam Fig, Copey, Cape, etc. Used for constructional work and furniture. S.G. ·9.

Clustered Column. A column formed of a number of small columns symmetrically grouped.

Cluster Pine. *Pinus pinaster*. See *Pine*.

c.m. Abbreviation for *caliper measure, q.v.*

Coach. **1.** A railway carriage consisting of several compartments for passengers. **2.** A heavy type of four-wheeled, horse-drawn, closed carriage ; especially applied to the type used for funerals. **3.** An apartment for the captain, near the stern, of old warships.

Coach Screw. A large screw with a square head and driven by turning with a spanner. See *Bolts*.

Coachwood. *Ceratopetalum apetalum* (H.). New South Wales, Australia. Pale brown or pinkish, darkening with exposure, smooth, lustrous, fragrant to odorous. Moderately heavy, soft, but tough and strong, not durable, shrinks greatly. Close, but rather short, grain. Requires slow and careful seasoning, easily wrought. Limited supplies. Used for coachbuilding, cabinet work, musical instruments, plywood, implements, spools, etc. S.G. ·62 ; C.W. 3.

Coak. A joggle or dowel in a scarfed joint to prevent one surface from sliding on another due to horizontal shear.

Coal-box. A domestic receptacle, or scuttle, for holding coal near to a fireplace.

Coal Tar. See *Creosote*.

Coaming. A stout piece bolted to a ship's deck to form a sill for deck houses, etc. A raised border to a cockpit, or to a hatch to keep it free from water. Vertical linings to deck openings : cockpits, hatchways, etc.

Coarse. See *Texture*.

Coarse Grain. Applied to wood with wide growth rings, denoting rapid growth. See *Grain*.

Coastal Sand Cypress. *Callitris columellaris.* Queensland cypress pine, *q.v.*

Coatings. Materials applied to the surface of wood as a preservative : cellulose, creosote, oil, paints, polish, varnishes. See *Preservation*.

Cobbling. Repairing casks.

Coble. A flat-bottomed fishing boat with large sail. A small flat-bottomed boat used for salmon fishing.

Coca or **Cacao.** *Erythroxylon sp.* (*H.*). Tropical America. Also called Aroba. Reddish brown. Very hard, heavy, strong, and durable. Used for wheelwright's work, cooperage, sleepers, poles, turnery. S.G. ·9. The name is also applied to species of *Sloanea*.

COCK BEADS

Coccoloba. See *Sea Grape*.

Coccos laricis. An insect very destructive to larch.

Cock. 1. A tap with spout 2. A weather vane.

Cock Bead. A bead standing above the surface of the material.

Cock-boat. A small boat.

Cockchafer Beetle. *Leucopholis.* A destructive forest pest in the East.

Cocking. 1. Fixing the hinges of a door so that the nose rises as the door opens. The knuckle of the bottom hinge projects more than that of the top hinge. This is necessary when the door frame is not plumb, or if the door is required to clear an obstruction. 2. Cogging or caulking.

Cock Loft. A small room, or space, in a roof just under the slates. It is generally used for storage purposes.

Cockne. Abbreviation for the name of the botanist Cockayne.

Cockpit. 1. An open chamber in the nacelle, fuselage or hull of aircraft, etc., for the pilot or crew. 2. Quarters, below deck, for junior officers or for wounded sailors. 3. An open steering well in a motor boat or small yacht.

Cockshead. A trefoil. See *Tracery*.

Cockspur Fastener. A casement fastener, *q.v.*, with a small bellcrank lever rotating on a plate and engaging with a slot on the frame.

Coco. See *Sapucaia*.

Cocoa. *Theobroma spp.* (*H.*). Widespread in the tropics. The tree is more valuable for its seeds than for its wood. The wood is light, soft, firm, with distinct rays, but it is of little commercial value. See *Sterculiaceæ*.

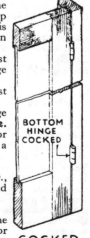

BOTTOM HINGE COCKED

COCKED HINGE

Cocoanut. *Cocus nucifera* (*H.*). Tropics. Indian palm. Also called Porcupine wood. Dark brown with black stripes. Hard, heavy, and very durable. Fine close grain. Small sizes. Used for decorative work and fancy articles. Stems used for sea-defence work. Important by-products : coir and copra. S.G. ·85.

Cocobolo. *Dalbergia retusa* (*H.*). Tropical America. Variegated, yellowish to rose or deep orange, irregular purple and black markings. Beautiful wood. Very hard, heavy, and durable. Interwoven grain. Fine uniform texture. Stable, and polishes well. Small sizes. Used for fancy articles, decorative work, and turnery. S.G. 1·1. INDIAN COCOBOLO, *D. cultrata*. Burma blackwood. Brown with purplish or black markings. Very hard, heavy, strong, and durable. Straight grain, coarse texture, dull, fragrant, polishes well. Difficult to season and very difficult to work when seasoned. Not so decorative as *D. latifolia* (Rosewood). Used for decorative work, turnery, etc. S.G. 1·1 ; C.W. 5. Also see *Sea Grape*.

Cocos, or **Cocus.** *Brya ebenus* (*H.*). Tropical America. W. Indies ebony. Yellowish to dark greenish brown, darker markings. Darkens with exposure. Very hard, heavy, and durable, but brittle. Roey grain, fine uniform texture. Small sizes, polishes well. Used for fancy articles, musical instruments, tools, door knobs, etc. S.G. 1·2 ; C.W. 5. See *Porcupine*.

Coelocaryon. See *Umokæze*.

Coelostegia. See *Punggai*.

Coffee Tree. *Gymnocladus dioicus* or *G. canadensis*. Kentucky coffee tree, *q.v.*

Coffeewood. *Coffea arabica*. See *Partridgewood*.

Coffer. 1. A small chest. 2. A deeply recessed panel in a ceiling or soffit.

Coffer Dam. A caisson, *q.v.* Two rows of piles and sheeting filled with clay to form a watertight compartment for underwater work.

Coffin Timbers. Chestnut, elm, hazel pine, oak, pitch pine, satin walnut.

Cog. 1. A key to prevent lateral movement in a carpentry joint. 2. A tooth in geared wheels. 3. The supports for lorry and motor bodies. 4. A small fishing boat. 5. Same as *chock*.

Coga. A patent automatic door catch.

Coggin. A wooden milk pail.

Cogging. A carpentry joint to prevent lateral movement between two pieces at right-angles, and where one supports the other, as a joist on a wall-plate or binder. See *Ceiling Joists*.

Cog Wheel. Usually implies a wheel with hardwood teeth, or cogs.

Cogwood. *Ceanothus chloroxylon* and *Zizyphus spp.* (*H.*). W. Indies. Buckthorn family. Yellowish green. Very hard, heavy, tough, and durable. Very dense and uniform. Used for rollers, cogs, machine parts, etc. S.G. 1·1. Also see *mistol*.

Cohucho. *Zanthoxylum cocoa* (*H.*). S. America. Light brown, close grain. Not exported. Used locally for furniture, carriage building. S.G. ·6.

Coir. The fibrous husk of the cocoanut.

Cola. See *Ogugu*.

Cold Frame. A wood frame with glazed top for hardening off young plants in gardens and nurseries.

Cold Press Resins. Urea-formaldehyde thermo-setting resins used for assembly work. They do not require heat for setting, but they are comparatively quick acting with a time interval of fifteen minutes. See *Hot Press* and *Plastic Glues*.

Coleoptera. See *Bark Beetle*.

Collapse. Irregular and excessive shrinkage during the drying of wood.

Collar. 1. The annulet on a column. 2. See *Collars*. 3. The lower end connections of the principal stays of a mast of a ship.

Collar Beam. The " tie beam " of a roof truss when at a higher level than the feet of the rafters, or above wall-plate level.

Collar-beam Roof. Similar to a couple-close, but with the tie-beam raised above the level of the wall-plate. See *Collar Tie* and *Hammer-beam Truss*.

Collar Joint. The joint between the rafter and tie in a collar roof. It is often a half-dovetailed halving and called the " carpenter's boast."

Collar Roof. One formed of close-couples or collar-beam rafters.

Collars. The topwood and limbs of trees after converting to logs.

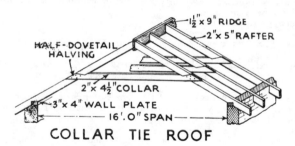

COLLAR TIE ROOF

Collar Tie. A collar-beam roof.

Collets. Spindle cutters.

Colombian Mahogany. *Cariniana pyriformis* (*H.*). C. America. Very similar to true mahogany, but more difficult to work as it soon dulls the tools. S.G. ·67.

Colonnade. A series of columns, bridged by lintels, and forming the side of a narrow covered path outside a building.

Colophony. Rosin, *q.v.*

Colorada. See *Fir* and *Quebracho*.

Colour. The colour of the woods included in the Glossary is given for the heartwood only. Colour in wood is due to the deposit of certain substances in the wood elements. Seasoning often changes the colour, probably due to oxidation of the mineral and chemical substances. Variation of colour due to decay may increase the value of the wood for decorative purposes. The difference in colour between the sapwood and heartwood of hardwoods is usually very marked.

The sapwood is lighter and usually whitish grey. In some trees it is hardly distinguishable and the amount of sapwood varies considerably with different species. See *Discoloration, Heartwood, Sapwood, Blue Goods, Dyewoods, Polychromatic,* and *Stain.*

Colubrina. *Colubrina rufa (H.).* C. America. Buckthorn family. Also called Snakewood. Yellowish red, darker markings. Very hard, heavy, strong, and durable. Fine texture. Rather difficult to work. Used for cabinet work, shipbuilding, sleepers. S.G. 1 ; C.W. 4. See *Alphitonia, Red Almond,* and *Snakewood.*

Columbian Pine. See *Douglas Fir.*

Column. A cylindrical post or pillar

Colura. See *Tester.*

Comb Cleat. See *Ship's Cleats.*

Comb Grain. The grain of quarter-sawn wood, *q.v.*

Combination. Applied to tools and machines that perform several different operations. See *Universal Plane* and *Tool Pad.*

Combination Beam. A flitched beam, *q.v.*

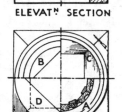

COLUMN CASING

Combretaceæ. The White Mangrove or Indian Almond Tree family, including Buchenavia, Calycopteris, Combretum, Conocarpus, Guiera, Laguncularias, Lumnitzera, Terminalia (Bucida).

Commercial Veneers. Plain rotary-cut or inferior veneers used for concealed parts of furniture, inner plies of plywood, etc.

Commifera. See *M'Kunguni.*

Commode. 1. A small sideboard or low ornamental cupboard. 2. A night-stool.

Commode Step. A step with curved, or bowed, riser, at the foot of a flight of stairs. See *Curtail Step.*

Commode Table. A console, or pier, table with marble top. It is a characteristic of Adam furniture. See *Console Table.*

Common. A term applied to many timbers to denote the best known or British species.

Common Furniture Beetle. *Anobium punctatum.* The cause of worm-holes in old woodwork. Also see *Furniture Beetles.*

Common Joists. The joists in a *single floor.* Bridging joists.

Common Pitch. 1. The pitch of a roof when the spars are about three-quarters of the span in length. 2. The pitch of fliers above and below winders, in stairs

Common Rafters. The rafters carrying the roof covering, as distinct from the principal rafter which is part of the framed truss. See *Box Gutter* and *King-post Truss.*

Communion Rail. An ornamental rail supported on short pillars surrounding an altar. (See p. 110).

Communion Table. A church altar at which Holy Communion is celebrated.

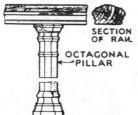

SECTION OF RAIL

OCTAGONAL
PILLAR

COMMUNION RAIL

Comocladia. See *Chicarron*.

Companion. 1. A skylight on the quarter-deck of a ship to light a cabin below. 2. A ladder or staircase leading from deck to cabins on board ship.

Companionway. A watertight shelter to a stairway on board ship. They have been superseded by deck-houses on modern ships.

Comparative Workability. A value comparing the labour costs involved in finishing joinery, cabinet work, etc. The wood selected as the unit for purposes of comparison is yellow, or white, pine. See *Working Qualities*.

Compartment. A part partitioned off from anything to form a smaller unit. One of the separated parts of a railway carriage.

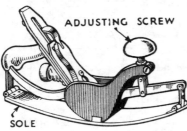

ADJUSTING SCREW

SOLE

COMPASS PLANE

COMPASS ROOF

Compass Plane. A smoothing plane with adjustable sole for circular work.

Compass Rafter. A curved rafter.

Compass Roof. 1. A semicircular roof. 2. A span roof. 3. A roof consisting of pairs of rafters with inclined ties which also serve as struts.

Compass Saw. A narrow saw for cutting circular arcs of small radii.

Compass Window. A semicircular bow-window.

Compass Work. Circular work.

Compo Boards. See *Building Boards*.

Compositæ. A family including Artemisia, Bæcharis, Bigelovia, Brachylæna.

Composite. 1. The last of the Roman orders of architecture. It is composed of parts of other orders. 2. A frame built of wood and metal.

Composite Carriage. A railway carriage with first-, second-, and third-class compartments.

Composite Door. A fire-resisting door made of wood faced with metal.

Composite Truss. A roof truss of wood and steel. The compression members are of wood and the tensile members of mild steel.

Compound Arch. An arch formed of a series of concentric arches placed successively within and behind each other.

Compound Beam. 1. A flitched beam, *q.v.* Also applied to one without an iron flitch. See *Beam*. 2. A built-up beam or girder.

Compound Shake. See *Shake*.

Compreg. A registered plastic-treated wood product.

Compression. A force tending to shorten a structural member by crushing the fibres longitudinally. The approximate compressive strength of timber, parallel to the grain, equals $SG \times 10^4$, when the moisture content equals 15 per cent.

Compression Wood. Timber from that part of a conifer that has been under compression, as under a branch or on the concave side of a bent tree. The cells are shorter than normally, thicker walled, and show spiral markings, and the wood is darker coloured and glassy.

Concave. Hollow. The reverse of convex.

Concentrated Load. A *point* load. A load resting on a very small portion of a beam. See *Distributed Load*.

Concentric. 1. Applied to circular arcs struck from the same centre but with different radii. Circles or spheres having different radii but a common centre. 2. Annual rings.

Concha Satinwood. Harewood, *q.v.* Also see *Satinwood*.

Conches. Shell-shaped roofs, or half domes.

Conchoid. The locus of a point originated by Nicomedes and from which is derived the entasis of a column, *q.v.*

Condensation Groove. A channel to collect the condensation on the inside of windows. A *weep-hole* allows the moisture to escape to the outside.

Conditioning. Adjusting the moisture-content of wood.

Cone. 1. The fruit, or seed vessel, of pines and firs. See *Coniferæ*. 2. A solid generated by a line (generator) revolving about an axis, at a fixed angle and point of intersection. Volume $= \frac{1}{3}\pi R^2 H$. Surface $= \pi RL$.

Confidante. A form of sofa with seats at the ends.

Conge. A cavetto joining the base or capital to the shaft of a column. See *Apophyge*.

Congowood. African walnut, *q.v.*

Conical Roof. A cone-shaped roof, such as a steeple with circular base.

Conic Groins. The groins produced by the intersection of conical vaults.

CONE

Conic Sections. Circle, ellipse, parabola, hyperbola.

Coniferæ. The important family of Gymnosperms that provides the bulk of our commercial woods. It is subdivided into seven families : Araucariaceæ, Cephalotaxaceæ, Cupressaceæ, Pinaceæ, Podocarpaceæ, Tax-

CONIC SECTIONS

CIRCLE   PARABOLA
ELLIPSE   HYPERBOLA

aceæ, Taxodiaceæ, *q.v.* The woods include cedar, cypress, fir, juniper, larch, pine, tsuga, yew. The trees are chiefly evergreen, cone-bearing, and needle-leaved. The timber is usually resinous, and is classified as softwood. There are 46 genera and 500 species. The wood tissue consists nearly entirely of tracheides with bordered pits, and has no vessels. The trees are profusely branched, and the leaves, or spines, are unbranched. See *Softwoods* and *Gymnosperms*.

Coniferous. Applied to woods from conifers : softwoods or non-porous woods.

Coniphora Cerebella. The cellar fungus. This disease requires very wet conditions. The fungus forms in strands and not in sheets, or cushions, and is yellowish brown to black in colour. The fructification is white in colour. The timber goes dark brown and splits longitudinally. See *Dry Rot*.

Conk or **Conky.** Applied to wood that has decayed due to a fungus.

Conoid. A solid or surface formed by a horizontal line which moves so that it always touches a vertical line at one end and a semicircle or semi-ellipse at the other. The solid is common in double-curvature work.

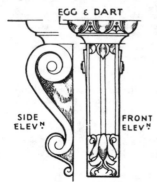

EGG & DART

SIDE ELEV?

FRONT ELEV?

CONSOLE BRACKET

Conoid Vaulting. Fan vaulting.

Conservatory. A greenhouse. A place for conserving tender or exotic plants.

Console. A large ornamental bracket, or corbel, usually with curves of contraflexure. A modillion. Also see *Ancone*.

Console Bracket. An ornamental bracket to the quarter gallery of a ship.

Console Table. A narrow table against the pier of a wall. It is supported by decorative brackets or curved legs and supports a large wall mirror, and is sometimes called a commode table, *q.v.*

Console Table Mirror. A large mirror fixed on brackets on a wall.

Constantine Measure. A New York method of measurement for logs of cedar and mahogany. The method allows for various deductions.

Constructional Work. Woodwork of a structural character, as the carcassing of buildings. Carpentry. The considerations for the selection of the wood are : cost, strength, durability. This Glossary distinguishes structural work, for purposes of selection, as heavy engineering work ; bridges, dock work, shoring, piling, etc.

Consuta Plywood. A stitched and glued plywood used in aircraft and shipbuilding. It is chiefly of cedar and mahogany and in sizes up to 60 ft. long, 8 ft. wide, and from $\frac{1}{8}$ in. to $\frac{3}{8}$ in. thick.

Container. A box in which goods are packed for transit. It may be of thin wood, plywood, or cardboard.

Content. See *Moisture Content*.

Continuous Beam. One beam continued over more than two supports.

Continuous Handrail. A handrail for geometrical stairs. The returns are formed by *wreaths* instead of by newels.

Continuous Impost. Applied to the supports of an arch in which mouldings on the arch are continuous to the ground.

Continuous Lights. A row of windows opened or closed in one operation by means of one piece of mechanism.

Contract Forms. Special agreement forms used between buyer and seller in the timber trade. The forms are generally known by code words such as *Albion, Finbob, Rufod, Uniform,* etc.

Contraction. A loss of volume. See *Shrinkage.*

Contraflexure. **1.** Contrary flexure. A continuous curve that reverses in direction, as in a cyma recta. **2.** The *points of contraflexure* in a fixed or continuous beam or column are the positions where the compressive and tensile stresses reverse.

Control Stick. The pillar with the mechanism for controlling aircraft.

Control Surface. The movable parts of wings and tail of aircraft, to control flight.

Converting or **Conversion.** Sawing timber into smaller sections. *Flat-* or *bastard*-sawn wood is cut tangentially to the annual rings. This produces handsome grain, due to irregular annual rings, in such woods as walnut and pitch pine, and is called flat grain. *Rift-* or *quarter*-sawn wood is cut radially, and gives edge, vertical, or straight grain. This method is best for floorboards and produces silver and

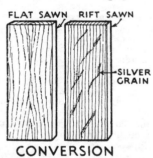

CONVERSION

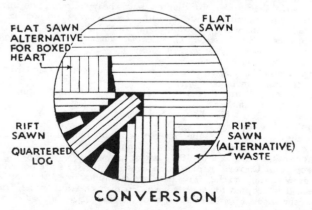

CONVERSION

ribbon grain, and the wood shrinks and warps less than when flat-sawn. *Rotary-cut* wood is sliced, or peeled, circumferentially for plywood. This produces unnatural grain, but is good for bird's eye and similar irregularities. *Slicing* is shaving a thin layer off a flat surface, and is the method generally used for veneers, *q.v.* See *Annual Rings* and *Shrinkage.*

Convolution. A fold, or revolution, of an expanding curve, as in a spiral, volute, or scroll.

Cooking. Same as moulding, *q.v.*

Coolibah. *Eucalyptus microtheca* (*H.*). Australia. Chocolate brown with white, wavy thread-like markings. Probably the hardest and heaviest wood known. Lustrous, interlocked grain. Difficult to work. Too hard for ordinary purposes. Excels lignum-vitæ for bearings, bushes, bowls, etc. S.G. 1·3.

Coom. 1. See *Camp Ceiling.* 2. Centring for bridges.

Coop. 1. A box, or cage, for birds or animals. 2. A cask or tub.

Cooper. A craftsman engaged in the making of casks. *Cooperage* is the art of making casks, and is usually classified as *wet* or *tight, dry,* and *white* cooperage.

Coopering. 1. Building up cylindrical surfaces with staves in cabinet work. 2. The craft of making casks.

Cooper's Joint. A mitred joint with tongue for cylindrical or polygonal work. See *Canted.*

Cooper's Shave. A curved draw-knife used in coopering.

Cooper-ware. The butt ends of ash poles, 16 to 18 ft. long, for cooperage, etc.

Copa. *Icaca panamensis* or *Protium spp.* (*H.*). C. America. Also called Bastard Cedar. Brown with golden lustre, smooth. Rather light and varies from soft to hard. Stiff, fairly strong, varies in durability. Straight grain, medium texture, distinct rays. Easily wrought. Little imported. S.G. ·5 ; C.W. 2.

Copaia. See *Fotui.*

Copaiba. *Copaifera officinalis* (*H.*). C. America. Reddish brown, dark streaks, lustrous. Moderately hard, heavy, strong, and durable. Fairly straight grain, medium texture. Easy to work and polishes well. Used for cabinet work, shipbuilding, turnery. S.G. about ·7 ; C.W. 3.

Copaifera. See *Bubinga, Copaiba, Copalwood, Kevazingo, Purpleheart, Pao d'oleo, Rhodesian Mahogany.*

Copal. See *Copa* and *Gumbo.*

Copalwood. *Copaifera coleosperma.* The suggested standard name for Rhodesian mahogany, *q.v.*

Cope. 1. The top, or cover, to a moulding box for casting metal. 2. Same as coping, *q.v.*

Copey. Cuban *Clusia rosea.* See *Clusia.*

Copie. See *Cupiuba.*

Coping. 1. Scribing the intersections of mouldings instead of mitreing. 2. A covering to protect from rain.

Copperas. Used as a dye for " silver greywood." See *Dyes.*

Coppice or **Copse.** A plantation in which the trees are not allowed to grow to timber size. The trees are cut and allowed to stool. The method is adopted for such as ash, Spanish chestnut, hazel, willow, etc., the shoots of which have a market value. See *Sprout Forest.*

Copsewood. Branches cut from a coppice, *q.v.*

Copying Lathe. A lathe that automatically turns the object like a copy, or *dummy*. Also called a Spoke Lathe.

Coquillos. The thick shell of the nut of the palm *Attalea funifera*. It is used for fancy articles, inlays, etc.

Corail. African padauk, *q.v.*

Coralline. Used as a dye for its rose colour. See *Dyes*.

Coralwood. *Adenanthera pavonina* (*H.*). India. Andamans. Brownish yellow, seasoning to a coral shade. Figure and texture similar to Spanish mahogany. Fine close grain and lustrous. Highly decorative wood. Used for cabinet work and turnery. S.G. ·8; C.W. 3·5.

Coramandel. Indian ebony. See *Calamander*.

Corbel. A projecting support on the face of a wall.

Cord. A timber measurement of 128 cub. ft. A volume of wood that fills a space 8 ft. × 4 ft. × 4 ft.

Cordia. *Cordia fragrantissima* (*H.*). Burma. Also called Kalamet, and Burmese Sandalwood. Rich reddish brown, darker streaks, fragrant. Moderately hard, heavy, and durable. Ray figure. Requires care in seasoning. Fairly difficult to work. A decorative wood suitable for cabinet work and superior joinery. S.G. ·75 ; C.W. 3·5. For species of the genus *Cordia* see *Babia, Bocote, Canalleta, Cyp* or *Cypre, Freijo, Hadang, Kalamet, Laurel, Muringa, Nopo, Omo, Peterebi, Princewood, Tepisuchil, Ziricote.*

Corduroy. See *Oak Walnut*.

Corduroy Road. A road formed of tree trunks halved longitudinally and laid crosswise in swampy ground.

Cordwood. Wood for charcoal burning. Firewood. Small-size wood, branches, etc., sold by the cord.

Core. **1.** The waste wood from a mortise. **2.** The base, or foundation, for veneer ; or the interior of plywood and laminwood. **3.** An iron bar along the underside of a handrail. **4.** The interior of a column faced with other material. **5.** The sand for the interior of a hollow casting. **6.** The heart of a log left after boxing. See *Box the Heart*.

Core Box. A box in which the cores are formed for moulding.

Core Box Plane. A patternmaker's plane in which the sole is in the form of a right angle for working a hollow.

Core Print. A projection on a pattern to make an impression in the sand for the insertion of a core. See *Pattern*.

Corf. **1.** A floating cage for fish. **2.** A large basket for hauling coal from a mine.

Corinthian. An architectural order, the chief distinguishing feature of which is the richly carved capital. See *Orders*.

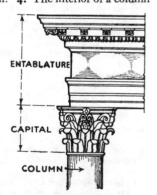

ENTABLATURE

CAPITAL

COLUMN

CORINTHIAN

Cork. The bark of the cork oak. *Quercus suber* and *Q. occidentalis.* It is used for insulation, floats, etc., and weighs about 15 lb. per cub. ft.

Cork-bark Elm. *Ulmus hollandica* or *U. major.* Dutch elm, *q.v.*

Cork Elm. *Ulmus thomasi* or *U. racemosa.* Rock elm, *q.v.* See *Elm.*

Cork Rubber. A piece of compressed cork, or wood faced with cork, round which the glasspaper is wrapped when finishing woodwork.

Corkscrew. A logging locomotive or stem-winder.

Corkwood. 1. *Pisonia obtusata* (*H.*). Tropical America. Dark grey. Moderately light and soft. Fairly straight grain, fibrous, ripple marks, coarse texture. Difficult to work smooth. Used for cooperage, cases, pulp. S.G. ·6. 2. *Anona glabra* (*H.*). Tropical America. Also called Pond Apple. The wood is of little commercial value. 3. *Leitneria floridana.* The lightest wood grown in the U.S.A. Not exported. S.G. ·2. 4. A whitewood, lighter than cork, from Uganda. The name corkwood is also given to several other tropical American woods, including balsa, blolly, bloodwood, and silk cotton tree. Also see *Rose Marara* and *White Basswood.*

Cornaceæ. A family of trees including Aucuba, Cornus, Marlea, Mastixia, Nyssa.

Cornel. See *Cornus, Dogwood,* and *Stonewood.*

Corner Cramp. A universal cramp. A cramp for mitred joints.

Corner Post. An angle post in framed structures.

Cornice. A large projecting moulding at the top of a piece of framing

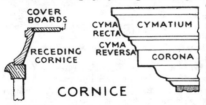

or at the junction of wall and ceiling. The top portion of an entablature. See *Capital, Order,* and *Blind.*

Cornice Bracket. Framed wood brackets forming a foundation for a cornice.

Cornice Moulding. A projecting moulding, but not so large as a cornice. Especially applied to the top of a piece of furniture such as a wardrobe. See *Bed Moulding.*

Cornish Elm. See *Elm.*

Cornus or **Cornel.** Dogwood. Cornaceæ family. Widely distributed. Trees cultivated chiefly for ornament. There is no demand for the wood in this country. Colour ranges from pinkish white to nearly black, and the weight varies from light to very heavy. There is great variation in different species. Grain straight or twisted and with deeply interlocked fibres. See *Nyssa* and *Dogwood.*

Coromandel. Variegated ebony. See *Calamander.*

Corona. 1. A wide, plane, vertical surface in a cornice, in which the bottom edge serves as a drip to throw off the water. See *Cornice.* 2. A chandelier or a circle of lights. 3. The apse to a church.

Coronel. *Krugiodendron ferreum.* Cuba. Also called Ironwood and Ebony.

Corpse Gate. See *Lych Gate.*

Corpsing. A shallow mortise in the face of the material.

Corrall. An enclosure. A pen for cattle or horses.

Corridor. A long communicating passage in a building giving access to rooms. A narrow passage from end to end of a railway carriage giving access to the compartments.

Corrugated. Wave-like or wrinkled. Regularly arranged into a series of longitudinal curves of contraflexure, or alternate ridges and hollows. The term is applied to an irregular wood surface due to collapse and shrinkage. See *Washboard.*

Corsican Pine. *Pinus calabrica, P. laricio,* or *P. nigra.* Austrian pine. Excellent softwood for general purposes. See *Pine.*

Cortex. That part of the bark of a tree between endodermis and epidermis. The outer covering of plants. See *Phloem.*

Corundum. An extremely hard mineral consisting of nearly pure crystallised alumina. Used as an abrasive.

Corves. Same as trams or tubs in mines. They are used for conveying the minerals.

Corvette. 1. A slow sailing ship for heavy cargoes. 2. A war vessel with flush deck and of lower rank than a frigate.

Corylaceæ. A family including Carpinus, Corylus, and Ostrya.

Corylus. See *Hazel.*

Corypha. See *Talipot.*

Cossonidæ. A family of wood-boring weevils. Their activities are similar to those of the common furniture beetle.

Costa Rica Mahogany. C. American mahogany, *q.v.*

Cot. A small bed or crib. A bed-frame suspended from the beams of a ship. A small cottage or hut.

Cot Bar. A curved bar, especially when in a semicircular sash for a sash and frame window with a circular head.

Cotchell. A parcel, or small quantity, of timber not sold through the usual trade channels.

Cote. A nest. A shelter for birds or animals. A sheepfold. A cottage or hut.

Cotinus. See *Chittamwood.*

Cotoneaster. N. India. A shrub-like tree allied to the hawthorn. Straw-coloured wood, not unlike boxwood.

Cotrets. Bundles of split wood.

Cottage Roof. A roof without principal rafters or trusses. It consists of common rafters, purlins, ridge, wall plates, and sometimes hips.

Cottered Joint. A joint for constructional woodwork consisting of strap, gibs, and cotters. It is often used to secure the tie beam to the king or queen post in a roof truss.

KING POST — CLEARANCE IN WOOD — GIBS — COTTERS — STRUT — CLEARANCE IN IRON — TIE BEAM — STIRRUP

ELEVATION   SECTION

COTTERED JOINT

Cotton Tree. See *Bombax* and *Silk Cotton Tree.*

Cottonwood. *Populus balsamifera*, *P. deltoides*, *P. monilifera* or Necklace Poplar, *P. tacamahaca* or Balsam Poplar, *P. trichocarpa* or Black Cottonwood, *P. heterophylla*, Swamp Cottonwood (*H.*).   N. America.   The name Cottonwood is often applied to tulip and cucumber trees.   They were often imported as American whitewood or Canarywood, *q.v.*   INDIAN COTTONWOOD, *Bombax malabaricum* (*H.*).   India.   Also called Semul.   White turning to pale yellow-brown on exposure.   Light and very soft, not durable or strong. Straight even grain, coarse texture.   Used for cases, tea boxes, plywood, matchboarding.   S.G. ·39.   See *Ceiba*.   NIGERIAN COTTONWOOD, *Ceiba pentandra*.   Also called Araba.   Dull oatmeal colour. Very coarse and open grain, loose fibred.   Only of local importance. Also see *Poplar* and *Ceiba*.

Cotylelobium.   See *Empata*.

Couch.   A piece of furniture, like a sofa, to accommodate more than one person, or on which one may recline.

Coulisse.   1. A grooved timber in which framing slides.   The grooved posts for a sluice gate or for the side screens on a theatre stage.   2. A channel or gutter.

Counter.   1. A table on which money is counted and goods displayed.   2. A dimension classification specially applied to mahogany.   3. Part of the stern of old wooden ships. See *Stern Timbers*.

Counter Batten.   1. A strong piece of wood fixed on the back of a wide board, or on a number of built-up boards, to prevent warping.   They should allow for free

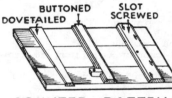

BUTTONED   SLOT
DOVETAILED   SCREWED

COUNTER BATTEN

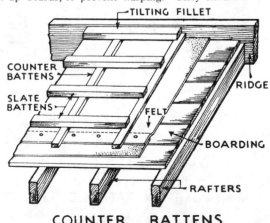

TILTING FILLET

COUNTER BATTENS

SLATE BATTENS

FELT

RIDGE

BOARDING

RAFTERS

COUNTER BATTENS

expansion and contraction, and are usually fixed by slot-screws or buttons, or they may be dovetailed into the boards. **2.** Vertical battens, under the slate battens, on a felted roof to allow any water to run to the eaves gutter.

Counter Brace. A brace to counteract change of stresses in a framed structure. The chief use is to resist wind pressure.

Counter Cramp. A cramp for securing end-to-end joints. Three battens are fixed lengthways and slotted to receive folding wedges which pull together the ends of the pieces to be jointed.

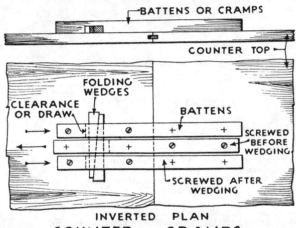

INVERTED   PLAN
COUNTER   CRAMPS

Counter Floor. The bottom layer of floor-boards when two layers are used ; or a layer forming a foundation for parquetry.

Counterfort. 1. An additional row of piles round a coffer dam to make more watertight and to give increased resistance against the pressure of the water. **2.** A pier or shore or tie at the back of a retaining wall.

Counter Lathing. See *Brandering*.

Counter Screen. A low screen on a counter.

Countersink. To form a conical sinking to receive a screw head.

Counter Skylight. A ceiling light or lay light.

Counter Timber. See *Stern Timber*.

Counter Veneer. The bottom layer in double veneers. It is used to prevent cracking of the face veneer.

Country Cut. A term used to denote *full* original sawn measure.

Coupé. 1. A four-wheeled closed carriage for two persons, with outside box for the driver. A single brougham. **2.** A small compartment at the end of a railway carriage, with seats on one side. **3.** A small two-seater automobile.

Couple. **1.** A pair of rafters. **2.** Two equal and parallel forces acting in opposite directions. The magnitude of a couple is the product of *one of the forces multiplied by the perpendicular distance between them.*

Couple Close. See *Close Couple.*

Couple Roof. A roof of small span consisting of pairs of rafters only.

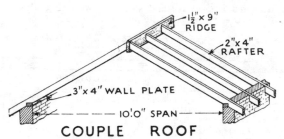

COUPLE ROOF

Coupling Ledge. See *Stops.*

Couratari. A genus of Lecythidaceæ, near *Cariniana,* q.v.

Courbaril. See *Locust.*

Course. A layer of boards in a timber pile. A row of shingles.

Court Cupboard. A form of cabinet of the Elizabethan period. It consists of a chest on legs with a recessed cupboard above.

Coussapoa. A genus of Moraceæ.

Cove or **Coving.** A hollow, or concave, moulding consisting of a quarter circle or ellipse. An inverted scotia.

Coved Bracket. A large shaped bracket carrying laths and plaster for a coved cornice.

Cover Board. A horizontal board covering rough timbers, pipes, cavities, etc. A board, or boards, forming a level surface on the top of the cornice to a cupboard. See *Blind, Cornice,* and *Curtain Rail.*

Covered Wagon. The name given to the fuselage unit of large aircraft.

Cover Flap. A hinged panelled cover to the boxings for window shutters.

Covering Board. One covering the heads of the ribs or timbers of a ship.

Covering Timbers. Boards used in a coal-mine as a protection against falling debris.

Cow-catcher. A strong frame in front of a railway engine to remove obstructions.

Cowdie or **Cowrie Pine.** Kauri pine, *q.v.*

Cowling. The panels enclosing the engine of an aeroplane.

Cow Tree. See *Massaranduba* and *Guayamero.*

c.p.d. Abbreviation for *charterers paying dues.*

Crab. **1.** A portable machine for hoisting purposes. See *Winch.*

2. An overhead crane moving in one direction only and called a log-turner. **3.** A raft with lifting tackle for handling logs in water.

Crab Apple. *Pyrus malus* or *Malus pumila* (*H.*). Great Britain. The wild apple. Brownish colour, hard, with fine grain. Difficult to season and work. Used for turnery, saw handles, mallet heads, cheeses or balls for game of skittles, cogwheel teeth, etc. S.G. ·7 ; C.W. 3·5.

Craboo. Nanche, *q.v.*

Crabwood. *Carapa guianensis* (*H.*). British Guiana. Mahogany family (*Meliaceæ*). Also called Andiroba and Carapa, etc. Reddish brown, lustrous. Moderately hard, heavy, tough, strong, and durable. Straight and ribbon grain, ray flecks. Slightly open grain, apt to pick up. Should be quarter-sawn. Checks easily with seasoning. Easy to work but difficult to finish for polishing. Fire resistant. Used for superior joinery and cabinet work. S.G. ·7 ; C.W. 4.

Cradle. **1.** A block, or frame, to steady small stuff for moulding on a spindle. **2.** Cylindrical centring, as for a culvert. **3.** Heavy framework running on a slipway, to receive a ship at low water. It is then hauled above high-water level for repairs to the ship. **4.** A framework of timber to form an ocean-going raft. **5.** A bed on rockers, for a child. **6.** A basket-like arrangement running on a line to rescue persons from a ship. **7.** A wood frame to keep the bedclothes off an injured limb. **8.** A movable adjustable scaffold for workmen repairing buildings. **9.** A wood box with sieve mounted on rockers, for washing minerals.

CRADLE FOR MACHINING SMALL STUFF

Cradle Bar. A cot bar, *q.v.*

Cradle Roof. See *Wagon Roof*.

Cradle Seat. The driving seat of a vehicle in a hammercloth.

Cradling. Rough brackets, or furrings, to carry linings or plasterwork round steelwork, etc. Also see *Cradle* (2).

Cradling Piece. A short joist between trimming joist and chimney breast to carry the ends of the floorboards.

Craft. **1.** A sailing ship. **2.** An occupation requiring manual dexterity and skill in the working of materials, together with expert knowledge of the particular craft or trade.

G-CRAMP

Cramp. **1.** An appliance for squeezing together two pieces of material, or for drawing the parts of a frame together. The kinds used in woodworking are : corner, flat-bar, flooring, G-cramp, matchboarding cramp, wooden, sash, universal or corner. Other cramping devices are cleats, clamps, handscrews. There are also various types of machines for the same purpose. **2.** A metal fixing for wood frames. The metal is fixed to the frame and the free end is bedded in the brickwork.

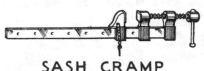

SASH CRAMP

Crane. A machine for lifting heavy weights. It usually consists of jib, mast, stays, tackle, and winding gear. There are many kinds : derrick, jib, travelling, wall, wharf, etc. ; and they may be operated by hand, steam, electricity, or they may be hydraulic. See *Derrick Crane*.

Cranked Bolt. One that is bent to shoot into a hole in advance of the back plate ; especially useful for pivoted sashes.

Crappo. Crabwood, *q.v.*

Cratægus. See *Hawthorn, Thorn, Whitethorn*.

Cratch. A fodder rack for horses and cattle.

Crate. An open packing case consisting of a light strong framework of rods and staves. It is flexible and allows for rough handling, and is used extensively in the pottery trades, etc. The types, for pottery, are distinguished as dishes, dumplings, humbugs, and sixes and sevens.

Cratoxylon. See *Gerunggang*.

Craupadine Door. A door hung on centre pivots so that it revolves about its vertical axis.

Credence. A small table at the side of an altar. A low cupboard with the top serving as a table. A side- or re-table.

Creepie. A milking stool.

Creeping Rafters. Jack rafters, *q.v.*

Cremorne Bolt. Espagnolette bolt, *q.v.*

Crenellated. Indented or notched. A moulding with regular sinkings on the top edge. See *Battlemented*.

CRENELLATED MOULDING

Crenelle. An embrasure or loophole. See *Battlemented*.

Creosote. A preservative and antiseptic obtained by distilling coal or wood tar. An important preservative for wood. See *Preservation*.

Crescentia. See *Calabash*.

Crescent Roof. A roof truss shaped like a quarter-moon. It consists of a semicircular outer member and a segmental lower member, or chord, connected together by struts and braces.

Crest. Ornamental work on the tops of buildings.

Crestings. Ornamental finishings to terminals, such as parapets, ridges, gables, etc.

Crib. **1.** A rack in a stable. **2.** A wood structure under a caisson. **3.** Circular ribs used as walings for excavated shafts. **4.** A small frame for a child's cot. **5.** A small boom or raft of logs. **6.** A bin of slat construction for storing materials of small size.

Crib Work. Layers of logs, in skeleton frame formation, sunk in water to serve as a dam.

Cricket Bat. See *Bat, Splice,* and *Willow.*

Cricket Seat. A folding seat to a vehicle.

Crimping. 1. Moulding or bending into corrugations. 2. An uneven, broad, curvature or corrugation of a wood surface, due to abnormal collapse or local shrinkage during the drying of the wood.

Cripple. 1. A form of bracket to rest on a roof, and slung from the ridge. It is used to support a slater's scaffold, or for hanging on the rungs of a ladder. 2. A board drawn up a chimney flue during construction to remove surplus mortar. 3. See *Jack Rib.*

Cripplings. Slender raking shores.

Crocket. An ornamental projection, at regular intervals, on canopies, windows, gables, etc. They are characteristic of Gothic architecture.

Crocodile Trees. Applied to S. American trees that have irregular rough bark due to excessive bleeding for rubber.

Crook. 1. A naturally curved, or crooked, timber specially useful for bends and knees in boatbuilding. See *Knees.* 2. Edge bend, *q.v.*

Crooked Grain. Cross grain due to irregular growth rings, intersection of branches, etc.

Crop. The handle, or stock, of a riding whip.

Crop Beam. See *Lock Gate.*

Crope. A finial, *q.v.*

Croquet Mallet. A long-handled wood mallet for striking the ball in the game of croquet.

CROCKET

Crosette. A right-angled projection at the top of a vertical architrave. Also called Elbows, Ears, or Ancons. See *Ancons.*

Crossbanding. A term used in plywood manufacture for a layer with the grain at right angles to that of the core, to resist shrinkage.

Cross-bands. The facings of laminboard.

Cross-bar. A horizontal piece across one or more other pieces, as for tying together the tops of goal posts.

Cross Chocks. Timbers fixed across the deadwood amidships, to strengthen the futtocks.

Cross-cut. 1. A handsaw, usually about 26 in. long and with six points to the inch, for cutting across the grain. 2. A two-handed saw for cutting heavy timbers and logs across the grain. 3. A pendulum circular saw with teeth specially shaped for cross-cutting. See *Saw Teeth.*

Crossfire. Figure extending across the grain, such as fiddleback, raindrop, finger roll, and mottle.

Cross Fracture. See *Upsett.*

Cross Garnet. See T-*Hinge.*

Cross Grain. Applied to wood in which the grain is not running lengthwise of the material or in one direction. The irregularity is due to interlocked fibre, or to uneven annual rings, or to intersection of branch and stem. Fibres presented endwise on the face of the wood. A term to denote an inclination of the annual rings on the face of a rift-sawn piece of wood. See *Grain*.

Crossjack Yard. The lower square yard of the mizzen mast of a sailing ship.

Cross Section. A section at right angles to the direction of the grain. A transverse section.

Cross Shakes. See *Shakes*.

Cross Tongues. Thin slips cut diagonally from a wide board and used as tongues in joints. This form of tongue is the strongest obtainable from ordinary wood, but it is more usual to use plywood. See *Angle Joints* (*f*).

Crosstrees. Pieces of hardwood, or spreaders, in the framework of the tops of masts, to support top-gallant rigging.

Crotch. 1. To cut notches in a log to provide a grip for dogs. 2. See *Dray*. 3. Ornamental grain, or feather, due to the struggle for direction of the fibres between forked branches and the trunk of the tree. Specially valuable in mahogany. 4. Naturally crooked timbers fixed on the keel, fore and aft, in wooden ships.

Crow Bar. A steel rod bent and tapered at one end, to use as a lever for moving heavy weights.

Crown. 1. The highest point of an arch, or the top of any ornamental feature. 2. The top of a tree including branches and foliage. 3. A shipping mark used in the grading of timber.

Crown Bar. The top timber carrying the boards in the centring for a tunnel.

Crown Cover. See *Saw Bench*.

Crown Cut. Sawn tangentially to annual rings. Flat-sawn. See *Conversion*.

Crown Piece. 1. A short bearing timber on a wall to carry the foot of a strut. 2. A ridge. 3. A sliding block used in centring for tunnels.

Crown Plate. A horizontal piece on the top of a post to provide a greater bearing surface.

Crown Post. A king post; especially applied to the post above the collar beam in a hammer-beam roof truss.

Crowntrees. 1. Small sleepers cut from about 8-in. diameter trees sawn longitudinally to produce two sleepers. 2. Pit props slabbed to be used horizontally against the roof.

Crow's Ash. *Flindersia australis* (*H.*). Australia. Lemon yellow to yellowish brown, oily, odorous. Very hard, heavy, strong, and durable. Resists insects. Sometimes called teak because of its excellent qualities. Close grain, uniform texture. Fairly difficult to work. Used for superior joinery and flooring, shipbuilding, wagons, exterior construction, etc. S.G. ·9 ; C.W. 3·5.

Crow's Nest. A small platform in an elevated position, for observation purposes.

Crow Stepping. An incline formed into a series of horizontal steps. Especially applied to a gable coping so formed.

Croze. 1. The cross groove at the end of a stave for a cask, in which the head is secured. 2. A cooper's router for preparing grooves.

Cruck. A bent, or forked, branch of a tree.

Cruck Construction. A primitive form of house or shelter. The framework consists of crucks connected together by a ridge tree.

Crush Room. The foyer to a place of entertainment.

Crustaceæ. A family of crustacean wood borers : Chelura, Linnoria, Sphæroma. See *Marine Borers*.

Crutch. 1. A V-shaped hardwood support for the main boom on a ship. A forked support. 2. Pieces of knee timbers to secure the cant timbers.

Crwth. A type of Welsh fiddle.

Cryptocarya. See *Beilschmiedia, Bolly Silkwood, Canary Ash, Oak Walnut, Queensland Walnut, Rose Walnut,* and *White Walnut.*

Cryptomeria. See *Sugi.*

Cryptophagidæ. A family of wood beetles that live in or just beneath the bark, or in decayed wood of logs, etc.

Crystals. Solidified salts in wood tissue. The most common are crystals of calcium oxalate.

Cuaba. *Amyris balsamifera.* Also called Amyriswood and Torchwood, *q.v.*

Cuanacaztle. Guanacaste, *q.v.*

Cuba. Exports to this country chiefly consist of cedar, cocuswood, degame, lancewood, lignum vitæ, and mahogany.

Cuban Pine. *Pinus caribæa,* syn. *P. heterophylla.* Honduras pitch pine or Caribbean pine. Similar to long-leaf pine. See *Pitch* and *Slash Pine.*

Cubby-hole. A small cabin, or snug place, to which one can retire for privacy.

Cube. A solid formed by six equal square sides.

Cubic. Applied to measurements of volume.

Cubicle. A small compartment, or room, for sleeping and dressing.

Cubing. Finding the volume or cubical contents.

Cucking Stool. A chair, or stool, to which delinquents were tied and exposed to public ridicule, and often ducked in water.

Cucumber. *Magnolia acuminata* and other spp. (*H.*). U.S.A. Closely resembles *Liriodendron.* S.G. ·48. See *Magnolia, American Whitewood,* and *Canary Whitewood.*

Cudgel. A heavy stout stick used in combat.

Cudgerie or **Cudgery.** *Flindersia schottiana* (*H.*). Australia. Also called Flindosy Beech and Silver Ash. Characteristics of ash and used for the same purposes. See *Flindersia.*

Cue. A long, tapering stick used in the game of billiards.

Cullenia. See *Durian.*

Cullis. See *Coulisse.*

Cull Lumber. Timber removed from a superior to an inferior quality. De-graded. *To cull* means to select. In forestry it means to select trees for felling. The term is also applied to rejected logs.

Culverhouse. A cote for birds.

Culvertail. A dovetail, *q.v.*

Cumamodendron. See *Canello.*

Cuneiform. Wedge-shaped.

Cuneoid. See *Conoid.*

Cunn. Abbreviation for the name of the botanist Cunningham.

Cunonia capensis. Red els, *q.v.*
Cunoniaceæ. A family including Ackama, Ceratopetalum, Geissois.

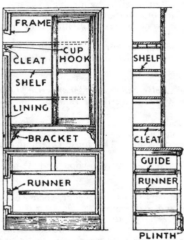

CUPBOARD FOR KITCHEN

Cunura. See *Tonca Bean.*
Cup. A conical receiver for the head of a screw. It is sunk and glued in the wood, for easy removal of the screw. See *Screws.*
Cupania. See *White Tamarind.*
Cupboard. A small room, or enclosed space, for storage purposes.
Cupboard Turn. See *Turn-button.*
Cupiliferæ. A family of trees under old classifications including alder, beech, birch, hazel, hornbeam, oak, sweet chestnut.
Cupiuba. *Goupia glabra* (*H.*). C. America. Light reddish brown ; darkens with exposure. Hard, heavy, strong, odorous, fairly durable. Irregular grain, uniform texture, difficult to work. Used for sleepers, paving, construction, furniture. S.G. ·85 ; C.W. 4.
Cupola. 1. A dome or spherical vault. Especially applied to the interior of a dome that is a separate structure from the outer framing. 2. A smelting furnace.
Cupping. See *Warping.*
Cupressaceæ. The Cypress family. A sub-family of Coniferæ, with three further sub-families : Cupressoideæ, Juniperoideæ, and Thujoideæ, *q.v.*
Cupressoideæ. A family that includes Cupressus and Chamæcyparis.
Cupressus. Cypress. Yellow cedar, *q.v. Cupressus obtusa*, Hinoki. *C. sempervirens*, common cypress. *C. thyoides*, white cedar. *C. macrocarpa*, Californian cedar. See *Cedar.*
Cuprinol. A proprietary preservative for wood. It is an organo-metallic salts in liquid form, easy to apply and with no offensive smell.
Cup Shake. A shake in wood due to lack of cohesion between successive annual rings, due to exceptional winds or lack of nutriment during a season of growth, or expansive force of frost. See *Shake.*
Curatella. See *Rough-leaf Tree.*
Curb, or Curb Plate. 1. A frame round the mouth of an opening or well. 2. A circular wall plate. 3. The plate at the change of inclination of a mansard roof. 4. A fender, or protection, in front of a fireplace. See *Kerb* and *Skylight.* 5. Frames used as foundations for tubbing or walling in a mine shaft.

Curb Rafter. The upper rafters of a mansard roof. See *Mansard Roof*.

Curb Roll. A roll used at the intersection of the two surfaces in a mansard roof.

Curb Roof. A mansard roof, *q.v.* A roof resting on a curb.

Curculionidæ. A family of wood-boring weevils.

Curf or **Kerf.** A saw cut partly through the material.

Curl. Figure in the grain produced, by suitable conversion, by the junction of a branch with the stem, or between two or more branches. It is specially valuable in mahogany and is also called Fan, Fork, Crotch, or Crutch.

Curly Birch. *Betula lutea.* See *Birch*.

Curly Grain. Ornamental figure in wood due to the fibres forming irregular curves or undulations. Large undulations produce " wavy " grain.

Curly Maple. *Acer saccharum.* See *Maple*.

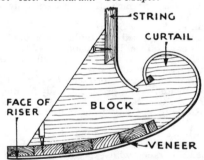

CURTAIL AND COMMODE STEP

Curtail Step. A step, at the foot of a flight of stairs, with the end in the form of a scroll. It is also sometimes applied to a semicircular, or drum, end.

Curtain. 1. Often applied to a wide horizontal board that serves as a screen, as the board on the front of a joiner's bench. 2. A fireproof screen dividing the stage from the auditorium in a theatre, or sections in a mine. 3. Suspended cloth, or other material, screens to windows, etc., as a furnishing.

Curtain Pole. A cylindrical rod from which curtains are suspended.

Curtain Rail. A shaped fascia serving as a rail in cabinet work. Also called a Curtain Piece or a Span Rail.

Curtisia faginea. Assegai, *q.v.* See *Lancewood*.

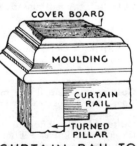

CURTAIN RAIL TO SIDEBOARD

Curupay. *Piptadenia spp.* (*H.*). Tropical America. Deep red ; black stripes. Very hard, heavy, tough, strong, and durable. Roey, smooth, feels oily. Difficult to work. Polishes well. Used for structural work, superior joinery, and cabinet work. S.G. 1 ; C.W. 4·5. Also see *Guayacan*.

Curved Filling. Planted pieces to give a curved outline or contour. See *Turret*.

Curved-rib Truss. A roof truss in which one of the main structural members is curved, as in a de Lorne, hammer-beam truss, etc.

Cushimucho. *Michelia sp.* Formosa. See *Champac*.

Cusp. A point where a curve stops and returns, so that the two branches, or arcs, have a common normal. It is usually ornamented in tracery work. See *Cincfoil*, *Tracery Work*, and *Chancel Screen*.

Cuspidated. Pointed. Terminated by a cusp.

Customary Measure. A method, established by long usage, of measuring logs, or balks, in which allowances are made for wane, etc. See *Measure* and *Actual Measure*.

Customary Square. See *Die Square*.

Customs Measure. Applied to timber measured by customs fund officials who issue official list of contents. Planks of hardwood are measured to $\frac{1}{4}$ ft. in length and $\frac{1}{4}$ in. in width and thickness, and calculated to the nearest $\frac{1}{2}$ ft. super.

Cut. Implies *bevel* in carpentry. Also see *Flat Cut* and *Deep Cut*.

Cut and Mitred Beads. The prepared beads forming the rebates for a rectangular pivoted sash.

Cut and Mitred String. A cut string mitred with the risers. See *Bracketed Stairs*.

Cut Bracket. A bracket moulded on the edge.

Cutch. *Acacia catechu* (*H.*). India. Also called Kath or Khair. Light to dark red ; darkens with exposure. Lustrous and smooth. Very hard, heavy, strong, and durable. Resists insects. Straight and close grain, coarse texture. Difficult to work due to secretion and hardness. Structural work, furniture, tools, wheel teeth, spokes, etc. S.G. ·98 ; C.W. 4.

Cut Lock. A lock for which the wood has to be cut away to receive it.

Cut Nail. A nail stamped from sheet metal, such as a floor brad. See *Nails*.

Cut Over. Applied to a forest in which the merchantable timber has been felled.

Cut String. One shaped to the outline of the steps. See *Bracketed Stairs* and *Strings*.

Cut Stuff. Market sizes of timber, such as deals, resawn into smaller sections.

Cutter. 1. A small one-masted vessel with deep keel, rigged like a sloop. 2. A boat to a warship. It is shorter, broader, and deeper than a pinnace, and usually has six oars.

Cutter Block. The block in a woodworking machine that carries the cutters or knives. There are many types, but for planing machines they must be circular, and they may have up to six knives. Recent research has proved that the angle at which the cutter is fixed in the block should vary for different woods, hence patent blocks with adjustable seatings for the knives are now used.

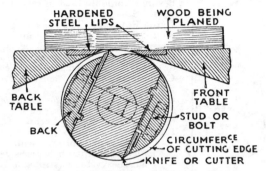

SECTION THRO' CUTTER BLOCK

Cutters. Blades, or knives, used in woodworking machines.

Cut-through. Applied to the door of a vehicle when the bottom is in front of the rocker.

Cutting Gauge. Like a joiner's marking gauge, but with a thin blade instead of a pin. It is used for cutting thin material instead of ripping or for forming small rebates.

Cutting Iron. See *Plane Iron.*

CUTTING GAUGE

Cutting List. A prepared list of the materials and sizes required for a woodworking job. Most firms require three copies, one each for setter-out, machinist, and office respectively. One copy accompanies the stuff throughout the various operations.

Cutting-off Chisel. A turner's chisel for parting the work after turning.

Cutwater. **1.** Heavy timber framing in triangular formation protecting the end of a pier or jetty. **2.** The built-up timbers forming the knee of the head of a ship. See *Stem and Head.*

c.v. Abbreviation for *centre V-jointed.*

Cycadales. One of the orders of Gymnosperms. The species are tropical woody plants resembling tree ferns or palms.

Cyclone. A mechanical collector of the dust and shavings from woodworking machines.

Cyclostemon. See *Banawi.*

Cydonia. See *Quince.*

Cylicodiscus. See *Okan.*

Cylinder. **1.** A prism of uniform circular section. *Volume* $= \pi R^2 L$. *Surface area* $= 2\pi RL$. **2.** A large vessel in which wood is placed for preservative treatment under pressure. See *Preservation.* **3.** A drum, or centre, round which veneers are bent for gluing on the blocks, for linings, etc., in circular work. (See illus., p. 130.)

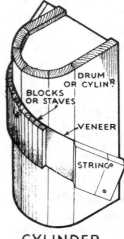

DRUM
OR CYLINR

BLOCKS
OR STAVES

VENEER

STRING

CYLINDER

Cylinder Saw. See *Barrel Saw*.
Cyma Recta. A curve of contrary flexure commonly used in mouldings. An ogee. See *Cornice*.
Cyma Reversa. The reverse of the cyma recta. The upper part is convex and the lower part concave. See *Cornice*.
Cymatium. The upper member of a cornice or capping. The cyma recta. See *Cornice*.
Cynometra. See *Muhimbi*, *Ping*, and *Zebrano*.
Cyp or **Cypre.** *Cordia gerascanthus* or *C. alliodora* (*H.*). C. America. Borage family. Also called Bocote, Canaletta, Princewood, Salmwood, Solera, and Ziricote. Variegated brown with irregular black markings, smooth, lustrous, feels oily. Very hard, heavy, tough, strong, and durable. Roey grain. Difficult to work, polishes well. Used for cabinet work, superior joinery, vehicles, veneers. S.G. ·85 ; C.W. 4.
Cypress. *Cupressus sempervirens* (*S.*). S. Europe and Asia Minor. Common cypress. Yellowish brown with reddish markings, darkening with exposure. Fairly soft, light, strong, and very durable. Fine uniform grain, close texture. Resists insects. Easy to work. Used for general purposes. S.G. ·5 ; C.W. 1·5. ALASKA CYPRESS. Yellow cedar, *q.v.* BALD CYPRESS. See *Louisianna Cypress*. CONGO CYPRESS. See *Vera*. HIMALAYAN CYPRESS, *Cupressus torulosa*. India. Light brown with darker streaks and mottles. Lustrous, smooth, fragrant. Fairly soft, light, and very durable. Straight even grain, moderately fine texture, easy to work. Useful wood for exterior or interior work. S.G. ·49 ; C.W. 1·5. HONDURAS CYPRESS. See *Yacca*. LAWSON'S CYPRESS. See *Cedar*. LOUISIANA CYPRESS, *Taxodium distichum* (*S.*). East U.S.A. Also called Bald, Black, Deciduous, Pond, Red, and Swamp Cypress. Yellowish red or golden brown. Very durable, but variable in other characteristics. Used for general purposes. NOOTKA SOUND CYPRESS. Yellow cedar, *q.v.* Similar to *Cupressus*. YELLOW CYPRESS, *Thuya excelsa*. U.S.A. Similar characteristics to common cypress. WHITE CYPRESS, *Callitris glauca*. New South Wales. Pale brown, fragrant, durable, short-grained and not strong. Resists insects. Not exported. S.G. ·65. See *Cedar*.
Cypress Pine. *Callitris glauca*, *C. calcarata*, and other spp. (*S.*). Queensland. There are about eight species of Queensland cypress, but only *C. glauca* (Western Sand Cypress) is of commercial importance. Buff colour, variegated streaks, fragrant. Fairly hard, heavy, strong, but brittle, very durable and resists insects. Numerous small firm knots. Used for joinery, flooring, piles, boats, construction. S.G. ·65 ; C.W. 2. See *White Cypress*.
Cyrtophyllum. See *Tembusa*.
Cytisus. See *Laburnum*.

D

d. Abbreviation for *dressed*.

Dabé. *Erythroxylum mannii* (*H.*). W. Africa. Also called Landa. Light brown with darker streaks and flecks. Moderately hard, heavy, and strong. Medium texture. Irregular stripe figure when rift-sawn due to oblique grain. Moderate sizes. Suitable for furniture and superior joinery. S.G. ·7 ; C.W. 3·5.

Dabema. Ekhimi, *q.v.*

Dacama. Angelin, *q.v.*

Dacrydium. See *Huon Pine*, *New Zealand Pine*, *Rimu*, and *Silver Pine*.

Dado. **1.** The plain part, between base and surbase, of a pedestal to a column. **2.** The lower part of a wall between skirting and dado rail.

Dado Base. The lowest members of a pedestal, *q.v.*

Dado Framing. Panelled framing, about shoulder height, fixed to the lower part of a wall.

Dado Head. An adjustable grooving saw. Cutter heads are inserted between two outer saws to give the required size of groove.

Dado Joint. The pattern-maker's name for a trench or housing. See *Housed Joint*.

Dado Moulding. See *Chair Rail* and *Surbase*.

PLASTER

GROUNDS PLUGGED TO WALL

PANEL MOULDING OR STOP CHAMFER

SKIRTING

DADO FRAMING

Dado Plane. A plane for grooving floors to receive the tongue on the bottom edge of a skirting or dado framing.

Dædalea quercina. A fungus that attacks oak, and sometimes other English hardwoods, and affects the wood in the same way as Merulius lacrymans.

Dagame. See *Degame*.

Dahat. *Tectona hamiltoniana*. India. Same genus as teak, but inferior and not imported.

Dahoma (Gold Coast). Ekhimi, *q.v.*

Dais. 1. A part of a floor raised above the level of the surrounding floor. A platform. **2.** A chair, or seat, with high back and canopy for one in authority.

Dakama. *Dimorphandra spp.* (*H.*). Guiana. Reddish brown. Very hard, heavy, and with coarse grain. Used for general purposes, sleepers, furniture. S.G. 1 ; C.W. 4.

Dak Ballow. *Parinarium sp.* (*H.*). Malaya. Very similar to merbatu, *q.v.* Extremely hard and difficult to work. Only used for heavy constructional work, piling, etc.

Dalbergia. Rosewood, *q.v.* Also see *Balsamo, Blackwood, Burma Tulipwood* or *Tamalan, Cocobolo, Indian Rosewood, Jacaranda, King-wood, Marnut, Poye, Sissoo, Yinzat.*

Dalz. Abbreviation for the name of the botanist Dalziel.

Dam. Any form of construction to obstruct the flow of water.

Damar or **Dammar.** A gum, or resin, used for transparent varnish. Soluble in turps. Also applied as an adjective to numerous E. Indian trees that produce commercial gums, especially of the Dipterocarpaceæ family.

Damar Hitum. See **Singga**.

Damar Laut. *Shorea utilis* and *Parashorea stellata* (*H.*). Malaya. Also called Selangan, Yacal, etc. Brown. Very hard, heavy, durable, with straight grain. Often substituted for Chengal. Used for structural work, piling, etc. S.G. ·95.

Dammar. See *Damar* and *Meranti*.

Dammara. See *Vatica sp.* and *Kauri*.

Dammar, Black. *Dipterocarpus indicus* (*H.*). India. Light red to greyish brown. Resin canals give lighter markings. Heavy, hard, moderately durable and strong. Straight and interlocked grain. Dull and feels rough to touch. Even, moderately coarse texture. Not difficult to work but requires care in seasoning, and liable to insect attack. Used for constructional work, planking, flooring, etc. S.G. ·78 ; C.W. 3.

Damper. A contrivance to check the vibrations of the strings of a piano.

Damping. Applying warm water to wood with woolly texture to raise the grain before final glasspapering. Applying water to the hollow side of a board to straighten the board.

Damson. *Prunus sp.* (*H.*). Britain. Bright yellowish red with variegated stripes. Hard with fine uniform texture. Used for fancy articles, turnery, etc. BITTER DAMSON. See *Simaruba*.

Dance Floor. Any form of resilient floor intended for dancing upon. There are several registered types.

Dancette. The chevron, or zigzag, moulding. See *Chevron*.

Dancing Steps. Balanced winders, *q.v.*

d. and h. Abbreviation for *dressed and headed*.

d. and m. Abbreviation for *dressed and matched*. There are numerous variations : *d.2s. and s.m.* means *dressed two sides and standard matched*.

Dandy. A small two-masted fishing vessel.

Daniella. See *Ogea*.

Danish Ash. *Fraxinus excelsior.* Common ash. Danish Beech. *Fagus sylvatica.* Common beech.
Danta. *Cistanthera papaverifera* (*H.*). W. Africa. Light brown, lustrous, striped. Moderately hard and heavy. Tough, strong, durable, and very flexible. Fine close grain and texture, but not easy to finish smooth; polishes well. Requires careful seasoning. A good substitute for ash and hickory. Used for implements, furniture, turnery, veneers, etc. S.G. ·75 ; C.W. 3·5.
Danzig Fir. *Pinus sylvestris.* See *Redwood.*
Danzig Oak. See *Oak.*
Dao. *Dracontomelum dao* (*H.*). Philippines. Light brown, darker markings. Moderately hard, heavy, and strong. Very durable, seasons well, and easily wrought. Fairly straight grain and fine texture. Resembles oriental walnut in grain and texture. Used for cabinet work, veneers, superior joinery and flooring. S.G. ·7 ; C.W. 3.
Daphnandra. See *Canary Sassafras.*
Daphne laureola. The spurge laurel. See *Laurel.*
Daphnis nerii. The oleander moth grub or chincona caterpillar. Very destructive to young trees.
Dapping. A term used for *notching* in heavy timber construction.
Dapple. Variegated figure, or fleck, in wood. Especially applied to mahogany.
Dar. See *Sedeng.*
Darby. A plasterer's long wooden floating rule.
Daru. *Urandu, Platea,* and other spp. (*H.*). Malaya, Borneo. Also called Dedaru. Yellow, variegated, lustrous, smooth. Very hard, heavy, and durable. Fine even grain, close texture, mottled and fine rays. Used for cabinet work, piling, etc. S.G. 1 ; C.W. 4.
Dash. The partition between horse and driver in a carriage, or between engine and driver in a motor car. It is the datum point from which the body dimensions are taken.
Dashboard. 1. A board to an aeroplane or motor car on which the indicators, gauges, etc., are fixed. 2. Any form of screen to the front of a vehicle to protect the occupants from mud splashes.
Dasher. The base of the dasher pole used in a butter churn.
Data. Conditions, dimensions, and other particulars for any particular piece of work.
Datiscaceæ. See *Tetrameles.*
Davenport. A small ornamental writing desk with drawers and hinged lid to desk. Named after the original designer. Also a large settee or divan.
Davits. Swivelled projections, with pulley tackle, to hoist or lower a boat over a ship's side.
Day. A division, or part, of a window between two mullions.
Day-bed. A couch with a drop end for full-length reclining.
Day-work. Work valued on the actual cost of materials and labour, with an agreed percentage to cover profit, plant, and supervision.
d.b.b. Abbreviation for *deals, battens, boards.*
D.C. Abbreviation for the name of the botanist A. de Candolle.
d.c.f. Abbreviation for *deal-cased frame.*
d/d. Abbreviation for *days after date.*

D. Don. Abbreviation for the name of the botanist David Don.

Dead. Applied to wood lacking its natural characteristics, probably due to over-maturity.

Dead Bolt. The bolt of a lock actuated by the key, to distinguish it from the bevelled bolt, or latch, operated by the knobs.

Dead Doors. Special doors on the outside of the gallery doors of wooden ships, for use in the event of damage.

Deaden. To kill a tree by girdling.

Deadener. A heavy balk with spikes arranged on a log-slide so that the spikes retard the speed of the logs passing under it.

Deadening. Sound-proofing.

Dead Eye. A small block of wood bored with three holes and used on board ship for extending the shrouds, etc.

Dead Freight. Empty space on board ship for which compensation is paid when the whole ship has been chartered.

Deadhead. **1.** A sunken, or partly sunken, log. See *Sinkers*. **2.** A wood block used as a buoy. **3.** A bollard, *q.v.* **4.** A tailstock.

Dead Knot. A knot in timber not firmly joined to the surrounding wood throughout. See *Knots*. The woodworker usually calls a decayed knot a dead knot.

Dead Light. A cover to protect the glass of a skylight on board ship.

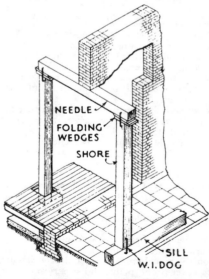

NEEDLE

FOLDING WEDGES

SHORE

SILL

W.I.DOG

DEAD SHORE

Dead Load. The forces acting on a structure due to gravitation. The weights of the materials. A steady load gradually applied.

Dead Lock. A lock actuated by a key only. One with a dead bolt only and having no latch. See *Stock Lock*.

Deadmen. **1.** Heavy timbers forming the anchor for a land-tie to quay walls, etc. See *Quay*. **2.** A strutted plank, on end, used by the bricklayer as a profile.

Dead Rollers. Applied to rollers used for moving heavy timbers that are not power driven.

Dead Shore. Vertical shores, supporting dead loads, as when supporting a needle. See *Shoring* and *Formwork*.

Dead Square. See *Die Square*.

Deadwood. **1.** The term is applied to wood from over-matured trees in European redwood; and, generally, to wood from trees that are dead before felling.

2. Timbers placed on the keel of a ship, especially fore and aft, to assist in building up and strengthening the framework. See *Keel* and *Stern Timbers*.

Deafening. Sound-proofing. See *Pugging* and *Deadening*.

Deal. A term applied to converted softwoods between 2 and 4 in. in thickness and 9 and 11 in. in width. The sizes vary at different ports. Deals are sold by the standard, *q.v.* See *White*, *Red*, and *Yellow Deal* and *Spruce*.

Deal Carrier. One who carries deals between *hick* and *stover*.

Deal Frame. Like a log frame, but with vertical feed rollers to keep the deal tight to the fence.

Deane. The name of the botanist H. Deane.

Deat. See *Dote*.

Death-watch Beetle. *Xestobium rufo-villosum*. Anobiidæ family. Very destructive to old interior constructional timbers. The insect is about ¼ in. long, dark brown in colour, with a mottled appearance due to patches of yellowish hair. The adults make a ticking noise, hence the name. The grubs burrow into the wood over a period of two years before emerging into the chrysalis stage. They are very difficult to eradicate when firmly established owing to the depth of the burrows. Fumigation by gas, carbon disulphide, etc., is usually effective. Several proprietary preservatives are also good, such as Cuprinol, Solignum, Hope's Destroyer, etc.

Debenture Timber. Imported mining timber, usually in the round.

Decagon. A rectilinear plane figure with ten sides. Area = (side)² × 7·694.

Decay. Disintegration of wood tissue due to action of organisms : bacteria, fungi. There is great variation in resistance of different woods to different organisms. All other things being equal, durability probably varies with density of wood ; but impregnation, to prevent decay, varies inversely with density. Chemical and mineral contents affect the resistance. Decay is often accompanied by discoloration, brittleness, decrease in density. Affected wood should be discarded for all structural and constructional work and fixed joinery, and carefully considered for any form of woodwork. In a few cases fungi attack increases the decorative value of wood. *Incipient decay* is an early stage of decay. See *Dote, Dosy, Foxy*, and *Brashy*. Also see *Durability, Disease, Defects*, and *Preservation*.

Decayed Knot. See *Knots*.

Deciduous. Applied to trees that shed their leaves annually. Nearly all the woods are classified as *hardwoods*, q.v.

Deciduous Cypress. Louisiana cypress. See *Cypress*.

Deck. A platform extending from side to side of a ship and serving as a floor. The distinguishing terms used are : flush, forecastle, gun, half, main or upper, middle, orlop, poop, quarter, and spar, decks.

Deck Cants. Timbers fixed on decks to receive wood framing. See *Cants* (2).

Deck Cargo. Cargo carried on the deck of a ship, which is restricted during certain seasons when rough weather is expected.

Deck Chair. A light, folding, portable chair, usually consisting of a folding frame and canvas. The canvas is fixed at the ends only so that it hangs freely to accommodate the body when resting.

Deck House. A structure on the weather deck of a ship to provide additional accommodation or protection.

Decking. 1. The working platform to a derrick tower that carries the derrick crane. 2. Shuttering for horizontal surfaces. 3. Timbers forming the floors of ship's decks, piers, jetties, wharfs, etc. See *Planking.* 4. The top covering of the fuselage of an aeroplane, or the upper surface of the hull of a flying boat.

Deck Planks. See *Decking* (3) and *Planking*.

Decorated Style. The second of the Gothic, or pointed, styles of architecture; also called Curvilinear, Geometrical, and Middle Pointed. It is rich in ornamentation and tracery work, including crockets, grotesques, diapers, etc. Fourteenth century.

Decorticate. To strip off the bark of trees.

Dedali. *Strombosia javanica* (*H.*). Malaya. Pale brown or yellowish. Fairly soft and light, but durable and fairly strong. Straight, very fine and open grain. Fine rays. Stable, and easy to work. Checks badly with careless seasoning. Little exported. S.G. very variable, about ·6 ; C.W. 2.

Dedaru. See *Daru.*

Deep Cutting or **Deeping.** Resawing timber parallel to its wider faces, as when converting deals to thin boards.

Deer Fence. A high fence round a park.

Defects. The principal defects common to timber are : bark pockets, burrs, callus, cross grain, gum veins, knots, pith, resin pockets, rind-galls, shakes stain, twisted grain, upsets, wandering heart, wane, warp, worm-holes. It is usually assumed that defects lessen the economic value of the wood, but some of them, such as burrs, increase the value considerably for decorative purposes, while others may only lessen the volume in conversion. Nearly all defects decrease the value considerably where strength is an important factor. See *Disease* and *Decay*.

Defibrator. Mechanical plant for disintegrating wood chips for fibreboard manufacture.

Deflection. Distortion or displacement of structural members due to the action of forces on the structure. Sagging or bending downwards. The formula for the deflection of a beam is $C \cdot \dfrac{Wl^3}{EI}$, where C is a coefficient depending upon conditions of loading and fixing. Refer to *Beams* (comparative strengths) where the values of C for the six examples are : $\frac{1}{3}$, $\frac{1}{8}$, $\frac{1}{48}$, $\frac{5}{384}$, $\frac{1}{102}$, and $\frac{1}{384}$ respectively. Hence the deflection for a supported beam, with central load $= \frac{1}{48} \cdot \dfrac{Wl^3}{EI}$ where W = load, l = length in inches, E = modulus of elasticity, and I = moment of inertia.

Degame. *Calycophyllum candidissimum* (*H.*). C. America. Variegated yellowish brown. Hard, heavy, strong, resilient, tough, and fairly durable. Straight and irregular grain. Fine uniform texture. Smooth, and polishes well. Easy to carve, and a good substitute for lancewood. Used for shafts, implements, turnery, cabinet work, superior joinery, etc. S.G. ·85 ; C.W. 3·5.

Dehaasia. See *Medang*.

Dehorn. To re-mark logs, for change of ownership.

Delcredere Commission. Additional premium ,charged by an agent to guarantee the solvency of buyer.

De l'Orme Truss. A roof consisting of built-up laminated semicircles, keyed together. Named after the designer. It is useful where only short lengths of material are available. Also called a Flitch-rib Roof, *q.v.*

Demerara Mahogany. Crabwood, *q.v.*

Demi Relievo. Sculpture in relief, in which half of the figure projects over the face of the material.

Demurrage. A fee charged by a railway or shipping company for undue delay in receiving goods, trucks, or ships. Payment for storage.

Dendrocalamus stricta. The strongest of Indian bamboos, and used in the earlier types of aircraft.

Dendroid. Having the form of trees.

Dendrometer. An instrument for finding the height of a tree.

Dennibloc. A registered blockboard.

Dennilam. A registered laminboard.

Density. Weight per unit volume. See *Specific Gravity*. Density of wood is influenced by rate of growth, percentage of summerwood, and, in specimens, the proportion of heartwood. The strength, toughness, hardness, durability, and resistance to cleavage usually increase with density. See *Weight*.

Density Rule. An approved specification for grading pitch pine. See *Grading*.

Dentate. Toothed. Serrated.

Denticulated. Applied to a moulding having dentils.

Dentils. Rectangular projections on the face of a moulding. They are usually formed by removing regular alternate parts of a fillet of the moulding, which is usually in two parts to facilitate removal.

Denya. Okan, *q.v.*

Deodar. *Cedrus deodara* (*S.*). Himalayas, N. India. Similar

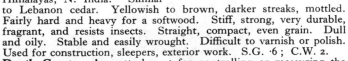

DENTILS

to Lebanon cedar. Yellowish to brown, darker streaks, mottled. Fairly hard and heavy for a softwood. Stiff, strong, very durable, fragrant, and resists insects. Straight, compact, even grain. Dull and oily. Stable and easily wrought. Difficult to varnish or polish. Used for construction, sleepers, exterior work. S.G. ·6 ; C.W. 2.

Depth Gauge. Any attachment for controlling or measuring the depth of a sinking. See *Bit Gauge* and *Twist Bit*.

Derby. See *Darby*.

Derm. A door post or threshold.

Derrick. A large pole with its foot connected to a vertical post or mast, and used for hoisting purposes on board ship.

Derrick Crane. A crane in which the jib has both horizontal and vertical movement.

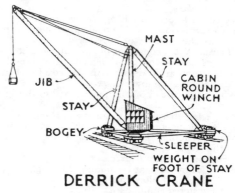

MAST

STAY

JIB

CABIN
ROUND
WINCH

STAY

BOGEY

SLEEPER

WEIGHT ON
FOOT OF STAY

DERRICK CRANE

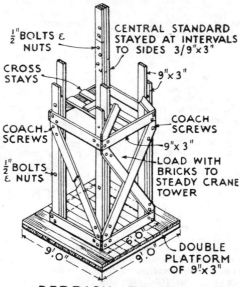

$\frac{1}{2}$" BOLTS &
NUTS

CENTRAL STANDARD
STAYED AT INTERVALS
TO SIDES 3/9"x3"

CROSS
STAYS

9"x 3"

COACH
SCREWS

COACH
SCREWS

9"x 3"

$\frac{1}{2}$" BOLTS
& NUTS

LOAD WITH
BRICKS TO
STEADY CRANE
TOWER

9'.0"

9'.0"

DOUBLE
PLATFORM
OF 9"x3"

DERRICK TOWER

Derrick Towers. The towers supporting the decking, or staging, for a derrick crane. There are three strong, braced, timber towers.

The king, or crane, tower supports the crane. The anchor, or queen, towers, tie down the sleepers of the crane by means of chains, to prevent the crane from overturning when loaded.

Desert banksia. *Banksia ornata* (*H.*). W. Australia. A small tree, but excellent wood for small cabinet work. See *Banksia* and *River banksia*.

Desert Gum. *Eucalyptus cliftoniana* (*H.*). W. Australia. Little commercial value. See *Gums*.

Desiccation. The artificial drying of timber in scientifically heated chambers. See *Seasoning*.

Desk. A table, usually with sloping top to write upon or read from. See *Lectern*. A writing desk usually has pedestals of drawers, and

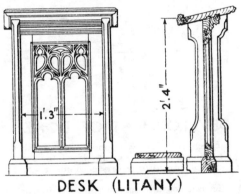

DESK (LITANY)

an enclosed portion with pigeon holes or stationery cabinet. The cover may be a rolltop, cylinder fall, or a hinged flap. In the latter case the flap, or fall, usually serves as the desk and is supported by metal straps or stays, or by slides. See *Kneehole Desk*.

Determa. *Ocotea rubra* (*H.*). C. America. Also called Surinam, Wana, etc. Light red and lustrous. Moderately hard, heavy, strong, but brittle, and fairly durable. Smooth, with distinct rays. Not difficult to work. Used for carriage construction, planking, furniture. S.G. ·7; C.W. 3.

Detrusion. The shearing of the fibres along the grain of the wood.

Development. The unfolding, or laying out, of a surface to show the true shape. The rebatment of a line into the planes of projection to show the true length and inclinations. See *Stretchout*. (See illus., p. 140.)

De Wild. Abbreviation for the name of the botanist De Wildeman.

dft. Abbreviation for *draft*.

d.h. Abbreviation for *double-hung*.

Dhaman. *Grewia tiliæfolia* (*H.*). India. Reddish brown, darker streaks and white spots. Fairly hard and heavy. Strong, flexible, tough, and very durable. Dull, smooth, odorous. Straight and wavy

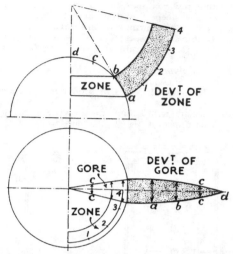

DEVELOPMENT OF DOME

grain, medium texture. Easily wrought and polishes well. The heart should be boxed in conversion. Used for carriage building, poles, shafts, joinery, furniture, sports equipment. S.G. ·72 ; C.W. 2·5.

Dhow. An Eastern coasting sailing ship. It has one mast, lateen sail, and broad stern.

Dhup. *Canarium euphyllum* (*H.*). India and Andaman Islands. Also called White Dhup and White Mahogany. Pale yellow to pinkish grey, lustrous. Light, soft, very smooth, but coarse texture. Interlocked grain. Not durable. Large sizes and easily wrought. Stains easily if air-seasoned. Very buoyant. Used for planking, packing cases, flooring, cabinet work. S.G. ·45 ; C.W. 2·5. *Canarium strictum* (Black Dhup) is very similar, but harder and heavier and more difficult to work. S.G. ·6 ; C.W. 3. *Vateria indica* is called White Dhup but is no relation to the above. Yellow to light brown. Lustrous but feels coarse. Moderately hard, heavy, not durable. Interlocked grain but fairly easy to work. Often sold as Malabar white pine. Uses as above. S.G. ·6 ; C.W. 2·5.

Dhupi. See *Sugi*.

Diabolo. A game in which a wood reel is whirled on a string held by two sticks.

Diaglyphic. Sculptured ornamentation below the face of the material.

Diagonal. A line, or member, crossing a figure, or frame, from corner to corner.

Diagonal Brace. See *Angle Brace.*
Diagonal Grain. Applied to converted wood in which the fibres do not run parallel to the axis of the piece, through faulty conversion or crooked tree. Also see Oblique Grain.
Diagonal Ribs. The ribs running from the angle between the walls to the crown, in a rib and panel vault.
Dialium. See *Ironwood, Jutahy, Kranji* or *Keranji.*
Diameter. A straight line passing through the centre of a circle or polygon, terminated at each end by the perimeter, and cutting the figure into two equal parts. A line passing through the centre of a sphere.

DIAMOND   FRET

Diameter Tape. A graduated tape that gives the diameter of a tree from the girth measurement.
Diametral Plane. A plane cutting a spherical or cylindrical solid along the diameter or axis.
Diamond Fret. An ornamental moulding shaped like a series of rhombuses, or diamonds, by means of intersecting fillets. Characteristic of Norman architecture.
Diamonding. A term applied to uneven shrinkage that causes a square to distort to diamond shape. It occurs in pieces in which the growth increments extend diagonally.
Diamond Match. See *Veneer Matching.*
Diamond Point. A turning chisel tapering to a point at the middle of the cutting edge.
Diamond-shaped Spout. See *Spouts.*
Diaper. Flat surfaces decorated by squares, or lozenges, and foliage.

DIAPER

The pattern is repeated in regular formation over an area.
Dibble. A tool for making holes in which to plant seedlings.
Dibetou. African walnut, *q.v.*
Dichopsis. See *Palaquium.*
Dicky. A small seat at the back of a horse-drawn carriage or two-seater motor car.
Dicorynia. Angelique, *q.v.*
Dicotyledons. Trees in which the seedling has two seed leaves. Deciduous, or broad-leaved, trees. Hardwoods, *q.v.* That branch of Angiosperms that supplies nearly all of the commercial hardwoods. It includes about 200 families, nearly 3,000 genera, and over 120,000 species.
Didelotia. See *Bubinga* and *Kevazingo.*
Didu. *Bombax insigne* (*H.*). Burma and India. White, seasoning to yellow-brown, with brown and black streaks due to sapstain fungi. Fairly lustrous. Very light and soft with straight, even, coarse

texture. Some ripple. Not durable. Easy to work. Used for packing cases, boxes, rotary veneers, etc. S.G. ·34 ; C.W. 2.

Didymopanax. See *Yagrume*.

Die. 1. The dado, or cubical part, of a pedestal. 2. A printer's block.

Diels. The name of the botanist L. Diels.

Die Square. Applied to timbers square in section and between quarterings and balks in size. *Dead square*, to distinguish from *customary square* which may include waney edge.

Diffuse-porous. A description of wood in which the pores are diffused, or scattered, throughout the annual ring so that there is no marked difference between those in the summerwood and in the springwood. The term is also applied to a gradual change in size and distribution throughout the growth ring. See *Ring-porous* and *Pores*.

Digesting. See *Wood Pulp*.

Diglyph. Similar to triglyph, *q.v.*, but with two channels.

Dihedral Angle. The angle between two intersecting plane surfaces, as at the hips and valleys of a roof.

Dillenia. *Dillenia pentagyna*. Several species (*H.*). India. Reddish grey, purplish, or reddish brown. White lines from chalky deposit. Moderately hard, heavy, strong, and durable. Interwoven and twisted grain, even and coarse texture. Liable to warp and feels rough. Attractive wavy figure when quarter-sawn. Fiddleback when tangential-sawn, but warps freely. Difficult to work when seasoned. Used for rotary veneers, superior joinery, sleepers, etc. S.G. ·63 ; C.W. 4·5. *D. Indica* is called Chalta. See *Simpoh* and *Catmon*.

Dilleniaceæ. A family of trees valued for their ornamental appearance rather than for their wood. Includes *Curatella, Dillenia, Doliocarpus*, and *Wormia*.

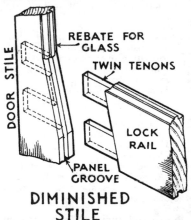

REBATE FOR GLASS

TWIN TENONS

DOOR STILE

LOCK RAIL

PANEL GROOVE

DIMINISHED STILE

Dilly. See *Bulletwood*.

Dimension Timber or **Stock.** Applied to wood cut to sizes required by different woodworking trades. An American term usually implying timbers from 6 in. × 6 in. upwards and from 10 ft. upwards. Sometimes applied to lumber larger than 1-in. boards. Also see *Small Dimension Stock* and *Furniture Dimension Stock*.

Diminished Stile. A door stile that is diminished in width at the lock rail to give a greater glass area above. A gun-stock stile.

Dimmers. Adjustable louvres, or shutters, to control the passage of sound.

Dimorphandra. See *Dakama, Mora*, and *Morabukéa*.

Dinette. A nook in a kitchen fitted with fixed seats and small table.

Dinghy. The smallest of a ship's boats. A small rowing or sailing boat.

Dinner Wagon. A framework of shelves on wheels, or castors, for conveying dishes, etc., in a dining-room. See *Wagon*.

Diospyros. Ebony, *q.v.* Also see *Calamander, Camagon, Ebony, Kaki, Kanran, Marblewood, Meribut* or *Arang, Persimmon, Uhu*.

Diplotropis. See *Sapupira*.

Dipterocarpaceæ. A family of trees including Anisoptera, Balanocarpus, Dipterocarpus, Dryobalanops, Hopea, Isoptera, Parashorea, Pentacme, Shorea, Vateria, Vatica.

Dipterocarpus. See *Apitong, Dammar* (*black*), *Eng, Gurjun, Hollong, Hora, Keruing, Kruin, Lauan, Yang*.

Dipteryx. See *Tonca Bean*.

Diptychs. Small folding doors, formed like shutters, especially when near an altar, and suitably painted or carved.

Direct Drive. An independent drive to a machine, which eliminates lines of shafting and long belting. The direct drive is safer, and better for high-speed machines and where the machines are not in constant use.

Disc or **Disk.** A thin cylindrical piece of wood. A quoit or discus.

Discoloration. Any change in the normal colour of wood. It may be due to fungi or chemical action. In softwoods abnormal colour, except " blueing," usually denotes decay. See *Colour* and *Stains*.

Discontinuous Growth Ring. One formed on one side of the stem only.

Disc Sander. A revolving disc covered with glasspaper for smoothing small and cylindrical work.

Diseases. The diseases in wood are decay, doatiness, druxiness, dry rot, foxiness, plethora, wet rot. Also see *Decay, Defects, Rot*, and *Watermark*.

Dished. The term applied to a hole chamfered or rounded on the edge, or sunk for the flange of a pipe.

Dishes. See *Crates*.

Dish Rack. See *Plate Rack*.

Dispatch Box. A box or case to hold documents.

Dispatch Money. A payment for loading a vessel in less than the arranged time.

DISTAFF

Display Cabinet. A show case.

Distaff. That part of a spinning wheel on which the yarn is wound.

Distance Piece. A wood block or strap to keep two members the required distance apart.

Distemonanthus. See *Ayan, Movingue, Satinwood, Yellow Satinwood*.

Distributed Load. A load spread over an area or length ; not concentrated. A distributed load may be uniform or otherwise.

Ditty Box. A box to hold needles, sewing cotton, etc.

Dividers. An instrument like a pair of compasses, but with both legs terminated by pin points. The instrument is used for dividing lines into a number of equal parts by trial and for stepping out or transferring equal distances.

Divi-divi. *Cæsalpinia coriaria* (*H.*). C. America. Chief importance for tannin.

Divining Rod. A forked twig, used by water diviners for discovering water, or metals, below ground.

dld. and **dly.** Abbreviations for *delivered* and *delivery*.

d.m. Abbreviation for *double moulded*.

Doatiness. See *Dote*.

Dock Dues or **Dockage.** Payment to the harbour, or dock, authorities for the use of the docks for loading or unloading. It is a timber trade custom for unmanufactured goods that the cargo receiver pays two-thirds of the dock dues.

Dock Gates. Large watertight gates to docks. In dry docks they also act as sluices.

Docks. 1. Artificial enclosures for ships while undergoing repairs

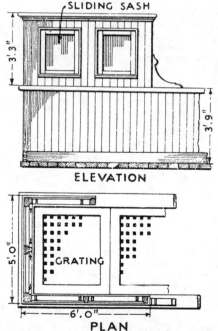

SLIDING SASH

3' 3"

3' 9"

ELEVATION

5' 0"

GRATING

6' 0"

PLAN

DODGER OR BRIDGE SHELTER

or loading or unloading. DRY DOCKS. Graving docks. The water
is pumped out when the gates are closed. TIDAL DOCKS. The level
of the water is the same both inside and outside. WET DOCKS. The
water level is kept constant by means of gates. **2.** See *Ramp* (3).
Dock Warrant. A receipt for goods warehoused at the docks.
Dock Work. The heavy constructional woodwork entailed in dock-
ing, harbours, jetties, piers, wharves, etc. In this country the timbers
normally used for this class of work include blue gum, elm, greenheart,
jarrah, pitch pine, turpentine.
Dodecagon. A rectilineal plane figure with twelve sides.
Area $=$ (side)$^2 \times 11 \cdot 196$.
Dodecahedron. A regular solid with twelve equal pentagonal faces.
Volume $=$ (linear side)$^3 \times 7 \cdot 663$. Surface $=$ (linear side)$^2 \times 20 \cdot 646$.
See *Polyhedra*.
Dodger. A bridge shelter on board ship.
Dog. **1.** A strong iron fastening for heavy timbers, especially in
temporary work. It consists of a length of
iron with the ends bent at right angles and
pointed. A small type is used for holding
together a glued joint until the glue sets.
2. Pressure mechanism to secure an object
in the required position. **3.** A strong piece
of steel bent and pointed at one end with a
ring at the other end, used in logging.

Dog Almond. Cabbage bark, *q.v.*
Dogcart. **1.** A high two- or four-wheeled carriage with seats back
to back. It should have space below to carry sporting dogs. **2.** A
loose term applied to several types of carriages.
Dog-legged. Applied to two parallel flights of stairs, continued in
opposite directions, with the strings in the same newel. The outer
strings of the flights are in the same vertical plane so that there is no well.

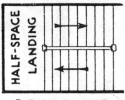

**DOG·LEGGED
STAIRS**

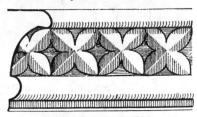

DOG TOOTH

Dog-leg Spout. See *Spouts.*
Dog Plum. See *Cape Ash.*
Dog Shore. **1.** A horizontal shore, distinguished from a flying
shore by having no braces. See *Flying Shore.* **2.** A heavy balk used
for supporting a ship under construction or repair.
Dog Tooth. An ornamental moulding with curved pyramidal projections.

Dogwood. 1. *Ichthyomethia piscipula* or *Piscidia piscupula* (*H.*). C. America. Yellowish brown, lustrous. Hard, heavy, very strong, and durable. Roey, medium texture. Polishes well. Used for piling, vehicles, shipbuilding, etc. S.G. ·87; C.W. 4. 2. *Cornus florida* (*H.*). Also called American Boxwood. Pinkish. Hard, heavy, very fine grain. Used for turnery, shuttles, inlays, etc. *C. sanguinea* (*H.*). Also called Cornel. Very hard and tough. Small sizes. Cultivated chiefly for charcoal and explosives. *C. nuttallii.* Pacific dogwood. Similar to above species of *Cornus*. Used for bobbins, turnpins, small pulleys, mallet and golf-club heads novelties, etc. S.G. ·7. See *Cornel* and *Cornus*.

Dolly. 1. A hardwood block placed on top of a pile to protect it from the pile hammer. 2. See *Puncheon*. 3. An upright roller for controlling moving logs. 4. A wood implement for swirling the clothes in a wash tub.

Dolly Tub. A wash tub.

Dolphin. 1. A group of piles for securing a boom or for protecting the piers of a bridge. A cutwater, *q.v.* 2. A short post serving as a bollard on board ship.

Dolphin Hinges. Used on the ends of the *falling* front of a secretaire.

Dolphin Striker. A spar, or boom, inclined downwards from the bowsprit cap to help to support the jib-boom.

Dombeya. See *Mukeo*.

Dome. A hemispherical roof. A cupola. The term is also applied

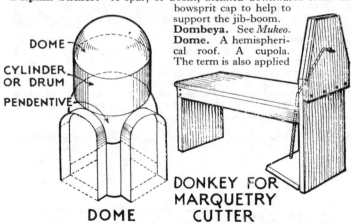

DOME

CYLINDER OR DRUM

PENDENTIVE

DOME

DONKEY FOR MARQUETRY CUTTER

to similar roofs with polygonal and elliptical plans. See *Gore*.

Dominant. A term in forestry applied to trees that form the main canopy. See *Suppressed* and *Wolf*.

Dominion Whitewood. See *Kahikatea*.

Don. The name of the botanist D. Don.

Donkey. 1. A wooden frame used as a stool when cutting veneers. 2. A kind of fret saw for marquetry.

Donkey's Ear. A mitre shoot that can be fixed vertically in a vice.

Donnacona. A registered wallboard, *q.v.*

Donsella. *Mimusops jaimique* (*H.*). C. America. Also called Acana and Almique. Several species from Mauritius. Reddish brown, variegated markings. Darkens with exposure. Feels oily but polishes well. Very hard, heavy, durable, and stable. Wavy and roey grain, fine uniform texture. Not difficult to work. Used for cabinet work, superior joinery, turnery. S.G. ·97 ; C.W. 3.

Dook. A wood plug.

Doona. *Doona spp.* (*H.*). Ceylon. Reddish brown, lustrous. Fairly hard and heavy. Strong and durable. Smooth close grain and distinct rays. Resembles teak in some respects.

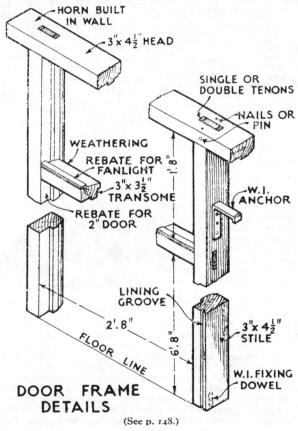

DOOR FRAME DETAILS

(See p. 148.)

Door Bar. See *Fall Bar* (2).

Door Case or **Casings.** Jamb linings rebated for a door.

Door Check. An appliance for controlling the closing of a door.

Door Cheek. See *Cheeks* (4).

Door Finishings. The ornamental features to a doorway ; architraves, moulded grounds, pediment, plinth blocks, etc.

Door Frame. A strong solid frame for a door, as distinct from door linings. It consists of stiles and head, and a transome if there is a fanlight. It is made from about 5 in. × 3 in. stuff. (See illus., p. 147.)

CORNICE

PANEL

DOOR HEAD

Door Furniture. The term is especially applied to handles, knobs, escutcheons, and fingerplates, but often used to include hinges, locks, latches, bolts, letterboxes, etc. The former group may be of wood, glass, metal, or any of the phenol or lactic products such as bakelite, erinoid, etc. Metal furniture may be obtained in over twenty different alloys and finishings.

Door Head. 1. A horizontal projection or decorative feature over a door. A canopy or hood. 2. The top rail of a door frame.

Door Holder. An appliance for securing an open door to the wall.

Door Linings. See *Door Casings.*

Door Nails. 1. Nails, or studs, with large heads to imitate bolts. They are intended to give an appearance of strength, in imitation of mediæval doors. 2. A nail, or stud, on which a door-knocker strikes.

MOULDED GROUND

DOUBLE FACED ARCHITRAVE

DOOR LININGS

Door Operator. A person engaged in the buying and selling of imported doors.

Door Plane. A coachmaker's plane for rebates and sinkings.

Door Posts. The stiles of a solid door frame. Door cheeks.

Doors. The various types of doors are : ledged, ledged and braced, flush-faced, framed and ledged, panelled, and solid. Panelled doors may have any number of panels and they are named according to the number of panels, or to some special feature as double-margin, gun-stock, Gothic, sash, etc., or to the particular use. The standard sizes are 6 ft. 4 in. × 2 ft. 4 in. to 6 ft. 8 in. × 2 ft. 8 in., in 2-in. rises. Special types of doors are : barred, cabinet, composite, double, dummy, dwarf, folding, fire-resisting, hatch, jib, revolving, sliding, stable, swing, trap, tambour, and warehouse. See *Barred, Dwarf, Flush, Gun-stock,* and *Panelled Door.*

Door Sill. A piece tying together the feet of door posts and usually serving as a threshold.

Door Spring. A spring arrangement that automatically controls the closing of a door. There are many patent types.

Door Stop. 1. The edge of the rebate against which a door closes. **2.** See *Floor Stop.*

Door Suite. A complete set of furniture for a door.

Door-way. The entrance to a building or to a room.

Doping. Coating fabric with dope to make it airproof and waterproof. Dope is chiefly of cellulose acetate or nitrate, with pigment if required.

Doric. The first and simplest in design of the three Greek orders

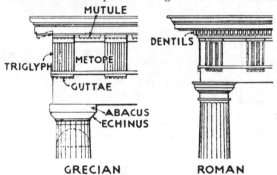

GRECIAN ROMAN

DORIC

of architecture. The order has no base, a plain capital consisting of a square abacus usually with echinus, and a fluted shaft. The characteristics are guttæ, metopes, and triglyphs. There are several variations in the Roman Doric.

Dormand, Dormant, or **Dormant Tree.** A sommerbeam or sleeper.

Dormer. A window with vertical casements in a sloping roof. See *Lucerne.*

DORMER

Dormitory. A sleeping apartment in schools, etc., accommodating a number of persons.

Dorsal or **Dorse.** 1. Wall hangings. 2. The back support for a canopy

Dory. A small boat.

Doryphora. See *Canary Sassafras*.

Dosy, Doty, or **Dozy.** Applied to wood that shows incipient decay due to wood destroying fungi. See *Dote*.

Dote. A general term used in lumbering for decay or rot in timber. Incipient decay. Localized patches of decay. It is generally denoted by discoloration, speckles, and an unusual odour, and is often due to lying in the round after felling.

Double-acting. Applied to a door that swings both ways. A swing door.

Double Bed. One to accommodate two persons.

Double-bellied. Applied to a turned baluster when both ends are alike.

Double-boxed. A boxed mullion prepared to receive two pairs of weights for sliding sashes.

Double Curvature. Circle-on-circle work, *q.v.*

DOUBLE DOVETAIL KEY

Double-cutting Band Mill. Wide band saws with two cutting edges to cut in either direction of feed.

Double Decker. 1. A ship with two decks. 2. A vehicle with roof seats.

Double Doors. A pair of doors meeting with rebates on the middle, or meeting, stiles. Sometimes called Folding Doors, *q.v.*

Double Dovetail Key. See *Dovetail Key*.

Double Faced. Applied to architraves moulded on both edges or to built-up architraves stepped to form two plane faces. See *Door Linings*.

Double Floor. A floor consisting of binders and common joists. See *Framed Floor* and *Binder*.

Double-handed Saw. A long saw with handles at each end for cross-cutting timber of large section.

Double Hung. Applied to *sash and frame* windows in which both sashes are hung by balance weights.

Double-knuckle Joint. A cabinetmaker's joint for screens so that the wings can swing in either direction. A specially shaped fillet is placed between the wings to make the screen draught-proof. See *Screen* and *Reversible Hinge*.

Double Laths. Specially thick plasterer's laths. See *Laths*.

Double Lean-to. A V-shaped roof. It consists of two lean-to roofs with a parallel or a V-gutter.

Double-margin Door. A wide door with the appearance of a pair of doors. It has four stiles, with the inner stiles keyed together with folding wedges. The rails are wedged to the inner stiles which are then keyed, and then the panels and outer stiles put in position.

Double Measure. Moulded on both sides.

Double Partition. A partition constructed with a cavity for sliding doors.

Double-pitched Roof. A mansard roof, *q.v.*

Double Quirked. See *Return Bead*.

Double Rebated. Applied to wide jamb linings that are rebated on both edges so that the door may be hung on either face of the wall. See *Door Linings* and *Finishings*.

Double Return Stairs. A staircase in which *return* flights branch to right and left; a return in two flights.

Double Ring. See *False Ring*.

Double Roof. Applied to roof trusses combining more than one type. A hammer-beam truss.

Double Sapwood. A defect in wood in which the sapwood has not been converted into heartwood, probably due to severe frost, but through which the cambium survived to continue future growth.

Double Skirting. A wide skirting built up in two widths. The top piece is set back. (See illus., p. 152.)

Double Step. A joint for heavy constructional timbers to avoid

FRIEZE RAIL

METAL BAR

FOLDING WEDGES

DOUBLE MARGIN DOOR

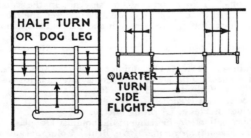

HALF TURN OR DOG LEG

QUARTER TURN SIDE FLIGHTS

DOUBLE RETURN STAIRS

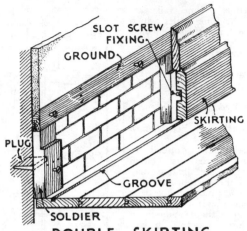

DOUBLE SKIRTING

weakening one of the pieces too much and to provide greater resistance against detrusion.

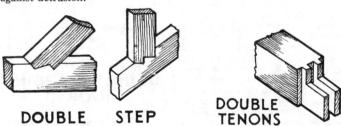

DOUBLE STEP   **DOUBLE TENONS**

Double Tenons. Two tenons side by side not in the same plane. Used on a lock rail to receive a mortise lock between the faces of the tenons or on thick stuff to give strength and prevent distortion. See *Twin Tenons*.

Double Tiered. Applied to a framed partition extending through two storeys.

Double Window. A frame containing two parallel sets of casements or balanced sashes. The space between is intended for insulation, but it must be carefully designed to be effective.

Double Wrack. See *Wrack*.

Doubling Fillet or **Piece.** A tilting fillet, *q.v.*

Dougl. Abbreviation for the name of the botanist Douglas.

Douglas Fir. *Pseudotsuga taxifolia* (S.). British Columbia and Western U.S.A. Also called Columbian Pine and Fir, Oregon Pine and Fir,

Puget Sound Pine, Red Fir, and Yellow Fir. Pink, toning to light reddish brown, resinous. Strong for its weight and fairly durable for a softwood. Very large sizes free from defects. Excellent timber for general purposes. Handsome grain due to contrasting colour and density in growth rings. Similar to redwood, *q.v.* Easy to work, stains and varnishes well, sometimes difficult to paint due to prominent grain. Used for constructional and structural work, piling, joinery, plywood, flooring, sleepers, wood blocks, etc. S.G. ·5 ; C.W. 1·75.

Doussié. See *Apa.*

Dovecote. A nesting box for pigeons or doves.

Dovetail. A joint for pieces at right-angles to each other and end to end. The pins, on one piece, are fan-shaped like a dove's tail, and fit in sockets or eyes in the other piece. The variations of dovetailed

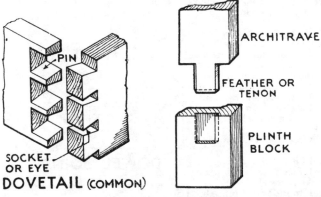

PIN

SOCKET OR EYE

DOVETAIL (COMMON)

ARCHITRAVE

FEATHER OR TENON

PLINTH BLOCK

DOVETAIL FEATHER

joints are : common, lapped, secret. The dovetail is also used for halving joints, etc., to prevent lateral movement. See *Housed, Sash, Shouldered, Slip,* and *Lapped Secret Dovetails.* The term dovetail is sometimes used to imply a good fit in a joint.

Dovetail Bit. A machine bit for machine dovetailing.

Dovetail Feather. A thin slip shaped like a dovetail to secure two pieces together as between architrave and plinth block.

Dovetail Key. A wood key with dovetailed section. See *Counter Battens.* A *double dovetail key* is used for end-to-end butt joints. See *Double Dovetail Key.*

Dovetail Key Tongue. See *Tray Frame Table.*

Dovetail Margin. Strips mitred round a flush door, with bevelled rebates to receive the edges of the plywood to prevent the plywood from springing. See *Edge Strip* and *Flush Door.*

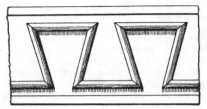

DOVETAIL MOULDING

Dovetail Moulding. A Norman moulding or triangular fret in the form of a continuous band of dovetails.

Dovetail Plane. A plane for forming bevelled sinkings.

Dovetail Saw. Like a tenon saw but smaller and with an open handle. It has about 14 points per inch.

Dovetail Socket. A movement used for a swivelled mirror, as in a dressing table.

Dovetail Tenon. See *Fox-wedging* and *Dovetail Feather*.

Dowel. A wood pin. A cylindrical piece of hardwood used in many types of joints, or in place of a mortise and tenon joint. Dowels should have a fine groove along their length to allow air and glue to escape. Metal dowels are used for fixing the feet of posts to stone, concrete, etc. See *Plate Dowel*.

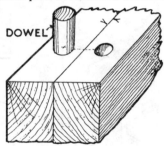

DOWEL JOINT

Dowel Bit. A short twist bit.

Dowelling Jig. An appliance to guide a bit when boring for dowels.

Dowelling Stock. A cooper's brace.

Dowel Pin. See *Nails*.

Dowel Plate. A steel plate for making dowels.

Dowel Rounder. A brace bit for preparing the end of a dowel to ease the entry into the hole.

Dowel Screw. A double-ended screw. See *Screws*. It is useful for making end-to-end joints of small stuff, and

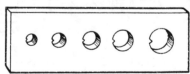

DOWEL PLATE

serves the same purpose as a handrail bolt.

Down Comer. A down spout, *q.v.*

Down Sizes. A term used to denote *exact* sizes off the saw.

Down Spout. A long square box, or closed trough, for conveying water from the eaves gutter to the ground. It is about 3 in. square inside and of 1-in. stuff, and used instead of cast-iron down pipes.

Dowsing Rod. Same as divining rod, *q.v.*

Doyle Rule. An official rule of the U.S.A. Hardwood Manufacturers' Association that allows for waste in conversion from the round log to inch board measure when calculating the contents of the log. The *Doyle-Scribner Rule* is a combination of two rules : the former is used for logs 27 in. diameter and under, and the latter for 28 in. upwards.

Dozy. See *Doty.*

d.p.c. Abbreviation for *damp-proof course.*

dr. Abbreviation for *door* and *debtor.*

d/r. Abbreviation for *dock return.*

Dracæna. Tropical liliaceous, palm-like plants, which include the dragon tree.

Dracontomelum. See *Dao, New Guinea Walnut, Paldao,* and *Walnut.*

Draft. 1. See *Draught.* 2. To prepare a preliminary sketch or scheme for anything to be constructed.

Drag. 1. The lower part of a moulding box, *q.v.* 2. A lumberer's term for collecting felled logs into the runway.

Drag Cart. A bummer, *q.v.*

Dragging Tie. A dragon tie. See *Angle Tie.*

Dragon Beam or **Piece.** The timber with the dragon tie carrying the foot of a hip rafter. See *Angle Tie.*

Dragon's Blood. The resin from the dracæna, or dragon, tree. It is red and used as colouring pigment.

Drag Sled or **Dray.** A single sled on which one end of a log rests when being dragged or hauled. It is also called a bob, crotch, scoot, sloop, or travois.

Drain Hole. A weep hole, *q.v.*

Draining Board. An inclined board, grooved to collect the water from draining crockery and to convey it back to the sink. The board is usually made of teak. See *Sink.*

Drammen Standard. A measure of timber, now seldom used, consisting of 120 pieces totalling $121\frac{7}{8}$ cub. ft. See *Standard.*

Drapery Panel. See *Linen Fold Panel.*

Draught, Draft, or **Draw.** 1. The clearance allowed in jib and cotter joints, and in draw-boring, to allow the joint to tighten. See *Gib and Cotter.* 2. The depth of a ship below water level.

Draughtboard. 1. A checkerboard for the game of draughts. A chessboard. 2. Applied to a design consisting of black and white squares.

Draught Check. Any arrangement in joinery to prevent draughts, especially for windows in exposed positions.

Draughting. Setting out a drawing in orthographic projection.

Draw. See *Draught.*

Drawback Lock. A lock in which a cranked lever operates the bolt from the inside. A key is used on the outside of the door in the usual way.

Draw-boring. Preparing the hole for a dowel, or pin, by boring the tenon and cheeks of mortise separately. The hole in the tenon is slightly in advance of that in the cheeks, so that driving the dowel tightens the joint. (See illus., p. 156.) See *Draw Pin.*

Drawbridge. A bridge that can be moved to allow for the passage

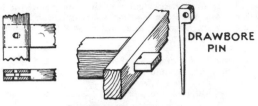

DRAWBORE
PIN

DRAWBORING

of vessels. It may be hinged to raise one end or it may swing horizontally.

Drawer. A sliding open box in a table, desk, cabinet, etc.

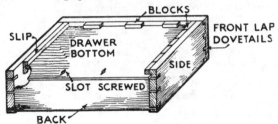

BLOCKS

FRONT LAP
DOVETAILS

SLIP

DRAWER
BOTTOM

SIDE

SLOT SCREWED

BACK

DRAWER CONSTRUCTION

Drawer Guides. Pieces planted on the runners to form a rebate, or guide, for a drawer. See *Guide*.

Drawer-lock Chisel. A small cranked chisel for preparing a mortise in a confined space. Also called a Bolting Iron, as it is used for bolting up, or preparing the mortise, for a drawer lock.

DRAWER-LOCK
CHISEL

Drawer Muntin. An intermediate rail for a long drawer to divide the bottom into two equal parts.

Drawer Pulls. Finger grips for the front of a drawer.

Drawer Slips. A planted grooved strip to receive a drawer bottom. It is used instead of ploughing the side of the drawer, *q.v.*

Drawer Stops. Small blocks glued on to the frame to prevent a drawer from sliding in too far. See *Guide*.

Drawing Appliances and Instruments. The minimum requirements for a draughtsman are : board T-square, set-squares, rules, pens, french curves, ink and pencil compasses, dividers, spring bows, protractor. There are many other aids to mechanical drawing.

Drawing Board. A thin board, clamped to prevent warping and squared round the edges, for mechanical drawing with T-square and set-squares. They are made in sizes to suit the drawing paper.

Drawing Paper. For ordinary work *Cartridge* paper is generally used. It is machine made, in over twenty different qualities. *Whatman* paper is more expensive and used for special purposes. Other kinds are Arnold, Causon, Creswick, David Cox, Joynson, Michallet, Van Gelder, Varley. The standard sizes in common use are : Emperor, 68 in. × 48 in. ; Antiquarian, Double Elephant, 40 in. × 27 in. ; Atlas, Columbian, Elephant, Imperial, 30 in. × 22 in. ; Half Imperial, 22 in. × 15 in., etc. *Bristol Board* is a stiff glazed paper used for special work. Tracing paper and cloth are transparent for copying drawings for reproduction, etc. See *Blue Print*.

Drawing-room. A withdrawing room. A parlour, lounge, or sitting-room to which people retire after dining and which is furnished with easy-chairs or drawing-room furniture.

Draw-knife. A cutting tool cranked at the ends for two handles. Also called a Shave, *q.v.* It is used for reducing the width of thin boards, chamfering, rounding edges, etc.

Draw, or **Draw-bore, Pin.** A tapered metal pin used for pulling

DRAW KNIFE

together a mortise and tenon joint where cramps are not available. The tenon projects through the mortise and a hole is bored through the tenon partly inside the mortise. The pin is driven in the hole and used as a lever to pull the tenon through the mortise. See *Drawboring*.

Draw-plate. See *Saw Bench*.

Draw-table. An extension table, *q.v.*

Dray-sled. A low cart for heavy loads.

Dress Circle. The lowest gallery of a theatre or a part of the gallery.

Dressed. Applied to wood that is surfaced or planed, on one or more sides. The term *dress up* means to finish off, or plane and sandpaper, a piece of woodwork.

Dressed and Headed. Applied to floorboards tongued and grooved on the ends.

Dressed and Matched. Applied to boards that are surfaced and tongued and grooved.

Dresser. 1. A kitchen fitment combining table, shelves, and drawers. It may be fixed in a recess or it may be self-supported. The upper shelves are usually enclosed by glazed doors. See *Cupboard*. There is a large number of registered designs called kitchen cabinets, *q.v.* 2. A boxwood, or hornbeam, tool used by the plumber for dressing sheet lead.

Dressing. Planing and finishing woodwork.

Dreyfusia. An insect specially destructive to silver fir.

Drift. 1. A large punch. 2. An appliance or a tapered piece of hardwood, for driving out the core after mortising on a hand-mortiser. 3. Applied to a saw cut that has run off the correct line of direction through faulty setting of the saw.

Drift-bolts. Iron or steel bars with one end pointed and the other end formed into a head. They are used as fastenings in heavy timber construction and serve the same purpose as spikes in smaller timbers. A common size is about 2 ft. 6 in. long and ¾ in. diameter.

Drifter. A fishing boat that carries drift nets, especially in herring fishing.

Driftwood. Wood cast up on the shore by the sea or tidal river.

Drill. A bit for boring, or drilling, holes in hard materials. Drills may be operated by hand, brace, bow, or machine.

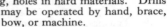

BOARDS

LEAD

2" TO 3"

CAPILLARY GROOVE

LEAD

DRIP

Drill Planting. Planting trees in rows.

Drimys. See *Winter's Bark*.

Drip. 1. A step in a lead gutter. 2. A projection on a moulding to break the flow of water, to prevent the water from running on to the wall, etc.

Drip Box. A cesspool to a roof gutter.

Drip Pan. A small gutter under a ship's window in an exposed position. A condensation groove.

Drip Shield. A projecting guard or narrow hood over a window to throw the water away from the casements.

Drive. A collection of logs being floated from point to point.

Driver. 1. A mechanical device on a lathe to revolve the work. 2. A cooper's tool for driving down the hoops on a cask. 3. A golf club with wood head.

Drone. A tube to a bagpipe that produces a deep vibrant note.

Drop. 1. An ornamental terminal at the bottom of a suspended post, as to a newel to an upper flight of stairs. A pendant, *q.v.* 2. Small cylinders or truncated cones used as a decorative feature. See *Guttæ*.

Drop Arch. A Gothic, or pointed, arch in which the radius of curvature is less than the span.

Drop Girt. See *Girt*.

Drop-handle. A drawer pull or door handle that drops flat to the face of the drawer or door when not in use.

Drop-leaf Table. A table with hinged leaves that hang vertically when not in use. When the leaves are in use they are supported by fly-rails. The drop-leaf is often used for tables fixed to walls, in this case hinged brackets are used as supports.

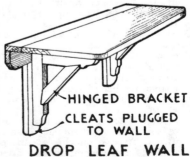

HINGED BRACKET

CLEATS PLUGGED TO WALL

DROP LEAF WALL TABLE

Drop Moulding. A panel moulding that lies below the face of the framing.

Drop Ornament. A split turned ornament characteristic of Jacobean work. A decorative detail like a husk. See *Festoon*, also see *Drop* (2).

Drop Print. See *Tail Print*.

Drop Siding. Weatherboarding for timber buildings. See *Sidings*.

Drum. 1. A temporary cylinder or centre used for veneers in circular work. The veneer is bent round the drum and blocked or staved on the back. See *Cylinder*. 2. The vertical support or stylobate for a dome or cupola. See *Dome*. 3. A cylindrical or hemispherical wood or metal frame with parchment or skins stretched over the ends. It is used as a percussion instrument in a band or orchestra. 4. Any large cylinder on which wires or cables are wound.

Drum Curb. A cylindrical lining to a well under construction. A circular curb or fence.

Drum End or **Head.** A semi-circular, or round, end to a step at the foot of a flight of stairs.

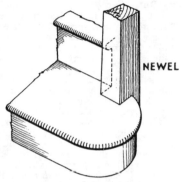

NEWEL

DRUM END STEP

Drumhead. The head of a capstan.

Drum-sander. Drums covered with sandpaper for finishing the surface of wood.

Drunken Saw. A circular saw running eccentrically so that the cut is greater than the thickness of the saw. Adjustable collars are used to give the eccentricity, or wood packings on each side of the saw may be used. The drunken, or *wobble*, saw is used for plough grooves, removing the waste between double tenons, open mortises, etc.

Druxiness or **Druxy.** A timber disease due to the attack of fungus on a wound in the cambium layer. The same as dote, *q.v.*

Drybilt. A prefabricated wood house.

Dry-cemented. Applied to plywood in which the veneers are dried before the adhesive is applied. See *Wet-cemented*.

Dry Cooperage. Casks for dry goods. See *Wet Cooperage*. The term is also applied to the converted wood for making the casks.

DRUNKEN SAW   **Dry Dock.** See *Docks*.

Dry-floated. Applied to Canadian softwoods that have dried after floating on rafts or that have been floated above water level.

Drying. Seasoning, *q.v.*

Drying Stresses. Stresses in timber due to a variation in shrinkage or expansion in adjacent layers, caused by a difference in moisture content.

Dry Joints. Applied to joints that are not glued.

Dry Kiln. A chamber for the artificial drying of wood.

Drymys. See *Winter's Bark.*

Dryobalanops. See *Keledan, Kapur, Teng Mang, Camphorwood.*

Dry Rot. Decay caused by contagious fungoid growth. There are several different forms but they are only distinguished by experts. The common and most destructive form is *Merulius lachrymans* or weeping fungus. The wood is covered with a yellowish or greyish blanket-like covering with darker patches, and there is an unpleasant odour. The disease spreads rapidly in moist stagnant air and reduces the wood to a cracked friable mass. Dry rot cannot occur with well-seasoned sound wood in a ventilated position. It is very difficult to eradicate dry rot as the spores get into the joints of brickwork, etc. All infected wood must be removed and burnt. The remaining timbers should be treated with a strong preservative such as corrosive sublimate (1 oz. per gal. of water, applied hot), creosote, sodium fluoride, sulphate of copper, hot lime, or a proprietary preservative, such as Cuprinol. The walls should be treated with a blow-lamp and the joints raked and re-pointed, and ventilation provided. Other forms are *Poria vaporaria*, which is recognized by the creamy fine strands ; *Coniophora cerebella* or cellar fungus, which has very dark thick strands ; and *Paxillus panuoides*, which is similar to *Coniophora.* These are usually found in cellars and mines, as they require more moisture than *Merulius.* See *Wet Rot.*

Dry-shined. A special finish to the interior of wardrobes, etc.

Dry Stock. Wood containing a normal moisture content after seasoning. See *Moisture Content.*

Dry-topped. Applied to a tree with a dead. or dying, crown, due to injury or disease.

Dry Wedging. Wedging-up framing temporarily without paint or glue.

D.S. Abbreviation for *drop siding*, and **d.s.** for *double sunk.*

d/s. Abbreviation for *days after sight.*

Duabanga. See *Lampati.*

Duala. African mahogany, *q.v.*

Duali. Rotary-cut veneers of Palosapis, *q.v.*

Dubbed-off. 1. Applied to a badly shaped or abrupt easing to a moulding. **2.** Removing arrises to facilitate easy entrance, as for a tenon to enter a mortise.

Dubin or **Dubini.** African mahogany. See *Mahogany.*

Dublin Standard. A measure of timber, now seldom used, consisting of 120 pieces 12 ft. × 3 in. × 9 in., or its equivalent of 270 cub. ft. See *Standard.*

Duboisia. See *White Basswood.*

Duck-board. 1. An inclined scaffold board with cleats at intervals for foothold. Also called a Duck-run or Roof Ladder. **2.** A footrun in a water-logged trench to provide a dry footway.

Ducking Stool. A long wood lever with seat at end to which offenders were tied and ducked in water.

Duck's-bill Bit. A spoon bit, *q.v.*

Dudri. *Machilus edulis* (*H.*). India. Silver grey, streaked and mottled, with fine rays. Moderately hard and heavy. Smooth. Similar to black chuglum in many respects. S.W. ·65 ; C.W. 3.

Duka. See *Tapirira.*

Dukalaballi. *Ficus spp.* 'Guiana. White or grey. Light and soft, but tough and strong for its weight. Used for boxes, etc. Little marketed. Also *Platymiscium spp.* Dark red. Hard and heavy.

Dulcimer. A percussion musical instrument consisting of a sounding board on which metal plates or strings are stretched and struck by leather-covered hammers.

Dumb Barge. A canal barge that depends on towing for movement.

Dumboard. A proprietary insulating board.

Dumb Piano. A keyboard for exercising the fingers. The keys can be adjusted for pressure. There are no strings, hence no musical sounds.

Dumb Sheave. A hole in a spar for a rope, or a half-sheave that does not revolve.

Dumb Waiter. A small lift for conveying food and crockery. A table used when serving dishes for meals. A dinner wagon. See *Wagon.*

Dummy. **1.** A pattern used in turning. See *Copying Lathe.* **2.** An imitation door to a wardrobe.

Dumoria. See *Makore.*

Dumpling. **1.** A cradle or wood block shaped like an inverted basin. It is used for moulding double-curvature work on a spindle. **2.** See *Crate.*

Dums. A local term for frames and casings.

Dune Cypress. *Callitris rhomboidea.* Queensland. See *Cypress Pine.*

Dungun. *Heritiera littoralis* (*H.*). Malay. Dark chocolate brown to purplish. Hard, heavy, strong, tough, and durable. Fine even grain, close texture. Very difficult to saw and work. Used for bowls, tools, piling, structural work. S.G. ·8 ; C.W. 5.

Dunnage. **1.** Brushwood or loose wood used as a support for cargo on board ship. **2.** Loose articles or wood packed amongst ordinary cargo to steady or separate cargo. **3.** Cull lumber, *q.v.*

Duodecagon. See *Dodecagon.*

Duodecimals. A variation of ordinary decimals in which sub-divisions of 12 are used instead of 10. They are used in costing building work, as they conform to feet and inches, and shillings and pence.

Dur. Abbreviation for the name of the botanist Durand.

Durability. The natural resistance of wood against disease, decay, and insect attack. Throughout this Glossary the term is used to imply suitability for exterior work. The durability of wood may be due to its structure, resin or oil or mineral or chemical content, hardness, and density, or to some peculiar property not understood ; but it is always increased by seasoning. There is great variation even in the same species, and only experience or scientific investigation can classify a wood as durable. Usually deep colour, strong natural odour ; infiltration with gums, resins, or oils, denote a measure of durability. See *Decay* and *Preservation*.

Duramen. Heartwood. The inner part of the tree that normally does not contain living cells. Usually it is darker and denser than the alburnum, or sapwood, but in some timbers there is little difference. See *Annual Rings* and *Heartwood*.

Durian Daun. *Durio spp.* (*H*). Malay. Dark red, odorous. Moderately hard and light to fairly heavy. Usually an excess of sapwood which is of little value. Coarse grain, shrinks and warps freely. Not strong or durable and subject to insect attack. Should be quarter-sawn and carefully seasoned. Used for cheap cabinet work, planking, temporary construction. S.G. about ·55.

Durian, Wild. *Cullenia excelsa* (*H*.). India. Pale pink to reddish brown, fairly lustrous, smooth. Medium even texture, straight grain. Light but hard. Not durable but easily treated. Very easy to work. Used for matchboarding, packing cases, carpentry, plywood. S.G. ·6 ; C.W. 2. Also see *Punggai*.

Durmast. *Quercus sessiliflora.* See *Oak*.

Duroi. The name of the botanist Du Roi.

Duromold. A patent process of building up veneered construction of double curvature.

Dust. Fine wood particles from wood-boring insects.

Dust Boards. 1. Horizontal panels between drawers. See *Guide*. **2.** Cover boards, *q.v.*

Dust Cart. One for the collection of household refuse.

Dustproof. Applied to joinery and cabinet work and shopfitting in which special joints are provided to keep out dust.

Dutch Elm. See *Elm*. The adjective *Dutch* also implies imported European wood that is similar to home-grown wood.

Dutch - elm Disease. *Graphium ulmi.* A fungus prevalent in N. Europe and which causes much damage to growing elm trees.

Dwangs. Stiffening pieces. Strutting between joists, and nogging pieces between studs.

Dwani. *Eriolæna candollei* (*H*.). Burma. Brick red, brown streaks. Hard, fairly heavy. Close even grain. Distinct rays. Not difficult to work but needs care in seasoning. Polishes well. Used for cabinet work, superior joinery, gunstocks, inlays, turnery. See *Salmon-wood*.

Dwarf Cupboard. One less than 3 ft. high.

Dwarf Door. A door less than 5 ft. 6 in. high. A screen door, *q.v.*

Dwarf Partition. A screen. A low enclosure or partition as used in offices to divide the private from the public sections.

Dwarf Walls. Sleeper walls supporting ground-floor joists.

Dyer. The name of the botanist Sir W. T. Dyer.

Dyera. See *Jelutong*.

Dyes. Wood is often dyed to imitate superior wood. It is usually in veneer form, and boiled in caustic soda to remove gum or resin,

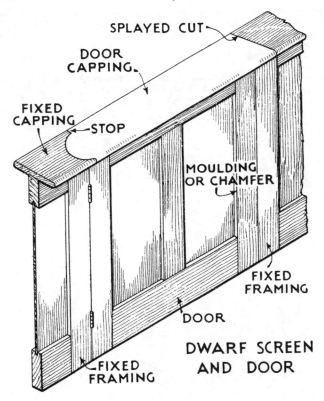

SPLAYED CUT

DOOR
CAPPING.

FIXED
CAPPING

STOP

MOULDING
OR CHAMFER

FIXED
FRAMING

DOOR

DWARF SCREEN
AND DOOR

FIXED
FRAMING

dried, and then immersed in the required dye for one day. The following solutions are used: BLACK. Immerse in hot logwood solution, dry, immerse in boiling solution of crushed nutgalls and copperas. BLUE. Indigo and sulphuric acid, with whiting to give required shade. BROWN. Equal parts of permanganate of potash and sulphate of magnesia diluted with water. GREEN. Dilute blue dye, add yellow to give required shade. GREY (Silver). Boiling solution of copperas (1·5 oz.) and water (80 oz.). This does not require previous treatment. LILAC. Archil in boiling water. RED. Carmine (1 oz.), liq. ammonia (2 oz.), water (40 oz.). ROSE. Coralline dissolved in boiling water with a little caustic soda. VERMILION. First dye black, then soak in warm solution of oxalic acid and water. VIOLET. Mix archil with blue. YELLOW. Picric acid (1 oz.), ammonia ($\frac{1}{2}$ oz.), water (80 oz.). See *Stains*, *Fuming*, and *Polychromatic*.

Dyewoods. Woods that are used as dyes commercially, because their colour is soluble. They are chiefly tropical woods such as fustic, logwood, brazilwood, camwood, barwood, quebracho, sappan, santal, yellow-wood, etc. There is a great variety of woods providing nearly every shade and colour, but most of them are not of commercial importance.

Dysoxylon. See *Kayatau, Miva Mahogany, Red Bean, Rose Mahogany, Rosewood,* and *White Cedar.*

E

E. A symbol for *modulus of elasticity*.

Eagle Square. A registered type of steel square, *q.v.*

Eaglewood. *Aquilaria agallocha* (*H.*). India and Burma. White to yellowish, often reddish tinge and dark bluish streaks due to sap-stain. Old trees have irregular masses of fragrant, darker, and harder wood probably due to fungus; this is the commercial eaglewood used for walking sticks, fancy articles, etc. It is lustrous and feels rough. The normal wood is very soft, light, and elastic, but not durable; fairly coarse and very subject to sapstain. S.G. ·36.

e. and c.b.1.s. Abbreviation for *edge and centre bead one side*. There are several variations.

e. and o.e. Abbreviation for *errors and omissions excepted*.

Earbreadth or **Earbed.** See *Headstock*.

Early English. The first of the Gothic styles of architecture. Thirteenth century.

Earlywood. Springwood, *q.v.* The term *earlywood* is more apt, as it applies to both tropical- and temperate-zone timbers.

Ears. 1. Crosettes, *q.v.* 2. Projecting lugs on pipes, for fixing.

Earth Board. That part of a plough that turns over the mould. A mould board.

Easel. A skeleton frame with a back strut, for displaying pictures, etc.

Easing. 1. Curving, or rounding, an angle to avoid an abrupt change of direction or sharp angle. 2. See *Easing Wedges*.

Easing Wedges. Folding wedges supporting a centre. When the arch is completed the wedges are *eased*, or slackened, a little, to allow the arch to settle. See *Centre*.

East Africa. Numerous woods exported to this country. Variety and quantity continually increasing : cedar, camphorwood, ebony, ironwood, mahogany, muvule, olive, pear, podo, sandalwood, satin-wood, yellow-wood.

East African Camphor. The standard name for *Ocotea usambarensis*. See *Camphorwood*.

East African Olive. The standard name for *Olea hochstetteri*. See *Olive*.

Eastern. A prefix distinguishing the timbers of Eastern N. America, as E. arborvitæ (white cedar), E. cottonwood, E. hemlock, E. white pine, etc.

East Indian Satinwood. The standard name for *Chloroxylon swietenia*. See *Satinwood*.

East Indies. Applied to the woods from the islands of the East Indian Archipelago, and often from India, Burma, etc., to distinguish them from similar species from other parts of the world. See *Amboyna, Apitong, Billian, Ebony, Kokko, Lauan, Paldao, Petaling, Rosewood, Satinwood, Serayah*, etc.

East London. See *Cape Box*.

Easy-chair. A well-padded lounge armchair.

Eaves. The bottom edge of a sloping roof where it meets or over-hangs the wall.

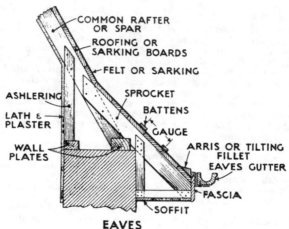

EAVES

Eaves Board. A feather-edged board serving as a tilting fillet, *q.v.*

Eaves Catch. A tilting fillet, *q.v.*

Eaves Fascia A board fixed, on edge, to the feet of the rafters. It usually carries the gutter, and often serves as a tilting fillet in cheap work.

Eaves Gutter. A horizontal trough to collect the rainwater from the roof, but with sufficient inclination, or fall, to convey the water to the down pipe. Sometimes called spouts.

Eaves Plate. A wood beam to carry the feet of rafters when there is no supporting wall for the rafters but only posts or piers. See *Wall Plate*.

Eaves Pole. A tilting fillet, *q.v.*

Eaves Soffit. The horizontal board, or surface, under projecting eaves.

Eba. Ekki, *q.v.*

Ebano. See *Ebony*, *Mastic*, and *Partridgewood*.

Eben or **Ebar.** Abura, *q.v.*

Ebenaceæ. The Ebony family, including Euclea, Diospyros, Maba, Royena.

Ebo. Ekki, *q.v.*

Ebonise. To stain or dye wood to imitate ebony.

Ebony. A very hard, heavy, tough, decorative wood, with fine smooth grain, and lustrous. There is considerable variation in the different species as it is widely distributed, and the colour ranges from dark green, with dark brown stripes, to black. It is usually sold by weight, and marketed in billets, averaging nearly 1 cwt. per billet. The logs are immersed in water for twelve months after felling ; the ends are then ringed with iron and wedged to prevent splitting. When the wood is very streaky it is usually called by some other name : arang, blackwood, calamander, camagon, kaki, marblewood, persimmon, thitpok, etc. S.G. about 1 ; C.W. 5. Ebony is used for fancy articles, inlays, shuttles, turnery, piano keys, bowls, etc. AFRICAN, or NIGERIAN, EBONY, *Diospyros spp.* The wood is usually named after the port of shipment : *Barutu* is coarse and inferior ; *Cameroon* is black, of good quality and fairly large sizes ; *Cape Lopez* is black with a proportion of grey, and fragrant ; *Gaboon* is black and of good quality ; *Ogowe* is similar to Gaboon ; *Old Calabar* is black and of good quality. AMERICAN EBONY, *Diospyros spp.* Tropical America. Black and variegated. Very hard, heavy, and strong, but brittle. The black heart is usually small, and the irregular yellowish sapwood makes a striking contrast for decorative work. S.G. 1·1 ; C.W. 5. Also see *Cocos.* ANDAMAN EBONY. See *Marblewood.* BROWN EBONY. See *Partridgewood* and *Wamara.* BURMESE EBONY, *Diospyros burmanica.* Inferior, because of small proportion of black heart. Also *Dalbergia cultrata,* called Yin Daik. Dark chocolate colour with black stripes ; good quality. CEYLON, or INDIAN, EBONY, *Diospyros ebenum, D. tomentosa,* and *D. embryopteris.* Usually black with lighter striations, grey sapwood with black streaks, and some wavy grain. The last-named species is inferior with very little black and is called Kaluwara or Speckled Ebony. S.G. ·9. E. AFRICAN EBONY, *Dalbergia melanoxylon.* Also called Mozambique Ebony and African Blackwood. Excellent wood, extremely hard and heavy. S.G. 1·4. E. INDIES EBONY. Ceylon ebony. GREEN EBONY, *Brya ebenus.* W. Indies. Greenish brown, variegated stripes. Very hard and heavy ; uniform and smooth. S.G. 1·1. GUIANA EBONY, *Swartzia sp.* Also called Bania. Purplish black, yellowish sapwood. S.G. 1·2. INDIAN EBONY. Ceylon ebony. MACASSAR EBONY. Unclassified. Celebes Isles. Variegated dark brown, black stripes. Used chiefly as veneer. MALAYAN EBONY, *Diospyros spp.* and *Maba spp.* Also called Kayu Arang. Average quality. Many species have no black heartwood. MOZAMBIQUE EBONY. See *E. African Ebony.* PHILIPPINE EBONY, *Maba buxifolia.* Chiefly variegated, but some supplies of black. W. INDIES EBONY. See *Cocos.*

Eccentric Load. A non-axial load. A force that does not act along the axis of a column, or through the centre of gravity of a section.

Echinocarpus. See *Lepcha.*

Echinus. An egg-shaped ornament. See *Egg and Dart.*

Eckebergia. See *Essenhout.*

Eclipse. A registered name for a telescopic scaffold board, and for a metal-faced plywood.

e.d. Abbreviation for *equivalent defects.*

Edge. The narrow surfaces of wood of rectangular section. The thin cutting side of a tool. An abrupt margin or border.

ELEVᴺ OF BOARD

PLAN

EDGE BEND

Edge Bend. The distortion of wood in season-ing in which the wood remains flat but bends edgeways in its own plane. Also called Spring.

Edge Grain. The grain produced on quarter-sawn wood. Also called Comb and Vertical Grain.

Edge Mark. A V-shaped mark to denote the better edge. See *Face Mark*.

Edge Nailing. Secret nailing of boarded surfaces.

Edger, or **Edging Bench.** A special saw-bench in saw-mills for squaring the edges of waney boards.

Edge Tools. Cutting tools that are sharpened on the oil-stone : chisels, gouges, plane irons, etc.

Edge Trimmer. A plane with the sole in the form of a right-angled recess, for planing the edges of small stuff square to the face.

Edging. 1. Straightening the edges of a board with plane or machine. 2. Narrow stuff used as a protective, or decorative, border. 3. Pre-paring the edges of flush doors for the edging strips.

Edging Machine. A special machine for preparing the edges of flush doors for the edging strips.

Edging Strips. Pieces fixed on the edges of flush doors to cover and secure the edges of the plywood. They are sometimes mitred over the top of the door. In cheap doors they are sometimes only on the shutting stile. They are sometimes called Bandings, Claddings, Clashings, Rail-ings, etc.

Edging Tool. A spade-like gardening tool with crescent-shaped blade.

e.e. Abbreviation for *errors excepted*.

Eel Grass. An effective insulator against sound. The grass is cured and laid at all angles to form a cushion full of cells of dead air.

EDGING STRIP

Effective Pillar Length. The length on which *the ratio of length to least radius of gyration* is calculated, in the design of columns.

Effective Span. The distance between the centres of bearing of a structural member, such as a beam or roof truss. See *Clear Span*.

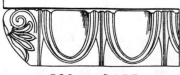

EGG and **DART**

Efforestation. See *Afforesta-tion*.

e.g. Abbreviation for *edgegrain*.

Egba. Mangrove, *q.v.*

Egboin. See *Eghoin*.

Egg and Dart. An ornamenta-tion consisting of a series of carved eggs, or echinus, sepa-rated by vertical anchors or darts. It is common in classic architecture, but subject to varied treatment. Also called Egg and Tongue, and Echinus. The carving is usually on an ovolo moulding.

Eghoin. *Terminalia superba* (*H.*). Nigeria. Also called Afara. Yellow, lustrous. Fairly hard and strong, but light. Mild, even texture, and easily wrought. Used for shelving, matchboarding, etc. S.G. ·5 ; C.W. 2. Also see *Idigbo*.

Ehretia. See *Roble* and *Silky Ash*.

Ehrh. Abbreviation for the name of the botanist Ehrhart.

Ekebergia. See *Cape Ash*.

Ekeing. 1. Lengthening a piece that is too short for its purpose, in shipbuilding. 2. A curved piece under the quarter-piece to the quarter-gallery of a ship.

Ekhimi. *Piptadenia africana* (*H.*). Nigeria. Also called Agboin, Dahoma, etc. Greyish brown, resembling plain oak. Moderately hard, heavy, strong, and durable. Interlocked fibres and coarse grain make it difficult to work. Liable to pick-up and warp. Polishes. Used for structural and constructional work, and cheap furniture. S.G. ·72 ; C.W. 4.

Ekki. *Lophira procera*, var. *alata* (*H.*). Nigeria. Also called Ebo, Kaku, African Oak, Red Ironwood. Reddish to purplish brown. Very hard, heavy, strong, and durable. Interlocked grain, white secretion. Difficult to work, but not difficult to season. Polishes. Used for structural work, sleepers, piling, wagon work, flooring, rollers, superior joinery. S.G. 1·1 ; C.W. 5.

Ekpaghoi or **Ekpakpogo.** *Berlinia acuminata* and *Canarium sp.* (*H.*). Nigeria. Also called Apado, etc. Reddish brown, darker markings, giving a " zebrano " effect. Fairly hard and heavy. Interlocked, fibrous. Difficult to work smooth, wandering grain liable to tear. Polishes. Used for superior joinery, panelling, veneers. S.G. ·72 ; C.W. 4. *Canarium schweinfurthii* is pinkish yellow, with fibrous, coarse woolly grain. Used for rotary-cut plywood. S.G. ·6.

Ekusawa. Opepe, *q.v.*

Elæocarpaceæ. *Tiliaceæ* or Linden family. Includes Elæocarpus, Echinocarpus, Muntingia, Sloanea, etc.

Elæocarpus. See *Blue Fig*, *Medang*, and *Quandong*.

Elæodendron. See *Mirandu*.

Elasticity. 1. That property of a material that allows it to regain its original condition after distortion. See *Modulus of Elasticity*. 2. Resilience. The reverse of rigidity.

Elastic Limit. The greatest stress that a material can be subjected to and retain its original property of elasticity. See *Permanent Set* and *Stress*.

Elastic Strain. Strain, due to stress, within the elastic limit.

Elbow Chair. One with rests for the arms. An armchair. See *Hepplewhite*.

Elbow Linings. 1. Linings to splayed window-jambs. 2. Linings between the boxings for shutters and the floor. (See illus., p. 170.)

Elbows. 1. Right-angle bends for pipes. 2. The returns of panelling in recesses. 3. See *Crosette*.

Elder. *Sambucus nigra* (*H.*). Britain. Also called Bourtree. Shrub-

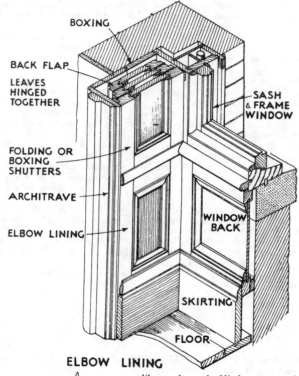

BOXING

BACK FLAP

LEAVES
HINGED
TOGETHER

SASH
& FRAME
WINDOW

FOLDING OR
BOXING
SHUTTERS

ARCHITRAVE

ELBOW LINING

WINDOW
BACK

SKIRTING

FLOOR

ELBOW LINING

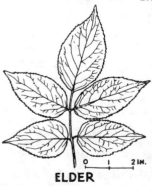

0   1   2 IN.

ELDER

like, and wood of little commercial importance. Yellowish brown. Varies in density. Uniform texture. Large pith, which is easily removed. Fine rays. Used for turnery, small fancy articles, combs, etc. Box ELDER, *Acer negundo*. Ash-leaved maple. See *Maple*.

Eleagnaceæ. A family including little timber of commercial importance. Includes Sea buckthorn.

Element. 1. A cellular unit of wood. 2. The simplest known constituents of any substance.

Elemi. A name applied to trees that provide gum-elemi, *q.v.*

Elephant. A moulding and recessing machine with an overhead spindle. An inverted spindle.

Elephant Beetle. The great weevil. An insect very destructive to young pines. It is dark brown with yellow spots and about ½ in. long.

Elevation. 1. The façade of a building. 2. The front view of an object drawn in orthographic projection.

Elgon Olive. See *Olive*.

Elizabethan. A transitional period of architecture between Gothic and Renaissance, prevailing during the reigns of Queen Elizabeth and James I. The best examples are at the older universities and in domestic architecture.

Ellipse. A section across a .cone not at right-angles to the axis. Area = πRr. See *Conic Sections*.

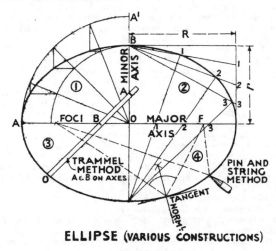

ELLIPSE (VARIOUS CONSTRUCTIONS)

Ellipsoid. A solid of which every plane section is an ellipse. See *Spheroid*.

Elm. *Ulmus spp.* (*H.*). Europe, N. America. Dull reddish brown. Moderately hard and heavy. Strong, very tough, and difficult to split. Excellent shock absorber. Handsome figure in plain-sawn wood due to contrast in growth rings. Very durable if kept *either* wet or dry, but should be treated for weathering. Shrinks · and warps freely unless correctly seasoned or reconditioned. Twisted grain. Difficulty of working is due to knots, irregular growth, and warping. Burrs used for veneers. Stains and polishes. Used for structural and constructional work, piling, weatherboarding, wheelwright's work,

ELM

furniture, coffins, etc. S.G. ·65 ; C.W. 3·5. AUSTRALIAN ELM.
See *Red Tulip Oak*. CANADIAN ELM. See *Rock Elm*. CHINESE ELM,
U. chinensis. Beautiful wood in colour and texture. COMMON ELM,
U. campestris or *U. procera*. English, or red, elm. Other English
species are usually included as common elm. CORK, or CORK-BARKED,
ELM. See *Rock Elm*. CORNISH, or JERSEY, ELM, *U. stricta* or
U. nii ms. English elm. Good quality. DUTCH ELM, *U. major*.
Common elm. ENGLISH ELM. Common elm. HICKORY ELM. See
Rock Elm. HUNTINGDON ELM, *U. vegeta*. English elm. INDIAN
ELM, *U.*, or *Holoptelia integrifolia*. Straw colour, dark streaks.
Cross-grained and hard. LOCK, or TIRLING, ELM, *U. minor*. English
elm. MEXICAN ELM, *U. mexicana*. Similar to common elm. RED
ELM. Common elm and slippery elm. ROCK ELM, *U. racemosa*.
N. America. Forms bulk of supplies on English market. Lighter
colour, straighter grain, finer texture, and easier to work than common
elm. All the properties of common elm, but not quite so good.
S.G. ·76. SCOTS ELM. Wych elm. SLIPPERY ELM, *U. fulva*.
U.S.A. Little timber value. SPANISH ELM. See *Laurel*. W. INDIES
ELM. See *Guacima*. WHITE ELM, *U. americana*. U.S.A. and
Canada. Also called Orhamwood, Grey, and Swamp, Elm. Fine,
close, silky grain and texture. Properties of rock elm. Most important
N. American elm. Used for veneers, furniture, bentwork, coffins,
cooperage, vehicles, implements, etc. S.G. ·7. WYCH ELM, *U.
montana*, *U. glabra*. N. Europe. Called Mountain, or Scotch, Elm
and Wych Hazel. Good timber, but sometimes crooked grain and
subject to insect attack. Used for chocks, packings, etc.
Els, Red. *Cunonia capensis* (*H.*). S. African elder. Excellent for
furniture, wagon, and wheelwright's work, turnery, etc. S.G. ·74.
WHITE ELS, *Platylophus trifoliatus*. S. Africa. Similar characteristics to red els and used for the same purposes, but inferior and easier
to work. S.G. ·62.
e.m. Abbreviation for *end matched*.
Embattlemented. See *Crenellated*.
Embellishments. Ornamentation, especially when in the form of
carved mouldings.
Emboss. To carve in relief. To raise in the form of carving. To
form bosses, *q.v.* Embossed mouldings are usually machine made.
They have the appearance of being carved and are used on cheap
furniture, coffins, etc.
Embothrium. Queensland satin oak, *q.v.*
Embrasure. 1. A narrow wall opening with splayed interior jambs.
2. The intervals, or crenelles, between the merlons, in crenellated
cornices or mouldings. See *Battlemented*.
Embuia. *Phœbe porosa* (*H.*). Brazil. Olive to chocolate brown ;
darkens with exposure, lustrous. Moderately hard, heavy ; strong
and durable. Straight and curly grain. Medium to fine texture,
smooth. Fairly easy to work. Used for cabinet work, superior
joinery, sleepers, etc. S.G. ·72 ; C.W. 3.
Emi-emi. See *Okweni*.
Empata. *Vatica sp.* (*H.*). Philippines. Resembles false acacia. Durable but not stable. Used for constructional work. S.G. ·8. See *Narig*.

Empennage. The rear unit of an aircraft. It consists of fin, rudder, tailplane, and elevator.

Empire. 1. Applied to woods from any part of the British Empire. **2.** Applied to a French style of decoration based upon the ancient Egyptian and Grecian.

Empire Andiroba. Crabwood, *q.v.*

Empty-cell Process. A term used in preservation to describe the method in which the surplus antiseptic is withdrawn from the cells. Also called Open-cell Process. When the wood is impregnated the tank is emptied and a vacuum created in the cylinder to withdraw the free preservative from the wood. Very often the amount of preservative to be retained per cubic foot of wood is stipulated. See *Preservation.*

Emry or **Emri.** Idigbo, *q.v.*

Emufuhai. Nigerian sycamore, *q.v.*

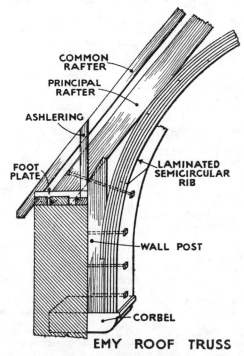

COMMON RAFTER

PRINCIPAL RAFTER

ASHLERING

FOOT PLATE

LAMINATED SEMICIRCULAR RIB

WALL POST

CORBEL

EMY ROOF TRUSS

Emy's Roof. A system of roofing in which the main member is a laminated semicircle.

Encarpa. A carved festoon of fruit or flowers. Often used to ornament a frieze, capital, mantelpiece, etc.

Encased. To enclose with a case or linings.

Encased Knot. A dead, or loose, knot partly or wholly surrounded by bark.

Encino. See *Roble*.

Enclosed Knot. One buried in the wood and not seen on the surface.

Enclosure. The framing enclosing a shop window, from window board to ceiling.

End Check. Seasoning checks at the end of a piece of wood.

Endecagon. A polygon with eleven equal sides. Area $=(\text{side})^2 \times 9\cdot366$.

End Grain. The grain shown on a cross-cut surface. A transverse section of a log. A section at right-angles, or approximately so, to the length of a piece of wood.

Endiandra. See *Orientalwood*, *Queensland Walnut*, and *Queensland Greenheart*.

Endive Scroll. A carved detail derived from a leaf, and characteristic of Chippendale and Louis styles.

Endl. Abbreviation for the name of the botanist Endlicher.

End Matched. 1. Applied to matched-boarding tongued and grooved on the ends. 2. Butt matched. Similar to book matched, but the ends are matched instead of the edges. When *book* and *end* matched are combined it is called a *four-way* match.

End Matching. A term used in the matching of veneers when the bottom ends of adjacent cuts are placed together. Specimens cut end to end.

Endodermis. The innermost layer of the cortex, *q.v.*

Endogens. Plants in which new growth takes place at the middle of the stem, such as palm trees, bamboos, etc.

Ends. 1. Short lengths of battens, deals, etc. 2. Roundwood, *q.v.*, converted into batten, board, and deal sizes.

End Split. A split at the end of a piece or log. See *Split*.

Eng. *Dipterocarpus tuberculatus (H.)*. India, Burma. Also called In. Resembles gurjun. Reddish brown, darkening with exposure, fragrant. Hard, heavy, strong, fairly durable. Difficult to season and work. Contains a gum secretion. Large sizes. Straight grain. Stains with corroding iron. Used for wagon and carriage work, superior joinery, utility furniture, flooring, constructional work. S.G. ·84 ; C.W. 4·5.

Engaged Column. A column forming part of a wall.

Engelm. Abbreviation for the name of the botanist Engelmann.

Engelmann. See *Spruce* and *Picea*.

Engl. Abbreviation of the name of the botanist Engler.

English. Specially applied to ash, elm, oak, and walnut. See *British Trees*.

Enneagon. A nonagon, *q.v.*

Enrichments. Embellishments, such as carvings, etc.

Ensonit. A registered wallboard, *q.v.*

Ensowal. A registered wallboard, *q.v.*

Entablature. The horizontal members supported by the columns in classic architecture. It includes cornice, frieze, and architrave. See *Orders* and *Corinthian*.

Entail. The delicate and elaborate parts of carvings.

Entandrophragma. See *Gedunoha, Mahogany, Sapele, Sipo,* and *Ubilessan*.

Entasis. The gradual swelling towards the middle of the shaft of a column. The swelling corrects the illusion of concavity, and increases the stability.

Enterolobium. See *Guanacaste, Keloba, Pacara,*

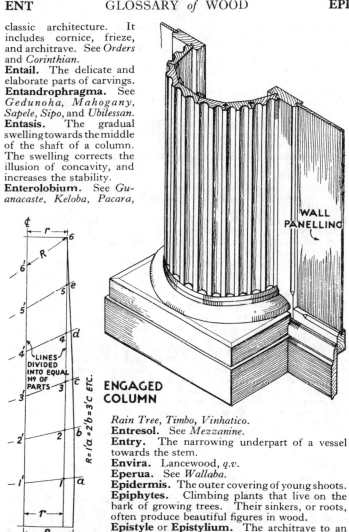

WALL PANELLING

ENGAGED COLUMN

ENTASIS OF SHAFT OF COLUMN

Rain Tree, Timbo, Vinhatico.

Entresol. See *Mezzanine*.

Entry. The narrowing underpart of a vessel towards the stem.

Envira. Lancewood, *q.v.*

Eperua. See *Wallaba*.

Epidermis. The outer covering of young shoots.

Epiphytes. Climbing plants that live on the bark of growing trees. Their sinkers, or roots, often produce beautiful figures in wood.

Epistyle or **Epistylium.** The architrave to an entablature.

Epitithydes. The upper member of the cornice to an entablature.

Equilateral Arch. A Gothic, or pointed, arch in which the arcs are struck with a radius equal to the span.

Equilateral Roof. A roof in which the spars are equal in length to the span, forming an equilateral triangle.

Equilibrium. A condition of rest produced by the action and reaction of a system of forces.

Equilibrium Frame. Double log-frames, to cut two logs at once. The frames balance each other which allows for greater speed and efficiency.

Equilibrium Moisture Content. The moisture content at which wood is stable and in equilibrium with the humidity of its surroundings. See *Moisture Content*.

Eremacausis. The gradual decay of wood due to exposure to the atmosphere.

Erica. See *Briar*.

Ericaceæ. A family that includes the rhododendron and strawberry tree. The timber is of little commercial importance.

Erimado. See *Okwen*.

Erin. *Polyodoa*, or *Picralinia, umbellata* (*H.*). Nigerian boxwood. Orange yellow. Resembles boxwood. Hard, heavy, and lustrous. Very close grain and uniform texture. Small sizes. Used for carving, turnery, etc. S.G. ·85.

Erinoid. A decorative multi-coloured casein product, obtainable in large sheets. Used for the same purposes as bakelite, for inlays, signs, etc.

Eriodendron. Ceiba or cottonwood, *q.v.*

Eriolæna. See *Dwani* and *Salmonwood*.

Eriostemon. See *Lancewood*.

Ernobius mollis. A species of furniture beetle.

Errol. Pynkado, *q.v.*

Erun. *Erythrophlœm guineense* (*H.*). Nigeria. Chestnut brown with lighter streaks. Very hard and heavy. Coarse, fibrous, and liable to pick up. Subject to shakes. Difficult to work. Used for structural and constructional work, flooring, etc. S.G. ·85; C.W. 5.

Erythrina. See *Ceibo, Grey Corkwood*, and *Mandat*.

Erythrophlœm. See *Erun* and *Ironwood*.

Erythroxylon. See *Coca* and *Redwood*.

Erythroxylum. See *Dabé* or *Landa*.

Escalator. A continuous moving stair, used instead of a lift.

Escallade. A ladder fixed vertically to a wall. A cat ladder.

Escallop. See *Scalloped*.

Escape. The apophyge between a column and its base.

Eschweilera. See *Manbarklac* and *Kakeralli*.

Escoinson. See *Scoinson*.

Escritoire. A writing desk with chest of drawers and pigeon holes often enclosed by hinged flap.

Escutcheon. 1. An ornamental plate covering a keyhole or carrying a door handle. See *Door Furniture*. 2. An armorial shield. 3. A boss in the centre of a vaulted ceiling. 4. The middle compartment in the stem of a ship in which the name of the ship was painted.

Esenbeckia. See *Larangeira* and *Satinwood*.

e.s.m. Abbreviation for *end standard matched*.

Espagnolette Bolt. A long bolt that secures a french window at the top and bottom, and sometimes middle, in one operation.

Espalier. Trellis work in gardens to carry climbing plants.

Espavé. *Anacardium excelsum* (*H*.). C. America. Also called Pesége Mahogany, Giant Cashew, Caracoli, etc. Variegated yellowish brown, lustrous. Distinct rays. Darkens with exposure. Moderately light and soft. Tough, strong, and fairly durable, but subject to insect attack. Roey with striped effect. Smooth with texture like cedar, and fairly easy to work. Used for vehicles, implements, carpentry, boxes. S.G. ·54 ; C.W. 3.

Espina de Corona. *Gleditsia amorphoides* (*H*.). C. America. Also called Espinillo. Reddish brown, lighter steaks. Hard, heavy, tough, strong, elastic, and fairly durable. Roey, medium texture, smooth, and not difficult to work. May be used for most purposes. S.G. ·9 ; C.W. 3.

Essenhout. Cape ash, *q.v.*

Essex. A registered wallboard, *q.v.*

Estrade. An elevated part of a floor. A low platform.

Estribeiro. *Luehea divaricata* (*H*.). C. America. Also called Acoita-cavallo. Grey to pinkish brown, variegated streaks. Moderately hard and heavy. Smooth, with uniform grain and texture, like beech. Sometimes roey. Fairly easy to work. Used for interior construction, furniture, woodware. S.G. ·6 ; C.W. 3.

Étagère. A tier of shelves supported by pillars ; like a *what-not*.

Eucalyptus. The most important Australian woods. Over 300 species, from shrubs to trees that rival the sequoia in size. They vary considerably in characteristics and properties, but the commercial woods are mostly very hard, heavy, tough, strong, durable, and fire-resisting. They should have the heart boxed in conversion. The following are important, and described alphabetically : Argento, Blackbutt, Bloodwood, Cadaga, Coolibah, Gimlet, Grey Box, Gum, Ironbark, Ironbox, Jarrah, Karri, Maiden, Mallee, Mallet, Marri, Merrit, Messmate, Morrell, Mountain Ash, Peppermint, Red Box, Red Mahogany, Redwood, Saligna, Santavera, Silvertop, Stringy-bark, Tallow-wood, Tasmanian Oak, Tingle-tingle, Tuart, Wandoo, White Ash, White Mahogany, Woollybutt, Wormwood, Yapunyah, Yate, Yellow Box, Yellow Cheesewood, York Gum, Yorrell, Yuba. Many others are given alphabetically in the Glossary.

Eucarya. See *Australian Sandalwood*.

Eucryphia. See *Ulmo* and *Leatherwood*.

Eugenia. Over 120 species in E. Indies, all known commercially as

ESPAGNOLETTE BOLT (TWO-WAY)

kelat, *q.v.* Several species in India and Ceylon, but little exported. See *Jaman.* Also see *Cherry Alder, Grumixaba, Makaasim, Rose Satinash, Water Pear.*

Euonymus. See *Spindle Tree.*

Euphorbiaceæ. The Spurge family which includes Alchornea, Aleurites, Bischofia, Bridelia, Buxus, Croton, Cyclostemon, Euphorbia, Gymnanthes, Hemicyclia, Hevea, Hieronymia, Hippomane, Homalanthus, Hura, Mabea, Mallotus, Manihot, Oldfieldia, Pera, Phyllanthus, Piranhea, Ricinus, Sapium, Trewia. Very important family for products other than wood, especially rubber.

Euphrœ. A long piece, with holes at regular intervals for the crowsfoot, for an awning on board ship.

European. A prefix to distinguish the origin of the wood. Commonly applied to ash, aspen, beech, birch, box, cherry, holly, hophornbeam, larch, lime, maple, redwood, spruce, walnut, yew.

Eusideroxylon. See *Billian* and *Kajœ.*

Euxylophora. See *Sateenwood, Satinwood,* and *Pau Amarello.*

Even Texture. A description of wood in which there is little contrast between springwood and summerwood. Uniform texture.

Evergreen. Applied to trees that continuously bear leaves, in contrast to deciduous trees, due to the fact that the leaves perform their functions longer than for one year. In temperate zones the term usually signifies *softwoods.*

Evergreen Beech. *Nothofagus cunninghami* (*H.*). Tasmanian myrtle. Pinkish colour. Characteristics and properties of English beech, *q.v.* Used for cabinet work, panelling, floors, carving, turnery, bentwork. S.G. ·8 ; C.W. 3·5.

Evergreen Magnolia. *Magnolia grandiflora.* See *Magnolia.*

Evergreen Oak. *Quercus ilex.* See *Oak.*

Everlasting Wood. Jarrah, *q.v.*

Evodia. See *Silver Sycamore.*

Evolute. See *Involute.*

Ewowo. Ayous, *q.v.*

ex. **1.** Implies *out of.* **2.** Abbreviation for *excluding.*

Excelsior. Woodwool, *q.v.*

Excœcaria. A genus chiefly shrubs. See *Gangwa* and *Sandaleen.*

Excrescence. An abnormal growth on a tree, as a burr.

exd. Abbreviation for *examined.*

Exell. The name of the botanist A. W. Exell.

Exfoliate. To scale off, or break away in thin layers, as occurs to the bark of certain trees. To separate into scales or laminæ.

Exit Holes. Worm holes, *q.v.*

Exocarpus. Same as santalum, *q.v.*

Exogens. Timber trees. Exogenous plants are those in which the new growth takes place in the cambium layer just under the bark. Outward growers.

Exostemma. See *Princewood.*

Expansion. An increase in the volume of wood due to increase of moisture content, *q.v.* Also see *Seasoning.* The volume of other building materials varies with temperature.

Expansion Bit. A patent extension bit in which the cutters can be adjusted to bore holes from $\frac{1}{2}$ to $1\frac{1}{2}$ in. and from $\frac{7}{8}$ to 5 in. diameter.

ex. 1st. Abbreviation for *extra firsts*, in graded timber.

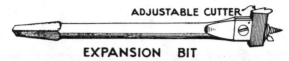

EXPANSION BIT

Extension Table. A dining-table that can be increased in length by means of sliding leaves, or by the insertion of extra boards. The

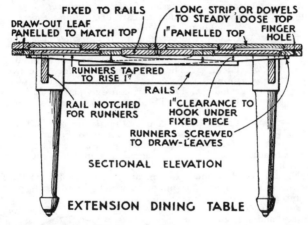

SECTIONAL ELEVATION

EXTENSION DINING TABLE

latter type is operated by mechanism that opens out the ends of the table. There are several registered types of extending mechanisms.

Extrados. The back, or outer curve, of an arch.

Extruded Metal. Bronze and other alloys formed into mouldings for shop fronts, fittings, etc. The hot metal is forced through dies to give the required shape, to about 12 B.W.G. in thickness. See *Kalamein*.

Eye. **1.** A small staple to receive a hook, to secure a door, etc. **2.** A general term for the centre of anything, as the centre of a volute, an opening at the crown of a dome, etc. See *Involute*. **3.** See *Cincfoil*. **4.** Inlays in bowls to provide bias. **5.** A dovetail socket.

Eyebrow. Applied to an upper window over which the eaves are raised in the form of a flat segment.

Eyelet or **Oilet.** **1.** A small eye. **2.** The end of a rod or rope provided with a hole or loop.

Eye Tree. The head of a spade handle, or of similar tools.

Eyne. See *Red Eyne*.

F

f. A symbol denoting the *maximum intensity of stress* in structural design. A small letter is added to show the kind of stress : $f_b =$ bearing, $f_c =$ compression, $f_s =$ shear, $f_t =$ tension. Also abbreviation for *free, first, full,* and *fathom.*

F. Abbreviation for *face, flat, factor of safety, Fahrenheit.*

Fabric. **1.** The frame, or structural part, of anything, especially of a building. **2.** Woven material, cloth.

f.a.c. Abbreviation for *feet average cube,* and for *free of address commission.*

Façade. The front, or elevation, of a building.

Face. **1.** The wide surface, or side, of converted timber. The better side of a prepared piece of wood. **2.** The lower, concave, part of a saw tooth.

Face Edge. The better narrow surface of converted timber.

Face Mark. A pencil mark put on the prepared face side of the material, near to the face edge.

Face Measure. Surface measure. The area of a face of a board. It is the same as *board measure* when the board is 1 in. thick.

Face Mould. A developed pattern, or templet, applied to the face of the material. It gives the outline to which the material must be cut, and is used in double-curvature work such as wreathed handrails, etc.

Face Plate. A flat plate chuck for a lathe.

Face Side. **1.** The better wide surface of material. The exposed face, which is prepared first and used as a basis of operations for further processes, moulding, etc. **2.** The side farthest from the heart in converted timber.

Facets or **Facettes.** The fillets between the flutings on a pilaster or column. See *Flutes.*

Face Veneers. Those used for the surfaces of plywood.

Facia. See *Fascia.*

Facing. **1.** Preparing the face side of the material. **2.** An architrave.

Facings. **1.** Thin wrought boards used to cover a rough or inferior

FACE

FACE MOULD FOR WREATH

surface. **2.** Finishings. **3.** Prominent parts on the main surface of a pattern for a casting.

Facing Up. Covering inferior material with thin superior wood.

Factor of Safety. *FS.* A number, by which the breaking load is divided to give the safe load, in structural design. Ultimate strength ÷ working stress = FS. The factor of safety for timber varies between 4 and 10 according to the quality of the wood and the conditions of use, but 6 is usual for permanent interior work and 8 for exterior work.

Factory Timber. Same meaning as shop timber, *q.v.*

Fagaceæ. A family of trees including Castanea, Castanopsis, Fagus, Nothofagus, Pasania (Lithocarpus), Quercus.

Fagara. *Zanthoxylum.* See *Bosuga, Knobwood, Munyenye, Okanhan, Okor,* and *Satinwood.*

Faggot or **Fagot.** **1.** A bundle of twigs, sticks, or small branches ; also a single stick. **2.** A bundle of pieces of steel of 120 lb. weight.

Fagræa. See *Anan, Malabera, Tembusa,* and *Yellowheart.*

Fagus. See *Beech, Clinker Beech, Cudgerie, Evergreen Beech, Myrtle, Queensland* and *Red Beech, Roble, Southland Beech,* and *White Beech.*

Fairing. **1.** A term used in aircraft and shipbuilding for additions to the structure that reduce head resistance. A streamlined cover. **2.** Removing the irregularities in the lines of a ship, in large-scale drawings. See *Laying-off.*

Fairlead. A sheave hole to lead a rope clear of an obstruction or to a more favourable position, on board ship.

Falcate Yellow-wood. *Podocarpus falcate.* S. Africa. Similar to Yellow-wood, *q.v.*

Faldstool. **1.** A folding stool. **2.** A litany stool, or low desk, at which the priest kneels.

Fall. **1.** The inclination of a flat surface or gutter to allow water to drain away. **2.** See *Writing Desk.* **3.** The free rope in pulley tackle. **4.** An elaborate tail to the vertical part of a carved bracket or console. **5.** Autumn, at the fall of the leaf.

Fall Bar. **1.** A pivoted bar for securing a ledged and batten door. It is operated by a finger-hole in the door, controlled by a keeper, and rests in a stop. **2.** A larger type, pivoted at the middle, is used for a pair of doors or gates, or for shutters. See *Locking Bar.*

Falling Axe. One used for felling trees.

Falling Cut. The *breaking down,* or first, cut of a log. There is an extra charge for this cut owing to labour entailed in packing and wedging, and to cover unforeseen difficulties.

Falling Home. A term used in shipbuilding for the curving inwards of the upper part of the top side.

Falling Line. The centre line of a developed wreath to a handrail.

Falling Mould. A templet for marking the depth of a handrail wreath, etc., after the stuff has been cut to the face mould. See *Face Mould.*

Falling Stile. The shutting stile of a gate with cocked hinge.

Fall of Timber. An area of trees marked out for felling ; or felled trees ready for removal.

Fall Pipe. A down spout.

Fall Shipment. Autumn shipment.
False Acacia. See *Locust* or *Robinia*.
False Amboyna. Maidu, *q.v.*
False Cedars. Cedrela, Cupressus, Juniperus, Thuya, etc.
False Ceiling. A second ceiling formed to provide cavities for pipes and wires, or for insulation.
False Ellipse. An approximate ellipse formed of circular arcs. See *Three-centred*.
False Heartwood. Dark innerwood due to disease or fungi.
False Keel. A member to strengthen or protect the keel of a ship.
False Plane. See *Sycamore*.
False Ring. An extra " annual ring " formed during a year's growth, or an incomplete growth ring due to interruption of growth during some part of the year. A false start in the season's growth. The phenomenon is due to abnormal weather conditions.
False Tenon. See *Inserted Tenon*.
Falsework. Centring or shuttering for concrete.
Family. A division in the classification of plants, and usually comprising a number of genera. See *Botany*.
Fan. 1. A sloping screen, or protection, to a scaffold. See *Mason's Scaffold*. 2. A revolving system of vanes used in artificial ventilation, or to exhaust the air in pipes, as to a cyclone, *q.v.*
f. and g. Abbreviation for *feathered and grooved*.
Fane. See *Vane*.
Fan Figure. A fan-like figure sometimes found in pollard oak.
Fang. Same as tang, *q.v.*
Fanlight. Originally applied to a semicircular sash, with radiating bars, over a door ; but now applied to any shape of light above the transome of a door. See *Door Frame*.
Fanlight Opener. A quadrant, *q.v.*
Fantail. 1. A short log of irregular shape, suggesting the name. 2. The arrangement for turning the cap of a windmill into the wind.
Fan Tracery. The elaborate arrangement of ribs and panels in a fan-vaulted ceiling, of the Perpendicular style of architecture.
f.a.q. Abbreviation for *fair average quality*.
Farthingale. An old type of chair with upholstered seat and back.
f.a.s. Abbreviation for *free alongside*.
F.a.S. Abbreviation for *firsts and seconds*.
Fascia. A long, horizontal, flat band between mouldings. A flat, wide, level board, standing on edge, as between the sash and cornice of a shop front. See *Blind* and *Eaves*.
Fascine. A bundle of twigs about 20 ft. long and 10 in. diameter. They are used as a foundation for a road in marshy ground, or as a protection against erosion. A faggot.
Fascine Building. One formed of logs and boarded surfaces.
Fashion Pieces. The timbers of a ship forming the shape of the stern. They are fixed to the stern post and wing transom. See *Stern Timbers*.
Fast. Permanent in colour. Applied to the colour of wood that does not fade with seasoning or exposure.
Fastenings and Fasteners. Bolts, coach screws, dogs, dowels,

holdfasts, nails, plate dowels, screws, staples, etc. Also applied to the various arrangements for securing doors and windows.

Fastigium. The pediment of a portico ; the ridge or gable of a building. A summit formed by sloping surfaces.

Fast Sheet. A fixed sash or stand sheet. A sash without frame ; or a sash fixed in a frame so that it will not open. See *Fixed Sash.*

Fathom. A timber measurement of 216 cub. ft. of stacked wood. A cube of 6 ft. or its equivalent. The volume often exceeds 216 cub. ft. for special purposes. See *r.c.fm.*

Fatigue. The deterioration of the resistance, or strength, of a material due to continual overstrain, as in an over-loaded beam.

Faucet. A tap to a barrel.

Faults. Defects, *q.v.*

Faun. A legendary figure combining man and goat. It was used as a decorative feature in the work of the Adams period.

Faurea. See *Terblanz.*

Fauteuil. 1. French for armchair. 2. The stalls of a theatre. 3. A faldstool.

Faux Satine Crotch. A trade name for figured veneers of cypress.

Favas. A form of diaper ornamentation resembling honeycomb cells.

Faveiro. *Pterodon pubescens* (*H.*). Brazil. Whitish to dark yellow. Used locally for exterior work, sleepers, etc. Not exported.

Faying. See *Snape.*

f.b.m. Abbreviation for *foot board measure.*

Fd. Abbreviation for *framed.*

f.d. Abbreviation for *free dispatch.*

f.e. Abbreviation for *feather edge.*

Feather. 1. A thin strip of wood used as a tongue. It may be of plywood or cut diagonally across a wide board. See *Slip Feather.* 2. The pendulum slip separating the weights for sliding sashes. See *Boxed Frame.* 3. Ornamental figure in wood, especially mahogany, due to the confusion of the fibres at the junction between branch and trunk or between two branches. 4. A fillet placed in the angle of a pattern to avoid a sharp interior angle of a casting, which is a source of weakness.

Featherboarding. See *Feather Edge* (1).

Feather Cone Fir. Noble fir, *q.v.*

Feather Edge. 1. Applied to boards that taper in thickness. See *Weatherboarding.* 2. The burr on a cutting tool after sharpening on an oil-stone. The burr is removed by rubbing the back of the tool flat on the stone. An over-sharpened cutting edge.

Feathering. Holding the scraper nearly parallel to the grain when finishing stringy hardwoods.

Featherings. The cusps in foliated tracery work. See *Tracery.*

Feather Tongue. See *Feather* (1).

Feed, Rate of. 1. The speed, in feet per minute, at which the wood goes through the machine or saw. 2. The linear length, in inches, cut by one revolution of the saw.

Feet Run. A term used when costing materials per foot of length.

Felling. The cutting down of trees by saw or axe. This should be done when the sap is at rest, in winter, and when the tree is at maturity.

If felled before maturity there is a greater proportion of sapwood. Many tropical hardwood trees are girdled about two years before felling.

Felling Shake. Upset, *q.v.*

Felloe or **Felly.** 1. The outer rim of the framework of a centre. 2. The rim, or a part of the rim, of a wheel.

Felt. 1. Silver grain, *q.v.* 2. Damp-proof material consisting of fibrous material impregnated with a waterproofer, usually bitumen. It is obtainable in rolls and in different widths. See *Counter Batten*.

Femerell. A louvre ventilator in a roof.

Femora or **Femur.** The spaces between the channels in a triglyph of the Doric order.

Fence. 1. A guard or protection. Fencing. 2. A guide on a machine or tool, as for a circular saw or plough. See *Fillester*.

Fencing. 1. An erection to enclose or protect a piece of ground or a building. It may be open or close boarded. A fence usually consists of posts, rails, and boards or lags, but there are many variations and registered types. 2. The materials for a fence or fencing. See *Interlaced Fencing*.

Fender. 1. Anything that defends or wards off. A horizontal balk protecting the foot of a scaffold from road traffic. See *Mason's Scaffold*. 2. A block of wood or cordage slung over the side of a vessel as a protection against contact with other objects, also called a Fender Beam. 3. A curb to a fireplace. 4. Heavy timbers protecting a wharf or quay wall. 5. A low rail in cowhouses. 6. See *Lock Gate*.

Fender Board. A board protecting the steps of a carriage from slush.

Fender Piles. An outer fence of piles to protect work from moving objects.

Fender Posts. Posts round a refuge, or " safety island," to protect pedestrians from road traffic.

Fender Stool. A fireside stool.

Fenestral. Belonging to, or like, a window. A small window, window blind, or shutters.

Fenestration. Arrangement of windows in a building.

Feng. *Liquidambar formosana* (*H.*). China. The tree is important for medicinal gum. The wood is not durable and used chiefly for tea chests.

Fen Pole. A pole used for jumping across ditches, used in the Fen districts.

Feretory. A tomb or shrine, or a chapel for same.

Fern Tree. Pink tamarind, *q.v.*

Ferolia. See *Satiné*.

Feronia. See *Wood-apple*.

Ferrairea. See *Sucupira*.

Ferrugo. A plant disease usually called rust.

Ferrule. A metal ring as round a chisel handle to prevent it from splitting. A thimble used for jointing pipes. Sometimes called Ferrel. See *Socket Chisel*.

Ferry Boat. Any type of boat for conveying persons and vehicles across rivers or narrow seas.

Ferule. A rod or cane or staff used by one in authority.

Festmeter. Cubic metre.

Festoon. A carved ornament in the shape of a garland, or loop, suspended from its two ends.

f.i.b. Abbreviation for *free into barge*.

Fiberlic. A proprietary building-board made from fibrous roots.

Fibreboards. Wall- or building-boards made from fibrous materials such as

FESTOON

bamboo, cane, straw, wood pulp, etc. They are used as a substitute for plastered surfaces and for their insulating properties. There is a great variety, with numerous finishes for the surfaces. See *Wall-boards*.

Fibres. Long, slender, thick-walled elements in broad-leaved trees. The longitudinal wood elements. Their function is to give strength. The weight and hardness of wood is usually proportional to the amount of wood fibre.

Fibre Saturation Point. The point at which the cell walls are fully saturated but the cells contain no free moisture. It is the point at which shrinkage commences with further seasoning, with moisture content usually about 30 per cent., but varies with different species. The physical properties of the wood do not change with change of moisture content when it is above *f.s.p.*

Fibre Stress. See *Stress*.

Fibrewood. *Laportea gigas* and other spp. (*H.*). Queensland. Brownish white. A very light soft fibrous wood, used as a substitute for cork, refrigerator linings, etc. S.G. ·25.

Fibril. The fine thread-like filament forming the cell walls in wood.

Ficus. See *Fig*.

Fid. 1. A conical hardwood pin for opening the strands when splicing a rope. 2. A square wood bar, shouldered at one end, for supporting the weight of the top-mast of a ship.

Fiddle. 1. A flat short piece of wood with two holes for securing guy ropes in tension. 2. See *Fiddle Rack*. 3. A violin, *q.v.*, but often implies viola, violincello, or double bass.

Fiddleback. Ripple. An undulating appearance of a smooth surface. Fine wavy grain common to sycamore and maple and used for the backs of violins. See *Figure* and *Grain*.

Fiddleback Maple. See *Maple*.

Fiddle Block. A ship's block shaped somewhat like a violin.

Fiddle-head. Carved scrollwork above the bow of a ship.

Fiddle Mottle. A combination of fiddleback and mottle.

Fiddle Rack. Portable guards or protecting edges, for fixing to tables on board ship in rough weather, to prevent articles from sliding off the table. See *Table Guard*.

Fiddlestick. A violin bow.

Fiddlewood. *Citharexylum, Vitex, Petitia, spp. Petitia* (*H.*). C. America. Light brown, dark stripes, lustrous. Hard, heavy, strong, and fairly durable. Straight, roey, and wavy grain. Rather fine texture, smooth, and polishes well. Rather difficult to work. Used for general purposes, cabinet work, superior joinery. S.G. ·9 ; C.W. 3·5.

Fielded Panel. A raised panel with a wide flat surface or one broken up into smaller panels.

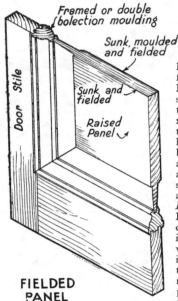

Framed or double
bolection moulding

Sunk, moulded
and fielded

Sunk and
fielded

Raised
Panel

Door Stile

**FIELDED
PANEL**

Field Maple. *Acer campestre.* See *Maple.*

Fife. A musical instrument somewhat like a flute, but smaller and more shrill in sound.

Fifie. A large carvel-built coastal fishing boat.

Fig. *Ficus spp.* About 600 species but only a few timber trees, and these are of little commercial importance. Great variation in woods, from very soft to hard. Whitish grey to light brown. Alternate bands of hard and soft tissue. Uneven texture and twisted fibres. Tough and strong for weight. Not durable and subject to sap stain. See *Balsam Fig, Banyan, Blue Fig, Jadi.*

Figure. Ornamental markings, or design, in wood. Normal figure is due to grain, rays, branches, variation of colour, irregularity and interlacing of fibres, pressure on the bark of growing trees. Abnormal figure is due to defects, parasites or climbers, fungi, insects, birds, wounds, stooling, pollarding, decay. Conversion is the important factor in obtaining figure. Much of the beautiful figure in wood is produced artificially in the growing tree. The following terms are described alphabetically in the Glossary : *Bird's Eye, Blister, Broken Stripe, Brown Oak, Burl, Burr, Callus, Crotch, Curl, Feather, Fiddleback, Finger Roll, Grain (curly, cross, interlocked, roey, spiral, oblique, wavy), Herring-bone, Honeysuckle, Lacewood, Lustre, Mottle, Oyster, Peacock's Eye, Plum, Pollard, Quilted, Ram's Horn, Ribbon, Raindrop, Ripple, Silver Grain, Stripy, Watermark.* Many of these terms are also used in combination. Also see *Conversion, Defects, Fungus, Grain,* and *Texture.*

Figurehead. A projecting ornamental, and often symbolical, figure at the stem of a ship. See *Stem and Head.*

Fijian Kauri. *Agathis vitiensis.* See *Kauri Pine.*

Filbert. The hazel tree, *q.v.*

File. An abrading tool for wood or metal. The various kinds are : parallel, half-round, round, rat-tail, triangular, float, knife, warding, pillar, cotter, riffler, feather-edge, and safe-edge. The *cut* may be rough, middle, bastard, second cut, smooth, or dead smooth, according to the number of cuts per inch, from 14 to 100 or more.

Filer. A mechanic who deals with the saws for a saw-mill. A *saw doctor*.

Filicium. *Filicium decipiens* (*H.*). India. Pale red to light brown, darker markings. Hard, heavy, strong, tough, fairly durable. Straight grain, fine even texture. Dull but smooth. Not difficult to work but requires careful seasoning. Used for wagon-construction, furniture, tools, turnery. S.G. ·9 ; C.W. 3·5.

Filler. A paste for filling the grain of wood before polishing. Whiting, plaster of Paris, or proprietary preparations are used, stained to the required colour. See *Stopping* and *Inert Fillers*.

Fillet. **1.** Strips of wood of small section. Narrow cleats to support shelves. **2.** A small flat moulding rectangular in section. The flat surface between two flutes of a column. See *Attic Base*. **3.** Slender joists bedded in a concrete floor, to which the floorboards are nailed. **4.** An angle or hollow in patternmaking. **5.** A cover, in aircraft, at a junction of the main structure to reduce interference drag. Also see *Wallboards*.

Filling Piece. Planted pieces of wood to produce a level, or plane, surface. See *Carling Sole*.

Fillister. An adjustable rebate plane.

Fillistered Joint. A rebated joint.

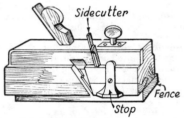

FILLISTER

Fin. An aerofoil vertically above the tail plane of an aeroplane. The fixed part of the tail unit of an aeroplane that contributes to lateral stability. It consists of fin post, ribs, and leading edge. See *Tail* (*Aeroplane*).

Finbob. See *Contract Forms*.

Fine Grain. See *Grain*.

Fine Set. See *Set*.

Fine Texture. See *Texture*.

Finfob. See *Contract Forms*.

Finger Board. **1.** The part of a stringed instrument, as a violin, on to which the fingers press the strings to give the pitch of the notes. **2.** The manual of an organ or the keyboard of a piano.

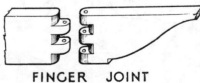

FINGER   JOINT

Finger-cone Pine. Western white pine, *q.v.*

Finger Joint. A joint for the fly-rail of a dropleaf table. It is a form of indented joint ; the projections on the fixed piece fitting in recesses in the end of the fly-rail to form a hinge by means of a strong wire. Also called a Knuckle Joint, *q.v.* Also see *Fly-rail*.

Finger Plates. Ornamental wood, metal, bakelite, or glass plates fixed on the shutting stile of a door to protect the paint.

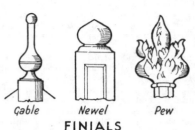

Gable   Newel   Pew

FINIALS

Finger Post. A directional sign post.

Finger Roll. Wavy figure in which the waves are about the width of a finger. See *Figure*.

Finger Slip. A small thin stone for sharpening gouges, etc. See *Oil-stone*.

Finial. An ornamental projection, or terminal, at the top or apex of a canopy, pinnacle, gable, spire, newel, etc. See *Turret*.

Fining Down. Implies *finishing* in certain specialist trades, or shaping an object towards its finished outline.

Finished Ground. One that is partly exposed and wrought.

Finishing Off. Preparing the finished surface of woodwork. See *Cleaning Up* and *Working Qualities*.

Finishings. The fixed joinery, as distinct from the carpentry, in a building. The non-structural part of anything to give a finished appearance.

Fin Keel. A thin deep keel as used on yachts.

Fink Truss. A type of trussed girder used in timber bridges. Also a type of steel roof truss. Named after the designer.

Fir. Applied loosely to numerous coniferous species, but correctly to the single-leaved conifers of the genus *Abies*. The wood is nearly white with yellowish markings, to pale brown, soft, light, rather weak and brittle, and not durable, and with numerous small hard knots. Fir is strong for its weight, very free from resin, and easily wrought, but it is inferior to spruce (*Picea spp.*). There is great variation in quality. The chief supplies are from N. Europe and Canada. Used

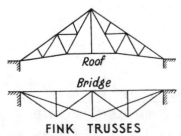

Backings — Grounds
Rebate
Stop — Architrave
Skirting
Panelled Jamb Linings — Plinth Block

FINISHINGS

Roof
Bridge

FINK TRUSSES

for kitchen fitments because of clean appearance, cheap joinery and interior constructional work, flooring, scaffold poles, pit props, etc. S.G. ·35 ; C.W. 1·25. ALASKA FIR. See *Western Hemlock*. ALPINE FIR, *Abies lasiocarpa*, W.N. America. AMABILIS FIR, *A. amabalis*, W.N. America. ARCHANGEL FIR. See *Red Deal* or *Redwood*. BALSAM FIR. See *Balsam Fir*. BRACTED FIR. See *Noble Fir*. CANADIAN FIR. See *Balsam Fir*. CALIFORNIAN, or COLORADO, WHITE FIR, *A. concolor*. CALIFORNIAN RED FIR, *A. magnifica*. Superior quality, like noble fir. COLUMBIA FIR. See *Douglas Fir*. DANZIG FIR. See *Redwood*. DOUGLAS FIR. See *Douglas Fir*. GRAND FIR. See *Lowland Fir*. GREY FIR. See *Western Hemlock*. HIMALAYAN SILVER FIR, *A. pindrow*, N. India. White to light brown, darker stripes and mottles, lustrous but dulls with exposure. Straight, even grain. Fairly durable, medium texture. Good supplies. S.G. ·5. LOWLAND FIR, *A. grandis*, W.N. America. MEMEL FIR. See *Redwood*. NOBLE FIR, *A. nobilis*, Western U.S.A, Harder, heavier, and superior in quality. Sometimes sold as larch. NORWAY FIR. See *Redwood*. OREGON FIR. See *Douglas Fir*. PETCHORA FIR. See *Redwood*. PRINCE ALBERT FIR. See *Western Hemlock*. RED SILVER FIR. See *Amabilis Fir*. RIGA FIR. See *Redwood*. SCOTS FIR. See *Redwood*. SILVER FIR, *A. alba* (*A. pectinata*). European silver fir or Swiss pine. C. and S. European highlands. Silver fir is applied to other firs, but discarded by the B.S.I. to avoid confusion. SPRUCE FIR. See *White Deal* or *Whitewood*. STETTIN FIR. See *Redwood*. WESTERN BALSAM FIR. See *Lowland Fir*. WESTERN SILVER FIR. See *Amabilis Fir*. WHITE FIR. This name is applied to several of the above species, but the B.S.I. suggest it be discarded. See *Whitewood, Redwood, Abies, Picea, Spruce*, and *Sitka*.

Fire. Variation in sheen or lustre in wood, due to reflection of light on non-parallel fibres, as in interlocked, roey, and curly grain.

Fire Doors. Fire-resisting doors. They may be of solid wood, composite, or metal.

Fire-resisting. Numerous timbers have a natural resistance against fire and the following are recognized as fire-resisting by the London County Council : Acacia, Beech, Crabwood, Douglas Fir, English Ash, Greenheart, Guarea, Gurjun, Hornbeam, Idigbo, Iroko, Jarrah, Karri, Keruing, Laurel, Maritime Pine, Meranti, Mora, Mukusu, Nigerian Walnut, Oak, Odoko, Okan, Padauk, Pyinkado, Secondi Mahogany, Seraya, Silver Greywood, Sweet Chestnut, Sycamore, Tasmanian My·tle, Teak, White Olivier, Yew. This list is added to as new woods are investigated. Timber may be made fire-resisting by chemical treatment with one of the following : Ammonium chloride or phosphate or sulphate, calcium chloride, borax, alum. The chemicals crystallize in the pores from which the air and moisture are first extracted. In some cases heat generates non-inflammable gases that drive out the air and so prevent combustion. In others, the chemicals fuse and glaze the cell walls which prevents access of oxygen from the air. Another effective method is to coat the surface with fire-resisting paint of which there are several proprietary makes on the market. These paints are usually sodium, or potassium, silicate with an inert filler, in the proportions of sodium silicate 1·1, kaolin 1·5, water 1. They should conform to specification BS/ARP 39 on Fire Retardants. All the methods aim to exclude oxygen, which is necessary to support combustion.

Firewood. A term used in the import trade for short lengths, or ends, of battens, boards, deals, and planchettes, up to 5 ft. 6 in. long. It is in demand for box- and case-making and is usually sold by the *piled fathom measure* of 216 cub. ft.

Fir Fixed. A term applied to unwrought timbers fixed by nails only.

Fir Framed. Applied to unwrought timbers with framed joints, as in roof trusses.

Firkin. A barrel of 9 gallons, or 1·44 c. ft., capacity. Modern usage implies any wood vessel of any capacity.

Firmer Chisel. The ordinary type of joiner's chisel as used for general purposes.

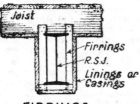

FIRRINGS

Firrings or **Furrings.** Fillets used to level up studs or joists to a plane surface to receive boarding. Cradling or rough brackets to carry lath and plaster, etc.

First Fixings. Plugs, grounds, etc., to which joinery is fixed.

First Floor. The floor next above the ground floor.

Firsts. A term used in the grading of timber to denote the best quality.

Firsts and Seconds. A term used in the grading of hardwoods, also called "prime quality."

Fisch. Abbreviation for the name of the botanist Fischer.

Fish. 1. See *Vesica*. 2. A piece lashed to a spar for strength.

Fishing. Lengthening structural timbers with the assistance of fish plates.

Fish Plates. Wood, or metal, plates used for securing together the ends of a lengthening joint in structural work.

Fissile. The property of wood that allows it to split or cleave. This depends upon the straightness of the grain in softwoods and the arrangement of fibres and rays in hard woods. Fissibility varies with moisture content.

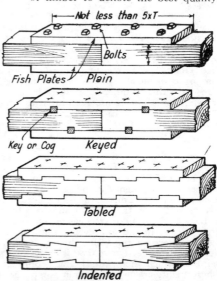

FISH PLATE JOINTS

Fistuca. The *monkey* of a pile-driver.

Fistulina hepatica. The " beaf-steak " fungus that forms *brown oak*. See *Oak*.

Fitments. Fittings of a loose, or semi-loose, character. The term is often applied to the registered types of kitchen furniture, such as cupboards and dressers, or cabinets, etc., common to modern domestic buildings. Built-in furniture.

Fitter's Screw. A patent double-screw for operating the frame of an extending table.

Fittings. Wood or metal furniture additional to the ordinary joinery finishings of buildings. They are distinguished according to their special uses as shop, office, bank, church fittings, etc.

Fit-up. A large piece of shuttering for concrete, framed so that it can be moved for repetition work.

Fitzroya. See *Alerce*.

Fixed Beam. A beam with the ends securely fixed, or adequately restrained in position and direction. This condition influences the bending moment, *q.v.*

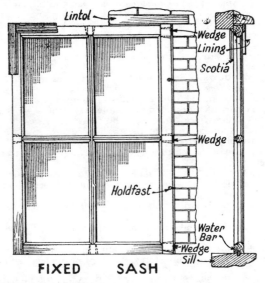

FIXED   SASH

Fixed Sash. See *Fast Sheet*.

Fixing Bricks. Bricks made specially to hold nails and screws, for fixing joinery, etc. See *Nogs*.

Fixing Fillet. A thin slip of wood, 9 in. × 4½ in., built in a joint of the brickwork to provide a fixing for joinery, etc. Also called a Pallet or Pad.

Fixings. Plugs, grounds, pallets, nogs, backings, soldiers, etc., to which joinery is fixed. See *Finishings* and *Double Skirting*.

Fixton Pad. Hardwood pads, used in place of hinged seats, for W.C.s.

Fixture. Anything attached to a building. Tenant's fixtures become the property of the landlord if their removal will cause material damage to the property. Landlord's fixtures are those provided by the landlord or by a previous tenant.

Flacourtiaceæ. Samydaceæ. A family with few trees of commercial importance for their timber : Casearia, Gynocardia, Homalium, Hydnocarpus.

Flag Staff. A pole on which a flag is displayed.

Flake. 1. A portable wooden fence. A field hurdle. 2. Silver grain, *q.v.* 3. Breaking away in thin layers or laminæ.

Flamard. An axe-like splitting tool used in certain woodland crafts.

Flamboyant. A Gothic, or Perpendicular, style of architecture in which wavy, or flame-like, tracery is a characteristic. The style predominated at the end of the fourteenth century.

Flame. 1. Applied to *plum figure* in mahogany, birch, etc. 2. Also applied to species of kurragong and she oak, *q.v.* 3. See *Fleam*.

Flange. 1. A projecting rim to a pipe. 2. A raised edge to the rim of a wheel. 3. The upper and lower parts of an " I " beam or rolled steel joist.

Flank. 1. The side, or wing, of a building. 2. Sometimes applied to the valley of a roof.

Flanking Windows. Small windows at the sides of an entrance doorway. See *Wing Windows* and *French Doors*.

Flanning. Splaying the interior jambs of an opening.

Flap. 1. A part of a desk top or counter top hinged so that it can be raised. 2. See *Back Flap*. 3. The folds, or leaves, of window shutters. 4. A hinged part to the trailing edge of the wing of an aeroplane, to control lift.

Flare. An outward slope to the hull of a flying-boat or to the sides of a vessel.

Flash. 1. See *Fire*. 2. *To flash* means to make watertight with sheet lead, or flashings, as at the intersection between chimney and roof.

Flash Board. A board or plank to control the outlet from a mill stream.

Flashing Board. A board to carry lead-work or flashings. See *Lear Board*.

Flat. 1. A domestic self-contained suite of rooms on one floor of a building of more than one storey. 2. A plane, or level and smooth, surface.

Flatcrown. *Albizzia fastigiata* (*H.*). S. Africa. Golden yellow, lustrous. Soft, light, compact and straight grain. Easily wrought. Used for general purposes, wheelwright's work, etc.

Flat Cut or **Flatting.** Sawing wood through its least dimensions or thickness. See *Deep Cut*.

Flat Grain. See *Flat-sawn*.

Flat Roof. A roof with less than 20° pitch.

Flat-sawn. Applied to timber cut tangentially to the annual rings, producing *flat grain*. The growth rings meet the face at an angle of 45° for at least half the width of the converted timber. Also called Plain-sawn, Slash-cut, or Bastard-sawn ; producing plain, slash, or bastard grain. See *Converting*.

Flayers. Men engaged in stripping the bark off felled trees.

Fleak. Flake (1), *q.v.*

Fleam or **Fleme Tooth.** Applied to a cross-cut hand-saw with teeth shaped like narrow isosceles triangles. The saw is very quick-cutting for softwoods but is not suitable for hardwoods. Also called a Peg-tooth, or Lance-tooth, or Needle-point Saw.

Fleche. 1. A tall spire with relatively small base. 2. A small spire-like ventilator on the roof of a building.

Flecks. Marks, spots, or dapples on the face of wood due to local irregularities in the grain or to extraneous matter. See *Pith Flecks* and *Ray Flecks*.

Fleur-de-lis. A carved ornament in the form of a conventional lily or iris.

Fleuron. Carved foliage, as in the abacus of the Corinthian capital.

Flexure. A curved outline. A bend, as in a loaded beam.

Flexwood. Very thin veneer, 0·3 mm. thick, mounted on fabric. It is waterproof, flexible, and may be obtained in sheets up to 10 ft. × 2 ft., and in twenty-five different woods.

Fliers. The steps in a straight flight of stairs. Parallel treads.

Flies. The stage wings in a theatre.

Flight. A series of fliers forming a complete unit of a stairs.

Flight Holes. Worm holes in wood.

Flimsy. Frail. Badly constructed. Anything not sufficiently strong for its purpose.

Flinching. See *Snape*.

Flindersia. See *Cudgerie, Crow's Ash, Hickory, Hickory Ash, Leopard Ash, Maple Silkwood, Queensland Maple, Rose Ash, Silver Ash, Teak, White Silkwood, Yellow-wood Ash*.

Flindosy Beech. Cudgerie, *q.v.*

Flintwood. *Eucalyptus pilularis*. See *Blackbutt*. Also applied to green satinheart or wilga, *q.v.*

Flitch. 1. A log, or part of a log, trimmed and prepared for conversion into veneers. 2. A part of a converted log suitable for further conversion, the rectangular cross-section of which is not less than 4 in. × 12 in. 3. The separate parts of a flitched beam. 4. A case of veneers consisting of about 500 pieces. See *veneers*.

Flitched Beam. A rectangular beam built up of two or more pieces bolted together. When there are three pieces the middle one is usually a wrought iron or steel plate. The wood flitches are obtained from a squared log cut up the middle. The outer faces of the log are placed together and one piece is turned " end for end " to equalize the strength.

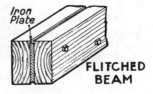

FLITCHED BEAM

Flitch Frame. A deal frame, *q.v.*

Flittern. A young oak tree.

Float. **1.** A raft of wood. **2.** The boards, buckets, or vanes of a water-wheel. **3.** A wood trowel used in plastering. **4.** A timber wagon. **5.** A single coarse-cut file. **6.** A measure of timber equalling 18 loads. **7.** Anything providing buoyancy to floating objects. Seaplane floats are described as inboard, main, tail, and wing-tip floats. **8.** A horse-drawn cart with low floor for carrying heavy goods, as a milk float, etc. **9.** Pieces of cork attached to fishing nets.

Float Board. See *Float* (2).

Floated Deals. Applied to Canadian softwoods that have been conveyed on rafts to port of shipment, which often causes discoloration.

Floating. Applied to anything that rides on the water. A floating bridge is built of rafts, pontoons, or boats.

Floating Floor. A patent insulated floor in which rebated wood battens rest on rubber isolators. The battens are loaded with slabs to prevent springing, and are boarded in the usual way.

Floating Policy. A single insurance covering goods in different places.

Floating Rule. A long wood straight-edge used in plastering.

Flogger. A flat wood mallet used by coopers.

Flogging. Levelling the joints of floorboards after they are nailed in position. Cleaning-up floors by hand tools. This is now usually done by abrading machines.

Flooded Gum. *Eucalyptus rudis* (*H.*). W. Australia. Also called Wormwood. Pale brown. In little esteem because of short grain. *E. grandis.* New S. Wales. Pale red. Open straight grain. Easily wrought, but not very durable and held in little esteem. See *Gum*.

Flood Gates. The lower gate of a lock. A sluice. A gate in a stream or river to control the flow of water.

Floor. The bottom timbers crossing the keel of a ship. Also see *Floors*.

FLOOR CASE

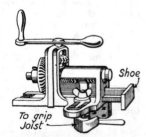

FLOOR CRAMP OR DOG

Floor Cases. See *Show Cases*.

Floor Cramps. Special types of cramps for squeezing up floorboards before nailing.

Floor Dog. **1.** A dog used with folding wedges, instead of a cramp. **2.** A machine for cramping up floorboards.

Floor Guide. Any form of groove in the floor to control sliding partitions, doors, etc.

Floor Hinge. See *Floor Spring.*

Flooring. Material for forming the surface of floors. See *Floors.* The factors for selection are cost, appearance, resistance to wear, and stability. A great number of woods are suitable, and the best qualities are rift-sawn. Slash-sawn boards should have the heart side

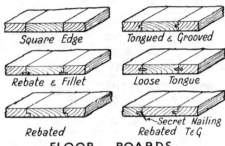

Square Edge   Tongued & Grooved

Rebate & Fillet   Loose Tongue

Rebated   Rebated T&G

Secret Nailing

FLOOR BOARDS

to the joists. Floorboards are sold by the *square, q.v.* The woods commonly used include redwood, whitewood, maple, oak, gurjun, Rhodesian teak, etc. Stock sizes are from 4 to 7 in. wide and $\frac{3}{4}$ to $1\frac{1}{4}$ in. thick.

Flooring Machine. A machine for preparing floorboards. The boards are planed on all sides and tongued and grooved in one operation. See *Four Cutter* and *Machines.*

Floor Joists. The timbers in a floor carrying the floorboards.

Floor Line. A line marked on the feet of door posts, etc., to denote floor level.

Floors. The horizontal divisions forming the storeys of a building. They are named in an ascending order : basement, ground, first, second, etc. ; and they may be of wood (blocks, boards, parquetry, plywood), wood and steel, concrete, reinforced concrete, or patented and registered types, of which there are many varieties. Wood floors may be single, double, or framed. See *Flooring* and *Mezzanine.*

Floor Spring. A boxed spring housed in the floor and controlling the movement of a swing door. See *Spring.*

Floor Stop. A rubber stop, fixed to the floor to prevent the door from banging on the wall or furniture.

Floor Strutting. Short pieces nailed between floor joists to stiffen them and to prevent *canting over.* The strutting may be herring-bone or solid. (See illus., p. 196.)

Floor Timbers. **1.** Those placed across a ship's keel and on which the bottom is framed. **2.** Beams, binders, and joists in timber floors.

Flor Amarilla. See *Lapacho.*

Floriated, Florid. A term applied to elaborate ornamentation.

Florida. A distinguishing prefix to certain species of pitch pine, *q.v.*, and denoting source.

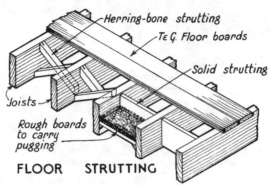

Herring-bone strutting

TE G. Floor boards

Solid strutting

Joists

Rough boards
to carry
pugging

FLOOR   STRUTTING

Florisia or **Floresa.** Cativo, *q.v.*

Flour. Wood dust, or fine sawdust, used with proprietary compositions for floors, synthetic compounds, etc.

Flourpaper. Very fine glasspaper, numbers oo and ooo.

Flower. Silver grain, *q.v.*

Flowered Satinwood. See *Satinwood*.

Flower Face. The inside face of tangentially sawn wood.

Flowering Dogwood. See *Dogwood*.

Flue. **1.** A shaft, pipe, or tube for the passage of air, gas, smoke, etc. **2.** The mouthpiece of an organ pipe.

Flueing or **Fluing.** Applied to twisted surfaces, as to a flush soffit for a geometrical stairs, or to the splaying of window jambs and linings.

Flue Pipe. An organ pipe, *q.v.*

Flue Work. The pipes of an organ.

Flume. A wood trough, or channel, conveying water. An inclined trough with running water for transporting logs. A water slide.

Fluorescence. A means of identifying some tropical woods, especially species of *Pterocarpus*. Shavings or sawdust are placed in a test tube with ethyl alcohol or water. After a day or two the solution is filtered and viewed in strong sunlight or beam of artificial light. The fluorescence is noted by the variable coloured luminosity, like the effect of light on shot silk.

Flush. Having the surface level with an adjacent surface. An unbroken surface.

Flush Bolt. A bolt housed into the wood so that it is flush with the surface.

Flush Box. A lead-lined tank or cistern for flushing a W.C.

FLUSH
BOLT

Flush Decked. Applied to a vessel with upper-deck level from stem to stern, and without raised obstruction.

Flush Doors. Doors with unbroken plane faces. Sometimes called solid doors. They may be framed and faced with plywood, or they may have an outer frame with fibre core, or made of laminwood; and they may be veneered. There are many different types and finishings. Various types of edging strips, *q.v.*, are used to cover the edges of the plywood facings.

Flush Panel. A panel that is level, or flush, with the face of the framing. See *Bead Flush.*

Flush Ring. A flush handle to operate the spindle of a lock. It is used instead of a door knob.

Flutes or **Flutings.** **1.** Cylindrical sinkings, as in pilasters, columns, etc. See *Engaged Column* and *Column.* **2.** A flute is a woodwind musical instrument consisting of a wood pipe with mouthpiece and a series of holes with keys controlled by the fingers. Also an organ stop with a flute-like tone. **3.** An old type of round-sterned sailing ship for stores and transport. It has three masts and is square-rigged.

Fluting Box. See *Moulding Box.*

Fly. **1.** See *Flies.* **2.** A form of two-seater carriage.

Fly Boat. A long light boat for passengers on a canal.

Flyer. See *Flier.*

Flying Boat. An aeroplane designed to enable it to take off from, or alight on, water.

Flying Bridge. **1.** The highest of several bridges on board ship. **2.** A floating bridge. **3.** A ferry boat attached by ropes to a bollard and swinging across a river by the action of the currents.

Flying Scaffold. A suspended scaffold.

Flying Screed. An overhead screed. One cantilevered from the shuttering, to allow for an unbroken surface in cement and plaster work.

Flying Shore. One fixed between walls, and without support from the ground. See *Shores.* (See illus., p. 198.)

Fly Leaf. A hinged leaf to a table. See *Gate-leg Table.*

Fly Rail. A rail that is drawn out or swung out to support a leaf to a table.

Fly Rod. A rod for fly-fishing.

Fly Wire. Fine woven mesh used to cover the joints of wallboards.

F.M. Abbreviation for *flush moulded.*

f.m. Abbreviation for *fathom* and for *festmeter.*

f.o.b. Abbreviation for *free on board.*

Foils. The inside arcs of tracery work, which is named according to the number of foils, as trefoil, quatrefoil, cincfoil, or multifoil. See *Tracery.*

f.o.k. Abbreviation for *free of knots.*

Fold. An enclosure for cattle, etc.

Lock Block

Edge Strip Banding or Railing

FLUSH DOOR

197

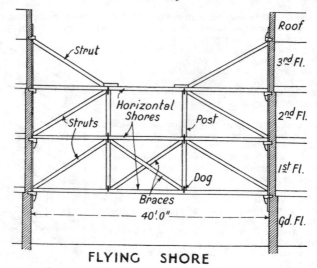

FLYING SHORE

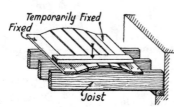

FOLDING FLOOR BOARDS

Folded Flooring. Several floor-boards sprung into position, between two fixed boards that serve as cramps. The method is used when cramps are not available.

Folded, or **Folding, Doors.** A pair of doors hung together so that one leaf folds over the other when open. They are often used as partitions between rooms, or for wide entrance doors. See *Double Doors*.

Folding Shutters. Shutters hung one to the other for folding into the boxing. See *Elbow Lining*.

Folding Trestle. See *Trestles*.

Folding Wedges. A pair of wedges sliding one on the other and driven in opposite directions to increase the distance between two bearing surfaces, as between a dead shore and needle or sole piece, or between a centre and its supports. See *Centring, Counter Cramps,* and *Dead Shore*.

Folha. See *Branca* and *Vermelha*.

Foliage or **Foliation.** Carvings in the form of leaves, flowers, etc.

Foliated. 1. Applied to tracery work with foils. **2.** Laminated. **3.** Leaf-shaped ornamentation.

Follow Board. A board used by the patternmaker to support a fragile pattern.

Follower. **1.** The bush for the spindle of a lock. **2.** A part of a machine actuated by another part.

Fomerell. A dome-like cover to a lantern.

Fomes. See *White Pocket Rot.*

Font. A vessel in a church for holding water for baptisms, etc.

Foocadie. *Terminalia buceras* (*H.*). British Guiana. An excellent wood with properties and characteristics of greenheart, but with coarser grain. S.G. ·9.

Foot. **1.** The leg of a chair. The *back foot* continues to form the frame for the back. **2.** The bottom end of a mast, post, rafter, etc.

Foot Block. A plinth block, *q.v.*

Footboard. **1.** An inclined board for the feet of the driver of a vehicle. **2.** A long step to the compartments of a railway carriage. **3.** A platform at the back of a carriage for the footman. **4.** A step to the door of a motor car.

Foot Bolt. A large tower bolt fixed vertically at the bottom of a door.

Foot Bridge. A bridge for pedestrians only.

Footing Beam. A tie beam, *q.v.*

Footing Piece. A sole piece, *q.v.*

Foot Pace. **1.** A landing to a stair. It is more usual to say quarter, or half, space. See *Pace.* **2.** An altar platform ; a dais.

Foot Plate. A piece connecting the foot of an ashler piece with a spar. It is often cogged on to a wall plate bedded in the wall.

Foot Stall. A plinth, base, or pedestal.

Footstick. A light spar for the foot of a jib or staysail.

Footstool. A low rest for the feet of a seated person.

Foot Waling. The lining, or inside planks, of a ship to prevent anything from falling amongst the floor timbers.

f.o.q. Abbreviation for *free on quay.*

f.o.r. Abbreviation for *free on rail.*

FOOT PLATE

Fore and Aft. Lengthwise of a ship.

Fore and Aft Sheer. The curvature of a ship's deck in the direction of its length.

Fore-cabin. Quarters for second-class passengers in the fore part of a ship.

Fore-carriage. The pivoted front part of a carriage that includes the front wheels.

Forecastle. The forward part of the upper deck of a ship.

Fore-deck. The front portion of a ship's deck.

Fore-edge. A projection at the front of a shelf on board ship. See *Gable.*

Fore Foot. **1.** The front curved member under a flying boat, etc. **2.** A piece terminating the keel at the fore end of a ship. See *Stem and Head.*

Fore-hook. Same as breast-hook, *q.v.*

Foremast. The lower section of the mast nearest a ship's bow. The term usually includes the fore-topmast.

Foresheet. The space in an open boat near to the bow.
Forest Grown. Applied to forest trees from self-sown seeds.
Forest Mahogany. See *Red Mahogany*.
Forest Oak. Australian beefwood, *q.v.*
Forest Red Gum. *Eucalyptus tereticornus.* Deep red. Hard, heavy, strong, tough, and durable. Straight and interlocked grain. Used for all purposes from structural work and piling to furniture. See *Gum.* S.G. 1.
Foretop. The platform at the head of the foremast of a ship.
Fore-topmast. The mast fixed at the top of the foremast.
Fork. 1. Figure in wood due to the fork, or intersection, between branch and trunk, or between two branches. See *Crotch.* 2. Anything with a bifurcated or Y-shaped end. Applied to numerous appliances consisting of handle and two or more prongs at the end, for digging, or for carrying hay, straw, etc., or for domestic use.
Fork Chuck. A forked lathe-centre.

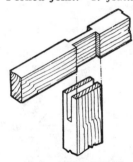

Forked Joint. 1. Joints formed with indentations like saw teeth. See *Indented Joint.* 2. A rod branched at the end to receive the end of another rod. The two ends are then secured by bolts or rivets. Also see *Knuckle Joint* and *Finger Joint*.
Forked Tenon. An open mortise sitting astride a tenon not at the end of the stuff. It is used to give the appearance of a stile over a continuous rail on wide doors, counter fronts, etc. A slight bevel on the

FORKED TENON

shoulders prevents the cheeks from opening when driven into position.
Forked Turn Screw. A screw-driver brace bit, with slot, for tightening the rivets in saw handles.
Forma. A channel for water.
Former. A temporary core round which layers of thin wood are bent, or built-up, to form a cylindrical strut, etc. See *Bentwood*.
Formeret. A rib lying next to the wall in groined vaulting.

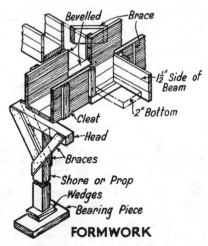

Bevelled — Brace
$1\frac{1}{2}''$ Side of Beam
$2''$ Bottom
Cleat
Head
Braces
Shore or Prop
Wedges
Bearing Piece

FORMWORK

Form Factor. A factor for obtaining the volume of the stem of a tree compared with a cylinder of the same height and mean diameter.
Formica. A proprietary plastic material, similar to bakelite. It is made in various colours.
Formica smaragdina. The red ant.
Forms. 1. Boxes, or troughs, in which concrete is cast. 2. Long low benches, with or without backs, to seat several persons.
Formvar Resin. A registered vinyl thermoplastic resin.
Formwork. Temporary timbering used in the casting of concrete *in situ*. Also called Shuttering or Falsework. See *Shuttering*.
Forssman Wood. Plywood from devitalised wood to make it stable against change of moisture content, to prevent cracking or blistering. Three-ply sheets are from ·04 in. thick and mounted on ordinary, or lower grade, plywood.

FORSTNER BIT

Forstner Bit. A patent brace bit for sinking holes that do not go through the material. The sharpened circumference fixes the position.
f.o.t. Abbreviation for *free on truck*.
Fotui. *Jacaranda copaia (H.).* C. America. Oatmeal colour, lustrous. Light, soft, not durable. Straight grain but not easily split, uniform texture, smooth. Easily wrought. Used for light construction, matches, cases, etc. S.G. ·44 ; C.W. 2.
Fou. See *Tortosa*.
Four-centred Arch. One in which the outline is struck from four centres. A Tudor arch.
Four Cutter. A large type of planing and moulding machine that prepares four sides of the material in one operation. The term four-cutter is in general use, but the machine may have six heads or cutter blocks.

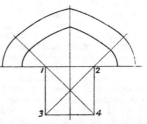

FOUR-CENTRED ARCH

Four-leaved Flower. A carved ornament characteristic of the Decorated style of architecture. It is usually placed in large hollow mouldings.
Four-way Match. See *End Matched*.
f.o.w. Abbreviation for *first open water*. Used in connection with shipments from ice-bound ports.

FOUR LEAVED FLOWER

Fowe. A hay-fork.

Foxiness or **Foxy.** Incipient decay, accompanied by a dull red stain. It is caused by fungi, and usually confined to trees grown in marshy soil.

Fox, or **Fox-tailed, Wedges.** Wedges in the end of a stub tenon. When the tenon is driven into the mortise the wedges open out the end of the tenon in the form of a dovetail.

FOX WEDGING

Foxw. Abbreviation for the name of the botanist Foxworthy.

Foyboat. A boat used in the mooring of ships.

Foyer. A large vestibule, or small hall, at the entrance to a public place of entertainment. It usually includes cloakrooms, payboxes, etc.

Fr. Abbreviation for the name of the botanist Fries.

Fracture, Cross. See *Upset.*

Fragrance. The pleasant scent in many woods due to the volatile chemical constituents. In some timbers, such as cedar, it is retained indefinitely, but in others it disappears with seasoning. Also see *Odorous.*

Frake. Afara, *q.v.*

Frame. 1. Usually applied to an assembly of pieces connected by halving, housing, mortise and tenon joints, or similar connections, and serving as a support or enclosure. See *Framing, Door Frame, Window Frame.* 2. A case, border, movable structure, etc. 3. To enclose in a border. 4. An open box, 6-ft. cube, for measuring a fathom of wood of small dimensions. 5. The timbers, or ribs, forming the cross-section or shape of a ship. 6. A box with glazed sloping top for forcing plants in a garden. 7. The essential structural elements of a building, etc., that support the remainder of the structure. 8. An inclined wood block used in dressing tin ore.

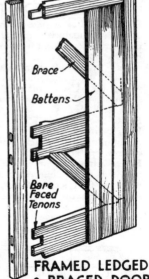

FRAMED LEDGED & BRACED DOOR

Frame Back. Framed and panelled backs to cabinet work. Used instead of plywood only.

Framed and Braced. A framed and ledged door with braces to prevent the nose from dropping.

Framed and Ledged. A door consisting of stiles, top rails, ledges, and battens. The battens, or boards, run the full length of the door from a rebate in the head. The ledges are thinner than the outer frame to allow for the thickness of the boards.

Framed Floor. A floor consisting of beam, binders, and joists. The binders are framed into the beam.

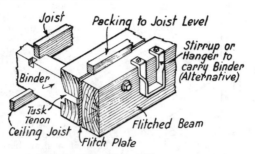

FRAMED FLOOR

Framed Grounds. Grounds, round openings, framed together and fixed as one piece.

Framed Linings. Panelled linings for interior doorways. See *Finishings*. Also see *Skeleton Linings*.

Framed Mouldings. Bolection mouldings framed together so that shrinkage is independent of the framing, and to prevent the mitres from opening. See *Fielded Panel*.

Framed Partition. **1.** A partition framed and braced to form a structural unit supported by the walls only. See *Partition*. **2.** Panelled framing serving as a partition.

Framed Roofs. Trussed roofs, *q.v.*

Framed Square. A term applied to panelled framing without mouldings.

Frame, or **Framed, House.** A timber house.

Framer. A maker of picture frames.

Frameri or **Framiré.** *Terminalia ivorensis* (*H.*). See *Idigbo*.

Frame Saws. Power saws in which a number of blades are secured in a frame, that moves in a vertical or horizontal direction, to make several parallel cuts in one operation. Also see *Bow Saw*.

Framing. **1.** The skeleton, or structural parts, of a frame. **2.** The act of constructing a frame.

Framing Timber. Wood suitable for the carcassing of buildings. Constructional timber.

Francis. Name of the botanist W. D. Francis.

Franking. Applied to a mortise and tenon joint in which a spur, or projection, is formed on the mortised member to fit in a recess in the shoulder of the tenoned member. It does not weaken the mortised member so much as haunching. It is usually described as a *reversed haunching*. The projecting spur need not be in line with the mortise. A variation is sometimes used for scribed sash bars, *q.v.*

Frank Jointing. A term used in functional cabinet work in which there is no effort made to hide the joints, but rather to emphasise them as a feature of the design.

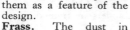

Frass. The dust in worm holes. The excrement of the insects.

Fraxinus. See *Ash*, *Olivewood*, and *Tamo*. Also see *Himalayan*, *Pumpkin*, *Red*, and *White Ash*.

Free Alongside. A term used when the price quoted is for delivery of goods unloaded.

Freeboard. The depth of a ship, between deck and waterline.

Free Moisture. Moisture in the cell cavities and intercellular spaces of wood; to distinguish from the moisture in the cell walls.

Free on Board. A shipping term used in the same sense as *free alongside*, *i.e.* goods placed within reach of ship's tackle without extra cost to purchaser. Also applied to other means of transport.

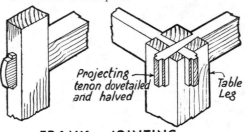

FRANKED JOINT

FRANK JOINTING

Free Overside. A term used when the purchaser is responsible immediately the goods leave the ship's side, and must provide craft to accept delivery.

Freijo. *Cordia goeldiana* (*H.*). Brazil. Called S. American Mahogany because of its similarity to *Swietenia sp.* S.G. ·75.

Frei-Jorge. *Cordia frondosa* (*H.*). Brazil. Greyish brown with darker streaks. Other characteristics as Freijo. S.G. ·65.

French Bit. A drill for very hard wood; similar to one for drilling metal.

French Casement. A large sash hinged and used as a door.
French Curve. A drawing appliance of varying curvature, and used for lining or inking freehand curves.

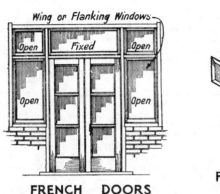

FRENCH DOORS

FRENCH HEAD

French Doors. A pair of large glazed sashes, hinged and used as doors.
French Fliers. Fliers round a rectangular well to a stairs.
French Head. See *French Spindle*.
French Polishing. The most common method of finishing interior hardwoods. The polish consists of shellac dissolved in methylated spirits. The pores of the wood are usually stopped with a filler, and the wood stained if required, before applying the polish by means of a pad or brush. There are many variations of finish. See *Fuming, Wax Polish, Spraying*.
French Spindle. A vertical‑spindle moulding machine. The cylindrical head, about $1\frac{1}{2}$ in. diameter, is slotted to receive the mild steel cutter, called *scraper*, or French, cutters. The stuff may run on the bare spindle, which makes it specially suitable for circular work of small radius. The cutters are secured by a long set-screw threaded down the head. This type of machine produces a well-finished surface due to the scraping action and perfect balance of the cutters, and to the speed, which is about 7,000 revs. per min.
Fresh Wood. A term applied to brittle and *short* wood, with little elasticity, due to poor lateral adhesion between the growth rings.
Fret or **Frette.** 1. A carved ornamentation consisting of projecting fillets in regular formation. See *Greek Fret*. 2. A grating. 3. Decorative perforated work. See *Jali Panel*.
Fret Saw. 1. A very fine saw for fret-work. 2. A jig saw, *q.v.*
Fret-work. Thin wood elaborately cut with a fret saw into intricate frets and perforations.
Friction Piles. Those depending upon the friction on the sides and having no firm bearing at the foot.

Frieze. 1. The top part of an interior wall between ceiling cornice and picture rail. 2. A pulvinated or flat member between the cornice and architrave in an entablature.

Frieze Panel. The top panel in a door with five or more panels, *i.e.* one with more than three rails. The upper panel in a panelled dado. See *Panelled Door*.

Frieze Rail. The rail immediately below the frieze panel.

Frigate. A light naval vessel with greater than normal speed.

Frithstool. A seat near the altar in a church. A chair of sanctuary.

frm. Abbreviation for *framing*.

Frog. A heavy timber at the mouth of a slide to control the direction of discharge of a log.

Fronton. An ornamental head to an entrance doorway, consisting of pediment, cornice, and consoles.

Frost Cracks. Fissures in trees caused by frost. The fissures close later and are covered with callus growth. They usually occur at the base of old trees, and form when the temperature is rising and not during the period of lowest temperature. They are a weakness in converted wood.

Frost Rib. A protruding ridge of callus due to repeated opening and healing of a frost crack, *q.v.*

Frost Ring. A circumferential brown layer in a growth ring due to injury of the cambium by frost.

Isometric View

FRUSTUM
HEX. PYRAMID

Frowy. A term applied to soft or brittle wood.

Frustum. The lower part of a solid when cut by a plane parallel to the base, as a pyramid or cone frustum.

f.s. Abbreviation for *fore sheer*. See *Fore and Aft Sheer*. Also for *free shorts*, in shipping.

f.t. Abbreviation for *full terms*.

Fucadi. Foocadie, *q.v.*

Fulcrum. 1. A prop or support. 2. The axis about which a lever rotates. 3. The foot of a couch.

Full. Slightly over size. The opposite to *bare*.

Full-bottomed. Applied to a ship with more than normal capacity below waterline.

Full-cell Process. See *Preservation* and *Empty-cell Process*.

Full Scale. Applied to log measurements in which no deductions are allowed for defects.

Fuming. Darkening oak by the fumes from ammonia, NH_3, before polishing. The wood is enclosed in an airtight chamber, or fuming box, and exposed to the fumes, and observed through glass until the required shade is obtained. Half a pint of ammonia in a shallow vessel will produce an olive-green shade in twelve hours in a chamber of 200 cub. ft. Fuming has no effect on the grain.

Fungicide. Any preparation that destroys fungi.

Fungus. The mushroom type of plant life that produces mould, mildew, etc. There is a great variety of fungi that attack trees and wood, and in most cases they are very destructive. In several cases they increase the decorative value, as in maple, oak, chestnut, eaglewood, etc. It is now suggested that bird's eye, peacock's eye, etc., are due to attack by fungus that retards growth in small localized areas. The depressions are filled by later growth. *Wood-destroying* fungi obtain their sustenance from the cellular structure and cause decay. *Wood-staining* fungi derive their sustenance from the cell contents and usually have no disintegrating effect on the wood substance. The growth of fungi depends upon suitable moisture content and temperature, sufficient oxygen and food. The preventive measures are seasoning to below 18 per cent. moisture content or treatment with toxic preservative. See *Decay, Dry Rot, Rot,* and *Preservation.*

Funnel. A smoke stack or ventilating shaft on board ship. It is usually a large cylindrical chimney projecting above the deck. A ship is often described according to the number of its funnels.

furn. Abbreviation for *furniture stock.*

Furniture. 1. Loose, or semi-loose, fittings to a building. It is described according to its special function : bank, house, school, or church furniture, etc. 2. The brasswork, or ironmongery, for doors, windows, cupboards, coffins, etc. 3. Frames round pages of type for printing. 4. Equipment, outfit, appendages.

Furniture Beetle. Anobiidæ family. The most widely distributed of boring beetles and very destructive to seasoned and mature woods. The common furniture beetle, *Anobium punctatum*, is the best known and is commonly found in old furniture, panelling, etc. It is closely related to the deathwatch beetle, but smaller. The treatment is to fumigate or to inject a strong insecticide into the flight holes, crevices, and joints to destroy the eggs and larvæ. It is most effective during late spring or summer. Several applications are necessary. The holes may then be stopped and the work repolished. Other species are : *Anobium denticolle, Ernobius mollis, Hedobia imperialis, Ochina hederæ, Priobium castaneum, Ptilinus pectinicornis,* and *Xestobium rufovillosum.*

Furniture Stock. Wood suitable for cabinet work.

Furniture Wood. A term often applied to imported wood that cannot be classified, but suitable for cabinet work. Very often the importer gives it a " coined " and confusing name.

Furrings. See *Firrings.*

Fuse Board. A board used in an electrical installation.

Fuselage. The structural framework or body of an aeroplane, to which the tail unit, main-planes, and undercarriage are attached.

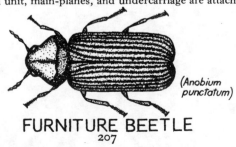

(Anobium punctatum)

FURNITURE BEETLE

Fust. 1. The shaft of a column. 2. The ridge of a roof. 3. A stick or staff.

Fustic. *Chlorophora tinctoria* (*H.*). W. Indies. Also called Amarillo, etc. Canary colour, darkening to brown. Very hard, heavy, tough, strong, and durable. Straight and interlocked grain. Uniform, fairly coarse texture. Distinct rays. Smooth and not very difficult to work. Used for cabinet work and as a yellow dyewood. S.G. ·86 ; C.W. 3·5.

Futchell. The provision for attaching the shafts or pole to a vehicle.

Futtocks. A crooked lower timber of a ship's framing, below the deck, especially one that crosses over the keel. The timbers towards the middle of a ship between the top-timbers and the floor. Ground futtocks are near the keel. The others are called lower futtocks.

Fyburstone. A registered wallboard, *q.v.*

Fygrin. *Tecoma peroba* or *Tabebuia sp.* Dutch Guiana. See *Peroba* and *Primavera*.

G

g. Abbreviation for *green, i.e.* unseasoned.

Gabbard or **Gabers Scaffold.** One built of square timbers bolted together.

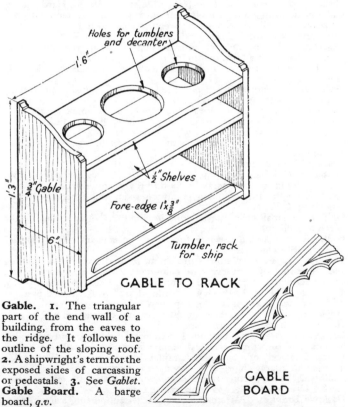

GABLE TO RACK

Gable. **1.** The triangular part of the end wall of a building, from the eaves to the ridge. It follows the outline of the sloping roof. **2.** A shipwright's term for the exposed sides of carcassing or pedestals. **3.** See *Gablet*.

Gable Board. A barge board, *q.v.*

GABLE BOARD

Gable Roof. A roof finishing against a gable and open to the rafters.

Gablet. A decoration shaped like a gable. A small gable. See *Gambrel Roof.*

Gable Window. A window with a triangular top; or one in a gable.
Gaboon Mahogany. *Aucoumea klaineana* (*H.*). French W. Africa.
Not a true mahogany. Also called Okoumé and Libreville Mahogany.
Other woods are often included in consignments, such as species of
Canarium and *Pachylobus*. It has some of the characteristics of
mahogany, but inferior, and is used extensively for plywood and
laminwood. Salmon pink to mahogany colour, lustrous. Light,
soft, not strong or durable. Irregular grain, broken stripe, fine rays,
coarse texture. Difficult to work because of woolly nature. Subject
to upsets. S.G. ·5; C.W. 4. See *Mahogany*.
Gading. *Hunteria corymbosa* (*H.*). Malay. Yellow. Hard, heavy,
with fine uniform grain. Finely distributed rays and glistening deposit
in pores. Small sizes due to irregular-shaped trees. Used as a
substitute for kemuning, *q.v.* S.G. ·85.
Gadroon. A decoration resembling large reeds or repetitive beads
with quirks. It is a form of nulling on turnery, friezes, etc.
Gaertn. Abbreviation for the names of the botanists J. and C. F.
Gaertner.
Gaff. A spar with a forked end used on sailing ships. It is usually
in a fore-and-aft direction and at the top of a sail. It is similar to the
boom which is at the bottom of the sail. **2.** A small spar. **3.** A
stick with iron hook for landing fish. A barbed fishing spear.
Gagil. Borneo white cedar or white seriah, *q.v.*
Gain. **1.** The lap in a lapped joint. **2.** A notch. **3.** The bevelling
of a shoulder.
Gaining. An alternative term for notching, *q.v.*
Galba or **Galaba.** *Calophyllum calaba* or *C. brasiliense* (*H.*). W.
Indies. Reddish. Moderately hard, heavy, and durable. Stable.
Used as a substitute for plain Honduras cedar and general purposes.
S.G. ·82; C.W. 2.
Galilee. A small gallery or balcony in a church. Also applied to the
porch of a church.
Gall. A nut-like excrescence formed by insects on certain species
of trees. See *Oak Gall*.
Gallery. **1.** A long platform projecting from the wall of a building
and supported by cantilevers or columns. **2.** A long narrow room,
or one serving as a passage of communication between other rooms.
3. A room used for the display of works of art. **4.** A covered passage.
5. A narrow projection at the top of a cabinet or sideboard. **6.** A
projecting platform, or frame, at the stern and quarters of a ship.
Gallesia. Pau d'alho. See *Pao*.
Galley. **1.** A low flat-bottomed boat with one deck and propelled
by oars, and sometimes sails. A sailing ship with lugsails for coastal
trade. The body of a boat. **2.** A captain's small boat. **3.** The cook's
quarters on a large ship. **4.** A tray used in printing.
Gallows. A supporting frame consisting of two posts and a cross-
piece. Used on trawlers to support the nets.
Gallows Bracket. A framed wood bracket supporting a load at the
outer end.
Galoche or **Galosh.** A wood shoe.
Gamari. The standard name for gumhar, *q.v.* Also see *Trewia*.

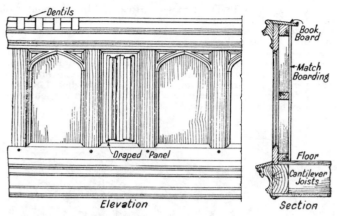

Elevation   *Section*

GALLERY   FRONT

Gamb. Abbreviation for the name of the botanist Gamble.

Gambrel or **Gambril Roof.**
The end of a roof partly hipped
and partly gabled.

Gambrel Vent. A triangular
louvre ventilator. See *Louvre.*

Gangau. *Mesua ferrea* (*H.*).
Burma. Also called Ironwood,
Mesua, Nahor, and Penaga.
Dark red with yellow secretion and streaks of dark
deposit. Very hard, heavy,
strong, durable, tough, and
elastic. Smooth, with medium
fine even texture. Straight
and interlocked grain. Very
fine uniform rays. Difficult to

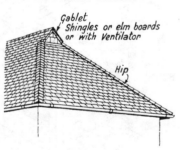

GAMBREL ROOF

work and season. Used for sleepers, structural work, gunstocks, tools,
walking sticks, etc. S.G. 1 ; C.W. 5.

Gangboarding. Boards used as a gangway, especially when in a
roof or on a ship. See *Gangway.*

Gang Mill. A machine in which several reciprocating saws, called
gang saws, cut at the same time.

Gang Mortiser. A mortising machine that cuts a number of mortises
in one operation.

Gang Mould. Formwork, or moulds, for concrete in which a number
of units are cast in one operation.

Gangplank. See *Gangway* (2).

Gangwa. *Excœcaria agallocha* (*H.*).   India.   Pale yellow, fairly lustrous. Light, soft, not durable. Straight, fibrous grain, with ray flecks. Medium texture. Easily wrought. Poisonous sap when worked green. Used for packing cases, toys, floats, etc. S.G. ·48.

Gangway. **1.** An incline for transporting logs from water to saw-mill. Also called Jack-ladder, Log-jack, and Slip. When it is mechanical it is called a Conveyor. **2.** A footway formed of one or more planks, for workmen. **3.** A narrow platform to board or leave a ship, or the opening in the bulwarks of a ship to receive the gangboard or gangway. **4.** The passage up the middle of a railway carriage, etc., with seats on each side. **5.** See *Stand*.

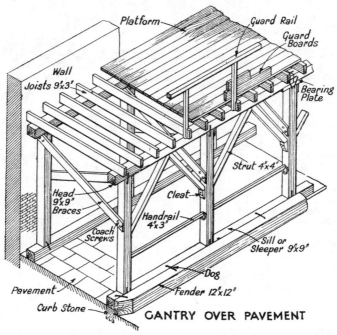

GANTRY OVER PAVEMENT

Gantry. **1.** A double staging of large square timbers carrying a travelling crane or a scaffold. Also see *Traveller*. **2.** Stagings for barrels in a cellar.

Ganua. See *Nyatop*.

Gap. A defect in the inner plies of plywood, showing a void on the edge.

Garage. A building to house one or more motor cars.

Garapa. *Apuleia leiocarpa*, syn. *A. præcox* (*H.*). Brazil. Golden brown. Appearance, characteristics, and uses of coralwood, *q.v.* S.G. ·9.

Garboard Strake. The first strake of planks in a ship next to the keel. The edge of the plank fits into a groove in the side of the keel.

Gardenia. *Gardenia latifolia* (*H.*). Indian boxwood. Several species. Yellowish white. Close compact grain. Very similar to European boxwood, but inferior. Small sizes. Also called Decamali, Orissa, Bihar, etc. S.G. ·8 ; C.W. 5.

Garlicwood. Pau d'alho. See *Pao*.

Garnet Hinge. See *Tee-hinge*.

Garnish Rails. Those forming the decorative, or face, parts of coachwork.

Garret. The top room of a building, immediately under the roof.

Garron. A large wrought nail with flat point and rose head.

Garuga. *Garuga pinnata* (*H.*). India. Reddish brown, with grey-black markings due to uneven infiltration, lustrous. Moderately hard, heavy, strong, and fairly durable. Straight and interlocked grain. Uneven coarse texture. Difficult to season but fairly easy to work. Liable to stain and insect attack. Used for furniture, planking, etc. S.G. ·64 ; C.W. 3. See *Kharpat*.

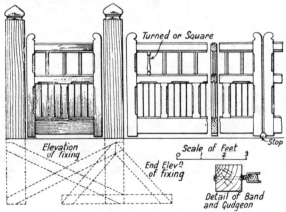

GATES TO CARRIAGE DRIVE

Gate. **1.** A framework of wood or metal controlling the entrance to an enclosure or the passage of water. A large batten door. Gates are described according to their particular use : carriage, crossing, dock, entrance, farm, field, garden, lock, lych, sluice, warehouse, wicket, yard. **2.** The passage through which the molten metal is poured for a casting. Ingates or runners.

Gateado. *Astronium graveolens.* See *Goncalo Alves*. Also *Swietenia humilis* (*H.*). Mexico. Golden brown, streaked with black. A species of mahogany. Harder, heavier, and denser than Honduras mahogany.

Gated Pattern. Small patterns fastened to the runners and gate fin for a multiple pattern in patternmaking.

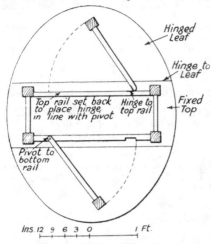

Hinged Leaf

Hinge to Leaf

Fixed Top

Hinge to top rail

Top rail set back to place hinge in line with pivot

Pivot to bottom rail

Ins. 12 9 6 3 0   1 Ft.

GATE LEG TABLE

Gate-leg Table. A table with a hinged leaf and a hinged leg to support the leaf.

Gate Posts. The strong uprights to a gate. They are called hanging and shutting posts according to their function. When they are of other material than wood or metal they are called piers.

Gauge. 1. A tool for marking parallel lines on wood. The various types are: butt, marking, cutting, grasshopper, mortise, pencil, roller. 2. A standard method of denoting the thickness of sheet metal and diameters of wire, or a tool for measuring the thickness or diameter. 3. See *Saw Tooth*. 4. See *Bit Gauge*. 5. See *Gauging Box*. 6. The distance between parallel rails, or the distance apart of the tails, or bottom edges, and nail-holes of consecutive rows of shingles, tiles, or slates. See *Eaves*. 7. Clock or dial gauges are used for measuring pressure, height, etc.

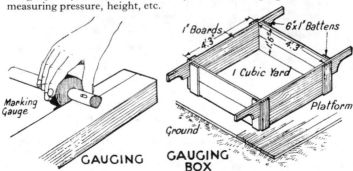

Marking Gauge

GAUGING

1" Boards

4'.3

6"x1" Battens

4'.3

1 Cubic Yard

Platform

Ground

GAUGING BOX

Gauge Piles. Guide piles, *q.v.*
Gauging. 1. Using a gauge. 2. Apportioning the materials for a plastic mix.
Gauging Board. A boarded surface, or platform, on which plastic materials are gauged and mixed.

Gauging Box. A box of known cubical capacity for measuring the quantities of dry materials for a plastic mix, such as concrete.

Gausan. *Schleichera trijuga* (*H.*). India. Ceylon oak or kosum. Reddish brown. Very hard, heavy, strong, and durable. Uniform grain but difficult to work smooth. Polishes well. Important for lac. Used for agricultural implements, wheelwright's work, etc. S.G. 1 ; C.W. 4·5.

Gavel. A small ornamental wood mallet, used by one in authority and as a symbol of office.

Gayac. See *Lignum-vitæ*.

Gazebo. A summerhouse in a garden, or a balcony or turret, commanding a wide view.

G-cramp. See *Cramps*.

Gean. *Prunus avium* (*H.*). Britain. Wild cherry. Red, hard, fairly heavy, strong. Fine compact grain. Easy to work and polishes well. Small sizes. Used for cabinet work, musical instruments, turnery, etc. See *Cherry*.

Gear. Applied to the tools, implements, tackle, etc., required to undertake a piece of work.

Gedge Bit. A twist bit with curved cutters specially adaptable for boring parallel to the grain. See *Twist Bit*.

Gedunoha or **Gedu Nohor.** *Entandrophragma angolense* and other species (*H.*). W. and E. Africa. Also called Gedu Lohor, Budongo Mahogany, Tiama, etc. Reddish brown, lustrous. Rather light and soft, moderately strong and durable. Interlocked grain liable to pick up, and rather difficult to work smooth. Fairly open and even texture. Liable to warp in seasoning unless quarter-sawn, which gives handsome stripe. Polishes well. Used for veneers, cabinet work, shopfitting, etc. S.G. ·5 ; C.W. 4.

Geelhout. *Podocarpus elongata* (*H.*). S. Africa. Yellowish. Large sizes. Not durable. Used for interior work only. Also called Yellowwood. *P. thunbergii* is a more durable wood with close even grain, and is brownish white. S.G. ·5.

Gee Throw. A strong wood lever with a curved iron point used in lumbering.

Geijera. See *Wilga*.

Geissois. See *Red Carrobean*, *Rose Marara*, and *Satin Sycamore*.

Gelam. See *Kelat*.

Gemmels. An obsolete name for hinges.

General Joiner. A woodworking machine that can be used for various operations. A universal woodworker, *q.v.*

Genipa. See *Jagua*.

Genizero. Kelobra, *q.v.*

Genoa Pine. A grading term used in the pitch-pine trade.

Genthi. See *Sedeng*.

Genus. A division in the classification of trees, etc., that has characteristics not found in other genera. Each genus is subdivided into species. See *Botany*.

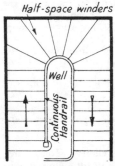

Half-space winders

GEOMETRICAL
STAIRS

Geometrical Stairs. A stair with a continuous string and handrail round a semicircular or semi-elliptical well.

Georgia Pine. Pitch pine, *q.v.*

Geronggang. *Cratoxylon spp.* (*H.*). Malay, E. Indies. Pale red or pink. Very soft and light. Not durable or strong. Sapwood of no value. Fine grain with faint rays. Liable to split. Little imported. S.G. ·34.

Gesso. A plaster surface prepared as a ground for painted decorations. A wood panel is generally used as a base for the plaster.

g.f.a. Abbreviation for *good fair average*.

Ghaut. A flight of stairs to a riverside landing stage.

Giac. Lignum-vitæ, *q.v.*

Giam. *Hopea lowii* (*H.*). Malay. Dark olive to very dark brown on exposure. Probably the hardest, heaviest, and darkest in colour of the Dipterocarps. Very durable and strong. Trees subject to beetle attack. Resinous secretion. Used where great strength and durability are required. S.G. 1·1.

Giant Arbor vitæ. Red cedar, *q.v.*

Giant Redwood. Sequoia, *q.v.*

Gib Door. See *Jib Door*.

Giblet Check. A rebate for an external door opening outwards. An external rebate.

Gibs. See *Cottered Joint*.

Gidgee. *Acacia spp.* (*H.*). Australia. Rich dark brown to nearly black, lustrous. Very dense and hard. Small sizes. Used as a substitute for briar. Excellent for turnery and carvings. S.G. 1·2. See *Myall*.

Gig. 1. A ship's boat with six or eight oars, and sometimes sails. It is a long narrow boat and usually reserved for the captain. 2. A high, light, one-horse carriage with two wheels.

Gig-stick. A radius rod used in setting-out, especially in plastering.

Gijam. Giam, *q.v.*

Gimlet. A small screw-pointed boring tool used by hand. A gimlet bit is a similar tool used with the brace.

Gimlet Gum. *Eucalyptus salubris.* Queensland. See *Gum*.

Gin. A tripod of three poles from which pulley tackle is suspended for hoisting purposes.

Gingkoaceæ. A family of Gymnosperms, but with only one species, *Gingko biloba.* It has the habits of a conifer but is deciduous and its leaves are like the maidenhair fern. See *Maidenhair Tree* and *Ischowood*.

Gin Wheel. A large pulley used for hoisting purposes. It has no mechanical advantage but only changes the direction of the effort.

Gippsland Grey Box. *Eucalyptus bosistoana* (*H.*). Victoria, Australia. Pale brown. Very hard, heavy, strong, durable, and tough. Close straight grain. Uses as Grey Box, *q.v.* S.G. 1·1.

Girandole. 1. A branched candlestick or bracket. 2. A mirror to a pier between windows.

Gird. A wooden hoop for a cask.

Girder. A large, or main, beam, but more especially applied to mild steel sections. See *Flitched Girder*.

Girdling. Cutting through the cambium layer round the circumference of a growing tree to prevent further growth or movement of sap. See *Bark Peeling*.

Girt. 1. Intermediate horizontal structural members, or rails, of framed wood buildings. A *drop girt* is one placed at a lower level to allow for better fixing in the corner post. 2. Same as *girth*.

Girth. The distance round a plane figure. The perimeter of a cross-section.

Girthing Tape or **Strap.** A specially graduated measuring tape for obtaining the volume of round logs. It

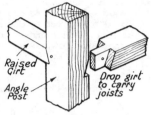

GIRTS

may be of steel, linen, or leather (strap). There are different varieties of tapes to give the volume according to some specified system of timber measurement. See *Quarter Girth* and *Measure*.

Girt Rope Cleat. See *Ship's Cleats*.

Give. 1. A term used to imply that a thing will accommodate itself, *i.e.* yield or bend without breaking. 2. Implies that a structural unit will fail when loaded.

Glass, or **Glazing, Bead.** Loose beads mitred and nailed round a rebate to secure a pane of glass. See *Barred Door*.

Glasshouse. An erection chiefly of glass, as a greenhouse, conservatory, studio, etc.

Glasspaper. Finely ground glass sprinkled on glued paper and used as an abrasive for finishing the surface of wood, etc. It is commonly called sandpaper. Waterproof glasspaper may be obtained for outside work. The usual strengths for woodwork are : oo, o, 1, 1½, fine 2, middle 2, and strong 2. See *Flourpaper* and *Sandpaper*.

Glasswood. See *Wilga*.

Glassywood. *Guettarda soleriana* (H.). W. Indies. Pale straw colour. Fairly hard, with straight close grain. Similar to rock elm. A useful wood, but little imported.

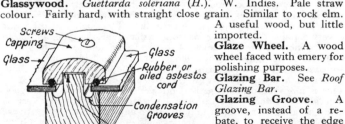

GLAZING BAR

Glaze Wheel. A wood wheel faced with emery for polishing purposes.

Glazing Bar. See *Roof Glazing Bar*.

Glazing Groove. A groove, instead of a rebate, to receive the edge of a pane of glass, as in the bottom meeting rail of vertically sliding sashes.

Gleas. Abbreviation for the name of the botanist Gleason.
Gleditschia. See *Honey Locust, Espina*, and *Tallow Tree*.
Gliders. 1. Non-mechanically driven aircraft with fixed wings.
2. Metal domes for the feet of furniture instead of castors.
Gliksten. A registered wallboard, *q.v.*
Globe. A sphere, *q.v.*
Glowood. A registered name for a processed plywood. Decorative woods are reproduced by photography on maple plywood.
Glue. An adhesive material of animal, mineral, or vegetable origin. Glue used in woodworking is chiefly of animal origin, from gelatinous substances as bones, hides, hoofs, fish refuse, etc. It is usually sold in cakes up to 6 in. × 6 in. × ⅜ in. ; as English, Scotch, Cologne, Medal (or Town), French, Fish, Russian, according to quality and process of manufacture. The cakes are broken into small pieces and soaked

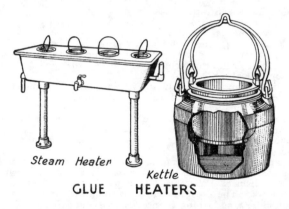

Steam Heater Kettle
GLUE HEATERS

in cold water for about twelve hours before being heated in a jacketed vessel to between 120° and 150° F. Overheating destroys efficiency, underheating prevents penetration. Dryness, warmth (70° F.), and hot glue are necessary for the best results. The quality of cake glue is very variable. Fresh glue is stronger than reheated glue, and clean utensils increase the efficiency. Unsoaked glue should not be mixed with the heated glue. POWDERED GLUE. Animal glue may be obtained ground in powder form. This form should be soaked for about an hour, otherwise the hints given for cake glue apply. LIQUID GLUE. This is more uniform in quality, but generally it is inferior to good cake glue. It does not require to be heated above 60° F. Various mediums are used to keep it in liquid form and to act as preservatives : carbolic acid, alum, alcohol, acetic acid, etc. COLD WATER GLUE. This is from dextrines or caseins, and is sold in powder form. Casein glues consist of casein with borax, lime, silica, or soda, etc., in proportions of 65 casein, 24 minerals, 1 petroleum. It is waterproof, fireproof, resists moulds and bacteria, and sets like cement. It is extensively used in plywood and aircraft, and must be used on the day it is mixed. The casein is obtained from milk by means of acids or rennet, which produces a curd. ALBUMEN GLUES, or cements, are from egg and blood sources, and are waterproof. MARINE GLUE. This consists of naphtha, indiarubber, and shellac in proportions of about 34, 4, 64. Another kind is made from resin, turpentine, and ordinary glue. WATERPROOF GLUE. To ordinary glue add 2 per cent. bichromate of potash, or chrome alum, or chromate of lime, or 1 per cent. formaldehyde liquid. FLEXIBLE GLUE. Add glycerine and sugar in proportions of 50, 46, 4 ; or 1 oz. of nitric acid to $\frac{1}{4}$ lb. of glue. FILM GLUE. This is made from synthetic resin and is waterproof. It is used for plywood, aircraft, tennis racquets, etc. Cellophane-like sheets of cold glue are placed between the wood laminations and the whole is heated in a press up to 260° C., which fuses the glue. RESIN GLUES. These are synthetic, and may be in two parts : a paste-like glue and a liquid hardener. When the two are placed together chemical action takes place and the glue sets hard. The time required depends upon the hardener. Other types are in powder form, consisting of resin and casein. Glue and hardener are mixed together just before use. Phenol-formaldehyde glues require heat and pressure. Urea-formaldehyde glues can be hot-press or cold-press. Resin glues have superseded other glues for aircraft, superior plywood, etc. See *Plastic Glues*. OIL-SEED GLUES. These are made from seeds : cotton seed, peanuts, soya beans, etc. Carbon bisulphide is added to increase water resistance. These glues are used extensively for plywood made from softwoods. VEGETABLE GLUES. These are in the form of pastes and obtained from flour, rice, starch, and from resins or gums. They are used for attaching other materials to wood.

Proprietary synthetic glues are now firmly established and fulfil the most exacting requirements. There is a great variety and new ones are constantly being introduced. Many of them require special presses, machines, etc., and some of them are chiefly used for veneering, such as the vacuum bag, " hot-water bottle," etc. The former consists of an airproof bag from which the air is extracted so that the bag collapses and squeezes the veneer to the core, which may be of any contour. In the " hot-water bottle " method the piece to be veneered is placed in the machine with the veneer glued and taped in position and resting in a long rubber bag. The whole is cramped in position by means of pressure bars and placed in the autoclave. See *Moulding* (3) and *Plastic Glues*.

Glued and Blocked. A term applied to angle joints that are glued and strengthened by wood blocks. See *Blockings*.

Glue Kettle. A receptacle consisting of an inner and an outer iron pot for heating glue. The water in the outer pot ensures that the glue in the inner pot will not attain a temperature above boiling point, otherwise the glue will burn. The inner pot should be of copper alloy or iron lined with tin. Iron will discolour the glue and stain many hardwoods. GLUE HEATERS. These may contain water, heated by gas, or they may be heated by steam. Usually there are several inner pots varying in size.

Glue Spreader. A machine for applying glue to large surfaces.

Glue Stains. These can usually be removed by wiping over with dilute oxalic acid.

Gluta. *Gluta spp.* (*H.*). India. Brownish red, orange to black streaks. Hard, heavy, durable, and stable, but brittle. Dull, with fairly coarse texture, but smooth. Interlocked grain, and difficult to work, but finishes well for polishing. Sometimes with fine chalky deposit. Seasons well. Liable to insect attack. One of the finest and most beautiful of Indian woods. Used for cabinet work, superior joinery, turnery, carving, etc. S.G. ·82 ; C.W. 4·5.

Glyphs. Channels, or grooves, used as a decorative feature. See *Triglyph*.

Gmelina. See *White Beech, Gamari,* and *Gumhar*.

Gnarl. A twisted knot.

Gnetum. Species of Gymnosperms with vessels.

Goal Posts. A frame consisting of posts and rail into which the ball is driven in the games of football and hockey.

Godroon. A carved ornament in the form of a cable or bead.

Going. The horizontal distance between the faces of consecutive risers in a flight of stairs, usually about 10 in., or rise × going = 66. The term is also applied to the horizontal distance between the first and last risers of a flight of stairs. See *Stairs* and *Step*.

Going Rod. A rod used when setting out the going of a flight of stairs.

Gold Birch. *Betula lutea.* See *Birch*. Also see *Queensland Greenheart*.

Gold Coast. Applied to several W. African woods to denote the source of origin. See *Sapele, Mahogany,* and *Walnut*.

Golden Chinquapin. *Castanopsis chrysophylla* (*H.*). U.S.A. Light brown, tinged or striped with pir..:. Fairly hard, moderately heavy, strong, and stable, not durable. Similar characteristics to birch and used for the same purposes. S.G. ·68 ; C.W. 2·5.

Golden Oak. A bright golden-brown colour in oak due to attack by the fungus *Eidamia catenulata*.

Golden Spoon. See *Spoon*.

Goldfields Blackbutt. *Eucalyptus clelandi* (*H.*). W. Australia. Pale brown. Very hard and strong, with straight grain. A useful timber but not exported. See *Eucalyptus* and *Blackbutt*.

Goldwood. See *Vinhatico*.

Goncalo Alves. *Astronium fraxinifolium* (*H.*). C. America. Called Zebrawood and Kingwood. Very varied in colour. Light to dark brown with very dark stripes ; becomes more uniform with exposure. Hard and heavy ; but considerable variation. Very durable and strong. Straight and roey grain, with fine texture and beautiful figure. Some specimens very difficult to work. Polishes well. Used for decorative work. S.G. ·9 ; C.W. 3·5 to 5. Also see *Aderno, Aræira, Urunday, Zebrawood*.

Gonioma. See *Kamassi*.

Gooseneck. **1.** A strong wood bar connecting two logging trucks. **2.** A curved iron at the bottom of a slide to check the speed of moving logs. **3.** A hinged device for attaching the boom to the mast of a ship.

Goose-pen. A large hole burned in a tree.

Gophor. A biblical wood identified as cypress or cedar. The name suggests a resinous wood.

Gordonia. See *Kelat*.

Gore. A wedge-shaped piece, as a lune for covering a dome. See *Development* and *Dome*.

Gorge. A cavetto, *q.v.*

Gossweilerodendron. See *Agba* and *Moboran*.

Gossypiospermum. See *Boxwood* (*Maracaibo*).

Gothenburg Standard. A measurement of 180 cub. ft., used in the pit-prop trade.

Gothic. Applied to Early English, Decorated, and Perpendicular styles of architecture prevalent from thirteenth to sixteenth centuries. An important characteristic is the pointed arch. Also see *Tudor*.

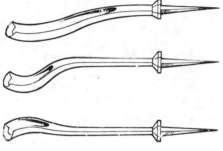

Gouge. A form of chisel with curved cutting edge for circular work. The various types are : bent, carving, firmer, paring or scribing, socket turning. They may be obtained in numerous sizes, like chisels. See *Slip*.

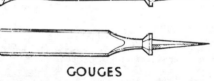

GOUGES

Goupia. See *Cupiuba* and *Kabukalli*.

Grabs. Tongs used in lumbering.

Grading. Arranging in order of size or quality. Sorting timber into different grades, or qualities. Most timbers have their own grading rules, or specifications, agreed upon by the timber associations concerned. See *Shipping Marks* and *Market Forms*.

Grain. The arrangement of the wood elements, especially on the surface of converted wood. The term is loosely used to denote many different characteristics, and the terms *grain* and *figure* overlap in the descriptions. The woodworker generally applies the term *figure* to unusually decorative wood, and *grain* to the normal figuration due to the growth rings. Many of the terms used are now standardised, and the following are described alphabetically in the Glossary : bastard, chipped, coarse, comb, cross, curly, diagonal, edge, end, even, feather, fine, flat, gnarly, interlocked, interwoven, mottled, narrow-ringed, oblique, plain, quarter, quilted, ribbon, rift, roey, short, silver, slash, spiral, straight, torn, twisted, uneven, vertical, wavy, wide-ringed. Also see *Conversion*, *Defects*, *Figure*, *Texture*.

Granadillo. A name given to several S. American woods : cocobola, cocuswood, crabwood, jutahy, partridgewood, quira.

Grand Bassam. See *Mahogany*.
Grand Fir. *Abies grandis.* N.W. America. See *Fir*.
Grand Lahou. See *Mahogany*.
Grand Piano. A large piano with horizontal strings. The plan is shaped somewhat like a harp to suit the different lengths of the strings.
Grandstand. An elevated structure, usually covered, on a sports ground, etc., from which a good view of the proceedings may be obtained.

GRAPE & LEAF ORNAMENT

Grape. See *Sea Grape*.
Grape and Leaf. An ornamentation characteristic of the Perpendicular style of architecture.
Grape Fruit. *Citrus grandis* (*H*.). U.S.A. and W. Indies. The wood is similar to lemonwood and orange.
Graphics or **Graphic Statics.** A method of solving problems in mechanics dealing with forces, moments, etc., by means of scale drawings instead of by calculations. It is based on the triangle, parallelogram, and polygon of forces, and the link polygon. See *Stress Diagram*.
Graphium. Dutch-elm disease, *q.v.*
Grapia. Ibera-pere, *q.v.*
Grapples. Two dogs connected by a chain for joining logs end-to-end for skidding.
Grasshopper Gauge. A pencil gauge for double-curvature work and for gauging over obstructions. See *Gauge*.
Grass Tree. Australian trees with grass-like foliage, named *Blackboy* and *Kingia*. They are of no value as timber trees, but provide useful by-products.
Grating. 1. A timber grillage, *q.v.*, for foundations on poor soil. 2. A framework of bars protecting an opening. 3. A lattice arrangement to keep a gutter free from snow, leaves, etc. 4. A number of struts arranged rather closely together between the beams of a ship. 5. A floor with square or rectangular perforations to allow for the free passage of water,

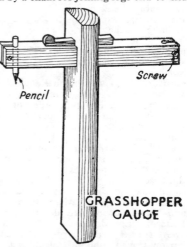

Screw

Pencil

GRASSHOPPER GAUGE

as used on board ship. It consists of rails checked across each other. See *Dodger*. 6. A continuous table round a greenhouse, for plants. The top is formed of 3 in. × 1 in. stuff spaced about 1 in. apart.
Gravel Plank. A strong board laid on edge along the bottom of a

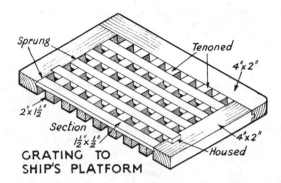

GRATING TO SHIP'S PLATFORM

wood fence to keep the ends of the boards, or palings, above the ground.

Graving Dock. See *Docks*.

Gravity. See *Centre of Gravity* and *Specific Gravity.*

Gray. The name of the botanist A. Gray.

Great Maple. *Acer pseudoplatanus.* Sycamore, *q.v.*

Great Sugar Pine. See *Sugar Pine*.

Great Weevil. The elephant beetle, *q.v.*

Grecian Fret. See *A la Grecque* and *Fret*.

Grecian Mouldings. See *Mouldings*.

Green. A term applied to unseasoned wood, *i.e.* containing free moisture in the cell cavities. Also see *Dyes* and *Polychromatic*.

Green Ash. *Fraxinus pennsylvanica lanceolata.* See *Ash*.

Green Ebony. *Brya ebenus.* Cocus, *q.v.*

Greenheart. *Nectandra rodiæi* or *Ocotea r.* (*H*.). British Guiana. Three varieties : white, black, and brown. Light to dark olive-green, often intermingled, lustrous. Very hard, heavy, strong, elastic, and durable. Very resistant to insect attack probably due to alkaloids (bebeerine). Straight grain, smooth, rather coarse, and porous on end grain. Easily split, and splinters are poisonous. Seasons readily, but inclined to check, and very free from defects. Probably the best timber for marine work. Used for marine structural work, shipbuilding, piling, brewer's stagings, stair treads, fishing rods, or where strength and durability are of primary importance. S.G. 1 ; C.W. 5. See *Musiga* and *Ekhimi*.

Greenhouse. 1. A house for the cultivation of plants. It consists chiefly of glass and wood bars. See *Conservatory*. **2.** The projecting gun turret of bomber aircraft.

Green Oak. An unnatural colour in oak due to the action of a fungus. The value of the wood is increased for decorative purposes.

Green-oil. A heavy grade of coal-tar creosote.

Green Room. A lounge or retiring room near the stage of a theatre, etc., for the use of the artistes.

Green Satinheart. See *Wilga*.

Grees. A staircase, or the separate steps of a flight of stairs.
Grevillea. See *Silky Oak*.
Grewia. See *Dhaman*.
Grey. An adjective distinguishing species of alder, elm, gum, iron-bark, poplar, etc.
Grey Birch. *Betula populifolia*. See *Birch*. Also *Bridelia exaltata* (*H.*). Queensland. Grey to chocolate, smooth, lustrous. Hard, heavy, moderately strong and durable, stable, fissile, somewhat like European birch. Straight grain, fine close texture, fine ray figure. Easily wrought and seasons readily. Suitable for construction, superior joinery, carriage building, turnery, etc. S.G. ·75 ; C.W. 3.
Grey Box. *Eucalyptus hemiphloia* (*H.*). Victoria and S. Australia. Also called Box Gum. Yellowish white to greyish brown. Very heavy, hard, strong, and durable. Interlocked grain. Resists insects. Used for structural work, piling, wagonbuilding, sleepers, etc. S.G. 1·1. *E. spenceriana*, W. Australia, is a similar wood, but darker and with white fibre-like markings. Used as a substitute for coolibah.
Grey Boxwood. *Hemicyclia australasica* (*H.*). Queensland. Also called Grey Marara, Yellow Tulipwood. Buff grey, lustrous. Close grain, even texture. Used for the same purposes as box (*Buxus*). S.G. ·85.
Grey Corkwood. *Erythrina vespertilio* (*H.*). Queensland. Very light and soft, but tough. Not strong or durable. Straight, stringy grain. Requires careful seasoning. Used for floats, surf boards, etc. Not exported. S.G. ·3.
Grey Fir. Western hemlock, *q.v.*
Grey Gum. *Eucalyptus propinqua* (*H.*). New S. Wales. Reddish. Very heavy, hard, strong, tough, and durable. Close grain. Gum pockets and shakes prevalent. Used for structural and constructional work, poles, sleepers, etc. S.G. 1.
Grey Handlewood. *Aphananthe philippinensis* and *Pseudomorus brunoniana* (*H.*). Queensland. Also called Native Elm and Rough-leaved Hickory. Creamy white or grey. Hard, heavy, very tough and elastic, not durable, smooth. Requires careful seasoning. Excellent wood for shafts, levers, clubs for games, mallets, etc. S.G. ·7.
Grey Pine. Jack pine, *q.v.*
Grey Poplar. *Populus canescens*. See *Poplar*.
Grey Teak. The Queensland name for white beech, *q.v.* It belongs to the same botanical family as Burmah teak.
Greywood. Figured sycamore dyed and used for cabinet work, etc. Also see *Silver Greywood* and *Harewood*.
Gribble. See *Marine Borers*.
Grid. 1. A room with open floor above the stage of a theatre, from which the lighting and scenery are controlled. 2. A framework of heavy timbers, above low-water level, to receive a vessel at high tide for repairs. 3. See *Grating*.
Grillage. A foundation consisting of layers of heavy timbers or mild steel joists laid at right angles to each other.
Grille. A grating, or lattice, protecting a window or other opening.
Grinding. Correcting the bevel of a cutting tool preparatory to sharpening. There are various types of revolving stones and wheels

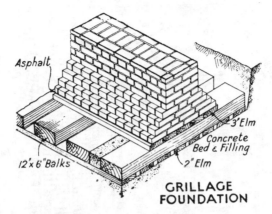

Asphalt

3' Elm

Concrete
Bed & Filling

12' x 6" Balks

2" Elm

**GRILLAGE
FOUNDATION**

for the purpose, and different abrasive materials are used, such as mill stone, emery wheels, carborundum wheels, etc.

Gripe. See *Forefoot* and *Stem and Head*.

Gripper. A spiked lever for gripping short logs to the chariot, or slide, for cross-cutting for firewood, etc., on a circular saw.

Griseb. Abbreviation for the name of the botanist Grisebach.

Griting. Teruntum, *q.v.*

Groin. The salient at the intersection of vaults, *q.v.*

Groove. A long narrow channel. See *Tongue and Groove* and *Matching*.

Gross Features. A term used in identification for the characteristics of wood that can be distinguished with little **GROOVES** laboratory assistance, *i.e.* weight, colour, odour, taste, grain, rays, resin content, etc. See *Identification*.

Ground Ash. American white ash. See *Cane Ash*.

Grounded Work. Joinery work fixed to grounds.

Ground Floor. The floor of a building near to the outside ground level.

Ground Mould. The invert mould for the concrete for a drain or tunnel.

Ground Plate. The lowest horizontal member of the framing for a timber building, into which the posts are framed.

Grounds. 1. Strips of wood fixed to a wall by means of plugs, etc., to which joinery work is secured. They are usually about 3 in. $\times \frac{3}{4}$ in., unwrought, and flush with the floating coat of plaster. Larger sizes are often framed. See *Double Skirting* and *Finishings*. 2. Backings, or cores, for veneers. (See illus., p. 226.)

Growth. The rate of growth varies considerably even in the same species of tree. The factors governing growth are : soil, position, and climate. There is also considerable variation at different periods

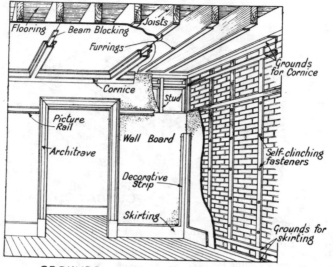

Flooring — Beam Blocking — Joists — Furrings — Grounds for Cornice — Cornice — Stud — Picture Rail — Wall Board — Architrave — Self-clinching fasteners — Decorative Strip — Skirting — Grounds for skirting

GROUNDS FOR WALL-BOARDS

in the life of the tree. In pitch pine, for instance, the growth rings may vary from 4 to 40 per inch.

Growth Rings. The rings shown on the cross-section of wood. They denote the periodic growth of the tree. See *Annual Rings*.

Groyne. A structure projecting into flowing water to hold it back, or to prevent scour, or to retain shingle.

Grub. The larva of an insect. See *Larva*.

Grub Holes. Holes in wood from $\frac{3}{8}$ to 1 in. diameter, due to larvæ or adult insects. They may be circular, oval, or irregular.

Grumixava. *Eugenia brasiliensis* (*H.*). Brazil. Yellow, with light reddish spots. A decorative wood for certain classes of cabinet work, but not yet imported.

g.s.m. Abbreviation for *good sound merchantable*.

Guacima. *Guazuma spp.* (*H.*). C. America. Pinkish. Moderately light, but tough and strong for its weight. Not durable. Fairly coarse texture, with silver grain. Suggests elm in some respects. Used for interior construction, dry cooperage, gunstocks, charcoal. S.G. ·6. Species of *Muntingia, Lætia, Prockia, Luehea* are also called Guacima.

Guaiacum. See *Lignum-vitæ*.

Guambo. *Phœbe spp.* (*H.*). C. America. Laurel family. Dark yellow-brown, lustrous. Moderately heavy and durable, and stable. Roey and silver grain. Rather coarse texture, with cross-grain, but smooth and fairly easy to work. Useful wood for most purposes. S.G. ·5 ; C.W. 3.

Guanacaste. *Enterolobium spp.* Eight species (*H.*). C. and N. America. Also called S. American Walnut and Rain Tree. Variegated walnut-brown, lustrous. Soft to moderately hard, fairly durable. Straight and roey grain. Medium to coarse texture. Hard specimens resemble walnut, and polish well. Easily wrought. Useful wood for cheap cabinet work, etc. S.G. ·36 to ·6 ; C.W. 2·5.

Guapaque. Paque, *q.v.*

Guapinole. *Hymenæa courbaril* and *Ebenopsis spp.* See *Locust.*

Guarabu. *Terminalia acuminata* (*H.*). Brazil. Light purple, lustrous. Hard, heavy, strong, durable. Very close grain with fine rays, and smooth. Similar to, and often marketed as, purpleheart, *q.v.* Rather difficult to work. Used for structural and constructional work, cabinetmaking, planking, etc. S.G. 1 ; C.W. 4·5.

Guaranta. Larangeira, *q.v.*

Guard. 1. An appliance on a machine or saw to protect the machinist against injury. There are many variations and patents that comply with Home Office regulations. See *Saw Bench.* 2. A protection to an opening on a scaffold. See *Gantry.* 3. See *Table with Guard.*

Guard Bead. See *Inner Bead.*

Guard Rail. Any form of rail that affords protection from machinery, falling from a scaffold, etc. A handrail to a landing or balcony.

Guarea. See *Marinheiro. Obobonekhui, Obobonufua, Trompillo.* Also see *Sapele.*

Guatacre. *Lecythis lævifolia* or *Chytroma sp.* (*H.*). Brazil. Light brown. Very hard, heavy, strong, durable, and resists insects. Difficult to season and work. Used for structural work, piling, sleepers. S.G. 1·1. Also called Guatequero.

Guatambu. *Aspidosperma tomentosum* (*H.*). C. America. Also called Lemonwood, Amarello, and Piquia Peroba. Canary yellow, lustrous. Moderately heavy to heavy, hard and strong, but not durable. Dense, with fine texture and straight grain. Easy to work and carve, and polishes well. Used for construction, flooring, cabinet work, turnery. S.G. ·7 ; C.W. 2·5. Also *Balfourodendron sp.*

Guatemalan Mahogany. C. American mahogany, *q.v.*

Guatteria. Lancewood, *q.v.*

Guayabi. *Patagonula americana* (*H.*). C. America. Variegated brown to purplish black. Very hard, heavy, strong, tough, durable, and resilient. Straight grain and rather fine texture. Fairly easy to work, and polishes. Used for implements, vehicles, furniture, superior joinery, turnery, etc. S.G. ·9 ; C.W. 3·5.

Guayabo. Nargusta, *q.v.* Several other species are called Guayabou.

Guayabou. *Mouriria guianensis* (*H.*). Venezuela. Like pauji, *q.v.*, but darker, harder, heavier, and more difficult to work.

Guayacan. *Cæsalpinia melanocarpa* (*H.*). S. America. Rich brown. Very hard, heavy, strong, and durable, with close grain. Used for paving, tools, wheelwright's work, etc. S.G. 1·1. The following woods are also called Guayacan : *Curupay, Urunday, Partridgewood, Lapacho, Lignum-vitæ,* and *Vera.*

Guayamero. *Brosimum columbianum* (*H.*). C. America. Whitish to greyish yellow. Hard, heavy, strong, but not durable. Straight grain, medium texture. Smooth, and not difficult to work. Suggests hickory, for which it is a substitute. Also called Cow Tree because of latex in bark. S.G. ·8 ; C.W. 3.

Guazuma. See *Guacima*.

Gudgeon. **1.** See *Band and Gudgeon*. **2.** A metal end to a wood axle. **3.** Irons with vertical holes to hold the pintles of a rudder.

Guelder Rose. *Viburnum opulus* (*H.*). Britain. The wood is hard and compact, but it has very little commercial value except for turnery.

Guettarda. See *Glassywood*.

Guiana. The important timbers exported to this country include Crabwood, Greenheart, Mora, Purpleheart, and Wallaba.

Guide Bead. See *Inner Bead*.

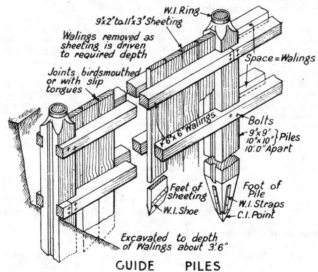

GUIDE   PILES

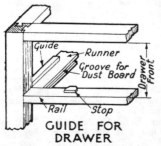

GUIDE FOR
DRAWER

Guide Pile. Piles placed at regular intervals and connected by walings or horizontal runners. The walings control, or guide, the sheet piling filling the spaces between the guide piles. The term *guide* is sometimes applied to those placed at salient points, and the intermediate piles called *gauge* piles.

Guides. An arrangement to control anything that slides, as a drawer or a lift.

Guijo. *Shorea guiso* (*H.*). Philippines. Lauan family. Brownish red, fairly lustrous. Hard, heavy, strong, but not durable. Handsome ribbon figure and fairly fine texture. Large sizes. Rift-sawn wood best for seasoning and figure. Stains and polishes well and not difficult to work. Used for cabinet work, superior joinery, flooring, vehicles, etc. S.G. ·8 ; C.W. 3·5.

Guill. Abbreviation for the name of the botanist Guillemen.

Guilloche. An ornament consisting of carved intertwining bands on large mouldings.

Guillotine. A trimmer, *q.v.* A large knife for trimming and squaring veneers.

Guinea Wood. New Guinea Walnut, *q.v.*

Gullet. The space between two consecutive saw teeth. See *Saw Teeth.*

Gulley Gum. *Eucalyptus Smithii.* Victoria, Australia. Difficult to work and of little commercial importance. See *Gum.*

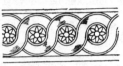

GUILLOCHE

Gum. 1. The name given to numerous smooth-barked species of Australian Eucalypti. They are mostly excellent woods and suitable for most purposes, especially where strength and durability are important factors. Gum veins detract from the value. They are generally very hard, heavy, durable, strong, resilient, and tough. When with wavy and interwoven grain they provide a valuable decorative wood. S.G. ·85 to 1·1 ; C.W. about 3·5. Blue Gum is probably the most valuable, but the following are all important. NEW S. WALES : Blue, Flooded, Grey, Murray Red, Spotted, and Sydney Blue Gums. QUEENSLAND : About forty different woods are called Gum, with a distinguishing adjective. Some of them are not Eucalypts. Those of commercial importance are described alphabetically. S. AUSTRALIA : Box (similar to Peppermint), and Red Gums. TASMANIA : Blue, Peppermint, Swamp, and White Gums. VICTORIA : Blue, Forest Red, Gully, Manna, Mountain Grey, Peppermint, River Red, Shining, Spotted Blue, Sugar, Swamp, and Yellow Gums. W. AUSTRALIA : Desert, Flooded, River, Ridge, Salmon, Spotted, Swamp, and York Gums. Also see' *Red Gum, Hog Gum,* and *Tupelo Gum.* 2. A juice that exudes from certain trees and that hardens on exposure, but is usually soluble in water. *Bush Gum* is from standing trees. *Fossil Gum* is dug from the earth.

Gum-anime. See *Locust* (*W. Indian*).

Gum Arabic. See *Babul.*

Gum, Black. *Nyssa sylvatica* (*H.*). U.S.A. Pale to greyish-brown. Fairly hard, heavy, tough, and strong. Inferior timber and little exported. Also see *Tupelo.* S.G. ·64.

Gumbo Limbo. *Bursera gummifera* (*H.*). C. America. W. Indian Birch or Gum-elemi. Light brown. Light, soft, fairly strong, not durable. Firm and tenacious. Straight and irregular grain. Medium texture and easily wrought. Used for general purposes. S.G. ·45 ; C.W. 2.

Gum Ducts. Intercellular canals in dicotyledons, especially of the Dipterocarpaceæ family. See *Resin Ducts.* The ducts in hardwoods contain various substances : gum, resin, oil, etc.

Gum-elemi. A fragrant stimulant resin used for ointments, etc. It is obtained from trees of the Burseraceæ family : *Boswellia, Canarium, Protium,* etc.

Gum Galls. A defect in wood consisting of concentric tubes of gum, and common to the Myrtaceæ family, especially Eucalyptus.

Gumhar. *Gmelina arborea* (*H.*). India, Burma. Now called Gamari. Also called Yamane. Light brownish yellow, lustrous, often variegated. Moderately light and soft, but durable. Fairly coarse grain, but smooth, and feels oily. Roe and mottle. Grain and texture similar to some mahoganies. Easily wrought. Suitable for cabinet work, superior joinery, vehicles, musical instruments, carving, etc. S.G. ·6 ; C.W. 3.

Gummer. A tool for *gumming*, or cutting out, the throats of a saw.

Gum Veins. An accumulation of gum or kino in the form of a vein or ribbon between the growth rings. It may be broken at intervals by wood tissue. The defect is common to many deciduous trees.

Gumwood. American Red Gum. See *Red Gum.*

Gun or **Gunning Stick.** A sighting device used for felling a tree in a required direction.

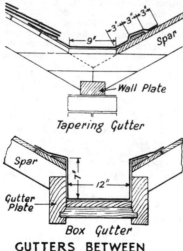

Tapering Gutter

Box Gutter

GUTTERS BETWEEN LEAN-TO ROOFS

Gun Stock. 1. Pieces of wood tapering in width and thickness. The term is used in the firewood trade. 2. The wood part of a rifle which is usually of walnut. 3. See *Diminished Stile.*

Gunwale or **Gunnel.** 1. The upper edge of the side of a ship. The timber often houses the top strake of planking. The line of junction between deck and side of a ship.

Gurjun or **Gurgan.** *Dipterocarpus turbinatus, D. alatus, D. costatus, D. incanus* (*H.*). India, Burma, Andamans. Dull reddish brown. Fairly hard, heavy, and strong. Durability varies with species, and it should be treated for exterior work. Handsome, close, even grain. Best when rift-sawn. Smooth, and not difficult to work. Large sizes. Polishing sometimes difficult due to exuding oil and resin, which forms an excellent polish of itself if rubbed well with a cloth wetted with spirits. Used for flooring, superior joinery, carriage-building, etc. S.G. ·75 ; C.W. 4.

Gürke. The name of the botanist M. Gürke.

Gussets. Angle stiffeners to prevent the distortion of framing.

Guttæ. Small conical-like drops used as ornaments. See *Doric* and *Drop.*

Gutta-percha. See *Sapodilla*.

Gutter. A channel for conveying water. The various types used on roofs are : box, eaves, parapet, saddleback, secret, tapering, trough, valley, and vee. They may be of wood, iron, lead, zinc, copper, or asphalt, etc. See *Parapet Gutter* and *Skylight*.

Gutter Bearer. Short pieces, about 2 in. × 2 in. section, supporting the gutter boards over which lead is laid. Any form of bearer carrying a gutter. See *Box Gutter*.

Gutter Boards. 1. See *Gutter Bearer* and *Box Gutter*. 2. See *Snow Boards*.

Gutter Plate. 1. A beam under a lead gutter. 2. The sides of a box gutter, carrying the feet of the rafters.

Guttiferæ. A family of trees including Calophyllum, Caraipa, Clusia, Kayea, Mammea, Mesua, Ochrocarpus, Platonia, Pœciloneuron, Rheedia, Symphonia, Vismia.

Guy. A rope stay, as to a pair of sheer legs or to a mast.

Gymnanthes. See *Crabwood* and *Sandaleen*.

Gymnasium. A hall equipped with appliances for physical exercises.

Gymnocladus. Kentucky Coffee Tree, *q.v.*

Gymnosperms. Trees bearing naked fruit or seeds. The commercial timbers in this classification are conifers, or softwoods, which include two families : PINACEÆ, which is divided into the sub-families *Araucariaceæ*, *Abieteæ*, *Taxodiaceæ*, and *Cupressaceæ* ; and the family TAXACEÆ which is divided into *Podocarpaceæ*, *Cephalotaxaceæ*, and *Taxeæ*.

Gyn. Same as sheer legs, *q.v.*

Gynocardia. See *Lemtan*.

Gyo. Gausam, *q.v.*

H

Hacienda. A farm or farmhouse.

Hack. 1. To cut or roughen irregularly or carelessly. 2. A grating on which things are dried. A rack for cattle. 3. See *Hack Cap*.

Hackberry. *Celtis occidentalis* (*H.*). U.S.A. Yellowish grey to light brown, with yellow streaks. Subject to blue sap stain. Moderately hard and heavy. Not strong but has good elastic and shock-resisting properties, rather fissile. Fairly stable and durable. Straight, some interlocked, grain, distinct rays. Not difficult to season or work. Usually marketed with elm or white ash. Used for furniture, sports equipment, boxes, crates, etc. S.G. ·5 ; C.W. 2·5. Other species from W. Indies are similar, but harder and heavier. Also see *Sugarberry*.

Hack Caps. Small roof-shaped wooden structures to cover hacks of drying bricks, to protect the bricks from sun and rain.

Hackia. *Siderodendron triflorum* and *Tabebuia spp.* (*H.*). S. America. Characteristics and uses as lignum-vitæ. S.G. ·95.

Hacking. Same as *Hack* (1).

Hackmatack. N. American larch or tamarack, *q.v.*

Hackney Carriage. A horse-drawn vehicle plying for hire.

Hadang. *Cordia macleodii* (*H.*). Indian. Light to olive brown, darkening with exposure, and darker veins. Hard, strong, tough, and durable. Fairly lustrous with smooth, oily feel. Distinct rays. Straight and wavy grain, with coarse uneven texture. Rather difficult to work. Used for furniture, wheelwright's work, tool handles. Suitable for superior joinery and cabinet work. S.G. ·7 ; C.W. 4.

Hæmatoxylon. See *Logwood*.

Haffit. 1. The end of a pew or church seat. 2. The fixed part of a cover to which the opening part, or lid, is hinged. See *Hall Seat*.

Haft. The handle, or shaft, of a striking tool, etc.

Hagg. Branches, brushwood.

Ha-ha. 1. A sunk fence. A fence in a ditch. 2. Punggai, *q.v.*

Hair Bracket. A moulding at the back of the figurehead of a ship.

Halban or **Haleban.** Leban, *q.v.*

Haldu. *Adina cordifolia* or *Nauclea c.* (*H.*). India and Burma. Light yellow to yellowish brown. Darkens with exposure. Moderately hard, heavy, strong, and durable, and resists insects. Straight or interlocked, fine grain, even texture. Grain liable to rise, should be damped before papering. Stains and polishes readily. Seasons fairly well and not difficult to work. Used for furniture, superior joinery, planking, turnery, carving, etc. S.G. ·65 ; C.W. 3·5.

Halesia. See *Snowdrop Tree*.

Half Assinie. W. African mahogany, *q.v.*

Half Blind Dovetail. A lap dovetail, *q.v.*

Half Butt. A billiard cue of intermediate length.

Half Deck. A deck over part of a ship only.

Half Lap. See *Halving*.

Halfordia. See *Saffron Heart*.

Half Pace. 1. See *Half Space*. 2. A part of a floor raised above the surrounding floor as to a fireplace, bay window, alcove, etc.

Half Pitch. Applied to a roof in which the rise is equal to half the span.

Half Rip Saw. A rip saw with slightly smaller teeth than usual.

Half Round. 1. A method of cutting veneers, similar to rotary cutting, but a wider sweep is obtained by means of a stay log. The method gives a greater variation of figure. See *Veneers* and *Rotary Cutting*. 2. A semicircular moulding.

Half Space. Applied to a landing or winders extending across two flights of stairs, or the full width of a staircase. See *Dog Leg* and *Geometrical Stairs*.

Half Stuff. See *Wood Pulp*.

Half Timber. A portion of a balk, or heavy square timber, cut longitudinally into halves.

Half Timbered. 1. Applied to timber-framed buildings in which the framing does not go through the full thickness of the wall, but is backed by an inner lining of brick or stone and which adds to the stability of the structure. See *Timber Framed*. 2. Framed timber buildings with the spaces, or panels, filled with other building materials.

HALF-TIMBERING

Half Timbers. The timbers in the framework of a ship towards the stem. They compare with the lower futtocks in the body of a ship.

Hall. An entrance lobby or vestibule. A large room for public meetings and entertainments. A collegiate dining-room. A large country house.

Halmilla or **Halmileel.** Trincomalee, *q.v.*

HALL SEAT

Halved. See *Halving*.
Halved and Mitred. An angle halving joint with the face mitred.
See *Bar Joints*.
Halved Laminated Joint. A cabinet-maker's bridle joint.
Halving. Removing half of the material from each piece in a cross
joint so that both surfaces are flush. There are many variations.
Hamamelidaceæ. The witch hazel family. Includes : Altingia,
Bucklandia, Corylopsis, Distylium, Fothergilla, Hamamelis, Liquid-
ambar, etc.
Hamillila. Trincomalee, *q.v.*
Hammer. **1.** A striking tool. The various types used by wood-
workers are : ball-pane, Canterbury, claw, cross-pane, Exeter, framing,
or heavy, pin, veneer, Warrington. There are many other types
peculiar to different trades. **2.** The mechanism that strikes the
wires in a piano. **3.** An auctioneer's mallet.

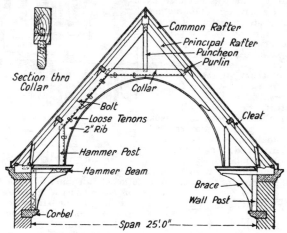

HAMMER-BEAM ROOF TRUSS

Hammer Beams. The short cantilevers at the feet of the principal
rafters in a hammer-beam roof truss. Also see *Angel Beam*.
Hammercloth. A decorative driving seat to a carriage.
Hammer Head Key. A hardwood key increased in width at the
ends to form a seating for wedges, to tighten an end-to-end joint.
It is commonly used in circular work instead of a handrail bolt.
Hammering. **1.** Pressing down veneers with the veneer hammer,
q.v. **2.** Correcting a saw that is buckled.
Hammer Marks. **1.** Distinguishing marks on timber. See *Shipping
Marks*. **2.** An accidental bruise on the face of wood. It can usually
be corrected by wetting the bruised area.

Hammer Post. The post springing from the hammer beam in a hammer-beam roof truss. See *Angel Beam* and *Hammer Beams*.

Hammock. **1.** Dense forest with undergrowth of creepers, etc. **2.** A canvas or string bed suspended by cords at the ends. It is used on board ship, in gardens, etc.

Hammock Chair. A deck chair, *q.v.*

Hamper. **1.** Applied to the top part of a tree that is of no use for timber. **2.** A strong basket with lid. **3.** Applied to things that obstruct or impede.

Hance or **Hanch.** Haunch, *q.v.*

Hand. A term used to denote on which side a door, etc., is hinged. The door is viewed from the side showing the knuckles of the hinges and named right or left hand hung accordingly. There is considerable confusion in the meaning of left-hand and right-hand lock, but the manufacturers of door furniture describe them as in the illustration. If the

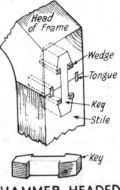

HAMMER HEADED KEY JOINT

door opens away from the operator with the lock on the right hand and the bevel of the bolt, or latch, facing the operator, it is a right-hand lock.

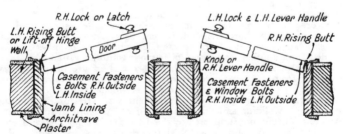

HAND OF LOCKS & FURNITURE

Hand-barrow. **1.** A small platform with handles at each end for conveyance by two men. **2.** Two light poles with rungs, like a ladder, for transporting bark, etc., in lumbering.

Hand-cart. A small two-wheeled cart pushed by hand.

Handed. Arranged in pairs, left and right hand.

Hand Feed. Applied to machines in which the stuff is fed by hand instead of by rollers.

Hand Hook. A special hook used for lifting one end of planks or deals when stacking, etc. See *Hicking*.

Handicraft. Skilled work performed by the hands, without machines. The term is usually applied to the educational training of children in manual dexterity and the use of tools.

Handlewood. See *Grey Handlewood.*

Hand Masts. Applied to poles and round timbers from the Baltic, between 24 and 70 in. circumference.

Hand Pike. A wood lever, about 7 ft. long, with metal point, for controlling floating logs, etc.

Hand-post. A directing sign-post.

Handrail. The top member of a balustrade, *q.v.* A guide for the hand in a staircase, etc. See *Newel.*

Handrail Bolt. A double-ended bolt for joining two pieces end-to-end, mitres, etc. See *Bolt.* The two pieces are bored endwise so that the bolt is half-way in each piece. Holes are cut, at right angles to the bolt, for the nuts, one of which is turned by a screwdriver, or similar lever, to tighten the joint.

Handrailing. The art of preparing handrails, especially when the work entails wreaths for geometrical stairs. Various methods are used : bevel cut, falling line, normal sections, tangent, etc. The work requires considerable skill and geometrical knowledge.

Hand Saw. Usually applied to the joiner's cross-cut saw, but may imply any kind of saw used by hand, as distinct from machine saws : block, bow, coping, dovetail, fret, keyhole, pad, rip, and tenon saws. See *Saws.*

Hand Screws. A form of wooden cramp operated by two threaded spindles. See *Caul* and *Shooting Boards.*

Handspike. A wood lever for moving heavy objects. The spoke of a capstan, etc.

Hangar. A shelter or building to house aircraft.

Hangers. 1. Strong vertical timbers to support the walings in shaft excavations. They are suspended from balks across the mouth of the excavation. 2. Any form of strap, by which one member is suspended from another, as ceiling joists from common rafters. 3. Iron shoes or straps by means of which cross members are suspended, as a binder from a beam, or bearers for shuttering, or eaves gutters from the rafters. See *Framed Floor.* 4. A wooded slope.

Hanging. 1. Hinging a door or casement. 2. Attaching the cords to a vertically sliding sash, clothes-airer, etc. 3. Applied to fitments fixed to a wall and not supported by the floor, as a *hanging cupboard,* etc.

Hanging Bar. A rail from which sliding doors are suspended.

Hanging Knee. One with its long arm downwards, in shipbuilding.

Hanging Sash. A vertically sliding sash. See *Sash and Frame.*

Hanging Stile. The stile of a door or casement receiving the hinges.

Hanging Wardrobe. 1. A single-carcass wardrobe, without drawer at the bottom. 2. A shelf with clothes-hooks, and with hanging drapery to protect the contents from dust.

Hanks. Wood rings fixed on the stays of a ship.

Hanoa. See *Uqueopokin.*

Hansom. An obsolete type of two-wheeled hackney carriage.

Haploromosia. See *Akoriko.*

Harbour Dues. The charges for anchoring a ship in a harbour.

Hard Beech. *Nothofagus truncata.* Clinker beech, *q.v.*

Hard Birch. *Betula lutea.* See *Birch.*

Hardboard. Wall-board of compressed wood fibre, etc. Pressure up to 2,000 tons is applied, which produces a hard, smooth, weather-proof board easy to cut, bend, and fix in large sizes.

Hard-grained. Wood with dense close grain.

Hard Maple. *Acer saccharum.* See *Maple.*

Hardness. The resistance of wood against indentation and abrasion. The former depends chiefly upon density, but the wearing properties depend also upon other factors : toughness, and the cohesion, size, and arrangement of the fibres. Hardness has little influence upon the durability or resistance to insect attack of wood.

Hard Pear. *Olinia capensis* and *Strychnos henningsi* (*H.*). S. Africa. Yellowish. Hard and tough, and used chiefly for wheelwright's work.

Hard Pines. The more resinous pines : long-leaf, short-leaf, Cuban, loblolly, and pitch pine.

Hard Rot. Brown rot, *q.v.*

Hardwickia. See *Anjan, Kolava*, and *Piney*.

Hardwood. A description applied to woods from deciduous and evergreen broad-leaved trees, or dicotyledons. The term does not infer hardness in its usual sense, but is a classification. Generally, the timber from broad-leaved trees is harder and heavier than that from needle-leaved trees, but there are many exceptions, and the lightest known woods are hardwoods. See *Specific Gravity*. Hardwoods are more complex in structure than softwoods, and consist of vessels, or pores, fibres, and parenchyma. They belong to the botanical group Angiosperms. Important British hardwoods : Ash, beech, birch, elm, chestnut, hornbeam, lime, oak, poplar, sycamore, willow. See *Softwoods, Conifers, Annual Rings, Growth, Light Hardwood*, and *Trees*.

Hardwood Flour. See *Flour*.

Hardy Catalpa. See *Catalpa*.

Harewood. 1. Figured sycamore or maple dyed for cabinet work, etc. The wood is chemically treated with ferrous sulphate or other salts, which react with the wood contents and produce a silver-grey hue, but it fades with age. 2. The name harewood is also applied to San Domingo satinwood, which is a lustrous yellow wood seasoning to silver grey, sometimes with dark green markings. It is very decorative, due to roe and mottle, and used for cabinet work, superior joinery, decorative articles, and veneers. S.G. ·8 ; C.W. 3.

Harmonium. A small organ in which the sounds are produced by reeds, through which air is forced from the bellows, operated by the feet.

Harms. The name of the botanist H. Harms.

Harness Casks. Teak barrels of special shape for holding salt beef on board ship.

Harpings. The fore parts of the wales, round the bow of a ship fixed to the stem.

Harpsichord. An old type of musical instrument from which the piano is derived.

Harpullia pendula. Australian tulipwood, *q.v.* *H. imbricata* is an Indian wood similar to inferior African mahogany, and of little commercial importance.

Harra. See *Myrobalan*.

Hart. Abbreviation for the name of the botanist Hartig.
Haskinizing. A proprietary method of wood preservation.

HASP & STAPLE

Hasp. A hinged and slotted plate, or link, to fit over a staple, and to secure a gate by means of a padlock.
Hat Block. A wood or metal block, or mould, on which a hat is shaped.
Hatch. 1. A cover to an opening in a floor or roof. A sliding trap door. An opening, or the frame and cover of an opening, in a ship's deck. 2. A dwarf door with an opening over it, or a half door. 3. See *Serving Hatch*.

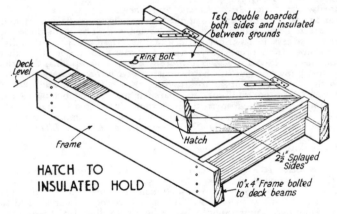

HATCH TO
INSULATED HOLD

Hatch Cover. A portable cover to a hatch in a ship's deck.
Hatched Mouldings. Applied to mouldings having two sets of intersecting lines carved on the surface.
Hatchet. A small axe with short shaft.
Hatchings or **Hachures.** Section lines on drawings.
Hatchway. An opening in the deck of a ship. See *Hatch*.
Hat Rack. A series of pegs or hooks for hats and coats.
Hat Stand. A piece of furniture with hooks or pegs for hats and coats. It may be a post with branched feet or a hall-stand.
Haunch. 1. A small projection due to reducing the width of a tenon. It is formed to allow for wedging near the end of a stile. See *Mortise* and *Tenon*. 2. An increase in the depth of a beam near to the supports, to resist shear. 3. The part of an arch midway between springing and crown.
Haunching. The recess formed to receive the haunch of a tenon.
Hawk. A small board, with handle underneath, for holding plastic materials when pointing, plastering, etc.

Hawse Bolster. Strong pieces round the hawse-holes of a ship. Also called naval hoods.
Hawse-holes. Cylindrical holes on each side of the stem of a ship for the cables.
Hawse Pieces. The foremost timbers in a ship, parallel to the stem.
Hawthorn. *Cratægus oxyacantha* (*H.*). Europe. Also called Whitehorn and May. Yellowish white. Hard, fairly heavy, tough, smooth, with fine close grain. Similar structure to pear. Small sizes. Used as a substitute for boxwood. S.G. ·65 ; C.W. 3·5.

HAWTHORN

Hay-box. An airtight box containing hay or cork dust as an insulation against loss of heat. It is used for retaining the heat in semi-cooked foods so that the process of cooking continues
Hay Loft. A loft above a stable.
Hayne. The name of the botanist F. G. Hayne.
Haywain. A farm wagon for carting hay.

Hazel. *Corylus avellana* (*H.*). Europe. Also called Nutwood. Pinkish white to pale reddish brown, with dark striations. Fine, uniform, straight grain. Soft, elastic, not durable. Of little commercial value. Butts valued for veneers. Branches very tough and pliant and used for cask hoops. Used chiefly for charcoal. S.G. ·62.
Hazel Pine. American Red Gum, *q.v.*
h.b. Abbreviation for *hollow backed*.

HAZEL

H., B., and K. Abbreviation for the names of the botanists Humboldt, Bompland, and Kunth.
h.b.s. Abbreviation for *herring-bone strutting*.
Head. **1.** A horizontal member resting on props, or shores, and supporting a centering. **2.** The top horizontal member of a frame. **3.** A machine cutter-block for a spindle or four-cutter. **4.** The ends of a cask or barrel. **5.** The top or highest part of anything, as *head of stairs*. **6.** The front or foremost part of anything, as *pier head* or *head of ship*. **7.** See *Figure Head* and *Stem and Head*.

Square head or spindle block

Paper packing

Irons or cutters

Bolt

Strut

HEAD ᴏʀ **CUTTER BLOCK**
(See p. 240.)

HEAD BLOCK
(See p. 240.)

239

Head Block. 1. A block bolted to a structural timber to resist a thrust from a strut, etc. The block may be keyed or tabled. **2.** A log under the front ends of the skids in a skidway. (See illus., p. 239.)

Head Board. A board forming the head, or top end, of a bed. A board near a manger to which an animal is tied.

Head-boom. A ship's jib-boom.

Head Chair. One with a head rest.

Headers. Nogs in stud partitions to take the top edges of wall-boards.

Head Gate. A lock gate that admits water.

Heading. Wood prepared for the ends of casks, or wood suitable for the purpose.

Heading Joint. An end-to-end joint, especially for floorboards.

Head Nailing. Nailing shingles near to the top instead of near the middle.

Head Piece. The capping piece of a quartered partition.

Head Post. The post near the manger in a stable partition between two stalls.

Head Rails. The curved rails at the head of a ship. See *Stem and Head*.

Head Room or **Headway. 1.** The clearance above the steps in a flight of stairs. It should not be less than 6 ft. 6 in., vertically over the face of a riser, above the tread. **2.** The clearance above a person's head when passing under any obstacle, as an arch, etc.

Head Saw. The principal breaking-down saw in a saw-mill.

Head Stays. The stays to the foremast of a ship.

Head Stocks. 1. The parts of a lathe between which the material to be turned is fixed. They are distinguished as *fast* and *loose*. **2.** The

Post — Head Tree or Bolster

HEAD TREE

end bars, or rails, in the bottom framing for wagons, etc. Same as ear-bed or ear-breadth. **3.** A wood bearer, usually elm, to which a bell is hung.

Head Tree. 1. A short bearing piece on top of a post. Also called *Bolster*. **2.** A tree to which the cable is attached, in lumbering.

Hearse. A carriage for conveying a coffin.

Heart. 1. The central core of a tree or log. It usually infers the pith with a few inches diameter of heartwood. See *Box the Heart*. **2.** A wood block grooved on the outside for a stay-rope and pierced by a large hole serrated on the bottom edge. It is used on board ship.

Heart Boxed. See *Box the Heart*.

Hearth. The fireproof part of a floor adjacent to a fireplace. The front hearth extends 6 in. beyond the opening on either side, and 1 ft. 6 in. in front, and is usually tiled. The back hearth is immediately beneath the fire and behind the face of the breasts.

Heart Hickory. Shagbark Hickory. See *Hickory*.

Heart Shake. A shake starting at the heart of a log. See *Shakes*.

Heartwood or **Duramen.** The inner part of exogenous trees that

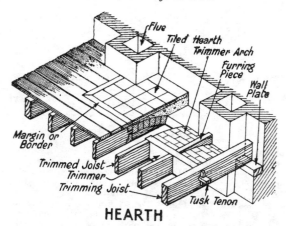

HEARTH

normally does not contain living cells. Also called *true wood*. Heartwood is usually harder, heavier, less permeable, and more durable than sapwood, and differs in infiltration content. Its chief function is to provide support to the tree and does not participate in the growth, although the amount of heartwood increases with the growth. The colour of our commercial woods ranges through every shade of yellow, orange, red, and brown, in addition to black and white. The amount of heartwood varies considerably with different species, and varies in the same species, and in some cases is difficult to distinguish. The number of rings in the sapwood, *q.v.*, may vary from one or two in some species to 200 or more in other species. See *Annual Rings*, *Duramen*, and *Ripewood*.

Heck. A local term for a door latch.

Heel. 1. The back end of a plane. **2.** The bottom of the hanging stile of a door. **3.** See *Saw Teeth*. **4.** Sometimes applied to the *cyma reversa* moulding. **5.** The lower end of the stern post of a vessel and the junction with the keel. See *Stern Timbers*. **6.** The lower end of a mast.

Heel Plane. A compass plane, *q.v.*

Heel Post. The post farthest from the manger in a stable partition. Also called *kicking post*. Also see *Lock Gate*.

Heel Print. See *Tail Print*.

Heel Strap. A wrought-iron strap to tie down the foot of the principal rafter in a roof truss. (See illus., p. 242.)

Heim. The name of the botanist F. Heim.

Helical. In the form of a helix or spiral. Applied to twisted surfaces in *ramp and twist work*, skew arches, and double curvature work generally. The centre line of a wreath to a handrail.

Helical Hinge. A hinge for doors swinging both ways. (See illus., p. 242.)

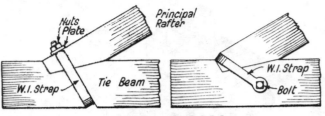

HEEL   STRAPS

Helicopter. An air machine that can ascend and descend vertically under control, and can move forward under its own power.

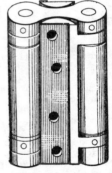

Heliocarpus. See *Amor seco* and *Jonote*.

Helioscene. A sun-blind consisting of small folding hood-frames.

Helix. 1. A line lying on the surface of a cylinder such that the increase of rise is constant with increase of angular advance, as in a screw thread. 2. A volute or spiral.

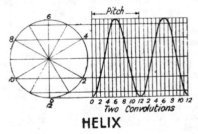

HELICAL HINGE

HELIX

Helm. That part of a ship in which the steering is controlled, and which contains rudder, tiller, wheel, etc.

Helve. The shaft of an axe or similar tool.

Hemicellulose. One of the important chemical constituents of wood substance, *q.v.*

Hemicyclia. See *Grey Boxwood* and *Wira*.

Hemlock, Eastern. *Tsuga canadensis* (*S.*). Eastern N. America. Also called White and Canada Hemlock, etc. Often called spruce. Light buff, with reddish brown tinge. Soft, light, brittle, splintery, and subject to shakes. Coarse and cross grain. Inferior to western hemlock, and held in little esteem. S.G. ·42. WESTERN HEMLOCK. *Tsuga heterophylla* or *T. Albertiana*. Western N. America. Also called Alaska Pine or Fir, Prince Albert Pine or Fir, B. Columbia Hemlock. Light yellowish brown, lustrous. Light, soft, strong. Durable if kept wet or dry, but subject to fungi attack in moist conditions unless treated. Splits easily when seasoned. Contrast between spring and summer wood but not so much as in eastern hemlock. Seasons well with care, and easily wrought. Non-resinous, with straight, even grain. Useful wood and in increasing demand. Used for joinery, construction, flooring, shelving, enamelled furniture, formwork, stable fittings as it repels vermin. S.G. ·45 ; C.W. 2.

Hemsl. Abbreviation for the name of the botanist Hemsley.

Hen Coop. A nesting place for poultry. A cote or roost.

Hendui. Ekki, *q.v.*

Henry. The name of the botanist A. Henry.

Hepplewhite. A cabinet-maker of the latter part of the 18th century. The furniture he designed was usually of inlaid mahogany and the characteristics were lightness and graceful curves. The backs of chairs were often shield-shaped.

Heptagon. A seven-sided polygon, *q.v.* Area=square of side × 3·634.

Heraklith. A proprietary insulating material and board.

Heritiera. See *Dungun* and *Sunder* or *Sundri*.

Hermes. A pillar with a finial in the form of a human bust.

Herminiera. See *Ambatch*.

Herring-bone. Timbers arranged obliquely in opposite directions. Also see *Floor Strutting*.

Herring-bone Figure. Formed by matching silver grain so that it radiates in opposite directions in adjacent pieces of wood. Also see *Veneer Matching*.

HEPPLEWHITE CHAIR

HERRINGBONE PANELS

Heterogeneous. Applied to woods in which the rays vary in size and shape.

Heterophragma. See *Karenwood, Petthan, Waras*.

Hewn Timber. 1. Squared or levelled with axe or adze. 2. Timber in the balk.

Hexagon. A six-sided polygon. Area=square of side ×2·598.

H-Hinge. A parliament hinge, *q.v.*

Hibiscus. See *Mahoe* (*blue*).

Hicking. Lifting the end of a plank or deal, usually by means of a hand hook, for the shoulder of the carrier.

Hickory. *Carya* or *Hickoria spp.* (*H.*). Canada and U.S.A. Related to walnut. Also called Butternut. About twelve species but the important species are shagbark, shellbark, mockernut or whiteheart, and pignut. White to yellowish brown with darker striations. The wood is very hard, tough, elastic, and strong, but not durable. Pignut is the heaviest and strongest. Straight, coarse, smooth grain, Very valuable for work to resist shocks. Similar characteristics to ash and superior in some respects. Used chiefly for sports equipment, implements, shafts, spokes, bentwork. S.G. ·7 to ·8 ; C.W. 4. BUTTERNUT, *H.* or *C. cordiformis*; MOCKERNUT, *C. tomentosa* or *H. alba*; NUTMEG, *H.* or *C. myristicæformis* ; OVAL PIGNUT, *H.* or *C. ovalis* ; PIG-NUT, *C. porcina* or *H. glabra* ; QUEENSLAND HICKORY, *Flindersia Ifflaiana*. Also called Cairn's Hickory. Little resemblance to *Carya spp.* Yellowish brown. Hard, heavy, tough and strong, with close grain. Used for heavy constructional work. S.G. ·85. SHAGBARK, *C. alba* or *H. ovata* ; SHELLBARK, *C. sulcata* or *H. laciniosa* ; WATER HICKORY, *H.* or *C. aquatica*. Also see *Pecan Hickory*.

Hickory Ash. Queensland Hickory. See *Hickory*.

Hickory Elm. Canadian Rock Elm, *q.v.*

Hiern. The name of the botanist W. P. Hiern.

Hieronymia. See *Tapana* and *Urucurana*.

Highboy. See *Tall-boy*.

High Humidity Treatment. Temporarily raising the humidity of the circulating air in a kiln when drying wood requiring special treatment.

Hill Tamarind. See *Jutahy*.

Hill Toon. *Cedrela serrata*. A species of toon, *q.v.*, but inferior and not so decorative.

Himalayan. Applied to woods from N. India. They are usually difficult to transport to the plains, hence little is exported except fir and spruce. HIMALAYAN ASH, *Fraxinus excelsior*. See *Ash*. H. CYPRESS, *Cupressus torulosa*. See *Cypress*. H. MAPLE, *Acer Campbelli*, *A. oblongum*, and *A. pictum*. See *Maple*. H. POPLAR, *Populus ciliata*. Similar to *Populus spp.* H. SILVER FIR, *Abies pindrow*. See *Fir*. H. SPRUCE, *Picea morinda*. The last two are usually marketed together indiscriminately. See *Picea* and *Spruce*.

Hingebound. Applied to faulty hinging in which the door is too closely set in the rebate so that it binds either on the stop or the rebate. The fault is usually due to insufficient allowance for paint.

Hinges. The types in common use are : black-flap, Bommer, box, butt, card-table, centre, continuous, counter-flap, cross garnet, dolphin, floor spring, garnet, H, helical, kneed, knuckle joint, lift-off, Mitchell, necked-centre, parliament, pew, piano, pin, quadrant, reflex, rising butt, rule-joint, screen, spring, stopped butt, strap, and tee. They may be of various metals and finishes. See *Cocked Hinge* and *Tudor Door*.

Hinoki. *Cupressus*, or *Chamæcyparis*, *obtusa* (*S.*). Japan. Straw colour, darker streaks. Light, soft, tough, elastic, durable. Fragrant and lustrous. Similar to yellow pine in texture but often with wavy grain. Easily wrought, polishes well. Excellent wood for general purposes. S.G. ·4 ; C.W. 1·5. Also see *Sugi*.

Hip. The salient angle at the intersection of two inclined roof surfaces. The angle between the surfaces is greater than 180°. See *Roofs* and *Jack Rafters*.

Hip Knob. A finial at the intersection between ridge and hips.

Hipped Roof. A roof with inclined ends instead of with gable. See *Hip*.

Hippocastanaceæ. A plant family including *Æsculus*.

Hippomane. See *Man-chineel*.

Hip Rafter. The rafter, in a roof, forming the hip.

Hip Roll. A long piece of circular section, with a V-recess on the underside, to cover the intersection of

the materials at the hips of a roof, or to carry and secure the lead flashings.

Hitch. A particular form of knot, or noose, used with rope lashings, when hoisting. See *Knots*.

Hitcher. A large boat-hook on a long ricker and used by men engaged as rafters.

Hive. A box or receptacle for housing honey-bees.

Hnaw. Haldu, *q.v.*

h.n.w. Abbreviation for *head, nut*, and *washer*.

Hoarding. A temporary fence to protect a building site during the erection of a building. Close fencing for the display of advertisements, or special erections for the purpose.

Hochst. Abbreviation for the name of the botanist Hochstetter.

Hockey. A field game in which a ball is driven by means of a strong stick with curved end ; hence, *hockey stick, club, posts*, etc.

Hod. A box-shaped tray with long handle. It is carried on the shoulder, for conveying bricks, mortar, etc.

Hog. A strong timber in a yacht, etc., placed over the keel when the planks are in position. In some ships it is bolted to the keel to provide a bearing surface for the timbers. A keelson. **2.** *To hog* means to bend in the form of a segment of a circle. **3.** A refuse grinder.

Hogback Girder. A built-up girder with a segmental top chord. (See illus., p. 246.)

HOGBACK GIRDER

Hog Frame. A trussed frame to a ship's deck to prevent distortion.

Hoggery. A piggery or pigsty.

Hogging. Cambering a horizontal structural member, *i.e.* raising it at the middle of its length.

Hog Gum. *Symphonia globulifera* (*H.*). Tropical America. Also called Boar Wood and Osol. Straw colour to greenish brown, lustrous. Fairly hard, heavy, strong, and durable. Straight, coarse grain, apt to splinter. Distinct rays. Not difficult to work. Used for construction, sleepers, etc. A similar species of *Symphonia* is exported from Madagascar. S.G. ·68 ; C.W. 3.

Hog Piece. See *Hog.*

Hog Plum. *Ximenia Americana* (*H.*). Tropical America. Reddish yellow. Very hard, heavy, with fine grain and fragrant, but of little economic importance. Sometimes used as a substitute for sandalwood. S.G. ·9. Also *Spondias mangifera.* India. Grey. Soft, light, not durable. Dull, feels rough, with very coarse texture. Distinct rays, straight grain. Easy to work, but not smooth. Subject to stain, and of little commercial value. Used chiefly for packing cases. S.G. ·37.

Hogs. Machines for breaking up small stuff and wood waste.

Hogshead. A barrel of 48 to 54 gallons capacity.

Hoist. A lift, or elevator, for goods rather than passengers.

Hoisting. Raising or lowering materials by means of mechanical appliances.

Hold. A space between keel and lowest deck for the storing of cargo on board ship. Any interior space for stowing cargo.

Holdfast. 1. An appliance for clamping wood on the bench for ease in working. See *Bench.* 2. A metal fastening driven into a joint of a wall to secure framing, etc. See *Raking Shore.*

HOLDFAST

Holding up. A term used when the stuff is large enough to finish to the required dimensions.

Hole. See *Pin-hole, Shot-hole, Worm-hole.*

Holigar. *Holigarna arnottiana* (*H.*). India. Nearly white, with yellowish cast, lustrous. Light, soft, not durable, but fairly strong and elastic. Smooth, coarse texture, straight grain. Easy to season and work, stains easily. Distinct rays ; of little commercial value. Used chiefly for packing cases. S.G. ·38.

Hollock. *Terminalia myriocarpa* (*H.*). India and E. Indies. Also called Panisaj. Greyish brown with beautiful dark striations, lustrous. Moderately hard, heavy, durable, and strong ; very stable. Smooth, with straight and wavy grain, but coarse texture. Seasons well, but rather difficult to work. Used for furniture, construction, wheelwright's work, plywood. S.G. ·7 ; C.W. 4.

Hollong. *Dipterocarpus pilosus* (*H.*). India. Reddish brown. Moderately hard, heavy, strong, but not durable unless treated.

Fairly straight with some interlocked grain. Dull, feels rough, with even, coarse texture. Distinct rays. Not difficult to work; smooth, and polishes. Present uses: constructional and structural work, sleepers, plywood for tea boxes, etc. S.G. ·7; C.W. 3. Also see *Gurjun*.

Hollow. **1.** A joiner's plane to form a convex, or rounded, surface. **2.** A concave moulding. See *Cove*. **3.** A sinking, or depression, showing on the surface of plywood due to a defect in the inner plies.

HOLLOW BACKED T&G. FLOORING

Hollow Backed. Boards with grooves on the back to provide a better seating, or ventilation, or to lesson the weight. Often applied to hardwood floor-boards.

Hollow Chisel. See *Mortiser*.

Hollow-horning. Same as honey-combing, *q.v.*

Hollowing Knife. A cooper's drawknife.

Hollow Partition. A wood partition with staggered studs, for sound insulation. A double partition with space between for sliding door.

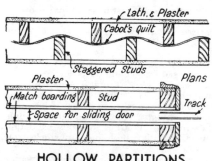

HOLLOW PARTITIONS

Hollow-wood Construction. Deal framed cores faced with plywood in place of panelled framing. See *Flush Door*.

Holly. *Ilex aquifolium* (*H.*). Europe. Also called Holm. Greyish white, sometimes greenish tinge. Hard, fairly heavy, very close fine grain with distinct rays. Warps, and liable to become spotted in seasoning. Used for inlays, stringings, woodware, veneers, turnery, and dyed as a substitute for ebony. S.G. ·76; C.W. 4·5. American Holly, a similar wood, is *Ilex opaca*.

Holm Oak. *Quercus ilex* (*H.*). Mediterranean. Called Holly Oak or Evergreen Oak. The wood is somewhat similar to English oak.

Holocalyx. See *Ibira-pepe* and *Uria-pepé*.

Holoptelea. See *Ayo* and *Elm* (Indian).

Holt. A grove or plantation. The word implies wood or timber.

Homalium spp. (*H.*). Philippines. Called Aranga. From yellow to pale chocolate colour, lustrous. Very hard, heavy, strong, and durable. Straight and interlocked grain, fine texture. Requires slow and careful seasoning. Difficult to convert but not difficult to work. Used for piling, structural and constructional work, sleepers, etc. S.G. ·83. Also see *Lancewood* (*Moulmein*), *Selimbar*, *Kalado*.

Homasote. A prefabricated wood house.

Homogeneous. Applied to woods in which the rays are similar in size and shape.

Honduras. The principal woods exported are : Bignonia, Bullet-tree, Cedar, Ceiba, Dogwood, Greenheart, Ironwood, Locust, Logwood, Mahoe, Mahogany, Rosewood.

Honduras Walnut. *Metopium sp.* (*H.*). Also called Black Poison-wood. Somewhat like ebony. Greenish brown, lustrous, smooth. Hard, heavy, with firm close grain and texture, distinct rays. Little exported.

Honeycombing. Checks in the interior of wood not seen on the surface. They are probably due to case-hardening in seasoning.

Honey Fungus. *Armillaria mellea.* A fungus very destructive to hardwoods. It attacks standing trees as well as converted wood.

Honey Locust. *Gleditsia triacanthus* (*H.*). U.S.A. Light red to reddish brown. Heavy, very hard and strong, fissile, durable, stable, gum deposits. Distinct rays and growth rings. Often confused with the wood of the Coffee Tree, *q.v.* Used for furniture, joinery, vehicles, construction, etc. S.G. ·7 ; C.W. 3. Also see *Locust* and *Mesquite*.

Honey-suckle. *Banksia marginata* (*H.*). Tasmania. Pinkish brown. Rather soft with reticulated appearance. Used for cabinet-work, turnery, etc. Also see *Rewarewa*. *Queensland spp.* are called Wallum and Honeysuckle Oaks.

Honing. Sharpening, *q.v.*

Honoki. *Magnolia hypoleuca* (*H.*). Japan. A variegated wood, like canarywood. Easy to work and used for general purposes.

Hoobooballi. *Loxopterygium sagotii* (*H.*). Tropical America. Also called Snakewood, *q.v.*, and included in imports of Zebra wood. Yellowish to reddish brown with darker streaks and flecks. Moderately hard, heavy, durable. Fairly straight close grain and medium texture. Smooth and easily wrought. Polishes, but oily exudations cause stains. Used for furniture, boatbuilding, gunstocks, etc. S.G. ·7 ; C.W. 2·5.

Hood. **1.** A cover or canopy, especially to a door or window opening. It is usually supported on cantilever brackets for a door opening. See *Canopy*. **2.** A structure, with door, above the companion hatch of a ship. Also see *Hawse Bolster* and *Stormproof Window*.

Hood Moulding. A label moulding, *q.v.*

Hook. Abbreviation for the names of the botanists W. J. and Sir J. D. Hooker.

Hook. **1.** See *Band* and *Gudgeon*. **2.** See *Saw Teeth*. **3.** A term used by the wood-worker to denote the projection of the cutting iron through the sole of a plane. The amount of hook varies with the hardness of the wood and the quality of the surface required.

HOOK & EYE

Hook and Eye. **1.** A gate strap-hinge with a hook to engage in a gudgeon shaped like a circular eye, to allow for the easy removal of the gate. **2.** A pivoted hook and a fixed eye, or staple, used as a fastening for hinged doors, casements, etc.

Hooker. A Plymouth sailing boat, like a cutter.

Hook Hole. A perforation in veneer due to the implements used in moving logs.

Hook Joint or **Rebate.** An S-shaped rebate used for casements, sash meeting rails, airtight show cases, etc. Also see *Tongues* (*Double*).

Hook Ring. An eye for a cupboard or cabin hook. See *Hook and Eye*.

Hoom. **1.** *Polyalthia spp.* (*H.*). India. Yellow, lustrous. Moderately hard, heavy, not durable. Straight even grain. Fine to medium texture. Stable and smooth, with distinct rays. S.G. ·55. **2.** *Saccopetalum tomentosum.* Also called Karri. Olive yellow to brown. Similar characteristics to Polyalthia spp. Both species are used for cabinet work and carvings.

Hoop Ash. *Fraxinus sambucifolia.* See *Ash*.

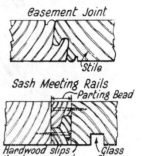

HOOK JOINTS

Hoop Pine. *Araucaria Cunninghamii* and *A. bidwilli* (*S.*). Australia. Also called Moreton Bay and Queensland Pine. Cream or ivory colour. Light, fairly soft, strong, tough, but not durable. Non-odorous. Close even texture. Smooth, easily wrought, and stains readily. Used for joinery, cheap cabinet work, construction, boxes, plywood. S.G. ·56 ; C.W. 2.

Hoops. **1.** The rings holding together the staves of a cask. A *catch hoop* is an iron hoop used temporarily for securing the ends of the staves, or chimes, while the cask is being made. See *Stave*. **2.** A large light wood ring used as a child's toy.

Hoopsticks. Roof carrying members in a vehicle.

Hopea. *Hopea parviflora* (*H.*). Several species. India. Dark reddish brown with purplish cast when seasoned and white lines. Hard, heavy, very durable, and strong. Dull, feels smooth, fine even texture, interlocked grain. Resists white ant. Fairly difficult to season and work. Polishes. Used for structures, construction, sleepers, etc., but too valuable for these purposes. An excellent wood where strength and durability are required. S.G. ·8 ; C.W. 4. Also see *Chengal, Giam, Meranti, Merawan, Manggachapui, Selangan batu, Selimbar, Thingan, Yacal*.

Hop Hornbeam. *Ostrya virginiana* (*H.*). U.S.A. Also called Ironwood. Whitish to light brown, reddish tinge and white specks. Very hard, heavy, strong, and stiff. High shock resistance. Shrinks greatly. Often confused with birch. Used for the same purposes as hickory : levers, handles, axles, woodware, novelties, etc. S.G. ·8 ; C.W. 3·5. EUROPEAN, *Ostrya carpinifolia*. Similar to above. See *Hornbeam*.

Hopper. **1.** A large funnel-shaped container for granular materials such as coal, grain, etc. **2.** A draught-preventer at the side of a casement hinged at the bottom.

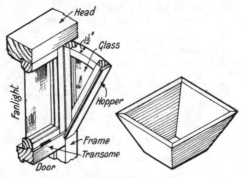

Head

1½"

Glass

Fanlight

Hopper

Frame

Transome

Door

HOPPERS

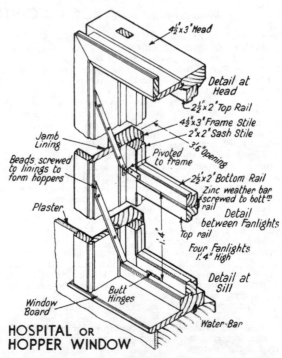

4½"x3" Head

Detail at Head

2½"x2" Top Rail

4½"x3" Frame Stile

2"x2" Sash Stile

3'.6" Opening

Jamb Lining

Pivoted to frame

Beads screwed to linings to form hoppers

2½"x2" Bottom Rail

Zinc weather bar screwed to bott^m rail

Plaster

Detail between Fanlights

Top rail

1'.4"

Four Fanlights 1'.4" High

Detail at Sill

Window Board

Butt Hinges

Water Bar

HOSPITAL or HOPPER WINDOW

Hopper Barge. A barge in attendance on a dredger.

Hopper Window or **Light.** A window formed of a series of fanlights or horizontal casements, rotating on rockers, or pivots, on the bottom rails. They are provided with hoppers unless between deep reveals, and are also called hospital windows, *q.v.*

Hopple. A block attached to a rope tied to the leg of a horse to impede movement.

Hop Poles. Vertical poles up which hops are trained.

Hoppus. 1. A system of measurement for round logs. The length is taken to the nearest $\frac{1}{2}$ ft. and the *quarter girth* measured at the middle to $\frac{1}{4}$ in. Then $(\frac{1}{4}$ girth$)^2 \times$ length $\div 144$ is the volume in c. ft., stated duodecimally. This system allows about 27 per cent. for waste in conversion. 2. The name of the author of Hoppus's Timber Measurer.

Hora. *Dipterocarpus zeylanicus.* Ceylon gurjun, *q.v.*

Horco cebil. *Piptadenia communis (H.).* S. America. Colour of light mahogany. Very hard, heavy, dense, and smooth, but not durable. Also called *Horco molle.* Used for cabinet work, etc. S.G. 1.

Horizontal. Applied to band and frame saws that cut horizontally.

Horizontal Shore. See *Flying Shore.*

Hormigo. Macawood, *q.v.*

Hornbar. The hind bar in the top carriage of a vehicle.

Hornbeam. *Carpinus betulus (H.).* Europe. Sometimes called White Beech. Yellowish white. Hard, heavy, strong, and tough, fairly durable. Difficult to split. Close grain with faint rays. Shrinks greatly. Used for cogs, implements, tools, turnery, etc. S.G. ·75 ; C.W. 4·5. CANADIAN, *Carpinus caroliniana.* Also called Blue Beech. Similar to above.

HORNBEAM

Horns. Projecting ends to the outside members of framing. They are intended to protect the corners until the frame is fixed, to allow for secure wedging, and to strengthen and assist the fixing of the frame. See *Joggle, Door Frame,* and *Panelled Door.* Also see *Jaws.*

Horn Timbers. Inclined timbers in the framing of a ship to form the projecting stern.

Horoeka. See *Lancewood.*

Horse. 1. A trestle, *q.v.* 2. A thwacking frame for bending pantiles. 3. A perforated board for fixing the damper rod of a brick kiln. 4. The board, or slipper, carrying the parts of a horsed mould for running plaster cornices mouldings. 5. Any form of framing used as a temporary or intermittent support. See

Zinc Profile

Zinc shoes or slippers

Stock

Horse or slipper

HORSED   MOULD

Clothes Horse. **6.** Gymnastic apparatus used for vaulting, etc. **7.** A timber across the deck of a ship along which a block or ring to a sail can slide.

Horse Boat. One towed by a horse along a river or canal.

Horse Box. **1.** A cage for transferring animals aboard ship. **2.** A vehicle or railway van for conveying horses. **3.** A stall for a horse.

Horse-cassia. Cana fistola, *q.v.*

Horse Chestnut. *Æsculus hippocastanum* (*H.*). Europe. Yellowish with pinkish tinge. Fairly light, soft, and not durable. Close even grain with fine ripple marks. Held in little esteem in this country. Used for turnery, cooperage, charcoal, etc. S.G. ·6 ; C.W. 2. BUCK-EYES. *Æsculus glabra* and *Æ. flava.* N. America. Soft inferior wood. Used for woodware, artificial limbs, etc. INDIAN HORSE CHESTNUT. *Æsculus indica.* Pinkish brown. Fairly soft and not durable. Dull, smooth. Straight, wavy and interlocked grain. Fine texture. Requires careful seasoning to avoid degrading. Easily wrought and polishes. Used for cheap furniture, woodware, cooperage, turnery. S.G. ·54 ; C.W. 2·5. JAPANESE HORSE CHESTNUT, *Æsculus turbinata.* Similar to Indian.

Horse Dam. A temporary dam in a stream, formed by large logs.

Horseflesh Mahogany. See *Moruro.*

Horsewood. See *Sea Grape.*

Horsing. Strutting up centering from a solid base.

Hospital Doors. Flush doors, *q.v.* They are usually provided with a small glass panel at the top.

Hospital Windows. Windows designed to provide ample ventilation without down-draughts. The name is specially applied to the hopper window, *q.v.* .

Hot-house. A greenhouse artificially heated.

Hot Press Resins. Urea-formaldehyde thermo-setting resins that require " cooking " in an autoclave. They allow for a time interval of several days for assembly of the components. See *Plastic Glues* and *Cold Press Resins.*

Houbooballi. See *Hoobooballi.*

Hounds. **1.** Two shaped pieces of wood, fixed to the mast of a ship, through which ropes are threaded. **2.** Side pieces to strengthen the body of a vehicle.

House-boat. A heavy type of stationary barge with superstructure. It is used for privileged spectators at river events, or as a summer residence.

Housebote. Wood that a tenant on an estate is allowed so as to keep the property in repair for which he is responsible.

Housed. See *Housing.*

Housed Dovetail. A dovetailed housing to secure the ends of shelves, rails, etc.

Housed String. . A stair string with housings for the treads and risers. The alternative to cut string, *q.v.* Also see *Handrail.*

Housing. **1.** A trench or sinking in a piece of wood to receive another piece, as the end of a shelf, stair tread, rail, etc. A *housed mortise and tenon* joint is one in which the end of the rail is let into the stile, as well as the tenon, so that the rail can carry a transverse load. **2.** A term used in schemes for the erection of dwelling-houses and flats.

Howdah. A framework to accommodate several persons on the back of an elephant.

Howe Truss. A wood roof truss for large spans. The illustration shows king, queen, and princess posts, but the number of auxiliary posts depends upon the span, which may be up to 80 ft.

Hoy. A small coasting vessel or barge. A Thames barge with sprit-sail.

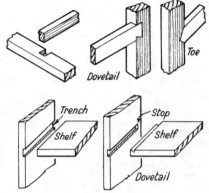

HOUSING JOINTS

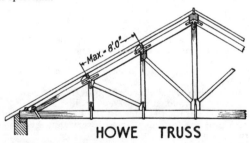

HOWE TRUSS

Hsiang. Chinese oak.

Hua. Chinese birch.

Huai. Chinese ash.

Hua-li-mu. Like blackbean, *q.v.*, but lighter in colour.

Huang yang. Chinese box.

Hub. Abbreviation for the name of the botanist Huber.

Hub. **1.** The centre part, or stock, of a wheel from which the spokes radiate. The nave of a wheel, *q.v.* **2.** A brake-block to prevent a wheel from turning.

Hububalli. See *Hoobooballi*.

Hudoke. Sucupira, *q.v.*

Huds. Abbreviation for the name of the botanist Hudson.

Hulk. **1.** A dismantled ship unfit for voyages. **2.** A heavy or unwieldy ship.

Hull. **1.** The frame or body of a ship. See *Stern Timbers*. **2.** The outer covering of anything.

Humbugs. See *Crate.*

Humiria. Bastard Bullet wood, *q.v.* Also see *Bullet wood* and *Tauroniro.*

Humiriaceæ. A family of trees including Humiria, Sacoglottis, Vantanea.

Hundred. See *Standard.*

Hung-li. Chinese Magnolia, *q.v.*

Hung Sash. A vertically sliding sash balanced by weights. See *Sash and Frame.*

Hung Tou. *Ormosia hosiei (H.).* China. The red bean tree. Rich red. Hard and heavy. The wood is beautifully marked, and esteemed for high-class cabinet work and fittings.

Hunteria. Gading, *q.v.*

Hunting Box. A small house generally used only during the hunting season.

Huntingdon Elm. See *Elm.*

Huntonit. A registered wall-board, *q.v.*

Huon Pine. *Dacrydium Franklinii (S.).* Tasmania. Also called Macquerie Pine. White. An excellent softwood, durable, fragrant, easily wrought, stable, and resists insects. Supplies difficult to obtain. Used for cabinet work, joinery, boatbuilding, etc. S.G. ·55 ; C.W. 2.

Hur. See *Arr.*

Hura. See *Possum Wood.*

Hurdle. A slender framework of cleft branches or twigs, used as a fence in agriculture and as an obstacle in sports.

Hurl Bat. A club used in the game of hurling.

Hurricane Deck. An elevated deck on a steamship, especially when over the saloon.

Hut. A small cabin or shelter. A light building of a temporary character, usually of wood, for workmen, soldiers, etc.

Hutch. Abbreviation for the name of the botanist Hutchinson.

Hutch. 1. A bin, box, or chest. 2. A low wagon used in coal mines. 3. A box for rabbits.

Hydroxylin. A registered wood substitute consisting of compressed sawdust and synthetic resin.

Hydulignum. A registered compressed wood, *q.v.*

Hygroscopic. Applied to materials that have the property of absorbing moisture from the atmosphere.

Hylecœtus. A genus of wood beetles very similar to *Lymexylon sp.,* *q.v.,* but they also attack coniferous trees.

Hylesinus. An insect that attacks unhealthy and felled trees. The ash bark beetle is the most common. It is black or dark brown, nearly ¼ in. long, and resembles Bostrichus.

Hylobius. The pine weevil. Very destructive to young coniferous trees.

Hylodendron. See *Akessi.*

Hylotrupes. See *Longhorn Beetle.*

Hymenæa. See *Locust* (W. Indies), *Courbaril* and *Guapino.*

Hymenodictyon. See *Kusan* and *Kuthan.*

Hymenolobium. See *Angelim.*

Hyperbola. See *Conic Sections.*
Hypericaceæ. A family of trees including Cratoxylon.
Hypertherum. A projecting cornice over the architrave of a
door.
Hyphæ. The hollow thread-like tubes of mycelium, *q.v.*
Hypsometer. An instrument for measuring the height of a tree.

I

Ibira. Applied to numerous tropical American woods : boxwood, bloodwood, canafistola, degame, guayacan, viraru.

Ibira-nira. *Bumelia obtusifolia* (*H.*). C. America. A species of buckthorn. Yellowish brown, greenish streaks, lustrous. Very hard, heavy, brittle, and not durable. Fairly straight grain, with fine smooth texture. Not easily wrought. Used for furniture, tools, constructional work, etc. S.G. 1 ; C.W. 4.

Ibira-pepe. See *Uria-pepé*.

Ibira-peré. *Apuleia leiocarpa* (*H.*). Tropical America. Also called Amarello. Golden yellow to brownish, lustrous. Very hard, heavy, strong, elastic, and durable. Fine uniform texture with some roey grain. Not difficult to work, and polishes well. Used for structural work, flooring, shafts, posts, etc. S.G. ·9 ; C.W. 3.

Ibiraro. *Pterogyne nitens* (*H.*). C. America. Light brown, sometimes black stripes, darkens with exposure. Hard, heavy, strong, durable, stable. Irregular grain and roey. Polishes well but difficult to work smooth. Used for cabinet work, construction, cooperage, turnery, sleepers, spokes. S.G. ·76 to 1 ; C.W. 4·5.

Ibo. 1. *Poga oleosa* (*H.*). Nigeria. Reddish brown, lustrous. Light and soft. An attractive wood with pronounced silky oak figure when quarter sawn, and easily wrought. The wood is under investigation and is expected to be in demand for interior fittings when its properties are better known. **2.** A prefabricated wood house.

Ibus Board. A proprietary wall-board.

Icacinaceæ. A family of trees including Calatola, Cantleya, Platea, Urandra, Villaresia.

Icaquito. See *Chozo*.

Ice Birch. *Betula sp.* Highly figured birch.

Ice Boat. One fitted with runners and sails for running over ice.

Ichthyomethia. See *Dogwood*.

Icing. Sprinkling water on a road during frost to assist the movement of logs.

Idagbon. *Pausinystallia spp.* (*H.*). Nigeria. Also called Nikiba. Resembles Haldu. Light orange brown with pink stripes. Hard, heavy, with characteristics of boxwood. Straight grain with patches interlocked that are liable to pick-up. Under investigation. Used for fancy articles, turnery, carving. Substitute for boxwood and pearwood. S.G. ·7.

Idaho White Pine. Western white pine. See *Pine*.

Identification. The identification of an unknown wood is usually very difficult except by scientific investigation. Specimens are prepared for microscopic examination. The following are some of the features to be examined. Presence of pores (hardwoods), absence of pores (softwoods). Ring or diffuse porous. Size, number, and arrangement of pores. Size and distribution of rays. Gum and resin ducts. Distribution of wood parenchyma. Tyloses in the pores. The various features in combination usually enable the expert to identify the wood botanically, especially in conjunction with weight, colour, hardness, fluorescence, lustre, scent, and taste. See *Photomicrograph* and *Gross Features*.

Idigbo. *Terminalia ivorensis* (*H.*). Nigeria. Also called Egboinnebi, Emri, etc. Pale lemon yellow to light oak colour, with darker markings, and lustrous. Moderately hard and heavy, stable, fissile. Should be boxed, as interior wood is inferior. Outer part of trunk provides fairly strong durable wood with uniform grain and texture. Easy to work smooth, with sharp arrises. Polishes well. Used for cabinet work, superior joinery, carriage work, construction, sleepers. S.G. ·62 ; C.W. 3.

Ijebo. Nigerian Sapele, *q.v.*

Ilex. See *Holly* and *Kakatara*.

Imago. An insect in its last or perfect stage of development.

Imbira Quiaba. *Sterculia sp.* (*H.*). C. America. Yellowish brown, lustrous. Very light and soft, not durable. Very strong for its weight. Straight grain, coarse but smooth texture. Used for light construction, cases, matches. S.G. ·25.

Imbirussu. *Bombax spp.* (*H.*). C. America. Brownish grey. Light, soft, tough, and strong for its weight. Straight grain, coarse texture. Difficult to work smooth. Used for purposes requiring lightness and strength. S.G. ·25.

Imbricate. To overlap, as with shingles or tiles.

Imbuia or **Imbuya.** *Phœbe porosa* (*H.*). Brazil. Also called Brazilian walnut. Varies from yellowish to chocolate brown, lustrous, fragrant, smooth. Fairly hard and heavy, durable, not strong, moderately stable. Straight, wavy, or curly grain, some beautifully figured. Easily wrought ; dust an irritant ; polishes well. Increasing in favour for superior joinery, cabinet work, etc. S.G. ·6 ; C.W. 3.

Impact Test. A hardness test. A steel ball is dropped from a recorded height, and the diameter of the indent is a measure of the hardness of the wood specimen.

Impale. To enclose, or fence, with stakes or palings.

Imperfect Manufacture. The term used for manufactured goods having defects due to faulty conversion or machining. See *Chipped Grain*, *Chip Marks*, *Machine Burn*, *Mis-matching*, *Skips*, *Tool Marks*, *Torn Grain*, *Variation*, and *Washboard Effect*.

Implement. A tool or appliance to perform some form of work, as agricultural or gardening implements, etc.

Impost. The cap or top of the support from which an arch springs.

Impreg. A registered name for plastic-treated wood products. See *Improved Wood*.

Impregnation. See *Preservation* and *Penetration*.

Improved Wood. A term applied to wood that has been treated to prevent movement due to variation of moisture content. Plies of very thin veneers specially treated are used for one type of " improved " wood. Another method is to impregnate with synthetic resin and then apply heat and pressure to the wood, which greatly increases the strength, stability, and durability of the wood. It is very difficult to work with ordinary tools as it is more like metal than wood. It is used for insulators, rollers, brake linings, pulleys, and in aircraft. The process of manufacture varies with the uses of the prepared wood. Many registered names are given to plastic-treated woods, and modern developments are providing many wood substitutes by the use of paper, sawdust, etc. : Benalite, Compreg, Hydroxylin, Hydulignum, Impreg, Jicwood, Papreg, Permali, Plyscol, Uraloy, etc.

In. See *Eng.*

Inbark. Bark enclosed in the stem of the tree by later growth, and exposed by conversion.

Inboard. The interior of a ship.

Incense Cedar. See *Cedar.*

Incipient Decay. An early stage of fungal decay in wood. It is usually accompanied by slight discoloration. See *Dote, Dosy, Foxy,* and *Rot.*

Incised. Cut or carved. Applied to carvings lying below the surface of the material.

Incising Machine. Used on timbers to give a more uniform penetration of the liquid when pressure creosoting. See *Preservation.*

Included Sapwood. Irregular areas or streaks of light-coloured wood embedded in the darker heartwood and having the appearance of sapwood. It is often found in Western Red Cedar.

Increment. The volume of wood produced during a given period of growth, or the increase in value of a stand. See *Accretion.*

Increment Borer. See *Borer.*

Incubator. A box-like apparatus, artificially heated, for hatching eggs and rearing chickens.

Incurve. To curve, or bend, inwards.

Indent. 1. A small hollow or depression. 2. A V-shaped notch. 3. An inward curve or recess.

INDENTED SPLAYED SCARF

Indented. Cut into the form of saw-teeth. Pointed or toothed.

Indented Joint. A fished joint in which both plate and beam are cut, or indented, to prevent the sliding of one surface on the other due to shear or tension. See *Fish Plates.*

India. The most important woods exported to this country include : Chuglum, Dhup, Ebony, Elm, Eng, Gurjun, Haldu, Kokko, Laurel, Padauk, Pyinkado, Pyinma, Rosewood, Satinwood, Silver Greywood, Sissoo, Teak, Thitka, Toon, White Bombway. Also see *Blackwood, Cedar, Chestnut, Cocobola, Deodar, Eaglewood, Gardenia, Himalayan, Horse Chestnut, Juniper, Laburnum, Olive, Palu, Piney, Poon, Poplar, Red Sanders, Sal, Soymida, Willow.* Many other Indian woods are included in the Glossary.

Indian Alder. *Alnus nepalensis* (*H.*). Brownish grey, lustrous. Light, spongy, smooth, with straight grain. Easily wrought. Used for matchboarding, tea boxes, etc. S.G. ·32 ; C.W. 3.

Indian Almond. *Terminalia catappa* (*H.*). Several species. Light brownish red, variegated streaks, lustrous, moderately hard, heavy, and strong. Twisted and interlocked grain ; fairly coarse texture. Difficult to season but not difficult to work. Polishes. Used for decorative work and general purposes. S.G. ·65 ; C.W. 3.

Indian Boxwood. Gardenia, *q.v.*

Indian Clubs. A pair of bottle-shaped wood clubs used in physical training.

Indian Laurel. *Terminalia alata, T. coricea, T. crenulata.* Often classified as *T. tomentosa.* See *Laurel.*

Indian Red Wood. See *Red Eyne.*

Indian Rosewood. *Dalbergia latifolia.* See *Blackwood* and *Rosewood.*

Indian Silver Greywood. *Terminalia bialata.* See *Silver Greywood.*

Indian Stones. Oil-stones made of corundum and alundum. See *Oilstones.*

Indigenous. Applied to woods that are native to a particular country. See *British Trees.*

Inf. Abbreviation for *inferior.*

Infiltration. Extraneous organic matter in wood that is infiltrated but forms no part of the wood substance proper.

Inga. See *Rain Tree* and *Waikey.*

Ingartu. *Cassia kotochyaua* (*H.*). Sudan. Pale salmon colour, smooth, lustrous. Hard, heavy, with close even grain. See *Cassia.*

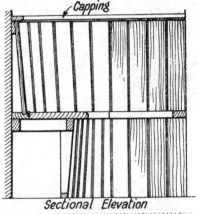

Sectional Elevation

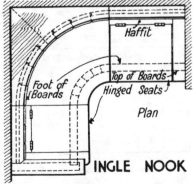

INGLE NOOK

Ingle-nook. An enclosed fixed seat near a fireplace.

Ingyin. *Pentacme suavis* (*H.*). India and E. Indies. Mahogany colour. Hard, heavy, strong, and very durable. Close cross grain, distinct rays. Requires slow seasoning and rather difficult to work.

Similar to Sal. Used for structural and constructional work, shafts, poles, etc. S.G. ·85 ; C.W. 4.

Initial Absorption. The amount of preservative absorbed before pressure is applied in wood preservation, *q.v.*

Initial Vacuum. The vacuum formed in the cylinder before the preservative is introduced in wood preservation, *q.v.*

Inkstand. Equipment for a writing desk to hold ink-wells and pens.

Inlay. Ornamentation formed by removing part of the surface and inserting a different material in its place, or a different kind of the same material. See *Bandings* and *Stringings*.

Inner Beads. The beads mitred round the inside of a sash and frame window or a Yorkshire light, to guide the sliding sashes. Also called guide, guard, and staff bead. See *Box Frame* and *Sash and Frame*.

Inner Linings. The inside linings of a sash and frame window that form the boxes for the weights.

Inner Plate. The inside wall plate when two plates are used. See *Eaves*.

Inodorous. Applied to wood without odour or scent. This is an important quality of wood for special purposes such as food containers, casks, cases, boxes, etc. See *Fragrance, Malodorous, Odorous, Scent, Caskwood.*

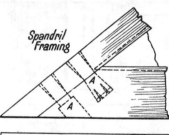

Spandril Framing

A = Hardwood Tenons

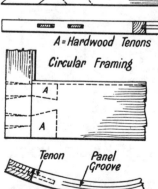

Circular Framing

Tenon　Panel Groove

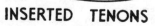

INSERTED TENONS

Inodorous Felt. An asphaltic flax felt used as underlining, etc.

Inoi. Ibo, *q.v.*

Insecticide. Any preparation that will destroy insects.

Insects. There is a large variety of ants, beetles, bugs, caterpillars, grubs, weevils, etc. Many only attack the growing trees, others only attack young plants, but many attack the wood after conversion and use, in buildings, furniture, piling, etc. The important insects destructive to trees and wood are described alphabetically. Also see *Ants, Beetles, Borers, Marine Insects.*

Inserted Bit. A tooth in an inserted tooth saw. See *Saws.*

Inserted Tenons. Hardwood tenons inserted in the rails of circular and oblique work in which the solid tenon would be too weak because of cross grain. Also applied to loose tenons that only prevent lateral movement.

Insulating. Applied to non-conducting materials that resist the passage of heat, sound, or vibration of any kind. Insulation is obtained chiefly by means of mass, trapped air in porous materials, or by scientific construction.

Insulite. A registered wall-board, *q.v.*

Insulwood. A proprietary building board.

Intagliated. Carved or engraved below the surface of the material. Sunk ornamentation.

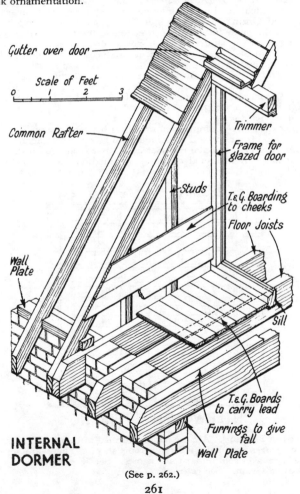

INTERNAL DORMER

(See p. 262.)

Intake. The reverse of offset, but the practical result is the same in both cases. See *Offset*.

Intarsia. Inlays forming pictorial effects. Pictures formed by inlays of differently coloured woods.

Intercostal Girders. Cross frames to strengthen the deck of a ship.

Interjoist. The space between two joists.

Interlaced. 1. See *Interwoven Fencing* and *Wovenboard*. 2. Applied to grain in which the fibres are interwoven in an irregular manner.

Interlocked Grain. Similar to spiral grain, *q.v.*, but with periodic changes in the pitch and sense of direction of the fibres. The spiral arrangement of the fibres is in opposite directions in successive growth increments. The arrangement produces ribbon grain when quarter-sawn, and the wood is difficult to work and split.

Internal Dormer. Vertical casements set back within the plane of a sloping roof. The arrangement entails a lead flat in front of the casements. The cheeks of the opening may be covered with lead or tiles. A glazed door or a sash and frame window may be used instead of casements. See *Dormer*. (See illus., p. 261.)

Inter-plane Struts. Compression members between the main planes at the rear and front spars of an aeroplane to give correct alignment.

Intertie. 1. The horizontal member in a trussed partition just above the doorways. A horizonal member in structural framing between sill and head. See *Partition*. 2. A short rail to stiffen two posts.

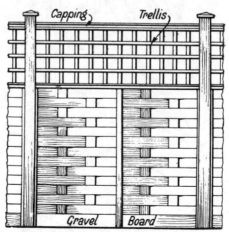

INTERWOVEN FENCING

Interwoven Fencing. A close fencing formed of thin material interlaced, or woven, to make a large strong sheet. There are several proprietary makes sold under trade names. See *Wovenboard*.

Interwoven Grain. See *Interlocked Grain*.

In the Open. A term applied to timber stacked in the open air, without cover.

In the Round. Applied to felled trees or logs before squaring and conversion.

In the White. Applied to finished furniture ready for polishing or other treatment.

Intrados. The soffit, or underside, of an arch or vault.

Intsia. (Afzelia.) See *Merbau* and *Ipil*.

Invert. The lower part of a sewer.

Inverted Arch. A segmental arch placed upside-down between the feet of piers to distribute the pressure.

Inviraro. A West Indies wood similar to oak but harder, heavier, and more tenacious. Used chiefly for wheelwrights' work.

Involute. A curve formed by unrolling the perimeter of a plane figure. The plane figure is the evolute, or eye.

Ionic. One of the architectural orders, *q.v.* The voluted capital is one of the chief characteristics.

Ionic Volute. See *Volute*.

Ipe. See *Peroba*.

i.p.f. Abbreviation for *intaken piled fathom*. See *Fathom*.

Ipil. *Intsia bijuga*, or *Afzelia spp.* (*H.*). Philippines. Yellowish, turns dark brown with age. Secretion forms black spots. Very hard, heavy, strong, durable, stable, and resists insects. Moderately fine texture with fine rays. Cross-grained but smooth, and polishes. Large sizes. Not difficult to season but fairly difficult to work. Used for furniture and cabinet-work sleepers, and where strength and durability are important. S.G. ·9 ; C.W. 4. Also see *Merabau*.

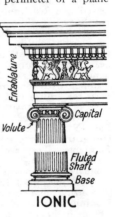

Iranian Box. Persian Box. See *Box*.

Irish Bog Oak. See *Bog Oak*.

Irish Standard. A measurement of timber consisting of 270 cub. ft. See *Standard*.

Iriya. Chuglum, *q.v.*

Iroko. *Chlorophora excelsa* (*H.*). W. Africa. Also called Odum, Muvule, Ulono, Oroko, etc. Golden to nut or dark brown, with dull streaks, lustrous. Moderately hard, heavy and stable, very strong and durable, fire-resisting, and resists insects. Even, coarse texture, with interlocked grain. Requires careful seasoning and fairly difficult to work due to picking-up. Smooth, and polishes well. Large sizes. Used for superior joinery, cabinet work, carriage work. A substitute for teak. S.G. ·7 ; C.W. 4.

Ironbark. Applied to several varieties of Eucalyptus. Excellent woods and large sizes. Pale brown, darkening with exposure. Very hard, heavy, strong and durable. Gum secretion. Wavy and close interlocked grain. Difficult to work. Used for structural work, carriage and wheelwright's work, piling, sleepers, flooring. S.G. 1·1 ;

C.W. 4. GREY, or WHITE, IRONBARK, *Euc. paniculata.* N.S. Wales and Victoria. Considered the best. BROAD-LEAVED IRONBARK, *Euc. siderophloia.* N.S. Wales. Inferior to Grey, and more red in colour. NARROW-LEAVED IRONBARK, *Euc. crebra.* Similar to broad-leaved. RED IRONBARK, *Euc. sideroxylon.* Also called Mugga. N.S.W. and Victoria. Softer, deeper colour, and inferior to Grey. RED IRONBARK, *Euc. leucoxylon.* Victoria. Equal to Grey. SILVER-TOP IRONBARK, *Euc. sieberiana.* Queensland. Called Gippsland Mountain Ash and White Ironbark. Used for general purposes. Not so durable as other ironbarks. Western Australia has excellent ironbarks, *Euc. melanophloia* and *E. terminalis*, but little is exported as the areas await development. Tasmania has also ironbarks similar to the mainland, but they are not exported. Queensland distinguishes about twelve different species. The most important are Grey, Broad-leaved, Narrow-leaved, and Red Ironbarks.

Ironbox. *Eucalyptus spp.* (*H.*). Australia. About eight different species are distinguished in Queensland alone. The most important are Black, Coolibah, Grey, and Yellow. BLACK, *E. raveretiana.* Light brown to brownish black, white veins and speckles. Very hard, heavy, strong, durable, and stable. Close even texture. Similar to lignum-vitæ and used as a substitute. Seasons well and works readily. Used for turnery, turnpins, bowls. Too hard for ordinary purposes. Also used for sleepers, structural work, etc. S.G. 1·1; C.W. 4. COOLIBAH, *E. microtheca.* Similar to above, but often harder and heavier. See *Coolibah.* GREY, *E. hemiphloia.* See *Grey Box.* YELLOW, *E. melliodora.* See *Yellow Box.*

Irongum. The name applied to several species of Queensland Eucalypts. They are so called to distinguish them from the softer varieties of gum-tree woods and because they are somewhat like Ironbark in properties and characteristics. The most important are Grey (*E. propinqua*), Lemon (*E. citriodora*), Red (*E. tereticornis*), River (*E. rostrata*), and Spotted (*E. maculata*), *q.v.* Also see *Gum.*

Ironing Board. A shaped board, covered with cloth, on which clothes are ironed in laundry work.

Ironmongery. Hardware. Metal fittings and fastenings used on woodwork, such as doors, windows, etc.

Irons. The cutters of planes and machines. See *Head* and *Jack Plane.*

Ironwood. Applied to nearly a hundred different woods throughout the world. They are hard, heavy, tough, dense, durable, and difficult to work. Many of them are species of acacia. AMERICAN IRONWOOD. Blue Beech and Hop Hornbeam, *q.v.* BORNEO IRONWOOD. See *Billian* and *Kajœ.* BURMA IRONWOOD. See *Gangau* and *Pyinkado.* C. AMERICAN IRONWOOD. See *Ibira, Lapacho, Mesquite, Palo diablo, Quebracho, Snakewood, Wamara.* E. AFRICAN IRONWOOD. See *Olivewood.* INDIAN IRONWOOD. See *Pyinkado.* MALAYAN IRONWOOD. See *Arang, Merbau, Penaga, Ru, Tembusu, Tempinis.* NIGERIAN IRONWOOD. See *Ekki.* QUEENSLAND IRONWOOD. *Erythrophlœum Labourcherii.* Reddish brown, lustrous. Resists insects. Used for turnery, sleepers, etc. S.G. 1·1. S. AFRICAN IRONWOOD (Black), *Olea laurifolia.* Very hard, heavy, durable, and close-grained. Used for wagon building, etc. (White), *Toddalia lanceolata*. Very hard, heavy, and tough, but not durable. Close grain. Used for spokes, tool handles, etc. Also see *Melkhout*. TASMANIAN IRONWOOD, *Notalea ligustrina*. Dark brown. Resembles lignum-vitæ and used for similar purposes. Polishes well. UGANDA IRONWOOD. See *Muhimbi*. Also see *Carrol, Lemon, Red*, and *Scrub Ironwoods*, and *Rata*.

Ironwood Box. *Syncarpia subargentea* (*H.*). Queensland. A very hard and heavy fine-grained wood, too scarce to be exported. Used for mallets, turnery, etc. It is too hard for ordinary purposes. S.G. 1.

Irul. *Xylia xylocarpa* (*H.*). India. Reddish brown, darker stripes. Very hard, heavy, and durable ; strong and tough. Resists insects. Interlocked grain. Difficult to season and to work smooth. Heart should be boxed. Used for structural work, sleepers, piling, etc. S.G. ·92 ; C.W. 5.

Irwin Bit. A good type of twist bit for boring in wood ; clean cutting, easily sharpened and in sizes from $\frac{1}{4}$ to 1 in. rising by $\frac{1}{16}$ in.

Iryanthera. See *Kirikowa*.

Ischo Wood. *Gingko biloba* (*S.*). Japan. Used extensively for lacquered furniture. Not exported.

Isolators. Rubber pads, on which the wood joists rest, for a floating floor. See *Floating Floor*.

Isometric. Pictorial projection in which the axes form angles of 120° with each other, with one of them vertical. All measurements

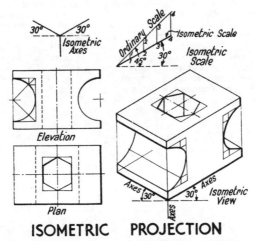

ISOMETRIC PROJECTION

are made along the axes. The isometric scale is not applied in ordinary practice.

Isonda Gutta Tree. East Indies. Produces the best quality of gutta-percha. The wood is not marketed in this country.

Isoptera. See *Selangan Batu, Sengkawang*, and *Yacal*.

Itako. *Celtis soyauxii* (*H.*). Nigeria. Also called Ohia, Ita, and Celtis. Light yellow, lustrous. Hard, heavy, with slightly interlocked grain and liable to pick up. Smooth, and polishes. Under investigation. Suggested for cabinet work and superior joinery. S.G. ·85 ; C.W. 4.

Italian Beetle. *Hylotrupes sp.* A dark brown insect $\frac{5}{8}$ in. long, very destructive to converted softwoods, especially if untreated, as in roofs.

Italian Poplar. See *Poplar*.

Italian Walnut. See *Walnut*.

Itapicura. *Goniorrhachis marginata* (*H.*). C. America. Deep yellow with black streaks. The yellow darkens to purple with exposure. Very hard, heavy, strong, and durable. Smooth, with roey grain. Rather difficult to work. Used for heavy construction, sleepers, etc. S.G. 1·1 ; C.W. 4.

Itchwood. Thitya, *q.v.*

Itrac. Yacal, *q.v.*

Ituri. See *Wallaba*.

Ivory Birch. *Baloghia lucida* (*H.*). Queensland. Ivory hue, smooth. Hard, heavy, tough. Fine even texture, straight grain. Not difficult to work. Used as a substitute for box : carving, inlays, turnery. S.G. 72.

Ivory Coast Mahogany. W. African Mahogany, *q.v.*

Ivory Wood. *Siphonodon australe* (*H.*). Queensland. Ivory white. Very hard and heavy. Fine straight grain and even texture. PINK or RED IVORY, *Rhamnus zeyheri*. S. Africa. Reddish colour. Very hard, heavy, strong. Fine even grain. Used for turnery, inlays, carving, instruments, chessmen, etc. S.G. ·8. *Balfourodendron riedelianum*. C. America. Ivory white to yellow, with fine darker streaks. Hard, heavy, tough, strong, but not durable. Fine, smooth, straight, and wavy grain. Fairly easy to work. Used for cabinet work, turnery, implements, oars, etc. S.G. ·75 ; C.W. 3.

Ixora. Indian silver greywood, *q.v.*

Izod. An impact testing machine for wood

J

Jacaranda. The name applied to several tropical American woods resembling rosewood. See *Rosewood (Brazilian)*. The genus *Jacaranda* is not associated with the woods of that name. See *Fotui* and *Caroba*.

Jacaruba. *Calophyllum brasiliense*. See *Santa Maria*.

Jack. 1. An appliance for raising heavy loads through small heights. It may be hydraulic, pneumatic, or operated by a screw thread and

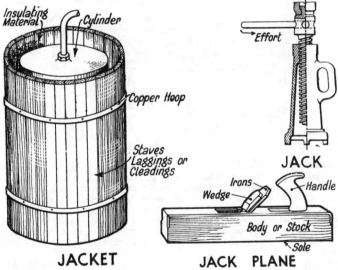

JACKET　　**JACK PLANE**

lever. **2.** Implies anything ordinary or common, or able to withstand rough usage.

Jack Chain. An endless chain with spikes for moving logs to the sawmill.

Jacker. A pond man, *q.v.*

Jacket. 1. An outer coat to protect insulating material. See *Cleadings*. **2.** The facing veneers of laminboard, *q.v.*

Jack Ladder. See *Gangway*, or *Log Haul-up*.

Jack Pine. *Pinus banksiana (S.)*. East and Central Canada. Also called Banksian, Grey, and Princess pine.

Jack Plane. A joiner's plane used for preliminary and rough work, or wood from the saw. The body, or stock, is about 17 in. × 3 in. × 3 in.

Jackpot. An obstruction, or snag, in lumbering.

Jack Rafters or **Timbers.** The short rafters, in the same plane as the common rafters, between hip and wall plate or between valley and ridge. See *Roof Timbers*.

Jack Rib. A curved jack rafter, as used in a turret roof with curved or domical outline. Also called a cripple or crippling.

Jack Slip. See *Log Haul-up*.

Jackstay. A bar on the top of a yard for the sail of a ship.

Jackwood or **Jak.** *Artocarpus integrifolia* (*H.*). India. Yellowish brown, darkens with exposure. Moderately hard, heavy, strong, and durable. Resists insects. Coarse variable texture and grain that makes it difficult to work smooth. Used for cheap cabinet work, interior joinery, turnery. S.G. ·6 ; C.W. 4.

Jackyard. A small spar.

Sideboard   *Table Leg*

JACOBEAN

Jacobean. A transitional style of architecture prevailing between the years A.D. 1603 to 1649. The characteristics were banded or twisted columns and Gothic features with Renaissance details. Spirally twisted columns, diamond-shaped panels, split turnings, and simple frets, were extensively used, especially in oak furniture.

Jacob's Ladder. A ladder for ascending the main mast of a ship. A vertical ladder, usually of wood treads with rope sides.

Jacq. Abbreviation for the name of the botanist Jacquin.

Jacqueria. Jackwood, *q.v.*

Jagged. Uneven or torn. Cut roughly ; or formed with indentations.

Jagua. *Genipa americana* (*H.*). C. America. Also called Genipa and Jacua. Brownish grey, yellowish hue. Hard, heavy, tough, elastic, strong. Many of the properties and uses of ash. Fairly straight grain, rather fine texture. Not difficult to work smooth. S.G. ·85 ; C.W. 4.

Jali Panel. A richly carved, perforated, reticulated panel, characteristic of Indian architecture.

Jalousie. A Venetian blind. An exterior blind or shutter framed with slats in the form of louvres. A jalousie screen is a louvre screen, *q.v.*

Jam. A congestion of logs in running water due to an obstruction.

Jamaica. Applied to several West Indies woods to denote the source of origin : ebony (cocus), mahogany (Cuban), lignum-vitæ, satinwood.

Jamaican Ironwood. *Sloanea jamaicensis.* See *Ironwood*.
Jaman. *Eugenia jambolana* (*H.*). India. Brownish grey, darker streaks, fairly lustrous but feels rough. Moderately hard, heavy, strong, fairly durable, and resists insects. Irregular and wavy grain, some interlocked. Medium coarse texture. Difficult to season without warping but not very difficult to work. Medium sizes. Used for structural and constructional work, sleepers, piling, selected wood for panelling and superior joinery. S.G. ·7 ; C.W. 3·5.
Jamba. Irul, *q.v.*
Jamb Linings. See *Linings* and *Finishings*.

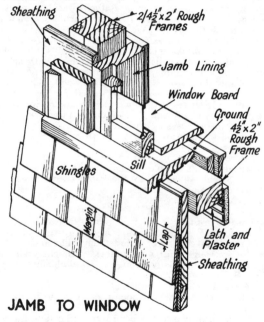

JAMB TO WINDOW

Jambs. **1.** The sides of wall openings. **2.** The stiles of framing fixed in wall openings. **3.** The sides to a fireplace.
Jambu. Pyinkado, *q.v.*, and Kelat, *q.v.*
Jangkang. See *Merawan* and *Simpoh*.
Janker. A two-wheeled log carrier.
Japanese. Applied to numerous woods, imported from Japan : Alder, ash, camphorwood, cedar, chestnut, Korean pine, larch, oak, sugi, walnut. See *Kuro matsu, Larch, Oak, Walnut.*
Japanese Ash. Several varieties. *Tamo* is excellent wood with beautiful grain and figure. It is heavier than English ash, strong

and durable, and used for superior joinery and veneers. *Sen* is not a true ash and is whiter and softer than *Tamo*. Both varieties are esteemed for cabinet work.

Japanese Beech. *Fagus sylvatica.* Very similar to British, but softer and milder. See *Beech.*

Japanned. Applied to iron fittings that are coated with a hard black varnish to protect the iron and prevent corrosion. Also see *Lacquer.*

Jarana. *Chytroma jarana* (*H.*). C. America. Light reddish brown, variegated streaks, darkens with exposure. Very hard and heavy ; durable, dense, strong, and odorous. Flexible but rather fissile. Straight grain, smooth, distinct rays. Rather difficult to work. Used for exterior construction, sleepers, decorative articles, walking sticks, fishing rods, etc. S.G. ·95 ; C.W. 4.

Jardinière. A pedestal or box for flowers.

Jarrah. *Eucalyptus marginata* (*H.*). W. Australia. Called Australian mahogany. Reddish or purplish brown, darkening with age. Hard, heavy, very strong and durable, fire- and insect-resisting. Very large sizes, and probably the most useful Australian wood. Usually straight grain. Interlocked and wavy grain is valued for decorative work, as are the butts. Even texture, moderately coarse. Subject to gum veins. Requires care in seasoning. Not difficult to work and polishes well. Used for all purposes : structural, piling, carriage and wagon work, cabinet work, superior joinery, flooring, paving, turnery, etc. S.G. ·9 ; C.W. 3·5.

Jarul. See *Pyinma.*

Jarvis. A wheelwright's double-handed plane for rounded work.

Jasil Bar. A patent water-bar to check the entry of water at the bottom of casements opening inwards.

Jatay. See *Jetay.*

Jati. Teak, *q.v.*

Jatia. *Phyllostylon brasiliensis.* San Domingo Boxwood, *q.v.*

Jaul. *Alnus sp.* (*H.*). C. America. Light brown, pinkish tinge ; darkens to bronze on exposure. Moderately light and soft ; tough and strong for its weight but not durable. Straight grain, smooth, with rather fine texture. Easily wrought. Used chiefly for carpentry and joinery. S.G. ·48 ; C.W. 2.

Jaw Box. A wood sink lined with lead.

Jaws. The horns at the ends of booms and gaffs to fit them to the mast of a ship.

Jear Block. One with two or three sheaves for hoisting the main- or fore-yards of a ship.

Jebu. Plain figured sapele, *q.v.*

Jeffry Pine. *Pinus jeffreyi.* Similar to and marketed as Ponderosa pine, *q.v.* See *Pine.*

Jelutong. *Dyera spp.* (*H.*). Malaya. White, very soft and light, weak and brittle, but stable. Subject to discoloration and insect attack, and not durable. Fine grain and rays, and easily wrought. Used for pattern making, drawing boards, matches, cases, match-boarding, etc. S.G. ·38 ; C.W. 2.

Jenisero. Kelobra, *q.v.*

Jenny. **1.** A travelling winch. **2.** A gin wheel, *q.v.* **3.** A type of spinning machine.

Jequitiba. *Cariniana spp.* (*H.*). Brazilian mahogany. Reddish to deep red brown, with some darker streaks. Very variable in weight and hardness. Tough and strong for its weight, and durable. Straight grain, medium texture, with distinct rays. Smooth, and easily wrought unless woolly. Another species is called Albarco or Columbian mahogany. Used for carpentry, superior joinery, veneers, etc. S.G. ·5 to ·7 ; C.W. 2·5.

Hip

Gable

JERKIN HEAD

Jerkin Head. A roof that is partly gabled and partly hipped. The feet of the hips are usually about halfway up the triangular gable.

Jesse. An ornate piece of sculpture or a church candlestick with branches representing the " Tree of Jesse."

Jesting Beam. A beam not serving any structural purpose. It is introduced for appearance or symmetry.

Jetay. W. Indies Locust. Also called Algarroba, Courbaril Plum, Jatay, Jutahy. See *Locust*.

Jettied Storey. A projecting upper storey of a half-timbered building.

Jetty. A landing stage. A structure, on piles, at the water's edge.

Jewel Blocks. Two small blocks suspended at the end of the main- and fore-top yards by means of an eye-bolt.

Jewel Drawers. Very small drawers on top of a dressing table for trinkets, etc.

Jib. The main member of a crane from which the load is suspended. See *Derrick Crane.*

Jibboom. A continuation of the bowsprit on a sailing ship. A spar rigged beyond the bowsprit.

Jib Door. A door flush with a wall and designed to be inconspicuous, and to form part of the surrounding surface. It is sometimes pivoted at the top and bottom instead of being hinged.

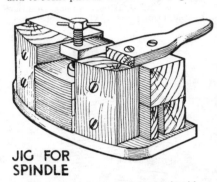

JIG FOR SPINDLE

Jig. **1.** A guide for shaping the material mechanically on a machine. Also called a jigging block, or dummy. **2.** A templet or guide for controlling tool operations. **3.** An arrangement to hold small stuff on the spindle moulding machine. A workholder.

Jigger-mast. The fourth mast of a sailing ship. A small mizzen mast. See *Masts.*

Jigger Router. A coach-maker's side router.

Jigger Saw. A reciprocating saw. A fret, or scroll, or jig saw. It is used for pierced and tracery work, and such work as is outside the scope of a band saw.

Jigger, To. To move a log by horse power over a level part of a slide.

Jigging Block. See *Jig.*

Jig Saw. A jigger saw, *q.v.*

Jiqui. *Bumelia spp.* or *Pera bumelifolia* (*H.*). C. America. Colour of rosewood. Darkens and hardens with age. Very hard, heavy, durable. Compact fine grain. Difficult to work. Used for piling, posts, sleepers, cabinet work. S.G. 1 ; C.W. 5.

Jobillo. Goncalo alves, *q.v.*

Jobo. *Spondias lutea* (*H.*). Mexico. See *Hog Plum.*

Jockey Block. A specially shaped block to carry curved stock over the spindle when machining.

Jocote. *Vitex longeracemosa* (*H.*). C. America. Yellowish brown, lustrous. Moderately hard, heavy, and durable. Strong but rather brittle. Smooth, with straight and irregular grain. Fairly easy to work. Used for general purposes, construction, etc. S.G. ·7 ; C.W. 3. The name Jocote is also applied to Fiddlewood, *q.v.*, and other species.

Jodelite. A proprietary wood preservative.

Jogged. Applied to squared logs that are stepped in their length to form a series of prism-like portions, instead of tapering.

Joggle. **1.** A small projection on the end of a framed member to strengthen mortised and other forms of angle joints. It is often ornamented as on the stile of a sliding sash. In structural work it allows for housing, and provides a better bearing resistance. **2.** See *Jogged.* **3.** A form of cranking of a beam or frame to receive deck plating or side plating, in shipbuilding, to provide a seating for the plates. In some cases the plates are joggled. **4.** See *Stop* (6).

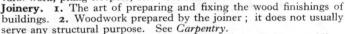

Joggled Piece. A post with a shoulder to provide abutment for a strut.

Johore Teak. *Parinarium oblongifolium* (H.). Malay. It is not teak. Now called Balau or Ballow. Yellow, seasoning to dark brown. Hard, heavy, strong, durable, and resists

JOGGLES

insects. Used for structural work, piling sleepers, etc. See *Merbatu*.

Joinery. 1. The art of preparing and fixing the wood finishings of buildings. **2.** Woodwork prepared by the joiner ; it does not usually serve any structural purpose. See *Carpentry*.

Joint Board. A board on which a pattern is rammed in moulding.

Joint Connectors. See *Timber Connectors*.

Jointer. 1. A plane longer than a try-plane for straightening or shooting long edges when making butt joints. **2.** A special machine for the same purpose. **3.** A cooper's plane for shaping the edges of staves.

Joint Fastener. A piece of thin corrugated metal used as a permanent fastener to tighten and secure glued butt joints. They are usually driven in the ends, but in cheap work they are sometimes driven in the face. Also see *Timber Connectors*.

Joint Gluer. A machine for jointing veneers without taping.

JOINT FASTENER

Jointing. 1. Preparing the joints of woodwork, especially butt joints. **2.** Sometimes applied to breasting, *q.v.*

Joint Mould. See *Section Mould*.

Joint Paper. Used between temporarily glued joints to simplify the processes ; the joint is easily broken when the work is completed. Also used to cover the joints of veneer until the glue has set, to prevent the edges from lifting. See *Taping*.

Joints. Prepared connections between two or more pieces of material. Woodwork joints are classified as (*a*) Angle Joints : blocked, bridle, butt, cogged, dovetailed, dowelled, feathered, franked, gun-stock, halved, housed, mitred, mortised, notched, rebated, scribed, tenoned, trenched. (*b*) Lengthening Joints : cramped, fished, halved, indented, keyed, lapped, scarfed, splayed or spliced, stepped, tabled. (*c*) Increasing Width : butt, clamped, dowelled, grooved, matched, rebated, secret screwed, tongued. (*d*) Bearing or Cross Joints : birdsmouth, bridle, cogged, halved, housed, notched, tenoned, tusk.

(*e*) HINGING AND SHUTTING JOINTS : bevelled, hook, knuckle, lapped, rebated, rule, and variations called air-tight, wa er-tight, storm-proof, etc. There are many combinations and variations of the above, and joints may be *secret* or *frank*. The cabinet-maker uses similar joints to the joiner, but there are several peculiar to the work, as : coopered, fall, finger, knuckle, laminated, rule, secretaire joints, etc. Other trades also vary the names of many of the above joints, but there is little difference in the construction. They are described alphabetically in the glossary. Woodwork joints are strengthened by battens, buttons, bolts, clamps, coach-screws, cramps, dowels, feathers, nails, screws, and by various types of fasteners.

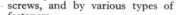

JOIST HANGER

Joists. Horizontal timbers bridging or spanning openings and carrying floorings and/or ceilings. See *Floors* and *Ceiling Joists*.

Jolli Panel. A jali panel, *q.v.*

Jolly Boat. A ship's small boat. It is like a ship's cutter but has four oars.

Jonote. *Heliocarpus americanus*. Amor seco, *q.v.*

Joy Stick. The control column of an aeroplane.

Jube. 1. A choir screen. 2. A rood loft over the rood screen in a church.

Juca. *Cæsalpinia ferrea* (*H.*). Brazil. Purple brown. Very hard, heavy, and durable, but liable to insect attack. Similar to some of the ebonies. S.G. 1·2.

Jucaro Negro. *Bucida buceras* (*H.*). Cuba. Dark brown, like black walnut. Very heavy and hard, strong, tough, elastic, and durable. Fine smooth grain. Not very difficult to work and polishes well. Used chiefly for structural work, piling, shipbuilding. S.G. 1·1.

Judas. An aperture or peep-hole in a door, etc.

Judas Tree. *Circis siliquastrum* (*H.*). Asia. Of little timber value.

Judio. Like Jonote, *q.v.* Mexico. Resembles sugar pine. Fast growing, straight grain. Used for veneer cores, pulp, etc.

Juglandaceæ. Walnut family. Includes Carya or Hickoria, Engel-hardtia, Juglans, Platycarya, Pterocarya.

Juglans. See *Walnut, Nogal, Butternut*.

Jujube. *Zizyphus jujuba* (*H.*). China. Similar to Buckthorn. Also called Baer and Rengha. Light walnut colour. Fairly hard and heavy. Similar to straight-grained black walnut for which it is a substitute.

Jumbo Bolt. A bolt operated by the thumb, instead of by a key, in a rim lock. It can only be operated from the inside of the room.

Jumper. See *Swage*.

Jump Saw. A circular saw that can be raised or lowered. It is used for cross-cutting.

Jungle Cypress. *Callitris macleayana*. See *Queensland Cypress Pine*.

Juniper. *Juniperus communis* (*S.*).   Britain.   Common juniper. Chiefly shrubs, and of little commercial value.   INDIAN JUNIPER, *Juniperus macropoda*.   Reddish brown, purplish cast, fragrant, lustrous, and smooth.   Light, soft, fairly durable.   Straight close grain, even texture.   Fine markings and mottled.   Easily wrought but subject to knots.   Transport difficulties limit supplies.   Used for pencils chiefly.   S.G. ·43.   AMERICAN JUNIPER.   Red cedar, *q.v.*

Juniperus.   Numerous species.   *J. virginiana*, Virginian cedar.   *J. barbadensis* or *J. silicicola*, similar to Virginian, Southern U.S.A. *J. communis*, see Juniper.   *J. procera*, see African cedar.   *J. chinensis*, Chinese juniper, or Tzu sung.   *J. bermudiana*, similar to Virginian. Also see *Australian Cedar* and *Juniper*.

Junk.   1. An eastern sailing ship.   2. A collection of useless things.

Jury.   An adjective implying " temporary."

Jury Strut.   A strut providing intermittent or temporary support, as to the folding wings of an aeroplane.

Juss.   An abbreviation for the name of the botanist Jussieu.

Jutahy.   *Dialium divaricatum* (*H.*).   C. America.   Reddish brown, darkens with exposure.   Very hard, heavy, tough, strong, durable, and dense.   Rather fine texture and interwoven grain.   Fairly difficult to work.   Used for structural work, piling, sleepers, spokes, implements.   S.G. ·9.   *Apuleia leiocarpa* and *Hymenæa courbaril* are also called Jutahy.

Jutili.   *Altingia excelsa* (*H.*).   India.   Reddish brown, fairly lustrous. Hard, heavy, strong, durable, odorous.   Wavy and interlocked grain. Moderately fine even texture and smooth.   Difficult to season and work.   Used for structural work, sleepers, wagon work.   S.G. ·77.

K

Kaba. *Betula sp.* Japan. Plain birch.

Kabbes. See *Sapupira*.

Kabukalli. *Goupia spp.* (*H.*). B. Guiana. Reddish brown, odorous until seasoned. Hard, heavy, durable. Used for furniture, general purposes, sleepers, etc. Also see *Cupiuba*.

Kadam. *Anthocephalus cadamba* (*H.*). India. Creamy white, lustrous, smooth. Fairly light and soft. Moderately strong, not durable. Straight grain. Medium even texture. Easy to season and work. Used for packing cases, matchboarding, cheap turnery. S.G. ·53 ; C.W. 2·5.

Kahikatea. *Podocarpus dacrydioides* (*S.*). New Zealand white pine, or Dominion whitewood. Yellow, large amount of white sapwood which is subject to attack by borers. Light, soft, moderately durable but subject to insect attack. Straight, even grain. Easily wrought. Large sizes free from defects. Characteristics of yellow pine. Used for cases, butter-boxes, veneer cores, joinery. S.G. ·48 ; C.W. 1·5.

Kaikawaka. *Libocedrus bidwillii* (*S.*). N. Zealand mountain cedar or Kawaka. Dark red. Soft, light, rather weak, durable, fire-resisting. Local demands equal supplies. Used for poles, shingles, pencils, joinery. S.G. ·47.

Kaim. *Stephegyne,* or *Nauclea,* and *Mitragyna, parvifolia* (*H.*). India. Greyish brown. Moderately hard and heavy, not durable. Very similar to Binga. Straight, interlocked, and wavy grain. Fine even texture. Fairly difficult to work because liable to pick up. Polishes well. Used for construction, cheap furniture, turnery, carvings, woodware, boxes. S.G. ·63 ; C.W. 3·5.

Kaita-da. Chaplash, *q.v.*

Kajatenhout. *Pterocarpus angolensis* (*H.*). S. Africa. Now called Muninga or Kajat. Also called Kejaat, Bloodwood, Sealing wax tree, and Transvaal teak. Dark brown, some variegated. Hard, heavy, strong, very durable but sapwood subject to borers. Seasons and works well. Good figure and esteemed for furniture, superior joinery, flooring, etc. S.G. ·75 ; C.W. 3·75.

Kajoe Besi. *Eusideroxylon zwageri* (*H.*). Borneo Ironwood. Brown. Very hard, heavy, strong, and durable. Used for structural work, piling, etc. S.G. ·9.

Kakaralli. See *Manbarklak*.

Kakatara. *Ilex sp.* (*H.*). C. America. Greyish white. Moderately hard, heavy, but not durable. Straight grain, medium texture. Fairly easy to work. Inferior to holly. S.G. ·8.

Kaki. *Diospyros kaki* (*H.*). Japanese ebony. Black, variegated streaks, or dark grey with black streaks. Hard, fairly heavy, odorous until seasoned. Highly decorative and esteemed for fancy articles. It has a very smooth cold surface, like marble. See *Ebony*.

276

Kakra. *Bruguiera sp.* (*H.*). India, Burma. Light red to reddish brown on exposure. Very hard, heavy, strong, and durable. Straight grain, fine even texture. Used chiefly for structural work. S.G. ·95.
Kaku. Ekki, *q.v.*
Kalabau. Merbau, *q.v.*
Kala Chuglum. Black chuglum. See *Chuglum.*
Kalado. *Homalium alnifolium* (*H.*). Nigeria. Yellow. Very hard, heavy, strong, durable. Straight and interlocked grain, but does not pick up. Seasons readily, fairly difficult to work. Under investigation. Suggested for sports equipment, decking, superior joinery, turnery. Subject to shakes in the log. S.G. 1.
Kalamein. A registered type of metal-faced mouldings, etc., used for shop fronts and fittings. Sheet bronze, from 12 to 22 gauge, is

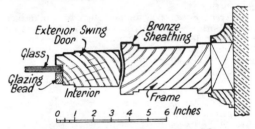

KALAMEIN SHEATHING

drawn on to a hardwood core up to 50 ft. in length. See *Metal Faced Joinery.*
Kalamet. *Cordia fragrantissima* (*H.*). Burma. Also called Burmese Sandalwood. Rich reddish brown, darker streaks, fragrant. Moderately hard, heavy, and durable. Ray figure. Requires careful seasoning. Fairly difficult to work. A decorative wood suitable for cabinet work, superior joinery, etc. S.G. ·75 ; C.W. 3·5. Also see *Cordia.*
Kalantas. Calantas, *q.v.*
Kalmia. See *Laurel.*
Kalopanax. See *Sen.*
Kalumpit. *Terminalia spp.* (*H.*). Philippines. Yellowish or greyish brown, lustrous. Moderately hard, heavy, and strong, but not durable. Wavy and cross grain, fairly fine texture. Characteristics of white lauan, *q.v.* Used for flooring, match-boarding, joinery, cheap cabinet work. S.G. ·6 ; C.W. 3·5.
Kalunti. White lauan. See *Lauan.*
Kaluwara. Indian ebony, *q.v.*
Kamahi. Towai, *q.v.*
Kamap. *Strombosia rotundifolia* (*H.*). Malaya. Light olive to purplish brown, with large proportion of yellow sapwood. Very hard, heavy, with close compact grain. Heartwood very durable. Medium sizes. Difficult to work. Similar to Petaling, *q.v.* Used chiefly for fancy articles, walking sticks, etc. S.G. ·84 ; C.W. 5.

Kamassi. *Gonioma Kamassi* (*H.*). S. Africa. Knysna boxwood. Characteristics, colour, and properties of boxwood. Used chiefly for weaver's shuttles and turnery.

Kambai. Simpoh, *q.v.*

Kambala. Iroko, *q.v.*

Kanooka. *Tristania laurina* (*H.*). Victoria and Queensland. A very hard, compact, decorative wood, used for fancy articles, pipes, furniture, etc., but little exported. Similar to Brush Box, *q.v.*

Kanran. *Diospyros mespiliformis* (*H.*). Nigerian ebony. Produces grey streaked ebony of very good quality. Centre of log is black. Difficult to work, as fibrous grain is liable to pick up. Used for decorative work and fancy articles. Outer pinkish wood for carving, turnery, etc. S.G. ·85.

Kanti. *Acacia ferruginea* (*H.*). India. Olive brown, darker markings, dull but feels smooth. Very hard, heavy, strong, and durable. Straight even grain, coarse texture. Requires careful seasoning and fairly difficult to work. Used for heavy construction, wagon work, brake blocks, turnery. S.G. ·98 ; C.W. 4.

Kanuka. See *Kanooka*.

Kanyin. Gurjun, *q.v.*

Kapar. Keruing, *q.v.*

Kapok Tree. Silk cotton tree, *q.v.*

Kapor or **Kapur.** Borneo camphorwood, *q.v.*

Kapp Balk. A Norwegian balk with two opposite sides hewn flat and two sides left round but free from bark.

Karani. Wild Durian, *q.v.*

Karanji. Indian Elm, *q.v.*

Karapa. Crabwood, *q.v.*

Karelian Birch or **Burl.** Figured birch.

Karen Wood. See *Waras* and *Petthan*.

Karri. *Eucalyptus diversicolor* (*H.*). W. Australia. Reddish brown. Resembles Jarrah. Very hard, heavy, strong, dense, tough, fire-resisting, but not so durable as Jarrah. Interlocked, with some wavy, grain. Moderately coarse texture, but smooth. Requires care in seasoning and fairly difficult to work. Polishes well. One of the largest trees. Used for structural and constructional work, wagon work, ship building, bent-work, implements, flooring, cabinet work, superior joinery, veneers ; also for poles, sleepers, and piles after treatment. S.G. ·92 ; C.W. 4.

Kartang. *Centrolobium sp.* (*H.*). B. Guiana. Also called Pau Rainha. Orange, variegated markings. Hard, heavy, durable. Used for constructional work, shingles, fencing, etc. S.G. ·75.

Karum-Kali. Indian ebony, *q.v.*

Kashi. Japanese oak, *q.v.*

Kata. Berangan, *q.v.*

Kath. Cutch, *q.v.*

Katpali. Mohwa, *q.v.*

Katsura. *Cercidiphyllum japonicum* (*H.*). Japan. Light nut brown, lustrous. Soft, light, with close, even grain. Very smooth and easily wrought. Retains sharp arrises. Fairly large sizes. Characteristics of Kauri pine. Used for furniture, lacquered work, delicate mouldings, cigar boxes, carving, engraving. S.G. ·34 ; C.W. 1·5.

Katus. *Castanopsis hystrix.* A species of Indian chestnut, *q.v.*

Katydid. Logging wheels, used in lumbering.

Kauri Pine. *Agathis australis,* or *Dammara a.* (*S.*). N. Zealand. *A. vitiensis,* Fiji Islands. *A. robusta, A. Palmerstonii,* and *A. microstachya,* Queensland. Pale yellow to light brown, lustrous. Fairly light and soft. Strong and elastic for its weight, and fairly durable. Large sizes free from defects. Less absorbent than other softwoods, hence difficult for treatment. Close, uniform, straight grain, and fine texture. Some wavy and mottled. Liable to warp. Easily wrought and retains sharp arrises. Polishes well. Smooth and non-splintering. Supplies limited. Used for joinery, cabinet work, vats, stills, carriage work, slides, decking, etc. S.G. ·56 ; C.W. 2.

Kauta. *Moquilea sp.* (*H.*). B. Guiana. Brown. Hard, heavy, durable. A useful wood but not exported at present. S.G. 1.

Kauta Balli. *Licania sp.* (*H.*). B. Guiana. Dark brown. Very hard, heavy, and strong, but not durable. Straight close grain. Used for heavy constructional work. S.G. 1·2.

Kawaka. Kaikawaka, *q.v.*

Kawanari. Simiri, *q.v.*

Kaya. *Torreya nucifera* (*S.*). Japan. Yellow, lustrous. Light, fairly hard, with smooth, close grain. Similar to Port Orford cedar. S.G. ·5. *Minusops elengi* (*H.*). Burma. Also called Bullet wood, Bukal, etc. Plum red. Hard, heavy, strong, and tough. Close compact grain, wavy and distinct rays. Used for cabinet work. S.G. ·9 ; C.W. 4·5.

Kayatau. *Dysoxylum turczaninowii* (*H.*). Philippines. Reddish yellow, lustrous, fragrant. Hard, heavy, strong, stable, durable, and resists insects. Fine, even texture, with straight grain. Some wavy and slightly crossed. Seasons and works well. Used for furniture and superior joinery. S.G. ·9 ; C.W. 3.

Kayea. *Kayea assamica* (*H.*). Assam. Also called Sia Nahor. Reddish brown, fairly lustrous, smooth. Very hard, heavy, but not durable. Interlocked grain, fine texture. Fairly difficult to work. Characteristics of Mesua, *q.v.* Plentiful supplies and under investigation. S.G. ·9.

Kayu. See *Mersawa, Malayan ebony,* and *Ru.*

k.d. Abbreviation for *kiln dry,* and *knocked down.*

Keaki. *Zelkova acuminata* or *Kelkova keaki* (*H.*). Japan. Elm family. Golden brown, lustrous. Hard, tough, elastic, durable, and very strong for its weight. Highly decorative wood with delicate bird's eye with outline of very fine fringes, and distinct rays. Easily wrought and highly esteemed for cabinet and decorative work. Plain wood used for structural work, shipbuilding, etc. S.G. ·65 ; C.W. 3.

Keekar. Babul, *q.v.*

Keel. 1. The central longitudinal structural member of a ship. It runs from stem to stern at the bottom of the ship's framing, for which it forms the support. 2. The snaped end of a log. 3. A coal

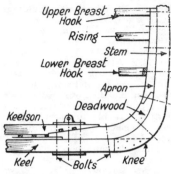

Upper Breast Hook

Rising

Stem

Lower Breast Hook

Apron

Deadwood

Keelson

Keel

Bolts

Knee

KEEL OF BOAT

barge. **4.** A fillet or projection forming part of a roll or scroll moulding. **5.** A Yorkshire river cargo boat.

Keel Blocks. Piles of wood balks for carrying the weight of a ship during construction or in dry dock. Similar blocks at the sides are called bilge blocks.

Keel Moulding. A Gothic moulding formed of two ogees meeting with a sharp edge, like the keel of a ship.

Keelson. A second interior keel over the keel, *q.v.* It is usually the chief longitudinal structural member or girder. See *Stem* and *Head*.

Keeper. A recessed piece of metal to restrict the movement of the fall-bar of a Norfolk latch, *q.v.* Anything that controls a moving or sliding object. A hollow metal box screwed to a door frame and into which the bolts of a lock shoot. A staple. See *Tower Bolt*.

Keeper Plate. A striking plate, *q.v.*

Kejaat. See *Kajatenhout.*

Keladan. A species of *Dryobalanops*, similar to Borneo Camphorwood in all respects but less durable.

Kelat. *Eugenia spp.* (*H.*). Malaya. Numerous species. Greyish to chocolate brown. Hard, heavy, strong, fairly durable. Rather coarse grain. Fine rays. Moderate sizes. Used chiefly for heavy constructional work. Small stuff used for tool handles, etc. S.G. ·61 to ·85.

Keledang. *Artocarpus lanceæfolia*, and other species (*H.*). Malaya. Dark yellow, but turns brown with exposure. Moderately hard and heavy, durable, strong. Coarse grain, distinct wavy rays. Difficult to work due to latex. Polishes. Large sizes. Little imported at present. Suitable for heavy construction and cheaper furniture. Valued for Chinese coffins. S.G. ·58 ; C.W. 4.

Kelempening. Mempening, *q.v.* '

Kelobra. Guanacaste, *q.v.*

Kemap. Petaling, *q.v.*

Kempas. *Koompassia malaccensis* (*H.*). Malaya. Orange red. Very hard and heavy, strong. Subject to insect and fungi attack but can be treated effectively. Distinct fine rays. Inclined to split. Difficult to cut and work because of hardness. Large sizes. Suitable for construction and furniture. S.G. ·92 ; C.W. 4·5.

Kemuning. *Murraya exotica* (*H.*). Malaya. Yellow, with olive or black mottles. Very hard and heavy. Very fine even grain, and rays. Requires careful seasoning. Small sizes. Good substitute for boxwood. Used for decorative articles, walking sticks, etc. S.G. ·9.

Kenmore. A registered wall-board, *q.v.*

Kennel. A box-like shelter for a dog.

Kentucky Coffee Tree. *Gymnocladus dioicus* and *G. canadensis* (*H.*).
U.S.A. Sometimes called mahogany. Cherry red to brown. Fairly
hard and heavy ; strong and durable. Coarse texture. Little exported.
Used for furniture, joinery, structural work, sleepers, etc. S.G. ·56.
Kenya. The woods exported include camphor, cedar, greenheart,
mahogany, mlange, muvule, olive, podo, satinwood, white pear. Also
see *Mueri, Muna, Mukeo, Muringa, Muhuhu.*
Kepong. Meranti, *q.v.*
Keranji. *Dialium spp.* (*H.*). Malaya. Dark brown, lustrous, odorous
before seasoning. Very hard and heavy ; strong, durable, and tough.
Fine grain and rays, with ripple. Large proportion of pale yellow
sapwood of little value. Fairly large sizes, but supplies limited.
Difficult to work ; polishes. Used for furniture, heavy construction,
etc. S.G. ·76 ; C.W. 4·5.
Kerb. See *Curb.*
Kerf. 1. A saw-cut only part way through the stuff. A saw groove.
2. The width of a saw-cut.
Keruing. *Dipterocarpus spp.* (*H.*). Malaya. Numerous species.
Same as Apitong, Eng, and Gurjun. Greyish or brownish red.
Moderately hard and heavy, to very heavy, strong and stiff. Experience
suggests that it is fairly durable in this country, but it should be
treated. Coarse grain, sometimes wavy, with good figure. Distinct
rays. Resin canals. Large sizes. Difficult to work because of resin
and hardness. Used for structural and constructional work, sleepers,
when treated, cheap furniture, flooring. S.G. ·6 to ·94 ; C.W. 4.
Ketch. A two-masted sailing ship. Similar to a yawl but with a
larger after-mast.
Ketenggah. *Murraya caloxylon* (*H.*). Malaya. Very similar to
Kemuning, *q.v.*
Kevazingo. *Copaifera arnoldiana, C. tessmanni, C. demeusei.* See
Bubinga.
Kevel. A nautical name for pegs and cleats, usually in pairs, to
which ropes are belayed.
Key. 1. A wedge, in carpentry. 2. A piece of hardwood inserted
in a joint to prevent movement of the adjacent surfaces. See *Fish
Plates, Double Dovetail,* and *Hammer Head Key.* 3. An appliance
for operating the bolt of a lock. See *Mortise Lock.* A *pass key* operates
a number of similar locks ; a *master key* operates all the locks of a
building ; a *servant,* or *journey,* key only operates the lock for which
it is intended. 4. A dovetailed batten, see *Counter Batten.* 5. Same
as Cotter, *q.v.* 6. A spanner for turning a nut to a bolt. 7. Levers
that control the hammers striking the strings of a piano, or the valves
of an organ, etc., and operated by the fingers. 8. Applied to numerous
things that fasten or secure one object to another, as railway keys, etc.
9. A plaster surface roughened to receive another coat. 10. See
Slip Feather. 11. See *Wreathed String.*
Keyaki. Keaki, *q.v.*
Keyboard. The series of keys to a piano or the manual of an organ.
See *Key* (7).
Keyed. Secured by a key, or wedge. Structural joints with keys
to prevent movement. See *Fish Plates* and *Key.*

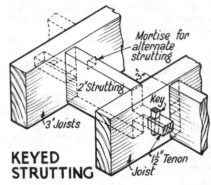

Mortise for alternate strutting

2" strutting

3" Joists

Key

KEYED STRUTTING

1½" Tenon

Joist

Keyed Strutting. Strong pieces tenoned through joists, etc., and wedged like a tusk tenon, to stiffen the joists.

Keyhole Plate. A metal plate to a keyhole to protect the surrounding wood. See *Escutcheon*.

Keyhole Saw. A pad-saw, *q.v.*

Khair. Cutch, *q.v.*

Khaja. *Bridelia retusa* (*H.*). India, Burma. Numerous vernacular names. Greyish brown. Similar to Butternut. Hard, heavy,

Setscrews to fix saw

Saw not in use

KEYHOLE SAW

with close and rather firm grain. Difficult to work smooth. A second-class wood and not imported.

Kharpat. *Garuga pinnata* (*H.*). India. Reddish brown. Fairly hard, durable. Even grain with distinct rays. Little imported. Used for planking, etc. S.G. ·65.

Khasia Pine. *Pinus khasya* (*S.*). Burma. Also called Tinya. Similar to *Pinus sylvestris*.

Khaya. See *Mahogany* (*African*).

Kiaat. See *Kajatenhout*.

Kiabooka. Amboyna, *q.v.*

Kibara. See *Yellow Beech*.

Kick. 1. The projection on the stock-board to form the frog in a brick. **2.** The throwing back of the wood in sawing and machining due to careless handling, defects in the wood, closing of saw-cut, etc. Also called back-lash. See *Riving Knife*.

Kick Back. 1. A term used in wood preservation for the surplus antiseptic that is released from the wood when the pressure is released. **2.** See *Kick* (2).

Kicker. A rail across the top of a carcase for drawers to prevent a drawer from dropping at the front when open.

Kicking Plate. A wide metal plate, usually stainless steel, on the bottom rail of a door, to protect the wood.

Kicking Strap. A heel strap that does not take any thrust but only ties the timbers together.

Kick Stamp. A machine for splitting wood to fuel size.

Kijaat. Kajatenhout, *q.v.*

Kilderkin. A barrel of 18 gals. capacity.
Kilhig. A strong pole used to lever a tree in the required direction when felling.
Kill. See *Knotting*.
Killesse. 1. A local term for a dormer or for a hipped roof. **2.** See *Coulisse*.
Killing Knots. Coating knots with *knotting* to prevent the resin from penetrating the paint.
Kiln Dried Saps. A grading term for southern yellow pine, U.S.A.
Kiln Dry. A description of wood that has been kiln-dried to a moisture content of less than 12 %.
Kilns. 1. Chambers for the seasoning of wood in which the temperature and humidity are mechanically and scientifically controlled.

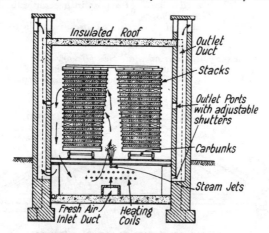

KILN FOR SEASONING

There are numerous types but the overhead internal fan compartment kiln is an efficient and economical form for normal purposes. It should accommodate two stacks, or piles, in width, 9 ft. high by 6 ft. wide, and it may be of any length up to 50 ft. It is provided with instruments for automatically controlling the temperature, and steam sprays for the humidification. Good circulation of the air is essential, and smoke is injected to test the currents. The heating surface should be in the proportion of 1 sq. ft. per 10 c. ft. volume of kiln. Sample pieces should be tested for rate of drying and moisture content. See *Progressive* and *Natural Draught Kilns*, *Seasoning*, and *Moisture Content*. Also see *Preservation*. **2.** Chambers for the manufacture of bricks, cements, lime, etc. There are numerous types : Hoffman, continuous, English, intermittent, Manchester, Scotch, etc. **3.** See *Autoclave*.

Kiln Seasoned. A description of wood that has been kiln seasoned but not to any particular percentage moisture content, unless it has been specified.

Kina. See *Oleo*.

Kindal. *Terminalia paniculata* (*H.*). Indian Almond Tree family. Greyish brown to rich brown, lustrous. Fairly hard and heavy; strong and durable. Straight grain, medium texture, distinct rays, and smooth. Seasons fairly well but not easily wrought for polishing. Good substitute for teak. Appearance and characteristics of black walnut. Used for structural, constructional, cabinet, and wagon work, etc. S.G. ·75; C.W. 4·5.

Kindling Wood. Firewood.

King. The name of the botanist Sir George King.

King Bolt. A bolt used instead of a post in a composite, or king-bolt, roof truss. See *King Post*.

Kingia. The grass tree of Australia. Fibres valuable for brooms and brushes.

Kingiodendron. See *Batete*.

King Nut Hickory. *Carya sulcata*. See *Hickory*.

King Post. 1. The vertical member at the middle of a king-post roof truss, which is the usual type of wood roof truss for spans between

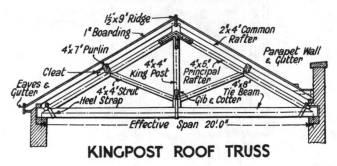

KINGPOST ROOF TRUSS

20 ft. and 30 ft. See *Roof Trusses* and *Cottered Joint*. 2. A compression member, with bracing wires, used in aircraft to limit the deflection of other members. 3. Short strong masts on cargo boats to serve as masts for derricks, etc., to facilitate loading and stowing.

King Tree. Mora, *q.v.*

King William Pine. *Athrotaxis selaginoides* and *A. cupressoides* (*S.*). Also called Red and King Billy pine. Tasmanian pine. Pinkish. Soft, light, fairly strong, durable, and stable. Straight, fine, and even grain. Open, varied texture. Easily wrought. Used for cheap cabinet work, joinery, pattern making, etc. S.G. ·36; C.W. 2.

Kingwood. *Dalbergia spp.* (*H.*). Tropical America. Species of Rosewood. Also called Violet Wood. Rich dark brown to almost black, variegated, lustrous, smooth. Rather small sizes. Used for decorative cabinet work, veneers, etc. S.G. 1; C.W. 4. See *Rose-*

wood and *Goncalo alves*. The name Kingwood is also applied to other highly decorative S. American woods, especially Zebrawood, *q.v.*

Kinked. Buckled ; undulating.

Kino. An astringent gum from eucalyptus trees.

Kiosk. 1. An open pavilion or summer house supported by pillars.

KIOSK

2. A small wood erection, like an elaborate glazed sentry box, but with serving hatch, for the sale of popular commodities.

Kirb. Curb, *q.v.*

Kirballi. See *Vayila*.

Kiri. *Paulownia tomentosa* (*H*.). Japan. Nut brown. Very light, soft, and stable. Fine even grain, and easily wrought. Used for linings of superior cabinet work, musical instruments, thin veneers, floats, etc. S.G. ·3 ; C.W. 1·5.

Kirikowa. *Iryanthera hostmanni* (*H*.). B. Guiana. Light brown, moderately smooth and lustrous. Moderately hard and heavy, elastic, and fairly strong. Straight grain. Not difficult to work. Little used in this country as yet. S.G. ·55 ; C.W. 3.

Kirkia. See *Wit Sering*.
Kissing Gate. A gate with limited movement and without fastenings

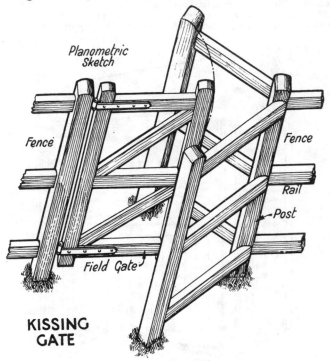

Planometric Sketch

Fence

Fence

Rail

Post

Field Gate

KISSING GATE

for a level crossing or field. It is arranged to prevent the passage of animals or vehicles.
Kist. A large box-shaped linen chest.
Kit. A set of tools. An outfit. The equipment necessary for a craftsman. **2.** A box or bag holding equipment or a set of tools. **3.** A small wood tub. **4.** A small fiddle.
Kitchen Fitments or **Fittings.** Cupboards, dressers, racks, shelves, tables, etc., used for the storing and preparation of food, etc. There are many patent and registered designs, with numerous labour-saving devices.
Kite Winder. The centre winder of quarter-space winders, *q.v.* It is so called because of its resemblance to a kite or trapezium. See *Geometrical Stairs*.
Klainedoxa. See *Akwekwe, Alukonraba,* and *Odudu*.
Knar or **Knarl.** A knot in wood. See *Gnarled*.

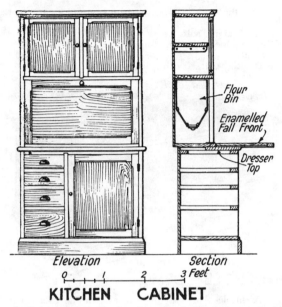

Elevation   Section

0   1   2   3 Feet

KITCHEN CABINET

Knee. **1.** A curved brace or bracket. **2.** A naturally crooked timber, cut from the crotch of a tree. **3.** Part of a sawmill carriage head-block

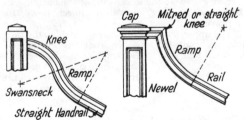

KNEE TO HANDRAIL

that carries the dogs to secure the log. **4.** A vertical convex curve in a handrail. It may be curved or straight and mitred. **5.** See *Crook* and *Keel*. **6.** Angle braces on a ship : beam-knee, carling-knee, etc. **7.** A bent piece of iron serving as a corbel, etc.

Knee Bolter. A machine consisting of a circular saw and small travelling carriage for preparing shingle bolts.

Knee Brace. An angle brace, from the tie beam of a roof truss to the support for the truss. It resists distortion due to wind pressure. See *Roofs*.

Kneed Bolt. A crank bolt, *q.v.*

Kneed Hinge. One with a right-angled bend for the knuckle. Used for skylights, etc.

Kneehole Desk. A desk with side pedestals.

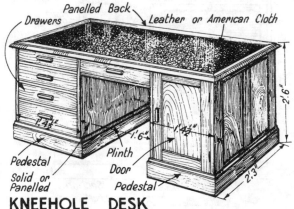

KNEEHOLE DESK

Kneeler. A simple plain type of low desk, similar to a Litany desk, *q.v.* The name is usually applied to a temporary or movable one.

Kneeling. A term applied to a handrail-easing leading to a lower level. A concave curve at the lower end of a handrail. See *Knee*.

Kneeling Boards. Sloping boards to church pews, for kneeling upon during prayers.

Knee Part. The upper part of a chair leg into which the rail is tenoned.

Knick-knack. A small, light, fragile ornament or article of furniture.

Knife. See *Bench* and *Riving Knife*. Also see *Veneers*.

KNIFE (BENCH)

Knife-board. 1. A board used, with emery powder, for cleaning cutlery. 2. A division between longitudinal back-to-back seats on top of an old type of omnibus.

Knight Heads. Strong vertical timbers near the stem of a ship for the bowsprit.

Knightia. See *Honeysuckle* and *Rewa-rewa*.

Knittles. Small pieces of wood by which the sail of a boat is bent to the yard.

Knob. A hard rounded protuberance. A boss. A rounded terminal

to anything, as to the spindle operating the latch of a door, etc.
See *Hand*.

Knob-kerrie. A club with heavy head used as a weapon.

Knob-stick. A walking stick with the head in the form of a knob.

Knobthorne. *Acacia pallens* (*H.*). S. Africa. A hard, heavy, strong, and durable wood used for guide blocks, fence posts, pit props, etc.

Knobwood. *Fagara davyi* (*H.*). S. Africa. A hard, heavy, strong, elastic wood, with close grain. Used for pick handles, yokes, etc.

Knocked Down. Dismantled, or taken apart, for easy conveyance. A term applied to the parts of a box before assembly.

Knot. 1. A section through a branch embedded in the trunk of a tree due to the natural growth. They are classified according to

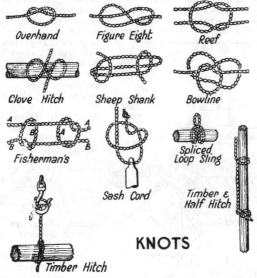

Overhand   Figure Eight   Reef

Clove Hitch   Sheep Shank   Bowline

Fisherman's   Spliced Loop Sling

Sash Cord   Timber & Half Hitch

KNOTS

Timber Hitch

size, quality, occurrence, and form, as follows—*Branched :* two or more branching from a common centre. *Dead :* not joined throughout to the surrounding wood. *Decayed :* one infected with fungus. *Encased :* a dead knot with bark included, or surrounded by pitch. *Enclosed :* one not visible on the face of the wood. *Large :* a knot over $1\frac{1}{2}$ in. diameter. *Live :* one firmly joined to the surrounding wood. *Loose :* a shrunken dead knot. *Oval :* one cut obliquely to the axis of the branch. *Pin :* a small knot less than $\frac{1}{2}$ in. diameter. *Pith :* a sound knot with pith hole. *Rotten :* in an advanced stage of decay, less hard than the surrounding wood. *Round :* one cut at right angles to the axis. *Spike* and *Splay :* a knot sawn lengthwise.

Sound : a live knot. *Standard :* one less than $1\frac{1}{2}$ in. diameter. *Tight :* a sound knot securely fixed. **2.** A length of sash cord of 12 yds. **3.** A fastening made by looping a cord on itself. There are numerous recognised forms used in scaffolding, etc., such as clove hitch, timber hitch, loop, spliced loop, sling.

Knot Hole. A hole in converted wood due to the removal of a loose knot.

Knot Saw. A small circular saw for cutting defects from shingles. A shingle jointer.

Knotted Over. Woodwork with the knots coated with knotting preparatory to priming.

Knotting. 1. Various compounds, usually shellac and methylated spirits, used to coat or kill knots and resin before applying paint. It is also used to cover coatings of tar products, etc., before painting. **2.** Adzing off the ends of branches of a tree trunk after felling.

Knuckle. 1. That part of a hinge containing the pin or pivot. See *Parliament Hinge.* **2.** The abrupt change or break in the line of a ship in the projecting after-end over the rudder.

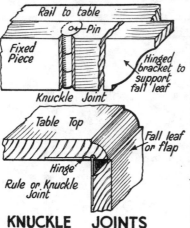

KNUCKLE JOINTS

Knuckle Joint. 1. A rule joint, *q.v.* **2.** Same as *finger joint*, but sometimes distinguished as being more elaborate. The fly-leaf and fixed piece are shaped to form a "butt-hinge," and bored for a wire to form the pin. **3.** A joint for lengthening an iron tension rod. One end is forked to receive the end of the other rod, and the joint is secured by a bolt or rivet in the form of a pin. See *Double Knuckle Joint.*

Knuckle Timbers. The foremost of the cant timbers of a ship.

Knur. A wood ball used in the game of knur and spell.

Knurled. Applied to a roughened surface, to provide a grip for the hand.

Knysna. S. African boxwood, *q.v.*

Koa. *Acacia koa* (*H.*). Hawaiian Islands. Red, black lines, lustrous. Hard, heavy, with fine close grain and texture. Wavy and curly grain. Characteristics of Australian Blackwood. Used for musical instruments, veneers, cabinet work. S.G. ·8 ; C.W. 4·5.

Koch. The name of the botanist K. H. E. Koch.

Koen. Abbreviation for the name of the botanist Koenig.

Kokko or **Koko.** *Albizzia lebbek* (*H.*). East Indies walnut, or the Siris tree. Dark walnut brown, variegated, lustrous. Moderately hard, heavy, and durable. Brittle, and requires careful seasoning to avoid checks. Interlocked grain and even coarse texture. Figured wood esteemed for decorative work. Fairly large sizes. Not easy to work, due to grain and woolly nature. Polishes well with careful filling. Used for cabinet work, superior joinery, veneers, flooring, turnery, carriage work, etc. Burrs very valuable. S.G. ·65 ; C.W. 3·5. NIGERIAN KOKKO. *A. ferruginea.* Also called Ayinre. Brown, lustrous stripes. Resembles Iroko but heavier and coarser in grain and easier to work. Spiral and open grain. Liable to pick up. S.G. ·72 ; C.W. 3. *Albizzia zygia.* Also called Ayinreta in Nigeria. Mahogany brown, darker streaks, lustrous. Reddish deposit. Moderately hard and heavy, very tough but not durable. Fairly straight grain. Cracks and warps unless carefully seasoned. Liable to pick up due to loose fibrous texture. Works to a smooth lustrous surface, but inclined to be patchy. S.G. ·68 ; C.W. 3. Both species are used for cabinet work, superior joinery, veneers, etc. Also see *Siris.*

Kolavu. *Hardwickia pinnata (H.).* India. Called E. Indian mahogany. Reddish brown. Fairly hard, heavy, and durable. Firm close texture with ray flecks. Exudes gum. Polishes well. Used for cabinet work and superior joinery. S.G. ·75 ; C.W. 4. Also see *Anjan.*

Koompassia. See *Kempas* and *Tualang.*

Koord. Abbreviation for the name of the botanist Koorders.

Koordersiodendron. Amugis, *q.v.*

Korean Pine. Siberian pine, *q.v.*

Korob. Sapodilla, *q.v.*

Korth. Abbreviation for the name of the botanist Korthals.

Kosterm. Abbreviation for the name of the botanist Kostermans.

Kosum. Gausan, *q.v.*

Ko tiao. Chinese walnut.

Kowhai. *Sophora tetraptera,* syn. *Edwardsia microphylla (H.).* New Zealand. Pale brown, lustrous. Somewhat like laburnum. Heavy, fairly hard, strong, tough, elastic. Coarse grain, ray flecks. Small sizes. Used for shafts, implements, cabinet work. S.G. ·78. *Sophora japonica.* China, Japan. Similar to above.

Kraftwood. A registered decorative type of plywood for interior wall surfaces. The face is specially treated with a mass of fine straight grooves to modify the appearance of rotary cut wood, and resin sealed for stability.

Krala. African mahogany, *q.v.*

Kranji. Keranji, *q.v.*

Krug. The name of the botanist L. Krug.

Krugiodendron. See *Palo diablo.*

Kruin or **Kruen.** Keruing, *q.v.*

Ktze. Abbreviation for the name of the botanist Otto Kuntze.

Kuhlm. Abbreviation for the name of the botanist Kuhlmann.

Kulim. *Scorodocarpus borneensis (H.).* Malaya. Chocolate brown, sometimes variegated bands. Disagreeable odour before seasoning. Very hard and heavy. Durable and resists insects except white ants. Fine grain, sometimes wavy. Moderate sizes. Used for structural work, piling, etc. S.G. ·8.

Kulscarp. Inferior Swedish mining timber.

Kumara. Tonca bean, *q.v.*

Kumbi. *Careya arborea* (*H.*). India. Light to dark brownish red with darker streaks. Moderately hard and heavy. Strong and durable. Straight grain. Medium coarse texture ; dull but smooth. Shrinks and warps considerably. Difficult to season and convert but not difficult to work. Used for furniture, superior joinery, etc. S.G. ·7 ; C.W. 3·5.

Kumbuk. *Terminalia glabra* (*H.*). Ceylon. Brown to nearly black. Hard, heavy, with close grain and distinct fine rays. Requires careful seasoning and not easy to work. S.G. ·8 ; C.W. 4·5.

Kumus. *Shorea ciliata* (*H.*). Malaya. Very similar to Resak, *q.v.*, but a little harder and not so straight grained. S.G. ·78.

Kungkur. *Pithecolobium confertum* and *Albizzia sp.* (*H.*). Malaya. Also called Medang Buaya. Similar to Rain Tree. Reddish or nut brown, lustrous. Moderately hard, very variable in weight. Durable and stable with fine grain and rays. Seasons readily. Used for structural and constructional work, suitable for cabinet work. S.G. ·4 to ·64.

Kural. *Bauhinia retusa* or *B. racemosa* (*H.*). India, Burma. Numerous vernacular names. Brown with very dark streaks and patches, variegated. Dull but smooth. Hard, heavy, and tough. Variable grain and texture, fine rays. Little imported. S.G. ·85 ; C.W. 4.

Kurana. *Cedrela sp.* (*H.*). B. Guiana. Light brown. Fairly light and soft ; durable. Coarse open grain, easily wrought. Used for furniture, cigar boxes, general purposes. S.G. ·5 ; C.W. 2.

Kuraru. Angelin, *q.v.*

Kurma. *Machilus macrantha* (*H.*). India. Brownish grey, lustrous. Fairly light and soft, with rather coarse grain. Difficult to work smooth. Used for general purposes, boat-building. Better qualities for cabinet work, superior joinery, etc. S.G. ·6 ; C.W. 3·5.

Kurokai. *Protium Schomburgkianum* (*H.*). B. Guiana. Pale brown, fairly lustrous and smooth. Moderately hard and heavy, fairly strong. Straight grain. Easily wrought. Little imported. S.G. ·5 ; C.W. 2·5.

Kuro-matsu. *Pinus thunbergii* (*S.*). Japan. Also called Omatsu. See *Matsu.*

Kurrajong. *Brachychiton spp.* or *Sterculia spp.* (*H.*). Queensland. BLACK KURRAJONG, *B. populneum*, *B. diversifolium*, *S. candata*, *S. diversifolia*. Creamy hue. Light, soft, not strong or durable. Smooth, lustrous, with mottle. Used for interior fittings. S.G. ·4. WHITE KURRAJONG. *B. discolor* or *S. discolor*. Similar to above. Used for interior fittings, excelsior, pulp, containers, but does not hold nails well. S.G. ·45. FLAME KURRAJONG. *B. acerifolium* or *S. acerifolia*. Similar to the above but inferior and of little commercial value except for excelsior.

Kurumi. Japanese walnut, *q.v.*

Kurz. The name of the botanist S. Kurz.

Kusam. *Schleichera trijuga* (*H.*). India. Also called Kusan, Kusum, Kosum, Gausan, Ceylon Oak, etc. Reddish brown. Very hard, heavy, strong and tough, but not durable. Interlocked grain, even medium texture, dull but smooth. Very difficult to season and convert and to work smooth. Used for wheelwright and wagon work, heavy construction, turnery. S.G. ·92 ; C.W. 5.

Kussia or **Kusiaba.** Opepe, *q.v.*

Kuthan. *Hymenodictyon excelsum* (*H.*). India. Light brownish grey, lustrous. Fairly light and soft ; not strong or durable. Straight grain. Medium even texture, feels rough. Some characteristics of beech. Seasons well, and easily wrought. Stains and varnishes but difficult to polish. Used for cases, planking, slack cooperage, cheap furniture, match-boarding, etc. S.G. ·5 ; C.W. 2·5.

Kwadwuma. *Guarea cedrata.* See *Obobonufua.*

Kyabooka. Kiabooka, *q.v.*

Kyanising. Preserving wood by impregnating with corrosive sublimate or bichloride of mercury solution (1 lb. to 5 gals. water).

Kydia. *Kydia calycina* (*H.*). India. Purplish grey. Very light and soft ; not durable, and subject to stain. Straight grain, even coarse texture, dull, and feels rough. Characteristics of Bombax, but harder and stronger. Difficult to season but easily wrought. Used for cheap interior work, match-boxes, splints, etc. S.G. ·3 ; C.W. 2.

L

l. Abbreviation for *linear* and *length*.

L. Abbreviation for the name of the botanist Linnæas and for von Linné.

lab. Abbreviation for *labours*. The term is used when estimating for the time expended on the materials for a job.

Label or **Label Mould.** A hood moulding over a window or door opening. A weather moulding.

Labill. Abbreviation for the name of the botanist Labillardiere.

Labourdonnaisia. See *Donsella* and *Almique*.

Laburnum. *Laburnum vulgare*, *L. anagyroides*, *L. alpinum*, or *L. cystisus* (*H.*). Europe and N. America. Brown to dark green, well-marked figure. Hard, heavy, with fine straight grain and medium

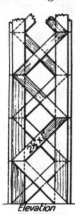

texture. Distinct rays. Difficult to work and small sizes. Used for oyster figure, cabinet work, turnery, inlays. S.G. ·82 ; C.W. 4·5. INDIAN LABURNUM. *Cassia fistula* (*H.*). Yellowish red to reddish brown, with darker streaks. Seasons to dark purple brown. Fairly lustrous. Hard, heavy, very strong, tough and durable. Straight smooth grain, medium texture. Difficult to season and fairly difficult to work. Polishes well. Trees girdled, or converted when green. Used for cabinet work, turnery, or where strength and durability are important. S.G. ·78 ; C.W. 4·5. Also see *Locust*.

Labyrinth Fret. See *Jalousie Panel*. A fret with many involved turnings ; reticulated.

Lac. A natural resin produced by the insect *Laccifer lacca* in the bark and twigs of certain Asiatic trees, and used in lacquer, varnish, etc., etc. See *Shellac*.

Elevation

Lace. A timber tying, or lacing, together other timbers. A brace or tie.

Lacebark. *Plagianthus betulinus*. New Zealand. Mallow family. Very decorative wood, similar to silky oak, *q.v.* Not exported.

Laced Column. Like a slender derrick tower, and used in temporary buildings such as exhibitions. It consists of four strong timbers held together by diagonal braces on all four sides.

Plan

LACED COLUMN

Lace Wood. The wood of the plane tree, quarter sawn and selected for ray figure and wavy grain. See *Buttonwood*. Also see *Silky Oak*.

Lacing. Same as taping, *q.v.*

294

Lacings. The vertical timbers in a coffer dam that tie together the horizontal struts.

Lacquer. 1. A varnish for metals to prevent tarnishing or oxidation. 2. The juices of certain Eastern trees, with suitable pigments and spirits of wine, used as a varnish for cabinet work and fancy articles, especially in China and Japan.

Lacquer Tree. *Rhus vernicifera.* Japan. Excellent ornamental wood, but the tree is valued more for its juice and wax for lacquer.

Lacunar. A deeply sunk panel in a ceiling. A coffer, *q.v.*

Ladder. A portable appliance used for climbing. It may be of wood, iron, or rope. The usual type consists of two long side pieces, or stringers, connected together by rungs at regular intervals. See *Step Ladder* and *Saddle Scaffold.*

Ladder Back. Applied to late seventeenth century chairs of which the backs are filled with several horizontal pieces somewhat like a ladder.

Ladder Rounds. The rungs, or staves, for a ladder.

Ladderwood. See *Machilus.*

Lading. See *Bill of Lading.*

Ladkin. A boxwood tool for opening the cames in lead lights, etc.

Laffatten. Sleepers from N. Europe cut from small trees. They are slabbed on two sides only, with rounded edges.

Lagerstrœmia. See *Banaba, Bongor, Jarul, Lendia, Leza, Nana, Pyinma.*

Lagging. 1. Building up patterns with narrow strips of wood. 2. Covering cylinders for insulation. 3. Small timbers driven behind the main timbering in mine shafts or drifts.

Laggings or **Lags.** Narrow strips of wood for building-up curved surfaces, broken plane surfaces, etc. See *Centre, Snow Boards, Jacket.*

Lagos. See *Mahogany.*

Lag Screws. Similar to coach screws but with conical points. They are used in heavy timber construction.

Laguna. Nicaraguan mahogany. Similar to Honduras mahogany.

Laguncularia. See *Mangrove (white).*

Lahou, Grand. African mahogany, *q.v.*

Laid Folding. See *Folding Floor.*

Laid-on. Planted, *q.v.*

Laid Yard. A trade term for the amount of tongued and grooved boarding required to cover 9 sq. ft.

Lakooch or **Lakuch.** *Artocarpus lakoocha (H.).* India. Yellowish to golden brown, to dark brown on exposure. Often darker streaks and white chalky lines. Lustrous but dulls with age. Feels rough. Moderately hard, heavy, strong, and durable, and resists insects. Straight and interlocked grain, even texture but very coarse. Seasons well with quick conversion and not difficult to work. Good appearance when quarter sawn, and large sizes. Difficult to polish due to resinous content. Used for piles, construction, furniture, boat-building, etc. S.G. ·56 ; C.W. 3·5.

Lalchini. *Calophyllum spectabile.* See *Nicobar Canoe.*

Lam. Abbreviation for the name of the botanist Lamarck.

Lamao. White lauan, *q.v.*

Lamb. Abbreviation for the name of the botanist Lambert.

LAMB'S TONGUE

Lamb's Tongue. A flat ogee moulding often used on sash bars.

Lamella Roof. A patent system of wood roof construction that allows for large unsupported spans without trusses. The network of framing resembles the diamond-shaped pattern of expanded metal, and consists of about 10 ft. × 8 ft. × 1½ in. for spans of about 70 ft. The units are prefabricated and delivered ready for erection for spans up to 165 ft.

Laminated. Applied to members that are built up to the required sizes by several thin layers or parallel sections. The method is often applied to circular work. See *Emy Roof*.

Laminboard or **Laminwood.** Thick compound board built up of a core with veneered faces. The core usually consists of small strips, up to ·28 in. thick, cemented or glued together, running at right angles to the grain of the face veneers which may be up to ¼ in. thick. In some cases the core also consists of veneers. There are numerous proprietary makes with varying arrangement of the cores. They are all arranged to give a stiff, stable, board. The boards may be obtained up to 20 ft. long by 8 ft. wide and 2 in. in thickness, and are extensively used in place of panelled framing. See *Blockboard* and *Battenboard*.

LAMINBOARD

Lampatia. *Duabanga sonneratioides* (*H.*). India. Grey, tinted or streaked with light brown, fairly lustrous, feels rough. Light, moderately hard and strong. Not durable except under water or protected. Straight, wavy, and interlocked grain, coarse texture. Seasons and works readily, but trees should be girdled. Ornamental with flecks, when quarter sawn. Used for match-boarding, tea chests, furniture, rotary veneers for plywood, light construction, etc. S.G. ·45 ; C.W. 2·5.

LANCETS FOR TENONER

Lancet Arch. A Gothic, or pointed, arch in which the radius of curvature is greater than the span. See *Equilateral* and *Drop Arches*.

Lance-tooth. A cross-cut saw with slender long teeth, like narrow isosceles triangles. It is quick-cutting for softwoods, but the teeth are too weak for hardwoods.

Lancets. The serrated shouldering irons to a machine tenoning block.

Lancewood. *Oxandra lanceolata* (*H.*). C. America. Sapwood lemon colour, heartwood nearly black but very small. Lustrous. Hard, heavy, strong, flexible, stable, but not durable and splits easily. Straight fine grain. Not very difficult to work and polishes well. Used where strength and flexibility are important : fishing rods, billiard cues, bows, tools, turnery. S.G. 1 ; C.W. 3·5. There are

other woods called lancewoods because they have similar properties
and used for the same purposes. See *Degame* and *Bulletwood*. CAPE
LANCEWOOD, *Curtisea flaginea*. S. Africa. Called Assegai wood.
Light red. Hard, heavy, durable, and tough, with fine close grain.
Used for shafts, spokes, implements, turnery, etc. S.G. ·9 ; C.W. 5.
MOULMEIN LANCEWOOD, *Homalium tomentosum (H.)*. Burma. Greyish
white, seasoning to reddish brown, lustrous. Very hard, heavy,
strong, elastic, and fairly durable. Dull but smooth. Fine even
texture and straight grain. Very difficult to season and work, polishes
well. One of the best woods to resist shear and tension. Used for
shafts, masts, furniture, and as *Oxandra sp*. S.G. ·93 ; C.W. 5.
NEW ZEALAND LANCEWOOD, *Pseudopanax crassifolium*. Light brown,
lustrous. Hard, heavy, with compact even grain. RED LANCEWOOD,
Albizzia basaltica. A small tree, producing extremely hard and
heavy wood. It is variegated and highly decorative. Used chiefly
for turnery. S.G. 1·2. TASMANIAN LANCEWOOD, *Eriostemon squameus*.
Small tree. Wood used for shafts, tools, handles, etc. W. INDIES
LANCEWOOD, *Guatteria virgata*. Characteristics and uses as
Oxandra sp.

Landa. See *Dabé*.

Landau. A carriage to hold four persons. It may be opened or
closed.

Landaulette. A brougham-shaped vehicle, similar to a landau.

Landing. 1. A wide resting place at the top of a flight of stairs, or
between two flights. See *Quarter Space* and *Half Space*. **2.** A platform
round which logs are collected for loading on to vehicles. **3.** A term
used in ship-building for *edge-lines*.

Landing Board. A projection to a scaffold for the convenience of
men stepping on to the scaffold or for depositing materials. See
Mason's Scaffold.

Landing Stage. A platform for convenient loading and unloading
of ships. It is often constructed to rise and fall with the tide. A
projecting platform to a scaffold for landing materials.

Landscape. Same as Bird's Eye, *q.v.*

Landscape Panel. A lay panel, *q.v.*

Lanete. *Wrightia sp*. Philippines. Creamy white. Very fine
texture and used for delicate carvings.

Langgong. See *Meranti*.

Lantern Light. A roof light with side lights. An erection on a flat
roof to provide light and usually ventilation. The plan may be
rectangular, polygonal, or circular. (See illus., p. 298.)

Lanza blanca. *Rapanea spp*. or *Myrsine sp*. S. America. Light
brown. Characteristics and properties of lancewood but lighter and
softer. S.G. ·75.

Lapacho. *Tecoma spp. (H.)*. C. America. Olive brown, darker
streaks, appears oily. Very hard, heavy, durable, and tough, but
splintery. Straight, irregular, and roey grain, with ripple marks.
Fine texture, and polishes. Contains lapachol. Difficult to work.
Characteristics of greenheart. Used for implements, tools, boat-
building, structural work, etc. S.G. 1·1 ; C.W. 4·5. Also *Tabebuia sp*.

Lapachol. A secretion in certain tropical trees that sets extremely

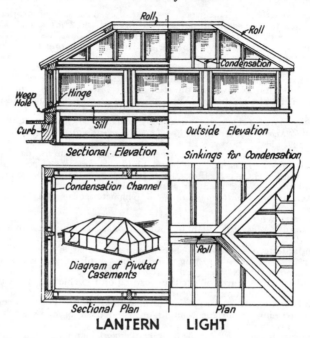

LANTERN LIGHT

hard and makes the wood difficult to work. It forms a yellowish powder that turns deep red when moistened with sodium carbonate or ammonia and is used as a dye.

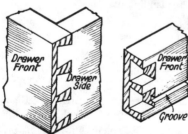

LAPPED DOVETAIL JOINT

Lap Board. A board resting on the lap of a person while he performs certain operations such as cutting out or ironing in tailoring, etc. Also see *Lapping Boards*.

Lap Dovetail. A dovetailed joint in which end grain is seen on one side only. It is used for drawer fronts, and sometimes called a half-blind dovetail. Also see *Halving Joints* and *Stopped Lap Dovetail*.

Lap or **Lapped Joint.** A joint in which one piece overlaps the other. Also see *Halving Joints*.

Lap or **Lip Mitre.** An angle rebated joint with a small

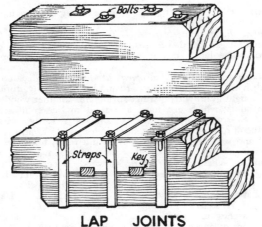

LAP JOINTS

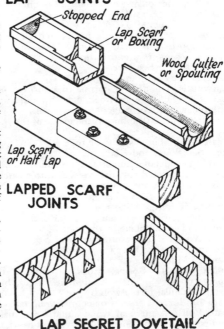

portion mitred at the face to avoid showing end grain. See *Lip Mitres*.

Laportea. See *Nettle* and *Fibrewood*.

Lapping Boards. Thin boards on which rolls, or bolts, of cloth are wound.

Lap Scarf. A joint used for lengthening wood gutters. The joint is secured by screws and made good by white lead and paint. It is also called a box scarf or boxing.

Lap Secret Dovetail. An easily made substitute for the secret dovetail.

Lap Sidings. Weatherboards that overlap each other and in which there is no preparation of the joints. See *Sidings* and *Clap Boarding*.

LAPPED SCARF JOINTS

LAP SECRET DOVETAIL

Lapwood. Debris left in the forest after felling and logging.
Larana. Obeche, *q.v.*
Larangeira. *Esenbeckia febrifuga* (*H.*). Brazil. Similar to Orange-wood. Small sizes. Used as a substitute for boxwood. See *Guarantan.*
Larboard. The left hand or port side of a ship, as opposed to starboard.
Larch. *Larix decidua,* syn.*L. europæa*(*S.*). Europe. A deciduous coni-fer. Reddish brown. Moderately hard and heavy. Strong, durable, and tough. Straight grain with distinct spring and summer wood. Fairly free from knots, which are hard and tend to loosen in seasoning. Resinous. Seasons easily but shrinks and warps badly and requires care. Difficult to impregnate but suitable for paint or varnish. Easily machined and wrought except for knots and resin, but gives trouble in conversion. Used for constructional work, fencing, piling, founda-tions, flooring, wheelwright's work, pitwood, sleepers, rustic work, furniture, planking, keelsons, etc. S.G. ·65 ; C.W. 2. The following species have the same characteristics and uses as European : Japanese or Red Larch (*L. Kæmpferi* or *L. leptolepis*) ; Black Larch (*L. americana*); Western Larch or W. Tamarack (*L. occidentalis*). American larches are also called tamarack and hackmatack, *q.v.* Several species of fir are sometimes wrongly called larch.
Larder. A domestic food store.
Laredo. Capomo, *q.v.*
Large Knot. One over 1½ in. diameter. See *Knots.*
Large-toothed Aspen. *Populus grandidentata.* Canadian Poplar, *q.v.* Also see *Aspen* and *Cottonwood.*
Laricio Pine. See *Pines.*
Larix. See *Larch* and *Tamarack.*
Larmier. The drip or corona to a cornice or entablature.
Larva. The stage between egg and pupa in insect development : caterpillar, grub, or maggot. See *Insects.*
Lashing Out. Cleaning up the inside of the frame of a tennis racquet, etc.
Lashings. Short lengths of rope for tying together scaffold timbers, etc. See *Whips* and *Knots.*
Last. A model of a human foot on which shoes are made and repaired.
Latch. A pivoted bar that engages with a catch to keep a door closed. A fastening that does not require a key. See *Norfolk Latch* and *Locks.*
Latchet Door. A wicket gate, *q.v.*
Lateral Bracing. Stays or struts to prevent racking due to wind pressure. Diagonal bracing on the backs of the principal rafters of a roof and in the plane of the roof.
Late Wood. The summerwood of a growth ring, *q.v.* The term is also applied to the later rings of old trees.
Latex. Milky juice in plants.
Latex Canals. Special tubes, or cells, in the ray tissue, for the storage of latex. They occur in few trees, but when large they are a serious defect in the wood.
Lath Binder. A frame for compressing laths for bundling.
Lath Bolt. A piece of wood suitable for making laths. The wood is imported in part-round billets without heart, and sold by the fathom.

Publisher's Introduction

FOR ten years the "Glossary of Wood" appeared serially in the monthly publication WOOD, a magazine, now ceased publication, dealing with all aspects of timber from the tree to the finished products, but chiefly the multifarious uses and fabrication of wood.

Due to much demand the Glossary was published in book form in 1948 and was at the time extensively enlarged and revised. The edition quickly sold out but sadly was never republished despite the obvious continued need for such a work. The same demand still exists to the present day and it is for this reason that the present publisher has re-issued this unique and comprehensive book.

The Glossary contains explanations of about 10,000 terms, more than 1,000 of which are also illustrated by clear line diagrams. They cover every significant term pertaining to trees, their location, the properties of their timber, uses, and botanical names; to structural uses of timber in building, engineering, ship and aircraft fitting; to woodworking machinery and tools, methods of woodwork construction, joints, finishes and polishes; to pests and diseases affecting trees and timber, and preservation and treatment of wood; to general timber trade and forestry practices, conversion of timber, and much more.

There are, inevitably, when a book is reissued in much the same form as when originally published, some terms which will apply in detail only to the items when first published. Examples of such terms are those referring to the use of wood where new materials for construction are now widely used. This is particularly relevant in the aircraft and shipbuilding industries. Another example is the author's use of British Standards and Government Regulations and readers should acquaint themselves with current codes of practice and regulations where necessary.

For those terms expressed with imperial units of measurement we have, for the benefit of readers, included a small table of conversion to the metric system. We have also included a table of geographical name changes appertaining to those used in the book.

GEOGRAPHICAL NAME CHANGES

As listed in the Glossary	*New Name*
British Guiana or Guiana	Guyana
British Honduras	Belize
Belgian Congo	Zaire
Cameroons	Republic of Cameroons
Ceylon	Sri Lanka
Dutch Guiana	Surinam
Formosa	Taiwan
French West Africa	Mauritania/Mali/Upper Volta/Niger
Gold Coast	Ghana
Indo China	Republic of Vietnam
Siam	Thailand
Tanganyika	Tanzania

IMPERIAL TO METRIC CONVERSION TABLES

Linear Measure
1 Inch = 25·400 Millimetres
1 Foot = 0·30480 Metre
1 Yard = 0·914399 Metre
1 Mile = 1·6093 Kilometres

Square Measure
1 Sq.Inch = 6·4516 Sq.Centimetres
1 Sq.Foot = 9·2903 Sq.Decimetres
1 Sq.Yard = 0·836126 Sq.Metre
1 Acre = 0·40468 Hectare
1 Sq.Mile = 259·00 Hectares

Cubic Measure
1 Cu.Inch = 16·387 Cu.Centimetres
1 Cu.Foot = 0·028317 Cu.Metre
1 Cu.Yard = 0·764553 Cu.Metre
1 Petrograd Standard (165 cu.feet) = 4·672 Cu.Metres

Liquid/Capacity Measure
1 Pint = 0·568 Litre
1 Quart = 1·136 Litres
1 Gallon = 4·546 Litres

Weight/Mass
1 Ounce = 28·3495 Grammes
1 Pound = 0·45359 Kilogramme
1 Stone = 6·350 Kilogrammes
1 Hundredweight = 50·802 Kilogrammes
1 Ton = 1·01605 Tonne

Abbreviations : H., *hardwood;* S., *softwood;* S.G., *specific gravity;*
C.W., *comparative workability;* *q.v., which see.*

A

a.a. Abbreviation for *always afloat.*
a.a.r. Abbreviation for *against all risks.*
Aaron's Rod. A long vertical moulding decorated with scrollwork
and foliage.
Abachi. Obeche, *q.v.*
Abaciscus. Any flat member. A rectangular plain surface.
Abacus. The crowning member of the capital to a column, or the
top of a pilaster. It forms the seating
for the horizontal members or a
springing for arched members. In
the Corinthian and Composite orders
of architecture it curves inwards and
has a central rose. In the Tuscan,
Doric, and Ionic it is rectangular.
Abassian Box. *Buxus sempervirens.*
See *Box.*
Abatement. Waste of material to
obtain the required sizes.
Abatis. A fence consisting of felled
trees with the branches pointing away
from the enclosure.

CORINTHIAN

DORIC
ABACUS

Abat-vent. The roof of a tower with small inclination.
Abele. *Populus alba,* or white poplar, *q.v.*
Abiache. Opepe, *q.v.*
Abies. The botanical name for fir. *Abies alba,* whitewood, or
white deal, Europe ; *A. amabilis,* red, or western, silver fir, W.N.
America ; *A. arizonica,* alpine fir, N.W. America ; *A. balsamea,* balsam
fir, N. America ; *A. concolor,* Californian white fir ; *A. fraseri,* balsam
fir, Virginia ; *A. grandis,* grand, or white, fir, W.N. America ; *A.
lasiocarpa,* alpine, or mountain, fir, W. America ; *A. magnifica,*
Californian red fir ; *A. nobilis,* noble, bracted, or white fir, W.
U.S.A. ; *A. pectinata=A. alba; A. pindrow,* Himalayan silver fir,
N. India. Refer to the timbers for other commercial names. Also
see *Picea* and *Fir.*
Abietineæ. Sub-order of Coniferæ. Includes pines, firs, spruces,
larch, cedar.
Abo. See *Pyinkado.*
Aboudikro. *Entandrophragma spp.* Sapele, *q.v.* Also see *Gedunohor*
Abrading Tools. Rasps and files. Also see *Sandpaper.*
Abrasion. See *Tests* and *Wear.*
Abura. *Mitragyna stipulosa* or *M. ciliata* (*H.*). W. Africa.
Also called Bahia, Eben, and Subaha. Light brown, pinkish tinge,

smooth. Little distinction between heartwood and sapwood. Rather light and soft, acid resistant, retains sharp arrises, shrinks considerably, not strong or durable. Fairly straight, interlocked grain, with rather fine even texture. Easily seasoned and wrought, finishes and polishes well. Used for carpentry, joinery, fittings, cheap cabinet work, batteries, oil-vats, mouldings, turnery, etc. S.G. ·52 ; C.W. 2.

Abutment. A support resisting a horizontal thrust from a curved or inclined member. The supports for a bridge. The pier from which an arch springs.

Abutment Cheeks. The surfaces on each side of a mortise receiving the thrust from the shoulders of a tenon.

Abut, To. To adjoin at the end. Usually abbreviated to *But.*

Acacia. *Acacia spp.* (*H.*). Widespread. About 450 species, but only about eight are imported. *Acacia arabica*, Indian Babul ; *A. catechu*, Cutch, Khair, etc. ; *A. formosa*, also called Sabicu ; *A. heterophylla*, Mauritius ; *A. homalophylla*, Australian Myall ; *A. koa*, Koa, Sandwich Islands (Hawaiian I.). See *Angica, Blackwood, Cutch, Gidgee, Kanti, Knobthorne, Koa, Myall, Raspberry Jam Wood, Talane, Thorn, Wattle.* FALSE ACACIA, *Robinia pseudoacacia.* The wood is called Robinia in this country, and Locust wood in America, *q.v.* Yellow to golden brown, with darker markings. Hard, strong, tough, elastic, and very durable. Fairly straight grain, rather coarse texture, but smooth. Requires careful seasoning, and difficult to work. Used for fencing, wheelwright's work, cabinet work, veneers, etc. S.G. ·71 ; C.W. 4. Also see *Queensland Acacia.*

Acajou. *Anacardium occidentale* (*H.*). C. America. Yellowish with red markings, beautiful figure. Not durable but resists insects. Only imported as veneers. S.G. ·5. Also see *Spanish Cedar.* Acajou implies mahogany, but it is usually accompanied by a distinguishing adjective.

Acana. Donsella, *q.v.*

Acanthaceæ. The Acanthus family. It also includes Bravaisia and Trichanthera, but is of little timber importance.

Acanthopanax. See *Sen.*

Acanthus. See *Sancho.*

Acanthus Leaf. The leaf of the bear's breech plant, which is imitated in carved ornamentation, and in the Corinthian and Composite capitals.

Acapú. *Vouacapoua americana* (*H.*) C. America. Chocolate brown lighter stripes, smooth. Hard, heavy, strong, and durable. Roey grain, fairly coarse texture. Difficult to work, polishes well. Used for superior joinery, shipbuilding, flooring, construction. S.G. ·9 ; C.W. 4. Also see *Sucupira.*

ACANTHUS LEAF

Acara. See *Araca.*

Acaricoára. *Minquartia guianensis* (*H.*). C. America. Greyish brown, smooth. Very hard, heavy, and durable. Straight grain ; small sizes. Little exported. S.G. ·97.

Accolade. An early decorative cresting over door and window openings.

Accommodation Bridge. A temporary bridge.

Accommodation Ladder. A swing ladder to allow passengers to embark on board ship or to give access to a pier, etc. See *Platform*.

Accra Mahogany. *Khaya ivorensis.* See *Mahogany*.

Accretion. Increase in height or diameter of a tree. See *Increment*.

Aceraceæ. Botanical family including maple, plane, and sycamore. *Acer campestre*, common maple, Britain; *A. campbellii*, Himalayas; *A. caudatum*, India; *A. dasycarpum=A. saccharinum*; *A. macrophyllum*, Pacific, Bigleaf, or Oregon maple; *A. negundo*, box-elder, or ash-leaved, maple; *A. nigrum*, black maple; *A. pennsylvanicum*, striped maple; *A. platanoides*, Norway maple; *A. pseudoplatanus*, sycamore, Britain; *A. rubrum*, soft, or red, maple; *A. saccharinum*, soft, or silver, maple; *A. saccharum*, rock, or sugar, maple. Also see *Harewood*.

Achi. Agba, *q.v.* See *Moboran*.

Achras. See *Sapodilla* and *Zapota*, or *Sapote*.

Ackama. See *Rose Alder*.

Acle. See *Akle*.

Acoita-cacalo. See *Estribeiro*.

Acorn. 1. The fruit of the oak tree. It is often imitated in carved ornamentation. 2. A cone-shaped ornament on the top of the spindle above the vane on the masthead of a ship.

Acrocarpus. See *Mundani*.

Acroter or **Acroterium.** A small pedestal to a pediment to carry an ornament or statue. An ornamental finish at the apex of a building.

act.std. Abbreviation for *actual standard*.

Actual Measure. The measure of timber in full, according to the system of measurement used. Usually opposite to customary measure, *q.v.* The dimensions of manufactured stock after machining.

Actual Merchantable. See *Used Length*.

a.d. Abbreviation for *air dried*. See *Air Seasoned*.

Adam. Applied to ornamentation characteristic of the work of the Adam brothers, especially Robert Adam. The important work was during the period 1760-1792.

Adansonia. W. Africa. The calabash, baobab, or monkey-bread tree.

Adenanthera. See *Coralwood*.

Aderno. *Astronium gracile* (*H.*). C. America. Yellowish red or brown, variable streaks, appears oily. Hard, heavy, durable, strong. Fairly straight and uniform grain. Not difficult to work and polishes well. Used for cabinet work, structural work, sleepers, axles, turnery, etc. S.G. ·9; C.W. 3.

Adina cordifolia. See *Haldu*.

Adze. A carpenter's tool used for squaring logs, levelling horizontal surfaces, etc. It has a thin arched blade at right angles to the shaft, as compared with the axe which has the blade parallel to the shaft.

Adzing Machine. A machine for preparing the seatings for the chairs for railway sleepers. It is usually combined with a boring machine for the bolt-holes.

3

Ægle. See *Bael*.

Aerofoils. The wings or planes of aircraft.

Æschynomeme. See *Solah*.

Æsculus. See *Buckeye* and *Horse Chestnut*. *Æsculus californica*, Californian Buckeye ; *Æ. glabra*, Ohio Buckeye ; *Æ. hippocastanum*, Horse Chestnut ; *Æ. octandra*, Yellow Buckeye. Also see *Tochi*.

Ætoma. The tympanum of a pediment.

Afara. Eghoin, *q.v.* Afara is the suggested standard name for this wood, which is also called Limba and White Afara. BLACK AFARA, idigbo, *q.v.*

a.f.b. Abbreviation for *allowance for bark*.

Afforestation. Converting ordinary land into forest by planting young trees, or saplings.

Affreightment. The chartering of ships to carry cargo.

African. Applied to the numerous woods imported from Africa. In the following it is included in the standard name. AFRICAN BLACK-WOOD, *Dalbergia melanoxylon*. See *Blackwood*. AFRICAN EBONY, *Diospyros crassiflora*, *D. dendo*, *D. mespiliformis*, *D. piscatorius*. See *Ebony*. AFRICAN MAHOGANY, *Khaya spp.* See *Mahogany*. AFRICAN OLIVE, *Olea hochstetteri*. See *Olive*. AFRICAN WALNUT, *Lovoa klaineana*. See *Nigerian Walnut*. Also see *Turtosa*.

Afrormosia. See *Satinwood*.

Aft. Towards the stern of a vessel.

Afzelia. See *Apa* and *Ipil*.

Aga. *Musanga smithii* (*H.*). Nigeria. Cream to yellow. A very light and soft fibrous wood. Interlocked grain. Rather difficult to work smooth and of little timber value. Useful for floats, insulation, paper pulp, etc. S.G. ·3.

Aganokwe. Makore, *q.v.*

Agathis. See *Kauri Pine*.

Agba. *Gossweilerodendron balsamiferum*. Nigerian cedar or Moboron, *q.v.* Agba is the standard name for this wood.

Agboin. Ekhimi, *q.v.*

Agents. All timber shipments pass through the hands of shipper's agents who sell to timber importers.

Aghako. See *Zylopia*.

Agoho. E. Indies. Hard, tough, elastic timber. Difficult to split. Used for tool handles, etc.

Agonis. See *Peppermint*.

Agromyza. A genus of fly, the larvæ of which produces pith fleck, *q.v.*

Aguano. A species of *Swietenia* from Brazil and Peru. See *Mahogany*.

Ahun. *Alstonia congensis* (*H.*). Nigeria. Also called Alstonia. Similar to Obeche, *q.v.* Cream. Light, moderately hard, lustrous. Gum ducts prevalent. Substitute for softwood. Useful for plywood cores, carving, matches, etc.

Aiele. See *Canarium* and *Dhup*.

Ailanto or **Ailanthus.** Tree of Heaven, *q.v.*

Aileron. Control flaps to the wings of aircraft. They are described as differential, floating, rotary, etc.

Ailon. Magnolia, *q.v.*

Aini. *Artocarpus hirsuta* (*H.*). India. Not imported. Similar species are Chaplash, Jacqueria, Keledang, *q.v.*
Air Box. A wood ventilating tube in a mine.
Aircraft. Any form of flying machine.
Air Doors. Doors in mines to regulate air currents.
Airer. See *Clothes Rail* and *Horse*.
Airframe. The structural part of an aeroplane.
Airscrew. A propeller for aircraft.
Air Seasoned. Applied to timber that is naturally seasoned in the

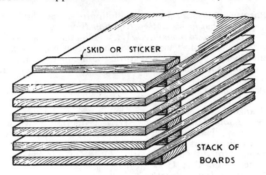

SKID OR STICKER

STACK OF BOARDS

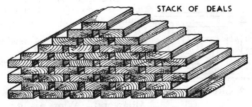

STACK OF DEALS

AIR SEASONING

open air, but usually protected from sun and rain. The timber is laid horizontally in stacks or vertically in steers. See *Seasoning*.
Aisle. Literally meaning a wing, but usually applied to the lateral division of a church.
Ait. Abbreviation for the name of the botanist Aiton.
Aité. *Gymnanthus lucida.* See *Crabwood*.
Aiye. Orodo, *q.v.*
Ajelera. *Canthium glabrifolia* (*H.*). Nigeria. Cream, smooth. Hard, moderately heavy. Mild grain, easily wrought. Substitute for hickory. Used for shafts, handles, rotary plywood, etc. S.G. ·6.
Akeake. *Olearia aricenniælobia* (*H.*). New Zealand. Similar to muskwood, *q.v.*, but lighter in colour.

Akessi. *Hylodendron gabunense* (*H.*). Nigeria. Cream to yellow or brown. Very hard and heavy. Interlocked fibrous grain. A substitute for persimmon. Under investigation. S.G. ·9.

Akle. *Albizzia acle* (*H.*). Philippines. Dark brown, like black walnut, odorous. Hard, heavy, durable, and resists insects, stable, fairly strong. Crossed and curly grain, irregular ribbon figure, fairly fine texture. Forms lather when rubbed in water. Easily wrought and seasons readily. An excellent cabinet wood. Used for musical instruments, gun stocks, superior joinery, vehicles, etc. S.G. ·7 ; C.W. 3.

Akleng Parang. *Albizzia procera.* Philippines. See *Siris*.

Akomu. *Pycnanthus kombo.* See *Nigerian Boxboard*.

Akoriko. *Haploromosia monophylla* (*H.*). Nigeria. Chocolate brown, variegated stripes, dull. Very hard, heavy, strong, and durable. Slightly interlocked grain, close texture. Difficult to work, especially when seasoned. Suitable for hard-wearing floors, turnery, handles, etc. S.G. ·85 ; C.W. 4·5.

Aku. Okwein, *q.v.*

Akume. Bubinga, *q.v.*

Akun. Uapaca, *q.v.*

Akwekwe. *Klainedoxa grandifolia* (*H.*). Nigeria. Fawn to brown, smooth, lustrous. Hard, heavy, strong, fairly durable. White deposit in pores. Straight and interlocked fibrous grain, liable to pick-up, even texture. Not difficult to work. Used for joinery, wagon construction, etc. S.G. ·85 ; C.W. 3.

a.l. Abbreviation for *all lengths*.

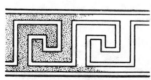

A LA GRECQUE

A la Grecque. A carved ornamentation having projecting fillets in continuous rectangular formation. The Greek Fret.

Alamo. A South American name for aspen, fig tree, and plane tree.

Alaska. See *Cedar* and *Hemlock*.

Alaus speciosus. The black carpenter beetle.

Albarco. Jequitiba, *q.v.*

Albawood. White oak, *q.v.*

Albespyne. Whitethorn, *q.v.*

Albizzia. See *Akle, Akleng, Kokko, Lancewood* (*Red*), *Siris*, and *White Siris*.

Alburnum. The outer part, or sapwood, of a tree. See *Annual Rings*.

Alcove. A deep recess in a room, usually containing a seat.

Alder. *Alnus glutinosa* (*H.*). Britain. Same family as birch. Whitish brown when cut, but the surface turns deep red and fades to pale brown colour when seasoned. Very durable if kept wet or dry. Will not stand alternate wet and dry. The wood is tender and homogeneous. Knotty timber and burrs valuable for curl and figure. Shrinks

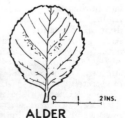

2 INS.

ALDER

6

greatly, nearly one-twelfth of its bulk. Used for piles, foundations, cooper's and cartwright's work, clog soles, rollers, hat-blocks, and furniture. S.G. ·64 ; C.W. 3. FORMOSAN ALDER, *A. maritima.* Excellent wood, but little imported. RED ALDER, *A. rubra.* Pacific coast of N. America. Similar to above and used for similar purposes, carving, turnery, etc. Also see *Cherry, Grey, Indian, Rose,* and *White Alder.*

Alecrin. See *Ibira-pepe.*

Alectryon. See *Titoki.*

Alerce. *Fitzroya patagonica* or *P. cupressoides* (*S.*). S. America. Red, variegated, purplish blue on exposure. Similar to cypress in grain, weight, etc. It is used locally for shingles because of its durability. S.G. ·45. *Libocedrus tetragona* is also called Alerce in S. America.

Aleurites. See *Wood Oil Tree* and *Candlenut Siris.*

Algaroba. Mesquite, *q.v.*

Aligna. Apa, *q.v.*

Alintatao. E. Indies (*H.*). A decorative timber for cabinet work. Yellow colour.

All., Fr. Abbreviation for the name of the botanist Allemão, Fr.

Allacede. An unidentified Philippine wood used as veneer for superior joinery and furniture. Reddish brown, darker brown broken stripes, some mottle. Rather coarse grain.

Alligator. 1. A boat used when handling floating logs. 2. A fork of a tree cut into a support for one end of a log to assist skidding on swampy ground. 3. See *Timber Connectors.*

Alligator Pear. See *Avocado.*

Alligator Tree. Red gum, *q.v.*

Alligatorwood. See *Marinheiro.*

Almery. See *Ambry.*

Almique. Donsella, *q.v.*

Almon. *Shorea eximia.* A pale red lauan, *q.v.*

Almond. *Amygdalus communis* or *Prunus c.* (*H.*). Cultivated for beauty of tree rather than timber. Hard, durable, and handsome timber. Used for general purposes in the East. See *Indian Almond.*

Almondwood. Chickrassee, *q.v.*, and Nargusta, *q.v.*

Almury. Same as ambry, *q.v.*

Alnus. See *Alder* and *Jaul.*

Aloe. Eaglewood or Paradisewood (*H.*), *q.v.* Laurel family. Fine grain and fragrant. Polishes well.

Aloe lactinea. A black caterpillar very destructive to young trees.

Alona. Nigerian walnut, *q.v.*

Alongside Delivery. The term implies delivery within reach of the ship's tackle, either on to craft or quay.

Alpenstock. An iron-pointed staff for use in mountain climbing.

Alphitonia. See *Red Almond.*

Alpine Ash. *Eucalyptus gigantea.* See *Mountain Ash.*

Alpine Fir. See *Fir.*

Alstonia. See *Chatiyan, Ahun,* and *White Cheesewood.*

Altar. An elevated place or structure for sacrificial purposes. The communion table in a Christian church.

Altar Piece. A decorative screen, or reredos, placed behind the altar.

Altingia. See *Jutili* and *Nanta-yop*.

Alto relievo. High relief in carving. Sculptured figures projecting at least half of their thickness from the background.

Alukonraba. Odudu, *q.v.*

Alupag. E. Indies (*H.*). Dark brown to black. Very heavy ; fine, even texture ; difficult to split. Used for tools, bowls, etc.

Ama. See *Pyinkado*.

Amamanit. *Eucalyptus deglupta* (*H.*). Philippines. Light red, lustrous. Fairly hard and heavy, strong, moderately durable and resists insects. Interlocked ribbon grain, moderately coarse texture. Requires careful seasoning, liable to warp. Easily wrought and polishes well. Resembles mahogany. Used for furniture, superior joinery, cabinet work, construction. S.G. ·6 ; C.W. 3.

Amaranth. *Peltogyne paniculata*. Also called Bois Violet. See *Purpleheart*.

Amarello or **Amarillo.** A C. American name applied to several woods, especially to species of Centrolobium and Aspidosperma. See *Aracá, Arariba, Fustic, Brazilian Satinwood, Vermelha, Vinhatico*, and *Waika Chewstick*.

Amargosa. See *Angelim*.

Amari. See *Amoora*.

Amate. *Ficus spp.* Mexico. Resembles basswood, *q.v.* Used for joinery, pianos, etc.

Ambatch. *Herminiera elaphroxylon* (*H.*). Substitute for balsa. Used in aircraft for insulating purposes.

Ambáy. *Cecropia adenopus* (*H.*). C. America. Similar species are called Trumpet Tree, Yagrume, etc. Pinkish-grey to light brown, smooth, stains easily. Light, soft, tough and strong for weight, not durable. Straight grain, coarse texture, distinct rays. Easily wrought. Used chiefly for packing cases, pulp, etc. S.G. ·44.

Amber. Fossilised resin. Gum exuded from trees and buried for long periods.

Ambit. A circuit. The perimeter of a figure or structure.

Ambo. A reading-desk or pulpit in early Christian churches.

Amboyna. *Pterocarpus indicus* (*H.*). E. Indies. Also called Vryabuca and Kiabooka. Light reddish brown to orange. Mottled, curled, and with burrs. Decorative timber for veneers, furniture, etc. S.G. ·7 ; C.W. 4. FALSE AMBOYNA, *Maidu*, *q.v.*

Ambrosia Beetles. Pinhole borers, *q.v.*

Ambry. A cupboard near to the altar to hold sacred utensils.

American. Applied to numerous woods from the U.S.A. : ash, beech, birch, black cherry, black walnut, chestnut, elm, holly, hop-hornbeam, lime, plane, red gum, oak, whitewood, etc.

American Whitewood. See *Canarywood, Tupelo, Magnolia*.

Amoora. *Amoora rohituka* and *A. wallichii*. Also called Amari, Thitni, etc. (*H.*). India, Burma. A second-class wood and not imported. It is deep reddish brown, fairly lustrous, with interlocked uneven grain, coarse texture, and visible rays.

Amor seco. *Heliocarpus americanus* (*H.*). C. America. Whitish. Very light and soft, stringy and spongy, not durable. Fairly straight grain, coarse texture, distinct rays. Difficult to work smooth. Bark

8

valuable for its strong fibres. Used for construction, cases, pulp, veneer cores, fire stops. A substitute for cork.

Amoteak. Sucupira, *q.v.*

Amphitheatre. An open space, or arena, surrounded by tiers of seats.

Amugis. *Koordersiodendron pinnatum* (*H.*). Philippines. Reddish brown, lustrous. Hard, heavy, strong, not durable, resists insects. Interlocked grain, wavy, fine texture. Seasons and works readily, polishes well. Used for interior construction, flooring, superior joinery, cabinet work, musical instruments, etc. S.G. ·8 ; C.W. 3·5.

Amygdalaceæ. The Almond family. Includes Amygdalus, Chrysobalanus, Hirtella, Licania, Moquilea, Parinarium, Pygeum.

Amygdalus. The almond tree.

Amyriswood. *Amyris balsamifera* (*H.*). C. America. Also called W. Indies Sandalwood, Torchwood, etc. Yellowish on exposure, oily, odorous, smooth, and lustrous. Very hard and heavy, brittle, durable. Fairly straight grain, fine uniform texture. Easily wrought, polishes well. Important for oil extract. See *Sandalwood*. S.G. 1 ; C.W. 3.

Anacardiaceæ. The Sumach family. Includes Anacardium, Astronium, Buchanania, Campnosperma, Canarium, Comocladia, Cotinus, Euroschinus, Dracontomelum, Gluta, Koordersiodendron, Loxopterygium, Mangifera, Melanorrhœa, Metopium, Mosquitoxylum, Odina, Parishia, Pistachia, Poupartia, Rhus, Schinopsis, Schinus, Spondias, Swintonia, Tapirira.

Anacardium. See *Caracoli, Cashew Nut, Espavé.*

Anaglyph. An embossed ornament in low relief.

Anamomilla or **Ananmwen.** See *Walnut.*

Anan. *Fagræa fragrans* (*H.*). Burma. Yellow or light brown, whitish streaks, lustrous, smooth. Hard, heavy, strong, durable, resists insects. Fine grain and texture, beautiful mottle. Used for cabinet work, veneers, etc. S.G. ·9 ; C.W. 4. Also see *Tembusa.*

Anchor. A carving representing a series of anchors or arrows. It usually alternates with the egg to form *egg and dart* enrichment to an ovolo moulding.

Anchoring. Tying down structural timbers to prevent lateral movement or uplift.

Anchor Towers. The towers for a derrick crane, *q.v.*, to which the stays are anchored.

Ancon. An angle or elbow. See *Crosette.*

Ancona. A figured walnut used as veneers.

Ancones. Carved ornaments on the sides of door casings, window openings, etc. Also called Trusses and Consoles.

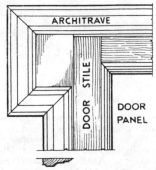

ANCON TO DOORWAY

Andaman Islands. See *Padauk, Pyinma, Satinwood, Bulletwood,* and *Marblewood.*

Andira. See *Angelin, Cabbage Bark, Macaya, Moca.*
Andiroba. Crabwood, *q.v.*
Angel-beam. A hammer-beam continued past the hammer-post and carved in human form.
Angelin or **Angelim.** *Andira inermis* (*H.*). C. America. Also called Cabbage Bark, Partridgewood, Macaya, etc. Reddish brown, odorous. Hard, heavy, tough, strong, elastic, very durable, resists

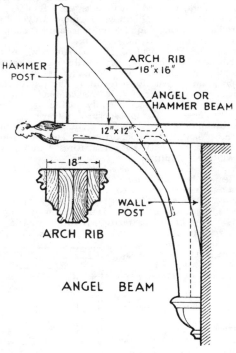

HAMMER
POST

ARCH RIB
18"x 16"

ANGEL OR
HAMMER BEAM

12"x 12"

18"

ARCH RIB

WALL
POST

ANGEL BEAM

insects. Wavy and roey figure and burr, coarse texture, distinct rays. Not difficult to work. Used for all purposes, and veneers. S.G. ·85 ; C.W. 3. ANGELIM AMARGOSA, *Andira vermifuga.* Similar to above, but lighter in weight and streaked with lighter coloured bands. S.G. ·75. ANGELIM ROSA, *Hymenolobium excelsum.* Also Pereira, *q.v.*
Angelino. *Ocotea caracasana* (*H.*). C. America. Yellow to dark olive, smooth, lustrous. Resembles olive. Moderately hard and heavy. Strong and very durable. Roey grain, medium texture. Not difficult to work and polishes well. Used for construction, cabinet work, etc. S.G. ·75 ; C.W. 3. Also see *Tuque.*

Angelique. *Dicorynia paraensis* (*H.*). C. America. Also called Bastard Locust, Angelica, and Nutwood. Olive brown, reddish patches, odorous. Hard, heavy, tough, strong, stable, elastic, very durable, and resists insects. Corrodes iron. Straight and wavy grain, ripple, medium texture, smooth. Not difficult to work. Does not take a high polish. Used for superior joinery, cabinet work, veneers, structural and constructional work, piling, etc. S.G. ·8 ; C.W. 3.

Angelisia. See *Merbatu.*

Angico. *Piptadenia rigida* (*H.*). C. America. Pale reddish brown,

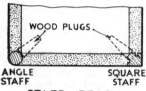

STAFF BEAD

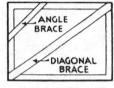

ANGLE BRACE

smooth, lustrous. Similar to mahogany. Hard, heavy, durable, strong. Roey grain, medium texture. Fairly difficult to work. Polishes well. The wood is not greatly appreciated and used chiefly for construction. S.G. ·95 ; C.W. 3·5. The name is sometimes applied to *Cana fistola, q.v.*

Angiosperms. Deciduous trees. Hardwoods. See *Trees.*

Angle Bar. The vertical bars at the angles of windows.

Angle Bead. A bead fixed to the salient angle of a wall. It serves as a screed for plastering and prevents injury to the corner. Also called angle staff and staff bead. It may be of wood or metal.

Angle Block. See *Blocking* and *Angle Joints* (*e*).

Angle Brace. A tie to strengthen the angles of framing. A diagonal brace divides rectangular framing into two equal parts.

Angle Bracket. A bracket forming the mitre of a built-up cornice.

Angle Joints. The jointing of timbers not in the same straight line or in

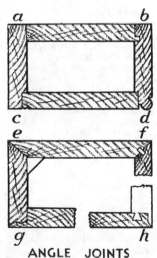

ANGLE JOINTS

the same plane. There is a large variety : halving, mortise and tenon, dovetailed, housing, bridle, cogging, keying, etc. The illustration shows (*a*) butt, (*b*) tongue and groove, (*c*) rebate with quirk bead, (*d*) rebate with return bead, (*e*) mitre with glued block, (*f*) mitre with loose tongue, (*g*) rebate and lip mitre, (*h*) lip mitre and dovetail as used on the carcase for drawers.

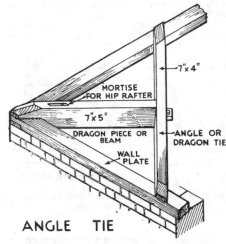

7" x 4"

MORTISE
FOR HIP RAFTER

7" x 5"

DRAGON PIECE OR
BEAM

WALL
PLATE

ANGLE TIE

←ANGLE OR
DRAGON TIE

Angle Post. The corner post in half-timbered structures, etc.

Angle Rafter. A hip rafter.

Angle Rib. A curved piece forming the mitre for a coved ceiling.

Angle Staff. See *Angle Bead*.

Angle, or **Dragon, Tie.** The roof timber tying together the wall plates and carrying one end of the dragon beam.

Angophora. A genus with several species in Queensland, but of little commercial importance. Two species, Applebox and Apple-gum, resemble spotted Irongum. Not exported.

Angsena. See *Narra* and *Sena*.

Aniba. Zebrawood, *q.v.*

Anisoptera. See *Mersawa* and *Polosapis*.

Anjan. *Hardwickia binata* (H.). India. Brick red to dark brown with black streaks. Hard, very durable, strong. Used for general purposes and cabinet work, piles and sleepers. S.G. ·9 ; C.W. 4.

Ankar. A proprietary wallboard, *q.v.*

Annex. An addition to a building. A supplementary building.

Annual Rings. The concentric rings of wood added annually to the growing tree in temperate zones. Tropical trees grow continuously. New growth takes place in the cambium layer just under the bark. Each ring consists of cells or tracheids, arranged vertically.

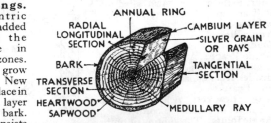

ANNUAL RING

RADIAL
LONGITUDINAL
SECTION

CAMBIUM LAYER

SILVER GRAIN
OR RAYS

BARK

TANGENTIAL
SECTION

TRANSVERSE
SECTION

HEARTWOOD
SAPWOOD

MEDULLARY RAY

ANNUAL RINGS ETC.

Medullary rays, radiating from the centre of the tree and arranged horizontally, interweave with the rings, giving strength and

Lath Bolter. A machine with a number of small circular saws to prepare the bolts for making laths.

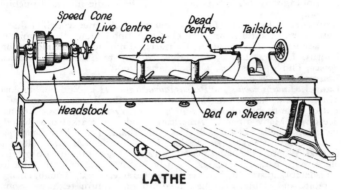

LATHE

Lathe. A machine for turning wood, etc., into circular form. See *Turning Machine.*

Lathridiidæ. A family of wood beetles. They attack the bark of trees or the decayed wood of logs and felled trees.

Lath River. One who prepares plasterer's riven laths ; or a special tool for preparing the laths.

Laths. Strips of wood of small section. They are generally named according to their particular uses : plasterer's, slater's, blind, etc. Plasterer's laths are from 3 to 4 ft. long, and riven or sawn from Baltic fir. Sawn laths are about 1 in. wide and $\frac{3}{16}$, $\frac{1}{4}$, or $\frac{3}{8}$ in. thick, and called single, lath and a half, or double laths, respectively. They are sold in bundles of 450 ft. Slater's laths, or battens, vary from $1\frac{1}{2}$ in. $\times$ $\frac{5}{8}$ in. to 2 in. $\times$ 1 in. Laths, or battens, to carry shingles are of greater section and depend upon the spacing of the rafters.

Lattice. An open network, or trellis, formed of narrow strips crossing each other. The open spaces, or meshes, may be rectangular or rhomboidal. See *Interwoven Fence* and *Mesh.*

Lattice Roof. One built up of laths in lattice formation like a Belfast truss, *q.v.*

Lattice Window. A window in which the spaces are filled with panels of lattice-work instead of glass ; or one glazed with diamond-shaped panes. See *Mesh* and *Dormer.*

Lauan. The name given to numerous Philippine species of the Dipterocarpaceæ family : almon, apitong, bagtikan, guijo, manggachapui, mayapis, narig, palosapis, red lauan, tangile, tiaong, white lauan, and yacal. They vary considerably in properties and characteristics from moderately soft to very hard, and from pale yellow to dark brown. They have some of the characteristics of mahoganies and are used for similar purposes. See *Red* and *White Lauan.*

Launch. **1.** Applied to numerous types of boats used for short voyages, either for pleasure or business. **2.** A large ship's boat for transferring stores and men to and from ship. It has superseded the long boat and carries more oars and has a flatter bottom.

Launching-way. Timbers supporting the cradle for the launching of a ship.

Launder. A local name for wood spouting. A water trough or flume.

Lauraceæ. Laurel family. Includes Actinodaphne, Beilschmiedia, Cinnamomum, Cryptocarya, Dehaasia, Endiandra, Eusideroxylon, Laurus, Lindera, Litsea, Machilus, Micropora, Nectandra, Ocotea, Persea, Phœbe, Sassafras, Umbellularia, etc.

Laurack. Acle, *q.v.*

Laurel. *Laurus nobilis* and *Prunus laurocerasus* (*H.*). Europe. Pinkish white. Hard, heavy, tough, and durable. Small sizes, rough grain, and of little timber importance. See *Bay Tree*. AMERICAN LAUREL. See *Cyp.* Many other C. American woods are marketed as laurel : avocado, canella, litsea, sassafras, silverballi, etc. See *Lauraceæ*. CALIFORNIAN LAUREL or Pacific Myrtle. See *Myrtle*. INDIAN LAUREL WOOD, *Terminalia tomentosa* (*H.*). Light brown to brownish black, with darker markings. Fairly lustrous and smooth. Varies considerably in weight, hardness, and characteristics. Tough, strong, durable, and stable. Straight medium grain, firm texture. Some interlocked and wavy grain and mottle. Highly figured wood very valuable. Difficult to season and work but very variable. Polishes well, especially with egg-shell or wax finish. Used for cabinet work, superior joinery, electric casings, veneers, etc. S.G. about ·9 ; C.W. 4 to 5·5. S. AFRICAN LAUREL, *Ocotea bullata*. Called Black Stinkwood. Golden yellow to dark brown. Hard, heavy, strong, lustrous. Moderately tough and elastic. Difficult to season and work. Polishes well. Used for high-class joinery and cabinet work. S.G. ·9 ; C.W. 5·5. Pacific myrtle is called Californian laurel, especially when burl figure. Cherry laurel is *Prunus laurocerasus*.

Laurelia. See *Pukatea*.

Laurus. See *Lauraceæ* and *Laurel*.

Laurustinus. *Viburnum tinus* (*H.*). S. Europe. Pinkish white. Hard, heavy, close grain. Similar to boxwood.

Lavoa. Nigerian Walnut, *q.v.*

Lawson's Cypress. See *Cedar* (*Port Orford*).

Lay Bead. A horizontal bead.

Lay or **Layer Board.** A board to level up the battens, and over which the lead is dressed, in roof valleys and gutters. Boards carrying the lead in roof gutters.

Lay Days. An agreed number of days allowed for loading or unloading a vessel.

Laying Down. Setting-out full size in ship-building, for fairing and to obtain the templets and moulds. *Laying off* has the same meaning. *Laying out* means setting-out, *q.v.*

Lay Light. A ceiling light, *q.v.*

Lay Out. 1. The arrangement of a building and its parts. 2. A pattern-maker's full-size drawing.

Lay Panel. One in which the grain runs horizontally.

Lay Stall. A separate stall in a byre.

Lazaret. 1. A hospital or vessel for quarantine of persons suffering from infectious diseases. **2.** A compartment, for stowing provisions, above the after peak of a ship.

l.c.l. Abbreviation for *less than car load lots*.

ld. Abbreviation for *load*.

Leaching. 1. The gradual reduction in the amount of preservative in timber due to the action of water. **2.** Leaking.

Leaders. Scribing cutters or shoulder lancets. See *Lancet*.

Leading Edge. The structural member forming the forward edge of a stream-lined body, as in aircraft, etc.

Leads. Back boards used in picture framing. Very thin boards sawn from deals.

Leadsman's Platform. A suspended grating over the side of a ship from which soundings are taken. See *Grating*.

Leaf. 1. A movable part of a table top. It may be hinged, sliding, or loose. See *Drop* and *Extension Table*. **2.** One piece, or fold, of folding screens, doors, etc. **3.** See *Grape* and *Leaf*.

Leaf Wood. Hardwoods, *q.v.*

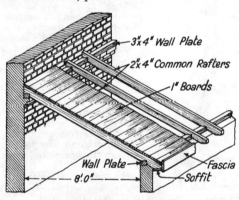

3″x 4″ Wall Plate
2″x 4″ Common Rafters
1″ Boards
Wall Plate
8′ 0″
Fascia
Soffit

LEAN-TO ROOF

Lean-to. A roof consisting of one sloping surface. A shed, or half-span roof. It is often formed against the side of a building which supports the top ends of the rafters.

Lear Boards. Layer boards, *q.v.*

Leather Jacket. A local name for Coachwood, *q.v.*

Leatherwood. *Euchryphia billardieri*. Tasmania. A flexible, tough, durable, and decorative wood suitable for shafts, handles, bentwork, etc. Small sizes and little exported. S.G. ·7. See *Ulmo*.

Leban. *Vitex spp.* (*H.*). Malaya. Also called Haleban. Colour varies from greyish to brown. Very hard, heavy, and durable. Fine grain with fine close rays and glistening deposit. Rather small sizes, crooked, and large knots. Used for boat-building, wagon work, implements, etc. S.G. ·8.

Lebanon Cedar. See *Cedar*.
Lecanium. Black and brown, or scaly, bugs. They are closely allied to the lac insect and very destructive to deciduous trees, from which they extract the sap.

Leche. See *Waika*.
Leche de Maria. *Calophyllum spp.* Mexico. See *Santa Maria*.
Lectern. A reading desk in a church.
Lecythidaceæ. Monkey Pot family. Includes: Allantoma, Bertholletia, Cariniana, Chytroma, Couratari, Couroupita, Eschweilera (Jugastrum), Grias, Gustavia, Lecythis, etc.
Lecythis. See *Guatecare, Jarana, Kakaralli, Sapucaia*, and *Monkey Pot*.
Ledeb. Abbreviation for the name of the botanist Ledebour.
Ledge. 1. The top surface of a projecting member. A narrow shelf. 2. The horizontal members holding together the battens in a ledged door.
Ledged Door. One consisting of battens, or vertical boards, and horizontal bars, or ledges. They are also called barred or batten doors and are usually strengthened by braces. See *Framed and Ledged*.
Ledger Board. See *Ribbon Board*.
Ledgers. 1. The horizontal timbers, in a scaffold, tying together the standards. See *Bricklayer's Scaffold*. 2. Pieces between the beams and under the deck of a ship.
Leeboard. 1. The front of a ship's berth. See *Berth*. 2. A wood frame fixed over the lee side of a ship to prevent it from losing leeway under certain conditions. It is used in a flat-bottomed ship.

LECTERN

Leeward. The side of a ship away from the wind.
Left Hand. See *Hand*.
Leg. 1. A support, longer and stronger then a prop, used in coal mines. 2. Any slender support, especially when inclined or one of a pair.
Lega. Kokko, *q.v.*
Leggotts. A fanlight opener operated by an endless cord with worm and wheel and rack. See *Quadrant*.
Legno. Applied to many C. American dark-coloured woods.
Leguminosæ. A very large family of trees with three subdivisions : Cæsalpinieæ, Mimoseæ and Papilionaceæ, and including the following genera : Acacia, Acrocarpus, Afzelia, Albizzia, Amphimas, Andira, Apuleia, Arthrocarpum, Baphia, Bauhinia, Belaira, Berlinia, Bow-

dichia, Brya, Butea, Cæsalpinia, Canavalia, Cassia, Castanospernum, Centrolobium, Cercis, Copaifera, Cynometra, Cytisus, Dalbergia, Dalea, Danilea, Detarium, Dialium, Dicorynia, Diphysa, Dipteryx, Distemonanthus, Eperua, Erythrina, Erythrophlœum, Genista, Gleditschia, Glyricidia, Gourliea, Gymnocladus, Hæmatoxylon, Hardwickia, Herminiera, Holocalyx, Hymenæa, Hymenolobium, Ichthyomethia, Indigofera, Inocarpus, Intsia, Kingiodendron, Koompassia. Laburnum, Lonchocarpus, Machærium, Macrolobium, Martinsia, Melanoxylon, Millettia, Myrospermum, Myroxylon, Olneyna, Ormosia, Ougeinia, Oxystigma, Parkia, Peltogyne, Phylloxylon, Pictatia, Piptadenia, Pithecolobium, Plathymenia, Platycyamus, Platypodium, Pœppigia, Pongamia, Prioria, Prosopis, Pseudocopaiva, Pterocarpus, Pterogyne, Pueraria, Robinia, Sindora, Sophora, Spatholobus, Swartzia, Sweetia, Tamarindus, Tipuana, Torresia, Wallaceodendron, Wistaria, Xylia, Zollernia.

Leiteira. *Brosimopsis spp.* (*H.*). C. America. Yellowish grey. Hard, heavy, tough, and strong, but not very durable. Straight coarse grain. Silica secretion makes it difficult to work when dry. Splinters easily. Used chiefly for protected structural work. S.G. 1.

Leitneria. See *Corkwood*.

Lemon. See *Citrus*. S. African wild lemon, *Xymalos monospora*, is an excellent ornamental wood for furniture and superior joinery but is not exported.

Lemon Ironwood. *Backhousia citriodora* (*H.*). Queensland. Pale red. Very hard and heavy. Close even grain and texture. Difficult to work. Small sizes and scarce. Used for mallet heads and turnery. S.G. ·95.

Lemonwood. See *Degame, Guatambu, Movingue,* and *Synara*.

Lemtan. *Gynocardia odorata* (*H.*). India. Also called Chaulmugri. Light brown, lustrous. Light, with smooth interlocked grain, and tough. Fine texture. Difficult to work when seasoned and subject to fungal stain. Used for construction, planking, etc. S.G. ·46; C.W. 4.

Lemusu. Nemesu, *q.v.*

Lendia. *Lagerstrœmia parviflora* (*H.*). India. Brownish grey to yellowish brown, fairly lustrous, smooth. Fairly heavy; hard, strong, durable. Straight and twisted grain, wavy grain very attractive, medium coarse texture. Very difficult to season but not difficult to work. Polishes well with careful filling. An excellent wood but for difficulties of seasoning. Used for construction, joinery, cooperage, implements, etc. S.G. ·75; C.W. 3.

Lenggadai. *Bruguiera parviflora* (*H.*). Malay. Dull to brownish yellow. Similar to Bakau and usually marketed as such, but it is inferior. S.G. ·9.

Lengthening Bar. An extension piece for a cramp.

Leningrad Standard. See *Petrograd Standard*.

Lentinus. A fungus, *L. lepideus*, very destructive to softwoods, especially in mines. It is one of the few fungi that attack creosoted wood.

Lenzites. A fungus, *L. sepiaria*, that attacks softwoods used for exposed structural work, fencing, etc. *L. abietina* is very similar. There are other species of lesser importance.

Leopard Ash. *Flindersia collina* (*H.*). Queensland. Very Similar to Crow's ash and Yellow-wood ash, *q.v.*

Leopard Wood. See *Letterwood*.

Lepcha. Hollock, *q.v.*

Leptospermum. See *Manuka* and *Tea Tree*.

Letheridge. A mixture of glue and litharge used as a stopping.

Letterbox. A box or receptacle behind an exterior door to hold letters pushed through a slot in the door.

Letterwood. *Piratinera guianensis*, syn. *Brosimum aubletii* (*H.*). C. America. Also called Snakewood, Leopard Wood, Tortoise-shell Wood. Several species. Dark red with irregular darker markings and speckles, suggesting snake skin. Extremely heavy and hard like horn. Very strong and durable but brittle. Straight fine grain, splits easily, difficult to work. Small sizes, sold by weight. Used for fancy articles. S.G. 1·2 ; C.W. 6.

Leucomb. A projecting cover to a crane, gin wheel, or other lifting tackle.

Levelling Pegs. Pointed wood pegs driven into the ground and used as a guide for a horizontal plane surface, or for measuring the depths of excavations. See *Sight Rails* and *Profile*.

Levelling Staff. See *Sopwith Staff*.

Lever. A rigid rod rotating about a fixed point, called the fulcrum, by means of which a resistance at one end is overcome by an effort applied at the other end. There are two other types, or *orders*, in which the position of the fulcrum is changed with regard to effort and resistance, but in all cases the *algebraic sum of the moments of the forces* about the fulcrum=zero ; or $E \times a = R \times b$. When the rod is bent it is called a bell crank lever.

LEVERS
(Three Orders)

Lever Boarding. Similar to louvres but the boards are pivoted at the ends and can be operated like a Venetian blind.

Lever Cap. Used, instead of a wedge, to fix the cutting iron of a plane. See *Smoothing Plane*.

Lever Lock. A lock in which several levers are raised by the key to shoot the bolt.

Lews. Hack caps, *q.v.*

Leza. *Lagerstrœmia tomentosa* (*H.*). India. Grey to greyish brown, lustrous, and smooth. Heavy, hard, and strong. Straight grain and ripple, fine even texture. Moderately difficult to season but fairly easily wrought. Used for flooring, panelling, furniture, wheelwright's work, tools, and implements. S.G. ·65 ; C.W. 3.

L.f. Abbreviation for the name of the botanist Linné (*filius*).

L'Hérit. Abbreviation for the name of the botanist L'Héritier.

Li. Chinese chestnut.

Lianes. A climbing woody vine. Climbing plants that attach themselves to trees in tropical forests.

Liber. Bast, *q.v.*

Libocedrus. See *Alerce, Cedar*, and *Kaikawaka*.
Libreville. See *African Mahogany*.
Licania. See *Chozo, Icaquito*, and *Kauta balli*.
Licaula. A species of palm tree.
Lich Gate. See *Lych Gate*.
Lid. **1.** The cover to a box, coffer, etc. It may be hinged or loose. The top of a coffin, which may be square-edged, plain, raised, double-raised, or panelled. **2.** Same as cap in mine timber-ing.

Lieblein. Name of botanist.

Lierne Ribs. The ribs in lierne, or fan, vaulting that do not spring from an abacus. The short ribs between the intermediate groin ribs to give stability and design to the panels.

Lifaki. *Entandrophragma sp.*, from the Belgian Congo. See *Mahogany*.

Lifeboat. **1.** A very buoyant boat, specially constructed, for life-saving. **2.** A ship's boat for use in case of shipwreck.

Life Cycle. The time required for the development of an insect from egg to egg.

Lift. **1.** An elevator or hoist. It may be operated by hand or electricity, or it may be hydraulic. **2.** A ring, hook, or handle, for raising sliding sashes, etc.

Lifting Shutters. Window shutters that slide vertically like sliding sashes. Also see *Shutters*.

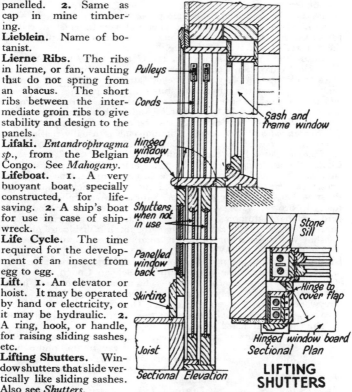

LIFTING
SHUTTERS

Lifting Tackle. Any form of mechanical appliance for hoisting materials, but specially applied to pulley tackle, *q.v.*
Ligas. See *Rengas*.
Light Back Saw. A small dovetail saw 5 in. long and with 30 teeth per inch.
Lightening Planer. A light type of machine in which short boards

are scraped by a fixed plane iron over which they are fed by a rubber-covered revolving drum.

Lighter. A heavy type of boat for conveying goods from vessel to wharf.

Light Hardwoods. Applied to tropical hardwoods having a specific gravity of less than ·5.

Lights. 1. The subdivisions of a window. The openings between mullions. 2. An opening in a wall or roof to admit light. A window.

Light-ship. A stationary ship bearing a light at the mast-head as a guide for shipping.

Lignasan. A proprietary sap-stain preventive.

Lignin. One of the important chemical constituents of cellular tissue. It has a hardening and binding, or cementing, function in the cell walls. See *Wood Substance.*

Lignum-vitæ. *Guaiacum officinale* and *G. sanctum* (*H.*). C. America. Also called **Guayacan**, **Congo Cypress**, etc. Olive brown to greenish black, feels waxy. Very hard, heavy, strong, tough, and durable. Fine uniform interwoven grain. Smooth, and polishes well ; very difficult to work. Several inferior species. A secretion forms a natural lubricant. Sold by weight. Used for bearings and machine parts, bowls, fancy articles, tools, castors, etc. S.G. 1·2 ; C.W. 5·5. See *Santowood* and *Verawood*. Lignum-vitæ is usually marketed to show the source of origin : Bahamas, Cuban, Jamaican, Mexican, Nicaraguan, Porto Rico, San Domingan.

Lilac. *Syringa vulgaris* (*H.*). Britain. Pale yellow to pinkish, with faint stripes. Hard, heavy, with close firm grain and ray flecks. Small sizes. Similar to tulip-wood. Used for fancy articles, inlays, etc. PERSIAN LILAC. *Melia azedarach.* India. Also called Tamago. Reddish brown, variegated, lustrous, gum deposits. Moderately hard and light, durable, and resists insects. Straight grain, coarse uneven texture, ray flecks. Easily wrought but liable to pick up. Polishes well. Used for furniture, rotary veneers, sport requisites, toys, etc. S.G. ·6 ; C.W. 3.

Limba or **Limbo.** Afara or Eghoin, *q.v. Limba noir* is the name given to the dark figured heartwood from the Belgian Congo, and *Limba blanc* to plain light-coloured wood. The amount of coloured heartwood varies considerably in different trees, as with Indian Silver Greywood. Other names are Frake and Offram.

Limber. 1. Limbs, or branches, of trees. 2. Flexible, easily bent. 3. The carriage attached to a gun.

Limbers. Conduits or gutters on each side of the keelson of a ship to form a waterway to the pump well. The short coverboards are called limber boards, and holes in the boards are limber holes.

LIME   0   1   2 *Ins.*

Lime. *Tilia vulgaris, T. cordata, T. parvifolia, T. platyphylla.* Also called tiinden. Yellowish white with reddish Lnge, fragrant. Soft, light, stable, not durable. Fine close grain with faint rays and ripple. Easily wrought. Used for carving, muscial instruments, artificial

limbs, turnery, etc. S.G. ·52 ; C.W. 1·75. See *Basswood* and *Citrus*.
Wild Lime. See *W. Indies Satinwood*.
Limed Oak. Applied to oak furniture in which the finish is obtained
by rubbing a paste of chloride of lime and water into the grain. Other
materials may be used to give the same effect. The wood is then
given a dull or egg-shell or wax polish.
Limnoria. See *Marine Borers*.
Limonia. *Limonia acidissima* (*H.*). India. A small tree of the box-
wood class. The wood is lemon yellow in colour and very similar
to boxwood but is in little demand. S.G. ·96.
Limousine. A carriage or motor car with closed body and roof
over driver.
Linaloe. Gumbo-limbo, *q.v.*
Lind. Abbreviation for the name of the botanist Lindley.
Linden. See *Lime*.
Lindera. See *Medang*.
Linear Measure. A measurement of length only.
lin. ft. Abbreviation for *linear foot*.
Linen Fold Panel. One moulded and carved to imitate the draping
or folding, of cloth.
Line of Nosings. A line, on the string board
of a flight of stairs, touching the intersections
of the faces of riser and tread of each step.
See *Nosing Line*.
Liner. 1. A passenger vessel for long sea
voyages. 2. A packing piece in shipbuilding.
Lines. 1. Narrow inlays. See *Bandings* and
Stringing. 2. The curves of the hull of a
vessel.

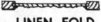

Lingoa. Amboyna, *q.v.* Also see *Padauk*.
Lingue. Apa, *q.v.*

**LINEN FOLD
PANEL**

Lining Out. 1. Levelling-up to give a plane
surface. 2. Marking out the stuff preparatory to sawing.
Linings. 1. Thin boards or wall boards over cavities or rough
surfaces to provide a finished surface, or for insulation. 2. The
rebated frame, or casings, for an interior door. See *Casings, Finishings,
Inner Linings, Sash* and *Frame, Furrings,* and *Wallboards*. 3. Interior
finishings to wardrobes, etc.

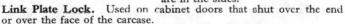

Thumb Moulding

LINING UP

Lining Up. Planting a narrow piece
under the edge of thin material to
give an appearance of greater thickness and to stiffen the edge.
Link Dormer. A dormer between
the roof and some other part of
the structure, such as a chimney
stack or an outside wall. The lights
are in the sides.
Link Plate Lock. Used on cabinet doors that shut over the end
or over the face of the carcase.
Linn. Abbreviation for the name of the botanist Linnæus.
Linociera. See *Olive*.

Linseed Oil. Raw linseed oil is produced from the seed of the flax plant. The raw oil is heated with a drier, such as litharge, for boiled linseed oil. Both are used as vehicles for paint, etc.

Linstock. A staff with a crotch, or fork, at one end.

Lintel or **Lintol.** A horizontal beam across an opening, usually to support the wall above. See *Fixed Sash*.

LINVILLE or **'N' TRUSS**

Linville Truss. A lattice girder with N-shaped bays in which the diagonals are in tension and the vertical members in compression.

Lip. An edge or border. A small thin projection. Also see *Cutter Block*.

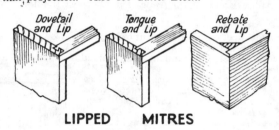

Dovetail and Lip   Tongue and Lip   Rebate and Lip

LIPPED MITRES

Lip Mitre. See *Lap Mitre*.

Lip Piece or **Block.** A short timber nailed over a joint in temporary timbering to secure the joint. It is often used for the joint between strut and waling. See *Timbering*.

Lipping. 1. Facing an exposed edge with superior wood, as the edges of shelves or laminboard. 2. A joint with one piece prominent to form a rebate.

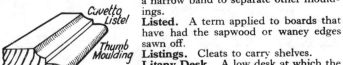

LIPPING

Liquidambar. See *Red Gum* and *Feng*.

Liriodendron. See *Canary Whitewood*, *Poplar*, and *Tulip*.

List. A cabinet-maker's term for a stay in a chair.

List or **Listel.** A square fillet in mouldings. It is generally used as a narrow band to separate other mouldings.

Listed. A term applied to boards that have had the sapwood or waney edges sawn off.

Cavetto
Listel
Thumb
Moulding

LISTEL

Listings. Cleats to carry shelves.

Litany Desk. A low desk at which the priest kneels when reciting the Litany. See *Desk*.

Litharge. Oxide of lead, *PbO*. It sets very hard when mixed with oil and is used as a drier for paints, mastics, stoppings, etc.

Lithocarpus. Tanbark oak. See *Oak*.

Litsea. There are numerous species of this genus, chiefly in S. Asia and E. Indies, but most of them are not appreciated in this country and have no market names. Yellowish grey to olive brown, fairly lustrous. Moderately soft and light, not durable and subject to attack by insect and fungus. Difficult to season but fairly easily wrought. See *Baticulin, Bollywood, Laurel Medang,* and *Queensland Sycamore.*

Little Joiners. Small inlays to remedy defects such as bruises, knot holes, etc.

Little Sugar Pine. Western white pine, *q.v.* See *Pine.*

Live Centre. The centre on a lathe that rotates the stuff. See *Lathe.*

Live Knot. A sound and firm knot. See *Knots.*

LITTLE JOINERS

Live Load. A moving load or variable force acting on a structure. See *Dead Load.*

Live Oak. *Quercus virens* and *Q. virginiana.* See *Oak.*

Live Rollers. Power-driven rollers to machines, etc.

Liverpool String Measure. A method of measuring the cubic contents of round timbers such as logs. The formula is the same as the Hoppus String measure, *q.v.,* but the allowances give results about 5 per cent. less.

Livery Cupboard. A cupboard to store bread to be distributed to the poor as alms.

Livistona. A species of palm trees.

Lizard. A drag sled, *q.v.*

Lizard Wood. See *Jocote.*

Lloyd. A registered wallboard, *q.v.*

lng. Abbreviation for *lining.*

Load. **1.** A volume of timber of 50 c. ft. hewn softwood, or 40 c. ft. hardwood or unhewn softwood, or 600 ft. super of 1 in. thick. A load should be about 1,680 lb. in weight. **2.** The external forces acting on a structure. See *Dead Load* and *Live Load.* Also see *Car Load.*

Loading Jack. See *Landing* (2).

Loading Way. A covered yard or vanway to a warehouse, etc.

Load Line. **1.** The water line along a ship's side when it is loaded. **2.** A scale diagram showing the sum of the several forces acting on a structure. See *Stress Diagram.*

Loam Board. A board with which a loam mould is *swept up,* in moulding.

Lobby. A passage, waiting room, or small hall in a building, serving as a means of communication between the various rooms or apartments.

Lobe. Same as foil, *q.v.*

Loblolly Pine. *Pinus tæda.* Florida. Southern yellow pine. It is marketed in this country as pitch pine, *q.v.* It has the characteristics and properties of long-leaf pine, *q.v.,* but is not so good.

Lock. **1.** A basin in a canal in which the height of the water is controlled by gates. See *Lock Gates*. **2.** A metal fastening operated by a key and used for securing doors, drawers, etc. The various kinds are : box, cupboard, desk, door, drawer or till, and wardrobe. They may be barrel, beak, cut, cylinder fall, dead, latch, lever, link plate, mortise, mortise-box, night latch, pad, rim, sliding door, stock, straight, cupboard, two-bolt, and Yale.

Lock Block. A muntin or block in the framing of a flush door to provide a fixing for the lock. See *Flush Door*.

Lock Corners. The corners of boxes prepared on a corner-locking machine or square dove-tailer. The pins are similar to dovetails but parallel or rectangular.

Locker. A small cupboard for a person's private property in schools, barracks, ships, etc.

Lock Gate. The gate to a canal lock, to control the height of the water. In this country lock gates are made up to 48 ft. deep by 38 ft. wide and usually of greenheart. The main framework consists of three vertical posts called heel, mitre, and centre posts respectively, and about nine horizontal members called crop beams. There are also intermediate vertical members called fenders. The members are built-up to the required section and keyed and bolted together. The spaces, or panels, are filled with 4-in. planking. Specially made machinery is used for the fabrication of the gates.

LOCK CORNERS

Locking Bar. A pivoted bar to secure a pair of gates. It engages with hook plates and staple and is secured by means of a padlock.

Staple for Padlock

LOCKING BAR

Locking Stop. A projection to limit the turning of the front carriage of a vehicle.

Lock Nut. A nut to a bolt screwed in position and fixed so that it will not work loose with vibration. There are many kinds and patents.

Lock Rail. The middle rail of a door that usually carries the lock. See *Panelled Door* and *Diminished Stile*.

Locust. **1.** *Hymenæa courbaril* or *Prosopis spp.* (*H.*). C. America. Also called W. Indian Locust, Algarrobo, Jatay, Courbaril, Gumanime, Simiri. Orange to dark brown with darker streaks. Darkens with exposure. Hard, heavy, tough, strong, stable, and fairly durable. Straight and irregular close grain and smooth. Difficult to work and polish. The tree provides copal. Used for construction, ship-building, cabinet work, etc. S.G. ·8 to 1 ; C.W. 4. BLACK or YELLOW LOCUST, *Robinia pseudacacia*. Europe and N. America. Also called False Acacia and White Laburnum. Now called Robinia. Yellow to brownish with varying shades of red and yellow. Hard, heavy, tough, and durable. Splits easily. Used for furniture, bentwork, sleepers, posts, wheelwright's work, etc. HONEY LOCUST, *Gleditschia tria-canthus*. U.S.A. Characteristics of Robinia. Also see *Acacia, Angelique, Mesquite,* and *Zebrawood*. **2.** A large insect, allied to the grasshopper ; very destructive to plantations.

Locust Wood. *Dicorynia paraensis.* See *Angelique.*

Lodge. 1. A hut or temporary residence. **2.** A small house sub-ordinate to a larger one. **3.** A room serving as a regular meeting place for certain societies.

Lodge-pole Pine. *Pinus contorta.* W. Canada and U.S.A. Pale brown. Soft, light, uniform with fine grain. Slightly resinous. Little exported. S.G. ·4. See *Pine.*

Lodget. Same as tabernacle on certain types of ships.

Loft. 1. A storage space immediately below the roof of a building. **2.** A gallery in a church or in a large apartment, such as a music loft.

Log. The stem of a tree felled and prepared for conversion. The term is also applied to short lengths of the stem, or to large branches. See *Balk* and *Squared Log.*

Loganiaceæ. A family of few timber trees, but including Antho-cleister, Cyrtophyllum, Fagræa, Logania, Norrisia, and Strychnos.

Log Band Mill. See *Log Saw.*

Log Deck. A platform for a collection of logs in a sawmill.

Log Dog. 1. A metal plate with spuds on an endless chain for elevating logs to the sawmill ; also called log-bracket, chair, saddle, shoe, and spur. **2.** Machine-operated jaws for securing a log whilst it is being sawn.

Log Dump. A platform to a sawmill on which logs are unloaded from vehicles or wagons. Also called a rollway.

Log Frame. See *Frame Saw.*

Loggia. A covered path or gallery on the outside of a building. A roofed verandah. A gallery behind a colonnade or open arcade.

Logging Wheels. A pair of wheels for transporting logs. They are about 10 ft. in diameter.

Log Haul-up. An inclined trough for drawing the logs up to the sawmill. Also called gangway, log-hoist, log-chute, jack-ladder, or jack-slip.

Log Jack. The mechanism for hauling logs to the sawmill.

Log Kicker. A lever mechanism for throwing the logs from the haul-up to the deck.

Log Lift. Mechanical contrivances for lifting logs from tidal waters to the log deck.

Log Rule or **Scale. 1.** A table showing the amount of timber that can be obtained from logs of given dimensions. **2.** A graduated stick giving the number of *board feet* in a log.

Log Run. A term applied to all merchantable grades of softwoods as they come from the saw. The term is a little more restricted when applied to hardwoods but covers all the better grades the full run of the log.

Log Saddle. Same as log dog, *q.v.*

Log Saw. A saw for converting logs. It may be a frame or band saw, *q.v.* Heavy types are also called log band mills. Also see *Machines*.

Log Turner. Mechanical devices for turning logs on a sawmill carriage or log deck.

Log-way. Same as log haul-up, *q.v.*

Logwood. *Hæmatoxylon campechianum* (*H.*). C. America. Also called Campeachy. Bright red, some darker streaks, darkens with exposure. Very hard, heavy, and durable. Strong but rather brittle. Irregular and interwoven fine grain. Smooth, and polishes ; rather difficult to work. Chief importance as a dye-wood.

Lombardy Poplar. See *Poplar*.

Lonchocarpus. See *Cabbage Bark* and *Rosa Morada*.

London Plane. *Platanus acerifolia*. See *Plane*.

London Standard. A measure of timber now little used. See *Standard*.

Long-arm. A long pole with hook for operating fan-lights, sliding sashes, etc.

Long-boat. The largest of a ship's boats. It is usually fitted with sails and often armed.

Long-bow. A bow used in archery, from 5 to 6 ft. long. The name implies a bow drawn by hand as distinct from a cross-bow.

Longeron. The main longitudinal member in the fuselage or nacelle of aircraft. See *Fuselage*.

Longhorn Beetle. *Hylotrupes bajulus*. A large insect about $\frac{5}{8}$ in. long, with grub nearly 1 in. long. The latter is very destructive, especially to roof timbers. It is not common in this country but is on the increase. Other species of the same family, *Cerambycidæ*, may be up to 3 in. long. They normally live just under the bark of trees or in green timber.

Long Jack. See *Yellow-wood Ash*.

Long-leaf Pine. *Pinus palustris*. Southern U.S.A. Light red or orange colour. Hard and heavy for a softwood. Very stiff and strong, and durable. Very resinous, with the summer wood well defined. Marketed in this country with inferior species such as pitchpine, *q.v.*

Long-pod. A spade gouge used in carving.

Long Timbers. Those from the deadwood to the top of the second futtock in the cant-bodies of a ship.

Long Tom. A wood sluice, in mining.

Long-tooth Gauge. A grasshopper gauge, *q.v.*

Longwood. A veneer term applied to veneers cut from the trunk of the tree.

Lonicera. Honeysuckle, *q.v.*

Lookum. See *Leucomb*.

Loom. 1. Part of the shaft of an oar between handle and rowlock. 2. A machine for weaving cloth.

Loophole Door. A trap door, *q.v.*

Loose. A term applied to band saws for places that fail to form a straight edge.

Loose Box. A large stall in a stable in which the horse is allowed free movement.

Loose Flange Butts. Door hinges that allow for the removal of the door without disturbing the screws.

Loose Knots. See *Knots*.

Loose Mouldings. Glazing mouldings or glass beads.

Loosened Grain. Slash figure check. The raising of fibres along the rings as they break out on a flat-sawn surface.

Loose Pieces. The detachable parts of a pattern for moulding that can be withdrawn separately from the main part of the pattern.

Loose Side. See *Tight Side*.

Loose Tenons. See *Inserted Tenons* and *Hammer-beam Roof*.

Loose Tongue. See *Tongue*.

Lopers. Slides for an extension dining table, *q.v.*, or for a bureau fall.

Lophira. Ekki, *q.v.*

Lophopetalum. See *Banate* and *Yamane*.

Lophora. Red Ironwood, *q.v.*

Lopped. Hewn or trimmed. Applied to a tree from which the top or branches have been chopped or sawn off.

Loppings. Lopped off branches and tops of trees.

Loquat. *Eriobotrya japonica* (*H.*). E. Asia. Red. Hard, heavy, strong, durable, elastic. Used for rudders, agricultural implements, etc.

Lorcha. An Eastern sailing vessel.

Lorry. A strong, low, open truck or wagon for heavy goods. It may be power- or horse-drawn.

Lorymer. Larmier, *q.v.*

Losonia. *Dysoxylum spp.* (*H.*). India. Resembles Poon. Reddish. Moderately hard, heavy, and stable, with fairly close grain. Easily wrought and polishes well. Used as a substitute for mahogany. S.G. 1·65 ; C.W. 3.

Loud. Abbreviation for the name of the botanist Loudon.

Louisiana Cypress. See *Cypress*.

Louis Style. An elaborate flamboyant ornamentation prevalent in France during the reign of Louis XIII to Louis XVI, from A.D. 1610 to 1793. The chief characteristic was elaborate scrolls, foliations, and shells in gilt stucco-work and without regard to symmetry.

Lounge. A room provided with easy chairs, couches, etc.

Loup. New Guinea walnut, *q.v.*

Louro. *Cordia spp.* (*H.*). India and C. America. Reddish brown, dark streaks, fragrant, and lustrous. Very variable in properties and characteristics. Distinct rays. Easy to season and work, and polishes well. S.G. ·56 to ·85 ; C.W. 3. See *Laurel* and *Vermelho*.

Louro Vermelho. *Octoea rubra*. (Uncertain.) (*H.*) Tropical S. America. Also called Determa in some areas. Brownish colour. Moderately hard and heavy. Strong and stable. Rather coarse texture. Roey figure, rays, and stripe. Requires careful seasoning, rather difficult to work. Polishes well with careful filling. Used for superior joinery, coachwork, boat-building, cabinet work, masts, turnery, etc. S.G. ·65 ; C.W. 3·5.

Louvre. A turret with sides formed of a series of inclined boards or sheets of glass, metal, or slate, called louvres, for ventilation. A louvre door, or window; has louvres instead of panels or glass.

Louvre Boards. The inclined boards or slats to a louvre ventilator or shutter. They may be hinged or pivoted to open or close as required.

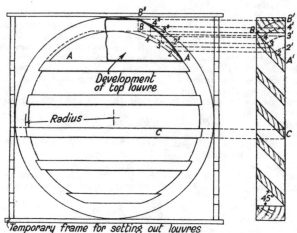

Temporary frame for setting out louvres
LOUVRE VENTILATOR

Lovely Fir. Amabilis fir. See *Fir*.
Lovoa. See *Nigerian Walnut*.
Lower Breast Hook. See *Keel*.
Lowland Fir. Grand fir. See *Fir*.
Lowry Process. An open-cell process in preservation, *q.v.* The excess creosote is removed by a final vacuum.
Lowside Window. A small window or opening in the South chancel of old churches.
Loxopterygium. See *Hoobooballi*.
Lozenge. **1.** A decoration on mouldings in the form of diamond-shaped carvings. It is one of the characteristics of Norman architecture. **2.** Small diamond-shaped, or rhomboidal, panels or panes of glass.
l.r. Abbreviation for *log run*.
Lubber's Hole. A hole in the platform of a ship's top, for easy access.
Lucarne. A vertical window in a parapet wall. It is like a dormer, but the latter stands back from the face of the building. The term is also applied to a dormer on the face of a spire.
Lucombe Oak. *Quercus lucombeana.* See *Oak*.
Lucuma. See *Mucuri, Red Silkwood,* and *Vaquetero*.
Ludat. *Avicennia officinalis* (*H.*). Malay. Also called Api-api. Sapwood pale grey, heartwood olive brown but seldom present.

Hard, fairly heavy, brittle, not durable, and very coarse grain. Difficult to split, season, and work. The wood is of little commercial value but sometimes included in shipments of Bakau.

Luehea. See *Estribeiro*.

Luffer Boards. Louvres, *q.v.*

Luffing. Swinging the jib of a crane or adjusting the helm of a ship.

Luff Tackle. One with a single and double block. There should be a hook for the tackle to each block.

Lug. A projection, generally to serve as a fixing, as an ear to a pipe, etc. See *Shelf Fittings*.

Lugger. A small sailing vessel with two or three masts and lug-sails for coastal trade.

Lumbayau. *Tarrieta javanica* (*H.*). Philippines. Light to dark red, lustrous. Moderately hard, heavy, durable, and tough, but not strong. Smooth, with straight grain, but rather coarse texture. Distinct rays. Rather difficult to season and convert, but fairly easily wrought and polishes well. Best if rift sawn. Used for cabinet work, superior joinery, planking, and as a substitute for mahogany. S.G. ·68 ; C.W. 3.

Lumber. An American term for converted wood ; also for felled trees prepared for the sawmill. Timber split or sawn for use in building. Hence a lumber mill is a sawmill.

Lumber Gauge. A tool for checking the accuracy of the machine processes as the stuff leaves the machine.

Lumber Jack. A tripod or stand used as a fulcrum for jacking timber to the top of a pile.

Lumber Transfer. A mechanical device for transferring converted wood from the live rollers, usually into another direction of travel.

Lumen. The cavity of a cell.

Lumnitzera. See *Teruntum*.

Lumpers. A gang of men, controlled by a stevedore, that contracts to do the work of unloading, etc., for a lump sum.

Lunes. Anything in the shape of a crescent or quarter moon. The gores in the development of a dome are sometimes called lunes.

Lunette. 1. An aperture in a concave ceiling for the admission of light. The space formed by a wall intersecting a vaulted roof, especially when it forms a window opening. A vertical arched opening intersecting an arched ceiling or vaulted roof. 2. A crescent-shaped ornament.

Lunumidella. *Melia dubia* (*H.*). Also called Ceylon cedar or mahogany. Reddish white. Characteristics and properties of Toon, *q.v.* Used for cabinet work and superior joinery.

Lup. New Guinea walnut, *q.v.*

Luster. *Syncarpia laurifolia*. Turpentine, *q.v.*

Lustre. Caused by reflection of light on the surface of the wood, and due to the undulating fibres and cell contents. If the light is absorbed the surface presents a dull appearance. Lustre is often a means of identification, and if a naturally lustrous wood loses its lustre it is a sign of incipient decay. Waxy or oily contents detract from the lustre. See *Fire* and *Figure*.

Lute. A musical stringed instrument like a guitar.

L.W. A registered wallboard, *q.v.*

Lych Gate. A covered entrance to a churchyard, but often used as a decorative feature to garden gates.

Lyctus. See *Powder Post Beetle.*

Lying Light. A ceiling light, *q.v.*

Lying Panel. A lay panel, *q.v.*

Lymexylonidæ. A family of wood beetles that attack green wood and standing trees. *Lymexylon navale* is specially destructive to oak trees.

Lysicarpus. See *Mountain Oak.*

Lysiloma. See *Sabicu.*

Lythraceæ. The Loosestrife family. Includes *Crypteronia, Duabanga, Lagerstrœmia, Physocalymma, Sonneratia,* etc. Few trees of importance for timber but several species valuable as dye-woods.

M

m. Abbreviation for *matched* and for *thousand*. See *Mille*.

m.1.s. Abbreviation for *moulded one side*. **m.2.s.** means *moulded two sides*.

Maba. See *Ebony* and *Persimmon*.

Macacahuba. *Platymiscium spp.* (*H.*). Brazil. Rich red, with variegated veining, smooth, lustrous. Very hard, heavy, strong, and durable. Ribbon grain. Not very difficult to work, and polishes well. Useful for most purposes where strength and appearance are important, but little imported. S.G. ·7 ; C.W. 3.

Macadamia. See *Nut Oak*.

Macaroni. A tool, shaped like an angular flat U in section, used in carving.

Macassar Ebony. *Diospyros macassar*. Celebes Islands. Brown with nearly black stripes. Used chiefly as veneers in this country for superior joinery and cabinet work. The wood has not yet been classified with certainty, probably because several species are sold under this trade name. See *Ebony*.

Macawood. Macacahuba, *q.v.*

Macaya. *Andira inermis*. Mexico. See *Angelim*.

Mace. A Jacobean ornamentation consisting of a turned, and often carved, ornament. It is split lengthwise and planted on sideboards, etc. Also called a split baluster or cannon. See *Jacobean*. **2.** An ornamental staff used as a symbol of office. **3.** A stick used in the game of bagatelle.

Machærium. See *Jacaranda, Negrillo, Rosewood, Tigerwood, Tipa*.

Machilus. *Machilus spp.* (*H.*). India. Four important species. Also called Ladderwood. Grey to light reddish brown, some darker streaks, lustrous and smooth. Indistinct heartwood. Moderately light, soft, strong and durable. Straight or twisted grain, even texture. Somewhat like chestnut. Should be girdled and quickly converted. Easily wrought and polishes well. Used for furniture, construction, etc. S.G. ·5 to ·65 ; C.W. 3. Also see *Dudri* and *Kurma*.

Machine Burn. Dark charred patches on machined wood due to over-heating of the machine cutters.

Machine Feeder. One who feeds the wood into the machine.

Machines. Woodworking machines may be obtained for nearly every specialist process, and the manufacturers will design them to client's own specification if desired. Improvements and variations are continually taking place in quality and tempering of materials, ball bearings, heavier and bigger types, automatic and portable types, canted tables and spindles, variable speeds, gear boxes instead of cone pulleys, safety devices, specialist types, etc. Modern machines are now motorised and large machines may have twenty motors for the different operations, with a push button start for each motor. Speed and rate of feed vary with the machine and kind of wood. Some machines have a speed of over 20,000 revs. per minute, and surfaces may have a feed rate of 450 ft. per minute. The following are standard types of machines from manufacturers' catalogues : blocking, boring (multi), branding, checking, chucking, combing, copying, corner locking, cramping, dowelling, dovetailing, drilling, edging, elephant, four-cutter, general joiner, glue spreader, grinders, guillotine, handle, jointer, joint gluer, lathe, matcher, mortiser, moulding, nailing, panel shaper, planing, plugging, pointing, recessing, rounding, router, sander, screwing, shive, slicer, slotting, spindle, squaring off, surfacer, taping, tenoner, thicknesser, trenching, turning, universal, veneer cutter and press, wire stitcher, and special types for plywood, barrel-making, etc. There are numerous specialist attachments to many of the machines and a great variety of grinders, sharpeners, etc. They are described alphabetically in the Glossary. Also see *Saws* and *Veneers*.

Machine Shop. A special department for the machines, as distinct from sawmill and assembly- or work-shop.

Machine Tailer. One who removes the wood as it comes from the machine and grades it.

Maclura. See *Osage Orange*.

Maco Gauge or **Templet.** An adjustable templet that can be quickly adjusted to obtain the outline of irregular contours of mouldings, etc.

Macquerie Pine. Huon pine, *q.v.*

Macrolobium. See *Zebrawood*.

Madhuca. A genus of the Sapotaceæ family. See *Betis* and *Nyatoh*.

Madrona. Strawberry tree, *q.v.*

Maftex. A proprietary insulating board.

Maga. *Thespesia grandiflora* (*H.*). C. America. Rich chocolate brown, variegated. Hard, heavy, strong, durable. Roe and ripple figure, fine texture. Not difficult to work and polishes well. Used for cabinet work, musical instruments, heavy construction, etc. S.G. ·8 ; C.W. 3.

Magadam. *Mimusops elengi.* See *Bulletwood*.

Magnolia. *Talauma sambuensis* and *Magnolia grandiflora* (*H.*). C. America and U.S.A. Variable yellowish to purplish, often with dark streaks. Firm, soft, straight grain. Resembles Canary whitewood, *q.v.*, and has the same characteristics and uses. MOUNTAIN MAGNOLIA, *M. acuminata.* Cucumber tree, *q.v.* EVERGREEN MAGNOLIA, *M. grandiflora.* Also see *Champac*, *Honoki*, and *American Whitewood*.

Magnoliaceæ. Magnolia family. Includes Drimys, Liriodendron, Magnolia, Manglietia, Michelia, Talauma, Trochodendron, Tetracentron, and Zygogynum.

Maho. *Sterculia pruriens* (*H.*). B. Guiana. Pale brown, lustrous. Moderately hard and heavy, not strong or durable. Straight grain. Easily wrought but liable to pick-up. Little imported. S.G. ·52 ; C.W. 3.

Mahoborn "Teak." *Dryobalanops aromatica.* Borneo camphorwood, *q.v.*

Mahoe. *Hibiscus elatus (H.).* C. America. Called Blue Mahoe and Mountain Mahogany. Varies from greenish to purplish, with variegated streaks, fragrant, lustrous. Moderately hard, heavy, strong, and durable. Straight, rather coarse grain, with distinct rays and ripple. Resembles black walnut. Easily wrought and polishes well. Used for cabinet work, superior joinery, veneers, vehicles, implements, etc. S.G. ·7 ; C.W. 2·5.

Mahogany. Over 60 different kinds of wood are sold under this trade name, because of the popularity of Cuban mahogany. Hence there is great variation in quality, and price, which may vary from 3d. to 30s. per ft. super. They are usually distinguished by the source of origin, and marketed as plain or figured. They are generally reddish brown in colour, moderately light and soft, with interlocked grain, highly figured, and of large sizes. Mahogany is particularly subject to upsets, *q.v.* It is used for cabinet work, superior joinery, veneers, and for all purposes where appearance and stability are of chief importance.

There are only three genera recognised as mahogany : *Entandrophragma, Khaya,* and *Swietenia,* and some claim the last named as the only true mahogany. The most valuable is Cuban, but supplies are scarce, and W. Africa supplies most of the mahogany for the English market. AMERICAN. See *Cuban* and *Honduras.* AFRICAN. See *E. African, S. African,* and *W. African.* AUSTRALIAN, *Dysoxylum sp.* Also called Rose Mahogany or Rosewood, *q.v.* An excellent wood but not a mahogany. Several species of Eucalyptus were marketed as mahogany. See *Jarrah,* and *White, Red, Miva, Rose,* and *Swamp-Mahogany.* BORNEO, *Calophyllum spp.* See *Penaga Laut* and *Seraya.* BRAZIL. Mostly not mahogany. See *Crabwood, Jequitiba, Santa Maria,* and *Vinhatico.* BRITISH GUIANA. See *Crabwood.* BURMA. See *Thitka.* CAPE LOPEZ, French Congo. Usually inferior to W. African, *q.v.* CEYLON. See *Lunumidella.* CHERRY MAHOGANY. See *Makoré.* COSTA RICA. Similar to Cuban, *q.v.* CUBAN, *Swietenia mahogoni.* Also called W. Indies, San Domingo, and Spanish Mahogany. Rich brown-red, darkens with age, lustrous, chalky deposit. Hard, heavy, durable, and very stable. Interlocked and ribbon grain, with roe and ripple. Firm texture. It is perhaps the most appreciated wood for cabinet work, superior joinery, pattern making, etc., because of its beautiful figure, ease of seasoning, working, and polishing, and for its stability. S.G. ·8 ; C.W. 3·5. E. AFRICAN. Species of *Entandrophragma* (Miovu, Mukusu, Mufumbi), and *Khaya* (Munyama). Very similar to W. African, *q.v.,* and marketed as Budongo and Uganda mahogany. GUATEMALAN. Like Cuban. HONDURAS, *Swietenia macrophylla.* C. America. The same genus as Cuban but from the mainland. It is golden red in colour, softer, and lighter than Cuban, and was originally distinguished as baywood. S.G. ·6 ; C.W. 3·25. INDIAN. See *Toon* and *Dhup.* MEXICAN. Like Honduras. NICARAGUAN. Like Honduras. NIGERIAN. See *W. African.* PANAMA. Like Honduras. PERUVIAN. Similar to Cuban. PHILIPPINE. See *Lauan.* RHODESIAN, *Copaifera coleosperma.* A very stable and hard-wearing wood used for superior flooring, etc. It is not a Mahogany but is a decorative wood that finishes well for polish, and is approved as fire-resisting. S.G. ·8; C.W. 4. SPANISH. See *Cuban*. S. AFRICAN, *Entandrophragma caudatum*. Called Berg mahogany. Similar to E. African species but with more ripple, and odorous during conversion. S.G. ·72. SURINAM. Ciròuaballi, *q.v.* SWAMP. See *Swamp Box*. TOBASCAN. Better qualities of Mexican; like Honduras. W. AFRICAN, *Khaya spp.* Nigerian or Gold Coast mahogany. There are several species and they are usually named after the district or port of shipment : Axim, Assinie, Bathurst, Benin, Dubini, Grand Bassam, ·Grand Lahou, Half-Assinie, Lagos, Sassandra, Secondé, and Sudan. Very variable in properties and characteristics. Pinkish, seasoning to reddish brown, lustrous. Better qualities are moderately hard, heavy, durable, and stable. Soft and spongy logs are common and very subject to shakes. Large sizes and often highly figured, and very similar to Honduras, but not so uniform in quality, and cheaper. S.G. ·65; C.W. 3·6. Also see *Cedar, Chinese and Colombian Mahogany, Espavé, Gaboon, Jequitiba, Makore, Mahoe, Moboran, Okoumé, Prima Vera, Sapele, Seraya*, and *Vinhatico*. Many other trade names are given in the Glossary.

Mahogany Birch. Richly figured American black birch, *q.v.*

Mahogany Gum. *Eucalyptus botryoides.* Australia (Victoria). Bright red ; resembles mahogany. Very hard, heavy, strong, tough, durable. Close firm grain. Difficult to season and work. Used for construction, carriage work, sleepers, etc. S.G. 1; C.W. 4·5.

Mahogany Measure. A method of measuring especially applied to mahogany. It allows for variation of thickness, width, and length, and for defects. See *Measure*.

Mahot. A prefix applied to several C. American woods : bocote, mahoe, mora, sapucaia, etc.

Mahua or **Mahwa.** Butter tree, *q.v.*

Mai. A prefix to most Siamese woods. The woods are described under their Malayan or Philippine names. Also see *Matai*.

Maiang. Nyatoh, *q.v.*

Maid. Abbreviation for the name of the botanist Maiden.

Maiden. A clothes horse, *q.v.*

Maiden Ash. Young ash wood from seed.

Maiden Gum. Spotted Blue Gum. See *Gum*.

Maiden Hair Tree. *Gingko biloba* or *Salisburia sp.* (*S.*). A Chinese tree with fern-like foliage but of little timber value. Also known as the living fossil, one of the few survivors of vegetation of the Carboniferous period.

Maiden Wood. Usually applied to young ash trees, suggesting greater suppleness or elasticity.

Maidou. *Pterocarpus pedatus.* Indo China. Very similar to Padauk and Amboyna, *q.v.*

Mailcart. 1. A light carriage for a baby. A form of perambulator. 2. A vehicle for conveying mail by road. A mail van.

Main-deck. The principle deck of a ship.

Main-mast. See *Masts*.

Main Planes. The aerofoils that give lift to an aeroplane. See *Fuselage*.

Main Tram Piece. A sole bar, *q.v.*

Main-yard. See *Yards.*
Mai Payung. Indian Blackwood or Rosewood, *q.v.*
Maire, Black. *Olea cunninghamii (H.).* New Zealand. Resembles European olive. Deep brown, black streaks. Very hard, heavy, tough, strong, and durable. Close grain. Difficult to season and work. Used for general purposes where strength and durability are important. S.G. 1 ; C.W. 4·5.
Maiyang. See *Yang.*
Majagua. Mahoe, *q.v.*
Makaasim. *Eugenia spp. (H.).* Philippines. Greyish brown or pale red. Odorous owing to large tannin content. Hard, heavy, strong, and durable. Twisted and curly grain. Difficult to season. Shrinks and warps freely if plain-sawn. Fairly difficult to work. Polishes well. Used for shipbuilding, piling, structural and constructional work, cabinet work, implements. S.G. ·83 ; C.W. 4.
Makai. *Shorea assamica (H.).* India, Burma. Greyish to light brown, white lines due to secretion, lustrous, but feels rather rough. Moderately hard and heavy and strong but not durable. Even, coarse texture. Fairly straight grain. Seasons fairly well and not difficult to work. Used for construction, planking, furniture, and for sleepers when treated. S.G. ·5 ; C.W. 3.
Makaka and **Makandoa.** Mangroves, *q.v.*, from Tanganyika.
Making Good. Repairing woodwork, etc., after alterations to any part of a building.
Makore. *Mimusops,* or *Dumoria heckelii (H.).* Nigeria. Cherry mahogany. Also called Aganokwe, Baku, etc. Reddish brown, lustrous. Moderately hard, heavy, durable, and stable. Straight, uniform, and interlocked grain, with roe and mottled figure. Fine compact texture and retains sharp arrises. Not difficult to work but requires care in seasoning ; polishes well. Fine dust is an irritant. Used for cabinet work, superior joinery, veneers, etc. S.G. ·75 ; C.W. 3.
Malabar Nim Wood. *Melia composita (H.).* India. Pinkish, seasoning to pale russet brown, lustrous. Soft, very light, not strong or durable. Straight coarse grain, uneven texture. Fine fleck on radial surface. Seasons well but subject to stain and shrinks considerably. Easily wrought. Used for packing cases, match-boarding, tea boxes, suitable for plywood. S.G. ·33 ; C.W. 2.
Malabera. *Fagræa fastigiata* or *F. crenulata (H.).* Malaya. Yellowish. Slight odour and feels greasy when dry. Moderately hard, heavy, durable, and resists insects. Coarse grain, distinct rays. Seasoning very slow and difficult. Limited supplies. Used chiefly for wharves and piling. S.G. ·66.
Malacca. A species of bamboo from Malaya.
Malaut. *Hopea anomala (H.).* Malaya. Sold as Resak, *q.v.*
Malenite. A registered name for a decorative plywood with *woven-wood* face.
Mall or **Maul.** A large mallet, or beetle, for levelling sets, driving small piles, etc.
Mallee. Applied to several species of Australian Eucalyptus in which several stems branch from the roots, or in which the tree has a dwarfed trunk, or to the shrub-like varieties. The woods have

little commercial value. The term is also applied to the white
cypress, *q.v.*

Mallet. A striking tool with hardwood head, usually beech, for
driving chisels. A smaller type has a metal head with hardwood
insets for the faces. Also with long handles for polo, croquet, etc.

Mallet-bark. *Eucalyptus occidentalis.* Australia. A hard, heavy
wood. Interlocking grain. Little imported.

Mallet's Spiral. A practical method of describing a spiral on turned
work. It consists of a light rectangular frame with a grooved diagonal
piece. In the groove is a cord coated with lamp black. When a
cylinder is rolled over the frame the lamp black leaves a spiral mark
on the cylinder. The diagonal is reversed for a left-hand spiral.

Mallet Wood. *Rhodamnia argentea* (*H.*). Queensland. Called
white myrtle, blackeye, etc. Pinkish. Hard, heavy, tough, strong,
durable. Fine uniform texture like boxwood. Used for tools, wheelwright's work, etc. S.G. ·82. INDIAN MALLET WOOD, *Schleichera
trijuga.* Light reddish brown. Very hard, heavy, strong, and durable.
Seasons well but difficult to work. Used for implements, tools,
furniture, etc. S.G. 1 ; C.W. 4·5. There are several Mallets from
W. Australia but their chief importance is for tannin, and the wood
is not exported. BROWN MALLET WOOD, *Rhodamnia trinervia* ; BLUE
MALLET, *Eucalyptus Gardneri* ; and RED MALLET, *Euc. astringens*, are
used chiefly for poles and the bark for tannin.

Malodorous. 1. Having an offensive odour. 2. A prefix applied to
several odorous Eastern woods.

Maloh. Chuglum, *q.v.*

Maltese Wood. Harder qualities of Mahoe, *q.v.*

Malugai. *Pometia pinnata* (*H.*). Philippines. Pinkish to light red,
lustrous. Hard, heavy, tough, but not strong or durable. Straight
and wavy fine grain. Requires careful seasoning and difficult to work.
Polishes well. Used for construction, masts, spars, furniture, cabinet
work, tight cooperage. S.G. ·84 ; C.W. 4.

Malus. See *Apple.*

Malvaceæ. The Mallow family, which includes Bastardiopsis,
Bombax, Bombycidendron, Cullenia, Gossypium, Hibiscus, Kydia,
Ochroma, Thespesia.

Mamee Apple. Zapote Mamey, *q.v.*

Manao. New Zealand silver pine, *q.v.*

Manbarklac. *Eschweilera spp.* (*H.*). Tropical America. Several
varieties. Olive or reddish brown, variegated streaks. Very hard,
heavy, tough, strong, and durable, and resists teredo. Smooth with
distinct rays. Difficult to season and work. Used for the same
purposes as greenheart. S.G. 1·2 ; C.W. 5.

Manchineel. *Hippomane mancinella* (*H.*). W. Indies, C. America.
Light yellowish brown, fragrant, lustrous. Hard, heavy, durable.
Close grain and beautiful figure. Poisonous sap. Used for cabinet
work, superior joinery, and exterior work. S.G. ·8 ; C.W. 3·5.

Manchurian Pine. Siberian pine. See *Pine.*

Mandania. *Acrocarpus fraxinifolius.* See *Mundani.*

Mandioqueira. *Qualea spp.* About 30 spp. from the Amazon
region, S. America. Yellowish brown. Hard, heavy, strong, fairly

durable. Interlocked grain. Not difficult to season but difficult to work, due to picking-up and fibrous nature. Also needs care in polishing. Useful for structural work, interior joinery and carpentry, flooring, etc. S.G. ·6 to ·8 ; C.W. 4.

Mandolin. A musical stringed instrument of the guitar type.

Mandong. Mempening, *q.v.*

Mandril. 1. A cylindrical piece of hardwood used for straightening lead pipes, etc. 2. A spindle, as on a lathe. A saw arbor.

Mangeao. *Litsœa calicaris* (*H.*). New Zealand. Small sizes. Used chiefly for handles.

Manger. A box, or trough, from which horses and cattle feed.

Manger Board. A strong bulk-head on the fore part of the deck of a ship to keep the water from the hawse-holes.

Manggachapui. *Hopea acuminata* and other species (*H.*). Philippines. Greyish yellow to golden brown on exposure, lustrous. Hard, heavy, strong, fairly durable. Straight grain, slightly interlocked, fine texture. Large sizes. Easily wrought and seasoned. Used for joinery, cabinet work, high-grade construction, etc. S.G. ·88 ; C.W. 3.

Mangifera. See *Mango* and *Pahutan.*

Manglé. See *Mangrove.*

Mangle Rollers. The rollers for a wringing machine, etc. They are usually made of sycamore or birch, and, for special cases, of lignum-vitæ.

Mango. *Mangifera indica* (*H.*). C. America, India. Numerous species. Chiefly greyish brown sapwood with yellow markings, with a little light chocolate brown heartwood. Rather hard, heavy, tough, resilient, and stable, but not durable except in water. Coarse texture with straight grain, and lustrous. Slightly interlocked and some curly grain. Subject to sap-stain and insect attack. Best qualities suggest ash, and used for same purposes, cheap furniture, tea chests, cases, etc. S.G. ·6 ; C.W. 2·5.

Mangona. African mahogany, *q.v.*, from the Cameroons.

Mangrove. *Rhizophora spp.* (*H.*). Numerous species growing on the seashore and in tidal swamps in tropical countries. There is great variation but the woods are generally reddish brown with darker stripes. Very hard, heavy, strong, durable. Often very irregular fine grain with distinct rays. Splinters and checks easily. Difficult to season and work, polishes well. Used for piling, sleepers, heavy rough construction. Important for tannin. S.G. 1·1 ; C.W. 5. BLACK MANGROVE, *Avicennia nitida.* C. America. Brown to nearly black ; lapachol present. Similar to Rhizophora spp. but unlignified laminations make wood of little value. S.G. ·9. WHITE MANGROVE, *Laguncularia racemosa* (*H.*). C. America. Also called Buttonwood. Yellowish brown with darker streaks. Hard, heavy, strong, with close grain and even texture. Not yet imported. The name White mangrove is also applied to lighter coloured wood of *Avicennia nitida* and to species of *Terminalia.* GREY MANGROVE, *Avicennia officinalis.* Queensland. Greyish brown, smooth. Hard, heavy, strong, and very tough. Not difficult to work considering hardness. Used for naves, knees, table legs, mallets, etc. S.G. ·85.

Mangwe. African Yellow-wood, *q.v.*

Manifest. A ship's document giving a detailed statement of the cargo, for dock and custom-house purposes.

Manihot. See *Tapioca.*

Manilkara. See *Nkunya.*

Maninga. Muninga, *q.v.*

Manna Gum. *Eucalyptus viminalis.* See *Gum.*

Manoa. New Zealand pine, *q.v.*

Mansard Roof. A roof of double pitch on both sides. The object is to provide space in the roof. It is also called a curb roof.

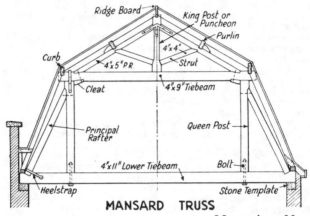

Ridge Board — *King Post or Puncheon* — *Purlin* — *4″x4″* — *Curb* — *4″x5″ P.R.* — *Strut* — *Cleat* — *4″x9″ Tiebeam* — *Principal Rafter* — *Queen Post* — *4″x11″ Lower Tiebeam* — *Bolt* — *Heelstrap* — *Stone Template*

MANSARD TRUSS

MANTEL

Mansonia. *Mansonia altissima* (*H.*). Nigeria, Gold Coast. Also called Ofun and Aprono. Purplish, fading on exposure to greyish brown, variegated markings, lustrous. Hard, moderately heavy, durable, fissile, and stable. Straight grain and smooth, with even, fairly close texture. Easily wrought and seasons well. Tends to irritate the skin. Resembles black walnut and used for similar purposes, fine cabinet work, propellers, etc. S.G. ·6 ; C.W. 3.

Mantel or **Mantelpiece.** The ornamental

front to a fireplace. It consists of pilasters, or jambs, and mantelshelf, and it may have panelling or mirror above the shelf.

Mantel Tree. A lintel to a fireplace.

Manual. A keyboard of an organ.

Manufactured. Applied to imported wood that is machined : matching, flooring, plywood, and all types of joinery, etc.

Manuka. The tea tree, *q.v.*

Maple. *Acer spp.* (*H.*). Chiefly from Europe and N. America, but widespread. The most important species on the English market is Rock maple. There is considerable variation in the different species but the wood is generally faint brownish white, hard and compact, with straight fine grain, and lustrous. It is moderately heavy and strong but not durable. It is used for flooring in strips and wood blocks, kitchen fitments, plywood, furniture, superior joinery, cooperage, woodware, rollers, implements, and for purposes where sharp arisses are required. Figured wood, especially bird's eye, is generally used as veneer. S.G. ·8 ; C.W. 3. AUSTRALIAN MAPLE. See *Maple Silkwood*. BIRD'S EYE MAPLE. Highly figured rock maple, *q.v.*, and chiefly used as rotary cut veneers. BLACK MAPLE, *A. negrum.* N. America, similar to rock maple. BLISTER MAPLE. Bird's Eye maple. CANADIAN MAPLE. Rock maple, *q.v.* EUROPEAN MAPLE, *A. campestre.* Britain. Also called Field or Common Maple. Little timber importance. Roots beautifully veined. Used for turnery, fancy articles, etc. *A. pseudo-platanus,* European great or false plane. See *Sycamore. A. platanoides,* Norway maple. Excellent wood and used for the same purposes as rock maple. HARD MAPLE.

MAPLE

Rock and Black Maple, *q.v.* HIMALAYAN MAPLES. Several species ; excellent wood and used as rock maple. JAPANESE MAPLE, *A. palmatum.* Excellent wood and used as rock maple. OREGON or BIGLEAF MAPLE, *A. macrophyllum.* Softer than rock maple, with attractive figure and burl. ROCK MAPLE, *A. saccharum.* N. America. Also called Hard or Sugar Maple. Brownish white or light reddish brown, darker lines showing growth rings. Hard, heavy, strong, stiff, not durable. Seasons, works, and polishes well. Fine compact grain, lustrous, hard wearing. Highly figured wood is called Bird's Eye, Blister, Curly, and Fiddleback Maple. RED MAPLE, *A. rubrum.* N. America. Similar to silver maple. SILVER or SOFT MAPLE, *A. saccharinum.* Also called Swamp or River Maple. Inferior and softer than rock maple. STRIPED MAPLE, *A. pennsylvanicum.* SUGAR MAPLE. Rock maple, *q.v.* U.S.A. MAPLE. Usually marketed as *hard* or *soft*. Hard maple includes *A. saccharum,* rock or sugar maple, and *A. nigrum,* black maple. Soft maple includes *A. rubrum,* red maple ; *A. saccharinum,* silver maple ; *A. macrophyllum,* Oregon maple ; and *A. negundo,* ash-leaved maple.

Maple Silkwood. *Flindersia brayleyana* (*H.*). Called Australian and Queensland Maple or Silkwood. No relation or resemblance to maple but belongs to the Satinwood family. Pinkish, lustrous, slightly

fragrant. Moderately hard and heavy, not durable, but strong and tough. Interlocked and wavy grain, very highly figured with moire and ripple figure when quarter-sawn ; sometimes with bird's eye. Requires care in seasoning and rather difficult to work owing to disturbed grain. Should be quarter-sawn. Polishes and fumes well. Butts valued for veneers. Used for cabinet work, superior joinery, carriage work, veneers, aircraft, turnery. S.G. ·6 ; C.W. 3·5.
Mara. E. Indian walnut or Kokko, *q.v.*
Maracaibo. Vera wood, *q.v.*
Maracaibo Boxwood. *Gossypiospermum*, or *Casearia præcox.* Venezuela. Also called W. Indian Boxwood and Zapatero, *q.v.*
Maranggo. *Azadirachta integrifoliola* (*H.*). Philippines. Nearly impossible to distinguish from good mahogany. It has all the characteristics and qualities of genuine mahogany and used for the same purposes.
Marara. See *Rose Marara* and *Grey Boxwood.*
Marbleboard. Proprietary wall-boards with surface to imitate marble.
Marblewood. *Diospyros Kurzi, D. marmorata,* and *D. oocarpa* (*H.*). Andamans. Ebony family. Also called Zebra Wood and Thitkya. Brown to nearly black, with variegated streaks. Hard, heavy, strong, especially in shear, durable. Properties very varied. Little lustre but smooth. Straight grain, fine even texture. Difficult to season and work, polishes well. Should be converted quickly, as for ebony. S.G. ·95 ; C.W. 5. Japanese Marblewood, *D. kaki.* Similar to above, with black and white stripes. Queensland Marblewood, *Olea paniculata* and *Acacia Bakeri.* Yellow with darker stripes. Hard, heavy, tough, and difficult to work. S.G. ·9. W. African Marblewood, or Bosse, similar to Andaman. Malayan Marblewood, or Buey, similar to Andaman. Also see *Macassar Ebony.*
Margin. 1. The projection of a stair string above the nosing line. 2. A narrow strip, or border, mitred round the hearth of a fireplace. See *Hearth.* 3. The exposed faces of stiles and rails in panelled framing, etc. 4. A variation, above or below, from a stated amount in a contract in the timber trade.
Margin or Marginal Lights. Narrow panes of glass round a big pane, to a sash or gun-stock door, etc., due to the introduction of bars. See *Ceiling Light.*
Marginal Mould. A moulded lining or fascia scribed to the wall or ceiling and covering the space between frame and wall. It is scribed to leave a parallel margin on the stiles and head of the frame.
Margined Door. Similar to double-margined, *q.v.*, but the central stile is in the form of a muntin with a central bead to give an impression of double stiles. The top and bottom rails run through.
Margosa. *Melia indica* (*H.*). India and Burma. Also called Neem. Golden brown, fragrant. Hard, heavy, strong, and very durable. A bitter secretion resists insects. Close fibrous grain, mottled. Fairly difficult to work. Small sizes. Used for furniture, chests, etc. S.G. ·8 ; C.W. 4.
Maria. See *Santa Maria, Jocote, Brauna, Roble.*
Marigold Window. A rose, or Catherine wheel, window.
Marine Borers. Crustaceans and Molluscs that are destructive

to wood. The best known are the gribble, *Limnoria lignorum*, and the shipworm, *Teredo navalis*. Very few timbers can resist either of these borers and in some waters the wood must be treated and sheathed in muntz or monel metal. Woods with silica content are most resistant but no wood can resist all kinds of borers. The shipworm, which is found up to 4 ft. long and ½ in. diameter, honeycombs the wood with holes up to 1 in. diameter. The gribble attacks the surface and the action of the water exposes fresh surfaces until the wood is eaten away. The *Chelura· terebrans* is similar to the gribble and is usually found with it. Full cell treatment with creosote, together with surface protection, is usually satisfactory. Other species : *Bankia* are molluscs like the teredo ; *Martesia* are molluscs resembling clams ; *Sphæroma* are similar to *Limnoria* but larger. British Guiana greenheart, Australian turpentine, opepe, and jarrah are very resistant.

Marinheiro. *Guarea spp.* (*H.*). C. America. Mahogany family. Also called Macaqueiro. Reddish colour. Moderately hard, heavy, tough, and strong. Durability very variable. Straight grain, medium texture, smooth. Easily wrought. Similar to Spanish cedar and used as a substitute for true mahogany. Needs careful seasoning. S.G. ·58 ; C.W. 2·5.

Maritime Pine. *Pinus pinaster*. S. and S.W. Europe. Used chiefly for pit props.

Marker. One who measures, marks, and tallies the wood as it comes from the saw.

Market Forms. The names given to the market forms of wood are balk, batten, board, casewood, deals, die square, ends, firewood, flitch, log, mast, pitwood, planchette, plank, planking, pole, quartering, scantling, sets, thick stuff, *q.v.* Wood is sold by board measure, cubic measure, standard, load, float, fathom, square, or by weight. See *Shipping Marks, Standard, Stock, Charter Party, Manufactured*.

Marking Awl. A pointed tool for setting-out the wood in cabinet-making.

Marking Hatchet. A special form of axe for blazing trees and stamping the blaze by means of a raised die on the head of the axe. A marking hammer is similar, for stamping proprietary marks on timber.

Marking Knife. A joiner's setting-out knife. It is usually about 8 in. long with a point at one end and a knife at the other.

MARKING KNIFE

Marking Out. Setting out, *q.v.* See *Rod*.

Marks. **1.** Imperfections in manufactured goods, as chip marks, tool marks, machine burns, etc. **2.** Customs *official's marks* denoting length and contents of logs. **3.** Letters or signs stamped on logs to indicate ownership. **4.** See *Shipping Marks*.

Mar. L. Abbreviation for *marginal lights*.

Marlocks. Australian eucalyptus shrubs.

Marmalade Tree. See *Zapote*.

Marnut. *Dalbergia spp.* or *Machærium spp.* (*H.*). Brazil. Rich violet brown with variegated markings. Like *Sissoo*.

Marquise. A large canopy, hood, or shelter, at the entrance to a shop, theatre, etc.

Marri. Australian tan-barks, or eucalypti, used chiefly for tanning and kino.

Marry, To. A craftsman's term meaning to pair or match.

Marsh. Abbreviation for the name of the botanist Marshall.

Martesia. See *Marine Borers*.

Martingale. A bar from the bowsprit cap of a ship to assist in supporting the jib-boom.

Marupa, Maruba. Simaruba, *q.v.*

Mascal Wood. *Vatica lanceæfolia* (*H.*). Assam, Burma. Light yellowish brown, with white resin canals, lustrous but dulls later. Moderately hard, heavy and strong, but not durable. Interlocked grain, ribbon streaks ½ in. wide. Even coarse texture. Distinct rays giving silver grain. Not difficult to season or work, but stains easily. Little used at present. S.G. ·63 ; C.W. 3.

Mashie. A golf club with iron head.

Mash Tub. A tub for the mash in a brewery.

Mask. A carving of grotesque form on keystones, friezes, panels, pediments, etc.

Masonite. Proprietary wood-fibre wall-boards.

Mason's Mitre or **Stop.** A mitre of a moulding carved on the solid as in masonry. It is characteristic of oak joinery in churches. See *Mitre* and *Bar Joints*.

Mason's Scaffold. A scaffold consisting of heavy squared timbers and independent of the wall for support.

Massaranduba. *Mimusops spp.* (*H.*). Surinam, Brazil. Light red to deep reddish brown, darkening with exposure, dull, but smooth. Extremely hard, heavy, strong, and very durable. Straight grain. Requires careful seasoning. Difficult to work, polishes well. Used chiefly for structural work, etc. S.G. ·98 ; C.W. 5. Also see *Bullet Wood* and *Mucuri*.

Sash Bars

Stop

MASON'S STOPS

Mast. **1.** The vertical post in a crane. See *Derrick Crane*. **2.** A guyed pole carrying pulley tackle for hoisting purposes. **3.** The vertical timbers carrying the rigging of sailing ships. They are distinguished as fore-, main-, mizzen-, and jigger-mast, with sections called lower, top, top-gallant, royal, and skysail. See *Spars*. **4.** A braced vertical structure for mooring an airship. **5.** A market term for round logs over 24 in. circumference.

Mast Cleat. See *Ship's Cleats*.

Mastic. *Sideroxylon mastichodendron* (*H.*). C. Amercia. Also called Wild Olive. Yellow to orange, lustrous, smooth. Very hard, heavy,

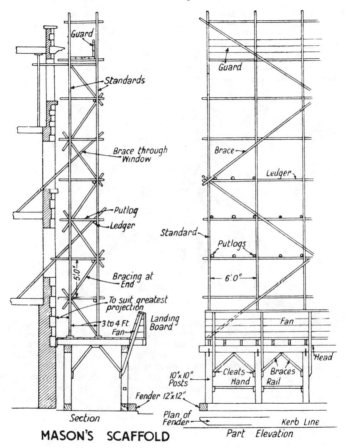

MASON'S SCAFFOLD

tough, strong, and durable. Straight, wavy, and roey grain. Fairly difficult to work, polishes well. Used for structural and constructional work, cabinet work, superior joinery, shipbuilding, sleepers, etc. S.G. 1 ; C.W. 4.

Mast Step. The footing for the bottom end of a mast to a ship.

Masur Birch. A trade name for specially figured birch.

Mata Buaya. *Bruguiera eriopetala.* Malayan Mangrove. Similar to Bakau, *q.v.*

Matai. *Podocarpus spicata* (*S.*). New Zealand black pine. Light yellowish brown, lustrous. Fairly hard, strong, and durable. Straight,

close, even grain and texture. Easily wrought, and polishes. Used
for exterior joinery, flooring, constructional work, containers, veneers,
etc. S.G. ·62 ; C.W. 2.

Matamata. Manbarklac, *q.v.*

Match-boarding. Matched boarding or matching. Thin tongued
and grooved boards with beaded or veed joints to hide shrinkage.
They are used to cover ceilings, walls, partitions, etc.

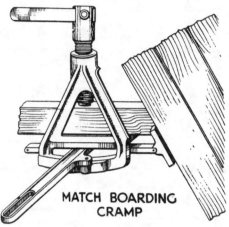

**MATCH BOARDING
CRAMP**

Matched. A term used in veneering for specimens cut from adjoining parts of the wood to give similarity, or symmetry, or continuance
of grain.   *Balance matched* has more than two pieces of uniform size
used in a single face.   *Book matched* means placing two adjacent
cuts together so that the second cut is turned over and placed alongside the first cut.   *Butt matched* is the same as end matched, *q.v.*
Random matched : unequal sized pieces used in a single face. *Side
matched* is applied to specimens placed side by side without turning.
Slide matched : the top sheet is slid into position and both faces
matched, similar to side matching. Both are common in quartered
striped wood. Also see *Veneer Matching.*

Matched Backs. The backs of furniture, such as bookcases, etc.,
formed of tongued and grooved boarding.

Matcher. **1.** One engaged in matching veneers. **2.** A machine for
preparing jointed boards.

Matching. See *Match-boarding* and *Matched.*

Matching or **Match Planes.** Pairs of planes for forming tongues
and grooves by hand.

Match Splints. Thin sticks for making matches.

Match Wood. Wood billets suitable for matches.

Matchwood. Wood splinters.

Maté. S. American holly.

Material Board. A cutting list, *q.v.*
Matopus. Penaga, *q.v.*
Matsu. Japanese pine. Also called Omatsu. AKA MATSU, *Pinus densiflora*. Red pine. Excellent mild wood, strong, durable, resinous, and easily wrought. It is superior to Kuro matsu, *q.v.*, and is used for general purposes.
Mattock. A form of grubbing axe.
Mattress Scantlings. Pitch pine, cut to sizes, for the frames of wire mattresses.
Matumi. *Adina galpini* (*H.*). S. Africa. Variegated yellow and brown, lustrous. An attractive cabinet wood but not exported.
Maturity. The age at which a tree attains its prime, which varies considerably with the kind of tree and conditions of growth, from seventy years for softwoods and a hundred years for hardwoods. After a period of maturity the heartwood begins to deteriorate.
Maul. See *Mall.*
Maulstick. A long thin rod used by artists as a hand rest.
Maxim. Abbreviation for the name of the botanist Maximowiez.
Maxlite. A patent system for sliding partitions and sashes.
Mayapis. *Shorea spp.* and *Parashorea spp.* A species of lauan, *q.v.* It is an excellent wood intermediate between red and white lauan. It is often marketed under its own name. Used for cabinet work, superior joinery, veneers, etc.
Mayeng. *Pterospermum acerifolium* (*H.*). India. Red. Fairly hard, heavy, strong, and durable. Not difficult to work and polishes well. Used chiefly for planking. S.G. ·6.
Mayflower. See *Roble.*
Maytenus. Tapir Wood, *q.v.*
May Tree. See *Hawthorn.*
Mazer. A bowl made of maple wood.
Mazzard or **Mazard.** *Prunus avium.* See *Cherry.*
M.b.m. Abbreviation for 1,000 ft. board measure.
Mchuur. A mangrove, *q.v.*, from Tanganyika. Other species are Makandoa and Makaha.
Md. Abbreviation for *moulded.*
Meal Girnal. A meal barrel, or bowie, of about 60 gals. capacity.
Mealy Bug. *Psenococcus.* An insect very destructive to tropical trees.

MEANDER FRET

Mealy Tree. Wayfaring Tree, *q.v.*
Meander. A complicated fret carving.
Measure and **Measurement.** There are many systems in use for calculating the volume of timber : standard, board, actual, customary, caliper, fathom, Hoppus', Customs', brokers' sale, Liverpool brokers', and surface, measure. Many important woods such as mahogany, teak, and pitch pine have their own particular rules. In most of the methods there are rules that allow for intermediate sizes, defects, irregularities, conversion, etc. Measuring is divided roughly into

four classes : converted wood, round and square logs, floated timber, and standing trees. Fancy hardwoods are sold by weight. Numerous tables, instruments, and appliances are available to assist in calculating, such as tapes, rules, calipers, etc., but considerable skill and experience are required, together with a knowledge of the peculiarities of the various systems and methods of marking the contents. Also see *Square, Flitch, Market Forms.*

Measuring Frame. A box-like frame without bottom and of known cubic capacity. It is used for measuring granular materials such as aggregates for concrete. See *Gauging Box.*

Measuring Tank. A tank for measuring the amount of preservative forced into the wood in the pressure process of preservation.

Meat Safe. A box or cupboard with perforated zinc sides for storing meat.

Mechanical Properties. See *Tests.*

Medallion. An elliptical or circular raised panel or tablet, with inscribed or carved surface.

Medang. The Malayan name for woods chiefly of the Lauraceæ family. There is great variation in the woods due to the different genera and species, and it is usual to include some qualifying term. See *Kungkur.*

Medang Tandok. *Micropora curtissi* (*H.*). Malaya. Resembles Meranti but without resin lines. Yellow, seasoning to olive, aromatic. Glistening deposit in pores and feels greasy. Variable, soft to moderately hard and heavy. Better grades fairly strong, durable, and stable. Fine uniform rays and grain. Easily wrought. Used for planking, cases, furniture, construction, carving. S.G. ·54 to ·72 ; C.W. 2·5.

Medic. Abbreviation for the name of the botanist Medicus.

Medieval. Applied to architecture of the fifth to fifteenth centuries. The term is usually confined to Norman and early Gothic.

Medium Pocket Rot. See *Pocket Rot.*

Medlar. *Mespilus*, or *Pyrus*, *germanica.* Britain. The tree is of the Rosaceæ family and is appreciated as a garden tree and for its fruit, but the wood has little commercial value.

Medulla. The pith of a tree.

Medullary Rays. Radial vertical sheets, or ribbons, of tissue formed across the growth rings of trees. They are also called rays (*primary* rays extend from bark to medulla, *secondary* rays are formed later in growth and do not reach the medulla), pith rays, or wood rays, and consist chiefly of parenchyma cells. In most woods they can only be discerned by means of a lens. They allow the radial transmission of sap ; and in many woods the outcrop of the rays on quartersawn wood, or when cut following the rays especially in symmetric trees, produces beautiful figure as silver grain, medullary spots, pith flecks, etc. See *Annual Rings.*

Meeting Rails. The rails of sliding sashes that meet when the sashes are closed. See *Sash and Frame Window.*

Meeting Stiles. The middle, or inside, stiles of folding doors and double casements. See *Barred Door* and *Hook Joints.*

Melaleuca. See *Paper-bark Tree* and *Tea Tree.*

Melanochyla. See *Rengas.*

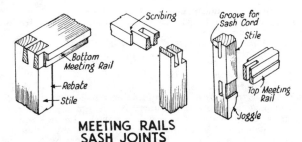

**MEETING RAILS
SASH JOINTS**

Melanorrhœa. See *Rengas* and *Thitsi.*
Melanoxylon. See *Brauna.*
Melantai. Meranti, *q.v.*
Melastomaceæ. A family that includes Bellucia, Miconia, Mouriria.
Melia. See *Lunumidella, Malabar Nim, Margosa, Lilac, Neem, Persian Lilac, Suradanni, Suryon,* and *Tulip Cedar.*
Meliaceæ. Mahogany family. Includes Amoora, Azadirachta, Cabralea, Carapa, Cedrela, Chukrasia, Dysoxylon, Entandrophragma, Guarea, Khaya, Lovoa, Melia, Sandorcum, Soymida, Swietenia, Trichilia, Xylocarpus.
Melica. Southland beech, *q.v.*
Melkhout. *Sideroxylon inerme.* S. African Ironwood. Brown with black streaks. Very hard, heavy, strong, tough, and durable. Very difficult to work. Little exported. S.G. 1·1 ; C.W. 5·5. See *Ironwood.*
Mellentia. See *Wenge.*
Melodeon. A form of reed organ, or an accordion.
Member. 1. An important piece in structural framework or timber framing. 2. An individual part of a moulding, or of an " order," or of anything built-up from a number of units.
Memel Fir. A name for *Pinus sylvestris.* See *Redwood.*
Mempening. *Quercus spp.* Malayan oak, of which there are about forty species. There is great variation in the different species. Many of them provide useful wood but with very large soft rays, which make the wood very fissile and difficult to season. The marketable woods are similar to *Q. rubra.* See *Oak.*
Mendong. Medang, *q.v.*
Mengkulang. *Tarrietia spp.* Malaya. Brick red, streaked, lustrous. Soft to moderately hard and heavy ; strong, not durable. Fairly difficult to work smooth due to variation of texture, polishes well. Suitable for cabinet work, superior joinery, wheelwright's work, etc.
Menispermaceæ. A family including Abuta, Anomospermum, Chonodrodendron, Cocculus, Disciphania, Hyperbænia, Pachygone, Pericamplylus, Tiliacora.
Menkabang Penang Tree. See *Chow.*
Mensa. The top of a church altar.
Mentangor. Bintangor, *q.v.*

Menzies Spruce. Sitka spruce, *q.v.*

Merabau. Malayan Ipil. Also called Merbau or Miraboo, Ironwood, Borneo " teak," etc. See *Ipil.*

Merago. *Adina rubescens* (*H.*). Malaya. Yellow. Hard, heavy, durable. Smooth with fine even grain. Excellent wood but small sizes.

Meranti. *Shorea spp.* and *Hopea spp.* (*H.*). Malaya. Philippine Lauan and Borneo Red Seraya. From pinkish yellow to dark red. Soft, light, not strong or durable and subject to insect attack. Resin canals conspicuous, fine straight rays, coarse grain. Requires care in seasoning, easily wrought, and polishes well. Large sizes. Substitute for mahogany, but some inferior species and other woods marketed as Meranti. Used for veneers, cabinet work, superior joinery, etc. S.G. ·31 to ·6 ; C.W. 3. Also see *Lauan* and *Nemesu.*

Merawan. *Hopea spp.* (*H.*). Malaya. Yellow to brown, darkening with exposure, dull. Fairly hard, heavy, and durable, but very variable in properties and characteristics. Medium rays, distinct resin canals. Used for furniture and constructional work. S.G. ·52 to ·8.

Merbatu. *Parinari spp.* and *Angelisia splendens* (*H.*). E. Indies. Also called Ballow, or Balau, but this name is now given to *Shorea* species. The wood is pale reddish, very hard and heavy, and durable. Small sizes, limited supplies, and difficult to work. Little exported. Used for piling, sleepers, structural work, etc.

Merbau. The suggested standard name for Ipil. See *Merabau.*

Merblanc. A charter used in the White Sea trade. See *Charter Party Forms.*

Merchantable or **Merch.** 1. Applied to specific grades of South and West U.S.A. lumber. Sound, free from shakes, large, loose, or rotten knots, or defects that materially impair the strength. Wood suitable for constructional purposes. Sap on corners up to one-third width and half thickness, or its equivalent, but sap not a defect on over 10 in. × 10 in. timbers. *Selected Merchantable* is a higher grade, and *Common Merchantable* a lower grade. See *Ukay* and *Selected.* 2. Applied generally to wood that is saleable for woodworking, or to that part of a tree that is of timber value.

Merchantable Lumber. The output of a sawmill, except mill culls. The term usually implies that it can be manufactured and sold with a profit.

Meristem. The tissue of living wood cells that are capable of repeated division.

Merlon. See *Battlemented.*

Merr. Abbreviation for the name of the botanist Merrill.

Merrit. *Eucalyptus Flocktoniæ.* Very similar to Salmon gum, *q.v.*

Mersawa. *Anisoptera spp.* (*H.*). Malaya. Yellow. Soft, light, not durable, and subject to insect attack. Coarse grain and rays. Contains large quantity of fluid resin. Large sizes and easily wrought. Used for furniture and interior work. S.G. ·48 ; C.W. 2·5. Also see *Palosapis.*

Mersida. A patent process for the treatment of hardwood veneers $\frac{1}{100}$ in. thick. The chemical treatment fills the pores and prevents movement so that the veneers can be used on bases other than wood.

Merulius. See *Dry Rot.*
Meshes. The spaces or openings in a grating, *q.v.* Like a net. See *Lattice.*

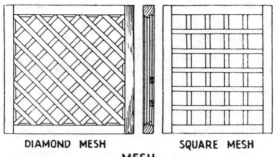

DIAMOND MESH   SQUARE MESH

MESH

Mespilodaphe. See *Canello.*
Mesquite. *Prosopis juliflora* (*H.*). C. America. Also called Honey Locust, Algaroba, Ironwood, etc. Rich dark brown, wavy darker lines, fragrant. Hard, heavy, tough, strong, very durable and stable. Irregular rather coarse grain. Fairly easy to work smooth but does not polish well. Used for construction, sleepers, vehicles, etc. S.G. ·8 ; C.W. 3.
Messmate. *Eucalyptus obliqua* and *Eucalpytus spp.* Stringy bark, *q.v.* Also called Tasmanian Oak and Peppermint.
Messua ferrea. See *Gangau, Penaga,* and *Nahor.*
Mesu. See *Nemesu.*
Mesur Birch. Russian birch with beautiful figure, supposed to be due to insect attack, known as pith flecks.
Metal Faced Joinery. Woodwork with metal facings. Stainless steel and bronze are commonly used, but there are numerous metals and alloys obtainable. See *Kalamein Sheathing* and *Plywood.*
Metal Windows. See *Surrounds.*
Metatracheal. See *Parenchyma.*
Metco. A registered wall-board, *q.v.*
Mete. 1. A conical or pyramidal figure. 2. A column or pillar. 3. A stockade or boundary wall.
Mether. An antique wooden drinking cup. It is usually square-shaped with four handles.
Metis. Betis, *q.v.*
Metoche. The spaces between dentils, *q.v.*
Metope. The slabs between the triglyphs in the Doric frieze, *q.v.*
Metopium. Honduras walnut, *q.v.*
Metric Foot. A compromise in the timber trade for some continental woods. It is a little over 13 in., or a third of a metre.
Metrosideros. See *Ohio* and *Rata.*

Me-tyl Wood. Veneered sheet metal. The metal is made porous for the bonding material to adhere. The sheets are waterproof.

Mews. A group of stables.

Mexican. Applied to numerous woods from Mexico, especially cedar, lignum-vitæ, and mahogany. Many other Mexican woods are described in the Glossary.

Mey. Abbreviation for the name of the botanist Meyer.

Mez. The name of the botanist K. Mez.

Mezzanine. 1. A low storey introduced between the levels of two main floors. 2. A floor beneath a theatre stage.

Mezzo-relievo. Middle relief in carving and sculpture. See *Relief*.

M.G. Abbreviation for *make good*.

M'gongo. *Ricinodendron rautanenii* and *R. africanum* (*H.*). Nigeria, Rhodesia, etc. Yellow. Very light and soft. Also called Corkwood, Mugongo, Mungongo. Substitute for balsa, *q.v.* Used for cores for flush doors and plywood, and for insulating purposes. S.G. about ·2.

Miall. See *Myall*.

Miang, Kapor. The hardest and heaviest of Borneo camphorwoods, *q.v.*

Miatico. Vinhatico, *q.v.*

Michelia. See *Champac*.

Michx. Abbreviation for the name of the botanist Michaux.

Micron. One millionth part of a metre. A unit of measurement in microscopic investigation of wood structure.

Micropora. See *Medang* and *Medang Tandok*.

Microscope. See *Photomicrograph* and *Identification*.

Microtome. A machine for cutting extremely thin sections of wood for microscopic examination.

Middle Cut. Portions cut from the middle of the length of a log. See *Butt Cut*.

Middling. A term used for middle grades, in several European gradings, such as Danzig timber.

Mid Feather. See *Parting Slip* and *Glazing Bar*.

Milkwood. *Mimusops obovata* (*H.*). S. Africa. Hard, heavy, tough, with close grain. Not exported.

Mill. Abbreviation for the name of the botanist Miller.

Mill. See *Sawmill*.

Milla. *Vitex altissima* (*H.*). India. Olive grey, often wide, darker streaks. Heartwood not distinct. Smooth but dull. Hard, heavy, strong, and durable. Straight, interlocked, and wavy grain, some mottle. Medium and even texture. Seasons fairly well, loses colour with water. Fairly difficult to work. Polishes well. Used for superior joinery, flooring, furniture, wheelwright's work, sleepers, etc. S.G. ·8 ; C.W. 4.

Mill Culls. 1. An American term for the poorest quality of wood from the mills, which is not exported. 2. Non-merchantable wood from the sawmills.

Mill Deck. Same as log deck, *q.v.*

Mille. 1. A measure of freightage for timber, equal to half of a Petrograd standard. 2. *One Thousand*. In the stave trade it implies

1,200, or ten " long hundreds." For boxboards, in London, it implies 1,200 pieces 28 in. by ¼ in. by 4 in.

Millentia. See *Wenge*.

Milletia. See *Thinwin*.

Mill Run. Usually implies all the saleable output of wood from a sawmill.

Mill-wheel. A wheel to turn the machinery of a water-mill.

Mill Work. 1. The manufacture of complete units of joinery : doors, windows, stairs, etc. **2.** Applied to units of joinery that are completely manufactured in the workshop, or factory, ready for fixing, especially when mass produced.

Mimber. A form of pulpit in a mosque.

Mimosa. See *Wattles*.

Mimosaceæ. A sub-family of Leguminosæ, *q.v.*

Mimusops. A genus with numerous species, widespread, and usually called bullet-wood. See *Balata, Beefwood, Bullet-wood, Kaya, Makore, Massaranduba, Milkwood, Mohwa Palu*.

Minar. A turret or lighthouse.

Minaret. A lofty slender turret with projecting balconies, characteristic of the Mohammedan religion and Arabian architecture.

Mindanao. White Lauan, *q.v.*

Mineral Streaks. Patches or streaks of discoloured and usually darker .wood that do not affect the strength and durability, *q.v.*

Mine Timbers. See *Chocks, Crowntrees, Props, Sleepers, Split Bars*, and *Splits*.

Mingris. Like Kempas, *q.v.*

Minolith. A grade of Wolman salts, *q.v.*, applied under vacuum and pressure. It gives protection against fungi and insects with a high degree of fire-resistance.

Minquartia. See *Acariguara*.

Minstrel's Gallery. A music loft. See *Loft*.

Miovu. *Entandrophragma cylindricum.* Uganda or Budongo Sapele, *q.v.*

Miq. Abbreviation for the name of the botanist Miquel.

Mirabeow. Merabau, *q.v.*

Mirandu. *Elæodendron glaucum (H.).* India, Burma. Red, smooth, fairly lustrous. Hard and heavy. Fine rays, close firm grain. Seasons fairly well and not very difficult to work. Suitable for cabinet work, etc., but little exported. S.G. ·85 ; C.W. 3·5.

MISERERE

Miro. *Podocarpus ferrugineus (S.).* New Zealand. Also called Bastard Black Pine. Brownish. Moderately light and soft, not durable, but very strong. Very fine even grain. Easily wrought and polishes well. Often sold as rimu because of resemblance. Used for joinery, constructional work, structural work, sleepers, veneers. S.G. ·6 ; C.W. 1·75.

Misereres. Small hinged seats in church choir stalls. They can be raised or lowered as required. Also called Misericorde, *q.v.*

Misericorde. A shelving projection on the under side of a hinged seat, providing a means of support to a person standing. Also applied to the whole seat. See *Subsellia* and *Miserere*.

Mismatching. 1. A misfit or imperfect alignment at the joints of matched boards, due to faulty machining. 2. Unsymmetrical matching of veneers.

Mistol. *Zizyphus mistol* (*H.*). S. America. A whitish, tough, flexible wood, with close grain. Small sizes are used chiefly for shafts and charcoal. S.G. 1·1. *Z. jujuba* and *Z. xylopyrus* are Indian species, but they are not imported and have no market names.

Mitragyna. See *Abura* and *Eben*. Also see *Kaim*.

Mitre. The intersection of two pieces or mouldings forming an angle. The joint, or mitre, bisects the angle, if the mouldings are of uniform section, so that the members of the moulding intersect on the line of the mitre. If the angle is not a right angle the mitre is referred to as acute or obtuse as the case may be. Other types are masons', bishops', curved, splayed, *q.v.* Also see *Angle Joints*, *Lipped Mitres*, *Knee*, and *Raking Mouldings*. Mitres may be plain, or they may have tongue or feather or spline, lip, stop, dovetail, double tongue, rebate, or butt, in combination.

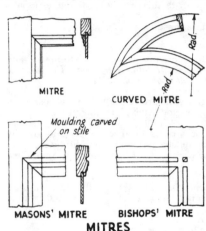

MITRE   CURVED MITRE

Moulding carved on stile

MASONS' MITRE   BISHOPS' MITRE

MITRES

Mitre Block Saw. A cabinet-maker's dovetail saw, *q.v.*

Mitre Box or **Block.** A long three-sided box or a rebated block in which mouldings are cut to the required mitre.

Mitred and Cut String. A stair string cut to the shape of the steps and mitred for the risers. Also see *Bracketed Stairs*.

Mitred Border. See *Margin* (2).

Mitred Cap. A cap to a newel into which the handrail is mitred. See *Knee*.

Mitre Dovetail. A secret dovetail, *q.v.*

Mitred Rebate. A joint for a canted corner comprising both mitre and rebate. The rebate holds the joint in position when assembling. See *Lipped Mitres*.

Mitred Tenons. Used on rails at right angles to each other, as in table legs.

Mitre Post. See *Lock Gates*.

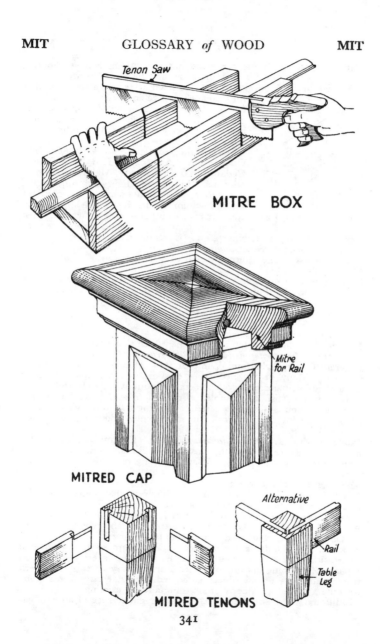

Tenon Saw

MITRE BOX

Mitre for Rail

MITRED CAP

Alternative

Rail

Table Leg

MITRED TENONS

341

Mitre Shoot. A frame for steadying a moulding whilst shooting the mitred end. See *Screw Mitre Shoot* and *Shooting Boards*.

Mitre Square. A bevel with a blade fixed at 45°, or a square with a mitred stock to give angles of 45° or 90°.

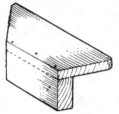

Mitre Templet. A rebated appliance for guiding the chisel when forming a small mitre, as when scribing small mouldings, etc. Also see *Templet*.

Miva Mahogany. *Dysoxylon spp.* (*H.*) Queensland. Also called Red Bean, Pencil Cedar, and Turnipwood. Mahogany colour, odorous. Moderately light and soft, fissile, not strong, but durable and stable. Straight grain, medium texture, handsome appearance. Requires care in seasoning ; irritant dust when sandpapered. Polishes well. Used for

MITRE TEMPLET

cabinet work, superior joinery, etc. G.S. ·6 ; C.W. 3.

Mixing Platform. See *Gauging Box*.

Mizzen Mast. The aftermost mast of a sailing ship.

Mkeo. See *Mukeo*.

M'kunguni. *Commiphora sp.* (*H.*). E. Africa. Greenish olive brown. Fairly hard and heavy, smooth, with close even grain. Suitable for cabinet work and superior joinery. S.G. ·65.

Mlange or **Mlanji.** S. African cedar. See *Podo*.

Mninga. Muninga, *q.v.*

Mo. Abbreviation for *moulded*.

Mobile. Applied to pitch pine shipped from that port.

Moboron. *Gossweilerodendron balsamiferum* or *Pterolobium sp.* (*H.*) Nigerian Cedar, Agba, and White or Pink Mahogany. Pale brown, lustrous. Many of the properties of Borneo cedar and resembles mahogany but paler in colour and rather gummy. Moderately hard, heavy, strong, stiff, and stable. Uniform fine grain and compact. Slightly resinous. Requires careful seasoning but works well. Large sizes. Used for cabinet work, superior joinery, flooring, etc. S.G. ·58 ; C.W. 2·5.

Moca or **Mocha.** *Andira jamaicensis* (*H.*). C. America. Yellowish brown, variegated, smooth, lustrous. Burls show large eyes which are very decorative. Used for veneers. Also see *Cabbage Bark*.

Mockernut. Hickory, *q.v.*

Mock Gable. The side of a carcase or pedestal that is against a bulkhead, or in a position where it is not seen, in ship joinery.

Mock Orange. *Prunus caroliniana* or *Philadelphus sp.* (*H.*). U.S.A. Also called Syringa. An ornamental tree, but the wood is not important and is not marketed in this country.

Models. 1. Buildings constructed in miniature so that a general impression may be obtained before the building is erected. Wood replicas of anything in miniature. 2. A preliminary attempt or pattern from which the actual object is copied. 3. Exercises for anyone learning a craft. 4. A copy of anything already existing.

Modilions. The enriched brackets under the cornice of an entablature of the Corinthian or Composite orders of architecture, or as a decorative feature to a cornice.

Modinature. The distribution and arrangement of the mouldings of an architectural order, or of a building.

Module. The radius of a column at the bottom of the shaft. A unit of measurement in classic architecture.

Modulus of Elasticity (E.) An imaginary stress necessary to stretch a piece of material to twice its length or compress it to half its length. A measure of the elasticity of a material, or of its power of recovery after being strained. Also called Young's modulus, $E = \text{stress} \div \text{strain}$. The value of E for ordinary structural timbers is between ·75 and 2·0 millions, in units of pounds and inches. An approximate formula for air-dried wood is $E = 2 \cdot 8 \times 10^6 \times S.G.$, but it can only be obtained by experiment. See *Stress and Strain*.

Modulus of Rupture. The equivalent fibre stress at maximum load. A constant used in structural design and obtained by experiment by loading numerous pieces of the material to destruction. It is suggested that modulus of rupture $= 2 \times S.G. \times 10^4$. Another formula is $C(S.G.)^2$, where $C = 50,000$ for certain pines, for 15 % moisture content. Bending moment at rupture = mod. of rupture $\times$ mod. of section $= f_r Z$. See f and *Modulus of Section*.

Modulus of Section (Z). The value of a structural member to resist a load, due to the size and shape of the cross-section. $Z = bd^2 \div 6$ for a rectangular section, such as a wood beam, or $Z = \text{moment of inertia} \div \frac{1}{2}d$.

Moench. The name of the botanist C. Moench.

Moeri. Red Stinkwood, *q.v.*

Moho. Balsa, *q.v.* Also *Belotia spp.*, British Honduras.

Mohwa. *Mimusops littoralis (H.)*. Andaman Isles. Also called Bullet Wood. Similar to Burmese kaya, *q.v.* Red. Hard, heavy, tough, strong, durable, but splits easily. Dense close grain, and wavy. Difficult to work. S.G. 1 ; C.W. 5.

Moisture Content. The percentage of moisture present in wood. The moisture may exist (*a*) free in the cell spaces, (*b*) in the cell walls, (*c*) in the protoplasm. It may be in all three conditions in sapwood but only in the first two in the heartwood. The strength and stiffness of wood change in almost inverse ratio to the moisture content change. A dry wood is about twice as strong as the same piece saturated and about 50 % stronger than air-seasoned. Wood is hygroscopic and should have a moisture content appropriate to the relative humidity of the atmosphere, or be *in equilibrium*, otherwise its stability is influenced and it either swells or shrinks. Moisture content

$$\text{Per cent.} = \frac{\text{orig. wt.} - \text{oven dry wt.}}{\text{oven dry wt.}} \times 100.$$

The moisture content of newly felled logs may vary from 40 to 200 % according to the species, etc. The following approximate values are important : green timber (heartwood), 40 % ; limit for fungoidal growth, 20 % ; air dried, 14 % ; roof timbers, 14 % ; exterior work, 16 % ; interior joinery (ordinary heating), 12 to 14 % ; (central heating), 8 to 10 % ; hence wood should have between 8 and 20 %

moisture content according to its surrounding conditions. When the moisture content is less than 12 % it is *kiln-dry*. See *Seasoning, Kilns, Shrinkage, Sorbitol*, and *Preservation*.

Moisture Content Meter. An instrument for automatically recording the moisture content of wood. Due to research new inventions are constantly being introduced. In a recent type a bell automatically rings whilst the wood is being machined if the fixed maximum moisture content is exceeded. The instruments are usually calibrated for one particular wood, and a chart is supplied with correction factors for other woods.

Moka or **Mokari.** *Schrebera swietenioides* (*H.*). India. Also called Ghant. Wide brownish grey sapwood, small claret purple heartwood. Irregular patches of false heartwood. Fairly lustrous, feels rough, but good surface. Hard, heavy, tough, strong, fairly durable but subject to insect attack. Widely interlocked grain, fine even texture. Seasons readily. Fairly difficult to work, polishes well. Used for shuttles, turnery, utensils. S.G. ·84 ; C.W. 4.

Molave. *Vitex littoralis, V. parviflora* (*H.*). Also called Philippine " teak." Yellowish brown. Hard, heavy, strong, and durable. Difficult to split. Straight and curly grain, fine texture. Seasons well. Difficult to work but finishes well. Used for shipbuilding, exterior work, high grade construction, flooring, sills, superior joinery, etc. S.G. ·8 ; C.W. 4. See *Teak*.

Molder. A moulding machine, *q.v.*

Mole. A breakwater.

Mollusc. See *Marine Borers*.

Molly. A mallet used in certain crafts. It is like an elongated mason's mallet.

Moment. The moment of a force is its turning effect about an axis. It is expressed in compound units. A force of 12 lb. turning about an axis 10 in. distant (perpendicular distance) has a moment of 12 lb. $\times$ 10 in.$=$120 lb. in.

Moment of Inertia (I). The *inertia areas* are the selected portions of the section of a structural member that are resisting the bending moments. If this sectional area is considered as built up of a very large number of very small areas, then I is the *sum of the second moments* of all the small areas about the neutral axis. If a_1, a_2, a_3, etc., are the areas and d_1, d_2, d_3, etc., their distances from the axis, then $I = a_1 d_1^2 + a_2 d_2^2 + a_3 d_3^2 + \ldots = \Sigma a d^2$ (the *sum* of ad^2). For a rectangular section, such as a wood beam, $I = bd^3 \div 12$.

Moment of Resistance (R). The moments of the internal forces acting in the resistance areas of a structural member. Bending moment$=$moment of resistance, or $M = R$.

Monel Metal. A white nickel-copper alloy used for sheathing timber as a protection against marine borers. Also used for laundry, chemical equipment, etc.

Monial. Mullion, *q.v.*

Monimiaceæ. A family of trees of little timber importance, except Daphnandra, Doryphora, and Laurelia.

Monitor. Same as Catamaran, *q.v.*

Monkey. Applied to both weight and slip-hook of a pile-driver.

Monkey Bread Tree. See *Baobab*.
Monkey Pot. *Lecythis sp.* (*H.*). B. Guiana. Also called Wadaduri. Light reddish brown. Very hard, heavy, tough, strong, and durable. Smooth but splintery. Close, even, straight grain. Difficult to work. Used for furniture, sleepers, structural work, etc. S.G. 1 ; C.W. 4·5.
Monkey Puzzle. Chili pine, *q.v.*

Monkey-tail. A vertical scroll to a handrail.
Monkey Tail Bolt. A type of tower bolt at the top of a tall door and having a long curved handle for easy operation from the floor.
Monkey Winch. A machine for grubbing up trees when clearing sites.

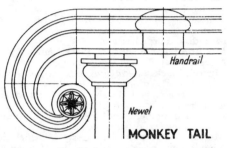

MONKEY TAIL

Monkey Wrench. A spanner with adjustable jaws.
Monocoque. A term used in aircraft construction in which the fuselage is built up in one single cell.
Monocotyledons. The subdivision of the Angiosperms in which the seedling has one seed leaf, or cotyledon. Includes non-banded trees, or inward growers : palms, bamboos, yuccas, rattans, etc. Compare with Dicotyledons.
Monoplane. An aeroplane with a single wing or plane.
Monstrance. A glass-faced shrine in which the consecrated Host is presented in Roman Catholic churches.
Monstrance Throne. A throne, usually with canopy, immediately behind the tabernacle.
Montant. A muntin, *q.v.*
Monterey Pine. *Pinus insignis* (*S.*). California. See *Pine*.
Moonshine Crotch. Swirling figure in crotch veneers.
Mooring Post. A mast for mooring, or securing, an airship. A post for mooring a ship.
Moorish or **Mosque.** Applied to Saracenic architecture. A characteristic is the arch, the soffit of which is struck from centres some distance above the springing. When only one centre is used the soffit is a little more than a semicircle and is called a horseshoe arch.
Mopstick. Applied to a handrail of circular section except for a small flat on the underside.
Moquilea. Oity, *q.v.* Also see *Kauta*.
Mora. *Dimorphandra mora* or *Mora excelsa* (*H.*). B. Guiana. Several species. Yellowish to reddish brown, variegated markings. Very hard, heavy, strong, and durable. Hard wearing and fire-resisting. Straight, coarse grain, but smooth, with ray and mottle. Large sizes. Difficult to work, polishes well. Substitute for greenheart. Used for shipbuilding, sleepers, structural work, paving, furniture, veneers. S.G. 1 ; C.W. 5. Also see *Fustic*.

Morabukea. *Mora*, or *Dimorphandra, gonggrijpii* (*H.*). B. Guiana. Brownish with darker streaks, smooth, fairly lustrous, odorous. Extremely hard, heavy, strong, and durable. Straight and ribbon grain. Difficult to work. Used chiefly for structural work, piling, etc. S.G. ·95 ; C.W. 4·5.

Moraceæ. Mulberry family. Includes Artocarpus, Bagassa, Broussonetia, Brosimopsis, Castilla, Cecropia, Chlorophora, Clarisia, Ficus, Helicostylis, Maclura (Toxylon), Morus, Olmedia, Perebea, Piratinera, Sloetia.

Morado. Purpleheart, *q.v.*

Moreton Bay Chestnut. Black bean, *q.v.*

Moreton Bay Pine. Hoop pine, *q.v.*

Morrel, Black. *Eucalyptus melanoxylon*. W. Australia. Dark brown with darker markings ; nearly black when seasoned. Very hard, heavy, strong, and durable. Interlocked grain. RED MORREL, *E. longicornis*. Similar to black, but with more gum veins. They are both good timbers and increasing in commercial importance, but they are only exported as veneers at present. S.G. 1.

Mortar. 1. A basin-like vessel in which substances are beaten or pounded into powder by a pestle. It may be of wood, porcelain, or metal. 2. A candlestick with bowl or saucer to catch the grease.

Mortar Board. A small flat board with handle underneath, for convenient handling of plastic materials, such as mortar.

Mortise. A recess, or hole, formed in one piece to receive a projection, or tenon, on the end of another piece. See *Tenon, Joints*, and *Mortise Joint*.

MORTISE
CHISELS

Mortise Chisel. A special chisel for forming mortises either by hand or machine. The hand tool has a very strong blade to withstand the leverage required to remove the core. The machine chisel may be reciprocating or hollow type. See *Mortising Machine*.

Mortise Gauge. A marking gauge with two spurs, or points, for scribing parallel lines on wood for mortises and tenons.

Mortise Joint. A joint used in framing, etc., in which the end of one member is prepared with a tenon to fit in a mortise in another member. The terms used in connection with mortise and tenon joints are : barefaced, box, chase, closed, double, dovetail, forked, foxtailed, franked, halved, haunched, housed, loose, notched, oblique, open, paired, shouldered, single, slot, stub, stump, sunk, tease, triple, tusk, twin. They are explained alphabetically in the Glossary. Also see *Tenon*.

Mortise Lock. A lock let into the edge of the door stile so that it is not visible on the face of the door. The lock cannot be removed when the door is closed.

Mortising Machine. A machine for cutting mortises and drilling holes. It may be operated by hand or power. The cutting is done by reciprocating chisel, hollow chisel, chain cutter, or twist bit. For

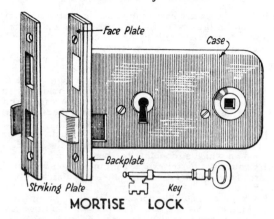

Face Plate

Case

Backplate

Striking Plate   Key

MORTISE   LOCK

large mortises the chain cutter is invariably used, and the mortise is
cut in one operation.   Multi-mortisers and borers will remove a
number of mortises or drill a number of holes in a single operation,
and may be obtained semi-automatic.   Multi- or gang-drilling machines
will bore up to eighty holes, vertically and horizontally and staggered,
through wood up to 6 in. thick.

Moruro.   *Cajoba arborea* and *Peltophorum adnatum* (*H.*).   C.
America.   Also called Horseflesh Mahogany and Bahama Sabicu.
Purplish brown with whitish markings, lustrous.   Hard, heavy,
durable, stable, and strong, but brittle.   Smooth roey grain.   Rather
difficult to work.   Polishes well.   Characteristics of mahogany and
used for same purposes : cabinet work and superior joinery.   It is
also used for structural work, sleepers, etc.   S.G. ·9 ;   C.W. 4.

Morus.   See *Mulberry* and *Nyakatuma.*

M.O.S.   Abbreviation for *moulded on solid.*

Motif.   The principal theme, or subject, prevailing in a design.

Mottle.   A term applied to figured wood that gives the impression
of an uneven surface, although smooth.   Broken wavy patches across
the face.   Twisted interwoven grain with irregular cross figure, which
is the mottle.   The effect is due to reflected light on the uneven
arrangement of the fibres.   There are several variations according
to the combination of effects obtained : bees'-wing, fiddle, peacock,
plum, ram, stop mottle, etc.   Mottle may be due to the wrinkling or
bending, in wave formation, of the growth rings due to the wind ;
uneven pressure of the bark ;   compression due to branches ;   big
climatic disturbance during growth ;   indentations in the surface of
the tree ; fungi ; etc.   The value of cabinet wood is increased greatly
when it is mottled.   BEES'-WING MOTTLE.   Very narrow stripes across
the face.   BLOCK MOTTLE.   Equal widths of mottle and space between
mottle.   FIDDLE MOTTLE.   The waves run across the wood in somewhat
regular formation.   PEACOCK MOTTLE.   In addition to fiddle or stop

mottle it has small eyes like bird's eye in maple. STOP MOTTLE. Like fiddle mottle but in broken formation. See *Figure* and *Grain*.

Mouchettes. Narrow flame-like shapes between the soufflets and the containing arch, in flamboyant tracery.

Mould. 1. A matrix, receptacle, or form, in which plastic materials are cast to the required shape. 2. A thin templet, or pattern, used to mark the outline of the material before it is wrought. See *Face* and *Falling Moulds*. 3. A fine vegetable growth, or mildew, that forms on wood in damp stagnant atmospheres. It is the least harmful type of fungus, and is usually confined to the surface of the wood. 4. Sometimes applied to a moulding or group of mouldings.

Moulded. 1. Applied to objects cast in a mould. 2. Applied to panelled framing with mouldings mitred round the panels, or to a piece of wood on which a moulding has been wrought.

Moulded Grounds. Partly exposed grounds moulded on the edge to form part of the architrave. Sometimes called architrave grounds. See *Door Linings*.

Moulded Plywood. Double curvature work built up of plies, as used in boatbuilding, aircraft, etc. The plies are built up on a mould, which is usually of wood, and either a male or female mould. Liquid or film synthetic resins are used as adhesives.

Moulded Stop. See *Stops*.

Moulding. 1. The process of shaping materials to recognised contours with tools or machines, or in a mould. 2. Applying mouldings to panels, etc. See *Mouldings*. 3. Applying heat and pressure, in an autoclave, to resin-bonded structures, plywood, etc., to complete the adhesive process. In some cases the work to be bonded, or " cooked," is placed in a watertight rubber bag, and hot water and compressed air are admitted into the autoclave as required, to set the synthetic-resin bonding agent. See *Hot Press*.

Moulding or **Fluting Box.** A box for fixing irregular objects such as table legs, columns with entasis, etc., whilst being moulded. A form of chuck, called a clock, is attached to the box for intermittent fixing as the object is rotated in the box. The clock is a circular disc bored at intervals on the circumference to engage with a dowel.

Moulding Machines. The various types of machines used for working mouldings on wood are : elephant or recessing machine, four-cutter, spindle, and universal or general joiner. See *Machines*.

Moulding Planes. The various planes used by the wood-worker for sticking, or shaping, mouldings by hand : beads, hollows, rounds, ovolos, etc. There is a great variety but they have been superseded by machines. Most workshops have a Stanley Universal plane, with which most mouldings can be stuck, if the quantity does not justify the setting-up of a machine. See *Mouldings*.

Mouldings. 1. Long lengths of wood, etc., of which the rectangular sections have been shaped to varied contours for ornamentation. 2. Classic mouldings, Roman and Grecian, include abacus, apophyge, astragal, cavetto, colarino, corona, cyma recta, cyma reversa, listel, ovolo, scotia, torus, trochilus. The contours of Roman mouldings are built up of circular arcs and the Grecian of the other conic sections : ellipse, parabola, hyperbola. Other mouldings are bead (quirk, sunk, return), chamfers, flutes, hollows, lamb's tongue, nosing, ogee (cyma recta), reeds, rounds, scotia.   There are other mouldings characteristic of different styles of architecture : billet, bird's beak, bowtel, cable, casement, chevron, diaper, dog-tooth, egg and dart, flower, keel, leaf, scroll, wave, zigzag, etc.   Mouldings are also named after the designer : Adam, Chippendale, Hepplewhite, Sheraton, etc. ; or according to the period : Queen Anne, Elizabethan, Georgian, etc. ; or according to the style of architecture : Gothic, Classical, Jacobean, etc.   Also see *Bolection, Cornice, Flush, Panel, Raking, Solid,* and *Sunk Mouldings.*

Mould Lines.   The lines on a mould, or templet, that coincide with the lines on the material when applying the mould. See *Face Mould.*

Mould Loft.   A room on the floor of which full size drawings are made to obtain the moulds in shipbuilding.

Mould Oil.   Applied to shuttering to prevent the concrete from adhering to the wood.

Moulmein.   Applied to timber from Lower Burma. See *Teak* and *Toon.*

Moulmein Cedar.   Toon, *q.v.*

Mount or **Mounting.**   To fix, or plant, on to the face of anything. A cardboard border to a picture.

Mountain Ash. *Eucalyptus regnans* and *E. delagatensis* (*H.*).   Victoria, Tasmania.   The two species are very similar woods and called Tasmanian or Australian Oak.   Pinkish or light brown.   Moderately hard, heavy, and durable.   Strong, resilient, elastic, and fissile.   Straight grain with fiddle-back.   Fairly coarse, even texture. Resembles plain-sawn oak.   Easily wrought and polishes well.   Seasons well if quarter-sawn but sometimes requires reconditioning.   Large sizes. Figured wood is esteemed for cabinet work, superior joinery, flooring, veneers, etc.   S.G. ·63 ;   C.W. 3·75.   *Euc. gigantea* is the N.S. Wales species and is usually called Alpine ash. Also see *Rowan.*

**ROWAN
MOUNTAIN ASH**

Mountain Cedar.   Kaikawaka, *q.v.*

Mountain Elm.   Wych elm, *q.v.*

Mountain Fir.   Alpine fir, *q.v.*

Mountain Mahogany.   Richly figured American black birch.

Mountain Marri. *Eucalyptus hæmatoxylon.*   W. Australia.   Little commercial value because of scattered distribution and difficulties of transport.

Mountain Oak. *Lysicarpus ternifolius* (*H.*).   Queensland.   Brownish-pink.   Also called Brown Hazelwood.   Moderately hard and heavy ; durable and strong but brittle ; lustrous, smooth.   Very fine even grain and texture.   Not difficult to work, and polishes well.   Used for cabinet work, superior flooring, carving, turnery, fancy articles. S.G. ·8 ; C.W. 3·5.

Mountain Pine. Western white pine, *q.v.*

Mountain Spruce. Engelmann spruce, *q.v.*

Mount Atlas Cedar. *Cedrus atlantica.* Europe, N. Africa. Similar to Deodar, *q.v.*

Mounting Steps. A block of two or three steps to assist in mounting horses.

Mouriria. See *Pauji.*

Mouth. 1. See *Remouthing Plane.* 2. The opening between the lips at the foot of an organ pipe, *q.v.*

Movement. A mechanical contrivance controlling the movement of anything.

Movingue. *Distemonanthus sp.* Nigerian Satinwood, *q.v.* Also called Ayan or Anyaran. See *Satinwood* and *Synara.* Also applied to species of Peroba, *q.v.*

Mowsteads. Dwarf partitions on a threshing floor.

Mozambique Ebony. African Blackwood. See *Ebony.*

M-Roof. A series of connected double-pitch roofs.

Mubugu. *Brachystegia spp.* See *Okwein.*

Mucuri. *Lucuma spp.* (*H.*). C. America. Reddish brown. Hard, heavy, tough, strong, and durable. Straight, rather fine grain, with distinct rays, and smooth. Similar to bullet wood. Fairly difficult to work. Used chiefly for heavy constructional work where durability is of first importance. S.G. ·9 ; C.W. 4.

Mudguard. The protective shield to the wheels of a vehicle.

Mudsill. A foundation timber resting on the ground. The bottom timber of a dam across a stream.

Muell. Abbreviation for the name of the botanist Mueller.

Muer. Pygeum, *q.v.*

Mueri. *Pygeum africanum* (*H.*). Kenya. Dull dark red. Hard, heavy, strong, durable, and tough. Fairly straight grain, even and fine texture. Difficult to season, and should be converted quickly to avoid waste. Fairly difficult to work. Polishes well. Used for construction, wheelwright's work, decking, sleepers. S.G. ·85 ; C.W. 3·5.

Mufumbi. *Entandrophragma utile.* Uganda. Budongo heavy mahogany, *q.v.*

Mugavu. *Albizzia coriaria.* Uganda Kokko, *q.v.*

Mugga. *Eucalyptus sideroxylon.* Red Ironbark, *q.v.*

Mugongo. See *Mgongo.*

Mugonyone. Kenya White Pear. See *Pear.*

Muhimbi. *Cynometra alexandrii.* Uganda Ironwood. Extremely hard, heavy, and durable. Only used where strength and durability are of primary importance. S.G. 1·1.

Muhuhu or **Muhuga.** *Brachylæna Hutchinsii* (*H.*). Kenya. Bright brown,˙ fragrant. Very hard, heavy, strong, durable, and fissile. Difficult to convert but not difficult to work. Resembles sandalwood. Used chiefly for local requirements at present. S.G. ·95.

Muiracoatiara. *Astronium le cointei.* Lower Amazon, Brazil. See *Goncalo alves* or *Putumuju.*

Muirapiranga. Satiné, *q.v.*

Mujua. *Alstonia congensis* (*H.*). Uganda. Also called Patternwood. A light, soft, even-textured wood with straight grain and easily wrought. Used for box-making, cores, etc. S.G. ·4.

Mukeo. *Dombeya Mastersii* (*H.*). Kenya. White, light, soft, tough, strong, and flexible, not durable. Straight grain, fine even texture, smooth. Requires careful seasoning but easily wrought. Used as ash and for match-boarding, boxes, crates, motor-car bodies, etc. S.G. ·52 ; C.W. 2·5.

Mukoge. Nyakatuma, *q.v.*

Mukusu. *Entandrophragma angolense.* Uganda. Budongo mahogany, *q.v.* See *Gedu nohor*.

Mulato. *Bursera simaruba* and *B. gummifera.* Mexico. See *Gumbo Limbo*.

Mulay Saw. A stiff pit-saw operated by one pitman. The upstroke is operated by some form of spring device.

Mulberry. The Mulberry, or Moraceæ, family includes many important hardwoods. They vary greatly from very soft and light to very hard and heavy. Many are too valuable for the silkworm industry to be marketed as timber. See *Bagasse, Berba, Fustic, Guayamero, Iroko, Leiteira, Letterwood, Oity, Poxa, Tunu, Satiné.* EUROPEAN MULBERRY, *Morus nigra* and *M. alba.* Yellowish to reddish brown. Distinct growth rings and faint rays. Coarse texture, irregular grain. Small sizes and of little commercial value. Used for cabinet work and fancy articles. INDIAN MULBERRY, *Morus alba.* Golden brown, darkening quickly with exposure, darker streaks, lustrous, smooth. Hard and heavy ; tough, elastic and stable, but not durable. Straight grain, distinct rays. Medium fine rather uneven texture. Requires careful seasoning but easily wrought. Substitute for ash. Used for sports equipment, implements, furniture, bentwork, turnery. S.G. ·63 ; C.W. 2·5.

Mule. A small fishing vessel, like a cobble, used on the Yorkshire coast.

Mule-ear Knot. A spike knot, *q.v.*

Mulga. *Acacia aneura* (*H.*). C. and S. Africa and Australia. Also called Mulga Spearwood. Brown with golden markings, lustrous, very smooth. Very hard, heavy, and strong. Dense close grain. Difficult to work, small sizes. Used for turnery, fancy articles, walking sticks, etc. S.G. 1·2 ; C.W. 5. The name is sometimes applied to stunted species of Eucalyptus, especially species of Gum, producing hard marble-like wood.

Müll. Abbreviation for the name of the botanist Müller.

Mulleting. Gauging the edges of panels to fit the plough groove in the framing. The edges of the panel are planed and tested by a temporary piece, or mullet.

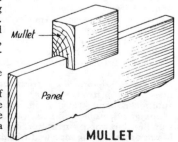

MULLET

Mullions or **Munnions. 1.** Vertical divisions in window frames and openings. **2.** A vertical sash bar is sometimes called a munnion.

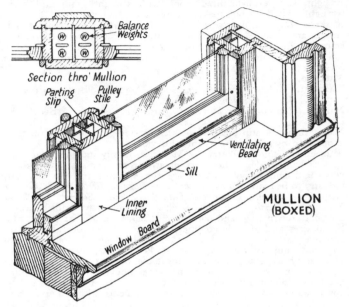

Section thro' Mullion

Balance Weights

Parting Slip

Pulley Stile

Ventilating Bead

Sill

Inner Lining

Window Board

MULLION (BOXED)

Multifoil. Tracery work with more than five foils. See *Tracery Panel.*

Multiple. Applied to the numerous machines that perform a number of the same operations at once.

Multiple Ring. A growth ring containing several false rings.

Multiply. Plywood built up of more than three veneers or plies. See *Plywood.*

Multiseriate. See *Rays.*

Mumble Bee. A small trawler used in the Brixham area.

Muna. *Sideroxylon sp.* (*H.*). Kenya. Pinkish, or light greyish brown. Light, soft, tough, not durable. Silica content is hard on saws and tools, otherwise easily wrought and polishes well. Used for cheap furniture, motor-car bodies, box-boards, etc. S.G. ·5 ; C.W. 3·5.

Mundani. *Acrocarpus fraxinifolius* (*H.*). India. Light red, darker lines, fairly lustrous, smooth. Moderately hard, heavy and strong. Interlocked and wavy grain, some mottled. Very coarse texture, resinous. Seasons well and not difficult to work. Used for planking, shingles, boxwood, furniture, plywood. S.G. ·68 ; C.W. 3.

Mungongo. Mugongo and Muhimbi, *q.v.*

Mungo River. See *African Mahogany*.

Muninga. *Pterocarpus angolensis*. The standard name for Kajatenhout, *q.v.*

Munnions or **Munions.** Vertical pieces between the lights on board ship. See *Mullions*.

Muntins. Vertical divisions between the rails of wood framing. Divisions between drawers and dustboards in cabinet work.

Muntz Metal. A malleable brass used for sheathing timber in underwater work. See *Marine Borers*.

Munyama. *Khaya anthotheca*. Uganda mahogany, *q.v.*

Munyenye. *Fagara angolensis* (*H.*). Uganda. Also called Silk Bark. Similar to English birch. Fairly hard and heavy, lustrous, with speckles and distinct rays. Straight grain, fine even texture, and smooth. Easily wrought. Rather small sizes. Little imported at present. S.G. ·68 ; C.W. 2·5. Also *Albizzia* and *Tetrapleura spp.*

Mura. Goncalo alves, *q.v.*

Mural. Attached or belonging to a wall.

Murbouw. Merabau, *q.v.*

Muringa. *Cordia Holstii* (*H.*). E. Africa. Light greyish brown with darker markings. Light, soft, very stable. Even grain and texture. Easily wrought and polishes well. Not difficult to season. Valuable wood, but little exported as yet. Used for furniture, superior joinery, panelling, decking, etc. S.G. ·5 ; C.W. 2.

Murraya. See *Andaman Satinwood*, *Kemuning*, and *Ketenggah*.

Murrayana. Lodgepole pine, *q.v.*

Murray Red Gum. See *Gum*.

Murtenga. *Bursera serrata* (*H.*). India, Burma. Reddish brown, fairly lustrous, smooth. Moderately hard and heavy. Straight grain, close, even texture. Suitable for cabinet work, carving, etc. S.G. ·75 ; C.W. 3.

Musanga. See *Aga*.

Musenene or **Musengera.** E. African podo, *q.v.*

Musgravea. See *Silky Oak*.

Musharagi. African olive. See *Olive*.

Music Stand. A light frame with ledge to hold music for an instrumentalist.

Music Stool. A seat without back for a pianist. It may be adjustable for height, or it may be framed as a box or small cupboard.

Musiga. *Warburgia ugandensis*. E. African greenheart. Similar to *Nectandra spp.* but darker and more decorative, with ripple and distinct rays ; also smoother, with closer grain and finer texture. S.G. ·82. See *Greenheart*.

Musk or **Muskwood.** *Olearia argophylla, Guarea spp.* (*H.*). Tasmania and Victoria. Smooth close grain. Difficult to season and work. Small sizes. Burrs very valuable for veneers. S.G. ·65 ; C.W. 4·5.

Musodo. Uganda corkwood. Similar to Mgongo, *q.v.*

Mutarakwa. African pencil cedar. See *Cedar*.

Mutari. *Panax sp.* (*H.*). E. Africa. Yellowish white. Very light, soft, tough, not durable. Characteristics of balsa, but heavier and harder. Little exported.

Mutirai. Ceylon satinwood, *q.v.*

Mutti. Species of Indian Laurel wood, *q.v.*

Mutule. A square block carrying the gutæ in the Doric cornice, *q.v.* A modillion rectangular in section.

Muvule. Iroko, *q.v.*

Muzaita. E. African camphor. See *Camphorwood.*

Muzzle. The fitting to the lower end of the sprit on board ship.

M.W. Abbreviation for mixed widths.

Myall. *Acacia homalophylla* and *A. pendula* (*H.*). Victoria and N.S. Wales. Also called Australian Acacia and Myall Spearwood. Dark brown to purplish, very smooth. Very hard, heavy, durable ; fragrant. Close texture with attractive figure, but rather conspicuous pores. Difficult to season and work. Small sizes. Used for cabinet and decorative work, fancy articles, turnery, carving, veneers. S.G. 1·2 ; C.W. 5.

Myaukchaw. *Homalium tomentosum.* See *Lancewood.*

Mycelium. Collection of hyphæ of fungi. See *Dry Rot.*

Myrobolan. *Terminalia chebula* (*H.*). India. Also called Harra, Panga, etc. Greenish grey sapwood with a little dark purple heartwood. Rather dull but smooth. Very hard, heavy, tough, strong, and durable, but not very stable. Narrow bands of interlocked grain, twisted and curly when rift-sawn. Medium fine texture. Difficult to season and work. Polishes well. Used for cabinet work, furniture, wheelwright's work, buffers, etc. S.G. ·8 to 1·1 ; C.W. 4·5.

Myrocarpus. See *Cabreuva.*

Myroxylon. See *Oleo Vermelho.*

Myrsinaceæ. A family with few woods of commercial importance : includes Aigiceras, Ardisia, Conomorpha, Jacquinia, Myrsine, Rapanea, Wallenia.

Myrsine. See *Beukenhout* or *Cape Beech, Canelon, Lanza.*

Myrtaceæ. Myrtle family. Includes a large number of important woods with very varied characteristics but usually very hard, heavy, durable, and strong. The most important is the Australian Eucalyptus. The family also includes Angophora, Backhousia, Barringtonia, Careya, Eugenia, Leptospermum, Myrtus, Pimenta, Psidium, Rhodamnia, Syncarpia, Tristania.

Myrtle. *Nothofagus cunninghamii* (*H.*). Tasmania Myrtle or Beech. Salmon pink to brownish. Hard, heavy, strong, very tough and hard wearing, fairly durable. Smooth, close grain, with feather and ripple. Resembles European beech. Not difficult to season or work, and polishes well. Used for furniture, cabinet work, superior joinery, turnery, flooring, veneers. S.G. ·75 ; C.W. 3·5. PACIFIC MYRTLE or Californian laurel, *Umbellularia californica* (*H.*). Also called Myrtle Burl and Baytree. Light brown, large proportion of paler sapwood. Hard, heavy, hard-wearing, and very resilient. Smooth, close grain, firm texture, distinct rays. Highly figured with bird's eye and blister. Polishes well. Figured wood esteemed for superior joinery, cabinet work, veneers, dance floors, etc. Also see *Satinwood (Andaman).*

N

N.A. Abbreviation for *neutral axis*.

Nacastillo. Nargusta, *q.v.*

Nacelle. The body of an aeroplane or the car of an airship, for the engine and crew. See *Fuselage*. The basket to a balloon.

Nacerda. See *Wharf Borer*.

Nagaed Wood. See *Rosewood (Honduras)*.

Nagal. *Premna tomentosa* (*H.*). S. India and Ceylon. Also called Naura, Kotokoi, etc. Difficult to distinguish from *Chloroxylon swietenia*. See *Satinwood*.

Nagari. Poon, *q.v.*

Nahor. See *Gangau*.

Nail-head Moulding. A Norman moulding having a series of diamond-shaped or pyramidal projections like nail heads.

Nailing Machine. Special machines for automatically nailing boxes, cases, etc., together. The nails are fed from a hopper and driven in the required positions.

Nail Pull. An appliance for extracting nails. (See illus., p. 356.)

Nails. A large variety of shapes, metals, and finishes. The common types are distinguished as wire (round or oval), cut, clasp, wrought. The largest types are called spikes, and the smallest brads, sprigs, tacks, or pins. Sizes rise by $\frac{1}{4}$ in. from 1 to $3\frac{1}{2}$ in., and then by $\frac{1}{2}$ to 6 in. long. Nails are often named according to their

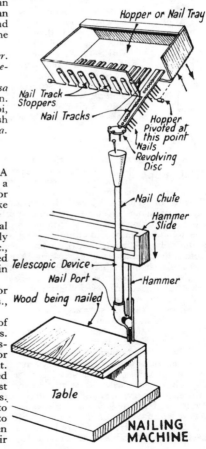

Hopper or Nail Tray

Nail Track Stoppers

Nail Tracks

Hopper Pivoted at this point

Nails Revolving Disc

Nail Chute

Hammer Slide

Telescopic Device

Nail Port

Wood being nailed

Hammer

Table

NAILING MACHINE

particular uses : box (smooth or barbed), casing, deck, felt, flooring, glazing, hurdle, mop, plate, etc., or according to the shape of the head or point. Special types used in wood-working are described alphabetically. Nails were described according to the cost per hundred as tenpenny (10d.), etc. ; the sizes ranging between threepenny (3d.) to sixtypenny (60d.). Another explanation is that " d " is a corruption of pound and 10d. infers that 1,000 nails weigh 10 lb.

Round Wire
Oval Wire
Cut Nail
Wrought Nail
Dowel Pin
Clout
Glaziers

NAILS

NAIL PULL

Nail Sets. Punches for driving nail-heads below the surface of the wood preparatory to stopping. See *Punches*.

Naito. Nyatoh, *q.v*

Najesi. Cuban Carapa, *q.v.*

Naked. 1. Applied to the structural timbers of a building, etc., before the finishings are applied ; as the flooring timbers before the floorboards are in position. 2. A continuous plane surface as opposed to the projections and ornamentation.

Nambar. American Cocobola, *q.v.*

Names of Woods. There is considerable confusion due to the variety of names given to the same species in many cases, probably due to the different national interests and sources of origin, but often due to indiscriminate naming of new woods. The botanical name is the only safe guide, and in a few cases the trade name is derived from the botanical name. Many Asiatic woods have over thirty vernacular names. The best-known alternatives are given for most of the woods described in the Glossary. The British Standards Institution has considered the names of woods used in this country and recommends a definite name for each wood, and the woods are described in the Glossary accordingly.

Nan. Chinese Medang, *q.v.*

Nana or **Nandi.** *Lagerstrœmia lanceolata* or *L. microcarpa* (*H.*). India. Also called Benteak. Light red to reddish brown, seasoning to light walnut brown. Darker markings on rift-sawn wood. Fairly lustrous and smooth. Moderately hard, heavy, and durable. Strong, elastic. Straight grain, coarse texture. Fairly difficult to season but not difficult to work. Should be girdled and kiln-seasoned. Used for furniture, carpentry, boat and carriage building, etc. S.G. ·7 ; C.W. 3·5.

Nance or **Nanche.** See *Spoon* (*Golden*).

Nan-mu-hua. *Phœbe sp.* China. Similar to camphor-wood, without fragrance. Highly figured, with fine close grain and rays, and useful for small cabinet work and fancy articles.

Naphthenates. One of the best preservatives for wood against fungi or insects. The salts may be copper or zinc. A 20 to 25 per cent. solution is usual. The results are permanent and there are no ill-effects, also the wood may be painted. The copper salts are soluble in alcohol which makes them suitable for unseasoned wood that will not absorb ordinary salts.

Nara. Onara, *q.v.*

Naranjo. *Terminalia sp.* (*H.*). C. America. Yellowish, lustrous. Moderately hard, heavy, and strong, but not very durable. Straight, curly, and roey grain. Not difficult to work and polishes well. Used for construction, cabinet work, sleepers, etc. S.G. ·7 ; C.W. 3.

Nargusta. *Terminalia amazonia*, syn. *T. obovata* (*H.*). W. Indies, B. Honduras. Also called Aromilla, Guayabo, Nacastillo, etc. Variegated olive brown colour. Fairly hard and heavy. Tough, durable, strong. Beautiful figure with roe and mottle. Fairly difficult to work and season. Used for decorative work, veneers, etc. S.G. ·62 ; C.W. 4.

Narig. *Vatica manggachapoi* (*H.*). Philippines. Also called Resak. Pale yellow, greenish streaks, turns dark brown with age. Very hard and heavy, strong and durable. Straight grain, slightly crossed, fine texture, distinct rays. Requires careful seasoning. Not very difficult to work. Very similar to Yacal and Molave, *q.v.*, and used for similar purposes. S.G. 1 ; C.W. 4.

Narra. *Pterocarpus indicus* and *P. vidalianus* (*H.*). Philippine Padauk, *q.v.* Marketed as red and yellow Narra, but both are products of either species. Also called Sena and Angsena. Light yellow to blood red, with faint characteristic smell. Hard, moderately heavy and strong. Durable and resists insects. Wavy, interlocked grain, with ribbon figure. Florid figure in plain sawn wood due to large vessel arrangement. Seasons well and fairly easy to work, polishes well. Shavings turn water fluorescent blue. Large sizes. Expensive because scarce. Used for cabinet work, high-class joinery, instruments, carriage building, etc. S.G. ·82 ; C.W. 3·5.

Narra Family. Leguminosæ, *q.v.*

Narrow-leaved Ironbark. *Eucalyptus crebra.* See *Ironbark.*

Narrow Measure. A term applied to an organ pipe of a narrow width compared with the length, to give a clear striking tone.

Narrow Ringed. Applied to slow-grown timber. See *Growth.*

Narthex. A part of early Christian churches for penitents. It was screened off from the main body of the church and was often in the form of a porch or vestibule.

Naseberry. Sapodilla, *q.v.*

Natalea. See *Ironwood.*

Native Box. *Bursaria spinosa* (*H.*). S. Australia. Excellent wood with characteristics of European Box.

Native Cherry. Australian Cherry. See *Cherry.*

Native Lime. See *White Cornelwood.*

Native Pear. *Pomaderris apetala* (*H.*). Tasmania. Small sizes and used for small cabinet work. Little imported. See *Pear.*

Nato. *Palaquium luzoniense* (*H.*). Philippines. Pale red, fairly lustrous. Variable weight and hardness, moderately strong, not durable. Straight grain, some wavy and slightly crossed, fairly fine texture. Seasons well and easily wrought, polishes. Used for the same purposes as Lauan : cabinet work, superior joinery, veneers, etc. S.G. ·7 ; C.W. 3. WHITE NATO. *Palaquium sp.* Creamy colour but has the same qualities as Red Nato. Many other species are called Nato.

Nattes. An interlaced decoration used in medieval architecture. It is somewhat like a diaper and resembles interwoven fencing.

Natural Draught Kiln. One in which the currents of air are not mechanically controlled, hence stagnant patches may occur. See *Kiln*.

Natural Seasoning. Converted wood stacked to allow the sap to harden naturally. The pieces are arranged so that the air will circulate freely round every piece, but they should be protected from sun and rain. This method does not impair the strength and colour, but it is slow and costly. See *Seasoning* and *Stacks*.

Nauclea. *Nauclea*, or *Adina*, *sessilifolia*. Similar to Haldu, *q.v.*, but harder and heavier. No market name because of limited supplies. Also see *Binga, Calamansaney, Kaim, Kadam, Stephegyne.*

Naval Hoods. Large pieces of thick stuff below and above the hawse-holes in a wooden ship. Also called hawse-bolsters, *q.v.*

Nave. 1. The hub of a wheel that carries the spokes and through which the axle passes. See *Wheel*. 2. The main body or central part of a church. The central aisle or division.

Nave Boring Machine. A machine for mortising the nave of a wheel for the spokes.

Nave Elm. English elm. See *Elm*.

Navel. A boss on a shield. A knob.

Neat-house. A shippon, *q.v.*

Neat Size. Net size. Exact size after preparation.

Neb. The lifting beam of a timber trailer.

Nebule Moulding. A wavy moulding or ornamentation.

Neck. 1. The piece connecting the body of a violin to the head. 2. A narrow connecting part.

Neck or **Necking.** 1. The connecting moulding between the shaft of a column and the capital, or the plain part between the mouldings and the shaft. 2. Small mouldings, astragal, etc., mitred or turned, around table legs, etc., in cabinet work. 3. See *Tabled Joints*.

Necked Bolt. A cranked bolt, *q.v.*

Necklace Poplar. Cottonwood, *q.v.*

Nectandra. A genus of the Laurel family. See *Canella, Cirouaballi, Tuque*, and *Embuia*.

Nedun. *Pericopsis mooniana* (*H.*). Ceylon. Deep brown, lustrous. Hard, heavy, strong, durable. Close firm grain and rays. A very attractive wood for cabinet work and superior joinery, but scarce. S.G. ·8 ; C.W. 4.

Needle Leaf. The name applied to coniferous or softwood trees, because of the spine-like leaves. See *Pine*.

Needle Point. See *Quarter-sawn*.

Needle Points. Needles, without eyes, used for delicate work, fractures, etc., instead of nails or pins. They are easily broken to the required length.

Needles. 1. Horizontal timbers passed through a wall to support the weight of the wall above during structural alterations. They are supported by dead shores, *q.v.* 2. A stout piece let into the wall to receive the thrust from a raking shore, *q.v.* 3. Thin rods of wood, metal, or bone, used in knitting and crochet work.

Needle Scaffold. One supported by cantilevers projecting from the face of a building.

Needle Wood. *Schima wallichii* (*H.*). India. Light red or reddish brown, with darker bands. Fairly lustrous and smooth. Moderately hard, heavy, strong, and durable. Not very stable. Even medium texture with twisted and irregular grain. Fairly difficult to season, shrinks considerably, should be girdled and kiln seasoned. Not difficult to work and polishes well. Used for joinery, tools, agricultural implements, sleepers, etc. S.G. ·72 ; C.W. 3·5. AUSTRALIAN NEEDLE-WOOD, *Hakea lorea* or *H. leucoptera.* Tasmania and Queensland. Rich reddish brown. Hard, heavy, tough. Close even grain, irregular rays. An excellent cabinet wood but small sizes and not exported. S.G. ·85.

Needling. 1. Dead shoring, *q.v.* 2. Needles supported by balks laid horizontally instead of by dead shores, as in under-pinning.

Needs. Wedges used in crate-making.

Neem. See *Margosa.*

Negrillo. *Machærium whitfordii* (*H.*). C. America. Chocolate brown to black, streaked and variegated, lustrous. Very hard, heavy, tough, strong, and durable. Roey grain, fine texture. Rather difficult to season and work, polishes well. Used for cabinet work and fancy articles. S.G. 1 ; C.W. 4.

Negrito. Simaruba, *q.v.*

Nellee or **Nelli.** *Phyllanthus emblica* (*H.*). India. Pinkish red. Fairly hard, heavy, tough, and durable. Close grain, uneven texture, smooth. Distinct rays and mottle. Difficult to season, warps freely. Easily wrought and treated. Used for furniture, turnery, construction. S.G. ·76 ; C.W. 3.

Nemesu. *Shorea spp.* (*H.*). Malaya. Also called Mesu and Dark Red Meranti. Also see *Oba suluk.* Dark red, with glistening resinous secretion. Very hard, heavy ; fairly durable. Distinct rays. Requires careful seasoning and fairly difficult to work. Large sizes but scarce and little exported. Used for heavy construction, furniture, boat building, etc. S.G. ·8 ; C.W. 4.

Neolitsea. A genus of the Lauracea family. See *Medang.*

Neonauclea. See *Nauclea.*

Neram. Keruing, *q.v.*

Nervures. The side ribs of a vaulted roof, as distinguished from the diagonal ribs.

Nest. 1. Applied to a set of similar things, usually diminishing in size to fit inside each other, of a small compact character : a nest of drawers or tables. 2. Any arrangement in which birds can lay and hatch eggs. (See illus., p. 360.)

Nettle Tree. *Laportea urtica* (*H.*). New South Wales. Grey, fairly smooth, and lustrous. Light, soft, moderately strong, and durable. Not exported. Also *Celtis spp.* and *Trema micrantha.*

NEST OF TABLES

Nettle Wood. *Celtis australis* (*H.*). India. Yellowish white seasoning to yellowish grey, darker streaks probably due to sap stain. Lustrous but dulls later. Fairly smooth. Moderately hard, heavy, and durable ; tough, elastic. Straight, fibrous-grained. Medium coarse, uneven texture ; distinct rays. Not difficult to season but should be quickly converted. Subject to insect attack. Resembles Indian elm but coarser. Used for bent-work, oars, tools, implements, carriage building, aircraft, etc. S.G. ·65 ; C.W. 3·5.

Neutral Axis. An imaginary line passing through the centres of gravity of the continuous cross-sections of a beam or other structural member. See *Neutral Section*.

Neutral Section. The longitudinal section of a beam or other structural member subject to bending where the stresses are neutral, *i.e.*, where the change from tension to compression takes place. It is a section normal to the plane of bending and along the neutral axis, and is usually called the neutral axis, *q.v.*

New Brunswick Spruce. *Picea mariana.* Black spruce. See *Spruce*.

Newel. 1. The post carrying the handrail to a flight of stairs. See *Handrail* and *Knee*. **2.** The central pillar carrying the inner ends of the steps in winding stairs.

Newel Cap. An ornamental top planted on a newel. See *Mitred Cap*.

Newel Joints. The joints between the handrail or string and the newel.

Newel Post. A newel, *q.v.*, but more especially applied to the one that is continuous from top of the flight of stairs to the floor, as a support.

Newel Stairs. Dog-legged stairs, *q.v.*

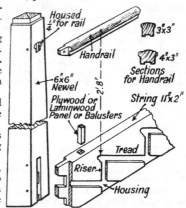

NEWEL HANDRAIL & STRING TO STAIRS

New England Hemlock.　Eastern hemlock, *q.v.*
New England Pine.　*Pinus strobus.*　White, or yellow, pine, *q.v.*
New Guinea Walnut.　*Dacontomelum spp.* (*H.*).　New Guinea.
Similar to Philippine Dao, *q.v.*
New Zealand.　Applied to the woods from that country.　SOFTWOODS :
Hoop Pine, Kahikatea, or white pine, Kaikawaka, Kauri, Matai,
Miro, Podo, Rimu, silver pine, Tanekaha, Totara.　HARDWOODS :
Beech, Black Maire, Honeysuckle, Puriri, Rata, Tawa.　They are
described alphabetically.　There are other woods but too scarce to
be of commercial importance.　Also see *New Zealand Beech.*
New Zealand Beech.　*Nothofagus spp.*　Sometimes wrongly called
birch.　There is variation in the different species but they are
similar to English beech.　BLACK BEECH, *N. solandri.*　Pale reddish,
often with black streaks, fairly durable.　Used chiefly for structural
work.　S.G. ·8.　HARD BEECH, *N. truncata.*　Pinkish.　Similar to red
beech but harder and more durable.　Used for same purposes.　S.G.
·8 ;　C.W. 4.　RED BEECH, *N. fusca.*　Pinkish red seasoning to light
brown.　Hard, strong, durable, fissile.　Difficult to season.　Used
for furniture, bentwood, etc.　S.G. ·7.　SILVER BEECH, *N. menziesii.*
Also called Southland Beech.　Pinkish, seasoning to light brown.
Lighter and softer than the other species and not so durable, but
strong.　It is in great demand for flooring, interior joinery, bentwood, furniture, implements, rifle stocks, etc.　S.G. ·56 ;　C.W. 3.
Ngeye.　Nigerian Kokko, *q.v.*
Niblick.　A golf club with small,
heavy round head.
Nicaraguan.　A prefix applied
to several Central American
woods :　cedar, lignum - vitæ,
mahogany, etc.
Nice Cargo.　One consisting of
sawn deals, battens, and boards.
Niche.　A small recess in a wall
for a statue or ornament.　It is
usually simicircular in plan with
a coved head—spherical, ellip-
soidal—but it may be polygonal
or square.　There is great
variety in the style and treatment.
Nick.　Indentations on a tally-
stock, *q.v.*　A small notch serving as a mark, guide, or for a
catch.
Nicking.　The incisions in an
organ pipe.
Nick-nack.　Knick-knack, *q.v.*
Nicobar Canoe Tree.　See
Canoe Tree.

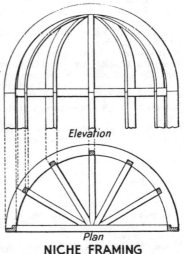

Elevation

Plan

NICHE FRAMING

Nieshout.　Sneezewood, *q.v.*
Nigerian.　Applied to the W. African woods from Nigeria.　All the
important woods are described alphabetically, including :　Ekki,

Iroko, Mahogany, Makore, Movingue, Mansonia, Obeche, Opepe, Sapele, Satinwood, Sycamore, Walnut, Whitewood.

Nigerian Box Board. *Pycnanthus kombo* (*H.*). Also called Akomu and Walile. Pinkish buff. Light, moderately hard, not strong or durable. Easily wrought but slightly woolly. Second-class wood with no special features. Suitable for rotary cut veneers for cheap plywood, shelving, cases, etc. S.G. ·38.

Nigerian Cedar. See *Moboron* or *Agba*.

Nigerian Cherry Mahogany. Makoré, *q.v.*

Nigerian Pearwood. See *Ofe* and *Obobonufua*. Standard name Guarea.

Nigerian Sycamore. See *Odoko*.

Nigerian Walnut. *Lovoa klaineana* (*H.*). Mahogany family. Also called African Walnut, Alona, Apopo, Dibetou, Goldon Walnut, Lavoa, Lovea, Sida, and Tigerwood. Goldon brown, variegated stripes, dark streaks due to gum veins, lustrous. Moderately hard, heavy, strong, and durable, and fire-resisting. Interlocked grain, roey, firm texture. Requires care in seasoning and fairly difficult to work due to spiral grain. Polishes well with careful filling. Used for the same purposes as black walnut and mahogany : superior joinery, cabinet work, propellers, gun stocks. Plain wood is used for flooring, etc. S.G. ·65 ; C.W. 3·5.

Nigger. A mechanical contrivance for turning a log when converting, etc.

Night Chair. A night stool. See *Commode*.

Night Latch. A spring latch operated on the inside by a knob and from the outside by a key.

Night Stool. A chamber utensil enclosed in a box-seat. A commode.

Nikiba. Idagbon, *q.v.*

Nim Wood. Malabar Nim Wood, *q.v.*

Nine-pin Block. One shaped somewhat like a nine-pin and fixed vertically under the cross-pieces of the forecastle bitts. It is used for a ship's ropes running horizontally.

Niri. Nyireh, *q.v.*

Nissen Hut. A semi-cylindrical form of hut named after the Arctic explorer.

Nkoba. *Lovoa brownii*. Uganda Walnut, *q.v.*

Nkunya. *Manilkara cuneifolia* (*H.*). Uganda. Rich red. Very hard, heavy, strong, and durable, with close compact grain. Difficult to convert and work through hardness. Used for structural work, paving, flooring, brake blocks, etc. S.G. ·95 ; C.W. 5.

Noble Fir. *Abies nobilis*. Also called Bracted, Silver, and White Fir, Oregon and Silver Larch. See *Fir*.

Node. **1.** The intersection of several members in structural framing. **2.** A point where two branches of a curve cross each other to form a loop. A double or triple point. **3.** The place on a stem which bears a leaf or leaves.

Noden Bretteneau. See *Preservation*.

Nog. **1.** A wood brick built into a wall for fixing purposes. **2.** A small block of wood. **3.** A projection on a log left from a sawn off branch. **4.** A projecting block built in a wall to carry a shelf.

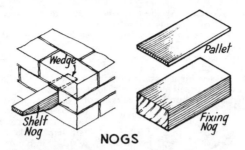

NOGS

Nogal. *Juglans australis* or *J. insularis* (*H*.). C. America. Walnut family. Variegated reddish or chocolate brown, lustrous. Faiıly light, soft, strong, and durable; stable. Straight grain. Easily wrought and polishes well. Substitute for black walnut, but lighter and softer. Used for furniture, superior joinery, carriage building, etc. S.G. ·6; C.W. 2·5.

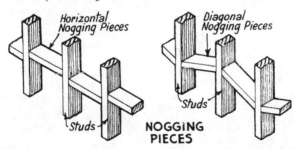

NOGGING PIECES

Nogging Pieces. Stiffening pieces between studs. See *Partition*.
Nogging Strips. Thin nogging pieces bedded between the courses of bricks to bond the bricks and stiffen the studs in a bricknogged partition.
N.O.M. Abbreviation for *Norwegian Official Measure*.
Nominal Measure or **Sizes.** The dimensions of stuff before machining.
Nom. Std. Abbreviation for *nominal standard*.
Nonagon. A polygon with nine sides. For a regular figure, area $=$ side$^2 \times 6 \cdot 182$.
Non-banded Trees. Inward growers, or endogens, *q.v.* See *Banded Trees*.
Non-conductors. Applied to materials that are poor conductors of heat or sound or electricity.
Nongo. *Albizzia zygia* (*H.*). Uganda. Light, soft, and not durable. Suitable for cheap furniture and interior joinery.
Non-inflammable. See *Fire-resisting*.

Non-porous Woods. Applied to woods that do not contain vessels or pores. Coniferous woods, or softwoods.

Nootka Cypress. *Chamæcyparis nootkatensis*. Yellow cedar, *q.v.*

Nootka Fir. Douglas Fir, *q.v.*

Nopo. *Cordia glabra* (*H.*). Mexico. Walnut colour. Rather light and soft, but tough and strong. Even grain with texture somewhat like red elm. Used for general purposes, veneers, etc.

Norfolk Latch. A fastening for a thin door, such as a batten door. It is operated by the thumb and consists of a fall bar raised by a latch lift, or sneck, and controlled by a carrier or keeper. The fall bar, or latch, engages with a catch on the frame. It is also called a Suffolk or thumb latch.

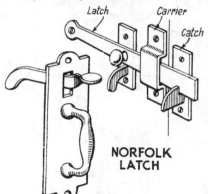

NORFOLK LATCH

Normal. A line at right angles to the tangent at a point on a curve. A perpendicular. The bed joints of an arch and joints in woodwork are usually normal to the curve.

Normal Section. A section along a normal. A method used in the preparation of hand-rail wreaths more especially for those of large radius.

Norman. A style of architecture prevailing in this country in the eleventh century. The important characteristics were richly decorated semicircular openings and low massive columns. Other features were arches with orders and chevron and zigzag mouldings. Also called English Romanesque.

North American Beech. See *Beech*.

North Carolina Pine. Loblolly pine and short-leaf pine. See *Pitch Pine*.

Northern. Applied to several woods, usually to distinguish the source of origin. See *Red Oak, Redwood, Spruce, White Cedar, White Pine, Whitewood*. Also applied to Queensland *Silky Oak*.

Northern Red Oak. American Red Oak. See *Oak*.

North Light Roof. A roof glazed along one side facing north, to obtain a uniform light not subject to alternate sun and shade. Also called a saw-tooth roof.

Norway. Applied to the woods from that country : fir, pine, spruce ; and to the maple, *Acer platanoides*, *q.v.*

Norway Pine. *Pinus resinosa*. U.S.A.

Patent Glazing

NORTH LIGHT TRUSS

and Canada. Very similar to *Pinus sylvestris*. See *Redwood*.

Nose. **1.** The bottom of the shutting stile of a door or casement **2.** A drip or a downward projection of a cornice. **3.** The rounded end of a log, for easier transport. **4.** The projecting moulding on a rib in vaulting. **5.** The front end of a spindle or mandrel.

Nose Bit. A brace bit like a shell bit, but with a turned-over cutting edge making it more suitable for boring in end grain.

Nosing. 1. A rounded projecting edge to a flat surface, such as a tread, window board, lead flat, etc. **2.** The projecting edge of a cornice, moulding, or drip. See *Wriggle*.

Nosing Line. An imaginary line touching the edges of the treads, in a flight of stairs, from which the margin is measured. See *Line of Nosings*.

Nosing Piece. A narrow piece with nosed edge to cover the ends of floorboards and to continue the nosing round the well of a stairs, etc.

Notalea. See *Ironwood*.

Notaphœbe. A genus of the Lauraceæ family providing a species of Medang, *q.v.*

Notch. 1. A groove, recess, or trench, in the side of one piece to receive the side of a cross-piece. When both pieces are notched it

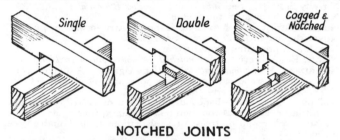

NOTCHED JOINTS

is called a double-notch. An alternative name is *gain* or *gaining*. The term is also loosely applied to trenching and housing. **2.** A V-shaped cut or slot. See *Tally*.

Notch Board. 1. A stair string. **2.** An appliance for gauging a stream of water.

Notched Tenons. Applied to tenons that are inserted from opposite sides of the same piece, and halved in thickness so that the ordinary size of mortise will serve for both tenons.

Nothofagus. Antarctic beech. See *New Zealand Beech, Roble, Tawhai,* and *Tasmanian Myrtle.*

Notice Board. A specially framed **NOTCHED** board for displaying announce-**TENONS** ments. It may be hung on a wall or fixed to posts. (See illus., p. 366.)

Novelty Saw. See *Plane Saw*.

365

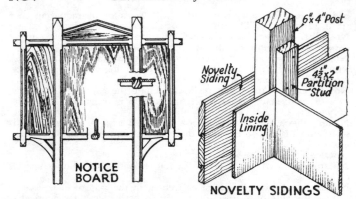

NOTICE BOARD

NOVELTY SIDINGS

Novelty Sidings. See *Sidings* and *Weather Boarding*.
Novingu. Nigerian Satinwood, *q.v.*
Noyer d'Afrique and **Noyer de Gabon.** African walnut, *q.v.*
Noyer du mayombe. Afara. See *Eghoin*.
Nozzle. A projecting spout or vent.
N.R.M.E. Abbreviation for *notched, returned*, and *mitred ends*.
Nsambya. *Markhamia platycalyx* (*H.*). Uganda. A tough, durable, insect-resisting wood used chiefly for poles and pit props. S.G. ·6.
N-Truss. See *Linville* and *Hogback*.
Nulling. A quadrant-shaped detail, turned or carved, in Jacobean oak furniture, etc. It is common on mouldings, friezes, etc.
Nursery. A group of trees, or a plantation, trained under expert supervision.
Nut. 1. The loose, or fixing, end of a bolt, *q.v.* It is a square, hexagonal, or round piece of metal pierced with a female screw for securing to the bolt. **2.** The fruit of many species of trees.
Nut Gall. An excrescence on oak trees due to insects. Also called oak galls or gall nuts.
Nutmeg. See *Stinking Cedar* and *Hickory* (*Hicoria myristicæformis*). Also W. African white cedar, *Pycnanthus kombo*. See *Nigerian Box Board*. The true nutmeg is *Myristica fragrans*.
Nut Oak. *Macadamia ternifolia* (*H.*). Queensland. Pale mauve-pink, finely veined. Very hard and heavy, tough but fissile. Straight compact grain, close texture. Used for turnery, fancy articles, etc. Too hard for ordinary cabinet work. S.G. 1 ; C.W. 5.
Nut Oil. Used in paints, but inferior to linseed and poppy oils.
Nutt. Abbreviation for the name of the botanist Nuttall.
Nut Tree. Hazel, *q.v.*
Nutwood. Locus wood, *q.v.*
N.W. Abbreviation for *narrow widths*.
Nyakatuma. *Morus lactea* (*H.*). Uganda. Bright yellow. Hard, heavy, strong, and durable. Seasons and works well. Little imported. S.G. ·8.

Nyatoh. *Palaquium* and *Payena spp.* (*H.*). Malaya. Local name Maiang. Often substituted for Meranti and Seraya. Reddish. Very variable in weight and hardness, fairly strong and stiff, not durable, and subject to insect attack. Rather fine grain and fine rays. Used for planking, cheap furniture, temporary construction. S.G. ·46 to ·71 ; C.W. 3. Also *Ganua spp.* and *Padhuca spp.* are called Nyatoh.

Nyatto. Betis or Belian, *q.v.*

Nyctaginaceæ. A family with a few woods of commercial importance. Includes Bougainvillea, Calpidia, Colignonia, Neea, Pisonia, Torrubia.

Nyireh. *Xylocarpus* (*Carapa*) *spp.* (*H.*). Malaya. Also called Niris, Neris, Nyiris. Similar to Mexican mahogany. Dull reddish brown, darker gum streaks. Hard, heavy, strong, durable, and very stable. Fine rays and ripple. Rather small sizes and wasteful in conversion. Used for cabinet work, superior joinery, tools, boat-building, etc. S.G. ·65 to ·8 ; C.W. 3·5.

Nyssa. *Nyssa sessiliflora* (*H.*). India, Burma. Yellowish white, seasoning to light brown. Fairly lustrous and smooth. No distinct heartwood. Moderately light, hard, and strong. Not durable and subject to insect attack. Straight grain ; even medium-fine texture. Seasons well and easily wrought. Liable to stain. Excellent wood but difficulties of transport prevent export. Used for joinery, tea boxes, turnery, cheap furniture. S.G. ·6 ; C.W. 2·5. Also see *Gum* (*Black*) and *Tupelœ Gum.*

O

Oak. *Quercus pedunculata* or *Q. robur*, and *Q. sessiliflora* or *Q. petræa* (*H.*). Europe. English oak. Rich light brown. Hard and heavy ; very strong and durable. Hardens and darkens with age. Large rays give beautiful silver grain when rift-sawn. Corrodes iron. Fumes, limes, and polishes well. English-grown oak is the best of the oaks, of which there are about 300 varieties, but good timber is expensive and it is difficult to work due to hardness and irregular grain. Requires careful seasoning. Oak is used for almost all purposes where strength and durability are important, superior joinery, external woodwork, furniture, flooring, structural work, cooperage, boat building, carving,

OAK LEAVES

etc. *Brown oak*, due to a fungus, and *Pollard oak* are in great demand for decorative woodwork, but it is usually in veneer form. S.G. ·8 ; C.W. 4. Many woods, not *Quercus spp.*, are wrongly called oak. AFRICAN OAKS. These are not oaks. Also called African " Teak " and Red Ironwood, *q.v.* See *Ekki, Turtosa*, and *Iroko*. AMERICAN OAKS. Numerous varieties, distinguished as red and white. They are not so hard, heavy, strong, or durable as English and easier to work, especially red oak. The red oak group includes : Red Oak (*Q. borealis* or *Q. maxima*) ; Shumard Red Oak (*Q. shumardii*) ; Black Oak (*Q. velutina*) ; Laurel Oak (*Q. laurifolia*) ; Southern Red Oak (*Q. rubra, Q. falcata*, and *Q. digitata*) ; Swamp Red Oak (*Q. rubra pagodæfolia*) ; Pin Oak (*Q. palustris*) ; Water Oak (*Q. nigra*) ; Texas Red Oak (*Q. texana*) ; Willow Oak (*Q. phellos*). Red oaks are more porous and less durable than white oaks. The white oak group includes : White Oak (*Q. alba*) ; Overcup Oak (*Q. lyrata*) ; Post Oak (*Q. stellata*) ; Swamp Chestnut or Cow Oak (*Q. prinus*) ; Swamp White Oak (*Q. bicolor* and *Q. platanoides*) ; Burr Oak (*Q. macrocarpa*) ; Chinquapin Oak (*Q. acuminata*) ; Chestnut Oak (*Q. montana*) ; Live Oak (*Q. virginiana*). APPALACHIAN OAK. From the mountainous regions of that name. One of the best U.S.A. white oaks, especially for decorative work, veneers, etc. AUSTRALIAN OAK. See *Mountain Ash* and *Silky Oak*. AUSTRIAN or EUROPEAN OAK. Also called Slavonian Oak. Similar to English, but milder and easier to work. It is usually quarter-sawn and known as wainscot oak. Only better qualities are imported. BOG OAK, *q.v.* BURR OAK. Specially valuable when English and from brown oak. See *Burrs* and *Brown Oak*. CALEDONIAN OAK. See *Queensland Oak*. CEYLON OAK, *Schleichera trijuga*. Used in aircraft. See *Gausan.* CORK OAK, *Q. suber.* S.W. Europe. Bark important for cork. DANTZIG OAK. As Austrian, *q.v.* DURMAST OAK, *Q. sessiliflora.* English Oak, *q.v.* EUROPEAN OAK. Same species as English. Usually named after source or port of shipment ; Austrian, Dantzig, Libau, Memel, Odessa, Polish, Riga, Slavonian, Stettin, Spessart, Volhynian. See *Austrian.* EVERGREEN OAK, *Q. ilex.* S. Europe. No pore rings. Very hard, heavy, and difficult to work. Also see *Cork Oak.* FOREST OAK, *Casuarina torulosa.* E. Australia. Properties and appearance of English oak. Coarse and harsh grain, with wide confused rays. Liable to insect attack and difficult to season. Used for cabinet work and veneers. S.G. ·9. FORMOSAN OAK, *Q. pseudo-myrsineæfolia.* Beautiful decorative wood both in figure and colour. S.G. 1·1. Also *Q. gilva,* similar to live oak, *q.v.* HIMALAYAN OAK, *Q. spp.* Properties and characteristics of English oak. Coarse fibrous grain. S.G. ·9. HOLLY OAK, *Q. morii.* Formosa. Salmon red, with darker streaks. Very hard, heavy, and smooth. Beautiful wood but little imported. S.G. 1. HOLM OAK. Evergreen Oak, *q.v.* INDIAN OAK, *Q. dilatata.* N. India. Similar to Live Oak. S.G. 1. Also *Q. lanceæfolia.* Evergreen. Brownish yellow wood. Other Indian species are *Q. incana, Q. lamellosa, Q. semecarpifolia, Q. semiserrata, Q. serrata.* They are variable in quality and usually inferior to European, hence little imported. JAPANESE OAK, *Q. grosseserrata, Q. acuta, Q. glauca, Q. crispula,* and *Q. mongolica.* Good quality with more even texture, lighter, and with finer rays than English. Used for superior joinery, cabinet work, etc. S.G. ·7 ; C.W. 3. LIVE OAK, *Q. virens.* U.S.A. Brown. Very hard, heavy, strong, durable, and stiff. A good cabinet and structural wood, but supplies reserved for naval work. S.G. ·9. LUCOMBE OAK, *Q. lucombeano.* W. England. A hybrid tree and inferior to English oak. MAIDEN OAK. Brown Oak, *q.v.,* with plain grain. POLISH OAK. See European Oak. QUEENSLAND OAK. See *Silky Oak.* ROCK OAK. Swamp chestnut oak. See *American Oak.* SESSILE OAK, *Q. sessiliflora.* English oak. SLAVONIAN OAK. See *Austrian* and *European Oak.* TASMANIAN OAK, *q.v.* TURKEY OAK, *Q. cerris.* Inferior to English, more like American red oak. VICTORIAN OAK. Mountain Ash, *q.v.* VOLHYNIAN OAK. Better qualities from Poland. See *European Oak.* WHITE OAK. Spanish Chestnut, *q.v.,* and American Oak, *q.v.* Also see *Mempening, Messmate, Pollard, Shea,* and *Silky Oak.*

Oak Walnut. *Cryptocarya corrugata (H.).* Queensland. Also called Washboard Tree and Corduroy. Pale cream, smooth. Fairly hard and heavy, tough, strong, durable. Firm fine grain, mottle, distinct rays. Requires care in seasoning, not difficult to work. Used for cabinet work, superior joinery, carving, general purposes. S.G. ·8 ; C.W. 3.

Oar. A pole with broad flat end, or blade, to propel a boat through the water. The oar works in a rowlock which serves as a fulcrum.

Oba-suluk. *Shorea polysperma* and *S. pauciflora.* The standard name for Borneo mahogany, *q.v.* See *Seraya* and *Nemesu.*

Obeche. *Triplochiton scleroxylon (H.).* Also called African and Nigerian Whitewood, Abachi, Arere, Ayous, Samba, Wawa, etc. Creamy white, lustrous. Soft, light, fairly durable and stable, not strong. Interlocked grain gives a stripy effect. Open uniform texture. Seasons and works easily, liable to warp. Rift-sawn wood liable to pick-up. Polishes well. Large sizes. Used for interior joinery, matchboarding, plywood, veneers, etc. S.G. ·35 ; C.W. 2.

Obelu. *Strombosia pustulata* (*H.*). Nigeria. Also called Atako. Characteristics of olive wood. Yellowish brown, irregular streaks, dark reddish brown watermarks. Hard, heavy, strong, and fairly durable. Interlocked and very compact grain. Difficult to work and requires careful seasoning. Polishes well. Under investigation. Suitable for furniture, flooring, etc. S.G. 1 ; C.W. 4.

Oblate. See *Spheroid.*

Oblique. Neither perpendicular nor parallel to some line, plane or solid, of reference. Arranged at an angle not a right angle. On the skew. Applied to a cone or cylinder when the axis is not at right angles to the base. See *Knots* and *Grain.*

Oblique Arch. A skew arch, *q.v.*

Oblique Grain. Spiral grain, *q.v.*

Oblique Joint. An angle joint not forming a right angle.

Oblique Mitre. One for an angle greater than 90°. See *Obtuse.*

Oblique Planes. A term used in geometry for auxiliary planes inclined to both planes of projection. See *Planes* and *Traces.*

Oblique Projection. Pictorial projection in which the elevation or plan or section is drawn in the usual way, and the other two sides shown by parallel projectors at an angle usually 30° or 45°. See *Axonometric* and *Planometric.*

Obobonekhui. Ofe, *q.v.*

Obobonufua. *Guarea cedrata* (*H.*). Nigeria. Now called Scented Guarea. Also called Pearwood, Bosse, and Obobo. Pale mahogany colour, lustrous. Moderately hard and heavy. Stiff, strong, stable, tough and shock-resisting. Close and interlocked grain, roe figure. Seasons well with normal care, and easily wrought. Figured wood liable to pick-up. Polishes well. Large sizes. Used for superior joinery, furniture, flooring, shop fittings, veneers, turnery. S.G. ·68 ; C.W. 3·5　Also see *Ofe.*

Oboe. 1. A musical wood-wind instrument. 2. An organ reed stop.

Observation Car. A railway coach specially designed for look-out purposes.

Obtuse. Applied to angles greater than 90°.

O.C. Abbreviation for *open charter.*

Occasional Table. A small general-utility table, usually of a decorative character and used in the drawing room of a house.

Occidental Plane. *Platanus occidentalis.* See *Plane.*

Occupational Bridge. A permanent foot-bridge across a river.

Ochanostachys. See *Petaling.*

Ochnaceæ. A family including *Cespedesia, Lophira, Tetramerista.*

Ochrocarpus. See *Ologbomidu* and *Bolo.*

Ochroma. See *Balsa* or *Polak.*

Ochrosia. India. A genus of the Dogbane, or Apocynaceæ, family The wood is of little commercial importance. Some species, as in the tidal forests of Madagascar, etc., produce a hard dense yellow wood, similar to boxwood, and used for shuttles, bobbins, etc.

Ocotea. Laurel family. See *Bois de Rose, E. African Camphor, Determa, Greenheart, Laurel, Louro vermelho, Muzaiti, Silverballi, Stinkwood*.

Octagon. A plane figure with eight sides. For a regular figure, area = square of side × 4·828.

Oculus. A round, or bull's-eye, window.

Odd Lengths. Applied to a quantity of sawn or manufactured stuff not of uniform lengths.

Oddments. Converted wood un-graded and of irregular sizes.

Odina. See *Wodier*.

Odoko. *Scottelia coriacea* (*H.*). Also called Nigerian Sycamore. Uniform cream to bright yellow, seasoning to light buff. Hard and heavy, but vari-able. Elastic, fairly strong, not durable.

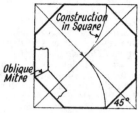

OCTAGON

Dense, close, straight grain. Fine texture, smooth. Liable to blue stain. Requires careful seasoning. Not difficult to work but some-times liable to pick-up. Polishes. Used for sports equipment, kitchen and school equipment, bobbins, turnery, carving, etc. S.G. ·7 ; C.W. 3.

Odour. Most woods have their characteristic odour, which is often a means of identification. If the smell is pleasant it is described as fragrant, *q.v.*, if very unpleasant it is malodorous, *q.v.* In the green wood it is due to the chemical constituents. In many cases it is due to decomposition of organic compounds on exposure to the air. See *Decay*.

Odudu. *Klainedoxa gabunense* (*H.*). Nigeria. Also called Alukon-raba. Dark greyish brown. Very hard and heavy. Close uniform texture. Interlocked grain. Difficult to work. Resists hard wear. Under investigation. S.G. ·85 ; C.W. 5.

Odum. Iroko, *q.v.*

Œcidium pini. Rust disease. A fungus that attacks growing pines.

Oerst. Abbreviation for the name of the botanist Oersted.

Ofe. *Guarea thompsonii* (*H.*). Nigeria. Now called Guarea. Also called Bosse, Pearwood, and Obobonekhwi. Rich cedar brown with glistening gum deposit. Hard and heavy. Milder and easier to work than Obobonufua, and not so much figure. Fibrous grain liable to pick-up. Exudes gum in seasoning and checks easily. Not difficult to work and polishes well. Similar to mahogany in strength and properties and used for similar purposes. S.G. ·75 ; C.W. 3. See *Obobonufua*.

Off Bearer. One who removes converted stuff as it comes from the saw.

Off Products. Remainder of tree after selection for timber. Used for packing cases, charcoal, and other by-products, *q.v.* Imported as rounds, ends, and short ends, for casewood.

Offram. Afara. See *Eghoin* and *Limba*.

Offset. 1. An automatic device for moving the carriage frame away from the saw-line, for the return of the carriage. **2.** A horizontal break in a wall to diminish the thickness. It is often used to carry a wall plate. **3.** A return in an eaves gutter or down-pipe to carry round, or over, a projection.

Off Sizes. The dimensions of converted wood " off the saw."

Ofun. Mansonia, *q.v.*

O.G. Abbreviation for *ogee*. Also suggested that it implies *old Grecian*.

Ogangwo. Lagos and Benin mahogany, *q.v.*

Ogea. *Daniellia ogea* (*H.*). Nigeria. Pinkish to streaked dark walnut brown. Moderately hard and heavy, with open fibrous or woolly texture. A substitute for mahogany but inferior. Liable to warp and rather difficult to finish for polishing. S.G. ·75 ; C.W. 4.

Ogechi. *Brosimum alicastrum* (*H.*). S. America. Also called Breadnut, Capomo, Laredo, etc. Yellowish tan colour. Fairly hard, heavy, and strong. Somewhat like maple. Close even grain, with ripple, and lustrous. Used for cabinet work, veneers, superior joinery, etc. S.G. ·65 ; C.W. 3.

OGEE MOULDINGS

Ogee. A moulding with contour of contraflexure. It consists of convex and concave arcs, like a narrow letter S inverted. It is the cyma reversa, but often applied to the cyma recta, as the ogee gutter.

Ogee or **Ogeval Arch.** A pointed arch consisting of two ogees.

Ogee Gutter. One with a cyma recta contour on the front. See *Lap Scarf*.

Ogugu. *Sterculia* or *Cola cordifolia* (*H.*). Nigeria. Buff colour with pinkish tinge. Moderately hard and heavy. One of the most stable woods. Open, straight, fibrous grain. Liable to pick-up. Little exported. S.G. ·65 ; C.W. 4. See *Orodo*.

Ohia. Itako, *q.v.*

Oil Bean Wood. See *Okpagha*.

Oillet. Same as eyelet, *q.v.*

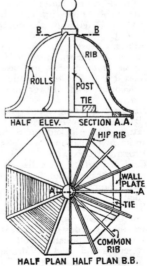

OGEE TURRET ROOF

Oils. Vegetable oils are from linseed, poppy, walnut, hemp, sunflower seeds, etc., and are used as a vehicle for paints, etc. Lubricating oils are usually of mineral origin. For oilstones, a mixture of neatsfoot and sweet oils is the most satisfactory. The stone should be periodically cleaned with paraffin oil.

Oil Seasoning. A method of boiling in oil, under a vacuum, to remove some of the moisture from green timber.

Oil Stain. Used for external woodwork, instead of paint, to show the grain of the wood.

Oil Stones. Used for sharpening woodwork tools. They may be natural or artificial stones. The favourite natural stones are : Arkansas,

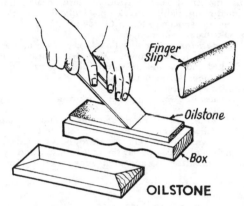

OILSTONE

Charnley Forest, Turkey, and Washita. Carborundum and Indian (corundum and alundum) are popular artificial stones, and may be obtained in various degrees of coarseness. Finger slips are made of the same materials and are of various shapes and sizes for sharpening gouges, etc.

Oilwood. M'gongo, *q.v.*

Oiti. *Moquilea tomentosa* (*H.*). C. America. Also called Cabraiba. Brown with purplish tinge. Very hard, heavy ; fairly strong and durable. Splinters easily. Straight grain, medium texture. Used for structural work; piling, etc., suitable for cabinet work. S.G. ·9. *Clarisia racemosa.* Also called Oiticia. Canary yellow to chocolate brown on exposure, lustrous. Fairly light and soft, not durable. Uniform grain and texture. Roey and with distinct rays. Difficult to work. Substitute for mahogany.

Ojoche. Ogechi, *q.v.*

Okan. *Cylicodiscus gabunensis* (*H.*). Nigeria. Also called Denya. Reddish and greenish brown with yellow tinge in pores, lustrous. Seasons to deep chocolate brown. Very hard and heavy, tough and strong. Coarse, with double spiral grain. Very difficult to work and requires careful seasoning. Used chiefly for structural work and superior flooring. S.G. 1 ; C.W. 5.

Okanhan. *Fagara unwinii* (*H.*). Nigeria. Also called Ata. Yellowish, variegated fawn and grey, lustrous. Moderately light, fairly hard, with fine compact grain. Easily wrought. Small sizes, suggested for rotary-cut plywood. S.G. ·6.

Okha. Honduras cottonwood. See *Ceiba*.

Okoko. *Sterculia oblonga* (*H.*). Nigeria. Yellow. Fairly heavy but easily indented. Mild, coarse grain. Some fibrous grain liable to pick-up. Not difficult to work. Little exported. Used chiefly for cases. S.G. ·75.

Okor. *Fagara macrophyllum* (*H.*). Nigeria. Also called Atagbe. Straw colour, darker streaks, lustrous. Hard, heavy, not durable. Fibrous inclined grain, liable to pick-up. Difficult to work. Suitable for superior joinery, veneers, etc. S.G. ·85 ; C.W. 4.

Okoume or **Okumie.** Gaboon, *q.v.*

Okpagha. *Pentaclethera macrophylla* (*H.*). Nigeria. Also called Oil Bean wood. Rich red brown. Very hard, heavy, and strong. Coarse interlocked and fibrous cross grain, liable to pick-up. Difficult to work. Used chiefly for structural work. S.G. 1.

Okpo or **Okopo.** Obeche, *q.v.*

Okwen. *Ricinodendron sp.* Nigerian Corkwood. See *M'gongo* and *Erinmado*.

Okweni or **Okwein.** *Brachystegia spicæformis* and *B. leonensis* (*H.*). Nigeria. Also called Bubinga, Emi-emi, and u (Ak*B. eurycoma*). Light to dark brown, dark markings. Hard, heavy. Coarse, double-spiral grain, liable to pick-up. Difficult to season and work. Not esteemed as yet, but attractive appearance for superior joinery, cabinet work, aircraft, etc. S.G. ·8 ; C.W. 4.

Olacaceæ. A family including Minquartia, Ochanostachys, Scorodocarpus, Strombosia, Ximenia.

Oldfieldia. See *Turtosa* or *African Oak*.

Old Woman's Tooth. A wood router, *q.v.*

Olea. See *Olive, Maire, Ironwood* (*black*).

Oleaceæ. A family including Alnus, Chronanthus, Forsythias, Fraxinus, Jasminum, Ligustrum, Linociera, Olea, Osmanthus, Schrebera, Syringas.

Olearia. See *Muskwood*.

Oleaster. The European wild olive. See *Olive*.

Oleo pardo. See *Cabreuva*.

Oleo vermelho. *Myroxylon toluiferum* or *Toluifera balsamum* (*H.*). C. America. Also called Balsamo, Cabreuva, and Quina. Variable reddish brown, purplish on exposure, fragrant. Resembles and a substitute for Cuban mahogany. Very hard, heavy, and durable. Straight and roey grain, medium texture. Fairly difficult to work ; polishes well. S.G. ·8 ; C.W. 3·5.

Olinia. See *Hard Pear*.

Oliv. Abbreviation for the name of the botanist Oliver.

Olive or **Olive Wood.** *Olea europa* (*H.*). S. Europe. Yellowish brown, variegated darker streaks, lustrous. Hard, heavy, very smooth. Small sizes. Used for decorative work, inlays, turnery, etc. S.G. ·9. *Olea hochstetteri*. E. African olive or Musheragi. Light brown, darker irregular markings. Handsome striking appearance. Very hard, heavy, durable, and strong. Close even grain and texture, slightly interlocked. Requires careful seasoning. Difficult to work due to hardness. Polishes well. Seldom over 10 in. wide. Used for flooring, cabinet work, carriage work, tools, turnery, etc. S.G.

·95 ; C.W. 4·5. Other Kenya olives : *Olea chrysophylla* (brown olive) and *Linociera welwitschii* (Elgon olive). *Olea laurifolia* (Black Ironwood), *q.v. Olea verrucosa* or *O. capensis*, is S. African or Cape olive. Variegated and lustrous. Very hard, heavy, durable, and strong. Used chiefly for wagon building and turnery. S.G. 1·1. *Olea ferruginea.* Indian olive. Very variable in colour from light brown with darker streaks to deep purple. Very hard, heavy, durable, elastic, and strong. Dull to fairly lustrous. Feels oily and smooth. Straight and shallow interlocked grain, fine even texture. Seasons fairly well. Not difficult to work but liable to pick-up. Polishes well. Excellent ornamental wood. Used for turnery, tools, veneers, fancy articles, etc. S.G. 1 ; C.W. 4·5.

Olive Ash. Applied to variegated brown heartwood of ash, *q.v.* Also see *Olivewood.*

Olive, Wild. *Ximenia americana* (*H.*). C. America. Reddish yellow, fragrant. Very hard, heavy, with fine texture. Substitute for sandalwood. S.G. ·9. Also see *Cabbage Bark* and *Mastic.*

Olivewood. A trade name for quarter-sawn American black ash, because of the brown heartwood.

Olivier. *Chuncoa obovata* (*H.*). C. America. Colour and properties of greenheart. S.G. ·8. Also *Terminalia aff. januarensis.* Yellowish brown, black mottles, lustrous. Fairly hard and heavy, durable. Like birch in texture and qualities, and used for similar purposes. S.G. ·76 ; C.W. 3. WHITE OLIVIER. Nargusta, *q.v.*

Ologbomidu. *Ochrocarpus africanus* (*H.*). Nigeria. Also called Bolo, *q.v.* Reddish brown. Moderately hard and heavy. Rather coarse, spiral grain, liable to pick-up. Little exported, but under investigation. S.G. ·7.

Oloko. Iroko, *q.v.*

Olon. *Turræanthus africana* (*H.*). Gold Coast. Now called Avodire. Also called Apaya. Dull white to golden cream, lustrous. Somewhat like sycamore in appearance. Moderately hard and heavy; stable. Not strong or durable. Strong figure with moire effect. Interlocked fine grain. Crotches produce beautiful feather figure. Requires care in seasoning. Easily wrought ; polishes well. Average sizes. Used for cabinet work, superior joinery, veneers, etc. S.G. ·6 ; C.W. 2·5.

Omatsu. See *Kuro matsu.*

Omnibus. A closed vehicle for numerous passengers, often with an upperdeck.

Omnibus Box. A box in a theatre to accommodate a large number of people.

Omo. *Cordia millenii* (*H.*). Nigeria. Yellowish brown to fawn, with darker markings, lustrous. Light, fairly soft, very resonant. Coarse, straight grain ; smooth. Easily wrought. Little exported, but under investigation. Good for sounding boards, etc. S.G. ·4.

Onara or **Nara.** Japanese Oak, *q.v.*

Oncosperma. Bayas Palm, Malay.

Ondo Mahogany. *Khaya ivorensis.* African Mahogany. See *Mahogany.*

Onion Wood. *Owenai cepiodora* (*H*.). New South Wales. Sold as bastard cedar. Strong odour which disappears with seasoning.

On-the-plank. A term used in carving and engraving for cutting along the grain.

Open-cell Process. A process of impregnating wood under pressure. The antiseptic is retained in the cell walls only, and the cells left empty. See *Preservation*.

Open Fencing. A fence with open spaces between the palings or boards.

Open Floor. One without ceiling, so that the joists are exposed to view.

Open Grain. Applied to hardwoods with large open pores. Coarse grain. See *Grain* and *Rings*.

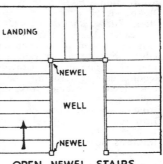

Open Mortise. A mortise at the end of the stuff so that the tenon cannot be fixed by wedging.

Open Newel Stairs. Applied to a staircase having a rectangular well and newels.

Open Roof. 1. A roof without ceiling so that the principals or trusses are in view. 2. A roof with trusses without tie beams.

Open Sheathing. See *Sheathing*, (2).

OPEN MORTISE

Open String. See *Cut String*.

Open Tank Process. Impregnating wood, at atmospheric pressure, by varying the temperature of the antiseptic in an open tank. See *Preservation*.

Open Up. To increase the tension of a log saw or band saw.

Open Valley. See *Valley* and *Shingles*.

Open Well. Applied to a staircase with large rectangular well. See *Open Newel Stairs*.

Opepe. *Sarcocephalus diderrichii* (*H*.). Nigeria. Also called Abiache, Bilinga, Badi, Kusia, Obiache, Ubulu, etc. Orange, yellow, seasoning to rich golden brown, slightly fragrant. Hard, heavy, strong, and durable. Coarse, fibrous, wavy, and interlocked grain. Requires careful seasoning. Not difficult to work but inclined to pick-up. Polishes well. Used for superior joinery, strip-flooring, structural work, piling, wagon construction, shipbuilding, cabinet work, etc. S.G. ·78 ; C.W. 3.

OPEN NEWEL STAIRS

Opopo or **Apopo**. African walnut. See *Nigerian Walnut*.

Opruno. Mansonia, *q.v.*

Orange. *Citrus aurantium* and *C. sinensis* (*H*.). Widespread. Yellowish white. Hard, heavy, tough. Fairly close and uniform grain and texture ; distinct rays. Not difficult to work. Small sizes. Used for

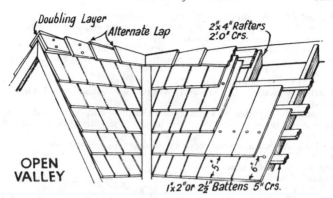

Doubling Layer
Alternate Lap
2"x 4" Rafters
2'.0" Crs.

OPEN
VALLEY

1"x 2" or 2½" Battens 5" Crs.

fancy articles. S.G. ·8. WILD ORANGE. W. Indies satinwood, *q.v.*
MOCK ORANGE. See *Andaman Satinwood.*

Orange Boxwood. *Celastrus dispermus* (*H.*). Queensland. Similar
to European boxwood, *q.v.*, and used for similar purposes. S.G. ·8.

Oratory. A small chapel attached to a church, or a private chapel.

Orb. A plain circular boss.

Orchestra. Part of a place of entertainment reserved for chorus and band, or
orchestra.

Orders. 1. Signifies an assemblage of
parts to certain established proportions.
Specially applied to classic styles of architecture : Doric, Ionic, Corinthian, together with Composite and Tuscan, in
which the base, shaft, capital, and entablature are in recognised proportions.
2. See *Family.*

Oregon Cedar. Port Orford cedar, *q.v.*

Oregon Larch. Noble fir, *q.v.*

Oregon Maple. Pacific maple. See
Maple.

Oregon Myrtle. Pacific myrtle, *q.v.*

Oregon Pine. Douglas fir, *q.v.*

Orelha. Vinhatico, *q.v.*

Orel Poles. Alder Poles.

Organ. A large musical instrument
in which pipes provide the sounds.
These are supplied with wind by
bellows and sounded by keyboards, or
manuals. Contact between keyboards
and pipes is made by means of stops.

Organ Case. The wood casing of an
organ.

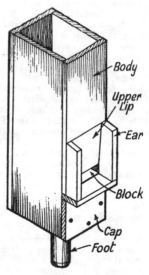

Body

Upper
Lip

Ear

Block

Cap

Foot

ORGAN PIPE

(See p. 378.)

377

Organ Loft. A gallery containing an organ.

Organ Pipes. The sets of pipes producing the musical sounds of an organ. See *Pallets.* (See illus., p. 377.)

Orham Wood. Canadian White Elm, *q.v.*

Oriel. A bow or bay window projecting from the wall and supported by brackets or corbels.

Oriental Plane. *Platanus orientalis.* See *Plane.*

Oriental Walnut. Queensland Walnut, *q.v.* Also called Oriental Wood.

Orientated. Applied to a church with chancel end pointing due east. Having an easterly direction, or having characteristics of Eastern architecture.

Orifice. A mouth, aperture, or opening, to a pipe or tube.

Orion. Guijo, *q.v.*

Orissa. See *Laburnum.*

Orites. See *Silky Oak.*

Orlop Deck. The lowest deck in a large ship.

Orme. The French name for elm.

Ormolu. A copper, brass, or gilded bronze alloy, used as a decorative feature on furniture.

Ormosia. See *Hung Tou.*

Oro. *Antiaris africana* (*H.*). Nigeria. Pale yellow, lustrous, stripes. Fairly light and soft. Fibrous, double spiral grain makes it difficult to work. Subject to insect attack. Held in little esteem. S.G. ·4. Also see *Tindalo.*

Orodo. *Sterculia rhinopetala* (*H.*). Nigeria. Also called Aiye. Reddish brown with lighter streaks. Hard, heavy, strong, tough, resilient, and hard wearing. Fibrous grain. Air seasons slowly without checking. Easily wrought. Used for flooring, sleepers, rollers, pick handles. S.G. ·8 ; C.W. 3.

Oroko. Iroko, *q.v.*

O.S. Abbreviation for *one side* and for *overside delivery.*

O.S. and W. Abbreviation for *oak, sunk and weathered.*

Osage Orange. *Maclura aurantiaca* or *Toxylon pomiferum* (*H.*). U.S.A. Also called Bois d'Arc. Bright orange, darkening with exposure, darker streaks, lustrous. Very hard, heavy, strong, flexible, tough, and durable. Coarse texture, irregular grain, shrinks greatly. Polishes well. Used for wagon framing, implements, bentwork, bows, wheelwright's work. Important for yellow dye. S.G. ·8.

Osanko. *Chrysophyllum prunefolium* (*H.*). Nigeria. Also called Osangbola. Resembles Avodire. Pale pink to yellow, lustrous. Moderately hard and light. Compact grain, even texture, ripple. Easily wrought. Under investigation. Substitute for light mahogany. Suitable for cabinet work, superior joinery, musical instruments, rotary plywood. S.G. ·6 ; C.W. 2·75.

Oscillating Saws. Frame Saws, *q.v.*

Osiers. 1. Tough flexible twigs used for various purposes but of little timber value. See *Willow.* 2. *Salix viminalis.* Usually shrub-like trees but sometimes up to 30 ft. high ; of little timber value. 3. PURPLE OSIER, *S. purpurea.* Shrubs, and of little value.

Osol. *Symphonia gabonensis* (*H.*). Nigeria. Light orange to brownish red. A straight grained wood, easily wrought, and suitable for constructional work. Under investigation. Also see *Hog Gum*.

Ostrya. Hop Hornbeam, *q.v.* See *Paque*.

Osun. Camwood, *q.v.*

Ottawa. A prefix to several Canadian pines, but chiefly *Pinus strobus*.

Ottoman. A box with cushioned top, like a sofa without arms or back. There are several types and shapes. Some are arranged as seats round a central pillar that serves as a back rest.

Otuto. *Cistanthera papaverifera* (*H.*). Nigeria. Light pink to reddish brown, lustrous stripes. Hard, heavy, strong, tough, and flexible. .Interlocked and fibrous grain, close texture, ribbon stripe. Seasons readily. Easily wrought but quarter-sawn wood liable to pick-up. Used in place of hickory and ash, and for cabinet work. S.G. ·8 ; C.W. 3.

Ougeinia. See *Sandan*.

Oui-ja. A board with figures and alphabet for use in a spiritualistic seance.

Oundy or **Undy.** A moulding with wave-like outline.

Outer Lining. The outside lining of a boxed frame for sliding sashes, *q.v.*

Outer String. The string farthest from the wall in a flight of stairs.

Outhouse. A small building detached from but belonging to a dwelling house ; or attached but without direct communication with the house.

Out of Round. A circular saw with its periphery not a true circle.

Out of Truth. Inaccurate, winding, or not square.

Out of Wind. A plane surface free from warp or twist.

Outrigger. 1. A cantilever beam projecting over the face of a building or side of a ship to carry a suspended scaffold or tackle for hoisting, etc. 2. The sleepers for a derrick crane. 3. A metal stay to carry a rowlock of a boat that projects over the side for increased leverage. 4. A projection from the splinter bar of a carriage to allow for an extra horse.

Out-to-out. 1. Applied to measurements taken from the extremes of framing, etc. Maximum size of anything. 2. See *Centre-to-Centre*.

Ova. An egg-shaped carving.

Ovaga. *Panda oleosa* (*H.*). W. Africa. Greyish white. Similar in texture and grain to a soft light-coloured oak. Used for veneers, etc. Under investigation.

Oval. Having the outline of an egg.

Oval Knot. See *Knots*.

Oven Dry. Applied to wood that does not lose moisture in a stated interval of time when placed in a ventilated oven at 100° C.

Over-all. Applied to dimensions that cover several smaller ones in the same plane. See *Out-to-out*.

Over-bark. A term used in the measurement of round logs when taken to the outside of the bark ; allowances are then made from prepared tables.

Overblown. Applied to an organ pipe that produces a sound higher than the correct one.

Overcup Oak. *Quercus lyrata*. American white oak. See *Oak*.

Overdoor. An ornamental finish at the head of a door. A pediment, *q.v.* Also see *Door Head*.

Overhanging Eaves. Common rafters continued over the wall plate to project over the face of the wall. See *Eaves* and *Sprocket*.

Overhead Canter. A log turner, *q.v.*

Overhead Screed. See *Flying Screed*.

Overlap. A ridge-like appearance on the surface of plywood due to overlapping of inner plies when joined plies are used.

Overlay. Applied to any thin covering, as strip flooring, etc.

Overmantel. The top part of the ornamental front to a fireplace. See *Mantel*.

Overrun. The difference between the actual output and the calculated output in conversion.

Overrunner. A large hoop used when assembling the staves for a cask.

Oversailing Floor. A floor projecting or cantilevered over the support, as is common for the first floor in the fronts of half-timbered buildings.

Overside Delivery. The process of unloading ships to a *lighter* or to a smaller vessel.

Overstorey. Clerestory, *q.v.*

Ovolo. A convex moulding in the form of a quarter circle or ellipse with listels.

OVOLO

Owenia. See *Rose Almond*.

Owowa. Obeche, *q.v.*

Oxandra. See *Lancewood*.

Oxford Frame. A picture frame in which the four pieces cross with halving joints, and project at the ends.

Oxter Piece. A piece of ashlering, *q.v.* A tie between ceiling joist and rafter in a cottage roof.

Oxydendrum. See *Sourwood*.

Oxylene. A process of chemically impregnating timber to make it fire-resisting.

Oylet. Eyelet, *q.v.*

Oyster Bay Pine. *Callitris tasmanica*. Tasmania. Not exported.

Oyster Shell or **Wood.** Thin oblique cross sections used as veneers in fancy cabinet work. The wood must be dense and of uniform texture, such as laburnum or yew. The complete growth rings are shown as inlays, hence they are usually cross sections of branches.

Oyster Trees. Mangrove trees in salt water to the roots of which molluscs attach themselves.

P

P. Abbreviation for *planed*; also used in combination, as *p and chd* for *planed and chamfered*.

Pa. Chinese alder. See *Alder*.

Pacara. *Enterolobium timbonva* (*H.*). S. America. Light brown. Moderately light and soft, not strong or durable. Used for joinery and furniture. Little imported. S.G. ·52.

Pace. A portion of a floor raised above the general level. A very low platform.

Pacific Maple. *Acer macrophyllum.* See *Maple* (*Oregon*).

Pacific Myrtle. See *Myrtle*.

Pacific Red Cedar. Western Red Cedar, *q.v.*

Pacific Walnut. New Guinea Walnut. See *Walnut* and *Dao*.

Package Kitchen. A compact unit for small kitchens, containing cooker, water-heater, refrigerator, sink, and storage shelves. It is made of wood or metal.

Packet Boat. A mail boat.

Packing. Material used for filling, levelling, or making solid. The act of packing or making solid.

Packing Cases. Large boxes in which goods are packed for transit. Small ones are called boxes or cases. Firewood, ends, and plywood are used for this type of work.

Pack Saddle Roof. See *Saddleback Roof*.

Pad. 1. A handle. 2. See *Tool Pad*. 3. See *Pressure Pad*. 4. A pannier used for measuring quantities of fruit.

Padal. Like Paral, *q.v.*, but not so strong.

Padauk or **Padouk.** *Pterocarpus dalbergiodes* (*H.*). Andaman I. Also known as Andaman redwood, Corail, Burmese rosewood, E. Indies mahogany, Lingoa, Yoma, etc. Deep crimson, seasoning to nearly purple, darker stripes. Hard, heavy, very strong and durable, elastic, stable, hard wearing, resists white ant. Interlocked grain; smooth. Seasons readily but must be girdled before felling. Rather difficult to work. Polishes well with careful filling. Sharp arrises should be avoided. Large sizes. A valuable decorative wood. Used for superior joinery, cabinet work, flooring, carriage construction, etc. S.G. ·75; C.W. 3·5. *Pterocarpus macrocarpus.* Burma. Very similar to Andaman, but slightly heavier, harder, darker, and stronger. Used for similar purposes and constructional work. S.G. ·8; C.W. 4. AFRICAN PADAUK, *P. soyauxii.* Similar to Andaman but slightly inferior. Also see *Bijasal* and *Macacahuba*.

Paddle. 1. A small spade-like implement with long handle for propelling a canoe, etc. A short broad-bladed oar used without rowlock. 2. One of the boards of a paddle wheel for propelling a ship. 3. A small sluice.

Paddle Wood. *Aspidosperma sp.* (*H.*). B. Guiana. Also called Yaruru. Yellowish brown. Hard, heavy, strong, and flexible. Open grain. Used for tools, paddles, etc. S.G. ·8.

Pader or **Padri.** Paral, *q.v.*

Padlock. A detachable lock used with hasp and staple for securing gates and doors from one side only.

Pad Saw. A saw with very narrow blade that is housed in the handle when not in use. It is used for small circular work. See *Keyhole Saw.*

Pagatpat. *Sonneratia apetala* (*H.*). Philippines. Also called Perepat. A mangrove tree. Reddish brown to almost black, with glistening deposit. Hard, heavy, dense, durable, and strong. Fine grain and even texture, fine rays. Salty secretion corrodes iron. Moderate sizes. Little imported. Suitable for heavy construction. S.G. ·7. A similar Indian wood is called Keowra.

Pagoda. A sacred building, temple, or tower, usually consisting of diminishing cylindrical storeys with top pyramidal in shape. They are characteristic of Eastern countries.

Pagoda Tree. See *Kowhai.*

Pahautea. New Zealand cedar. See *Kaikawaka.*

Pahudia. See *Tindalo* and *Ora.*

Pahutan. *Mangifera altissima* (*H.*). Philippines. Dark brown, variegated stripes, lustrous. Hard, heavy, but not durable. Straight and wavy grain, fine texture. Not difficult to season and work, and polishes well. Substitute for walnut. Used for cabinet work, veneers, etc. S.G. ·8 ; C.W. 3.

Pai. Chinese boxwood.

Pail. A round open vessel for carrying liquids. A bucket. It is slightly conical in shape and may be of wood, galvanised iron, or enamelled.

Paired. Applied to two similar things matched on opposite hands.

Pala. *Wrightia tomentosa* and other species (*H.*). India. White, seasoning to lemon yellow, fairly lustrous and smooth. Light, moderately hard, not durable. Straight and wavy grain, fine even texture. Stains easily ; small sizes. Easily seasoned and wrought. Esteemed for carving and turnery. S.G. ·49.

Paladian. Applied to the Renaissance style of architecture introduced by Inigo Jones about the year 1605.

Palafitte. A prehistoric N. Italian lake dwelling built on piles.

Palanquin. An Eastern conveyance consisting of a long box with opening sides and carried by four bearers by means of poles.

Palaquium. See *Nato, Nyatoh, Pali,* and *Taban.*

Paldao. Dao, *q.v.*

Pale Fence or **Palings.** Stakes, pickets, or open boarding forming a fence or enclosure. A pale is a narrow piece of wood.

Palette. A thin shaped board on which an artist mixes his colours.

Pali. *Palaquium ellipticum* or *Dichopsis elliptica* (*H.*). India. Light red to reddish brown, fairly lustrous, feels smooth. Moderately hard, heavy, and durable. Strong and elastic. Straight and wavy grain. Even medium texture. Rather difficult to season and resists treatment. Should be girdled and quickly converted. Not difficult to work. Liable to warp on re-sawing. Used for joinery, cheap

furniture, flooring, planking, cooperage, match-boarding. S.G.
·55; C.W. 2·75. See *Nyatoh*.
Palisade. A fence of pointed stakes or palings.
Palisander. Brazilian rosewood or Jacaranda, *q.v.*
Pallet. **1.** A thin slip of wood (9 in. × 4½ in. × ⅜ in.) built into the
joints of brickwork, etc., and to which joinery is fixed. See *Nogs*.
2. Thin boards for conveying newly moulded bricks to the hacks for

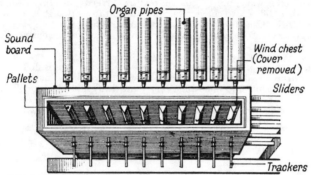

PALLETS TO ORGAN

drying. **3.** Flat wood blade with handle used by potters. **4.** Long
narrow triangular-shaped pieces, or valves, to the sound board of a
pipe organ. They act as valves and admit the wind to the pipes as
required.
Palmettes. Carved ornaments incorporating the Egyptian palm
leaf. The name is sometimes applied to carvings of the honeysuckle
flower. See *Anthemion*.
Palms. A great variety of tropical and sub-tropical trees. They
are inward growers or endogens, and of little timber value in this
country, but used for all purposes in countries of origin. The amount
of pith varies considerably in different species, but the wood is usually
very hard, durable, and sometimes highly decorative. Used for
marine defence work, fancy articles, veneers, inlays, substitute for
ivory, etc. The most important species are given alphabetically.
Palo Blanco. *Calycophyllum multiflorum* (*H.*). S. America. White.
Hard, heavy, strong, durable. Close even grain. The wood is used
for all purposes, especially furniture and turnery, but it is expensive.
Other varieties are Palo amarilla and Palo santo, but they are not so
good. S.G. 1 ; C.W. 4.
Palo Cruz. Not classified. (*H.*) S. America. Light yellow. A
hard, tough, flexible wood, like hickory.
Palo Diabolo. *Krugiodendron ferreum.* C. America. Called Black
Ironwood and Axemaster. One of the hardest and heaviest woods.
Small sizes and little imported. S.G. 1·4.

Palo Maria. Santa Maria, *q.v.*
Palo Morado. Purpleheart, *q.v.*
Palo Mortero. Tipa, *q.v.*
Palo Mulatto. Goncalo Alves, *q.v.*
Palosapis. *Anisoptera thurifera* (*H.*). Philippines. Also called Mayapsis, Mersawa, Bella Rosa, etc. Light yellow, sometimes with pink streaks that fade on exposure, dull. Moderately hard and heavy, fairly strong and elastic, stable but not durable. Straight grain, some crossed and wavy, fairly coarse texture Needs careful seasoning and rather difficult to convert but not difficult to work. Polishes well. Large sizes. Used for interior joinery, construction, planking, veneers, etc. Rotary-cut veneers are also known as Duali, and quarter-sliced as Bayott. S.G. ·6 ; C.W. 3.
Palta. Avocada, *q.v.*
Palu. *Mimusops hexandra* (*H.*). India. Similar to Bullet wood, but coarser. Red to light purplish brown, darker lines, darkens with seasoning. Dull but smooth. Very hard, heavy, strong, tough, and durable Straight, irregular, and interlocked grain. Fairly fine even texture. Distinct handsome grain due to variation in growth rings. Difficult to season, convert, and work. Girdled three years before felling. Polishes well. Used for presses, tools, structural work, rollers, brake blocks. Too hard and heavy for ordinary purposes. S.G. 1·1 ; C.W. 5.
Pametia. See *Riugan.*
Pampres. A carved ornamentation of vine and leaves.
Pan. Abbreviation for *panel.*
Panache. Pendentives formed by a domical roof over a square substructure. It is usually a spherical triangle.
Panaka. *Pleurostylia wightii* (*H.*). Ceylon, Madagascar. Also called Piyari. Reddish yellow, thin darker lines. Hard, heavy, tough, and strong. Fine close grain, smooth, decorative appearance, fine rays. Used for fancy articles. S.G. ·78.
Panama Mahogany. *Swietenia macrophylla.* Central American mahogany. See *Mahogany.*
Panax. See *Silver Basswood* and *Mutari.*
Panda. See *Ovaga.*
P. and S. Abbreviation for *planking and strutting.*
Pane. **1.** A panel. **2.** A square or sheet of glass cut to the required sizes. **3.** The sawn or hewn surface of a log in sided timber.
Panel. A surface distinct from or lying above or below a surrounding border. The recessed parts of framing. A thin wide piece fitted into the members of thicker surrounding framing. See *Beaded Panels, Chancel Screen, Dado, Dwarf Screen, Engaged Column,* and *Fielded Panel.* **2.** A grading of Cottonwood, Gum, and Poplar.
Panel Gauge. A marking gauge with long stem and wide stock, or fence, for gauging panels to width.
Panelled Door or **Framing.** Doors or framing consisting of stiles, rails, and probably muntins, mortised and tenoned together, and with the spaces filled with panels. See *Panel.*
Panelled Linings. Wide linings to door and window openings consisting of panelled framing. See *Finishings.*

Panelling. Panelled surfaces ; especially applied to panelled framing for the walls of rooms. See *Wall Panelling* and *Wainscoting*.

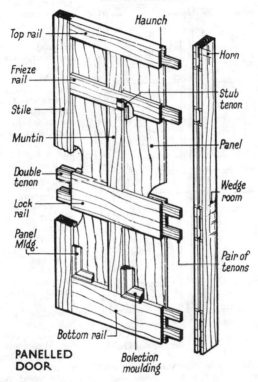

Top rail

Haunch

Horn

Frieze rail

Stub tenon

Stile

Muntin

Panel

Double tenon

Wedge room

Lock rail

Panel Mldg.

Pair of tenons

Bottom rail

PANELLED DOOR

Bolection moulding

Panel Moulding. Loose mouldings mitred round panels and lying below the surface of the framing. See *Elbow Lining* and *Panelled Door*.

Panel Pin. See *Pin* (1).

Panel Plane. The cabinet-maker's name for a metal try-plane.

Panel Planer. A thicknesser, *q.v.*

Panel Raising Machine. A machine for preparing raised panels.

Panel Saw. A hand saw, similar to, but smaller than, the cross-cut. It is about 22 in. long, with 10 to 12 teeth per inch, and is more convenient for cutting wide thin stuff than the tenon saw.

Panel Shaper. A special type of machine for shaped panels and for veneering shaped work.

Panga. Myrobolan, *q.v.*

Panic Bolts. Patent bolts on the special exit doors of public buildings, used in the event of fire or panic. The bolts are operated by pressure at the middle of the door.

Swing doors

Floor spring

PANIC BOLTS

Panisaj. *Terminalia myriocarpa* (*H.*). Assam. Also called Hollock, Lepcha, and Sunloch. Light brown, beautiful darker striations. Darkens with age ; lustrous, smooth. Fairly heavy, but hard and strong, not durable. Straight and wavy grain, fiddleback in rift-sawn wood, coarse texture. Seasons fairly well but liable to surface cracks. Easier to work when green. Similar to Laurel, *T. tomentosa*, q.v. Used for furniture, plywood, interior joinery and construction. S.G. ·75 ; C.W. 4.

Pannier. 1. A wicker basket carried on the back of a person or animal. **2.** A corbel.

Pansh. Abbreviation for the name of the botanist Panshin.

Pantechnicon. A large van for removing furniture.

Pantograph. An appliance consisting of four light wood rods pivoted together and used for reproducing a drawing to a different scale.

Pantry Window.

Pants. Same as spats, *q.v.*

Pao. A prefix to numerous Central American woods. Pao Amarello. Sateen wood, *q.v.* Pao Branco, *Auxemma gardneriana.* Chocolate to purplish brown, variegated. Fades a little on exposure ; lustrous, smooth. Fairly hard, heavy, strong, and durable. Straight grain, coarse texture. Easily wrought. Resembles black walnut and used for similar purposes. S.G. ·7 ; C.W. 2·5. Pao Concha. Also called Shellwood and Ropala, *q.v.* Pao d'Alho, *Gallesia scorododendron.* Called garlic wood because of odour. Greyish white. Fairly light and soft. Irregular grain ; coarse, uneven texture. Difficult to work smooth because of laminated structure. Held in little esteem.

S.G. ·58. PAO D'ARCO. Ipé or Lapacho, *q.v.* PAO D'OLEO, *Copaifera guianensis.* Red, lustrous, smooth. Hard, heavy, strong, durable. Distinct rays. Used for furniture, sleepers, masts. S.G. ·9. PAO FERRO, *Cæsalpinia sp.* Brazilian ironwood, *q.v.* Resembles African blackwood. PAO MARFIN. Ivorywood, *q.v.* PAO MULLATO. See *Degame.* PAO POMBO. Tapirira, *q.v.* PAO ROXO. Purpleheart, *q.v.* PAO SETIM. See *Satinwood.* PAO SANTA. Santowood, *q.v.* Also see *Pau.*

Pap. The outlet to an eaves gutter.

Papaw. *Carica papaya* (*H.*). Sub-tropical, widespread. Straw colour. Very light and soft. Substitute for Balsa but heavier. Important for fruit. S.G. ·35.

Paper-bark Tree. *Melaleuca leucadendron* (*H.*). Queensland. Hard, heavy, durable. Used for piles and underground work. See *Tea Tree.*

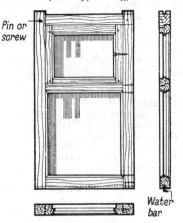

Pin or screw

Water bar

PANTRY WINDOW

Paper Birch. *Betula papyrifera.* See *Birch.*

Paper Joint. Placing paper in a glued joint, so that it is easily broken, for temporary work.

Paper Wood. Short lengths of round whitewood for wood pulp.

Papilionaceæ. Sub-family of Leguminosæ and includes Andira, Bowdichia, Brya, Butea, Centrolobium, Dalbergia, Diphysa, Diplotropis, Dipteryx, Erythrina, Eysenhardtia, Ichthyomethia (Piscidia), Machærium, Millettia, Myrocarpus, Myrospermum, Myroxylon, Ougeinia, Platycyamus, Platymiscium, Platypodium, Pongamia, Pterocarpus, Sweetia, Tipuana, Torresea, Vouacapoua.

Papita. *Sterculia campanulata* (*H.*). India. Ivory white, sometimes dark streaks due to a fungus, very lustrous. Light, soft, not durable, subject to fungi and insect attack. Moderately strong and flexible. Straight grain, even coarse texture, large rays. Conspicuous fleck on radial surface. Must be seasoned quickly. Suitable for matchboarding, cooperage, cases, etc. S.G. ·33 ; C.W. 2.

Papo. Ekpaghoi, *q.v.*

Papri. Indian Elm, *q.v.*

Papuan Walnut. New Guinea walnut. See *Dao.*

Paque. 1. *Ostrya guatemalense* (*H.*). C. America. Also called Guapaque. Brown. Very similar to hornbeam in properties and characteristics. 2. *Diospyros. spp.* Mexico. A species of ebony.

P.a.r. Abbreviation for *planed all round.*

Parabola. A section of a cone parallel to a side. See *Conic Sections.* Area $= \frac{2}{3}$ base $\times$ height.

Parabolic. Shaped like a parabola, as an arch or moulding.

Paraboloid. A solid generated by a parabola revolving about its axis.
Paradise Wood. Simaruba, *q.v.* Also called Parahyba.
Paral. *Stereospermum spp.* (*H.*). India, Burma. Also called Padri.
Numerous local names. Yellowish brown, darker streaks, lustrous.
Moderately hard, heavy, strong, durable, and elastic. Strong in shear
and shock resistance. Straight grain, mottled, variable texture.
Difficult to split. Seasons readily, easily treated, and resists insects.
Easily wrought before seasoning, but a hard crystal deposit, when
dry, dulls tools. Used for sleepers, flooring, plywood, cheap furniture.
S.G. ·65 ; C.W. 3·5.
Parallel Gutter. See *Box Gutter*.
Parallelogram. A quadrilateral, or four-sided plane figure, with
opposite sides parallel. Area = base × vertical height.
Parallelogram of Forces. The graphical method of finding the
components of a force, or the resultant of two forces, by means of a
parallelogram. See *Graphics*.
Parallel Strips. Winding strips, *q.v.*
Para Mahogany. Crabwood, *q.v.*
Parament. The furniture, etc., of a state room or apartment.
Parana Pine. *Araucaria brasiliana* (*S.*). S. America. Variegated
brown with occasional bright red streaks. Subject to sap-stain.
Variable in weight and hardness. Fairly strong but not durable.
Straight grain, uniform texture. Easily wrought. Requires careful
seasoning. Polishes well. Useful wood for general interior purposes.
S.G. about ·56, specimens up to ·8 ; C.W. 2.

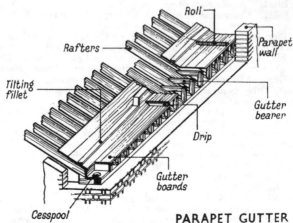

PARAPET GUTTER

Parapet Gutter. A gutter behind a parapet wall. See *Box Gutter*.
Parari. Paral, *q.v.*
Parashorea. See *Bagtikan, Damar Laut, Mayapis, Resak, Tavoy,
White Seraya*, and *White Lauan*.

Parasite. A destructive organism living on the tissue of another organism such as wood.

Parastas. A pilaster, *q.v.*

Paratecoma. See *Peroba*.

Paratracheal. See *Parenchyma*.

Para Wood. Brazilwood, *q.v.*

Parcel. A term loosely used in the import trade for quantities of softwood.

Parchment Panel. Same as linen-fold panel, *q.v.*

Parclose. **1.** A church screen with open tracery panels so that it only partly screens the enclosure. See *Chancel Screen*. **2.** A parapet round a gallery.

Pardillo. Bocote and Laurel, *q.v.* The name is applied to various species of *Cordia*, especially *C. alliodora*.

Pare or **Paring.** To trim or cut lightly. Reducing little by little.

Parenchyma. Wood tissue chiefly concerned with the distribution and storage of carbohydrates. It consists of thin-walled, brick-shaped, cells with simple pits, and may be axial or radial. The arrangement is a useful aid in identification. It is sometimes called soft tissue. Four types are recognised : *diffuse*, irregular single strands ; *terminal*, narrow layers occurring at the end of a season's growth ; *metatracheal*, in tangential layers independent of the vessels and appearing as lighter concentric bands, but they are sometimes broad and conspicuous ; *paratracheal*, associated with the vessels, with several variations known as *aliform*, *confluent*, and *vasicentric*. Also see *Rays*.

Parinari. See *Dak Ballow, Johore " Teak,"* Merbatu.

Paring Chisel. A long slender bevel-edged chisel. It is not intended to be used with the mallet because of the slender blade. See *Pare*, *Chisels*, and *Gouges*.

Parishia. *Parishia insignis* (*H.*). India, Burma, Andaman I. Also called Red Dhup. Light pinkish grey to pale brown, lustrous but soon dulls, due to fungal stain. Light and soft, not strong or durable. Straight grain ; fibrous, coarse, even texture. Seasons well and easily wrought. Suitable for plywood, match-boarding, etc. S.G. ·45 ; C.W. 2.

Park. Abbreviation for the name of the botanist Parker.

Parl. Abbreviation for the name of the botanist Parlatore.

Parliament Hinge. A hinge shaped like the letter H. The knuckle projects considerably over the face of the door when closed, and allows the door to clear a projection, such as an architrave, when open.

Paroba. *Achras gonocarpa* (*H.*). Brazil. Yellowish white. Heavy, fairly soft but strong. Close grain, even texture. Easily wrought. Useful wood but little imported. Used for general purposes, ship-building, etc. S.G. ·75. *Sapota speciosa*

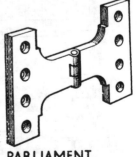

PARLIAMENT HINGE

resembles pencil cedar. Reddish, close even grain. Excellent wood, easily wrought. S.G. ·7 ; C.W. 2.

Paroto. Guanacaste, *q.v.*

Parquet or **Parquetry.** Surfaces formed of small pieces of varied coloured woods, and usually in geometrical designs. Often applied to floors.

Parrel or **Parral. 1.** The decorative parts of a chimney piece. **2.** A fitting to keep the yard close to the mast of a ship. It consists of ribs and trucks to form a kind of collar to enable yards to slide easily up and down the mast.

Parsipu. *Thespesia populnea* (*H.*). India. Salmon colour. Hard, heavy, strong, tough, durable. Cross grain, close texture, distinct rays.

Parsor. A gimlet, *q.v.*

Parted Pattern. A two-piece pattern for convenience of moulding. See *Pattern*.

PARQUET BORDER

Parting Bead. A thin bead separating sliding sashes. See *Sash and Frame*.

Parting Slip. A thin slip separating the weights for sliding sashes. See *Boxed Frame* and *Mullion*.

Parting Tool. A V-shaped gouge. A special chisel for cutting off the work when turning.

Partition Boards. Match-boarding dressed both sides.

Partitions. Interior division walls. They may be of wood, lath and

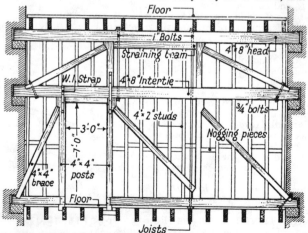

PARTITION TO CARRY FLOOR

plaster, glazed framing, bricks, patent blocks, concrete, or steel, and they may be bearing, collapsible, fixed, folding, or non-bearing.

Partners. Large pieces fixed in frames round masts, capstans, bowsprit, and pumps in large wooden ships.

Partridge Wood. *Cæsalpinia grenadillo* (*H*.). C. America. Also called Coffee Wood, Brown Ebony, and Mesquite. Variegated chocolate brown with fine lighter lines. Very hard, heavy, strong and durable. Interwoven grain, medium texture. Difficult to work. Used for structural work, superior decorative work, fancy articles. S.G. 1·1 ; C.W. 4·5. Also see *Angelin* and *Cabbage-bark*. The name partridge-wood is also applied to oak that has been attacked by the fungus *Stereum frustulosum*.

Party Fence. One separating adjacent properties or one standing on lands of different owners.

Pasania. *Quercus sp.* (*H*.). Formosa. Like chestnut but with medullary rays. Excellent wood for cabinet work and superior joinery. S.G. ·64 ; C.W. 3. Also see *Mempening*.

Patache. An early type of Mediterranean sailing vessel, somewhat like a brigantine.

Patagonula. Guayabi, *q.v.*

Patand. A sill, or plate, resting on the ground and supporting vertical timbers.

Patapsco. A trade name for a wavy mottled maple, with blister but not bird's eye.

Patch. 1. A rough area on veneer due to oblique or twisted grain. 2. An inlay to remedy a defect. See *Little Joiner*. 3. A repair to plywood.

Pateræ. Circular or elliptical flat ornamentation on friezes, furniture, etc., usually in the form of a cup or dish.

Patin. 1. A patand, *q.v.* 2. A plate used in religious ceremonies.

Patina. 1. Gloss on woodwork produced by age. 2. A bowl.

Patten. 1. The base of a column or pillar. 2. A patten-sole is a wood sole for footwear, very similar to a clog-sole.

Patter. A thick wood float for tamping and levelling cement surfaces.

Pattern. A wood or plaster model for a casting.

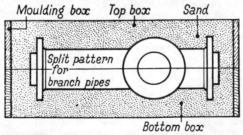

PATTERN

Pattern-shop. A workshop where the pattern-maker prepares wood models from which metal castings are made in the foundry.

Pattern Staining. Dark bands on plaster ceilings through moisture depositing dirt due to the unequal conductivity of the background, *i.e.*, joists and laths.

Patternwood. See *Mujua*.

Pau. See *Pao*.

Pau amarello. *Euxylophora paraensis* (*H*.). Brazilian satinwood. Clear yellow, darkening with exposure, fairly lustrous, smooth. Hard, heavy, strong, and durable. Straight and curly grain. Difficult to work. Little imported. S.G. ·8 ; C.W. 4. See *Satinwood*.

Pau Brazil. Brazilwood, *q.v.*

Pauji. *Mouriria pseudo-geminata* (*H*.). C. America. Variegated yellow brown, fine white lines. Hard, heavy, strong, but not durable. Straight and roey grain, medium texture. Not difficult to work, and polishes well. Used for furniture, construction, etc. S.G. ·78 ; C.W. 3.

Paulownia. See *Kiri*.

Pau roxo. *Peltogyne densiflora* (*H*.). Brazil. Dark brown, to purple on exposure, fairly lustrous, smooth. Very hard. Heavy, strong, and durable. Straight and wavy grain, ribbon effect. Not difficult to work ; polishes well. Little imported. S.G. ·8 ; C.W. 3.

Pausinystalia. See *Idagbon*.

Pav. Abbreviation for the name of the botanist Pavon.

Pavilion. A single-storey building for entertainments or for a sports field. A light ornamental building, often of wood, for the players on sports fields. The name is also applied to more pretentious buildings for theatres and for spectators on sports fields.

Pavilion Roof. A polygonal roof, *q.v.* Also applied to any form of roof curved in the direction of the pitch, especially when for places of entertainment.

Paving. The materials for or the act of forming a pavement or road. Numerous woods are used in the form of blocks for this purpose.

Paving-block Saw. A machine with a number of cross-cut saws spaced at equal distances apart for cutting paving blocks.

Pavisade. A partition or fence on a ship.

Paxillus. A dry rot fungus. See *Dry Rot*.

Payena. See *Belian, Betis, Nyatoh*.

P.C. Abbreviation for *prime cost*.

Pcs. Abbreviation for *pieces*.

P.e. Abbreviation for *plain edged*.

Peacock's Eye. Small localised circular markings in certain woods, like bird's eye in maple, that add considerably to the decorative value. It is now suggested that they are caused by fungi, but embryo buds, insects, and birds are also claimed to be the cause. See *Bird's Eye* and *Figure*.

Peak. See *Throat* (3).

Peak Joint. The joint at the apex of a roof truss.

Pear. *Pyrus communis* (*H*.). Europe. Wild pear or Choke pear. Yellowish brown. Like Lime but harder and tougher. Fine even grain and texture. Moderately soft, stable ; strong, not durable. Annual rings distinct. Large mottle figure. Small sizes. Easily wrought. Used for drawing instruments, carving, cabinet work, domestic utensils, etc. S.G. ·7 ; C.W. 2. ALLIGATOR PEAR, *Persea americana*. Also called Avocado and Butter Pear. Light reddish brown, fine texture. Easily wrought. S.G. ·65 ; C.W. 2. CHINESE PEAR, *Pyrus sinensis*. Resembles yellow boxwood. Used for engravings, fancy articles, etc. DOGWOOD PEAR, *Pomaderris apetala*. Tasmania. Light brown. A small tree, but burrs on the trunk produce beautiful figure for cabinet work. HARD PEAR, *Olinia cymosa*. S. Africa. A heavy, strong, durable wood used for structural work, fencing, etc. NATIVE PEAR, *Xylomelum occidentale*. W. Australia. Deep red, lustrous. Distinct rays. Beautiful cabinet wood but local demands prevent export. S.G. ·75. NIGERIAN PEARWOOD. See *Ofe* and *Obobonufua*. WHITE PEAR, *Apodytes dimidiata*. Kenya and S. Africa. Greyish brown, purplish streaks. Fairly heavy but hard and strong. Fine even close grain and texture. Large sizes. Little imported as yet. Used chiefly for wagon building. S.G. ·75. WILD PEAR. The service tree, *q.v.*

Peavy. Similar to a cant hook, but with a pike instead of toe ring and lip.

Pecan. *Carya pecan, C. aquatica.* See *Hickory*.

Pecky. Applied to timber showing symptoms of decay, especially to bald cypress. Wood with localised patches of decay.

Pedals. The part of a pipe organ operated by the feet of the organist.

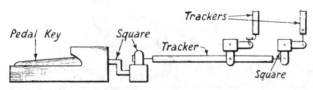

PEDALS TO ORGAN

Pedestal. 1. A base for a column, statue, or other ornament. See *urn*. 2. The framing supporting a counter or knee-hole desk, or table. 3. A bearing for a shaft in machinery. 4. A circular block above or below the wheel plate of a vehicle to build up the thickness.

Pedestal Desk. See *Kneehole Desk*.

Pedestal Sideboard. One with two side carcases connected by a middle portion.

Pediment. A triangular or segmental ornamental head to an opening, as a doorway ; or over the entablature at porticoes and ends of buildings in Classic architecture. See *Raking Mouldings*. (See illus., p. 394.)

Peduncle. 1. The main stalk that bears flowers or fruit of plants. 2. A stalk-like connection.

Pedunculate Oak. English Oak, *q.v.*

Peel or **Peeling.** 1. Removing the bark of logs and trees. 2. See *Veneers* and *Rotary Cutting*.

Peeler. A machine for cutting rotary veneers, *q.v.*

Peen. The wedge-shaped end of a hammer head.

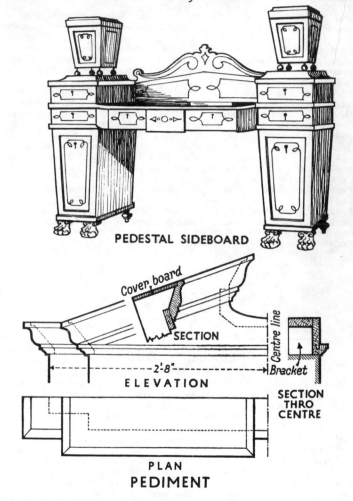

PEDESTAL SIDEBOARD

PEDIMENT

Peg and Cup. Brass dowels and sleeves, for joints in patterns.
Pegged. Same as pinned, *q.v.*
Pegged Out. A term used when a building site has been set out with pegs, *q.v.*
Peggy. Same as pecky, *q.v.*

394

Pegs. 1. Slender pointed pieces of wood for driving into the ground when setting out a site for building purposes, or with hook-shaped head for securing cords to tents, etc.
2. Pins for securing the joints of framing. 3. A wood hook or pin for hats and coats. 4. Pins on which strings are stretched in musical instruments. They are usually of rosewood or palisander for violins.
5. A two-pronged appliance for attaching clothes to a line or rope.
6. A tapered pin, or bung, for the hole in a cask.

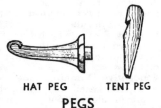

HAT PEG   **TENT PEG**

PEGS

Peg Stay. A casement stay with holes engaging with a pin to secure the casement when open.
Peg Tooth. See *Fleme Tooth*.
Pegui. *Canjocas brasiliensis* (*H.*). Brazil. Yellowish brown. Heavy, fairly hard, strong, and durable. Straight grain, porous. Used for the same purposes as greenheart. S.G. 1.
Pegunny. *Bauhinia hookeri* (*H.*). Queensland. Dark brown to black, variegated, creamy sapwood. Very hard, heavy, and tough. Fine close grain and texture. Difficult to work ; polishes well. Used for fancy articles, walking sticks, xylophones, etc. S.G. 1·2 ; C.W. 5.
Pehlitzu. Chinese beech.
Pele. A fortified tower.
Pelewan. *Tristania spp.* Malay. Usually marketed with Penaga, *q.v.*
Pellegr. Abbreviation for the name of the botanist Pellegrin.
Pellet. 1. A slightly tapered cylindrical plug used as a *little joiner*, *q.v.*, for covering sunken screws. 2. Small flat cylindrical or hemispherical projections used for ornamentation on mouldings, etc.
Pellet Moulding. One consisting of a series of small hemispherical projections.
Pelmet. An ornamental head, or valance, on the inside of a window, to hide the blind and curtain fittings.
Peltogyne. See *Amaranth, Guarubu, Purpleheart,* and *Pau roxo*.
Peltophorum. See *Cana fistola* and *Moruro*.
Pembroke. A small table with flaps on either side supported by hinged brackets.
Pen. A small enclosure for fowl, sheep, cattle, etc. A coop or fold.
Penaga. Gangau, *q.v. Mesua sp.* and *Calophyllum sp.* Malay.
Penak. Chengal, *q.v.*
Pencil Cedar. *Juniperus spp.* See *Cedar*.
Pencil Roe. Irregular fibre growth producing regular narrow alternate bands of different shades of colour in quarter-sawn decorative hard-woods.
Penda. *Xanthostemon spp.* (*H.*). Queensland. Yellow to pale red, odorous, very smooth and lustrous. Very hard, heavy, tough, and strong, but brittle and not durable. Fine dense grain and texture ; some interwoven and wavy grain. Difficult to work. Yellow penda is softer than the other species and more suitable for general purposes.

Used for rollers, dance floors, protected structural work. S.G. 1; C.W. 4 to 5.

Pendant. 1. An ornamental finish at the bottom of a suspended post as to a newel. Ornamentation suspended from a roof or ceiling. 2. Applied to either of a pair of statues or ornaments.

Pendant Post. The hammer post of a hammer-beam roof truss, continued below the hammer beam as a decorative feature. Also sometimes applied to the lower side or wall post of a hammer-beam roof truss, *q.v.*

Pendentive Dome or **Roof.** A dome or dome-shaped vault over a rectangular chamber in which the corners are filled with pendentives to provide a circular or polygonal seating for the dome.

Pendentives. The spherical triangles formed at the corners by the intersection of two vaults, or by a dome springing from a rectangular base. See *Dome* and *Vaults*.

Pendulum Saw. A machine cross-cut saw that is drawn across the stationary wood in the process of cutting, by swinging from the point of suspension like a pendulum. See *Swinging Cross-cut*.

Pendulum Slip. A parting slip, *q.v.*

Penetration. A term referring to the relative resistance of woods to impregnation by preservatives. Some woods are easily injected, others are very resistant, according to the structure. Usually the sapwood is more easily impregnated than the heartwood. Bark is very resistant and should be completely removed before treatment. Summer wood is usually more resistant than springwood, especially in ring-porous woods. When there is little difference between spring and summer wood the absorption is uniform. In a species the absorption varies inversely as the density but the relation does not hold for woods of different species. The condition and arrangement of the cells also governs the uniformity and amount of preservative absorbed. Only experience can determine the absorptive properties of a species. Douglas fir and some species of spruce are difficult to treat, while long-leaf pine, beech, and red oak are readily injected. A durable wood may resist impregnation whilst a non-durable wood may be easily treated and made more durable than the durable wood. See *Preservation*.

Penfold. A pound or enclosure for cattle.

Pengarawan. Merawan, *q.v.*

Peniophora. A fungus, *P. giganteum*, that often attacks felled soft-wood logs and the sapwood of stacked wood. The decay shows as streaks of yellowish brown starting at the ends of logs or timber.

Penkwa. Sapele, *q.v.*

Pennantia. See *Brown Beech*.

Penstock. A sluice over a water-wheel. A wood trough or conduit for conveying water.

Pent. A hood along a wall to protect the wall from driving rain. See *Pent Roof*.

Pentace. See *Thitka* and *Thitsho*.

Pentaclethera. See *Ogpagha*.

Pentacme. See *White Lauan, Ingyin,* or *Sal*.

Pentagon. A polygon with five sides. Area$=$side$^2 \times 1 \cdot 72$ for a regular figure.

Penthouse or **Pentee.**   A protecting hood to an opening.   Also same as pent, *q.v.*

Pent Roof.   Same as lean-to roof, *q.v.*

Pepperidge.   See *Gum (Black)* and *Tupeloe Gum.*

Peppermint.   *Agonis flexuosa (H.).*   W. Australia.   The willow myrtle.   Of little commercial importance except for by-products. Several species of Eucalyptus, chiefly second-class woods, are also called by this name.   BLACK PEPPERMINT, *Eucalyptus amygdalina.* Tasmania and Victoria.   Light brown.   Moderately hard and heavy. Fissile, fairly durable.   Fairly close, straight grain.   Used for joinery, construction, shingles, carriage building, etc.   S.G. ·83.   PEPPERMINT GUM, *E. odorata.*   Yellowish to brownish white.   Very hard, heavy, tough, and durable.   Small tree for a Eucalyptus and often with hollow trunk.   Used for wagon construction.   S.G. 1.

Pepperwood.   California myrtle.   See *Myrtle.*

Pequa.   Obobonekhui, *q.v.*

Pequia.   See *Piquia.*

Pera.   See *Jiqui.*

Perambulator.   A light carriage for a child, pushed by hand.

Perawan.   See *Meranti.*

Perch.   1. A bracket or corbel.   2. The understays between the axles of a vehicle.   3. A rail, with top edge rounded, on which fowl rest in pens and cotes.   4. A measure of length of 5½ yds.   5. A steer, *q.v.*

Pereira.   *Platycyamus regnellii (H.).*   C. America.   Variegated rose red, fading to yellowish brown.   Fairly hard, heavy, tough. strong, and durable.   Straight grain, coarse texture, smooth.   Rather difficult to work.   Used for structural work, sleepers, construction, cooperage, etc.   S.G. ·75 ;   C.W. 4.

Perepat.   Pagatpat, *q.v.*

Perfections.   Shingles, 18 in. long.

Pergola.   An arbour or covered walk formed of plants trained on trellis work.

Pericopsis.   See *Nedun.*

Perimeter.   The bounding line of a plane figure.   The distance round anything ;   the girth or circumference or the sum of the sides.

Period.   Applied to architecture and furniture characteristic of any particular past period, such as Tudor, Elizabethan, Jacobean, Queen Anne, etc.   The English periods are approximately as follows : Tudor, 1485-1603 ;   Jacobean, 1603-1649 ;   Cromwellian 1649-1660 ; Charles II, 1660-1685 ;   William and Mary, 1689-1702 ;   Queen Anne, 1702-1714 ;   Early Georgian, 1714-1727 ;   Chippendale, 1745-1780 ;   Adam, 1760-1792 ;   Hepplewhite, 1760-1790 ;   Sheraton, 1790-1810.   The French periods are :   Renaissance, 1515-1600 ; Louis XIII to Louis XVI, 1610-1774 ;   Directoire, 1795 ;   Empire, 1805 ;   Restoration, 1815.

Perkins Jig.   An appliance for housing staircase strings on a spindle machine.

Permali.   A registered name for impregnated wood.   See *Improved Wood.*

Permanent Set.   See *Elasticity* and *Strain.*

Pernambuco Wood. Brazilwood, *q.v.*

Peroba. A prefix to several Central and South American woods, but chiefly confined to two groups : Aspidosperma and Bignoniaceæ. The woods are used for nearly all purposes. PEROBA ROSA, *Aspidosperma polyneuron.* Variegated rose red to yellowish pink with purple streaks. Variable in weight and hardness. Strong but rather brittle. Varies in durability. Straight, cross, and roey grain, fine uniform texture. Excellent wood for construction, joinery, and furniture. Used for structural work. S.G. about ·8 ; C.W. 3. See *Guatambu.* IPE PEROBA, *Paratecoma*, or *Tecoma*, *peroba*. White peroba. Also called Fygrin, Peroba amarello, etc. Light brown or pale golden olive, variegated, lustrous. Hard, heavy, tough, strong, durable, and stable. Beautiful fine roey grain, medium texture, smooth. Fairly easy to work. Requires careful seasoning. Poisonous splinters. Used for veneers, superior joinery, and cabinet work. S.G. ·75 ; C.W. 3.

Perpendiculars. Vertical imaginary lines for calculating tonnage of a ship.

Perpendicular Style. The latest of the Gothic styles of architecture (fifteenth and sixteenth century). The special features are large window areas and perpendicular lines in the tracery. Also called *Late Pointed.*

Perr. Abbreviation for the name of the botanist Perrottet.

Perron. A staircase with large landing on the outside of a building. A platform at the entrance of a large building, approached by steps, especially when with double-approach flights.

Pers. Abbreviation for the name of the botanist Persoon.

Persea. Laurel family. See *Pear (Alligator)*, *Avocada, Nan.*

Persian Box. Common box. See *Boxwood.*

Persian Lilac. See *Lilac.*

Persienne. An external blind to a window, consisting of adjustable slats in a frame.

Persimmon. *Diospyros virginiana (H.).* C. America. Ebony family. The Virginian Date Palm. Brown to almost black, variegated. Hard, heavy, tough, strong, elastic. Close, roey and wavy grain, fine rays. Difficult to work. Small sizes. Used for shuttles, golf-stick heads, shoe lasts, turnery, cabinet work, variegated wood for veneers. S.G. ·8 ; C.W. 5. BLACK PERSIMMON, *D. texana.* Similar to above and used for similar purposes. CHINESE PERSIMMON. *D. kaki.* Similar to above but heavier, and black. GREY PERSIMMON, *Maba fasiculosa* and *Diospyros pentamera.* Queensland. Similar to American persimmon but lighter in colour.

Perspective. A pictorial drawing of an object by radial projection. The relative proportions and positions appear as in a photograph. See *Wheel.*

Peruvian. See *Mahogany* and *South American Cedar.*

Petaling. *Ochanostachys amentacea (H.).* E. Indies. Also called *Kemap* and *Mentati.* Brown, to light brown on exposure. Hard, very heavy. Strong but very fissile. Durable and resists insects but not all fungi. Fine grain and rays. Requires careful seasoning, but not difficult to work. Rather small sizes. Used for structural and constructional work, furniture, etc. S.G. ·92 ; C.W. 3·5. Also see *Kemap*.

Petchora Fir. *Pinus sylvestris.* See *Redwood.*

Peterebi. *Cordia spp.* (*H.*). C. and S. America. Laurel family, several species. Very variable in colour and weight. Fragrant and durable, with beautiful grain, and polishes well. Excellent woods, used for superior joinery, cabinet work, etc. S.G. about ·6 ; C.W. 4.

Peterlinium. A proprietary preservative for wood.

Petersburg or **Petrograd Standard.** A measure of softwoods based on 120 pieces 12 ft. × 11 in. × 1½ in., or 165 cub. ft., equal to 1,980 ft. board measure. See *Standard.*

Petir. Sepetir, *q.v.*

Petitia. Fiddlewood, *q.v.* Also see *Roble.*

Petrails. Strong beams in timber-framed buildings. See *Post and Petrail.*

Pet. Std. Abbreviation for *Petrograd Standard.*

Petthan. *Heterophragma adenophyllum* (*H.*). Burma. Also called Karen. Yellowish red with green tinge. Very hard ; heavy, strong, elastic, durable. Not very stable. Firm close texture, ray flecks. Properties and qualities of greenheart. S.G. ·8 ; C.W. 4. See *Waras.*

Petwun. Trincomalee Wood, *q.v.*

Pew Hinge. A parliament hinge, *q.v.*, with ornamental knuckle.

Pews. Enclosed seats in churches.

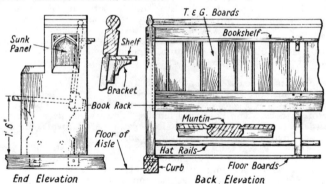

PEW

Phebalium. See *Satinwood.*

Phellinus cryptarum. A dry rot fungus that causes decay of oak in buildings.

Phialodiscus. See *Ukpenekwi.*

Philippine I. A large variety of woods, characteristic of the E. Indies, are exported from these islands. The important woods are described alphabetically.

Philippine Mahogany. See *Lauan.*

Philippine Teak. See *Molave.*

Phlebotaenia. See *Vera*.

Phloem. The inner bark of trees, or bast.

Phœbe. A genus of the Laurel family. See *Guambo, Imbuia* or *Imbuya, Medang*.

Pholas. A small marine shell fish, like a mussel, very destructive to timber.

Photomicrograph. A photograph of a small object as viewed under a microscope. Wood specimens require careful preparation to avoid tearing the very thin sections. The following describes the method usually adopted for scientific investigation. A ¾ in. cube is boiled in water to drive out air. It is then placed in a 50 % mixture of alcohol and glycerine to preserve and soften it. Very hard specimens are placed in a standard solution of hydrogen fluoride for several days to break down the lignin. The specimen is then washed in running water for two days and again soaked in alcohol and glycerine. Three sections are sliced by a microtome, two adjacent sides and one end grain. The slices may be as thin as ·01 mm. They are washed in distilled water and stained, usually with safranine solution, and again washed with alcohol. They are then treated with cedar-wood oil or xylol, and then mounted on a slide with a drop of canada balsam and protected by a cover glass. See *Identification*.

Phyllanthus emblica (*H.*). India. No market name as yet. Numerous vernacular names : ambal, daula, onra, osirka, nelli, etc. Dull red to reddish brown, purplish cast. Dull but smooth. Fairly hard, heavy, and durable. Straight and wavy grain, slightly interlocked. Large rays, fiddleback. Coarse uneven texture. Difficult to season but not difficult to work. Rift-sawn wood, very decorative. Under investigation. Suitable for furniture, superior joinery, etc. S.G. ·8 ; C.W. 3·5. There are other species, but they are of little commercial value.

Phyllocladus. See *Celery Top Pine* and *Tanekaha*.

Phyllostylon. See *Boxwood* (*San Domingan*).

Phymatodes. A species of Longhorn bettle, destructive to beech.

Phytolaccaceæ. The Pokeweed family. The woods, generally, are of little commercial value. Includes Agathestis, Barbenia, Ercilla, Gallesia, Phytolacca, Rhabdodendron, Seguieria.

Piano-stool. A seat without back or arms for a pianist. The seat may be adjustable for height or the lower part may be a music cabinet.

Picea. A genus including spruce and fir, *q.v.* *P. abies*, European spruce, Baltic whitewood. (See *Fir* and *Whitewood*.) *P. alba, P. canadensis, P. glauca*, N. American white spruce. *P. amabilis*, Silver fir. *P. ascendens*, W. China. Excellent wood. *P. balsamea*, Balm of Gilead fir. *P. engelmanni*, Engelmann or Mountain spruce. *P. excelsa*, see *Fir*. *P. Hondœnsis*, Manchuria. Like redwood. *P. mariana* or *P. nigra*, East N. American black spruce. *P. morinda* or *P. Smithiana*, Himalayan spruce. *P. omorika*, Serbian spruce. *P. rubens*, Red Spruce, E. Canada and U.S.A. *P. rubra*, Canadian red spruce. *P. sitchensis*, Sitka spruce. Also see *Roumanian Pine*.

Picho. Tipa, *q.v.*

Pickaroon. A piked pole used in logging on rivers. Also called a hookaroon.

Picket. Narrow pointed strips of wood used for fencing.

Picking Up. A term used when interlocked or cross grain plucks up or breaks away when the stuff is being wrought.

Pickled Oak. Limed Oak, *q.v.*

Pickling. 1. A term often used for the Bethel process of preservation. 2. Treating old roof timbers against insects and decay.

Pick-up Rod. A rod used in shipbuilding for recording heights.

Picræna. See *Bitterwood, Simarupa.*

Picralinia. See *Erin.*

Pictorial Projection. Any form of drawing that shows the three views of orthographic projection in one view : isometric, oblique, perspective, axiometric, and planometric.

Picture Rail. An ornamental rail round a room from which pictures are suspended. See *Grounds.*

Piece Mould. A mould consisting of several pieces that can be removed separately.

Pie-crust Top. The top of a small tea table with raised moulded edge. See *Pillar Table.*

Piedroit. A pilaster or pier without base and capital.

Pien-ch'ai. *Liquidambar sp.* China. See *Satin Walnut.*

Piend Check. A bird's-mouthed rebate, as in stone steps with flush soffit.

Piend Rafter. A hip rafter.

Pier. 1. A support of brick, stone, or concrete. Same as *post* in wood. 2. A projection on a wall to give additional support to a beam. 3. A landing place, jetty, or wharf projecting for some distance into the water. 4. Narrow solid work between openings in walls.

Pieræna. See *Bitterwood.*

Pier Glass. A long narrow mirror fixed on a pier between two openings in a wall.

Pier Table. One supporting a pier or wall mirror.

Pigeon Ash. New South Wales Mountain Ash, *q.v.*

Pigeon-hole. 1. A small space in a desk or other fitment for storing papers or small objects. 2. An opening to a cote for birds. 3. The spaces between stacked timbers, such as deals. See *Piling* and *Air Seasoning.*

Pigeon Wood. See *Sea Grape.*

Piggery or **Pigsty.** A place in which pigs are kept.

Pig Nut. See *Hickory.*

Pigwood. See *Hog Gum.*

Pike Pole. A long pole, up to 20 ft. long, with steel pike for controlling the logs in river driving.

Pike Staff. Similar to a pike pole, but lighter.

Pilaster. A thin rectangular pier projecting from the face of a wall. It is often of wood and only serves as a decorative feature, as in shop fronts, wall panelling, etc. See *Rolling Shutter.* Ornamental rectangular projections on the face of framing giving an appearance of strength and support as in counter fronts, etc. (See illus., p. 402.)

Pilaster Capping. An ornamental projecting top to a pilaster.

Pilcher's Stop Rot. A proprietary preservative for wood.

Piler. A stacker, *q.v.*

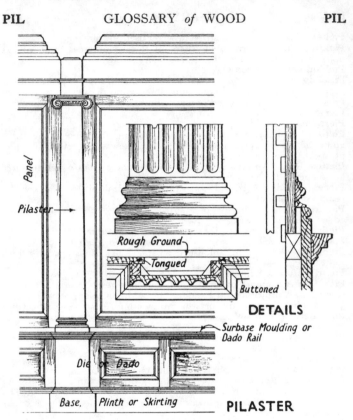

DETAILS

Surbase Moulding or Dado Rail

PILASTER

Piles. Long pointed supports driven into poor soil by a pile-driver to provide a foundation for a structure, or to protect an enclosure for under-water work, etc. Piles may be wood, steel, or concrete, and the last may be formed *in situ*. See *Bearing, Friction, Guide*, and *Sheet Piles*. Also see *Piling* and *Stack*.

Pilg. Abbreviation for the name of the botanist Pilger.

Piling. **1.** The act of driving piles, *q.v.* **2.** Arranging timber in stacks or piles, for storing and seasoning. The various types of stacks are as follows : *Bristol pile* is a large square pile of deals with alternate layers at right angles to each other. The pieces should be pigeon-holed. *Cabinet piles* are arranged so that the lengths gradually diminish toward the top. *Close piling*, without skids, is bad because of staining and decay. *Pigeon-hole piles* have the pieces laid with spaces between, for seasoning. *Stack piles* are about five feet wide with the pieces edge to edge. This does not allow for drying unless they are skidded.

Tier piles are only one piece wide and used for expensive cut woods of wide or large section, and for logs, to allow for easy removal of individual pieces. *Racking* is arranging converted wood on end in racks. The method is good for seasoning but there is little control against warping. See *Stacks*, *Skids*, and *Sanitation*.

Piling Sticks or **Strips.** See *Skids* or *Stickers*.

Pillar. 1. A detached support, or column, *q.v.* In the London Building Act, 1932, the term includes steel columns, stanchions, and struts. 2. Uprights in railway carriage construction and in the framing of vehicles. 3. The vertical structural supports in a ship.

Pillar and Claw. A table with centre pillar and claw feet.

Pillar Table. A small table with central support or pillar with branched feet.

Pillory. See *Stocks* (1).

Pillow. A template. A wood block used as a bearing or support. The support for an axle or shaft. A block supporting the inner end of the bowsprit of a ship.

Pillow Frame. A form of sawing stool, shaped like a tripod.

Pilot House. A small structure on board ship for the helmsman and navigating equipment.

Pilot Nails. Nails used temporarily when erecting shuttering.

Pimentiera. *Trichilia alta* (*H.*). C. America. Light red-brown. Hard, heavy, tough, strong, not durable. Straight grain, fine texture, smooth. Fairly easy to work, polishes well. Like birch and used for similar purposes. S.G. ·72 ; C.W. 2·5.

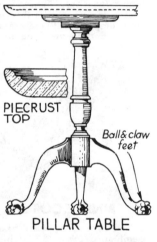

PIECRUST TOP

Ball & claw feet

PILLAR TABLE

Pimento. *Pimenta officinalis* (*H.*). W. Indies. Salmon colour. Very hard, heavy, tough. Firm close texture, fine grain. Warps badly and should be quickly converted. Small sizes. Used for fancy articles, walking sticks, etc. S.G. 1·1.

Pin. 1. A fine wire nail with very small head. 2. The steel wire through the knuckle of a hinge. 3. A cylindrical wood peg, or dowel, driven through a mortise and tenon joint, as at the corners of sashes. See *Clamps* and *Open Tenon*. 4. See *Dovetail*. 5. A small barrel of 4½ gals. capacity. 6. See *Band and Gudgeon*. 7. See *Clamps*. 8. *To pin* is to wedge, or fix, the end of a timber into a wall. 9. See *Knots*. 10. French for *pine*.

Pinaceæ. The pine family. A sub-family of Coniferæ. Includes *Abies*, *Agathis*, *Araucaria*, *Callitris*, *Cedrus*, *Cryptomeria*, *Juniperus*, *Keteleeria*, *Larix*, *Picea*, *Pinus*, *Pseudolarix*, *Pseudotsuga*, *Tsuga*. Also see *Cupressaceæ*, *Taxaceæ*, *Araucariaceæ*, and *Podocarpaceæ* for further subdivision.

Pinang. Meranti, *q.v.*

Pinaster Pine. Maritime pine, *q.v.*

Pincers. A tool for extracting nails consisting of two bent levers loosely riveted together. Also see *Nail Puller* and *Claw Hammer*.

Pinch Bar. A steel rod for levering or moving heavy objects.

Pinch Dog. A small dog for securing glued joints until the glue has set. See *Dog*.

Pindahyba. *Xylopia emarginata, X. frutescens* (*H.*). Brazil. Greyish white, darker streaks, lustrous, smooth. Close even grain and fine texture. Used for tool handles and for decorative woodwork. Little imported. The name Pindahyba is also applied to Brazilian lancewood.

Pin Dote. Small decayed spots at the ends of logs.

Pine. *Pinus spp.* There is a large number of important softwoods known as pine, with a distinguishing prefix. The sources of supply are widespread. There are about twenty-five species in N. America alone. Both in the trade and botanically there is much confusion in classification. Pines are chiefly light, soft, stable, and fairly durable. See *Pinus*. The woods are described alphabetically in the Glossary. In the following list the alternative name, for reference, is given in brackets. African pine (*African Yellow-wood*); Alaska p. (*Western Hemlock*); Alpine p. (*Swiss p.*); Arolla p. (*Swiss p.*); Austrian p. (*Corsican p.*); Banksian p. (*Jack p.*); Baltic p. (*Redwood*); Bhotan p.; Black p. (*Lodgepole p.*); Blue p. (*Bhotan p.*); Bosnian p. (*Corsican p.*); Brazilian p. (*Parana p.*); Brown p. (*Huon p.*); Bull p. (*Western Yellow*

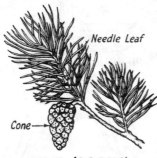

Needle Leaf

Cone

PINE (SCOTS)

p.; Bunya p.; Caribbean p. (*Cuban p.*); Celery Top p.; Cembran p. (*Swiss p.*); Chilian p.; Cluster p. (*Maritime p.*); Columbian p. (*Douglas Fir*); Corsican p.; Cowdie p. (*Kauri p.*); Cuban p.; Cypress p. (*Cypress*); Dantzig p. (*Redwood*); Douglas p. (*Douglas Fir*); Finger Cone p. (*Western White p.*); Florida p. (*Pitch p.*); Georgia p. (*Pitch p.*); Grey p. (*Jack p.*); Hazel p. (*American Red Gum*); Hoop p.; Huon p.; Idaho p. (*Western White p.*); Jack p.; Japanese p. (*Kuro Matsu*); Kauri p.; King William p.; Korean p. (*Siberian p.*); Larch or Laricio p. (*Corsican p.*); Loblolly p. (*Pitch p.*); Lodgepole p.; Longleaf p. (*Pitch p.*); Long-leaved p. (*Chil p.*); Macquarie p.; Manao p. See *New Zealand p.*; Manchurian p. (*Siberian p.*); Maritime p.; Moreton Bay p. (*Hoop p.*); New Zealand p. (*Kahikatea p.* and *Silver p.*); Norway p. (*Red p.*); Oregon p. (*Douglas Fir*); Ottawa p. (*Red p.* and *White p.*); Parana p.; Pinon p.; Pitch p.; Ponderosa p. (*Western Yellow p.*); Quebec p. (*Red p.* and *Yellow p.*); Queensland p. (*Hoop p.*); Red p.; Redwood; Scots p.; She p.; Shortleaf p. (*Pitch p.*); Siberian p.; Silver p. (*Western White p.*, *Siberian p.*, and *New Zealand p.*); Slash p. (*Cuban p.*); Southern p. (*Pitch p.*); Stone p. (*Swiss p.*); Sugar p.; Swiss p.; Tonowanda p.

(*Yellow p.*) ; Weymouth p. (*Yellow p.*) ; White p. ; Yellow p. Also see *Kuro Matsu, Matai, Miro, Podo, Rimu, Pinus,* and *Pinaceæ.* There are many other alternative names, but it is suggested that they are unnecessary and confusing.

Pine Sawyer. A beetle, genus *Monohammus,* that attacks the sapwood of logs, especially pine.

Piney. *Hardwickia pinnata* (*H.*). India. Light purplish red to dark red, darker lines, smooth, fairly lustrous. Moderately heavy, but very hard, strong, and elastic, fairly durable, subject to insect attack. Wavy grain, fiddleback, oily exudation. Requires care in seasoning and should be converted green with boxed heart. Not difficult to work ; polishes well. Suitable for cabinet work, superior joinery, turnery, veneers, constructional work, etc. S.G. ·6 ; C.W. 3·5.

Pinfold. A pound for straying cattle.

Ping. *Cynometra polyandra* (*H.*). India. Light to claret red, deeper fine lines, fading to pinkish grey, lustrous, smooth. Hard, heavy, very strong, fairly durable. Straight and curly grain, medium texture. Difficult to season, fairly difficult to work. An attractive wood suitable for cabinet work, rotary cut plywood, etc. S.G. ·88 ; C.W. 3·5.

Pin Hinge. Similar to a butt hinge but with loose pin that can be removed so that the door can be removed without disturbing the screws.

Pinho. Parana Pine, *q.v.*

Pinhole Borers. Beetles of Scolytidæ and Platypodidæ families. Very destructive to green timber.

Pinholes. Small holes in wood not more than $\frac{1}{16}$ in. diameter due to pinhole borers. Also see *Shot* and *Worm Holes.*

Pin Jointed. A term used in the design of pillars in which the ends are adequately restrained in position only. See *Fixed Ends.*

Pink. An old type of sailing vessel. The planking curves sharply to the stern post which gave rise to the term *pink-sterned.*

Pinking. Decorating thin material along the edge by piercing and scallops.

Pink Mahogany. Agba or Moboron, *q.v.*

Pin Knot. One not more than $\frac{1}{4}$ in. diameter ; but the size is up to $\frac{1}{2}$ in. with some American Lumber Association gradings. See *Knots.*

Pink Pine. New Zealand Silver Pine, *q.v.*

Pink Tamarind. See *White Tamarind.*

Pink Wood. *Beyeria viscosa* (*H.*). Tasmania. Rose colour. Hard, with close grain. Small sizes. Used for inlays, cabinet work, fancy articles. Also see *Tulipwood.*

Pinnace. A boat rowed with eight oars. A small vessel with oars and sails, and usually two masts. A ship's boat like a ship's barge but smaller for conveying junior officers ashore. A petrol or steam-driven boat used as a tender for a larger vessel.

Pinnacle. An ornamental terminal to a tower or roof, or a pointed turret.

Pinnate. A compound leaf. A single petiole with several leaflets attached. Shaped or branched like a feather.

Pinnatified. A single leaf nearly approaching a pinnate leaf in appearance, as an oak leaf.

Pinnay. Bintangor, *q.v.*

Pinned. Fixed securely. Wood framing secured at the joints by pins, usually in addition to wedges. See *Pin*.

Pinned Tenon. See *Pin* (3) and *Open Mortise*.

Pinoleum Blind. An outside blind that allows for light and air but protects from the rays of the sun.

Pinon. *Pinus edulis* (*S.*). Southern U.S.A. Yellowish, fragrant. Rather hard, and coarse grain. Fine texture. Small sizes and little exported. S.G. ·55.

Pin Rail. A wood rail with pegs for hats and coats. See *Pin*.

Pintle. A pin. The iron rod on which the rudder of a ship turns.

Pinus. Pine genus. See *Pine*. *Pinus austriaca* =*P. nigra*; *P. Banksiana*, Jack pine; *P. bungeana*, Chinese Sung, peculiar to China; *P. calabrica* =*P. nigra*; *P. caribæa*, Cuban pine; *P. cembra*, Siberian pine; *P. contorta*, Lodge-pole pine; *P. densiflora*, Kuro matsu; *P. echinata*, Shortleaf pitch pine; *P. edulis*, Pinon; *P. excelsa*, Blue pine or Bhotan; *P. halepensis*, Aleppo pine; *P. heterophylla* =*P. caribæa*; *P. Jeffreyi*, Jeffrey pine; *P. Khasya*, Khasia pine, like Chil; *P. koraiensis* =*P. cembra*; *P. lambertiana*, Sugar pine; *P. laricio* =*P. nigra*; *P. latifolia* =*P. contorta*; *P. longifolia*, Chil or Chir; *P. manchurica* =*P. koraiensis*; *P. merkusii*, like Chil; *P. mitis* =*P. echinata*; *P. monticola*, Western white pine; *P. Murrayana* =*P. contorta*; *P. nigra*, Austrian or Corsican pine; *P. palustris*, Longleaf pitch pine; *P. pinaster*, Maritime pine; *P. pinea*, Stone pine; *P. ponderosa*, Ponderosa pine; *P. resinosa*, Red pine; *P. rigida*, Pitch pine; *P. rubra* =*P. resinosa*; *P. sinensis*, Chinese horse-tail pine, similar to *P. densiflora*; *P. strobus*, Yellow pine; *P. sylvestris*, Scots pine and Redwood; *P. Tæda*, Loblolly pine; *P. thumbergii* =*P. densiflora*.

Pipal. *Ficus religiosa* (*H.*). India. Not exported because of native reverence for tree. The wood is of inferior quality. See *Fig*.

Piped Rot. Heart rot in oak due to the fungus *Stereum spadiceum*.

Pipe Hook. See *Wall Hook*.

Pipe Racks. Boards, about 4 in. above the sound-board of an organ. They are supported by rack-pillars and support the pipes.

Pipes. 1. Decayed channels, usually at the heart, in trees. 2. The channels or tubes forming the essential components of an organ for the musical sounds. There is a great variety, both wood and metal. They vary in length up to 32 ft. according to the pitch, and they may be from less than 1 in. to over 16 in. square, or they may be triangular or rectangular in section. Circular ones are of metal. A wood pipe consists of body, block, cap, and foot. The top may be closed or open. See *Organ Pipe*, *Stop*, and *Pallet*. 3. Tubes for conveying gases and liquids. 4. A barrel of 105 wine gallons capacity.

Holdfasts

Pipes

Hinged

PIPE COVER

Pipe Stave Oak. A standard size of quarter riven oak stave, used in cooperage.

Pipli. *Bucklandia populnea* (*H.*). India. Brown or greyish brown, fairly lustrous. Very similar to satin walnut. Moderately hard, heavy, strong, and elastic. Interlocked grain, very fine texture. Requires slow seasoning and to be converted green. Easily wrought, polishes well. Used for joinery, furniture, constructional work, etc. · S.G. ·63 ; C.W. 2·5.

Piptadenia. See *Angico, Carbonero, Curupay, Ekhimi, Horco.*

Piquia. *Caryocar villosum* (*H.*). C. America. Greyish brown, appears waxy. Hard, heavy, strong, tough, durable. Irregular, interwoven, and roey grain, rather coarse texture. Fairly difficult to work ; polishes well. Used for shipbuilding, piling, structural work, cooperage, wheelwright's work, etc. S.G. ·8 ; PIQUIA MARFIN, *Aspidosperma eburnea.* Brazilian satinwood. Like W. Indies satin-wood. S.G. 1. PIQUIA PEROBA, *A. tomentosum.* Brazil. Called Lemonwood. See *Guatambu.* Also see *Peroba* and *Satinwood.*

Piratinera. See *Letterwood* and *Snakewood.*

Pirn. A reel or bobbin.

Piscidia. See *Dogwood.*

Piscina. A sink, or a niche containing a sink, in a church.

Pisonia. See *Blolly.*

Pistacia integerrima (*H.*). India. Excellent decorative cabinet wood, but supplies limited and no market name. Native names : Kangar, Kakar, Masua, etc. S.G. ·45.

Pitch. **1.** The ratio of rise to span in a roof, etc. The inclination of roofs or stairs. See *Pitched Roof* and *Tread.* **2.** A thick dark resinous substance obtained by distilling tar. A by-product of pines and

Minimum Pitch	Materials
½	Thatch, Shingles
½₂·₆	Plain Tile
⅓	Small Slates
⅓·₅	Asbestos Tiles
¼	Ordinary Slates, Pan Tiles
⅕	Large Slates, Roman Tiles
⅛	Corrugated Sheets, Felt
½₈₀	Lead, Zinc, Copper

45°   37·5°   33·5°   30°   26·5°   22°   14°

PITCHES OF ROOFS

firs. **3.** The advance in one revolution of a helix or screw thread or propeller. The distance between the centres of adjacent wheel teeth or rivets. **4.** See *Saw Teeth.* **5.** The angle of inclination between the cutting iron and the sole of a plane. **6.** The forward movement of an aeroplane propeller in one revolution.

Pitch Blister. A pitch pocket breaking on the face of converted stuff.

Pitch Block. A temporary wood block cut to the pitch of a flight of stairs and used when shaping wreaths, etc.

Pitch Board. A triangular templet used when setting out stair strings preparatory to housing. The sides of the triangle represent rise, going, and pitch. A margin piece, or fence, is used to keep the pitch board the correct distance from the top edge of the string. The pitch board gives the faces of tread and riser, and separate templets are used to mark out the thickness.

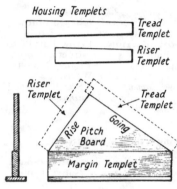

Housing Templets

Tread Templet

Riser Templet

Riser Templet

Tread Templet

Rise   Pitch Board   Going

Margin Templet

PITCHBOARD AND TEMPLETS

Pitched Roof. An inclined roof. When the inclination is less than 20 it is termed a flat roof for purposes of design. The minimum amount of pitch depends upon the kind of roof covering.

Pitching Block. Any form of block, or planted piece, to simplify the cuts and to provide a bearing for the heads of inclined struts. rafters, etc.

Pitching Piece. **1.** A rough bearer receiving the end of a carriage piece for a flight of stairs. The ends are fixed in the wall and newel, or in the wall only for geometrical stairs, where it acts as a cantilever. **2.** The timber carrying the tops of the rafters in a lean-to roof.

Pitch Pine. *Pinus spp.* (*S.*). Southern U.S.A. Includes Longleaf, Shortleaf, Loblolly, Slash, and Pond pines. Golden yellow with deep red markings due to darker and denser summer wood. Strongest and heaviest of pines ; very resinous, very stiff, durable. Uneven grain. Handsome figure when plain sawn. Curly wood esteemed for panelling. Used extensively for structural work, temporary timbering, interior joinery to be varnished. It is too resinous for paint. Seasons slowly and shrinks greatly. Not difficult to work except for resin, and on end grain. Weight very variable due to difference in rate of growth. Large sizes. The name pitch pine is confined to the species *P. rigida* in America and is inferior to other species. In the timber trade pitch pine is sold by the cubic foot. S.G. ·7 ; C.W. 2. See *Cuban, Florida, Lobolly, Longleaf, Pond, Shortleaf,* and *Southern Yellow Pine.* Also see *Pine* and *Pinus.*

Pitch Pockets. Defects due to accumulation of resin between the growth rings of coniferous woods. The pockets may vary from $\frac{1}{8}$ in. to several inches wide circumferentially, up to 1 in. deep radially, and up to 12 in. long.

Pitch Seam. Also called pitch shake or streak. Openings along the grain following the outline of the growth rings and containing resin.

Pith. 1. The middle core of a stem, consisting chiefly of parenchyma. It is the centre round which growth takes place and does not function after the sapling stage. Also called heart or medulla. 2. See *Solah*.

Pithecolobium. See *Camachile* or *Caumochil*, *Kungkur*, *Medang*, *Moruro*, *Rain Tree*, *Sabicu*, *Vinhatico*, *Zebrawood*.

Pith Flecks. Pith-like irregular discoloured streaks of tissue in wood, due to insect attack on the growing tree. A small callus due to injury of the cambium by insects. Wound parenchyma forms on the occluding tunnels, and cross-sections form the flecks on converted wood. There is no connection with the pith.

Pith Knot. See *Knots*.

Pith Rays. See *Rays*.

Pit Props. Small round timber used as struts to support the roof in a mine.

Pits. Parts of the cell wall in wood tissue left thin during the thickening of the walls. They serve as means of communication between the cells.

Pit Saw. A large, usually two-handed, saw used for ripping up logs, etc., in saw pits, *q.v.* Sometimes a mechanical contrivance dispenses with one sawyer, or the saw may be sufficiently stiff.

Pitt. Abbreviation for the name of the botanist Pittier.

Pitted Sap Rot. Decay caused by the fungus *Polystictus abietinus* in the sapwood of felled conifers. In the early stages it is difficult to detect except for a slight yellowish to tan discoloration, later this may develop into elongated cavities.

Pittosporum. See *Tallow Wood* and *White Hollywood*.

Pitt's Patent. A method of fixing door knobs to the spindle.

Pitwood. 1. Timber used in mines : bars, chocks, cogs, crowntrees, laggings, props, sleepers, splits, split bars, etc. 2. Timbers for propping the roofs of roadways in mines. See *Mine Timbers*.

Pityrantha. See *Vidpani*.

Pivot. A pin on which anything rotates. Also called centres. Pivots and sockets are fixed on sashes, just above the middle of the height so that the pivoted sash will swing easily to a closed position. Also see *Hospital Window*. (See illus., p. 140.)

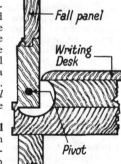

Fall panel

Writing Desk

Pivot

PIVOTED FALL JOINT

Pivoted Fall Joint. A joint in a bureau or secretaire for the flap when it is pivoted. See *Quadrant Stay*.

PIVOT AND SOCKET

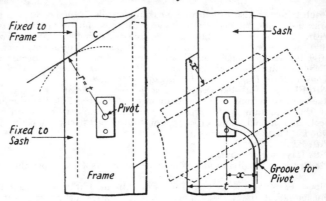

PIVOTS AND PIVOTING

Pix. An ornamental box for the consecrated Host in Roman Catholic churches.

Pix Canopy. An ornamental cover, usually in the form of a spirelet. It is similar to a font cover.

Pixpinus. A charter party form, *q.v.*, for the pitch pine trade.

Pky. Abbreviation for pecky, *q.v.*

Plafond. The soffit of the larmier to an entablature, or to projecting eaves. A ceiling.

Plagianthus. See *Lacebark*.

Plain Mouldings. A term used in coffin-making for mouldings that are not carved or embossed.

Plain Sawn. Flat sawn. Wood cut tangentially to the annual rings, when converting, *q.v.* It produces handsome grain when there is contrasting colour between spring and summer wood. Also called bastard, flat, and slash sawn. The term is applied to all sawn wood that is not classed as quarter sawn, *q.v.*

Plain Stripe. Alternating darker and lighter stripes running continuously along the length of a piece, due to cutting wood with definite growth rings on the quarter.

Plain Weather-board. Ordinary square-edged boards of rectangular section. See *Weather Boards*.

Plan. 1. The top view or horizontal section of an object drawn in orthographic projection. Anything drawn or represented on a horizontal plane, as a map or horizontal section of a building. 2. Any full-size drawing in shipbuilding may be called a plan.

Planceer. The soffit of an opening, cornice, or stairs. The soffit to the corona of a Gothic cornice.

Planceer Piece. Horizontal timbers carrying lath and plaster or boards to form a soffit to overhanging eaves.

Planchettes. Narrow boards under 4½ in. wide. See *Market Sizes*.

Planchonia andamanica. Red bombway, *q.v.*

Plane. **1.** A stock, or body, with cutting iron, for shaping the surface of wood. The following are used in woodworking, many of which may be obtained in wood or metal : badger, bead, Bismarck, block, bullnose, carriage-maker's, chamfer, chariot, coach-maker's, compass, dado, door, dovetail, edge, fillester, heel, hollows, jack, jointer, matching, mitre, moulding, ogee, old woman's tooth, ovolo, panel, plough, rebate, or rabbit, reed, roughing, router, rounds, scraper, scrub, scurfing, shave, shoulder, side rebate, smoothing, spokeshave, toothing, trenching, try, turning, Universal, veneer, *q.v.* There are numerous patent types : Bailey, Norris, Preston, Sargent, Stanley. There is a great variety of moulding planes, but they have been superseded by machinery. Also see *Bench Planes.* **2.** A flat surface ; one in which any two points lying in the surface may be joined by a straight line lying in the surface. **3.** An aileron of an aeroplane. See *Fuselage.* **4.** Scottish name for Sycamore, *q.v.* Eastern Plane, *Platanus orientalis.* Europe. Yellowish red. Confused with sycamore but softer, lighter, and with very broad rays. Used in cabinet and piano trades. S.G. ·6 ; C.W. 2·5. London Plane, *Platanus acerifolia.* Maple-leaf plane. Suggested to be a cross between Eastern and Western. Yellowish red. Distinct rays. Sometimes called Lacewood. Used for furniture and veneers. Western or American Plane, *P. occidentalis* or *P. racemosa.* See *Buttonwood.*

Planed. Surfaced. Wrought to a smooth surface by hand or machine. See *Plane* and *Machine.*

Plane, or **Novelty, Saw.** A circular saw that cuts a very smooth surface. It is hollow ground, with the teeth arranged in series of five. The advance, or clearing, tooth in each group is larger than

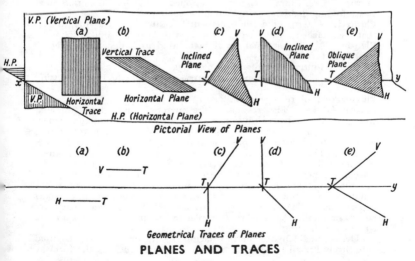

Pictorial View of Planes

Geometrical Traces of Planes

PLANES AND TRACES

the other four. Another type, the *planer saw*, is hollow ground and without set.

Plane Rest. A strip of wood on which bench planes rest to protect the cutting irons. See *Bench*.

Plane Sliced. See *Slicer*.

Plane Table. An instrument consisting of a drawing board on a tripod for plotting a survey.

Planing Machine. See *Surfacer, Four Cutter, Panel Planer*, and *Thicknesser*.

Planing Skips. Defects in the smoothness of the surface due to faulty machining.

Plank. 1. Square sawn softwood 2 to 6 in. thick and 11 in. or more in width. There is considerable difference in the limits in different markets. A plank in hardwood is from 1½ in. × 9 in. and 8 ft. upwards in length. See *Market Sizes*. 2. The piece from which a handrail wreath is shaped.

Planking. 1. A loose term often used for flooring and matching above normal thickness, decking for ships, scaffold boards, temporary floors of deals or planks, boarded linings of timber houses, etc. 2. Temporary timbering for excavations. 3. Outer skin of a wood vessel in shipbuilding.

Planning. The art of designing the lay-out of buildings.

Planometric. A pictorial representation of an object. The plan is drawn as in orthographic projection and oblique projectors represent the sides of the object. It is a form of oblique projection. See *Kissing Gate*.

Plant. A collective term for the tools, machinery, etc., necessary to carry on a business. Especially applied to builder's materials such as scaffolding, ladders, cranes, mixing machines, etc.

Planted. A term used for ornamental features, such as mouldings, that are attached, or applied, and not formed on the solid.

Plaque. A circular ornamental plate used as a decorative feature on a wall, furniture, etc.

Plash. To interweave branches or twigs.

Plaskon. The name applied to urea-formaldehyde thermo-setting resin.

Plastergon. A registered wallboard, *q.v.*

Plastic Glues. The resin-bonding materials used for laminated wood, plywood, stress-skin construction, and mass production generally. They are divided into several groups : (*a*) Thermo-setting resins, which include phenol-formaldehyde, urea-formaldehyde, soya-formaldehyde, melanine resin, etc. (*b*) Thermo-plastics, which include cellulose derivatives, acryl-polymers, vinyl polymers. (*c*) Casein plastics. (*d*) Natural resin glues. Their uses are nearly unlimited in cabinet making, building, aircraft, ship- and coach-building, etc., but they require expensive equipment and expert supervision. Thermo-setting resins are permanent when they are cured, or moulded. The thermo-plastic group can be made plastic again by re-heating.

Urea-formaldehyde thermo-setting resin, or plaskon, is used extensively for all types of laminated structures such as the fuselage of aircraft, boats, or any other form of monocoque construction. It is applied in liquid form, usually by machine spreaders, either as *hot-press* or *cold-press* resin. Hot-press resin requires " cooking," or moulding, in an autoclave. It allows for a time interval of two or three days between application and moulding, hence it is invaluable for complicated structures. The prepared work is clamped and placed in a watertight rubber bag, and the whole is placed in the autoclave. Hot water and compressed air are then admitted into the autoclave to provide the required heat and pressure. Cold-press resin is used for assembly work and does not require heat, but it only allows for a time interval of fifteen minutes before the pressure is applied. The time interval between the application of the bonding agent and moulding is controlled by the hardener which is added just before use. Dry resins may be applied in powder form or as a cellophane-like sheet placed between the laminations. The prepared work is placed in a heated press, which fuses the resin to make a permanent joint. Methods of curing by electric currents are being rapidly improved. Intensive research is proceeding on the application of plastics to wood construction involving many new methods. Also see *Glues, Moulding* (3), and *Improved Wood*.

Plasticity. A property of wood that allows it to retain its shape when bent. Opposite to elasticity. See *Bentwood, Pliable*, and *Urea*.

Plastic Wood. A paste for filling holes and open joints in wood. Also see *Improved Wood*.

Plat. A landing to a staircase.

Platanaceæ. The Plane tree family, the most important genus of which is *Platanus*.

Platanus. See *Plane, Buttonwood*, and *Sycamore*.

Plat-band. **1.** A door lintel. **2.** A flat fascia or a string course. Any flat rectangular moulding in which the width is much greater than the projection.

Plate or Platt. A horizontal timber on a wall to distribute the load from other timbers. See *Wall, Foot*, and *Pole Plate*, and *Gutters*. **2.** See *Tracery*.

Platea. Dedaru, *q.v.*

Plate Dowel. A plate that screws on to the face at the foot of a door post as a fixing. A projecting dowel is let into the stone or concrete sill. It is generally used when repairing the foot of the post.

Plate Lock. A stock lock, *q.v.*

Plate Rack. A rack, usually near a sink and over a draining board, into which washed crockery is placed.

Plate Shelf. A shelf, recessed near the back, on which plates are displayed. It is often fixed round a room at picture-rail height. See *Wall Panelling*.

Plate Staple. See *Striking Plate*.

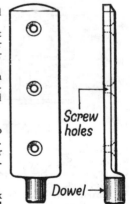

Screw holes

Dowel →

PLATE DOWEL

Platform. A part of a floor raised above the level of its surroundings. A projecting grating over the side of a ship. Also see *Derrick Tower*.

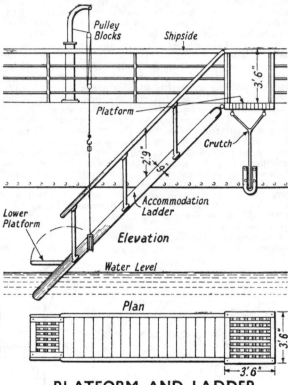

PLATFORM AND LADDER

Plathymenia. See *Vinhatico*.
Platonia. See *Bacury*.
Platycyamus. See *Pereira*.
Platylophus. See *Els* and *White Alder*.
Platymiscium. See *Quira*, *Macacahuba*, *Macawood*, *Vencola*, and *Roble*.
Platypodidæ. See *Pinhole Borer*.
Play. The same as clearance. The space between adjacent surfaces to allow free movement, as between the edge of a door and the rebate.
Play-house. A theatre.
Pleach. Same as plash, *q.v.*

Pleiococca. See *Silver* and *White Aspen.*
Pleiogynium. See *Tulip Plum.*
Plethora. A defect in wood due to uneven distribution of the sap during growth.
Pleurostylia. See *Panaka.*
Pliable. Easily bent ; flexible. The bending of wood is important for bentwood furniture, light-weight spars and struts, etc. Steaming is the usual method of making wood pliable, but recent research suggests that tanning agents are more effective. The wood is soaked for one or two days in a solution of about ·5 per cent. of the tanning agent, which may be natural or synthetic. Wood so treated is easily bent and may be folded on itself through 180° It retains the position when dry and loses none of its qualities. See *Bentwood* and *Urea.*
Pliers. Narrow-jawed pincers for bending and cutting wire, etc.
Plinth. **1.** A plain thin planted piece of wood at the bottom of a fitment or wall. A plain skirting board. **2.** A projecting surface at the bottom of a column, wall, etc. **3.** A flat plain square block supporting a pedestal or column. See *Pilaster* and *Cupboard.*
Plinth Block. A block at the foot of an architrave against which a skirting or plinth abuts. See *Finishings.*
Pln. Abbreviation for *plain.*
Plough. **1.** An adjustable plane, with usually eight different sizes of bits, for forming grooves. The spindles are sometimes threaded

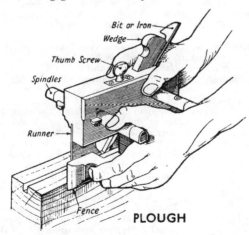

Bit or Iron
Wedge
Thumb Screw
Spindles
Runner
Fence   **PLOUGH**

and adjusted by hardwood nuts. **2.** An agricultural implement for turning over the ground. Also a similar implement for making a track through deep snow.
Plough Beam. The central shaft of a plough (2).
Ploughed. The same as grooved, *q.v.* A groove for a tongue or to receive a panel, and formed by a plough (1) or by machine.

Plough-staff. A form of spade with long handle.
Plucked Up. A term used when the fibres on the surface of wood have been torn up due to careless planing. The defect often occurs with interlocked grain.
Plug. **1.** A wedge-shaped piece of wood driven into a joint in brick-

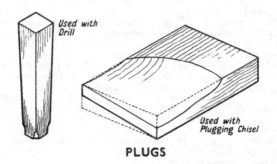

Used with
Drill

Used with
Plugging Chisel

PLUGS

work, etc., or into a drilled hole, as a fixing for joinery, etc. There are several patent types, of other materials than wood. The holes for the plugs are prepared by plugging chisels or drills. See *Double Skirting*. **2.** The stopper to a wood organ pipe. **3.** A repair to plywood.
Plug Centre Bit. A brace bit in which the " centre point " is like a cylindrical plug to fit into a previously bored hole so that it serves as a centre whilst boring a larger hole round it.
Plugging Machine. A machine for boring out and plugging defects in wood, such as knots, pitch pockets, etc. The machine bores, glues, and prepares the plug.
Plug Tenon. A stub tenon in the middle of the stuff to prevent

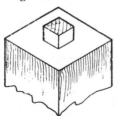

PLUG TENON

lateral movement and provide a good bearing surface for a post. Also called a spur tenon.
Plum. *Prunus domestica* (*H.*). Europe. Reddish brown, variegated stripes. Hard, heavy, with close firm texture and smooth. Small sizes. Valuable for turnery, woodware, small cabinet work. S.G. ·8. BLACK PLUM. See *Jaman*. Also see *Hog Plum*.
Plumb. Vertical. See *Plumb Rule*.
Plumb Rule. A parallel straight edge usually of pine and about 5 ft. × 4 in. × ¾ in. A heavy piece of metal called a plumb-bob is suspended by string from the top of the rule near the bottom. The rule is used to determine perpendiculars.
Plume. Figure due to crotch in mahogany.
Plum Figure. Irregular patches on the face of wood, produced by

indenting the growing tree. New growth follows the contour of the indent and on conversion the patches show variation in grain and lustre. See *Figure*.

Plum Mottle. Dark egg-shaped markings. See *Plum Figure* and *Mottle*.

Ply. A thin slice of wood, as used for plywood and laminwood. See *Plywood* and *Veneers*.

Plymax. A registered metal-faced plywood, *q.v.*

Plyscol. A proprietary " improved " wood, consisting of veneers or fibrous material and synthetic resin.

Plysyl. A resin-bonded plywood specially useful for formwork for concrete. Also made for wall linings and floors.

Plywood. Layers of veneer, or plies, glued together, with the grain of the plies alternating in direction. This produces a strong board and counteracts shrinkage if the glue is satisfactory. The best qualities are made with film glue that is resistant to moisture, splitting, peeling, or bubbling, by the dry-glue process, usually called resin bonded. See *Glue*. In some brands the plies are stitched together in addition. The number of plies varies according to the quality and thickness from 3 ply to as many as 36 alternating laminations. Plywood varies from $\frac{1}{32}$ in. to over 1 in. thick. The thickness is

Rule

Bob

PLUMB RULE

usually stated in millimetres, but the face measurement in inches or feet. Standard sizes are up to 120 in. long and 60 in. wide, with special sizes up to 192 and 96 in. respectively. The first dimension denotes the direction of the grain on the face. Special *cut sizes* are made to suit specialist trades. Three-ply is usually from 3 to 6 mm., 5-ply from 6 to 12 mm. There may be any number of plies when over 9 mm. thick. Ordinary plywood is from rotary-cut veneers. Better qualities may be faced with sliced or sawn veneers. European and Japanese plywood is chiefly of alder, birch, ash, oak, or redwood ; Australian of hoop pine, silky oak, maple, or walnut ; N. American of Douglas fir, gum, maple, or as European. Many other woods are used according to the local supplies. The Assam plywood mills use chiefly Hollong and Hollock, especially for the plywood for tea boxes. Methods of manufacture are continually improving.

Qualities are graded according to the knots and other defects in one or both faces as *AA*, *A*, *AA/BB*, *B*, *BB*. Gaboon plywood is graded as Crown, 1-, 2-, and 3-star. Special qualities may be faced with almost any kind of figured veneer, with a core of poplar, basswood, aspen, spruce, or pine, and is called *built-up stock*. Ordinary plywood veneered after manufacture is called *veneered-stock*. Metal-faced plywood may have facings of aluminium, galvanised iron, copper, or stainless steel, and is manufactured in panels up to 40 ft.

long. Bakelite and other phenol and lactic compounds are also used as facings. See *Veneers*, *Laminwood*, and *Plastic Glues*.

Pn. Abbreviation for *partition*.

Po. Chinese cypress.

Poachwood. Logwood, *q.v.*

Pocket. 1. The hole in the pulley stile of a sash and frame window for placing the weights in the box. 2. See *Pitch Pocket*. 3. A space in a wall to receive the end of a beam, etc.

Pocket Chisel. See *Sash Chisel*.

Pocket Piece. The removable piece forming the pocket of a sash and frame window.

Pocket Print. A drop print in pattern-making. A print continued above the position to be occupied by the core.

Pocket Rot. Decay in wood limited to pockets, or small areas, surrounded by apparently sound wood. In some cases the decay is so pronounced as to leave a hole. When the decayed part is less than $\frac{1}{2}$ in. long and $\frac{1}{4}$ in. wide it is called *medium pocket rot*.

Prepared for Pulley

Groove for Parting Bead

Pocket

Saw Cut

Pocket removed

Screw

Wedge

POCKET TO PULLEY STILE

ELEVATION

SECTION

POCKET SCREWING

Pocket Screwing. Gouging a thumb slot to receive the head of an inclined wood screw.

Pod Bit. A nose bit for a brace.

Podium. A continuous pedestal for a row of columns.

Podo. *Podocarpus gracilior*, *P. milanjiana*, *P. usambarensis* (S.). E. Africa. Also called Yellowwood, Musengera, etc. Cream to pale brown, lustrous. Moderately soft, light, strong, and durable; hard wearing, odourless, smooth. Straight grain, fine even texture. Easily wrought but requires care in seasoning to avoid checks. Stains and polishes well. Excellent softwood; large sizes free from defects. Used for joinery, furniture, butter crates, flooring, etc. S.G. ·5; C.W. 1·5.

Podocarpaceæ. The Podo family. A sub-family of Coniferæ. Includes Podocarpus, Dacrydium, Micrœachrys, and Saxegothæa.

Podocarpus. A genus providing the most useful softwood of the Southern Hemisphere. See *Brown Pine, Geelhout, Kahikatea, Matai* or *Mai, Miro, Mlange, Musengera, Thitmin, Totara, Upright Yellowwood, White Pine, Yacca,* and *Yellowwood.* There are also large quantities in S. America, but little exported : *Cobola, Chaquera, Mañio Pinheirinho, Pino,* etc.

Pœciloneuron. See *Vayila.*

Poga. See *Ibo* and *Inoi.*

Pogo. See *Spicey Cedar.*

Pohutukawa. See *Rata.*

Point. A tooth for an inserted-tooth saw. Also see *Points.*

Pointed Arch. A Gothic or Tudor arch.

Pointer. Same as tripper, *q.v.*

Pointing Machine. One for tapering the ends of cylindrical pieces.

Points. Saw teeth, *q.v.* See *Saws.*

Poir. Abbreviation for the name of the botanist Poiret.

Poison Wood. Several C. American woods are so called because of the irritating lactic from the bark of the tree, especially Possum wood, Crabwood, and Honduras walnut.

Poitrail. Same as Petrail, *q.v.*

Poker Work. Designs formed on wood by means of a red-hot needle or poker.

Polacre. A Mediterranean sailing vessel with sharp bow and light rigging.

Polak. *Ochroma bicolor* (*H.*). W. Indies. Properties and characteristics of balsa, *q.v.*, but heavier. The name is sometimes applied to balsa. S.G. ·22.

Pole. 1. A straight slender tree, in forestry, from 4 to 12 in. diameter, breast high. A small pole is 4 to 8 in. diameter and a large pole is 8 to 12 in. diameter. 2. A market form of slender tapering round timber from 24 to 35 ft. long and up to 6 in. diameter at the middle. They are used extensively for scaffolding, wireless, and flag poles, etc. See *Masts.* 3. Any long slender cylindrical piece of wood, such as curtain poles, etc. 4. A measure of length of 5½ yds. or of area of 30¼ sq. yds. 5. A long staff for impelling a punt through water. 6. A loose shaft of a two-horse vehicle.

Pole Plate. A horizontal timber, or beam, carrying the feet of rafters. It is stronger than a wall plate and carried by the ends of the tie beams. See *Box Gutter.* (See illus., p. 420.)

Poling Boards. Short vertical boards used in temporary timbering for excavations. See *Timbering.*

Polish. A prefix to several woods from Poland, especially ash and oak, to denote the source of origin.

Polishing. See *French Polishing.*

Pollard. Applied to woods, especially oak, from trees that have been pollarded.

Pollarding. Continuous lopping of the top, or poll, of a tree to encourage fresh growth. The succession of new shoots, or knurls, produces a highly decorative wood, like burrs, valued for veneers.

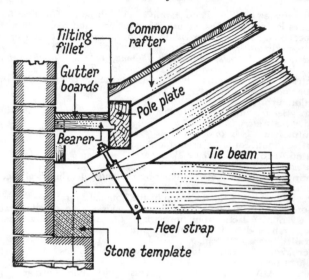

POLE PLATE

Polyalthia. See *Hoom* and *Red Wood*.

Polychromatic. Meaning many-coloured. Applied to decorative woods that are variegated, or in which the colour lends itself to decorative treatment.

Polygonaceæ. The Buckwheat family, including Coccobola, Ruprechtia, Triplaris.

Polygonal Roof. A pointed or dome-like roof with its base in the form of a polygon.

Polygons. Rectilineal plane figures with more than four sides : pentagon, hexagon, heptagon, octagon, nonagon, decagon, dua-decagon, *q.v.* In mechanics figures with more than three sides are considered polygons, as in " polygon of forces."

Polyhedra. Solids with many sides. There are five regular poly-hedra : cube, dodecahedron, icosahedron, octahedron, tetrahedron.

Polyporous. A fungus, *P. schweinitzii*, that attacks mature softwood trees and produces *brown cubical butt rot.* There are other species of less importance. See *White Butt Rot.*

Polystictus. Fungi very destructive to both softwoods and certain hardwoods. *P. abietinus, P. versicolor.* The decay is called *white*, or *pitted, rot.*

Poma. Toon, *q.v.*

Pomele. A trade name for small blister figure in mahogany.

Pometia. See *Malugia*.

420

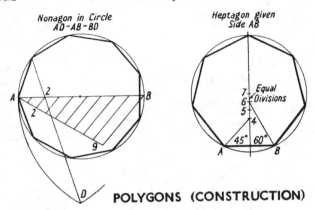

Nonagon in Circle
AD = AB = BD

Heptagon given
Side AB

POLYGONS (CONSTRUCTION)

Pommel. A spherical ornament, as a ball finial.

Pomponia. *Bombax buonopenzense* (*H.*). Nigeria. Buff colour, lustrous. Fairly easy to work. Second-class wood and little exported. Suitable for packing cases. S.G. ·6.

Pond. A rectangular enclosure at the docks for floating unloaded timbers.

Pond Apple. See *Corkwood*.

Pond Cypress. Louisiana Cypress, *q.v.*

Ponderosa or **Pondosa Pine.** *Pinus ponderosa* (*S.*). W.N. America. Also called Bull, Californian White, and Western Yellow Pine, *q.v.* It is similar to yellow, or white, pine, but more variable in density, texture, and resin content. See *Western Pine*.

Pond Pine. *Pinus rigida serotina*. See *Pitch Pine*.

Pongamia. See *Thinwin*.

Pontoon. 1. Applied to a bridge formed across a series of floating objects, such as air-tight cylinders or flat-bottomed boats. 2. Sometimes applied to a float for a sea-plane.

Pony Girder. A secondary girder supported by cantilevers only.

Poon. *Calophyllum inophyllum, C. elatum* and other *spp.* (*H.*). India, Burma, Andaman I. Reddish brown, darker streaks, lustrous, smooth. Fairly hard, heavy, strong and elastic, durable, not very stable. Interlocked grain, medium texture, faint ray flecks on a radial surface. Seasons well, with salt-water seasoning. Not difficult to work, polishes well. Used for shipbuilding, cabinet work, joinery, construction, etc. S.G. ·65 ; C.W. 2·75. Poon or Poonspar, *C. tomentosum*. Similar to above but superior, stronger, harder, and more uniform in weight. Sometimes sold as Malabar red pine. Used as above and for structural work. S.G. ·6 ; C.W. 3.

Poop. An erection on the top deck at the stern of a ship. A light deck above the main deck for a short distance forward from the stern.

Poplar. *Populus spp.* (*H.*). Willow family. Widespread. White to greyish brown, darker streaks, often lustrous. Fairly light and soft.

Good abrasive resistance and does not splinter. Shrinks greatly and not durable. Straight close grain, smooth, firm uniform texture. Easily wrought. Used for turnery, toys, sabots, matches, cooperage, cabinet-making, pulp, etc. S.G. ·5 ; C.W. 1. AMERICAN POPLAR. See *Aspen, Balsam,* and *Cottonwood.* BALM POPLAR. Canadian poplar. BAY POPLAR. Tupelo gum, *q.v.* BLACK POPLAR, *P. nigra.* Europe. Similar to grey poplar. BLACK ITALIAN POPLAR, *P. serotina.* Europe. CANADIAN POPLAR. The suggested name for N. American poplars, especially *P. tacamahaca* and *P. grandidentata.* GREY POPLAR, *P. canescens.* Europe. Similar to white poplar but rather superior and harder. HIMALAYAN POPLAR, *P. ciliata.* Inferior to Indian and not exported. INDIAN POPLAR, *P. euphratica.* Also called Bahan.

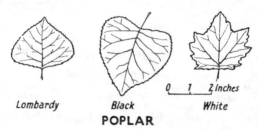

Lombardy · Black · White

POPLAR

Reddish brown, dark lines, producing tortoiseshell figure in bastard sawn wood. Lustrous, but dulls later. Some interlocked grain ; uneven texture. Light, fairly hard, durable in water. Seasons well and easily wrought. Used for well linings, planking, plywood, lacquer work, etc. S.G. ·47; C.W. 2. ITALIAN POPLAR. Black Italian poplar, *q.v.* LOMBARDY POPLAR, *P. fastigiata.* Little timber value. *P. deltoides* or *P. monilifera,* N. America, called Eastern Cottonwood or Necklace Poplar. Cottonwood is marketed as American whitewood in this country. *P. tremula,* European aspen, chiefly from Finland and Russia. Similar to white poplar. *P. tremuloides,* Canadian aspen. *P. trichocarpa,* N.W. America, called balm, black, and balsam cottonwood. SILVER POPLAR, aspen, *q.v.* WHITE POPLAR, *P. alba.* The best known European poplar, also called Abele. It often has beautiful figure due to the sinkers of mistletoe. YELLOW POPLAR, *Liriodendron tulipifera.* See *Canary Wood.*

Poppet. 1. A short piece of wood. **2.** The top frame supporting the pulleys for a hoist. **3.** The tail-stock of a lathe. **4.** Applied to numerous small mechanical objects, such as a *poppet valve,* etc.

Popple. *Populus tremuloides* and *P. grandidentata.* Northern U.S.A. Aspens, *q.v.*

Poppy Head. An elaborate carved ornament often found in old churches at the top of pew ends.

Populus. See *Poplar, Aspen,* and *Cottonwood.*

Porch. A roofed approach or shelter to a doorway.

Porcupine Wood. *Cocus nucifera.* Widespread in tropical and sub-tropical countries. A hard decorative wood from the cocoanut palm. It is reddish brown with variegated black and white stripes. Used for fancy articles, walking sticks, etc., and for construction in the country of origin.

Pored or **Porous Woods.** Hardwoods. Woods containing vessels, or pores, from broad-leaved trees.

Pores. The sections of the vessels on a transverse section of wood. In some woods the pores are very minute and can only be distinguished through a lens, in others they are easily seen. They may be uniformly or irregularly arranged and distributed, which is an aid to identification. See *Diffuse* and *Ring Porous* and *Vessels*.

Poria. See *Dry Rot* and *Rot*.

Port. 1. The left-hand side of the hull of a ship, or aeroplane, looking forward. An opening in a ship's side, including the small round scuttles. 2. A gate or gateway.

POPPY HEAD

Portable Machines. Saws and machines for use on the site. There is a big variety of planers, saws, sanders, etc.

Portal. 1. An entrance passage. 2. The smaller of two gates at the entrance to a building. 3. An arch to a gateway.

Portal Bracing. Wind bracing. A gusset bracing between column and beam in tall framed buildings.

Port Bars. Strong pieces of oak fitted in the port sills of a wooden ship. They carry the hooks for the port shackles.

Portcullis. A strong grating of wood or metal sliding vertically in coulisses to guard a gateway.

Port Hole. An opening in the side of a ship for light and air, or for discharging cargo, etc. A gun embrasure

Portico. A large porch supported by columns.

Port Lids. Doors, hinged at the top, to close the ports in wooden ships.

Port Orford Cedar. See *Cedar*.

Porto Rico. A prefix, usually to Lignum-vitæ and Cuban mahogany, to denote source of origin.

Port Riggle. See *Wriggle*.

Possi Possi. *Sonneratia acida* (*H.*). Malay. The River Willow. Not a true willow but has similar characteristics and properties.

Possum Wood. *Hura crepitans* (*H.*). C. America. Creamy white or yellowish brown streaks, and lustrous. Light and soft, like basswood. Ribbon grain. Easily wrought. Cheap substitute for Spanish Cedar. S.G. ·4 ; C.W. 2·5.

Post. A vertical timber acting as a support, or in constructional framework, or in fencing. See *Centres*. Often distinguished for a special purpose, as bed, door, goal, king, princess, queen, sign, soundpost, etc.

Post and Pan. Half-timbered buildings with panels, or *panes*, of lath and plaster or brickwork.

Post and Petrail. Same as *Post and Pan.*

Postern. A small gate or door serving as a private entrance, usually at the back of a building.

Postique. Applied to ornamentation added after the main work is completed. Superadded.

Pouch Table. See *Work Table.*

Poudresse. A lady's dressing table.

Poui. *Tabebuia serratifolia (H.).* W. Indies ebony. Two varieties, green and grey. See *Ebony.*

Pounce Box. One with a perforated under-lid for holding perfumes.

Pound. An enclosure for straying cattle.

Pouring Gate. An aperture for pouring the molten metal into a mould.

Powder Post Beetles. Lyctidæ and Bostrychidæ families. They are very destructive to the sapwood of certain freshly cut or partly seasoned hardwoods : ash, chestnut, elm, hickory, oak, sycamore, walnut, willow, etc. Fine textured woods are generally immune from attack. Careful inspection of stocks in March and October is recommended. Accumulation of waste wood encourages rapid infection of stocks. See *Sanitation.*

Powel-Wood. A process of preservation by impregnating the wood with a saccharine compound. See *Preservation.*

Poye. African Blackwood, *q.v.*

P.p.e. Abbreviation for *planed plain edged.*

Pram. 1. A perambulator. 2. A flat-bottomed barge used in the Baltic. 3. A boat in which the planks are gathered together to form a tapering bow. They end in a piece like a small transom.

Pratt Truss. The Linville or N-truss, *q.v.*

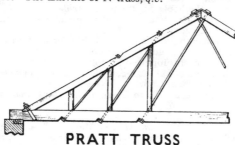

PRATT TRUSS

Predella. 1. A foot or kneeling stool. 2. The platform or step on which an altar rests, or the raised shelf at the back of the altar.

Premna. See *Nagal.*

Presbytery. The area at the east end of a church, near the altar, reserved for officiating clergy.

Presdwood. A registered wall-board, *q.v.*

Preservation. The treatment of wood to make it more durable. Many woods are naturally durable, others may be made durable by treatment. Oils or mineral or chemical solutions are forced into the cells to increase the resistance against fungi and insect attack. Preservatives are broadly divided into two groups : salts and tar-oil. Salts of arsenic, copper sulphate or chromate, mercuric chloride, naphthenates, tannin, sodium fluoride, zinc sulphate or chloride, sugar, and zinc meta-arsenite, are all effective. Arsenic and mercury are poisonous and are not in general use except in the tropics, but they can be painted over, which is an advantage. Many processes are named after the inventor : Boucherie (sulphate of copper), Burnett (zinc chloride), Blythe (carbolic and tar acids), Card (cresote oil and zinc chloride), Haskin's, Kyanising (mercuric chloride), René, Wolman (dinitrophenol and sodium fluoride). Noden-Bretteneau (electric), Powel-Wood (saccharine). See *Celcurising*.

The antiseptic should be cheap, permanent, easily applied, colourless, odourless, toxic to fungi and insects, non-corrosive to metals, non-poisonous to animals, and have no effect on the qualities of the wood. No antiseptic satisfies all these considerations. Perhaps the best pressure solutions for this country in order of merit are : creosote oil, Wolman salts (2 % solution), and sodium fluoride (4 % solution). The preservatives may be injected by capillary or by hydrostatic or pneumatic pressure. CREOSOTE. Pressure plants are best for creosote oil. The timber is run into the chamber on carbunks, air is exhausted to form a vacuum to extract the air from the cells, and the oil is forced into the wood under a pressure of 200 lb. per sq. in. In the *full-cell* process the oil remains in the cellular spaces, but it is expensive. See *Empty Cell Method*. It is usual to specify the amount of oil per cubic foot, *i.e.*, redwood sleepers have 10 lb. per cub. ft. Processes to control the amount of oil retained are : Bethel (full cell), Rueping (empty cell), Lowry (surplus extracted by vacuum), and Boulton methods.

Other preservatives may require some variation, but the above is a general description for pressure impregnation. Many woods resist impregnation, and absorption is assisted by incising, *q.v.* The most naturally durable woods are usually the most resistant against impregnation. Dry wood is more absorbent than green. Open tanks are used for *dipping*. The wood is steeped until the required amount has been absorbed. The preservative is gradually heated to about 180° F. This expels the air, and on cooling the preservative is drawn in to take its place. It is less effective than pressure, but cheaper. COATINGS : paint, varnish, creosote, tar (coal or Stockholm), oils, polish, cellulose, charring. Well-known proprietary preservatives are : Boracure, Carbolineum, Celcure, Cuprinol, Hope's destroyer, Jodelite, Microleum, Permolite, Peterlineum, Pilcher's Stop Rot, Presotim, Solignum, etc. Also see *Seasoning*, *Fire Resisting*, *Kilns*, *Penetration*, *Sanitation*, and *Urea*.

Press. **1.** Any apparatus for squeezing or enclosing. **2.** A machine used in the manufacture of plywood, flush doors, etc., for pressing the separate parts together after applying glue or resin. **3.** A cupboard with shelves for clothes, books, etc. **4.** A dwarf cupboard or wardrobe for linen. **5.** A set of bookshelves in a library.

Press Box. A space reserved for newspaper representatives on sports fields, etc.

Pressed-fibre Boards. See *Wallboards.*
Press Roll. The roller that presses the timber against the feed roller, for feeding the stuff through the machine.
Pressure Cabin. An aeroplane cabin in which ordinary atmospheric pressure can be maintained at high altitudes.
Pressure Pad. A pad operated by the machinist for controlling the pressure of the sanding belt on the wood.
Pressure Process. See *Preservation.*
Pricked Sizes. A term sometimes used in conversion for sizes slightly under the stated sizes. It infers $\frac{1}{16}$ or $\frac{1}{8}$ in. under size according to the method of conversion, *i.e.,* 1 in. pricked size $=\frac{7}{8}$ in.
Prick Post. 1. A light post between the main posts in a wood fence.
2. Same as Queen post. 3. A jamb post.
Prickwood. The spindle tree, *q.v.*
Prie-dieu. A low desk, with book shelf, for the priest during prayers.
Primavera. *Tabebuia Donnell-Smithii* (*H.*). C. America. Also called Roble, Apamate, or white mahogany. Yellowish to greyish brown. Light, moderately hard and fairly strong, not durable. Straight and interlocked grain, some wavy, roey, and ribbon figure, firm texture, smooth. Seasons well. Easily wrought and polishes well. Substitute for superior hardwoods for furniture, etc. S.G. ·5 ; C.W. 2·5.
Prime. Applied to the best quality in the grading of timber. See *Grading* and *Shipping Marks.*
Priming Coat. The first coat of paint on woodwork. It is a thin paint consisting of white and red lead, boiled oil, and turps, to penetrate the pores and crevices before stopping.
Prince Albert Fir. Western hemlock, *q.v.*
Princess Pine. Jack Pine, *q.v.*
Princess Post. Auxiliary posts in a queen-post roof truss. The truss itself is often called a princess-post truss.

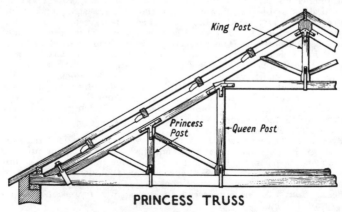

PRINCESS TRUSS

Princewood. *Cordia gerascanthus.* C. American laurel. Also called Canaletta, Solera, Ziricote, etc. See *Cyp.* Also *Exostemma caribæum.* W. Indies. Called Cerillo. Light brown, darker streaks. Very hard and heavy, smooth, with close grain. Small sizes. Used for turnery and small cabinet work. S.G. 1·1.

Principal Posts. The door posts in a wood framed partition, *q.v.*

Principals. The principal rafters of a roof truss. The inclined members that carry the purlins. The term is also applied to the complete truss. See *Roofs, Box Gutter,* and *Hammer Beam Truss.*

Print. 1. A projection that makes an impression as a guide for the insertion of a core in pattern-making. 2. See *Printing.*

Printing. The mechanical reproduction of drawings, usually called blue-printing. The method is to trace the drawing with Indian ink on transparent paper, and to place it in front of ferro-prussiate paper. Exposure to a strong light turns the ferro-prussiate paper blue, but the Indian ink intercepts the light, leaving the drawing in white lines. The paper is then immersed in running water to fix the chemicals. If the blue-print is over exposed (bluey lines), wash in oxidising solution (potassium or sodium bichromate) and again wash in water. With modern equipment semi-transparent paper may be used for the drawing, such as detailing paper, and special pencils used instead of Indian ink. Ordinary paper treated with 25 % diethyl phthalate and 75 % solax is sufficiently transparent for printing from pencil originals. *Ferro-gallic* paper gives black lines on white background. *Coralin* and *Ozalid* papers give reddish brown lines on white background. *B.W. Direct* and *Diazotype* or *Dyeline* processes produce a dyed line, red brown or black, on a white ground. *True to Scale* prints are black lines on white paper or cloth without variation in size. They are very similar to ordinary drawings, and may be coloured. There are other patented processes. Improvements are continually being introduced, both in papers and chemicals. Ferro-prussiate and ferro-gallic papers are developed in water, other papers are usually chemically developed. PHOTOSTAT. This is a photographic reproduction without negative plate, but it requires expensive equipment. HECTOGRAPH. This is the cheapest and simplest method of multiple reproduction of small drawings. The drawing is made with hectograph ink and then placed on a gelatine pad which absorbs the ink. Plain paper is then pressed on the pad to obtain the reproduction. Up to one hundred prints may be obtained by this method.

Prinzwood. A trade name for quartered elm veneer.

Prioria copaifera (*H.*). C. America. Called Cativo, Tabasaro, etc. Buff to light reddish tan, seasoning to brownish buff, with nearly black core, variegated streaks, lustrous. Fairly light, soft, not strong or durable. Straight, interlocked, and wavy grain, even medium texture, gum ducts, distinct rays. A useful cabinet wood and likely to be marketed in this country in the near future. S.G. ·44 ; C.W. 3.

Prising. Levering. Moving a heavy object by means of a lever.

Prism. A solid of uniform section throughout its length. Surface area = perimeter of end × length. Volume = area of end × length.

Prismoid. A solid that approaches the form of a prism, as a tree trunk. The volume $= \frac{1}{6}l(a_1 + 4a_2 + a_3)$, where $l =$ length and $a =$ sectional areas at the ends and middle.

Privet. *Ligustrum vulgare* (*H.*). Europe. Like olivewood, smooth, lustrous. Hard, heavy, tough, and strong. Small sizes. Used for inlays, small cabinet work, etc. CHINESE PRIVET, *L. lucidum.* Light yellow-brown, very smooth. Hard with close even grain and texture. Cultivated for wax insect. Wood used for walking sticks, handles, hay forks, etc.

Profile. 1. The outline or contour of anything. The side view silhouette of the hull of a ship. The cross section of a moulding, etc. **2.** A temporary wood guide used in the erection of brickwork.

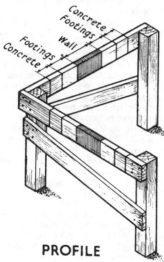

PROFILE

Profile Box. A box templet with square ends and the inside shaped to a moulding or hand-rail, for marking the shoulders of the rail.

Profile Plan. Same as shear plan, *q.v.*

Progressive Kilns. Long kilns in which timber is continuously passing through on carbunks. They are only suitable where large quantities of the same kind of wood, and of similar section, are to be seasoned, as the air is not under scientific control throughout the length of the kiln. The wet timber enters at one end and emerges dry at the other end. The hot air enters at the opposite end to the wood, hence it is cool and humid when it reaches the wet wood.

Projection. 1. See *Orthographic, Oblique, Isometric, Axiometric,* and *Planometric.* **2.** The overhang of a cornice, moulding, etc.

Promenade Deck. An upper deck on a liner where passengers may stroll about.

Prompt Box. A place near a theatre stage for the prompter.

Prop. A wood post. See *Centres.* A support, strut, or stay.

Propeller. 1. A screw-like attachment turned by power to move a vessel or machine. **2.** Airscrews for aircraft, are usually made of black walnut, mahogany, maple, or ash. They are laminated, and built up before shaping. A profile machine is used to shape them, but they are finished by hand. See *Fuselage.*

Properties. 1. See *Tests.* **2.** Things used as adjuncts to a play on a theatre stage.

Property Room. A room near to the stage of a theatre in which scenery, etc., is stored.

Props. Properties (2).

Propyleum. A vestibule, court, or portico at the gates or entrance of a building.

Proscenium. The stage of a threatre in front of the drop scene, or the frame to the curtain.

Prosopsis. See *Mesquite* and *Locust.*

Proteaceæ. Silky oak family. Includes : *Banksia, Cardwellia, Carnarvonia, Embothrium, Grevillea, Guevina, Leucadendron, Lomatia, Macadamia, Orites, Protea, Roupala, Stenocarpus.*

Protium. See *Kurokai.*

Protractor. An appliance for setting out angles. It is usually a semicircular piece of celluloid graduated in degrees.

Proud. Applied to anything slightly above its position, as a built-up surface not flush in places. Too full or too high.

Prow. The fore part of a ship.

Prunus. See *Almond, Blackthorn, Bullace, Cherry (Black), Gean, Laurel, Sloe, Mock Orange, Sakura, Wild Cherry, Wild Plum.*

p.s.e. Abbreviation for *planed and square-edged.*

Pseudopanax. See *Lancewood.*

Pseudotsuga. See *Douglas Fir.*

P.S.H. Abbreviation for *Petrograd Standard Hundred.*

Pterocarpus. See *Amboyna, Arisauru, Barwood, Bijasal, Bloodwood Kajatenhout* or *Muninga, Kejaat, Maidou, Narra, Padauk, Pudek, Red Sanders, Sangre, Sena, Vengai.*

Pterodon. See *Faveiro.*

Pterogyne. See *Ibiraro.*

Pterolobium. See *Moboran.*

Pterospermum. See *Mayeng* and *Welanga.*

Pteroxylon. See *Sneezewood.*

p.t.g.v.2s. Abbreviation for *planed, tongued, grooved,* and *veed two sides.* Numerous variations.

Ptilinus pectinicornis. A species of furniture beetle.

Pudek. Penaga Laut, *q.v.*

Pudlinks. Same as putlogs, *q.v.*

Puget Sound Pine. Douglas Fir, *q.v.*

Pugging. Non-conducting material to prevent the transmission of sound. It is often used between the joists of floors.

Pug Piles. Dovetailed piles.

Pukatea. *Laurelia Novæ Zelandiæ (H.).* N. Zealand. Pale brown, darker streaks. Good for cabinet work, superior joinery, etc., but supplies are scarce.

Pull. Any appliance to provide a convenient grip for the hand, as to a chain, drawer, door, etc.

Pulley. A grooved wheel revolving on a pin in a block. It is used for changing the direction of the pull on a rope for raising loads. See *Clothes Rail, Lifting Shutters, and Pulley Stiles.* Pulley blocks, or tackle, are movable blocks containing one to four pulleys. They are in pairs, with a continuous rope passing over top and bottom blocks in turn, and used where a mechanical advantage is necessary to raise a heavy load. The ratio of effort to load, neglecting friction, is $W = nE$, where n is the number of pulleys in the tackle.

Pulley Stiles. The vertical pieces in a sash and frame window that carry the pulleys and contain the pockets. See *Mullion, Pocket,* and *Sash and Frame.*

Pullman.   The name of the designer of a luxurious type of railway carriage and sleeping berth.

Pullman Sash Balance.   A patent spring in the form of a pulley that takes the place of weights for balanced sashes.

Pulp.   The production of pulp from waste wood and small round logs is an important industry for the manufacture of paper, patent compositions, cellulose, rayon, etc.   A pulp mill requires complicated machinery, including barking and chipping machines that reduce small logs to chips in a few seconds.   The chips pass to *digesters,* mixed with *liquor,* and *cooked* at very high temperatures, after which the liquor is drained off in a *diffuser,* to provide several by-products. The pulp is screened and washed, passed on a wire carrier to the drying machine, and then bundled.   Aspen, fir, hemlock, and spruce are chiefly used, but many other white-coloured woods, both hardwoods and softwoods, are also used by chemical pulp producers. Pulp may be produced mechanically by crushing in water, or chemically by breaking down in soda, sulphite, or sulphate solution.   The latter method is superior and more costly.

Pulpit.   An elevated enclosed stage for the preacher in a church. Also see *Sounding Board.*

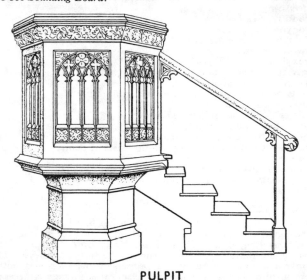

PULPIT

Pulvinated.   Applied to a convex frieze.   Anything convex or cushion-shaped.   A swelling.

Pump Handle. The bar connecting the body with the hind carriage of a vehicle.

Pumpkin Ash. *Fraxinus profunda.* U.S.A. The wood is from wet lowlands and is light, soft, and brittle. See *Ash.*

Pumpkin Pine. Sugar pine and White pine, *q.v.*

Pump Screwdriver. One that operates by pressure only and that springs back to the normal position when the pressure is released. Also a spiral screwdriver. See *Screwdrivers.*

Punah. *Tetramerista glabra* (*H.*). Malay. Brownish yellow or yellow, fragrant, forms a lather with water. Moderately hard and heavy, to heavy. Not durable. Very coarse grain, uneven texture, varied rays. Used for interior constructional work, flooring, etc. S.G. ·74.

Punch. 1. A steel tool for driving nails below the surface of the wood. 2. A small cranked chisel for tightening the nut of a handrail bolt. 3. A centre punch has a sharp point for marking metal.

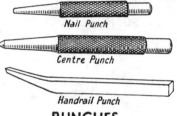

Nail Punch

Centre Punch

Handrail Punch

PUNCHES

Puncheon. 1. A short post, like a king post, in the upper part of a hammer beam, or queen post roof truss, etc., or in a trussed partition. It is sometimes called a quarter. 2. Short timbers supporting the horizontal frames in a box coffer dam. See *Timbering.* 3. A short post or block placed on a pile when the head of the pile is lower than the fall of the ram. 4. A large cask of varying capacity ; for spirits 120 gals. ; beer, 72 gals. ; wine, 84 gals.

Puncher. The engineer who controls the yarder, in lumbering.

Punggai. *Cælostegia Griffithii* (*H.*). Malay. Orange, darkening on exposure, odorous, glistening deposit in pores. Rather soft and light, not durable, flexible. Coarse grain and rays. Easily wrought. Large sizes. Little imported. Used for furniture, sabots, etc. S.G. ·54 ; C.W. 2·5.

Punky. Applied to wood showing signs of decay, or to soft spongy heartwood.

Punt. A flat-bottomed river boat propelled by a long pole.

Pupa. The chrysalis form of an insect.

Purfled. Having a decorated border. Inlaid bordering. An edge or parapet ornamented with crockets.

Purflings. Very fine inlays round the edges of the table and back of a violin to prevent splitting.

Puriri. *Vitex littoralis.* Called New Zealand teak. Greyish brown. Very hard, heavy, strong, and durable, but subject to insect attack. Attractive figure, close texture. Excellent wood but supplies very scarce. Used for structural work, sleepers, turnery, etc. S.G. 1·1.

Purlin Roof. One in which the purlins are carried by cross walls instead of by trusses.

Purlins. Horizontal beams lying on the backs of the principal rafters of a roof truss, between ridge and wall plate, to carry the common rafters. See *Roofs*. Also see *Belfast Roof* and *Centering*.

Purpleheart. *Peltogyne spp.* Much confusion with woods of *Copaifera spp.* (*H.*). B. Guiana. Also called Amaranth, Morado, Violet Wood, Palisander, Pau roxo, etc. Brown to purple on exposure. Very hard, heavy, strong, tough, durable, and elastic. Fine close grain, distinct rays, wavy and roey, very smooth. Seasons readily, difficult to work due to hardness, polishes well. Wax finish best to retain purple colour. Large sizes. Valuable for furniture, decorative work, turnery, marquetry, or where strength and durability are essential. S.G. 1 ; C.W. 4·5. Also see *Guarabu* and *Pau roxo*.

PUSH BLOCKS

Push Block. A shaped block to protect the hand when using the circular saw or spindle moulder.

Push Brace. Operated like a rachet screwdriver and used for light work in a confined space.

PUSH BRACE

Push Cart. A hand-cart or barrow.
Push Chair. A *bath-chair* for invalids or a *go-chair* for babies.
Pusher. 1. A mechanical device to complete the last few inches of a saw-cut on a circular saw, as a protection. 2. See *Stop Action*. 3. A propeller, of an aircraft, behind the engine.
Push Stick. A piece of wood with bird's-mouth end or steel point used for pushing the stuff through the circular saw or machine.

Pussur Wood. *Carapa moluccensis* (*H.*). India. Wine red, variegated streaks, smooth, feels greasy, dull. Heavy and hard, strong, very durable, resists insects. Heavy organic infiltration. Straight, slightly interlocked, grain ; fine even texture. Seasons well, and fairly easy to work for polishing. Used for cabinet work, boat-building, constructional work, tools, spokes, etc. S.G. ·77 ; C.W. 3.

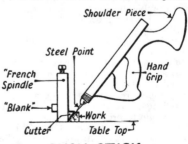

PUSH STICK

Putlogs. Short bearers used in scaffolding. In a bricklayer's scaffold, *q.v.*, one end of the putlog is carried by the wall and the other end by a ledger.   Imported birch putlogs are 6 ft. ×3 in. ×3 in.   Also see *Mason's Scaffold.*

Putter.   A club used for putting in golf.

Putty.   Plastic material that sets hard when used for stopping, pointing, glazing, etc.

Putumuju.   See *Kartang* and *Vermelha.*

Puzzle Fruit Tree.   *Carapa obovata.*   Similar to Pussur wood but lighter in colour and in weight and not so durable.   It also has less infiltration, and is more difficult to work smooth.   S.G. ·75 ;   C.W. 5.

Pycnanthus.   See *Nigerian Box Board.*

Pye Gun.   A high-frequency gluing gun for resin-bonding localised areas.   It is specially suitable for double curvature work.

Pygeum.   See *Red Stinkwood* and *Mueri.*

Pyinkado.   *Xylia dolabriformis* or *Mimosa xylocarpa* (*H.*).   India, Burma.   Also called Ironwood, Acle, Obo, Jambu, Errol, etc.   Reddish brown to rich red wine colour.   Very hard, heavy, strong, tough and durable.   Hard-wearing, non-fissile, and resists insects.   Not very stable.   Close wavy grain, fine rays, beautiful figure, smooth.   Difficult to season and work, but polishes well.   Difficult to glue due to gum secretion.   Used for all purposes :   structural, piling, lock gates sleepers, furniture, decking, keel blocks, etc.   S.G. 1 ;   C.W. 5.

Pyinma.   *Lagerstræmia flos reginæ* (*H.*).   India.   Also called Jarul.   Light red to reddish brown, lustrous, smooth.   Moderately hard and heavy.   Strong and fairly durable.   Resists treatment.   Straight and wavy grain, medium coarse texture.   Seasons and works readily for polish.   Used for superior joinery, furniture, turnery, shipbuilding, etc.   S.G. ·7 ;   C.W. 3.   ANDAMAN PYINMA, *L. hypoleuca.*   Similar to above but heavier, harder, stronger, more durable, and with coarser texture.   S.G. ·75.

Pylon.   1. A slender lattice tower.   2. The structural part of an auto-giro.   A pyramidal arrangement of struts in aircraft.   3. One wing or leaf of a double door or gate.

Pyramid.   A solid having a rectilinear base and its sides triangles, tapering to a point at the apex.   Volume⇒$\frac{1}{3}$ area of base × height.   See *Frustum.*

Pyrus.   See *Apple, Cherry, Hawthorn, Medlar, Mountain Ash, Pear, Rowan, Service Tree, Whitebeam.*

Pyx.   An ornamented box or vessel used in churches to contain the Host.

Q

Q.P. Abbreviation for *quartered partition*.
Qtd. Abbreviation for *quartered*.
Q.T. Window. A patent mechanical device for controlling sliding sashes.

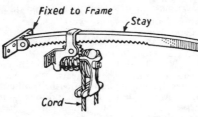

QUADRANT

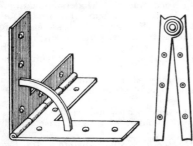

QUADRANT HINGES

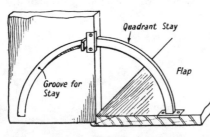

QUADRANT FLAP STAY

Quadra. **1.** A square border or frame. **2.** A plinth to a *podium*.

Quadrangle. **1.** A four-sided plane figure. A square or rectangle. **2.** A rectangular court surrounded by buildings, especially to schools and colleges.

Quadrant. **1.** A stay for a fanlight or casement hinged horizontally. **2.** Quarter of a circle; an angle subtending 90°. **3.** A surveying instrument.

Quadrant Hinge. A stay and hinge combined to support a flap.

Quadrant Stay. A metal stay to support a falling flap, as to a secretaire or bureau.

Quadrilateral. A four-sided plane figure: square, rectangle, rhombus, rhomboid, trapezoid, trapezium.

Quagginess. A defect in timber consisting of numerous shakes, usually from trees grown in loose soil.

Qualea. See *Mandioqueira*.

Quality. See *Conversion, Defects, Disease, Figure, Grain, Moisture Content, Preservation, Resistance, Selection, Shipping Marks,* and *Tests*.

Quamwood. *Schizolobium*

Kellermanii, S. parahybum (*H.*).   British Honduras.   Also called Qualm and Zorra.   Yellowish white, bluish stain.   Soft, woolley texture, distinct rays.   Not imported.

Quangdong. *Elæocarpus spp.* (*H.*).   Queensland.   Several species, distinguished as brown, white, grey, and silver.   Also called Blueberry Ash, Blue Fig, and Caloon.   White to brownish white.   Light, soft, tough, strong, fairly durable, and stable.   Easily wrought.   Useful wood for joinery, cheap cabinet work, bentwood, turnery, etc.   S.G. ·5 ; C.W. 2.

Quants.   Tapered handles for boat-hooks, etc.   They are 14 to 24 ft. long, with a top diameter of about 1¼ in., and usually marketed with poles and rickers.   Poles for propelling craft in waterways.

Quararibea.   See *Veroity*.

Quarrel.   A lozenge, or diamond-shaped, pane of glass or panel, or piercing in tracery work.   A quatrefoil in tracery work.

Quarter. 1. See *Quarterings*. **2.** See *Puncheon*. **3.** A square panel. **4.** The part of a ship's side towards the stern.   Also see *Quarter Deck*.

Quarter-butt.   A billiard cue of special length.

Quarter-deck.   The part of the upper deck of a ship abaft the mainmast.

Quartered.   Logs cut into four quarters for radial conversion, to obtain figure due to rays, ribbon figure, etc. *Quartered oak* is rift-sawn oak.   See *Quarter Sawn, Conversion*, and *Quarterings* (4).

Quartered Partition.   A partition formed of studs, or quarterings.

Quarter Girth Measurements.   A method of finding the cubic contents of a round log.   Quarter of the girth is assumed to be the side of an equivalent square, and the area of the square multiplied by the length equals the volume.   See *Measure*.

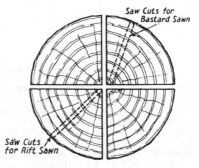

QUARTERED LOG

Quarterings. 1. Wood of small section used as studs in a plastered partititon.   Market sizes 3 in. by 3 in. to 4½ in. by 4 in. **2.** A deal or plank cut into four equal pieces. **3.** Small stuff, square in section, suitable for carpentry or first fixings. **4.** Veneering panels in four pieces as a decorative feature.   The veneers should be cut from four consecutive veneers from the log and carefully matched.   See *Matched*.

Quarter Pieces.   Pieces of wood at the after part of the quarter gallery, near the taffrail, in a wooden ship.

Quarter Rails.   Moulded pieces from the top of the stern to the gangway.   It is a fence to the quarter-deck.

Quarter Round.   An echinus moulding, or ovolo, *q.v.*

Quarters.   See *Quarterings*.

Quarter Sawn.   Converted stuff cut radially, which produces ribbon and silver grain. See *Conversion*. Edge grain, vertical grain, quarter-cut, rift-sawn, all mean the same as quarter-sawn. When the saw-cuts are inclined to the rays to produce pencil-like stripes, and to avoid large flakes, it is sometimes called bastard sawn, and the grain is also called *needle point* and *comb grain*.

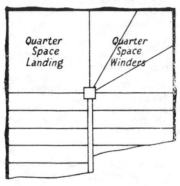

Quarter Sliced. See *Slicing*.

Quarter Space or **Pace.** Applied to a landing that is the width of one flight of stairs, or half the width of the staircase in a dog-legged stairs.

Quarter Staff. A stout wood staff, about 7 ft. long, formerly used as a weapon.

Quarter Turn. A wreath for a handrail turning through 90°, or one right angle.

Quaruba. *Vochysia spp.* (*H.*). Brazil. Pinkish brown. Fairly light, rather hard, moderately strong and durable. Varies considerably in texture and working qualities. Some woolly and somewhat like Sapele. Requires care in seasoning.

QUARTER SPACE

Polishes well with careful filling. A substitute for second-class mahoganies. S.G. ·6; C.W. 3. See *San Juan, Femeri*, and *Vochysia*.

Quassia Wood. Valuable for medicinal purposes but not for wood. Also see *Bitterwood*.

Quatrefoil. Tracery work with four cusps and foils. See *Tracery Panel*.

Quay. A place for the loading and unloading of vessels. A wharf.

Quebec. A prefix denoting the source of numerous Canadian woods. Especially applied to aspen, birch, elm, poplar, and white pine.

Quebec Standard. A measure of timber of 100 pieces 12 ft. × 11 in. × 2½ in., or 229¼ cub. ft. See *Standard*.

Quebracho. *Schinopsis spp.* (*H.*). C. America. Also called Iron-wood, Tanwood, and Red Lignum-vitæ. Cherry red, darkening with exposure. Very hard, heavy, strong, tough, and durable; brittle. Irregular grain and roey. Fine and uniform texture. Very difficult to work, polishes well. Important for tannin. Used for structural work, sleepers, cabinet work. S.G. 1·1; C.W. 5. WHITE QUEBRACHO, *Aspidosperma sp.* Yellowish to rose red, variegated. Fades with exposure. Similar properties and characteristics to *Schinopsis spp.*, but more resilient and not so durable. It is also softer and lighter, but used for similar purposes. S.G. ·95. The two varieties are known as Quebracho Colorado and Q. Blanco. Species of *Astronium, Brasilettia, Lysiloma* are also called Quebracho.

Queen Anne. A style of period furniture, 1702-1714. Marquetry, carving, cabriole legs, etc., were common features.

Queen Bolt. A bolt used instead of a post in a Queen post roof truss.

Queen Posts. The two vertical members in a Queen post roof truss,

QUEEN ANNE

or the two principal posts of any framed truss. Also see *King* and *Princess Posts*.

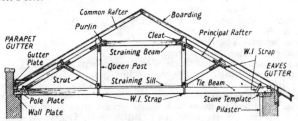

QUEEN POST ROOF TRUSS

Queensland.   A prefix to numerous woods from Queensland, Australia, many of which are valuable veneer woods : *Maple Silkwood, Oak, Pine, Red Tulip Oak, Rose Alder, Rose Walnut, Sandalwood, Satinay, Satin Sycamore, Silky Oak, Silver Ash,* and *Walnut Bean.* Many other Queensland woods are described in the Glossary.

Queensland Acacia.   Numerous species.   See *Gidgee, Marblewood, Mulga, Myall, Spearwood,* and *Wattle.*

Queensland Cypress Pine.   *Callitris spp.*   Marketed as Cypress Pine, *q.v.*   About seven important species, hence there is considerable variation in the wood, which is steadily increasing in importance. The wood has not been favoured because of its brittleness and the large number of firm small knots.

Queensland Greenheart. *Endiandra compressa* (*H.*). Also called Golden Birch and Whitebark. Olive hue, metallic lustre. Very hard, heavy, strong, tough, and durable ; elastic. Very compact with fine even grain. Difficult to work. Used for fishing rods, turnery, handles, golf sticks, etc. S.G. 1 ; C.W. 4.

Queensland Maple. See *Maple Silkwood*.

Queensland Oak. Caledonian Oak, *q.v.*

Queensland Pine. Includes Bunya, Hoop, and Kauri Pines, *q.v.*

Queensland Red Beech. Maple Silkwood, *q.v.*

Queensland Red Cedar. *Cedrela toona.* See *Toon*.

Queensland Sycamore. *Litsea spp.* (*H.*). Also called Brown Bolly-wood and She Beech. Pale brown, dull but smooth, faint reddish brown specks. Fairly soft and light. Firm, tough, and stable. Moderately strong and durable, but subject to borers. Straight grain, some wavy, rather woolly texture. Easily wrought and seasoned, large sizes. Useful second-class cabinet wood. Used for motor bodies, small boats, interior joinery, bent work, carving, plywood. S.G. ·6 ; C.W. 3.

Queensland Walnut. *Endiandra Palmerstonii.* Resembles American black walnut. Called Oriental Wood. Reddish brown to almost black, variegated, irregular black markings. Hard, heavy, strong, stable, not very durable. High insulation properties. Interlocked, wavy and curly grain. Odorous until seasoned. Contains silica which is hard on high-speed tools. Difficult to work and requires care in seasoning. Finishes and polishes well. Large sizes. Used for superior joinery, cabinet work, electrical appliances, propellers, veneers, etc. S.G. ·75 ; C.W. 5. Also see *Oak Walnut*.

Quercus. The botanical name of Oak, *q.v.* Also see *Mempening*, *Pasania*, *Holm Oak*, *Turkey Box*, and *Willow Oak*.

Quick Beam. The Rowan Tree, *q.v.*

Quickset. A hawthorn hedge.

Quick Sweep. Applied to circular work of small radius.

Quilted or **Quilting.** **1.** Applied to highly figured veneers that have the appearance of folds or waves, or blister-like, as often found in maple. **2.** Ridges on sawn wood, usually from reciprocating saws.

Quina. Applied to numerous C. American trees from which bitter medicinal extracts are obtained : Bitterwood, Cabreuva, Cinchona, etc.

Quince. *Cydonia vulgaris* (*H.*). India. Yellowish white, darker streaks. Fairly hard and heavy. Very close grain. Small sizes. Used chiefly for turnery. Other varieties are cultivated chiefly for their fruit. S.G. ·74.

Quira. *Platymiscium polystachyum* (*H.*). C. America, Panama Redwood or Vencola. Also called Roble and Macawood. Rich red, variegated streaks, lustrous, and smooth. Very hard, heavy, and durable. Fine roey grain. Fairly difficult to work, polishes well. Used chiefly for cabinet work. S.G. ·8; C.W. 3·5. Also *Andira spp.*

Quirk. A narrow flat groove, or sinking, forming part of a moulding. It is commonly placed along the back edge of a planted panel moulding to break the joint.

Quirk Bead. A semicircular moulding with a quirk. It is used at a salient corner, or to break the joint between two boards, as in match-boarding. See *Return Bead* and *Angle Joints* (*C.*).

Quirk Cutter. A tool for forming quirks, especially in circular work. Also see *Router*.

Quoin. 1. The external corner of a building. 2. A wedge ; especially the small hardwood wedges used for fixing the type in the formes in the printing trades ; or large wedges used to prevent a barrel from rolling.

Quoit Terminal. An inverted monkey tail or vertical scroll at the bottom end of a handrail to serve as a warning on arrival at the foot of a flight of stairs. It is a very unusual terminal to a handrail.

R

R. Abbreviation for *rebate, reaction,* and for *moment of resistance.*
r. A symbol for *radial section.*
Rabat or **Rebatment.** A term used in geometry when a surface is rotated about its trace into one of the planes of projection to show the true shape.
Rabbet. Same as rebate, *q.v.*
Rabbet Line. The line of contact between planking and keel of a ship.
Rabo Lagarto. Ruda, *q.v.*
Race. A water-course to a water-wheel or slide. A channel, slot, or groove in which something slides or runs.
Race Knife. See *Raze Knife.*

RAZE KNIFE

Rack. 1. Any arrangement of framing, shelving, hooks, bars, etc., for holding or storing things : book, fodder, hat, plate, pipe rack, etc. See *Gable to Rack* and *Pew.* 2. Any type of frame used for stretching or straining. 3. A rail with teeth to engage with wheel teeth, as for a pinion. 4. See *Piling.* 5. See *Pipe-rack.*
Racked. Temporary timbering braced to prevent distortion.
Racket or **Racquet.** An open frame, with handle, tightly strung with catgut for use in certain games : tennis, badminton, squash, etc. A battledore.
Racking. 1. Out of square. Frames distorted through strain. 2. See *Piling.*
Rack-pillar. See *Pipe-rack.*
Rack Saw. A log saw with travelling table.
Radial Bar. A radius rod, *q.v.*

Radial Sawn. Rift sawn, *q.v.*
Radial Section. The surface of wood cut along a plane containing the axis or the pith.
Radial Shake. One originating at the circumference of the log. See *Shakes*.
Radial Square. A tool for drawing radial lines on circular work. Also called a circular square.
Radial Step. A winder, *q.v.*
Radiator Dry Kiln. One in which the air is heated over steam pipes and then forced into the chamber containing the timber.

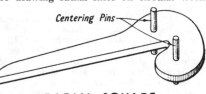

Centering Pins

RADIAL SQUARE

Radiogram. A combination of radio receiver and gramophone.
Radius of Gyration (K.). A property of the cross section of a column or strut, used in design, and denoting a relative measure of stiffness. It depends upon the sectional area and its distribution about an axis. $K = \sqrt{I \div A}$ where I = moment of inertia and A = sectional area.
Radius Rod. A rod equal in length to the radius of the required circle and used for testing or describing circular arcs. A trammel.
Raffling. Applied to a notched edge to carved foliage.
Raft. A flat floating structure. An assemblage of logs or timbers for floating down stream. When buoyancy has to be provided for a raft of heavy hardwoods it is often called a float.
Rafters. 1. The slender timbers, or common rafters, running from ridge to wall plate, and carrying the covering of a roof. Also called spars. See *Close Couple, Principal Rafter, Centering*, and *Jack Rafter*. 2. Dockers who unload timbers from ships and parcel them into rafts ready for sale.
Rafter Shoe. A cast-iron socket for the foot of the principle rafter of a roof truss.
Raft Foundation. 1. A site for a building in the form of a raft supported on piles. 2. A grillage foundation, *q.v.*
Raggled. Slotted or notched for a series of laths. See *Grating*.
Raglin or **Raglan.** A slender ceiling joist.
Rail. 1. A horizontal member of a frame, panelling, fence, etc. 2. A bar protecting an enclosure, staircase, etc. See *Handrail*. 3. A rod from which something is suspended, or runs by means of rings, hooks, pulleys, etc. 4. A piece fitted on the heads of the bulwark stanchions in shipbuilding. 5. Large ornamental mouldings planted on the upper parts of wood ships, but sometimes serving as a protection. They were variously described as drift, fife, rough-tree, sheer, or waist rails, etc. 6. Lengths of steel, called lines, shaped for flanged wheels to run along.
Railings. 1. A fence or open iron protection for an enclosure. 2. See *Bandings* (2) and *Flush Door*.
Railway Keys. Hardwood wedges to fix a railway line to the chair, which is in turn fixed to the sleeper.

Railway Roof. A cantilever or umbrella roof as used on station island platforms.

Railway Tie. A sleeper, *q.v.*

Raindrop. Figure in wood consisting of a combination of mottle and broken stripe, or fiddleback occurring in single or widely spaced waves. The waves in the fibres occur singly or in groups with big intervals between.

Rain Tree. *Pithecolobium*, or *Inga, saman*, or *Enterolobium spp.* (*H.*). Widespread. India, S. America. Also called Saman, Zorra, etc. Nut brown in colour. Somewhat like *Albizzia lebbek* but more uneven in texture and lighter and softer. The tree is valued for its beauty and shade rather than the wood. S.G. ·5. See *Gaunacaste* and *Kungkur*.

Rainwater Pipe. See *Down Spout*.

Raised Girt. See *Girt*.

Raised Lid. A term used in coffin-making for built-up lids. The usual method is to plant a raised panel. It may be single or double-raised. For a double-raised lid, the panel is cut out of the top and rests on a moulded strip, mitred and planted on the top, so that the lid shows three tiers in thickness.

Raised Panel. One thicker at the middle than at the edges, for strength and ornamentation. See *Fielded Panel*.

Raising Plate. A horizontal timber carrying the feet of other timbers. A pole plate.

Raizes. Buttresses, *q.v.*

Rajbrikh. Indian laburnum. See *Laburnum*.

Rake. 1. A long-handled implement with teeth for collecting hay, straw, etc., or levelling earth, or mixing plastic materials. 2. Slope or inclination, as to a ship's mast or funnel. 3. The set of saw-teeth. 4. The projection of the upper part of a wooden ship at the height of stem and stern beyond the ends of the keel.

Rake Board. Same as barge board. An inclined board.

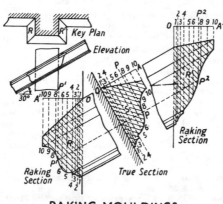

RAKING MOULDINGS

Raker. 1. See *Raking Shores*. 2. A wind brace, but more especially in steel construction. 3. Strong timbers fixed between purlins to carry a dormer.

Raking Balusters. Those with foot and head blocks shaped to the pitch of the stairs. Common in Jacobean architecture. Also applied to metal balusters that are not vertical.

Raking Moulding or **Cornice.** One inclined to the horizontal, and with horizontal returns. See *Pediment*. The illustration shows the method of obtaining the contours of the returns.

Raking Pile. An inclined pile.

Raking Risers. Applied to a step in which the riser slopes inwards to provide more foothold. More usually applied to the risers of winders, because of the narrow ends.

Raking Shores. Inclined struts supporting a defective wall or whilst it is undergoing structural alterations. Also see *Shores*.

Raking Surface. A twisted surface, as under the winders of a geometrical stairs.

Ralli-car. A light two-wheeled carriage of the dog-cart type. It accommodates four persons sitting sideways.

Ram. 1. Any heavy implement for ramming, pounding, or driving. The hammer of a pile-driver. **2.** To consolidate loose material with a pommel or rammer.

Rammer. A heavy implement for consolidating loose material, levelling sets, etc. It usually consists of a heavy base piece of wood or metal and a long vertical handle.

Ramon. Ogechi, *q.v.*

Ramp. 1. An inclined runway, floor, or road, usually curved in plan. **2.** A vertical curve, or large easing, in a handrail or moulding. A concave sweep from a lower to a higher level. See *Knee to Handrail*. **3.** A raised structure at a saw mill on which the sawn wood is stored and from which the wood can be loaded on to wagons or ships by gravity.

Ramp and **Twist.** Applied to surfaces that both incline and turn. A twisted or raking surface.

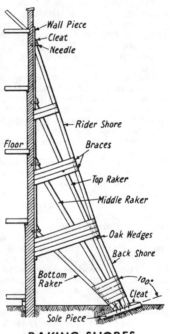

**RAKING SHORES
WITH RIDER**

Rampant. Applied to an arch or vault with springings at different levels, which requires a *rampant centre* for its construction.

Ram's Horn. Figure in wood that is a variation of ripple or fiddle-back. See *Figure*.

Rance. A shore or prop.

Rand. A border or margin. The waste piece left from straightening the edge of a board.

r. & b. Abbreviation for *rebate and head*.

Random Matched. Pieces of unequal sizes combined in one face. See *Matched* and *Veneer Matching*.

Random Widths. Especially applied to shingles in which there is considerable variation in the widths in each bundle.

r. & v. Abbreviation for *rebate and vee*.

Range Cleat. See *Ship's Cleats*.

Range Cypress. See *Queensland Cypress*.

Ranges. Kevels, *q.v.*

Ranging Bond. Brickwork with long horizontal wood strips built in the joints to serve as fixings.

Ranging Down. Correcting the tips of the teeth of a circular saw by means of a piece of York stone, applied whilst the saw is running.

Ranging Poles. Surveying poles for sighting, or ranging, levels. They are 6 to 10 ft. long with steel shoes and differently coloured sections for easy sighting.

Rangoon Teak. See *Teak*.

Rankine. The name of the engineer responsible for several important formulæ used in structural design.

Rank Set. Opposite to fine set. Applied to plane irons when the back iron is set well back from the cutting edge, for rough work.

Rantree. The Rowan Tree, *q.v.*

Rapanea. See *Canelon*.

Rap-rig. A patent type of scaffolding of light timbers easily assembled and adjusted.

Rasp. An abrading tool for wood and lead. It is like a rough file but with separate teeth instead of furrows.

Raspberry Jam Wood. *Acacia acuminata* (*H.*). W. Australia. Also called Violet wood. Crimson to violet in colour. Hard, heavy, strong, durable, and very fragrant. Excellent wood for inlays, turnery, small cabinet work, etc. S.G. 1.

Rassac. Empata, *q.v.*

Rata. *Metrosideros lucida* and *M. robusta* (*H.*). New Zealand Ironwood. Reddish brown. Hard, heavy, very strong and durable. Difficult to work. Little exported. Used for structural work, sleepers, wheelwright's work and shipbuilding. S.G. ·75. The tree is unique, as seeds can propagate in the branches of other trees. The roots, up to 3 ft. diameter, trail down and suckers crush the supporting tree to destruction.

Ratchet. A mechanical contrivance for converting circular movement into a to-and-fro movement, as in braces, screw-drivers, etc. A small projection, or pawl, engages with a toothed wheel to check movement.

Rate of Feed. A term used in machining, stating the speed in feet per minute at which the stuff passes through the machine. See *Machines*.

Rat Tail. A round tapered file.

Rauli. Southland Beech, *q.v.*

Rave. A rail in the side framing of a vehicle. It is distinguished as floating, head, middle, or top rave.

Ray Fleck. Part of a ray on rift-sawn wood.

Rays. Ribbons of tissue, chiefly parenchyma cells, that run radially through the wood at right angles to the axis of the tree. They are also called medullary rays, wood rays, and pith rays. In many woods they are not discernable except under the microscope, but when they are distinct they produce characteristic figure in rift-sawn wood, such as silver grain in oak, *q.v.* There are two types : medullary and vascular rays.

Raze Knife. A hooked tool used for scribing, or marking, the cubical contents of measured logs. Also called a scribe, or screeve knife, or scratcher.

r.c.fm. Abbreviation for *Russian fathom*, which is 7 ft. cub. or 343 cub. ft.

Rdm. Abbreviation for *random sizes*.

Reaction. The counteraction of a force, or forces, to produce equilibrium. The term is specially applied to the upward resistance of the wall against the downward pressure of the supported load. See *Equilibrium*.

Reading Desk. An inclined shelf for holding an open book and usually supported by a tall pedestal. See *Lectern* and *Desk*. In public

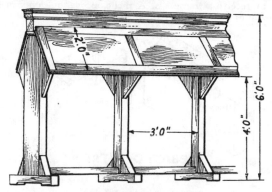

READING DESK

libraries it is in the form of a continuous framing and intended to hold a number of newspapers and periodicals.

Real Salie. *Brachylæna sp.* (*H.*). S. Africa. Also called Vaalbosch. A fairly hard, tough, light coloured wood with close even grain Used for fishing rods, golf sticks, etc.

REAMERS

Reamer or **Rimer.** A machine or brace bit for enlarging or tapering holes.

Rebate. A recess along the edge of a piece of material to receive

another piece, or a door, sash, or frame. Also called *rabbet*. See *Angle Joints*, *Finishings*, and *Floor Boards*.
Rebated Mitre Dovetail. A special form of secret dovetail, *q.v.*
Rebated Siding. See *Weather Boarding*.
Rebound. Same as kick-back, *q.v.*
Receding Cornice. One that sets back instead of projecting. See *Cornice*.

REBATE PLANE

Recepage. Cutting back young trees to produce clean straight trunks.
Recess. **1.** The rebate, or set-back, to the jambs and head of a wall opening to receive a frame. **2.** A niche, alcove, or cavity in a wall. **3.** A sinking in the surface of wood.

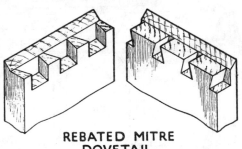

**REBATED MITRE
DOVETAIL**

Recessed Arch. An arch within and behind another arch. A compound or concentric arch.
Recessing. Forming sinkings on the surface to give an appearance of overlaid fretwork. Also see *Sunk Panel*.
Recessing Machine. See *Elephant*.
Reciprocating Saw. One operating up-and-down or to-and-fro, as distinct from circular or band saws.
Recorder. A wood-wind musical instrument shaped somewhat like a flute.
Red. A prefix applied to a large number of woods to distinguish the wood or the species. It is not always applied because of the colour but perhaps because the wood or bark is darker than in other species. The distinction may refer to either wood, heartwood, or tree.
Red Almond. *Alphitonia excelsa* (*H.*). Queensland. Also called Red Ash, White Myrtle, etc. Pale pink, darkening to brick red, variegated. Fairly hard and heavy, strong and durable. Firm grain and texture. Requires care in seasoning, not difficult to work, polishes well. Used for all purposes : small structural parts, fencing, flooring,

carpentry, coach and railway carriage work, fancy articles, turnery. S.G. ·6 to ·8 ; C.W. 3.

Red Ant. *Formica smaragdina.* Not so destructive as the White Ant, *q.v.*

Red Ash. *Fraxinus pubencens* (*H.*). N. America. Similar to white ash. Also *Alphitonia excelsa.* S. Australia. Resembles ordinary ash on conversion but darkens to deep red with age. Very decorative but little exported. S.G. ·85. Also see *Mountain Ash* or *Tasmanian Oak* (Australian), and *Ash.*

Red Bean. *Dysoxylon muelleri* (*H.*). N.S. Wales. Red. Substitute for cedar and soft mahoganies. Called Turnip wood because of the odour during conversion. Used for furniture and superior joinery.

Red Beech. *Fagus ferruginea* (*H.*). N. America. Similar to English beech but coarser in texture. Warps and splits easily. *Nothofagus fusca.* See *New Zealand Beech.* Also see *Maple Silkwood.*

Red Birch. *Betula lutea.* N. America. Applied to heartwood only. See *Birch.*

Red Bloodwood. *Eucalyptus corymbosa* (*H.*). Queensland. Red, lustrous, smooth. Hard, heavy, very strong and durable, resists insects and fire. Subject to concentric kino veins which makes the wood shelly. Requires care in seasoning and fairly difficult to work, polishes. The gum veins restrict its use for many purposes. Used chiefly for heavy structural work, sleepers, fencing, etc. S.G. ·9 ; C.W. 3.

Red Bombway. *Planchonia andamanica* (*H.*). Andaman I. Brownish red, smooth. A very hard, heavy, strong and durable wood, but not very stable. Little imported. S.G. ·98. See *White Bombway.*

Red Box. *Eucalyptus polyanthema* (*H.*). Victoria, Australia. Dark reddish brown. Very hard, heavy, strong and durable. Close interlocked grain. Difficult to work. Used for sleepers, paving, wagon building, etc. S.G. 1·1 ; C.W. 4·5.

Red Cabbage Bark. Angelin, *q.v.*

Red Camphor. Borneo camphorwood, *q.v.*

Red Carrobean. *Geissois Benthami* (*H.*). Queensland. Similar to Rose Marara. Reddish brown. Hard, heavy, strong and tough, but not durable. Close grain, even texture, fine brown rays. Easily wrought and polishes well. Requires care in seasoning. Used for flooring, panelling, brush stock, etc. S.G. ·7 ; C.W. 3.

Red Cedar. Pencil cedar. See *Cedar.* Also see *Western Red Cedar*, *Red Seraya*, and *Toon.*

Red Cypress. Louisiana cypress. See *Cypress.* Also applied generally to cypress of a deep colour.

Red Deal. **1.** Redwood, *q.v.* **2.** Redwood sawn into *deal* sizes.

Red Dhup. Parishia, *q.v.*

Red Elm. English Elm, *q.v.*

Red Els. *Cunonia capensis* (*H.*). S. Africa. Reddish brown, darker streaks. Hard, heavy, strong, and tough. Straight even grain with little figure. Close texture but fibrous. An excellent wood for plain turnery, furniture, and work requiring strength and durability, but little exported. S.G. ·75.

Red Eyne. *Soymida febrifuga* (*H.*). Indian Red Wood or Soymida.

Rich dark brown, lighter streaks, dull but smooth, feels oily. Very hard, heavy, durable, and strong, and resists insects. Close irregular twisted grain, coarse texture. Difficult to season and work, due to hardness, infiltration, and knots. Used for cabinet work, superior joinery, carriage work, carving, turnery, and also heavy constructional work. Rotary cut veneers are highly decorative. S.G. 1 ; C.W. 3·5.

Red Fir. Noble Fir. See *Fir*.

Red Gum. *Liquidambar styraciflua* (*H*.). U.S.A. American Red Gum. Also called Sweet Gum, Bilsted, etc. Reddish brown, variegated, lustrous, smooth. Moderately hard, heavy, and tough. Fairly strong, elastic, but not durable. Warps freely unless carefully seasoned. Uniform grain and texture, interlocked, beautiful figure. Easily wrought and polishes well. The heartwood, about 40 per cent. of the log, is called Satin Walnut. The sapwood is called Sap Gum or Hazel Pine, *q.v.* Used for furniture, veneers, superior joinery, etc. The sapwood is used for packing cases, etc., or stained for furniture. S.G. ·5 to ·65 ; C.W. 2. See *Gum*.

Red Heart. An incipient stage of red rot.

Red Hickory. The heartwood of hickory, *q.v.*

Red Ironbark. *Eucalyptus creba*. See *Ironbark*.

Red Ironwood. Ekki, *q.v.* Also see *Ironwood* (*Erythrophlœum Laboucherii*).

Red Ivory. *Rhamnus zeyheri* (*H*.). S. Africa. Red. Very hard and heavy, with fine even grain. Excellent wood for small decorative work, turnery, etc. See *Buckthorn*.

Red Lauan. *Shorea negrosensis*, and other spp. (*H*.). Philippines. The species are distinguished as red lauan proper, bataan, mayapis, tanguile, tiaong. See *Lauan*, *Meranti*, and *Seraya*.

Red Mahogany. *Eucalyptus resinifera* (*H*.). New South Wales. Rich red. Hard, heavy, strong, and durable. Hardens with age. Close interlocked grain, some roe and mottle. Used for constructional work, joinery, flooring, fencing, etc. S.G. ·9 ; C.W. 4. See *Eucalyptus* and *Mahogany*.

Red Maple. *Acer rubrum*. Light and soft wood. Little imported. See *Maple*.

Red Meranti. *Shorea acuminata*, *S. eximia*, *S. parvifolia*, *S. dasyphylla*, and other spp. See *Meranti*, *Lauan*, *Nemesu*, and *Yellow Meranti*.

Red Oak. *Quercus borealis*, *Q. rubra*, and other spp. See *Oak*.

Red Pear. *Scolopia ecklonii* (*H*.). S. Africa. Useful wood but scarce and not exported.

Red Pine. *Pinus resinosa* (*S*.). Canadian and U.S.A. red pine. Characteristics and properties between those of yellow pine and redwood. Specimens may be confused with either. Light red to reddish yellow, Uniform grain, fairly elastic and tough. Not durable unless treated. Used chiefly for interior work. S.G. ·5 ; C.W. 1·5. Also see *King William Pine*, *Kuro Matsu*, and *Rimu*.

Red Rot. A loose term describing incipient stages of fungi attack. Also decay in which the cellulose is destroyed leaving lignin. See *White Rot*.

Red Sanders. *Pterocarpus santalinus* (*H*.). India. Red Sandalwood or Coralwood. Orange red to almost black, fairly lustrous. Very hard, heavy, and durable. Properties similar to teak. Interlocked grain, ripple, medium fine texture. Exudes gum. Seasons well, but slowly. Difficult to work, polishes well. Used for structural work, implements, decorative work, carving, turnery, etc. The santalin content makes it a valuable dyewood. S.G. 1·1 ; C.W. 4·5.

Red Satinay. See *Satinay*.

Red Serayah. See *Seraya* or *Seriah*.

Red Silky Oak. *Stenocarpus salignus* (*H*.). Queensland. Also called Beefwood. Dark red, otherwise similar to Silky Oak, *q.v.*

Red Siris. See *Siris*.

Red Spruce. *Picea rubra* (*S*.). N. America. Similar to black spruce. Used chiefly for pulp. See *Spruce*.

Red Stringybark. *Eucalyptus macrorrhyncha* (*H*.). Victoria, Australia. Light brown. Moderately hard, heavy and durable. Close grain and fissile. Shrinks and warps badly if seasoned quickly. Not difficult to work. Used for structural and constructional work, flooring, sleepers, poles, etc. S.G. ·8 ; C.W. 3. See *Stringybark*.

Red Stripe. Caused by cracks in summer-felled wood that admit spores of fungi. The cracks close when floated to the mill and the spores germinate and cause decay.

Red Tingle Tingle. See *Tingle-tingle*.

Red Touriga. See *Touriga*.

Red Tulip Oak. *Tarrietia argyrodendron* and *T. peralata* (*H*.). Queensland. Several other varieties, but inferior. Variegated pale pink to reddish brown. Beautiful figure, wavy and mottled, and striking silver grain. Fairly hard ; heavy, stable, elastic and strong. Not very durable, and fissile. Straight open grain. Requires careful seasoning. Fairly difficult to work and does not take glue readily. Fumes and polishes well. Used for furniture, superior joinery, veneers, plywood, bent-work, flooring, electric fittings, fishing rods, etc. S.G. ·85 ; C.W. 3·5.

Redwood. *Pinus sylvestris* (*S*.). N. Europe. Baltic redwood. Also called red and yellow deal, Scots or Baltic fir or pine. Often named after the port of shipment. There are seventeen trade names for this species commonly used in this country. Pale reddish brown, distinct rings, moderately resinous. Light, soft, moderately strong and durable, easily treated. Straight grain. Easily seasoned and wrought. Maximum sizes 4 in. by 11 in. Used extensively for carpentry, joinery, flooring, sleepers, poles, piling, pitwood, etc. S.G. ·4 to ·65 ; C.W. 1·75. ANDAMAN REDWOOD. Padauk, *q.v.* BORNEO REDWOOD. Red Seraya, *q.v.* HONDURAS REDWOOD, *Erythroxylon affini* (*H*.). Brownish red, similar to the darker African mahoganies. Uneven, hard and soft texture in narrow stripes, fine rays. Suitable for cabinet work and superior joinery. Also see *Coca*. INDIAN REDWOOD, *Polyalthia cerasoides* (*H*.). Olive grey, yellowish cast, lustrous, smooth. Hard, moderately heavy. Straight grain, distinct rays, medium fine texture. Requires careful seasoning, easily wrought, polishes well. Suitable for joinery, turnery, matching, and planking. S.G. ·68 ; C.W. 3. ZAMBESI REDWOOD. Rhodesian " teak," *q.v.* Also see *Quira* (*Panama*), *Sappon* (*Indian*), *Satine* (*Brazilian*), *Sequoia*, and *Soymida*. The name redwood is sometimes applied to Eucalyptus spp., especially *E. resin-ifera*, and *E. transcontinentalis*, but little is exported.

Reeds. 1. A series of sunk beads, usually without quirks, on the face of the stuff. Thin narrow strips, side by side, on the surface of furniture. 2. Semicircular projecting mouldings on pillars, shafts, etc. 3. A vibrating tongue to certain musical instruments, organ pipes, etc.

REEDS

Reel. A spool or bobbin on which cotton sewing thread or silk is wound, or fishing lines or log lines, etc. A large cylinder on which cables, wires, etc., are wound.

Refectory. A room for refreshments in a scholastic or religious community.

Reflex Hinge. Used for reflex mirrors. It also serves as a rising butt.

Refrigerator. A chamber or box the inside of which is kept at a very low temperature for preserving perishable foods.

Refuse Grinder. A machine for converting saw-mill refuse into fuel, or pulpwood to chips. Also called a hog, edging grinder, or chipper.

Registers. Organ stops. See *Stops*.

Register, To. To indicate a particular part or point.

Reglet. A listel or fillet. A small rectangular moulding. Also a rectangular groove or *raglet*. A small flat moulding as an interlacing ornament or as a division between panels. Strips of wood used in formes in the printing trade.

Regula. The fillet under the triglyph in the Doric order of archi-tecture. Same as *reglet*.

Rejects. Pieces rejected because they do not conform to the required standard or specification. They are often marketed as such. See *Cull Lumber*.

Releasing Key. A tapered piece, or wedge, for easing shuttering for concrete work.

Relief. Prominent carved ornamentation is referred to as *in relief*. It may be high, half, or low relief (alto, mezzo, or basso relievo), according to the degree of prominence.

Relieving Arch. An arch to relieve a lintel from the load above. A rough arch.

Re-mouthing. Fitting a new mouth-piece in the sole of a wood plane to close the mouth. Metal plates may be obtained for the same purpose.

Renaissance. Applied to sixteenth century architecture that attempted to replace Gothic by the classic architecture of Greece and Rome.

Rendered Laths. Riven, or split, laths for plaster-work. See *Laths*.

Rengas. *Melanorrhœa spp.* (*H.*). Malay. Also called Borneo Rosewood, Straits Mahogany, or Black Varnish Tree. Variegated rich red. Moderately hard and heavy to heavy. Rather coarse texture. Slow in seasoning. Rather difficult to work, and irritates the skin. Used for superior joinery and cabinet work. S.G. ·7 ; C.W. 4.

Rengha. See *Jujube.*

Rengkong. Mersawa, *q.v.*

Rentokil. An insecticide that destroys insects in wood.

Reprise. 1. A return of a moulding for an internal angle. 2. A stool for a mullion or jamb.

Rere Arch. An arch with a level soffit over splayed jambs.

Reredos. An elaborately carved screen in the chancel of a church usually serving as a background to the altar. An altar piece.

Res. Abbreviation for re-sawn.

Resak. *Shorea* and *Vatica spp.* (*H.*). Malay. Brownish yellow to dark brown, very smooth. Very hard and heavy. Durable and strong, but subject to pin-hole borers. Straight grain, ripple, distinct rays, resin canals. Requires careful season-

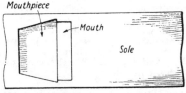

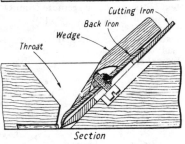

Section

REMOUTHING PLANE

ing. Not difficult to work. Used for furniture, superior joinery, structural work, sleepers. S.G. ·72 to 1 ; C.W.3.

Re-saw. To re-saw means to cut converted wood into smaller sections. There are numerous types of re-saws : gang, sash gang, band, etc.

Reset, To. Correcting the set of the teeth of a saw.

Resilience. The property of a material to recover from strain, due to its elasticity. The work done per unit volume in producing strain.

Resin. Solidified vegetable products : amber, dammar, gum anime, copal, etc. Yellowish brown and transparent or translucent. It exudes from trees and the moisture, or oil, is evaporated, leaving the solid substance which is used for varnish, lacquer, etc. It is soluble in ether, alcohol, or turpentine, but not in water.

Resin Bonded. See *Plastic Glues.*

Resin Canal, or **Duct.** Intercellular tubular cavities containing resin and surrounded by excreting cells. They are common to most softwoods—pines, firs, etc.—and they occur in many hardwoods, especially of the Dipterocarpaceæ family.

Resin Gall or **Pocket.** Excessive local accumulations of resin or gum. See *Pitch Pocket.*

Resinous. Applied to coniferous trees that contain resin : pines and firs.

Resistance. The qualities and properties of wood to be considered for different purposes include resistance against abrasion, bending, compression, conduction, fission, fungus, impact, insect, penetration, resonance, shear, tension, torsion, treatment, wear, and weather. See *Tests* and *Selection.* Also see *Moment of Resistance.*

Resolution of Forces. A term used in mechanics for finding the components of a force.

Resonance. The quality of reflecting sound, which makes wood especially useful for musical instruments. Silver fir and sugar pine are very resonant or vibrant.

Respond. A pilaster forming a pair, or matching another pilaster or column. A half-pillar.

Respond Newel. A half-newel fixed to the wall or dado to pair with the newel.

Rest. See *Lathe.*

Rest House. A hut, remote from other buildings, in which travellers may obtain shelter. Now applied to more pretentious buildings used as hostels.

Resultant. A force that replaces two or more other forces. The result of compounding a number of forces.

Resweld. A proprietary resin-bonded plywood, suitable for exterior work, formwork, etc.

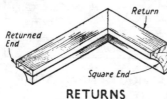

Return

Returned End

Square End

RETURNS

Retable. A back screen to a church altar.

Retaining Wall. A wall resisting the pressure of earth.

Reticulated Moulding. One with interlaced fillets on a prominent member of the moulding, and characteristic of Norman architecture.

Retinospera. Japanese cypress.

Return. 1. The continuation of a moulding or member on an adjacent surface, usually at right angles. 2. The end of a moulding shaped to the contour of the moulding.

Return Bead. A bead stuck on both faces of a salient corner. See *Angle Joints* (*d*).

Return End. See *Return* (2).

Reveal. The side of a door or window opening between the frame and the face of the wall, *i.e.*, at right angles to the face of the wall.

Reverse. A templet shaped to the reverse of a moulding to test the contour when working it to shape.

Reversal of Stress. Change from one kind of stress to another in structural members, as tension to compression. These conditions tend to produce fatigue in the material.

Reversible Hinge. A patent hinge for screens to enable them to fold in either direction.

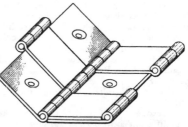

REVERSIBLE SCREEN HINGE

Revolution Marks. Machine burns, *q.v.*

Revolving Bracket. A piece removed and a corresponding piece fitted into the rail at the end of a table to support a flap. The bracket revolves on an iron pin the top of which is held in position by the false top of the table.

Revolving Doors. Doors, or wings, that revolve about a central axis like a turnstile. The doorway may have two or four compartments, with a corresponding number of wings, so that it is draught-proof.

Revolving Shutters. Narrow strips of wood strung on flexible metal bands, or hinged together, to form a continuous surface to protect an opening. The shutter winds round a cylinder in a box above the opening when not in use. There are several patent types. See *Blind, Rolling Shutter,* and *Tambours.*

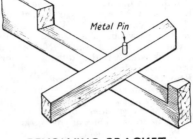

Metal Pin

REVOLVING BRACKET

Rewa-rewa. *Knightia excelsa* (*H.*). New Zealand Honeysuckle. Deep red, lustrous. Hard, heavy, strong, fairly durable, and fire-resisting. Beautiful figure with silver grain and mottle. Used for cabinet work, piling, etc. Little exported. S.G. ·8.

Rfg. Abbreviation for *roofing.*

Rhamnaceæ. The Buckthorn family. Includes *Alphitonia, Ceanothus, Colubrina, Condalia, Emmenospermum, Krugiodendron, Reynosia, Rhamnus, Zizyphus.*

Rhamnus. Buckthorn, *q.v.* Also see *Red Ivory* and *Ivorywood.*

Rhizophora. See *Bakau, Egba, Mangrove,* and *Rhizophoraceæ.*

Rhizophoraceæ. The Mangrove family. A small tropical family that includes

REVOLVING DOORS

Bruguiera, Carallia, and *Rhizophora.* The trees are chiefly confined to sea coasts and tidal marshes. The wood ranges through orange red to dull brown, and is hard, heavy, strong and durable. Rays produce beautiful figure on rift-sawn wood. The wood is of little importance in this country as yet.

Rhodamnia.   See *Malletwood.*

Rhodesian Balsa.   See *M'gongo.*

Rhodesian Mahogany or **Copalwood.**   *Copaifera coleosperma* or *Afzelia quanzensis* (*H.*).   The suggested standard name for this wood is R. copalwood.   Also called M'chibi.   Fairly heavy ; very hard, stable, and fire-resisting.   Not a mahogany, and with few of its characteristics.   Rather difficult to work, polishes well.   Used chiefly for superior flooring.   S.G. ·75 ; C.W. 4.

Rhodesian Teak.   *Baikiæa plurijuga* (*H.*).   Reddish brown.   Also called Zambesi Redwood, Mukusi, and Umgusi.   Very hard, heavy, durable, and stable.   Insect and fire resisting.   It is not teak but has many of its qualities and used for similar purposes.   Excellent for block flooring.   Difficult to work because of hardness.   S.G. ·85 ; C.W. 4·5.   See *Teak.*

Rhododendron.   *Rhododendron arboreum* and other species.   Chiefly shrubs, cultivated for ornament, but specimens up to 40 ft. by 8 ft. girth are found in Asia.   The wood is pinkish white to reddish brown, rather dull, but smooth.   Moderately soft, light, not durable.   Straight and irregular grain, very fine texture, ray flecks.   Shrinks and warps badly.   Easily wrought.   Used for turnery, utensils, suggested for rotary plywood.   S.G. ·5 ; C.W. 2.

Rhodosphæra.   See *Tulip Satinwood.*

Rhomboid.   A parallelogram with its angles not right angles and adjacent sides unequal.   A rhombus is similar but its four sides are equal.   Area = base × vertical height.

Rhone.   An eaves gutter.

Rhus.   A genus producing the lacquer trees of Japan.   See *Sumach, Lacquer Tree, Varnish Tree,* and *Tulip Satinwood.*

Rib.   **1.** A curved rafter.   See *Jack Rib* and *Ogee Turret.*   **2.** The

RIB CENTRE

curved framework of a centre, *q.v.*   **3.** An intersecting member of a vault.   **4.** Projecting ornamental bands on ceilings.   **5.** A shaped member to receive the covering of parts of aeroplanes, or of a ship.   **6.** The beam portion of a reinforced concrete T-beam.   **7.** See *Ribbing* (3).   **8.** Small pieces of wood pierced with two holes used as the trucks of a parrel on board ship.   **9.** Small strips for laminated circular work in cabinet-making.   See *Rim* and *Ribbing Up.*

Rib and Panel.   A term used in vaulting in which a framework of arched ribs carry thin panels, called cells, webs, or infilling.

Riband or **Ribband.**   **1.** Temporary wood stringers to hold the frames of a vessel in their relative positions in shipbuilding.   **2.** Longitudinal members in boatbuilding.   **3.** Timbers, in shuttering ior beams that run along the caps of the struts to steady the beam sides. **4.** See *Ribbon*.

Riband Back. The name applied to Chippendale chairs in which carved ribands, or ribbons, are a characteristic feature.

Ribbed Framing. Applied to timber framed buildings in which curved struts and ornamental timbers are added to the main timbers.

Ribbing. 1. The ribs for panel vaulting. See *Rib and Panel*. **2.** Applied to a corrugated surface due to variation in the shrinkage of spring and summer wood in seasoning. Also called *wash boarding*. **3.** The contoured side of a violin. Usually of maple.

Ribbing Up. Building up circular work by laminated ribs in cabinet work. See *Rim*.

Ribbon. 1. A decorative fillet, or narrow strip, especially if undulating. **2.** A distinctive painted moulding on the side of a ship. **3.** See *Girt* and *Ribbon Board*.

Ribbon and Stick. A carved detail representing a ribbon wound round a stick, characteristic of Louis XVI. design.

Ribbon Board. A kind of girt for a timber-framed building, or balloon framing, in which the studs run through and are notched for the ribbon.

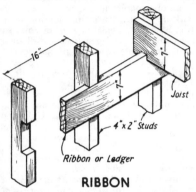

RIBBON

Ribbon Grain. Stripe figure or roe grain. Parallel zones of alternating grain, in rift-sawn wood, due to interlocking grain.

Ribbon Rail. A light metal rail, or core, tying together the tops of iron balusters. It is usually covered by a wood handrail.

Rib Centre. One for a segmental arch in which the curvature is formed in one piece. See *Rib*.

Riblets. Secondary ribs in aircraft construction.

Rib-ply. A registered reinforced plywood. Ribs are formed during manufacture on one or both sides, producing a rigid but light board.

Rich. Abbreviation for the name of the botanist *Richard*.

Ricinodendron. See *M'Gongo* or *Mugongo*, *Musodo*, and *Okwen*.

Ricker. 1. A long pole or sapling used for handling floating timbers at the docks. **2.** A small scaffold pole.

Rick Stand. A wood floor for a rick, or pile, of any kind of loose material.

Ride or **Riding.** Applied to a door when it touches any part of the floor as it opens ; or to a butt joint open at the ends, or *hard* at the middle.

Rider. 1. The top shore that rests on a short back shore, or springs from the back of a raking shore, in a system of raking shores, *q.v.* **2.** The band of a *band and gudgeon*.

Rides. Pins, or gudgeons, that carry the bands in *bands and gudgeons*.

Ridge. **1.** The highest part or apex of a roof, *q.v.* The horizontal timber to which the tops of the common rafters are fixed. Sometimes called a ridge board or ridge pole. See *Collar Tie*. **2.** The internal angle, or nook, of a vault.

Ridge and Valley Roof. Same as M-roof, *q.v.*

Ridge Gum. *Eucalyptus alba.* W. Australia. Of little economic importance and not exported. See *Gum*.

Ridge Roll. A rounded piece planted on the ridge and over which lead is dressed to form the ridge covering. Tingles are fixed under the roll to prevent the lead from lifting. See *Barge Board* and *Lantern Light*.

Riding Shore. See *Rider*.

Riffle. The lining for the bottom of a sluice, or a strip of wood used as an obstruction in a sluice.

Riffler. A bent rasp for shaping concave circular work.

Rift Saw. A special type of circular saw, usually with inserted teeth, for converting slabbed logs into flooring strips. It has four or more arms projecting from the plate to carry the cutting teeth.

Rift Sawn. Quarter sawn, *q.v.* See *Converting*.

Riga Last. A timber measure of 65 cub. ft. in the round, or 80 cub. ft. of squared timber.

Rig, To. To equip. A general term for providing or erecting the preliminary essentials for a job.

Rigger. **1.** A scaffolder. **2.** One who fits up the rigging of a ship. **3.** A craftsman concerned with the assembling of aircraft. **4.** Lumberers who attach cables, etc., for dragging felled logs.

Rigging. **1.** The cordage, etc., for the masts and sails of a sailing ship. **2.** Scaffolding. **3.** Same as *to rig*.

Right Hand. See *Hand*.

Rigid. Stiff. Able to resist bending or change of form.

Rigidex. A proprietary cold process glue.

Rilievo. See *Relief*.

Rim. **1.** The circumference of circular work. **2.** A raised border or margin. **3.** The sides and back of a drawer.

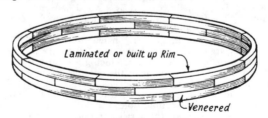

Laminated or built up Rim

Veneered

RIM FOR TABLE

Rimda. Thingan, *q.v.*

Rimer. See *Reamer*.

Rim Latch. A fitting for keeping a door closed. It consists of a spring latch in a metal case with a rim for fixing to the door, and operated by a spindle and knobs. It is fixed on the inside of the room and is usually provided with a sliding thumb bolt to secure the door.

Rim Lock. A combined latch and lock. Similar to a rim latch but it has a key to operate the bolt.

Rim Speed. The speed of the perimeter of a circular saw. See *Saw Teeth*.

Rimu. *Dacrydium cupressinum* (*S.*).
N. Zealand red pine. Light brown,
varied streaks. Moderately soft, light,
and durable. Strong and stable.
Straight even grain, often figured, fine
texture. Seasons well, easily wrought,
and polishes well. Excellent wood for
joinery, flooring, cabinet work, turnery,
plywood, cooperage, etc. Figured stuff
for veneers and superior joinery. Valuable for kraft pulp. S.G. ·6 ; C.W. 1·5.

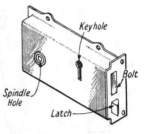

RIM LOCK

Rincer. A machine reamer.

Rind Galls. Surface wounds, as from torn-off branches, covered by later growth.

Ring. See *Growth Ring*.

Ringas. Rengas, *q.v.*

Ring Bolt. One with a drop-ring handle that falls flat when not in use. See *Hatch*.

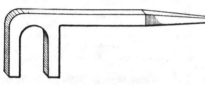

RINGER IRON

Ringer Iron. An implement, used in crate-
making, for bending and
twisting the staves.

Ring Fence. The narrow
circular fence used for
circular work on a spindle
machine.

Ringing. Girdling, *q.v.*

Ringing Machine. A small type of pile-driver operated by manual labour.

Ring Latch. A fall bar, or latch, operated by a drop-ring handle.

Ring Porous. Applied to woods in which there is a distinct contrast in size and number of the pores in different parts of a growth ring. The larger pores in the spring wood produce a well-defined zone. The change may be abrupt between spring and summer wood. See *Pores* and *Diffuse Porous*.

Ring Rot. Decay in logs following the direction of the growth rings.

Ring Shake. A cup shake, caused by lack of cohesion between adjacent growth rings. Also see *Shakes*.

Ring-tail. See *Boom*.

Rink. A place specially prepared for skating.

Rio Rosewood. Brazilian Rosewood, *q.v.*

Rip. To saw lengthwise, or *with the grain*.

Ripewood. A term applied to certain trees, as silver fir, that do not produce apparent heartwood. The colour is uniform throughout.

An intermediate stage between heartwood and sapwood in trees in which the change is slow and very gradual.

Ripper. 1. A rip saw. 2. A slater's tool.

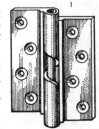

Rip Saw

Sawing Stool

RIPPING

Ripping. Using the rip saw. Sawing along the grain.

Ripple Marks. 1. Figure in wood caused by the buckling of the fibres due to compression in the growing tree. Also called fiddle-back, and characteristic of sycamore. Fine transverse markings, or striations, across the grain of tangential sawn wood, that have a tier-like structure of the wood elements due to storied rays and other storied elements. Also see *Roll* (5). 2. The name given by machinists to *machine marks*.

Rip Saw. A hand-saw, like a cross-cut, but with specially shaped large teeth (4 or 5 points per inch) for sawing along the grain. The name is also applied to special types of machine saws. See *Saw Teeth*.

Rise. 1. The difference in height between the ends of an inclined member. 2. The vertical distance between consecutive stair treads. 3. The height above the springing of the highest part of the soffit of an arch. See *Roof*, *Segmental Arch*, and *Step*.

Rise and Fall. A term applied to circular saws in which spindle or table can be adjusted for height to regulate depth of cut.

Risen Mouldings. Mouldings that project above the face of the framing. Like a bolection moulding but without the rebate.

Riser. 1. The vertical front piece of a built-up step. See *Handrail* and *Step*. 2. The framework, with the rocker-piece, forming the wheel-house of a vehicle.

Rising Butt. A hinge with a helical knuckle joint that causes a door to rise as it opens, to clear the floor coverings. It is more effective than cocking ordinary butts, as the door rises uniformly.

Riugan. *Pametia pinnata* (*H.*). Formosa. Cherry-wood colour, lustrous. Hard, heavy, with close wavy grain. Glistening deposit and fine rays. A useful cabinet wood, but little imported. S.G. ·85.

Rive or **Riven.** Same as cleft, or split. Specially applied to plasterer's laths, *q.v.*

River Banksia. *Banksia verticillata* (*H.*). W. Australia. Pale yellow. Beautiful grain and medullary rays make it a valuable furniture wood, but little exported. S.G. ·56.

RISING BUTT

River Gum. *Eucalyptus rostrata* (*H.*). W. Australia. Pinkish red. Hard, heavy, dense, fissile. Straight grain. Used for general building, poles, etc. Not imported. See *Gum*.

River Maple. Soft Maple, *q.v.*

Riving Knife. 1. A steel plate to open a saw-cut behind the saw to prevent the stuff from closing on the saw at the back, and kicking. See *Saw Guards* and *Saw Bench*. 2. A tool used for riving or splitting laths.

r.l. Abbreviation for *random lengths*.

Ro. Abbreviation for *rough*.

Robinia. *Robinia pseudoacacia*. Europe and U.S.A. The suggested standard name for this wood. See *Acacia* and *Locust*.

Roble. *Platymiscium polystachium* (*H.*). Venezuela. Rich reddish brown, some variegated, smooth, lustrous. Very hard, heavy, durable, strong, and stable. Some ribbon grain. Fairly difficult to work, polishes well. Used for all purposes. S.G. ·75 ; C.W. 3·5. *Tecoma pentaphylla*. C. America. Similar to Catalpa, *q.v.*, but lighter in colour. Greyish brown, fine brown stripes, smooth, lustrous. Moderately hard, heavy, and strong. Very durable and resists insects. Some wavy and roey grain, medium texture. Easily wrought. Used for superior joinery, flooring, sports and farming implements, construction, turnery, etc. S.G. ·62 ; C.W. 2·5. *Fagus batuloides*. S. America. Resembles American oak and used for similar purposes. S.G. ·56. *Nothofagus sp.* Called Chilean Oak or Beach, Antarctic Beech. Like English beech, but softer, lighter, and more durable. S.G. ·52 ; C.W. 3. Also see *Quira*. The name Roble (a Spanish name for oak) is applied to many Central American woods that have characteristics of oak, but usually there is a distinguishing prefix. The woods include species of *Catalpa, Ehretia, Ekmanianthes, Petitia, Platymiscium, Tabebuia*, etc. Many of them are of the Bignoniaceæ family.

Roble Colorado. Quira, *q.v.*

Rob-roy. A canoe partly decked in, and propelled by double-bladed paddles.

Rock Elm. *Ulmus racemosa* or *U. thomasi*. See *Elm*.

Rocker. 1. The side connection of standing pillars in vehicle construction, or a panel below seat line or body side. 2. Curved pieces fixed to a cradle or to the bottom of chair legs to give a rocking motion to the cradle or chair. 3. See *Hopper Window*.

Rocker Piece. The framework, with the riser, forming the wheelhouse of a vehicle.

Rocking Frame. A special form of mould for concrete. The rocking, wave-like, motion produces better and stronger concrete due to consolidation.

Rock Laths. Strong slater's laths for *stone slate* roof coverings.

Rock Maple. *Acer saccharum*. See *Maple*.

Rock Oak. See *Oak*.

Rock Saw. A special circular saw suspended from a long arm to remove a wide kerf on the upper surface of a log. The object is to detect any foreign matter liable to injure the head saw. The sawdust is removed by a suction hood. It is also called a Barking Saw.

Rock Wool. Slag wool.

Rococo. A style of decoration prevalent in the 18th century applied to European furniture and architecture consisting of ornamental scrollwork and shell motifs. Highly decorative and ornate. From the French 'rocaille'.

Rod. 1. A board on which a woodwork job is set out full size and from which the material is set out preparatory to construction. 2. A slender stick, cane, or twig. 3. Hazel twigs and branches used for filling in between the staves in crate-making. 4. A measure of length of 5½ yds. 5. A measure of brickwork of 272¼ sq. ft., 1½ bricks thick.

Rod Machine. A special machine for making cylindrical rods, such as broom handles, etc.

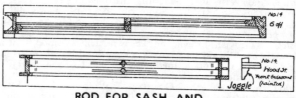

**ROD FOR SASH AND
FRAME WINDOW**

Roe or **Roey.** Short broken ribbon, or stripe, figure in quarter-sawn decorative hardwoods, due to spiral formation of the fibres, or inter-locked grain, in the growth rings. The irregular growth produces alternate bands of varying shades of colour and degrees of lustre.

Roll. 1. A cylindrical piece for ornamentation or for a machine. 2. A shaped piece over which lead is worked at the joints on roofs, etc. Also see *Ridge Roll*, *Parapet Gutter*, and *Lantern Light*. 3. A Gothic moulding often used as a drip, and consisting of a round

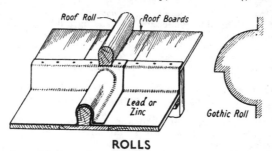

ROLLS

broken horizontally with the top half slid forward to project over the lower half. 4. The crossbar of a logging sled. 5. A bold form of ripple, *q.v.*

Roller. A cylinder revolving on its axis and used for crushing, smoothing, propelling, etc. Also see *Roller Feed*. A cylinder placed under a heavy object to facilitate movement, or one on which thin material runs or is wound, as a blind roller, towel roller, etc. A cylindrical rod for striking a measure, such as a bushel.

Roller Board. A board in some pipe organs carrying rollers of different lengths as part of the mechanism between the pedals and manuals.

Roller Feed. Applied to saws and machines in which the stuff is fed through the machine by power rollers.

Rolling Pin. A cylindrical piece of hardwood used as a kitchen utensil. It is usually of sycamore or beech.

Rolling Shutters. See *Revolving Shutters*.

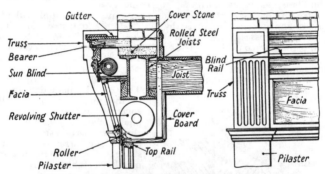

ROLLING SHUTTER
AND SUN BLIND

Rolling Stock. Railway stock : engines, carriages, wagons, vans, etc.

Roll Top. Applied to a desk with a flexible or tambour fall. It consists of narrow strips glued on canvas or held together by metal strips. The fall slides in grooves in the desk ends. See *Writing Desk*.

Rollway. See *Log Dump*.

Roman. Applied to the style of architecture practised by the Romans. It was founded on the Grecian but was characterised by its boldness and circular contours, and included Tuscan and Composite Orders. See *Doric, Mouldings*, and *Orders*.

Romanesque. Applied to the architecture derived from the Roman and prevailing before the introduction of Gothic.

Roman Mouldings. See *Mouldings*.

Rones. Gutters, *q.v.*

Rood. 1. A carved figure of Christ on the Cross. 2. A measurement of area, quarter of an acre.

Rood Arch or **Beam.** An arch or beam across the chancel of a church, between choir and nave, usually to support the rood.

Rood Loft. A small gallery over the chancel screen in a church.

Rood Screen. One separating the choir and presbytery from the nave.

Rood Steeple. A spire over the entrance to the chancel of a church.

Rood Tree. The Cross used as a religious symbol.

Roof Beam. A tie-beam.

Roof Capping. Trimming the top of a pole to throw off water.

Roofers. Boards under shingles.

Roofing. 1. The materials forming a roof. 2. The work entailed in forming a roof.

Roof Ladder. A long board with cleats nailed on at intervals to provide foothold. It is used on roofs to avoid breaking tiles, shingles, etc. See *Saddle Scaffold*.

Roof Lights. Any arrangement to provide light from the roof of a building : skylights, dormers, lantern lights, *q.v.*

Roofs. Roofs are classified as single (common rafters only), double (rafters and purlins supported by end walls), and trussed or triple-membered (trusses, purlins, and rafters). They are named according to the shape : conical ; curb or mansard ; domical ; flat ; gable, or span, or cottage, or double pitch ; gambrel ; Gothic ; hip ; hopper ; jerkin ; lean-to, or pent, or single pitch, or shed ; M-roof ; north light or saw-tooth ; ogee ; pavilion ; saddle-back ; V-roof ; etc.

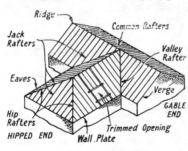

ROOF TIMBERS

They are also named according to the materials forming the covering : asbestos-cement, asphalt, boards, felt, iron, lead, shingles, slates, stones, tiles, etc. The arrangement and shape of the constructional members, which may be of wood, steel, or composite, also give name to the roof : arched, Belfast or bowstring, cantilever or umbrella, collar, collar beam, couple, couple close, French, hammer beam, king post or bolt, lattice, mansard, north light, open, purlin, queen post or bolt, scissors, square-to, and trussed rafter. The name of the designer is another means of identifying a roof : Cubitt's (king and queen bolts), de l'Orme, Emy, Fink (French or Belgium), Howe, Linville, Pratt, etc. Roofs are also distinguished by the type of building they cover, as church roof, etc. The above roofs are described alphabetically in the Glossary. Also see *Pitch*.

Roof Sticks. The curved ribs forming the top of a railway carriage.

Roof Trees. An obsolete term applied to several structural members of a roof.

Roof Truss. A triangulated framework of wood or steel for supporting the roof coverings. See *Roofs*.

Rooiels. Red Els, *q.v.*

Roost. A strong rod, or rail, on which fowl rest in cotes. A perch.

Rooster. See *Gooseneck*.

Root. 1. That part of a plant fixed in the earth and by means of which the sap is drawn from the soil. 2. The bottom of a tenon, nearest to the solid wood. 3. The bottom of the cut forming a saw tooth or screw thread.

Root Burr. A burr at the base of a tree. See *Burr.*

Root Spurms. Roots that appear above the ground and stretch some distance up the trunk. See *Buttress.*

Roove. A washer, especially one on which a nail is clinched.

Ropala. *Roupala brasiliensis* (*H.*). C. America. Silky oak family. Also called *Pao concha* and *shellwood.* Reddish to chocolate brown, black streaks. Very hard, heavy, strong, and durable. Straight fibrous grain, coarse texture, ray flecks. Fairly difficult to work, polishes well. Used for cabinet work, exposed structures, sleepers. S.G. ·85 to 1 ; C.W. 4. Also see *Beefwood.*

Rope Feed. Applied to circular saw benches on which the stuff is pulled through mechanically by a rope. Now superseded by roller feeds, *q.v.*

Rope Figure. Due to the twist of broken stripe figure being in one direction.

Rope Moulding. Cable moulding, *q.v.*

Rosa. See *Vermelha.*

Rosace. A rose window, or a rose-like ornament.

Rosaceæ. The rose family. Includes about one hundred genera and 2,000 species. The trees are important for decorative purposes, and for edible fruits and seeds : almond, apricot, apple, cherry, nectarine, peach, pear, plum, raspberry, etc. Many genera provide useful woods : *Angelisia, Amelanchier, Aronia, Cercocarpus, Chrysobalanus, Coupia, Cratægus, Hirtella, Licania, Moquilea, Parastemon, Parinari, Prunus, Pygeum, Pyrus, Rosa, Sorbus.* The sub-families are *Amygdalaceæ, Pomaceæ,* and *Spiræaceæ.*

Rosamay. Rose Mahogany, *q.v.*

Rosa Morada. *Lonchocarpus hondurensis* and *Tabebuia sp.* (*H.*). C. America. Light brown, smooth, lustrous, close grain, and somewhat like a pale walnut, mottled. Only obtainable as veneers.

Rose. 1. A decorative plate for door furniture. 2. A perforated nozzle to a pipe for spraying liquids.

Rose Alder. *Ackama Muelleri* and *A. quadrivalvis* (*H.*). Queensland. Also called Corkwood, Rose-leaf Marara, etc. Reddish mauve, smooth. Very light and soft, but tough and fairly strong ; not durable or stable or fissile. Close even texture. Easy to work, nail, glue, and stain. Used for shoe heels, cabinet work, electric fittings, interior joinery. S.G. ·52 ; C.W. 2.

Rose Almond. *Owenia venosa* (*H.*). Queensland. Mahogany colour, some variegated, fragrant, oily nature. Hard, heavy, tough, strong, resilient, durable. Straight, close, handsome grain. Requires care in seasoning. Fairly difficult to work and glue, polishes well, small sizes. Used for shafts, felloes, axe handles, wagon poles, screws, turnery, flooring, etc. S.G. ·95 ; C.W. 3·5.

Rose Ash. *Flindersia lævicarpa.* Very similar to Yellow-wood Ash, *q.v.*

Rose Bit. A brace bit for countersinking holes in wood or soft metals.

Rose Butternut. *Blepharocarya involucrigera* (*H.*).   Queensland.
Rose pink with flecks.   Light, soft, fissile ;   but tough, strong, and
stable ;   not durable.   Straight, woolly grain, some fiddleback.   Works
readily ;   dust an irritant.   Used for plain cabinet work, cooperage,
interior joinery.   S.G. ·6 ;   C.W. 3.
Rose Gum. *Eucalyptus saligna* and *E. grandis.*   See *Gum.*
Rose Mahogany. *Dysoxylum fraseranum.*   See *Rosewood.*   The
suggested standard name for this wood is Australian Mahogany.
Rose Marara. *Geissois*, or *Weinmannia*, *lachnocarpa* (*H.*).   Queensland.   Rose pink to mauve, smooth.   Hard, heavy, strong, stable,
but not very durable.   Straight close grain and fine even texture ;
fine rays.   Requires care in seasoning.   Not difficult to work.   Used
for carving, turnery, bearings, flooring, stocks of planes.   S.G. ·82 ;
C.W. 3.   See *Sweetbark.*
Rosemary.   Coarse fast-grown pitch pine, *q.v.*
Rose Nails.   Wrought nails with rose-shaped projecting heads.
Rose Satinash. *Eugenia francissi* (*H.*).   Queensland.   Rose colour,
lustrous, smooth.   Hard, heavy, tough,
durable.   Straight fibrous grain.   Requires slow and careful seasoning.   Not
difficult to work.   Used for flooring, tool
handles, barrows, etc.   S.G. ·78 ;   C.W. 3.
Rose She Oak.   See *Shea Oak.*
Rose Silkwood.   See *Silkwood.*
Rosette.   A carving with radiating lines,
circular arrangement of leaves, etc., or
rose-like.   A pateræ carved to imitate a
rose.
Rose Walnut.   See *Walnut.*
Rose Window.   A circular window with
radiating bars and usually decorative
tracery work.   Also called marigold and
catherine-wheel windows.

ROSETTE

Rosewood. *Dalbergia Spp.* (*H.*).   Widespread.   Generally distinguished by source or port of shipment.   The name is applied to
numerous other woods because of their rose-like fragrance.   Dark
purplish brown, variegated, fairly lustrous, smooth.   Hard, heavy,
durable, elastic, strong, stable.   Close uniform grain and fine texture ;
ribbon grain.   Heartwood often shaken and decayed.   Beautiful
wood and costly.   Seasons slowly and rather difficult to work ;   polishes
well.   Used for cabinet work, superior joinery, pianos, veneers, parquet
floors, carving, etc.   S.G. ·8 ;   C.W. 3 to 4.   AFRICAN ROSEWOOD.
See *Bubinga* and *Padauk*, AUSTRALIAN ROSEWOOD, *Dysoxylon fraseranum.*   Now called Australian mahogany.   Resembles Honduras
mahogany, fragrant.   Hard, heavy, stable, durable.   Not difficult to
work.   S.G. ·8 ;   C.W. 2·75.   BAHIA or BRAZILIAN ROSEWOOD, *Dalbergia nigra.*   See *Jacaranda.*   Also called Cabiuna and Palisander.
BORNEO ROSEWOOD, see *Rengas.*   BURMESE ROSEWOOD, *Pterocarpus
sp.*   See *Padauk.*   DOMINICA ROSEWOOD, *Cordia sp.*   See *Laurel.*
EAST AFRICAN ROSEWOOD, *Dalbergia melanoxylon.*   Also called Blackwood, *q.v.*   EAST INDIES ROSEWOOD, *Dalbergia latifolia*,   Indian Rose-

wood. Also called Blackwood and Shisham, *q.v.* FRENCH ROSEWOOD, Madagascar Rosewood. HONDURAS ROSEWOOD, *Dalbergia stevensonii.* Also called Nagaed wood. INDIAN ROSEWOOD, *Dalbergia cultrata,* *D. latifolia,* and *D. sissoo.* Also called Sissoo and Bombay Blackwood,

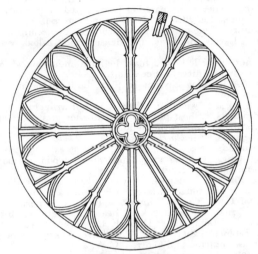

ROSE WINDOW

q.v. Excellent wood for high-class work. S.G. ·88. MADAGASCAR ROSEWOOD, *Dalbergia greveana.* Very variable in colour. Obtainable as veneers only. RIO ROSEWOOD. Same as Bahia. ROSETTA ROSEWOOD, East Indies rosewood. SEYCHELLES ROSEWOOD, see *Mahoe.* Also see *Bois de Rose, Cocobola, Kingwood, Palisander, Poye, Quira, Tulip Wood,* and *Yinzat.*

Rosin. A distillation of turpentine from pines. Also called colophony. See *Resin.*

Ross. A local term for accumulations of irregular growth on the bark of trees.

Rossing. Removing the bark of trees. The workman is called a *rosser.*

Rostrum. **1.** A platform or pulpit in a hall or place of worship. It usually accommodates several people. **2.** A curved end of the prow of a ship. **3.** Any curved projection resembling a beak.

Rosy. Applied to grain of wood that runs irregularly or overlaps.

Rot. Decomposition of wood due to fungi. It is distinguished as brown, dry, hard, piped, red, soft, wet, and white rot. White rot, produced by *Polystictus* and *Poria* species of fungi, is more destructive than brown or yellow brown rot, produced by *Coniophora, Lenzites,*

Paxillus, and *Trametes species*. Also see *Decay, Disease, Dry Rot, Sanitation, Brown Oak, Partridge Wood*, and *Figure*.

Rotary Cutting. Also called peeling. The steamed log is mounted in a large lathe and turned against the knife which peels the veneers in long sheets, usually from $\frac{1}{28}$ to $\frac{1}{8}$ in. thick. For half-round and back-cut veneers the part-log is mounted *off-centre* to modify the grain figure. See *Plywood, Veneers*, and *Converting*.

Rotary Figure. The grain of wood produced by peeling, *q.v.*

Rothholz. Layers of darker and denser wood called compression wood, *q.v.*

Rotten Knot. One less hard than the surrounding wood, or one in an advanced state of decay. See *Knots*.

Rotunda. A building circular in plan, both inside and out, and with domed roof.

Rough. Applied to wood left off the saw, not wrought.

Rough Brackets. Short pieces nailed to the sides of a carriage to a flight of stairs to support the middle of the treads.

Rough Carriage. See *Carriage Piece*.

Rough Grounds. See *Grounds*.

Roughing Out. Cutting away superfluous wood preparatory to forming a moulding by hand.

Roughing Plane. A single-iron wood plane used by the cabinet-maker for dirty and rough surfaces. Also called a Bismarck Plane.

Rough Leaf Tree. *Curatella americana* (*H.*). C. America. Reddish brown, variegated. Moderately hard, heavy, and durable. Interwoven coarse grain, distinct rays. Difficult to work. Used for general purposes, cabinet-making, etc. The leaves contain enough silica to be used as sandpaper. S.G. ·77 ; C.W. 4.

Rough Patch. A rough area on veneer, generally due to oblique or twisted grain.

Rough String. Same as carriage piece, *q.v.*

Roumanian Pine. *Picea excelsa*. A favourite wood for the sound boards of pianos because it is slow grown with fine uniform grain, and dense for spruce. See *Picea* and *Spruce*.

Round. **1.** A nosing. A semicircular moulding as on the edge of a stair tread. **2.** Same as rung, *q.v.* **3.** A joiner's plane to form a hollow.

Round Butt. A market name for a large long pole, about 6 in. diameter at the middle.

Round Disease. An infectious and very destructive fungus that attacks growing pines, especially *Pinus pinaster*.

Rounded Corner Joint. A corner joint used in cabinet work.

Rounded Steps. See *Bullnose, Curtail*, and *Drum End*.

Roundel. **1.** Same as astragal, *q.v.* **2.** A small disc or ring.

ROUND PLANE

Rounding Machine. A special machine for shaping broom and brush handles. It can produce rods of varying diameter, or curved.

Round Knot. One produced by cutting at right angles to the branch. See *Knots.*

Round Shave. A cooper's tool, like a bent draw-knife, for circular work.

Round Step. A step with semi-circular end. See *Drum End.*

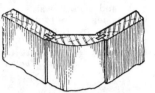

ROUNDED CORNER JOINT

Round Timber. Applied to round logs ready for conversion.

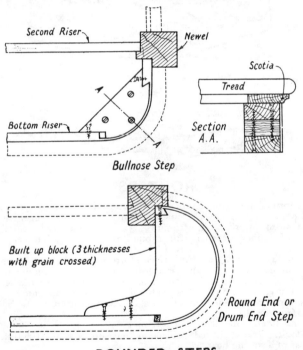

ROUNDED STEPS

Round-up. Same as camber, in shipbuilding.

Round Wood. Off products of trees, such as the tops and large branches. They are chiefly Baltic softwoods, in lengths from 5 to

9 ft., and 6 in. upwards in diameter. Used for packing case and box making. Also round pieces for pit props or pulp.

Roupala. See *Ropala*.

Rouping. See *Rueping*.

Router. **1.** A woodwork tool for levelling the bottom of a trench or groove. See *Old Woman's Tooth*. **2.** Special tools for shaping

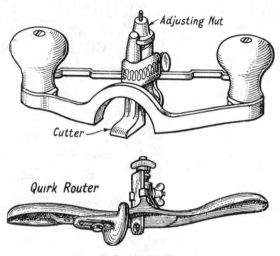

ROUTERS

and forming quirks on circular work, etc. The various types are : beading, boxing, carver's, circular, coachmaker's, fence, jigger, lining, ovolo, pistol, quirk, sash, stringing. **3.** A machine for forming the joints of rounded corners for cabinet framing. **4.** The elephant machine, *q.v.*, is called a heavy duty router.

Rowan Tree. *Pyrus aucuparia* (*H.*). Europe. Also called Mountain Ash, but no relation to ash. Reddish brown, darker summer wood. Moderately hard and heavy, tough, elastic. Close grain, some beautifully figured, with pith flecks. Discolours easily with glue, and of little commercial value. Small sizes. Used for furniture, tools, etc. See *Mountain Ash*.

Rowing Boat. One propelled by oars only.

Rowlock. The U-shaped projection that serves as the fulcrum for the oar of a rowing boat.

Roxb. Abbreviation for the name of the botanist Roxburgh.

Royals. Shingles 24 in. long.

Royena. See *Blackwood*.

R/S. Abbreviation for *resawn*.

Ru. *Casuarina equisetifolia* (*H.*). Malayan ironwood. Reddish to dark brown. Very hard, heavy, and strong. Fine grain, distinct rays. Rather small sizes and knotty. Little exported because of its importance as firewood. S.G. ·65 to 1.

Rubbing Pieces. Vertical timbers to protect piers and wharves from moving vessels, etc.

Rubbing Streak. A strong horizontal plank fixed to the side of a vessel as a protection.

Rubiaceæ. The Madder family. Includes Adina, Anthocephalus, Calderonia, Calycophyllum, Canthium, Cephalis, Cinchona, Coffea, Exandra, Gardenia, Genipa, Hymenodictyon, Morinda, Vauclea, Randia, Relbunium, Rubia, Sarcocephalus, Sickingia, Stephegyne. A great variety of trees, but more important for edible fruits and medicines than for wood.

Ruby Wood. Red Sanders, *q.v.*

Ruda. *Zanthoxylum* or *Vochysia spp.* (*H.*). Classification uncertain. C. America. Variegated like yellow poplar, and very similar in characteristics, but a better class wood. Used for shipbuilding, aircraft, veneers, cabinet work, etc. Little imported. The name is also applied to West Indian Satinwood, *q.v.*

Rudder. 1. A mechanism to steer a vessel, aeroplane, etc. A vertical flap for directional control of aircraft. It is hinged to the stern-post of the fuselage, *q.v.*, to control the yawing of the plane. See *Tail* (*Aeroplane*). 2. A paddle for stirring malt in a mash tub.

Rudder Post. A tapering post in the tail of an aeroplane.

Rudd's Table. An antique table with various contrivances for extending different sections, wings, and trays, forming a general utility piece of furniture.

Rudenture. Same as cable moulding, *q.v.* Shaped like a rope or staff, carved or plain, and used in hollows such as the flutings of a column.

Rueping. An empty cell method of preservation, *q.v.* Creosote is forced into the wood, which is enclosed in a cylinder, and then extracted.

Rufod. See *Contract Forms*.

Rule Joint. A table or knuckle joint, *q.v.*, as used for the hanging leaf of a table. Special hinges may be obtained for the joint.

Rule or **Ruler.** A flat strip of wood or metal graduated for measuring ; or a hardwood cylinder for ruling lines on paper. The woodworker's rule is 2 or 3 ft. long and is two- or four-fold.

Rules. Regulations framed by various lumber organisations for the measurement and sale of timber.

Rum Cherry. *Prunus serotina.* See *Cherry*.

Rum-tum. A light sculling boat.

Run. 1. A gangway. Planks, bridging gaps or rough ground, and used by workmen for transporting materials, especially by wheelbarrow or hod. 2. Linear measurement, as *foot run*, etc. 3. Applied to a saw-cut not true to a required line. 4. The horizontal distance between the ends of an inclined member.

Rundle. A rung, or ball, or ring.

Rundlet or **Runlet.** A small barrel.

Rung. 1. A stave of a ladder or one tying together the legs of a chair.

2. A floor timber of a ship. **3.** A radial handle to a steering wheel. **4.** A staff or cudgel.

Runlet. Obsolete term for a small wine cask.

Runner Fittings. Ball-bearing rollers for sliding doors.

Runners. 1. Bearers along which anything slides. The supports of a drawer. See *Guides, Drawer, Tracks,* and *Extension Table.* **2.** The bearers for an arch centre. **3.** Horizontal timbers supporting the bearers, or joists, for shuttering. **4.** See *Plough.* **5.** Vertical poling boards for loose ground. They are driven as the soil is excavated. See *Timbering.* **6.** A projecting deal at the end of a timber pile to carry gang planks for the men forming the stack. **7.** Metal rails for sliding partitions, etc., to run along. **8.** A carling sole, *q.v.* **9.** See *Sett.*

Running Planks. 1. Stout timbers along which the top centering for a culvert is drawn for a new length of construction after one part is completed. **2.** A gang-plank or runway.

Running Rules. Wood strips, fixed temporarily, and used as guides for a horsed mould. When narrow bands of plaster are used, instead of wood, they are called *running screeds.*

Runtree. The head to a quartered partition.

Runway. 1. A slide for logs. **2.** See *Run* (1).

Rupala. See *Ropala.*

Rupr. Abbreviation for the name of the botanist Ruprecht.

Ruprechtia. Viraru, *q.v.*

Rupture. Tearing of the fibres across the grain of wood. It is variously described as felling, transverse, or thunder shake, cross fracture, or upset.

Russel Jennings Bit. One of the best types of twist bit for the brace.

Russpruss. A Baltic charter party, *q.v.*

Rustic Work. Applied to garden equipment, summer-houses, etc., constructed of branches and light poles to give a rural appearance.

Rutaceæ. The Satinwood family. Important for edible fruits and for wood, and includes : Ægle, Amyris, Atalantia, Balfourodendron, Chloroxylon, Citrus, Esenbeckia, Euxylophora, Evodia, Fagara, Feronia, Flindersia, Helietta, Limonia, Merrilia, Murraya, Pilocarpus, Zanthoxylon.

Rutger's Process. A process of pressure preservation consisting of a mixture of oil and salt solution, usually creosote and zinc chloride.

R.W.B. Abbreviation for *rebated weather boards.*

S

S. Abbreviation for *surface, square, side*, and *shear*.

Sabia. Not classified. (*H.*) Brazil. Reddish brown. Very similar to Cuban mahogany. Hard, heavy, strong, and stable. Close even grain, fine rays. Suitable for cabinet work and superior joinery. S.G. ·95.

Sabicu. *Lysiloma sabicu* or *L. latisiliqua* (*H.*). C. America. Also called Cuban Sabicu and Wild Tamarind. Chestnut brown, darker stripes, lustrous. Rather hard and heavy ; strong but brittle ; very durable and stable. Roey, smooth, and polishes well. Not difficult to work. Properties of mahogany and used for similar purposes. S.G. ·85 ; C.W. 3·5. BAHAMA SABICU. See *Moruro*. AFRICAN SABICU. See *Satinwood*.

Sabot. 1. A wood shoe. 2. An iron skid or runner.

Sacarium. 1. The part of a church within the altar rails. 2. A piscina.

Saccoglottis. See *Atala*.

Saccopetalum. See *Hoom*.

Sacome. The profile of a moulding, or member of a moulding, etc.

Sacristy. A vestry to a church.

Saddle. 1. A block used as a seating for circular work on a spindle moulder. A cradle, drum, or cylinder. 2. A cleat, or block, fixed to a spar to receive the foot of another spar. 3. A thin piece under an interior door, and fixed to the floor, to allow the door to clear the floor coverings. 4. An appliance for marking the shoulders of ship framing. 5. A shaped piece for clamping serpentine work to the caul, when veneering. 6. A headtree, *q.v.*

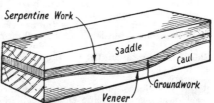

SADDLE FOR VENEERING

Saddle Back. Applied to cappings and copings that are weathered both ways.

Saddle-back Board. A raised and rounded floor board under a door to allow the door to clear the floor coverings. See *Saddle* (3).

Saddle-back Roof. 1. A span roof gabled both ends. Also called a Saddle Roof. 2. A tower roof with gables.

Saddle Bar. A metal bar to strengthen leaded lights.

Saddle Bead. A specially shaped glazing bead to serve for two adjacent panes of glass, especially when for a curved bar. See *Glazing Bar*.

Saddle Cramp. One with extra deep shoes for cramping-up over projections or curved work. A frame-work for cramping-up flights of stairs.

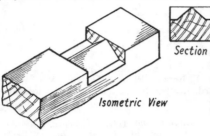

Section

Isometric View

SADDLE JOINT

Saddle Joint. A joint to provide a good bearing surface for the foot of a vertical timber and to prevent lateral movement.

Saddle Scaffold. A scaffold sitting astride a ridge and resting on the roof, for chimney work. See *Straddle Pole*.

Saddle Tenon. The projection in a bridle joint, *q.v.*

Saddle Tree. The wood frame of a saddle for a horse.

Safan. *Talauma Hodgsoni* (*H.*). India, Burma. Also called Balukhat, Boramthuri, etc. Brownish grey, fairly lustrous, feels rough. Rather hard and heavy, not durable. Straight and irregular grain, even medium texture, fine rays. Not difficult to season and work but liable to pick-up. Used for tea boxes, handles, etc. Suitable for match-boarding, etc. S.G. ·72 ; C.W. 3·5.

Safe. 1. A cupboard or box for preserving food. **2.** A lead-lined floor, with raised border, under a tank or cistern.

Safe Edged. Applied to a flat file with plain edge.

Safe Load or **Strength.** A proportion of the ultimate or breaking strength of a structural member. It depends upon the selected factor of safety, *q.v.*

Safety Devices. The many appliances used on woodworking machines to satisfy Home Office Regulations : cages, circular blocks, fences, grippers, guards, pushers or push sticks, riving knives, etc.

Safety Lintel. A substitute for a relieving or safety arch. It is used to relieve a stone lintel or soldier arch from the weight of the wall above, and may be of wood, steel, or reinforced concrete.

Safety Rail. A second hand-rail to a balustrade or stairs for the use of children.

Saffron Heart. *Halfordia scleroxylon* (*H.*). Queensland. Pale saffron, appears oily. Very hard, heavy, tough, strong, elastic, and durable ; fissile. Properties and characteristics of Greenheart. Straight close grain and fine texture, some cross grain and fiddleback. Requires careful seasoning, and difficult to work. Rather small sizes. Used for fishing rods, golf sticks, dance floors, etc. S.G. 1 ; C.W. 4.

Saffron Tree. *Chrysophyllum cainito* (*H.*). C. America. The Satin-leaf or Star-apple Tree. Reddish brown. Rather hard, heavy, strong, and durable. Very similar to Mucuri, *q.v.* Used for structural and constructional work. S.G. ·8.

Sag or **Sagging.** The downward bend at the middle of a horizontal or inclined member. The deflection of a member between the points of support.

Sagawoita. Kenya Satinwood, *q.v.*
Sageræa. See *Bow-wood* (*Andaman*).

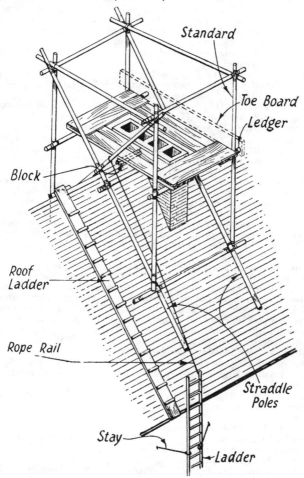

Standard

Toe Board

Ledger

Block

Roof
Ladder

Rope Rail

Straddle
Poles

Stay

Ladder

SADDLE SCAFFOLD

Sagger. A packing box used for annealing malleable cast-iron objects.
Sagowood. See *Threaded Box*.

Sag Rod. A hanger, *q.v.*

Sagwan. *Tectona grandis.* See *Teak.*

Sailing Ships. Those that depend on wind and sails for motion. In some cases they are assisted by auxiliary engines. The various types are : barque, brig, brigantine, clipper, cutter, ketch, lugger, schooner, scow, sloop, smack, xebec, yacht, etc.

Sail-over. Projecting or over-sailing, *q.v.*

Sain. Indian Laurel Wood, *q.v.*

Saka. Purpleheart, *q.v.*

Sakura Wood. *Prunus pseudo cerasus* (*H.*). Japan. Used for lacquered furniture. Not exported.

Sal. *Shorea robusta* (*H.*). India. Pale brown to reddish brown on exposure, dull, smooth. Hard, heavy, very strong and durable, moderately stable. Interlocked and cross grain, medium texture. Difficult to season and work, liable to pick-up. Excellent wood where strength and durability are required, equal to greenheart. Used for structural work, sleepers, piling, wheelwright's work, implements, etc. S.G. ·82 ; C.W. 4. Also see *Thitya.*

Salai or **Salia.** *Boswellia serrata* (*H.*). India. Yellowish to greenish brown, darker wavy lines, dull, feels rough, odorous. Fairly light and soft, not strong or durable. Straight even rough grain, rather coarse texture. Easily wrought, but requires careful seasoning. Used for cheap furniture, planking, cases, masts, toys, slack cooperage. Suggested for decorative plywood, panelling, etc. S.G. ·58.

Salicaceæ. The Willow family. See *Populus* and *Salix.*

Salie. See *Real Salie.*

Salient. Applied to a projection or to an external angle or corner. An angle less than 180°.

Saligna Gum. *Eucalyptus saligna* (*H.*). S. Africa. Pinkish red to reddish brown on exposure. Hard, heavy, strong, fairly durable. Straight grain with ripple and fiddleback. Difficult to season without checks but not difficult to work. Similar to mahogany in many respects. Little exported. S.G. ·8 ; C.W. 3·5.

Salisb. Abbreviation for the name of the botanist Salisbury.

Salisburya. See *Maiden Hair Tree.*

Salix. See *Willow* and *Sauce Colorado.*

Salle. A hall or large room. A public dining-room in an hotel.

Sallies. *Willow trees.*

Sallow. *Salix coprea, S. cinerea, S. aurita* (*H.*). Species of willow, of little timber value. The bloom is often called " palm." See *Willow.*

Sally. 1. The projecting tongue to a scarfing. See *Table Joints.* 2. See *Salient.* 3. Sometimes applied to a bird's-mouth in carpentry. 4. A timber with a hole for a bell rope.

Sally-wood. The Willow, *q.v.*

Salmon Gum. *Eucalyptus salmonophloia.* W. Australia. Red, darkens with exposure. Very hard, heavy, strong, and durable. Straight grain. Beautiful wood, and increasing in commercial importance. S.G. 1 ; C.W. 4. See *Gum.*

Salmon Wood. *Eriolæna candollei* (*H.*). India. Brick red to light reddish brown, variegated, darker streaks, fairly lustrous. Somewhat like acacia. Heavy, very hard and strong, fairly durable. Straight even grain, fine texture, ripple. Difficult to season, fairly difficult to work, polishes well. Used for heavy furniture, superior joinery, wheelwright's work, etc. S.G. ·78 ; C.W. 3·5.

Salmwood. Cyp and Peterebi, *q.v.*

Salon. A drawing- or reception-room.

Saloon. A large public room for receptions, dancing, or for the sale of liquor. A railway carriage for sleeping or dining. A large communal cabin on a ship.

Salt. Common salt is sprinkled over hardwood boards to expedite air seasoning and to prevent checks. The salt draws out the moisture but keeps the surface moist. Also see *Preservation*.

Salt Box. A wood container in which salt is kept in a kitchen.

Sam. Chaplash, *q.v.*

Samak. Kelat, *q.v.*

Saman. See *Raintree*.

Samar. Bajac, *q.v.*

Samba. Obeche, *q.v.*

Sambucaceæ. A family including Sambucus. See *Elder*.

Samohu. *Chorisia speciosa* and other *spp.* (*H.*). C. America. Brownish grey. Light, soft, fairly tough and strong, but not durable. Straight coarse and firm grain, distinct rays. Easily wrought. Used for cases, slack cooperage, etc. S.G. ·44.

Sampigi. See *Champac*.

Samydaceæ. See *Flacourtiaceæ*.

Sanai. Mersawa, *q.v.*

Sanara. Synara, *q.v.*

Sancho-arana. *Bravaisia floribunda* (*H.*). Central America. Greyish white, both sapwood and heartwood ; smooth. Light and soft, but firm, fairly strong and tough, not durable. Straight grain, medium textura, distinct rays. Easily wrought. Used for box boards, pulp, charcoal, etc. S.G. ·52.

Sandaleen. *Excœcaria lucida, Gymnanthus lucida, Spirostachys africana* (*H.*). C. America, S. Africa. Dark red, variegated, fragrant. Very hard and heavy, with close even grain. Used for small decorative work, turnery, inlays, etc. S.G. 1.

Sandalwood. *Santalum album* (*H.*). India. Light yellowish brown to dark reddish brown on exposure, fairly lustrous, feels oily, aromatic. Very hard and heavy. Straight and very close wavy grain, bird's-eye. Fine even texture, small sizes. Yields sandalwood oil. Seasons well but slowly, works readily and polishes well. Valuable for trinkets, fancy articles, cabinet work, especially linings, carvings, etc. S.G. ·9 ; C.W. 4. *Santalum cygnorum.* India and W. Australia. Similar to *S. album*. Parasite on roots of other trees, hence small sizes. Important oil extract. Used for turnery, Eastern joss sticks, and fancy articles. S.G. ·8. *Santalum spicatum.* W. Australia. Used chiefly for oil and incense. S.G. ·6. *Amyris balsamifera.* W. Indies. Also called Amyris Wood and Torchwood. Odorous, lustrous, smooth, easily wrought, and polishes well. Also see *Red Sanders*.

Sandan. *Ougeinia dalbérgioides* (*H.*). India. Numerous local names. Light golden brown, ageing to reddish brown, darker streaks, lustrous,

smooth, but feels rather rough. Gum and often white deposits.
Very hard and heavy. Strong, durable, and tough. Narrow interlocked grain, fairly coarse texture, very fine rays. Resembles Shisham.
Requires care in seasoning and should have heart boxed. Difficult
to work, due to hardness and interlocked grain. Polishes well. Used
for wheelwrights' work, implements, pounders, tools, cabinet work,
heavy furniture, etc. S.G. ·84 ; C.W. 4·5.

Sandbags. Bags of hot sand used when veneering shaped work.

Sand Cushion. See *Helmet*.

S. & E. Abbreviation for *surfaced and edged*. Numerous variations.

Sanden. See *Sandan*.

Sander. A machine for sandpapering woodwork. The machines are
made for all purposes, from small cylindrical work such as penholders
to finished joinery such as doors, and for floors. They are distinguished
as bobbin, disc, drum, belt (vertical or overhead), dowel, portable,
etc. Drum sanders may have up to eight drums for sanding both
sides of the work. See *Sandpaper* and *Glasspaper*.

Sanders. See *Red Sanders* and *Yellow Sanders*.

S. & M. Abbreviation for *sunk and moulded*, applied to skirtings.

San Domingan Boxwood. *Phyllostylon brasiliensis* (*H.*). C. America.
Yellow to pale brown. Hard, heavy, very fine uniform straight
grain. Used for carvings and turnery. S.G. ·95.

San Domingo. The port of shipment for several important woods :
Cuban mahogany, satinwood, lignum-vitæ, etc. The name often
signifies superior quality. See *Harewood*.

Sandpaper. The name applied to all kinds of abrasive papers used
by hand or machine, except emery cloth. The abrasive material may
be glass, flint, garnet, aluminous oxide, etc., and it may be mounted
on paper or cloth, in sheets or rolls, to suit different types of machines.
It may be waterproof. Glass and garnet are used for glasspaper for
handwork. Aluminous oxide, flint, and garnet are used for machine
papers. The abrasive is crushed and sifted to the required grade and
affixed to the paper by glue. The process consists of spreading,
beating, and re-heating to allow the abrasive to sink into the glue
and make contact with the paper. Kraft paper is used for handwork
and manilla paper or white duck cloth for machine work. Ordinary
glasspaper is graded as follows, o, 1, 1½, F2, M2, S2, 2½ and 3. Garnet
finishing paper for special work is graded from 7/0 to 1/0. Flint
paper for machines is from 3/0 up to 4, and special quality garnet
paper for machines is graded from 4/0 to 3. In both cases the grade
rises by ½ after 1/0. See *Glasspaper*, *Flour-paper*, and *Sander*.

Sandwich. A built-up " balk " of boards glued together to be sawn
for the cores of blockboard. When the glue is hard the balk is cut
into *core slices* to the thickness required for the blockoard. As many
as twenty-four wide boards are glued face to face to form a balk.

Sandwich Beam. A flitched beam, *q.v.*

Sandwich Construction. The term applied to light wood roof
trusses, etc., consisting of single and double members with the single
members sandwiched between the double members.

Sang. Chinese Mulberry, *q.v.*

Sangre palo. See *Tapsava*.

Sangre de Toro. *Maytenus* and other *spp.* (*H.*). Mexico. Very variegated in colour, greyish to green. Irregular grain and figure. Not imported.

Sanitation. The following are suggestions for keeping the timber yard in a healthy condition to prevent decay and staining of stacked wood. Masonry foundations for the stacks, or piles, sufficiently high above the ground to allow for good air circulation. Dry softwood stickers. Careful piling for quick drying. Frequent collection and burning of rubbish. Treatment against decay of permanent timbers such as sleepers, runways, etc. Treatment of sapwood with chemical solution. See *Sap Stain.* Removal of grass, weeds, etc. The yard should be on dry and high ground and well drained to avoid stagnant water.

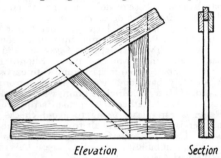

Elevation   *Section*

SANDWICH CONSTRUCTION

San Juan. *Vochysia guatemalensis* and other *spp.* (*H.*). C. America. Also called Quaruba. Light brown, golden lustre. Light, soft, brittle, but durable. Firm texture, roey, visible rays. Not difficult to work but rather woolly. Requires careful seasoning. Used chiefly for construction and interior joinery. S.G. ·5 ; C.W. 3.

Santalaceæ. A family including Exocarpus, Fusanus, and Santalum.

Santalum. See *Sandalwood.*

Santa Maria. *Calophyllum calaba* or *C. brasiliense* (*H.*). C. America. Also called Bari, Palo Maria, Leche Maria, Chijole mahogany, or Crabwood. Like Indian Poon. Red, dark stripes. Moderately hard and heavy. Very tough and strong. Elastic, stable, and durable, and resists insects. Some irregular grain, medium texture, ribbon and roe figure, smooth. Not difficult to work and polishes well. Good substitute for Honduras mahogany. Used for superior joinery, veneers, vehicles, and for structural work. S.G. ·75 ; C.W. 3. Also see *Galba.*

Santa Vera. *Eucalyptus spp.* California. Pinkish brown. Ribbon stripe and beeswing figure. Very decorative. Only marketed as veneers.

Santo Wood. *Zvllernia paraensis* (*H.*). C. America. Called Páu Santo. Variegated olive brown, greenish black, striped. Smooth, feels waxy. Extremely heavy, hard, and durable. Splintery. Fine, uniform, with some interwoven grain. Difficult to work. Used for fancy articles, cutlery handles, substitute for lignum-vitæ. Perfume extracted. S.G. 1·3 ; C.W. 5.

Sap. The circulating fluid or juices o n which the life and growth of a tree depends. See *Sapwood.*

Sapan. *Cæsalpinia sapan* (*H.*). E. Indies. Orange red, variegated

477

streaks, lustrous. Fine close grain, distinct rays. Like Brazil-wood. Used for dyewood, fancy articles, etc.

Sap Burrs. Those of no value for veneers.

Sapele. *Entandophragma cylindricum* or *E. utile* (*H.*). W. Africa. Also called Sapele Mahogany, Gold Coast Cedar, Acajou, Sapelli, Aboudikrou, Penkwa, Tiama, Sipa, etc. Sometimes includes species of Guarea. Two main species *Ogiekpogo* and *Ubilesan*, each of which has " man " and " woman " trees. Slightly heavier than Khaya spp. Light mahogany colour, darkening with exposure, pale pink sapwood. Moderately hard and heavy ; tough, strong, stable, fragrant, and lustrous. Interlocked close grain, with roe and mottle ; even texture, visible rays. Rather too regular stripy roe figure. Large sizes. Requires careful seasoning to avoid checks and warp. Should be quarter-sawn. Not difficult to work but liable to pick-up. Polishes well. Used for superior joinery, cabinet work, veneers, plywood, flooring, shipbuilding, etc. S.G. ·64 ; C.W. 4.

Sap Gum. See *Red Gum*.

Sapindaceæ. The Soapberry family. Includes Æsculus, Blepharocarya, Cupania, Diploglottis, Dilodendron, Filicium, Harpullia, Litchi, Magonia, Melicocca, Ratonia, Sapindus, Schleichera.

Sapindus. See *Soapnut Tree*.

Sapling. A young tree over 3 ft. high and less than 4 in. diameter at breast height. A large sapling is one over 10 ft. high.

Sapling Pine. White Pine, *q.v.*

Sapocarana. Not classified. (*H.*) Brazil. Bright yellow. Fairly hard, heavy, strong, and durable. A good substitute for satinwood. S.G. ·78.

Sapodilla. 1. Genera providing gutta-percha, balata, chicle, etc. See *Sapotaceæ*. Many species are too valuable for chicle (chewing gum) to fell for timber. The woods include : Achras, Almique, Bullet-wood, Ibira, Isonda, Mastic, Mucuri, Nispero, Paroba, Sapote, Sweetwood. 2. *Achras sapota* (*H.*). Honduras. Also called Chico, Zapota, etc. Reddish brown. Very hard and heavy. Fine close grain and texture. Excellent for turnery. S.G. 1·1.

Sapota. See *Paroba*.

Sapotaceæ. The Sapodilla family : Achras, Bassia, Bumelia, Calocarpum, Chrysophyllum, Ganua, Labourdonnaisia, Lucuma, Madhuca, Mimusops, Palaquium, Payena, Pouteria, Pradosia, Sideroxylon.

Sapote. *Calocarpum sp.* or *Sideroxylon sp.* (*H.*). C. America. Also called Mamey, Marmalade Tree, Zapote, etc. Light greyish red. Hard, heavy, strong, and durable. Distinct rays. Fairly difficult to work. Used for structural work, sleepers, cabinet work, vehicles, etc. S.G. ·9 ; C.W. 3·5. The name is applied to several species of Sapodilla, and to *Matisia*, *Grias*, *Achras*, and *Casimiroa spp*.

Sappan. See *Sapan*.

Sapper. A circular saw and a small travelling carriage operated by the knee. It is used chiefly for preparing shingle bolts. Also called a knee bolter.

Saps. Pieces of hardwood, chiefly sapwood.

Sap Stain. Stain in sapwood due to the action of fungi, probably through late felling or oxidation. In some cases it is restricted to the surface but in others it penetrates through the sapwood. The colour varies considerably according to the fungus. Dipping in solutions such as Lignasan, Dowicide, etc., are preventives. Ethyl-mercury chloride, chlorinated phenol, and sodium bicarbonate are amongst the chemical solutions proved to be effective.

Sapucaia. *Lecythis spp.* Monkey Pot, *q.v.* Also see *Kakeralli* and *Lecythis*.

Sapwood. The outer part of a tree trunk that contains the living cells in the growing tree ; also called alburnum. The amount of sapwood varies considerably with different species. Some have very little whilst others, such as red gum, may have 60 per cent. sapwood. Usually the proportion diminishes with age. In some species it is difficult to distinguish, but usually there is a big contrast between heartwood and sapwood. The colour in softwoods is generally bluish and in hardwoods greyish or dirty white. Specifications usually state *no sapwood*, but in carefully seasoned wood most of it is suitable for ordinary purposes. In some woods the sapwood may be more durable than the heartwood after treatment, because of its greater absorption. In some cases sapwood is preferred for special purposes, such as hickory for handles or shafts, persimmon for shuttles, etc. In polished hard-woods the difference in colour usually prohibits its use. Sapwood is less resistant to attack by insect and fungi than heartwood and it is lighter in weight, but it is more easily treated. See *Annual Rings*.

Sapwood Trees. Those from which the wood shows little difference in colour between heartwood and sapwood, such as ash, aspen, beech, holly, sycamore, white spruce, etc.

Saqui. *Bombacopsis sp.* (*H.*). C. America. Wild Cedar. Resembles Spanish cedar. Reddish brown. Light, soft, durable. Straight fine grain, rather coarse texture, smooth. Easily wrought and seasoned. Used as a substitute for cedar. S.G. ·52 ; C.W. 2.

Sarcocephalus. See *Bilinga*, *Opepe*, *Nauclea*, *Yellow* or *Canary Cheesewood*.

Sarg. Abbreviation for the name of the botanist Sargent.

Sargent Plane. A patent metal plane, but with wood sole.

Wood Sole

SARGENT PLANE

Sarking Boards. Close boarding to carry roof tiles, shingles, or slates. Thin boards used as a lining.

Sash. **1.** The separate lighter frame to a window, carrying the glass. It may be hinged, pivoted, sliding, or fixed. See *Fixed Sash, Casement*, and *Sash and Frame*. **2.** The frame in which gangsaws are stretched. It slides vertically in grooves.

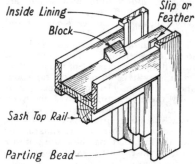

Sash and Frame. Applied to windows with vertically sliding sashes and boxed frame. Cast-iron or lead balance weights are attached to cords which pass over pulleys and are then nailed to the edges of the sashes. There are several patents to dispense with the cords which require periodic renewal. See *Boxed Frame*.

Sash Bar. See *Bar* (1) and *Lamb's Tongue*.

Sash Beads. Those forming the rebates between a pivoted sash and the frame, or that serve as a guide or stop for a sash. See *Inner Bead* and *Glazing Bead*.

Sash Bit. A Gedge pattern twist bit ; also called a railway carriage bit or long-nose bit.

Sash Centres. The centres of the pivots, or the pivots, for a pivoted sash. They are fixed a little above the middle of the height, or centre of gravity, of the sash, so that it closes of its own weight.

Sash Chisel. One for cutting the pockets in the pulley stile of a sash and frame window. It has a wide, very thin blade, sharpened on both sides.

Sash Cords or **Chains.** The cords attached to the sliding sash and

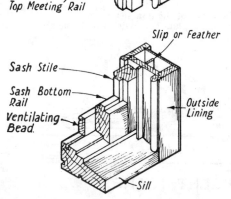

SASH AND FRAME WINDOW

balance weights of a sash and frame window. Chains are used for heavy sashes. See *Lifting Shutters*.

Sash Door. One with the top part glazed. See *Gun Stock Door* and *Panic Bolts*.

Sash Eye. A ring screwed to a vertically sliding sash to facilitate opening and closing.

Sash Fastener. An appliance fixed on the meeting rails of sliding sashes to secure them when closed.

Sash Gang. A type of resaw or gang mill. It is so called because it has an oscillating frame, or sash, carrying the saws, that allows the saws to swing free on the upward stroke.

Sash Linings. The inner and outer facings to the boxed frame of a sash and frame window, *q.v.*

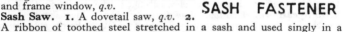

SASH FASTENER

Sash Saw. 1. A dovetail saw, *q.v.* 2. A ribbon of toothed steel stretched in a sash and used singly in a machine saw, where motive power is limited.

Sash Stuff. Prepared material for sashes. It is usually of stock sizes, moulded and rebated. Stiles and top rail 2 in. by 2 in., bottom rails, 2 in. by 3 in., meeting rails $2\frac{3}{8}$ in. by $1\frac{3}{4}$ in. or $1\frac{1}{2}$ in. See *Sash and Frame* and *Boxed Frame*.

Sash Weights. Cast iron or lead weights used to balance vertically sliding sashes.

Sassafras. *Atherosperma moschatum* (H.). Tasmania and Victoria. Colour very variable from light greyish brown to nearly black. Moderately heavy, but soft and not durable or strong. Straight grain, fine uniform texture, irregular rays, giving ripple effect. Easily wrought. Small sizes. Used for cabinet work, carving, turnery, woodware. S.G. ·65; C.W. 2·5. *Sassafras officinale.* U.S.A. Pale to dark brown, fragrant. Soft, light, fairly durable and flexible. Not strong but stable. Coarse texture, straight grain. Large sizes scarce. Sometimes sold as black ash. S.G. ·45. CANARY SASSAFRAS, *Doryphora sassafras.* Queensland. Not exported. The name Sassafras is applied to several C. American species of laurel, especially to Bois de Rose, and also to Gumbo Limbo, *q.v.* The name is sometimes applied to Queensland Camphorwood. See *White Siris*.

Sassandra Mahogany. *Khaya anthotheca.* African Mahogany. See *Penkwa* and *Mahogany*.

Sasswood. Erun, *q.v.*

Sateenwood. *Euxylophora paraensis* (H.). C. America. Also called Pao Amarello. Bright golden yellow, lustrous, smooth. Hard, heavy, strong, durable. Wavy and roey grain. Uniform medium texture. Large sizes. Fairly difficult to work. Excellent for contrasting with dark woods for interior fittings and cabinet work. S.G. ·8; C.W. 3·5.

Satinash. See *Rose Satinash*.

Satinay. *Syncarpia hilli* (H.). Queensland. Called Red Satinay. Like Satiné in figure and colour. Bright pink to light brown, lustrous, smooth. Fairly hard, heavy, very durable and strong. Resists insects and fire. Close texture; ripple. Beautiful variation when fumed. Large sizes. Requires care in seasoning; not difficult to work. Used for veneers, superior joinery, furniture, plywood, carving, flooring, etc. S.G. ·8; C.W. 3.

Satiné. *Brosimum paraense* or *Ferolia guianensis* (*H.*). C. America. Also called Satiné Rubane, Muirapiranga, Cardinal and Brazil wood, Satinee, etc. Red with golden lustre, variegated yellow and red stripes, feels waxy. Very hard, heavy, and durable; splinters readily. Straight grain; smooth, distinct rays. Not difficult to work and polishes well. Used chiefly for cabinet work. S.G. 1; C.W. 3.

Satinheart. See *Wilga.*

Satin Oak. *Embothrium wickhami* (*H.*). Queensland. Pink; lustrous. Very light and soft, fissile, fairly strong, very durable. Straight open grain, coarse texture, large red-coloured rays. Easily seasoned and wrought, but liable to pick up and difficult to glue. Polishes well. Used for superior joinery, shingles, cabinet work. S.G. ·5; C.W. 3.

Satin Sycamore. *Ceratopetalum virchowii* (*H.*). Queensland. Like Coachwood but more decorative. Fawn colour. Fairly soft, with loose texture, and woolly. Little imported. S.G. ·6z. Also *Geissois spp.* Queensland. Rose pink, lustrous, fawn markings. Fairly light and soft, not durable, but tough and strong. Fine even grain and texture. Requires careful seasoning, not difficult to work, polishes well. Used for cabinet work, flooring, superior joinery. S.G. ·6; C.W. 3.

Satin Walnut. See *Red Gum.*

Satinwood. There is a great variety of satinwoods, including species of Citrus, and they vary considerably in properties and characteristics. ANDAMAN S., *Murraya exotica.* Also called Chinese boxwood or Myrtle. Light yellow, darker streaks. Hard and heavy. A substitute for boxwood, but coarser and not so uniform. Small sizes. Used for cabinet work, fancy articles, instruments, turnery. S.G. ·83; C.W. 4. AUSTRALIAN S., *Phebalium billardieri.* Victoria. Similar to W. Indies satinwood. Excellent cabinet wood. BRAZILIAN S., *Euxylophora paraensis* or *Esenbeckia sp.* Similar to W. Indies satinwood but a little lighter, and easier to work but more brittle. See *Sateen Wood.* Also see *Piquia Marfin.* CEYLON S. Same as E. Indies. CONCHA S., *Zanthoxylum sp.* Also called San Domingan S. See *W. Indies Satinwood.* EAST AFRICAN S., *Fagara* or *Zanthoxylum sp.* Also called Ata, Sabicu, Sagawoita, etc. Excellent wood and similar to E. Indies Satinwood. EAST INDIES S., *Chloroxylon swietenia.* India, Ceylon. Allied to mahogany. Also called Flowered Satinwood, Behra, Buruta, Mutirai, etc. Golden yellow to brownish, darker streaks, lustrous, fragrant, smooth. Very hard, heavy, strong, durable, and stable. Subject to *borers* but resists *teredo.* Beautiful, fine, even, variegated grain and figure. Interlocked and twisted fibres, mottle, roe, silver grain. Even texture. Liable to gum veins and cup shakes. Requires careful seasoning. Difficult to work because it is liable to pick-up. Polishes well. Used for all purposes in India: furniture, veneers, structural work, piling. Only small stuff exported, for cabinet work and fancy articles. S.G. 1; C.W. 4·5. JAMAICA S. Same as W. Indies Satinwood. NIGERIAN S. See *Yellow Satinwood.* SAN

DOMINGAN S. Same as W. Indies Satinwood. SCENTED S. Coach-wood, *q.v.* WEST AFRICAN S., *Afrormosia sp.* Similar to E. Indies Satinwood. WEST INDIES S., *Fagara flava* or *Zanthoxylum elephantiasis*. Golden yellow, lustrous, fragrant, appears oily. Very hard, heavy, and stable ; fairly durable. Fine uniform grain and texture, wavy, roey, and mottled. Not difficult to work, unless oily or figured, and polishes well. Figured wood obtainable only as veneers. Used for cabinet work. S.G. ·85 ; C.W. 4. Species of *Euxylophora* and *Amyris* are also included in W. Indies Satinwood. See *Pau amarello* and *Amyris*.

Saturation Point. The fibre saturation point of wood. See *Moisture Content*.

Sauce Colorado. *Salix humboldtiano* (*H.*). Central America. Resembles *Salix nigra*. Reddish grey, sapwood ill-defined. Light, soft, but firm, tough and strong for its weight, not durable. Rather coarse texture, straight grain. Easily wrought. Not imported. S.G. ·44 ; See *Willow*.

Saugh. Willow, *q.v.*

Savin. *Juniperus virginiana.* The shrub of aromatic red cedar.

Savory. A bay in a vaulted ceiling.

Saw. A thin steel blade, disc, or band, with toothed edge. Saws may be machine or hand saws. Machine saws are circular, oscillating blades, or continuous bands. There are also numerous specialist types described alphabetically. The various types of *circular saws* are : crosscut, edger, and rip. Small types may be obtained with swivelled saw, rising or canted table, and they may be portable. Large types have roller feeds and drives. *Crosscut saws* may be straight line, mitre, pendulum, double, multiple, etc. *Oscillating saws* vary from the light fret, jigger, or scroll saws to the heavy frame saws used in conversion. *Band saws* vary from the small standard type to large types for re-sawing and the conversion of logs, called log band mills. The latter is roller driven with rolling table or carriages. There is a great variety of brazing, grinding, setting, sharpening, and swaging machines and equipment necessary for doctoring machine saws. HAND SAWS. The various types of hand saws are : back, band, block, bow, compass, coping, cross-cut, double-handled, dovetail, fleme-tooth, frame, fret, hack, half-rip, hand, keyhole, nests, pad, panel, pit, rat-tail, rip, Spearfast, and tenon. The types of saws and the terms used in sawing are described alphabetically in the Glossary. The size and shape of saw teeth vary according to the work both for hand and machine saws. See *Machines, Saw Teeth, Widia, Plane, Tipped, Tubular,* and *Veneer Saws*.

Saw Alive. Same meaning as *saw through and through*.

Saw Arbor. The shaft and bearings for a circular saw.

Saw Around. To convert a log by sawing on all the faces in turn, to obtain the best quality of the lumber.

Saw Bench. The cast iron frame carrying a machine circular saw together with the wood extensions. (See illus., p. 484.)

Saw Block. A low trestle for supporting the wood whilst sawing. See *Ripping*.

Sawbya. *Sterculia alata* (*H.*). India. White, to brownish on exposure, lustrous. Light, fairly hard and strong, fissile, not durable. Straight grain, distinct rays. Requires careful seasoning. Used for cheap planking, packing cases, plywood for boxes, etc. S.G. ·48.

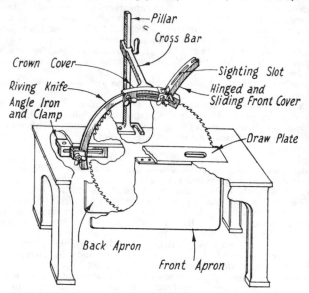

SAW BENCH AND GUARDS

Saw Chops. Any appliance for securing a saw whilst it is sharpened. See *Chops*.

Saw Doctor. A mechanic who keeps the saws in condition.

Saw Falling. A term applied to wood as it falls from a log converted according to the judgment of the sawyer, without regard to size and quality. Unselected or ungraded wood.

Saw Guards. See *Safety Devices*.

Saw Guide. Any arrangement to control the run of a circular or band saw.

Sawing Horse. Same as sawing stool or saw block, *q.v.* Two strong V-shaped frames for holding a log whilst it is being sawn.

Saw Kerf. See *Kerf*.

Saw Mill. The buildings and plant necessary to convert logs or lumber to smaller sections.

Sawn. Applied to wood from the saw without any other preparation. Converted timber. Specially applied to boards, deals, planks, etc.

Saw Pits. 1. Pits used for converting heavy timbers to smaller sections. The double-handed saw is operated vertically by a man in the pit and one above. Such pits are still in use where machines are not available. **2.** The space under a machine saw to receive the sawdust.

Saw Set. An appliance for correcting the set of saw teeth. See *Set*.

Saw Tailer. An off-bearer, *q.v.*

Saw Teeth. The type of teeth for machine saws depends upon the direction of the grain, and whether it is hard or soft wood. For the same kind of wood and same conditions the size of the teeth depends

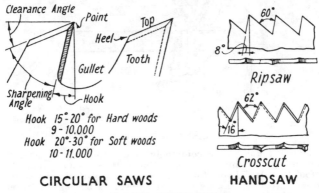

Hook 15°-20° for Hard woods
9 - 10.000
Hook 20°-30° for Soft woods
10 - 11.000

CIRCULAR SAWS   **Ripsaw**

Crosscut
HANDSAW

SAW TEETH

upon the rim speed, which is governed by the revolutions per minute and the diameter of the saw in a circular saw. The rim speed is usually about 10,000 ft. per minute. The type of teeth for hand saws are crosscut, fleme, peg, rip, raker, spearfast. Also see *Set*.

Saw Through and Through. A term used when converting a log with parallel cuts only.

Saw Tooth Rack. See *Shelf Fittings*.

Saw Tooth Roof. A north-light roof, *q.v.*

Saw Vice. A vice for holding a saw whilst it is sharpened. (See illus., p. 486.)

Sawyer. 1. One who controls the machinery required for converting logs and timber into smaller sections. The quality and quantity of the sawn wood depends upon his skill and judgment. **2.** A beetle destructive to softwoods.

Sayar. Semul, *q.v.*

Scab. A wound in a tree covered by later growth.

Scaffold or **Scaffolding. 1.** A temporary platform or stage, with its supports, for carrying men and materials during the erection or repairs of buildings, etc. The usual types are bricklayer's, cradle, derrick, flying, gantry, independent, mason's, saddle, stagings, straddle, suspended, travelling. The timbers used are called standards, ledges, braces, putlogs, and scaffold boards. Guards and fans are used for

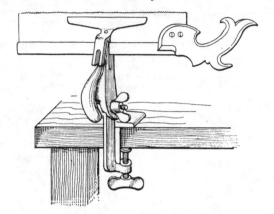

SAW VICE

protection. Fir poles are commonly used. **2.** A timber stage in a coal mine. **3.** A temporary erection for the execution of criminals.

Scaffold Boards. Those forming the working platform of a scaffold. They are usually 1½ in. by 9 in. and up to 12 ft. long. The ends should be bound with hoop iron.

Scaffold Cradle. See *Suspended Scaffold*.

Scaffold Trestle. See *Trestles*.

Scalariform. Shaped like a ladder. A term used in the description of vessels and cells of plants.

Scald. An injury to the cambium of a tree due to exposure to strong sunlight.

Scale. 1. A thin flake on a surface. **2.** The ratio representing the difference between the dimensions of a drawing or model and the dimensions of the actual object. **3.** A ladder. **4.** A leaf-bud before bursting. A bract of a pine cone, etc.

Scale Board. 1. A thin strip of wood used when setting type for printing. **2.** Thin slices of softwood, prepared by slicing, for such uses as hat boxes, match boxes, backing of pictures, etc.

Scaling. Determining the volume of logs.

Scaling Ladder. A fireman's ladder.

SCALLOPED

Scallage. A lych gate, *q.v.*

Scalloped. A decorative edge to thin material, carved or shaped in the form of a sinuous curve or small segments. Like the edge of a scallop shell.

Scaly Bug. See *Lecanium*.

Scamilla or **Scamillos.** A plain block, or sub plinth, to elevate a statue or column.

Scanfin. A charter party form, *q.v.*, now superseded by the Baltwood Charter.

Scano. A prefabricated wood house.

Scant. Applied to sawn stuff slightly under nominal or specified sizes. The same meaning as *short*. See *Tolerance*. Sometimes applied to waney goods.

Scantlings. **1.** Wood of small section and varied dimensions. Waste wood left from conversion. **2.** Redwood, white pine, and spruce, 2 to 4½ in. wide by 2 to 4 in. thick, and over 8 ft. long. Pitch-pine, Oregon, and Californian pine of any length, under 6 in. wide and from 2 to 5 in. thick are marketed as scantlings at present, but the British Standard Institution suggests that the sectional sizes given for redwood should be generally applicable. **3.** A timber frame to support a cask.

Scape. The apophyge to a column. That part of a column where it springs from the base. See *Apophyge*.

Scarf or **Scarph.** **1.** A lengthening joint, in structural timbers, that

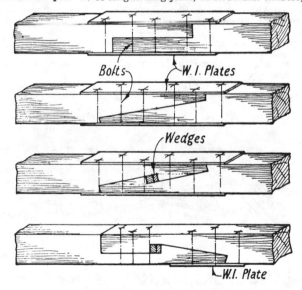

Bolts   *W. I. Plates*

Wedges

W.I. Plate

SCARFED JOINTS

does not increase the cross-sectional area. Also see *Tabled Joints*.
2. A serrated joint to increase the glued area.

Scarfing. Preparing scarfed joints.

Scarlet Oak. *Quercus coccinea*. U.S.A. Classified as red oak. See *Oak*.

Scene Dock. The place in a theatre for storing scenery.

Scenery. The painted scenes used as a background on a theatre stage.

Scenography. Drawing in perspective.

Scented Guarea. *Guarea cedrata.* See *Obobonufua.*

Scented Mahogany. Sapele, *q.v.*

Scented Satinwood. Coachwood, *q.v.*

Schau. Abbreviation for the name of the botanist Schauer.

Scheff. Abbreviation for the name of the botanist Scheffer.

Schima. See *Needlewood.*

Schinopsis. See *Quebracho.*

Schinus. See *Arœira.*

Schizolobium. See *Quamwood.*

Schizomeria. See *Cherry Birch.*

Schlage Lock. A patent barrel lock operated by a push button in the knob.

Schleichera. See *Kusan* or *Kusum, Gausan,* and *Mallet Wood.*

Schneid. Abbreviation for the name of the botanist Schneider.

Schomb. Abbreviation for the name of the botanist Schomburgh.

Schooner. **1.** The most common type of sailing vessel, rigged fore-and-aft, usually with two masts with gaffsails. If it has more than two masts it is named accordingly. **2.** A covered wagon.

Schrebera. See *Moka.*

Schum. Abbreviation for the name of the botanist Schumann.

Sciagraph. A sectional drawing showing the interior, especially of a building.

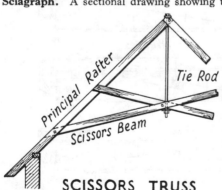

SCISSORS TRUSS

Scissors Truss. A simple type of truss for an open roof in which the strut and lower chord are in a straight line, or one member.

Sclerosis. Hardening of the cell walls of wood by lignification.

Scoinson. See *Sconcheon.*

Scoinson Arch. A squinch arch, *q.v.*

Scollop. See *Scalloped.*

Scolopia. See *Red Pear.*

Scolytidæ. A family of bark-beetles that includes pin-hole borers, *q.v.*

Sconce. **1.** A protecting screen or partition. **2.** A bracket carrying a light or reflector. **3.** See *Squinch.* **4.** A chimney seat or ingle-nook, *q.v.*

Sconcheon. **1.** Same as squinch. **2.** A splayed inside jamb. The interior edge of a window jamb.

Scoop. A large ladle. A coal scuttle.

Scoop Wheel. A water-wheel with buckets, or vanes.

Scoot. 1. See *Drag*. 2. An American term for scantlings (1) in hardwoods. 3. An American grading term for hardwood lumber of an inferior quality.

Scooter. A toy runabout consisting of a board on two wheels and a steering handle.

Scop. Abbreviation for the name of the botanist Scopoli.

Scope. A bundle of twigs.

Score, To. To make a mark or notch ; to keep count.

Score Board. Used in games to show the progress of the game.

Scorodocarpus. See *Kulin*.

Scotch. See *Scots*.

Scotch Bracket. Short pieces of laths bedded in plaster screeds for small plaster cornices.

Scotching. Removing a shaving of bark the whole length of the *heads* for a crate. The shavings are removed with a scotching iron, similar to a cooper's round shave.

Scotia. A concave moulding. See *Trochilus* and *Attic Base*. The contour may be formed of two circular arcs or it may be elliptical. See *Mouldings*.

Scots Elm. Wych Elm, *q.v.*

Scots Pine or **Fir.** *Pinus sylvestris*. See *Redwood*.

Scottelia. See *Odoko*.

Scow. A large square-ended, flat-bottomed boat used as a barge or lighter.

Scraper. A thin steel blade, usually about 5 in. by 3 in., used for finishing the surface of

SCOTIA

hardwoods before sandpapering. The corner of the edges are turned over to form the cutting edges by means of a piece of cylindrical steel or a gouge. To sharpen the scraper, first file the edge straight and square, and then rub on the oilstone, as shown, to remove the file marks. Wet the edge and rub the sharpener flat on the sides to remove any burr. Then turn over the corners with the sharpener applied at an angle of about 80° as shown. Sometimes the blade is mounted in a holder or in a special type of plane. There are several patented types. Scrapers with curved outline are used for mouldings, etc.

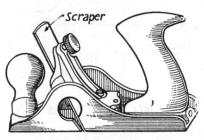

SCRAPER PLANE

Scraper Machine. A machine with a stationary cutter for finishing the surface of wood. The action is similar to that of a hand scraper.

The boards are fed through the machine by roller feed gear, and the scraper cutter removes the planing machine marks. The scraper is from 24 to 60 in. long.

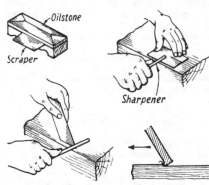

Scraping Tools. Wood-turning chisels, including diamond points or side tools.

Scratch Bead. One formed by the head of a screw projecting at the required distance from a piece of wood that serves as the fence. It is a useful expedient when a bead plane is not available.

Scratcher. 1. A hook-like gouge for marking trees. Also called a bark blazer or tree scribe. 2. A fine router for stringings in cabinet work. 3. A hooked tool for handling heavy cases.

Preparing Cutting Edge   Scraper in action

SCRAPER SHARPENING

Scratching. Forming or scraping small mouldings on hard woods or around curves with a scratch stock.

Scratch Stock. 1. A tool like a cutting gauge used to form grooves for inlays. 2. A specially made appliance to grip a thin cutting iron to shape small mouldings.

Screeds. 1. Guides for the rule when plastering. 2. An alternative name for grounds.

Screen. 1. A protection or shelter. A partition for privacy. It may be fixed, or loose and folding. See *Chancel* and *Dwarf Screens*, and *Parclose*. 2. A large rectangular sieve for grading sand and gravel. 3. A sight board on a cricket field.

~Cutters

SCRATCH STOCKS

Screen Door. See *Dwarf Door*.

Screen Hinge. See *Reversible Screen Hinge*.

Screeve. Same as scribe, *q.v.*

Screw. Anything helical or spiral-shaped, so that it advances as it revolves. Cylindrical metal fastenings with a helical thread to engage with the material to which it is screwed. The head of a wood screw may be flat, raised, or round, and it is slotted so that it can be rotated with a screw-driver. Wood screws may be of iron, brass, gun-metal, copper, or alloys, and they may be galvanised, japanned, oxidised, chromium, or nickel-plated. Sizes range from $\frac{1}{8}$ to 7 in. long, in gauges from o to number 30. The various types of screws are : coach, dowel, grub, lag, nail, set, and wood screws. There are many patent types.

SCREEN, FOUR FOLD

Screw-box. A device for cutting screws out of wood.

Screw-drivers. There is a large variety in both sizes and types : brace-bit, London pattern, cabinet or spindle pattern, pump, ratchet, spiral, and wheel. See *Screwing Machine* (2). (See illus., p. 492.)

Screwing. See *Pocket-*, *Secret-*, and *Slot-screwing*.

Screwing Machine. **1.** A machine for cutting screw threads and tapping, in wood. **2.** A machine for inserting wood screws.

Screw Jack. See *Jack*.

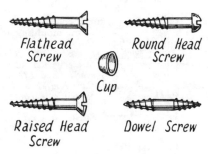

Flathead Screw

Round Head Screw

Cup

Raised Head Screw

Dowel Screw

SCREWS FOR WOOD

Screw Key. A spanner, *q.v.*

Screw Nail. A spiral nail that revolves when driven with the hammer, to give greater security.

Screw Rollers. Live rollers with coarse spiral threads to roll the stuff to one side as it comes from the saw or machine.

Screw Shackle. See *Tension Sleeve*.

Scribed. Applied to a joint between intersecting moulded members in which the moulding on one piece is fitted on to the moulding of the other piece. This type of joint is used in place of a mitre. The

LONDON PATTERN

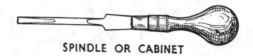

SPINDLE OR CABINET

SPIRAL

RATCHET

SCREWDRIVERS

moulding is first mitred, as shown, and then cut square to the outline

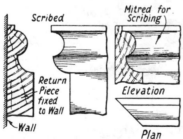

SCRIBED SKIRTING BOARD

of the moulding as shown. See *Bar Joints.*

Scriber. **1.** A pointed piece of steel used for setting out the stuff. See *Marking Knife* and *Raze Knife.* **2.** A swivelled arm carrying a router for circular inlays.

Scribing. **1.** The act of shaping material to an irregular surface. Preparing a scribed joint. See *Meeting Rails.* **2.** Marking or incising with a scriber. Also see *Spiling.*

Scribing Cutters. The irons on a tenoning machine for preparing the shoulders of rails and muntins to fit on to a moulding. Also see *Shoulder Lancets.*

Scribing Iron. A raze knife, *q.v.*

Scrieve Board. A large board or specially prepared floor for setting-out full size in shipbuilding.

Scrim or **Scrimp.** Coarse canvas. Any form of woven fabric for covering the joints of wall-board, reinforcing plaster work, etc.

Scrog. Brush-wood or branches.

492

Scroll. A carved spiral ornament. A continuous curve consisting of circular arcs of increasing radii. Also see *Monkey Tail.*

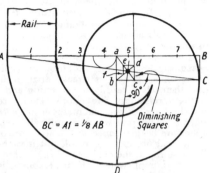

SCROLL TO HANDRAIL

Scroll and Casement. A moulding characteristic of Gothic architecture.

Scroll Foot. A characteristic of the carved feet of furniture of the Queen Anne period. See *Chippendale.*

Scroll Moulding. Same as roll moulding, *q.v.*

Scroll Saw. A machine fret-saw.

Scroll Step. A curtail step, *q.v.*

Scrowl. Pieces bolted to the knee of the head of a ship in place of a figure.

Scrubbing Board. A corrugated board used when washing clothes. A washboard.

Scrub Ironwood. See *Ironwood Box.*

Scrub Plane. A metal plane with single iron for roughing off.

Scrub Wood. A small stunted tree.

Scuffled. Applied to a roughly sawn surface that has had the loose fibres removed but still shows a sawn surface.

Scull. A short spoon-bladed oar.

Sculler. A small boat propelled by a pair of sculls.

Sculpture. The art of carving figures in wood, stone, etc.

Scunchion or **Scuntion.** See *Sconcheon.*

Scuppers. Deck drains. Outlets to allow water to escape from the deck of a ship.

Scurfing Plane. A jack plane with single iron, for use on rough and dirty surfaces.

Scutcheon. See *Escutcheon.*

Scuttle. **1.** A small aperture in a roof or ceiling. A trap door. **2.** A hatchway of a ship, or a hole or opening. **3.** A shallow basket. **4.** A small receptacle for coal to stand near a fireplace.

Scuttle-butt. A water cask with a hole for dipping in a cup, etc.

Scythe Stock. The long wood handle to a scythe.

Sd. Abbreviation for *seasoned.*

Sdg. Abbreviation for *siding.*

Se. Chinese maple. See *Maple.*

S.e. Abbreviation for *square-edged.*

S.E. Abbreviation for *stopped end.*

Sea Grape. *Coccoloba uvifera* (H.). Central America. Many local names, including pigeon wood, horsewood, yarua, cocobolo, etc. Dark brown, tinged with red or violet, sometimes streaked. Very hard, heavy, and compact. Strong but brittle, durable. Irregular grain, fine texture. Not difficult to work considering hardness, and polishes well. Sometimes confused with commercial cocobolo wood. Used for cabinet work and furniture. S.G. ·96 ; C.W. 3·5.

Seam. 1. Same as seasoning check, *q.v.* **2.** Joints in carvel-built boats, ship's decks, etc., that need caulking.

Seaplanes. Aircraft that can take off from the surface of water. Hydroplanes.

Seasoning. Reducing the moisture content of wood to suitable proportions. There is great variation in different woods. Many lend themselves readily to seasoning and are naturally stable, others are very refractory and require expert treatment to avoid warping and checking. The usual methods are *natural*, *water*, and *kiln* seasoning, *q.v.* Different woods require different methods. Some tropical woods are best seasoned in the sea. Air or natural seasoning is slow, costly, and not under control, but it is the best for some woods and does not impair the qualities. Water seasoning consists of submerging logs in running water. Kiln seasoning is the best for most woods as it is quick and under control so that the vagaries of the particular wood can be dealt with. The chief considerations are heat, humidification, and circulation of air. Careless seasoning causes checks and distortion, especially on further conversion, due to case-hardening. Newly felled trees contain acids, gums, resins, tannin, sugar, and various mineral and chemical constituents, in an emulsified active condition. The wood is stronger and more durable and stable if these constituents are inert, which requires considerable time, even after the moisture is evaporated. Science and experience are yet uncertain whether maturing to produce the chemical inactivity should take place before or after kiln seasoning. Wood loses 15 to 40 per cent. of its weight and shrinks considerably with seasoning. *Second seasoning* is the further drying or stoving of framing, etc., before it is wedged up or prepared for fixing, so that the wood will acquire the equivalent moisture content of the atmosphere in which it has to be fixed. See *Air Seasoning, Kilns, Moisture Content, Preservation, Salt, Shrinkage, Strength,* and *Urea.* Also see *Girdling.*

Seasoning Checks. Those that occur on the surface during seasoning. See *Checks* and *Shakes.*

Seat. 1. The portion or area of a plate carrying another member, such as a rafter. **2.** Anything on which a person rests in a sitting position, as a stool, chair, couch, etc. **3.** The base line for the development of the face mould, in hand-railing.

Seating. 1. A pad or template to provide a level and firm support for a structural member, such as a beam. **2.** A prepared bed for something to rest on. **3.** Applied collectively to a number of seats in a building.

Seat Rail. A cross-piece between the legs of a chair.

Seco. A prefabricated wood house, with plywood structural elements.

Secondary Beam. A binder, *q.v.*

Second Fixings. Applied to joinery that is fixed after plastering : architraves, skirtings, picture rails, handrails, etc. See *Fixings.*

Second Growth. Natural forest growth after felling or fire. In some cases the growth is from the stumps of felled trees.

Secondi. *Khaya ivorensis.* African or Nigerian mahogany, *q.v.* Also see *Mahogany.*

Seconds. See *Shipping Marks* and *Quality.*

Second Seasoning. See *Seasoning.*

Secretaire. A writing desk with pigeon holes and drawers. A bureau or escritoire.

Secretaire Drawer. A drawer to a secretaire in which the front falls level when the drawer is open, to form a flat table or desk.

Secretaire Fall Joint. A special joint for the falling front of a secretaire. See *Pivoted Fall Joint.*

Secretaire Fall Panel. The flap of a secretaire or bureau that falls flat to serve as a desk. See *Pivoted Fall Joint.*

Secret Dovetail. A dovetailed joint having the appearance of a mitred joint when assembled. See *Dovetail.*

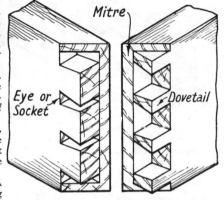

SECRET DOVETAIL

Secret Fixing. A method of fixing polished or varnished joinery without showing nail holes or screws. See *Secret Screwing.*

Secret Gutter. A valley, gable, or chimney gutter hidden by shingles, tiles, or slates.

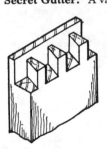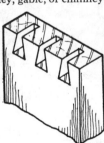

Secret Lapped Dovetail. Similar to a lapped dovetail, *q.v.*, but the sockets are stopped and the dovetails hidden. See *Stopped Lapped Dovetail.*

Secret Mitred Dovetail. See *Secret Dovetail.*

Secret Nailing. 1. Skew nailing through the tongues of boarding so that the nail holes are not seen. **2.** Raising a narrow thick shaving from the face of the stuff

SECRET LAPPED DOVETAIL

before driving a nail. The shaving is raised by chisel or special tool. The strip is then glued in position over the nail hole. See *Floor Boards.*

Secret Screwing. A method of strengthening butt joints and fixing superior joinery. The screws project about $\frac{3}{8}$ in. from one piece and are inserted into corresponding holes in the other piece. They are then driven along parallel slots slightly larger than the shanks of the screws. Also see *Slot Screw*.

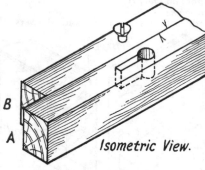

Isometric View.

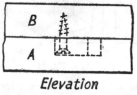

Elevation

Section

Plan

SECRET SCREW JOINT

Sectioning. Denoting the kinds of material by variation of the section lines in a sectional drawing. The alternative method is to use uniform section lines and print the names of the materials.

Section Modulus. See *Modulus of Section*.

Section Mould. A cardboard, plywood, or zinc templet cut to the required contour. It is marked or fixed on the end of the stuff, which is shaped accordingly. See *Moulds*.

Section Rod. Same as setting-out rod.

Sections. The parts of a mechanical drawing representing the interior, or section, of anything. The cross-section of a moulding, etc.

Sector. 1. A part of a circle bounded by an arc and two radii. 2. A mathematical instrument.

Sedan. 1. An enclosed chair carried on two poles. 2. A handbarrow.

Sedeng. *Boehmeria regulosa* (*H.*). India. Also called Dar and Genthi. Light red, fading to dull reddish brown, smooth. Light, fairly soft and durable, strong. Straight grain, even coarse texture, distinct rays. Easily seasoned and wrought. Limited supplies, small sizes. Used chiefly for carving and turnery. S.G. 5 ; C.W. 2·5.

Sedilia. Recessed seats in the choir of a church, usually for the priests. See *Misericorde*.

Brick	*Stone*	*Concrete*
Timber	*Excavation* *Earth*	*Plaster*

SECTIONING

496

Seed Lac. See *Shellac*.

Seem. Abbreviation for the name of the botanist Seeman.

Seepage. Leakage. To trickle or ooze out.

See-saw. A plank hinged on a short central post that serves as a fulcrum so that the ends of the plank move up and down. It is used by children at play.

Segl. Abbreviation for *segmental*.

Segment. 1. A part of a circle bounded by an arc and a chord. Area of small segments $= \frac{2}{3} hc$, where h = height and c = chord. The radius of the circle $= \frac{1}{2}\left[\frac{(\frac{1}{2}c)^2}{h} + h\right]$. 2. A part of a sphere cut by a plane. Surface area $= \pi(r^2 + h^2)$. Volume $= \frac{1}{6}\pi h(h^2 + 3r^2)$. When the plane is diametral the segment is a hemisphere. 3. Short pieces for building up circular work.

Segmental Arch. An arch consisting of a circular arc less than a semicircle.

Segment on Segment. Double curvature work consisting of segments of circles in plan and elevation. Same meaning as *circle-on-circle*.

Circle   Sphere

SEGMENT

Moving Pencil marking outline of Elevation

Block

Ribs

Rise

Radius

Laggings

90°

Span

SEGMENT ON SEGMENT CENTRE

Seize. 1. To bind, stick, or jam, especially through heat or friction. 2. To lash together with ropes.

Sekondi. Secondi, *q.v.*

Selangan Batu. Sometimes called "Borneo teak." See *Ipil, Merabau*, and *Teak*.

Selans. Australian Silky Oak, *q.v.*

Selected or Selects. 1. Applied to the best qualities in the grading of wood. 2. A specific grade in hardwoods. See *Shipping Marks*.

Selected Merchantable. A special grade of Douglas fir and other N.W. American softwoods. It is a superior grade to *Merchantable, q.v.*

Selection. The factors governing the selection of wood for a particular job, are cost, fitness for its purpose, and supplies. Cost includes labour in working. See *Comparative Workability*. The quality of wood depends upon its position in the log, and condition of growth : climate, soil, and surroundings. There is considerable variation even in the same species. Other factors controlling its fitness for a particular job are given alphabetically. See *Quality, Resistance, Strength, Supply, Tests*, and *Working Qualities*.

Self-acting Saw. One with a mechanical feed motion.

Self Log. A term applied to logs sold singly because of their superior quality.

Self-supporting Partition. One supported by the walls and independent of any support from the floors. See *Partition*.

Selimbar. *Hopea* or *Shorea spp.*, and spp. of family Flacourtiaceæ. Malay. Usually marketed as Balau, *q.v.* See *Resak*.

Sel. Struc. Abbreviation for *select structural*, a grading term for pitch pine.

Semal. See *Semul*.

Semi. Abbreviation for *semicircular*.

Semicircular Head. The top part of a frame shaped to half of a circle. The curved part may be cut from the solid or built-up or laminated.

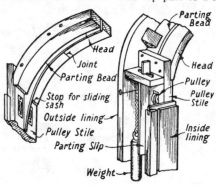

SEMI-CIRCULAR HEAD

Seminai. Betis, *q.v.*

Semul. Cottonwood, *q.v.*

Sen. *Acanthopanax recinifolius* or *Kalopanax r.* Resembles and often marketed as Japanese Ash, *q.v.*

Sena. *Pterocarpus indicus* (*H.*). Malay. Also called Narra, *q.v.*, and Angsena. Rosewood yellow to blood red. Often streaked and fragrant. Large proportion of whitish sapwood. Moderately hard, heavy, and durable. Ripple and faint rays. Large sizes. Excellent wood for cabinet work. S.G. ·75 ; C.W. 3·5.

Sénécl. Abbreviation for the name of the botanist Sénéclauze.

Senegal Mahogany. See *African Mahogany*.

Sengkawang. *Isoptera borneensis* (*H.*). Malay. Often marketed as Balau and Resac. Brown. Very hard, heavy, durable, and strong. Straight and crooked grain. The latter is very difficult to work. Does not polish easily. Used for structural work, furniture, boat-building, planking, etc. S.G. about ·85 ; C.W. 3 to 5.

Sentry Box. A small wood hut sufficiently large to hold a man sitting or standing. It is intended to provide shelter against the weather for a sentry or a watchman.

Separators. Packings between structural timbers that are bolted together. Cast-iron spools are preferred, when exposed to the weather, as they do not retain the moisture so much as wood packings.

Sepetir. *Sindora spp.* (*H.*). Malay. Borneo. Brownish, oily. Large proportion of whitish sapwood. Light, soft, stable, not strong or durable. Fine grain, distinct rays and growth rings. Not difficult to work. Used for light construction and furniture. S.G. ·44 ; C.W. 3.

Sepira. Greenheart, *q.v.*
Sept. An enclosure or railing.
Sequoia. *Sequoia sempervirens* and *S. gigantea* (*S.*). Californian or Giant Redwood. The largest known trees. Reddish brown. Soft and light, but variable. Stable and durable, brittle and fissile. Strong for its weight. Resists insects and chemicals. Grain variable from rather fine to coarse ; and from uniform to uneven texture. Not appreciated in this country. Burr wood is highly figured and called Vervona. Easily wrought, except end grain. Used for panels (but now superseded by plywood), shelving, shingles, sleepers, pipe-lines, greenhouses, tanks, and for structural work because of large sizes without defects. In America it is used for nearly all purposes. S.G. ·4. C.W. 1·75. See *Redwood* and *Vervona*.
Seraya or **Seriah.** *Shorea macroptera* and other *spp.* (*H.*). Some-times called Borneo Cedar or Mahogany. Malayan Seriah is sold as Red or Yellow Meranti. Pink to dark red. Distinguished as white and red Seriah. The latter is moderately hard, heavy, and strong, but variable in properties, not durable, distinct rays, fine grain. S.G. ·38 to ·6 ; C.W. 4. WHITE SERAYA. *Parashorea malaanonan* and other species. Also called Borneo White Cedar. Lighter colour, softer, coarser grain, and not so variable in properties and characteristics. S.G. ·6 ; C.W. 4. Both varieties require care in seasoning, work readily, and polish well, and obtainable in large sizes. Used as a substitute for mahogany : furniture, superior joinery, vehicles, veneers, cigar boxes, etc. See *Lauan* and *Meranti*.
Serbian Spruce. *Picea omorika*. Jugoslavia. Not distinguished from Whitewood, *q.v.*, in marketing.
Serentang. Same as Terentang, *q.v.*
Sering. See *Wilde Sering* and *Wit Sering*.
Serpentine. Applied to a sideboard in which the front has a sinuous curve in plan. See *Saddle*.
Serrated. Toothed. Shaped like the cutting edge of a saw.
Servant Key. See *Key*.
Servery. A place for preparing trays of food in a restaurant.
Service Tree, Wild. *Pyrus torminalis* (*H.*). Europe, N. America. Wood similar to Rowan Tree. Yellowish red, lighter streaks. Moderately hard and heavy. Close even grain, small sizes. Little marketed. S.G. ·65. TRUE SERVICE. *P. sorbus*. Also called Cormier. Like pearwood, and used for similar purposes. Fawn colour. Hard, with fine grain and some beautiful figure. Used for cabinet work. See *Wild Service*.
Serving Hatch. A small aper-ture, with door, for the serving of food, drinks, crockery, etc.

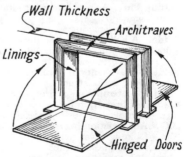

SERVING HATCH

It is in the division wall between kitchen and dining-room of a house. The door may be hinged or pivoted, or it may slide or revolve. There are several registered designs.

Serving Table. A sideboard on which food is placed ready for serving.

Sessile. Without stalk. Also see *Peduncle.*

Sessile Oak. *Quercus petraea* or *Q. sessiliflora.* English oak, *q.v.*

Set. 1. A nail punch. **2.** The projection of the teeth on alternate sides of a saw to provide clearance when sawing. See *Saw Set.* **3.** The non-elastic deformation of wood. See *Permanent Set.* **4.** An assortment of similar things. **5.** To harden : as glue, cement, plaster, etc. **6.** See *Setting.* **7.** The parts to build up a coffin : sides, ends, bottom, and lid. **8.** A complete unit of framing in mine timbering.

Set Screw. One with parallel shank to engage with metal which is tapped to receive the screw.

Set Squares. Thin triangular drawing appliances, usually of wood or celluloid, for drawing vertical and inclined lines. The angles are 90, 45, 60, and 30 degrees.

Sett. A term applied when more than one wagon or truck is required to transport a long log. The second truck is called a bolster or runner. A *sett* of timber is a load of timber conveyed by such a double conveyance.

Settee. 1. A long seat with low arms and back. A sofa. **2.** A type of small sailing vessel with long prow.

Settee Bolt. A special type of bolt for securing fitments to the deck of a ship.

Setting. 1. Drying out of steamed and bent wood. **2.** Fixing the back iron correctly to the cutting iron of a plane. **3.** Correcting the teeth of a saw to give the necessary clearance. See *Saw Set.* **4.** Drying, or hardening, of glue, or any plastic materials. **5.** Placing a thing in its correct position.

Setting Out. Preparing full-size drawings on rods by the setter-out, from which the dimensions are transferred to the stuff preparatory to machining and construction. See *Rods.*

Setting Pin. An appliance for straightening rake handles, etc., after steaming.

Setting Up. Erecting.

Settle. A hall seat with the lower part framed as a box, or receptacle. A long wood seat, or bench, with high back and arms, to accommodate three or four persons.

Settlement. Uneven sinking of an erection causing distortion, cracks, etc., in the structure.

Set-up. A temporary fitting up of a portable sawmill.

S.4.S. Abbreviation for *surfaced four sides.*

S.ft. Abbreviation for *superficial feet.*

S.G. Abbreviation for *specific gravity* and *slash grain.*

Sha. Chinese silver fir.

Sha-chu. Chinese walnut.

Shack. A rough hut, or shanty.

Shackle. A link or staple. See *Tension Sleeve.* Devices for coupling and fastening.

Shadbolt. A patent stay for a fanlight.
Shade. Any arrangement to give protection from the sun's rays.
Shade Pine. Sugar pine, *q.v.*
Shade Roller. A roller for a screen or blind.
Shaft. 1. Anything long and slender, as the handle of a striking implement or tool, the stem of a tree, the bars, or poles, of a horse-driven vehicle, etc. 2. The cylindrical part, or body, of a column, between base and capital. 3. The part of a chimney above the roof. 4. A vertical excavation leading to a mine. 5. A ventilating pipe or conduit. See *Entasis* and *Entablature*.
Shaft Log. A block through which the steering shaft of a ship passes.
Shagbark. *Hicoria* or *Carya alba*, *C. sulcata* or *C. ovata*. White hickory, *q.v.*
Shagreen. A fish skin used for inlays.
Shaitan Wood. Chatiyan, *q.v.*
Shake. A cleavage, or split, in wood. A separation between adjacent layers of fibres. Shakes are described as compound, cross, cup or ring, heart, radial, shell, or star shakes, *q.v.* Felling or thunder shakes, are ruptures, or upsetts, *q.v.* Heart and star shakes are radial, and due to shrinkage through seasoning or old age. Cup shakes are circumferential between ad-
jacent layers of growth rings and may be due to damage of the cambium layer, or lack of nutriment, or termites. Shakes are usually caused by wind, frost, lack of nutrition during growth, or by faulty seasoning. Also see *Check*.

Shakes. Hand riven ½-in. shingles, longer than normal, and often staggered for special effect.
Shales. Same as shingles, *q.v.*

SHAKES

Shallop. A sloop, *q.v.*
Sham Beam. A decoration on a ceiling to imitate a beam. It is often used when panelling a ceiling with wall-boards.
Shank. 1. The part of a tool connecting the handle to the working part. A long connecting part to any appliance. 2. The shaft of a column. 3. The straight part worked on the wreath of a handrail to receive the bolt for the joint. 4. A device for locking the teeth in an inserted-tooth saw.

SHARK'S JAW

Shanty. A temporary wood building. A hut or mean dwelling.
Shaped Work. Curved work.
Shapings. Curves in cabinet making.
Share Beam. That part of a plough in which the share is fixed.

Shark's Jaw. A special tool used for turning handrail bolts.
Sharpening. See *Oilstone*, *Saws*, and *Set*.

Sharpening Horse. A frame for securing a circular saw whilst being sharpened. Also see *Saw Chops*.

Sharpie. A straight-sided sailing boat with pointed stem. It may be flat- or V-bottomed.

Shaves. See *Spokeshave, Cooper's Round Shave*, and *Draw Knife*.

Shaw. A small plantation, coppice, or wood.

Shaw Guard. A registered type of guard for a spindle moulding machine.

Shay. A low four-wheeled carriage. A chaise.

Sh.D. Abbreviation for *shipping dry*.

Sheal or **Shealing.** A shelter for sheep. A hut or shelter for fishermen or shepherds.

Shea or **Shee Oak.** *Casuarina quadrivalvis* (*H*.). Tasmanian Oak. Dull brownish to flame-red. Hard, heavy, tough. Beautiful figure for cabinet work, due to medullary rays. Rather small sizes and difficult to work. Polishes well. Also used for implements, veneers, shingles, fancy articles, etc. *C. Fraseriana*, W. Australia, is a similar wood. S.G. ·83. BULL OAK. *C. suberosa*. Similar to above but darker and more durable. There are several good varieties from Queensland. They have a distinguishing prefix as red, black, flame, rose, river, etc.

Shear. 1. A tendency for one part to slide over another. To cut as with a pair of scissors. 2. See *Sheer*.

Shear Boom. A boom for guiding floating logs in the required direction. A fender, or glancing, boom.

Shear Force. The resultant force acting on a structural member

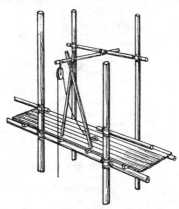

and tending to produce shear, either horizontal or vertical. At any section of a beam it is the algebraic sum of the forces acting on one side of the section.

Shear Legs. Two or three poles secured together at the top and spread apart at the feet, like a tripod. Pulley tackle is suspended from the top for hoisting purposes. A guy rope is often used instead of the third pole.

Shear Lines. The profile (or side elevation) of the hull, etc., of a ship. See *Sheer*.

Shear Plan. A longitudinal elevation, in shipbuilding, to show the sheer of the decks.

Shears. A long wood bed for a lathe, but sometimes applied to an iron bed. See *Lathe*.

SHEAR LEGS

Sheathing. 1. A thin protective covering, as muntz metal coverings to wood piles, or the bottom of a ship. 2. Close boarding on spars or studs used as a base for the final coverings. It is often distinguished as open (square edged) or tight (matched). Especially applied to the diagonal boarding under the sidings of timber buildings. **3.** Ship's planking or decking. Also see *Weather Boards, Jamb,* and *Sidings.*

Sheave. 1. The wheel of a pulley. **2.** A sliding escutcheon to a keyhole.

Shed. A wood shelter or outhouse.

Shed Roof. A lean-to roof, *q.v.*

Sheer. 1. See *Shear.* **2.** Fore and aft or longitudinal curvature of a ship's deck.

Sheer Strake. The strake in the topside under the gunwale of a ship.

Sheet. Sometimes applied to a sash, especially when it is fixed.

Sheeted. A shipwright's term for a surface built up of tongued and grooved boards.

Sheeting. 1. Close horizontal timbering for excavations. See *Guide Piles.* **2.** Shuttering, *q.v.* **3.** Match-boarded surfaces on a ship.

Sheet Piling. Close piling forming a panel between large piles. They are usually pointed and shod with metal for easy driving. See *Guide Piles.*

Shelf. A horizontal flat board, framed in a cupboard, etc., or fixed to a wall, and used for storing articles. A ledge. See *Housing Joints.*

Shelf Fittings. Any arrangement for fixing or holding shelves in position. It is usual to fix the shelves by means of cleats, nogs, or brackets. In recesses, in cheap work, the shelves often rest on nails and the plastering fixes the shelves in position. There are numerous patent and other devices used in cabinet work for adjusting the height of the shelves as required.

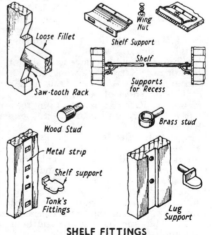

Shelf Nog. A projecting piece of wood supporting a shelf. It is built into the wall, like a cantilever, and is a substitute for a bracket. See *Nogs.*

Shelf Piece. A long timber running the length of a vessel under the deck beams.

Shell. Same as *Carcase.*

Shellac. A resinous gum produced from seed lac.

SHELF FITTINGS

It is used with methylated spirits for french polish, and for certain kinds of varnish.

Shellbark. See *Hickory.*

Shell Bit. A semicylindrical brace bit for boring small holes.

Shell Shake. A cup or ring shake showing on converted wood. See *Shakes*.

Shellwood. Ropala, *q.v.*

Shelly. Applied to wood in which shell shakes appear.

Shelter. 1. See *Dodger*. 2. A decorative shed-like structure, open on the sides and provided with seats. Any form of hut, cabin, or shed to provide shelter against the weather. 3. Any form of strong structure to give protection against air raids.

Shelves. 1. Plural of shelf. 2. To shelve means to incline slightly.

Shelving. 1. Same as shelves. 2. Fixing shelves. 3. To slope or incline.

She Oak. See *Shea Oak*.

Sheraton. A designer of domestic furniture (died 1806) whose

SHERATON

work is still greatly appreciated. It is usually of mahogany with decorative inlays, and has similar characteristics to Hepplewhite furniture, but simpler and more severe. He is sometimes styled the master of cabinet-makers, and he introduced numerous ingenious contrivances used in cabinet work.

Sheriff. A patent fanlight opener.

Shet. Same as joint.

Shides. Shingles, *q.v.*

Shield. 1. An escutcheon. 2. Anything that defends or protects. A movable screen.

Shield Back. Applied to a Hepplewhite chair in which the back has the outline of a shield.

Shieling. A roughly constructed shelter. See *Sheal*.

Shikhuma. Kenya Satinwood, *q.v.*

Shimer Head. A cutter head used for tongues and grooves, or *matching* the edges of boards.

Shims. 1. Packings used to bring the surfaces of battens to a plane surface to carry wall-boards. 2. Long narrow patches glued into plywood panels.

Shingle Bolt. See *Bolt*.

Shingle Press. A special machine for bundling shingles.

Shingles. Wood "tiles." Thin pieces of wood, somewhat like slates, used for covering roofs and exterior walls. They may be cleft, especially if of oak, but the modern type is rift sawn, tapered, and of Western red cedar. The ordinary size, 5X, is 16 in. long and of random widths, but longer shingles may be obtained: Perfections, 18 in., Royals, 24 in. The thickness tapers from $\frac{5}{10}$ in. at the butt to about $\frac{1}{16}$ in. They can be steamed and bent for swept valleys. Shingles should be ventilated underneath and not laid on felt or close boarding. The margin for ordinary sizes, 5X, should not exceed 5 in. for roofs or $7\frac{1}{2}$ in. for walls. See *Barge Boards*, *Jamb*, and *Shakes*.

Shingle Weaver. One who makes shingles.

Shining Gum. *Eucalyptus nitens*. Victoria, Australia. See *Gum*.

Ship-lap. A dovetailed halving joint for the corner of ship's coamings.

Ship-lap Boards. Rebated weather boards, *q.v.* See *Sheathing*, *Sidings*, and *Weather Boarding*.

Shipmast Locust. *Robinia pseudoacacia.* See *Locust*.

Shippen. A cow-house or stable.

Shipper's Marks. See *Shipping Marks*.

Shipping Culls. Applied to wood that is culled in shipping. It is a grade above mill culls.

Shipping Days. An agreed period allowed for loading a vessel.

Shipping Dry. A term applied to wood that is sufficiently seasoned to prevent fungi attack or deterioration on board ship. The moisture content is equal to that of air-dried wood.

Shipping Marks. Private trade marks of timber exporting firms to denote the quality of the wood. Different combinations of letters, signs, and colours are used for different sizes and qualities, and they are painted or stamped or branded on each piece. The marks have no significance to the user unless he has had experience with the exporting firm. Thousands of different combinations of marks are used for European timbers alone. See *Grading*, *Quality*, and *Selection*. The illustration shows the marks used by exporting firms of five different countries. The Swedish and Norwegian firms export six grades, but the Canadian and Japanese only three grades. Most timbers have their separate grading rules or specifications, agreed upon by the timber association concerned.

Shipping Ton. A volume of wood of 42 cub. ft.

Ship Plane. An aeroplane adapted for landing on a ship's deck.

Shippon. See *Shippen*.

Ship's Cleats. See *Cleats*. On board ship cleats are distinguished as belaying, comb, girt-rope, mast, range, shroud, shoulder, sling, stop, thumb, etc. (See illus., p. 506.)

NORWAY Redwood

FᵢB   FᵤB   FᵢₗₗB   FᵢᵥB   FᵥB   Fᵤ/ₛB

SWEDEN Redwood

1ˢᵀ   2ᴺᴰ   3ᴿᴰ   4ᵀᴴ   5ᵀᴴ   Unsorted

P+H   P++H   P·H   P··H   P✱H   P⊕H

CANADA Whitepine

1st. Qual.
2nd. Qual.
3rd. Qual.

U.S.A. Hardwoods

Prime   Selects   №1 Common   №2 Common

BAN   BAN   BAN   BAN

JAPAN Hardwoods

Prime   1st. Quality   2nd. Quality

★★★★   ★★★   ★★

SHIPPING MARKS

Ship Spikes. Heavy, nail-like fastenings, forged from square bars of iron, with a wedge or chisel point.

Ship-way. See *Slip-way*.

Ship-worms. Marine insects of the *Teredo* and *Bankia* genera that attack wood. No wood can resist all kinds of marine insects.

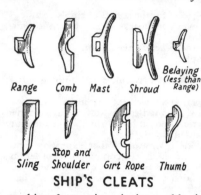

Range   Comb   Mast   Shroud   Belaying *(less than Range)*

Sling   Stop and Shoulder   Girt Rope   Thumb

SHIP'S CLEATS

Those containing silica are most resistant. See *Teredo* and *Marine Borers*.

Shipwright. A ship's carpenter. A shipbuilder.

Shira or **Shiron.** Japanese oak, *q.v.*

Shisham. Sissoo, *q.v.* See *Rosewood (Indian)*.

Shittim Wood. Lebanon cedar, *q.v.*, or *Acacia seyal*. A Biblical wood not definitely classified.

Shive. A cooper's term for a piece of wood for a bung for a cask.

Shive Machine. Special machines for turning, ringing, and boring shives for barrels.

Shiwan. Also called *Gumhar, Gamari, Yemane*, etc. See *Gumhar*.

Shlp. Abbreviation for *ship-lap*.

Shoe. 1. Any form of iron foot for a wood member. A metal point for the foot of a pile. 2. A bent piece at the foot of a down-pipe to turn the water away from the wall. 3. The seating for the heel of a swing door. See *Floor Spring*. 4. A metal slipper in the bottom edge of a sliding door or sash to engage with the runner. 5. An adjective applied to numerous parts of and appliances for footwear : shoe-tree, -last, -form, -heel, -filler, etc. Also see *Spring Hinge*.

Shooks. Casewood cut to the required sizes. A bundle of sawn or split pieces sufficient to form a barrel, or a set of boards for a box. Boxboards or caseboards, *q.v.*

Shoondul. Ipil, *q.v.*

Shoot. 1. An inclined slide or trough. See *Chute*. 2. To straighten the edge of a board with a plane. 3. A new growth on a plant, either from a bud or from the roots.

Shooting Board. A framed board for steadying a piece of wood whilst shooting the edge with a plane. It may have an attachment for shooting mitres. See *Mitre Shoot*.

Shooting Box. A small country house used in the shooting season.

Shooting Plane. An extra long try-plane.

Shop Board. A bench, used especially by tailors.

Shop Timber. Wood to be cut or re-converted for further manufacture. The wood is graded as the percentage of the face measure that will produce a certain number of cuts of the required quality and size.

Shore. A wood prop or support. Temporary timbers to support defective buildings, or those undergoing structural alterations, or ships under construction or repair, or the sides of large excavations. *Dead* shoes are vertical. *Flying* shores are horizontal, or have no support from the ground. *Raking* shores are inclined from the ground. See *Dead, Flying,* and *Raking Shores,* and *Formwork*.

Shorea. A genus with many species and including a large number of valuable tropical woods : Almon, Balau, Cangu, Guigo, Damar laut, Kalunti, Kumus, Mayapis, Lauan, Makai, Manggasinora, Meranti,

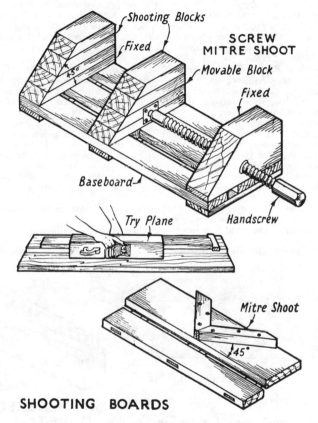

SHOOTING BOARDS

Oba-suluk, Resak, Sal, Selangan batu, Selimbar, Seraya, Talura, Tanguile, Thitya, Tiaong, Yakal.

Shoring. 1. A system of shores. 2. The preparation and fixing of shores. See *Shore*.

Short Column. Applied to struts or columns that are sufficiently short to fail by direct crushing only, as distinct from one that may also fail by bending. The maximum length depends upon the least diameter, but varies for different materials. See *Long Column*.

Short Ends. Wood in deal, batten, or board sizes, up to 5½ ft. long.

It is usually described as firewood and used for box-making. See *Ends*.

Short Grain. Applied to carroty or cross-grained wood in which the fibres fracture with little splintering. A piece that will fail easily under a transverse load. The defect may be due to brittleness of the fibres, or to diagonal or sloping grain, *i.e.*, the direction of the grain is not parallel to the axis of the stuff, possibly due to faulty conversion. See *Slope of Grain*.

Shortleaf Pine. *Pinus echinata* (*S.*). Southern U.S.A. Also called Slash pine, Carolina pine, Shortleaf yellow pine, but marketed in this country as pitch pine. Similar to longleaf pitch pine but softer, lighter, less durable, not so strong, and smaller sizes. See *Pitch Pine*, *Loblolly Pine*, and *Longleaf Pine*.

Shorts. A term applied to certain hardwoods, especially mahogany, less than 6 ft. long. Applied generally, in grading, to pieces shorter than the standard length.

Shot. Applied to the edge of a board that has been planed straight. See *Shoot* (2).

Shot Holes. Small holes in wood due to boring insects, called shothole borers. The holes are between $\frac{1}{16}$ and $\frac{1}{8}$ in. diameter. See *Pinhole* and *Wormhole*.

Shoulder. A projection to give strength and support to anything, as the projection at the root of a tenon, *q.v.*

Shouldered Dovetail. A joint often used between rails and legs, or posts.

Shouldered Dovetail Housing. A joint in shelving in which the

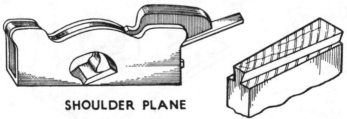

SHOULDER PLANE

dovetail tapers in its length. See *Housing Joints*.

Shouldered Tenon. A tusk tenon, *q.v.*

Shoulder Plane. A metal rebate plane for cleaning up shoulders to tenons.

SHOULDERED DOVETAIL HOUSING

Shovel-board. A board marked out for a game with discs or half-pennies.

Show Board. The inside stall board of a shop window, or the window bottom. The show space on the inside of a shop window. Sometimes called the stall-board.

Show-board Framing. The *window back* of a shop window.

Show Case. An air-tight glazed case for displaying goods. They are distinguished as floor, counter, and wall cases.   See *Silent Salesman* and *Floor Case*.

Show Wood. Applied to upholstered chairs in which part of the frame is exposed.

Shragged. Applied to poles that have been cleaned of side branches.

Shread Head. A jerkin head, *q.v.*

Shreddings. Short light timbers used as bearers under a roof in old buildings.

Shrine. A case in which sacred things are deposited. An altar.

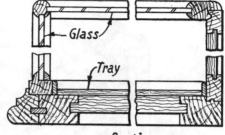

Section

SHOW CASE

Shrinkage. The decrease in the dimensions or volume of wood due to decrease of moisture content. It is usually expressed as a percentage of the original dimensions or volume : linear or volumetric shrinkage. *Directional* shrinkage refers to one of the grain directions : tangential, radial, and longitudinal. The last is usually

SHRINKAGE EFFECTS

insignificant in practice. The approximate variations due to seasoning are : longitudinal, 1 per cent. ; radial, 2 to 7 per cent. ; tangential, 6 to 15 per cent. There is great variation in different species but the tangential is about twice the radial shrinkage. The illustration shows the effects of shrinkage exaggerated. See *Seasoning* and *Moisture Content*.

Shroud Cleat. See *Ship's Cleats*.

Shrub. A woody plant, smaller than a tree, and with several separate stems growing from the roots or near the ground.

Shrubbery. A number of shrubs, usually bordering a walk in a garden.

Shuffle Board. See *Shovel-board*.

Shumard Oak. *Quercus shumardii*. American red oak. See *Oak*.

Shuting. An eaves gutter.

Shutter Bar. A long pivoted bar for securing folding shutters.

Shutter Blind. See *Jalousie*.

Shuttering. The formwork, falsework, or sheeting used for concrete work, especially *in situ*. Also see *Formwork*. (See illus., p. 510).

Shutters. 1. Protections for window or other openings. They may be boxing or folding, hinged, loose, rolling, or sliding shutters, and they may be on the inside or outside. Exterior types are often

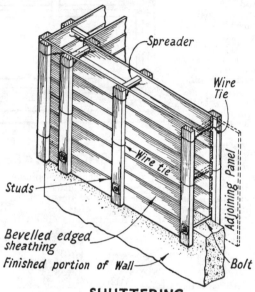

SHUTTERING

louvred to allow for light and ventilation. See *Boxing, Elbow Lining, Ventilating*, and *Lifting Shutters*. **2.** Panels or boards and ledges for shuttering. **3.** A device in the swell-box of an organ to regulate the volume of sound.

Shutting Joints. The various types of joints used for the shutting stiles of doors and casements : rebates, bevels, single and double lap, hook, round, etc. See *Joints*.

Shutting Stile. The stile on the opposite edge of a door or casement to the hinges. It carries the fastenings.

Shuttle. An implement used in weaving. It is made of extremely hard wood to withstand the wear due to shooting to and fro with the thread of the weft.

Shuttle Blocks. Pieces cut to size for shuttles.

Siamese. See *Mai*.

Siberian. Applied to the woods, chiefly redwood and yellow pine, exported from N. Russia and Manchuria.

Siberian Larch. *Larix sibirica.* See *Larch*.

Siberian Pine. *Pinus cembra.* See *Pine*.

Sickingia. See *Arariba*.

Sicupira. See *Sucupira*.

Sida. African or Nigerian walnut, *q.v.*
Side. The wide surface or face of converted wood in woodworking, but applied to any of the four faces in the timber trades, especially by Customs' officers.
Side Axe. An axe ground on one side only.
Side-bar. A piece of side framing of a cart or wagon.
Sideboard. A side table. A piece of dining-room furniture of a decorative character for holding dishes, etc. They are named according

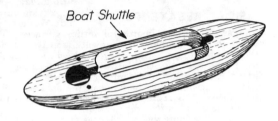

Boat Shuttle

Stick Shuttle

SHUTTLES

to the period of the design or designer : Adam, Elizabethan, Jacobean, etc. ; or according to some special form of construction : tambour, bow centre, pedestal, loose pedestal, single carcase, etc.
Sideboard Table. A combined sideboard and table for small rooms, flats, etc. The table slides into the sideboard when not in use.
Side-car. 1. A small car attached to the side of a motor-cycle. 2. A jaunting car.
Side Cut. Applied to a piece sawn so as not to include the pith.
Side Dresser. A tool to regulate the degree of spread of a saw tooth when using a swage.
Side Flights. See *Double Return Stairs*.
Side Grain. The grain shown on any longitudinal surface, as distinct from end grain.
Side Gutter. A small lead gutter down the slope of a roof, as to a chimney or gable.
Side Hook. A bench hook, *q.v.*
Side Light. See *Margin* and *Wing Light*.
Side Matched. See *Matched* and *Veneer Matching*.
Side Posts. Princess posts, *q.v.*

Side Rebate. A small metal plane for cleaning up plough grooves, etc. They are usually in pairs, right and left hand.

SIDE REBATE PLANE

Siderodendron. See *Hackia*.

Sideroxylon. See *Mastic, Melkhout, Muna, Sapote, Sweetbark, Yellow Boxwood, Yellow Butterwood*.

Side-table. A small table used for carving during a meal.

Side Wavers. Purlins, *q.v.*

Sidings. **1.** Weather boards for vertical surfaces. There are several market forms : feather-edge, novelty, rebated, ship-lap, *q.v.* Also called drop sidings. See *Weather Boarding*. **2.** A N. American term for boards. **3.** Slabs left from squaring a log.

Sidol. A proprietary wood preservative.

Sieb. Abbreviation for the name of the botanist Siebold.

Sight Board or **Screen.** A large white screen used on cricket fields to help the batsmen to sight the ball.

Sighting. Deciding the straightness or position of anything by means of the eye and judgment only, without instruments or mechanical aids.

Sight Line. **1.** A datum line. **2.** The inside edge of a member of a piece of framing.

Sight Rail. Any form of fixed level rail for sighting levels and gradients, especially when excavating.

SIGHT RAILS

Sigmoid. Shaped like the letter S.

Sign. A board, or other prominent object, to indicate the existence of a particular occupation or business, or to point out the direction and distance of a place.

SILENT SALESMAN

Signal. Any apparatus for conveying information implying some recognised instruction. A railway signal consists of a post with a lever that is operated from a signal box.

Sign Board. A wide fascia board over a shop front, or one fixed to a post or suspended from a wall, describing a business and ownership.

Sign Post. A directional sign. It usually points the direction and states the distance of another place.

Silent Salesman. A show case, about 6 ft. high, standing on the floor of a shop or store.

Silica. A flinty mineral secretion in wood that makes the wood very resistant to insect attack, but not to decay, and adds to the difficulty of working.

Silk Cotton Tree. *Ceiba pentandra* and *C. occidentalis* or *Eriodendron anfractuosum* (*H.*). C. America. Also called Ceiba, Cottonwood, Odoum, and Kapok. Same family as Balsa. Brownish white. Light, soft, strong and tough for its weight. Not durable, very buoyant. Irregular coarse grain, distinct rays. Easily wrought. Used for cases, slack cooperage, etc. S.G. ·4 ; C.W. 2.

Silkwood. See *Queensland Maple* or *Maple Silkwood*.

Silky Ash. *Ehretia acuminata* (*H.*). Queensland. Light fawn, smooth, lustrous. Moderately soft and light. Firm, tough, and strong for its weight. Similar to white ash in grain and texture, distinct rays, and mottle. Easily wrought, polishes well. Used for interior joinery, cabinet work, implements, churns, turnery, etc. S.G. ·6 ; C.W. 2·5.

Silky Beech. *Villaresia moorei* (*H.*). Queensland. Resembles English beech. Also called Maple. Silver grey, lustrous. Fairly hard and heavy, tough, fairly strong, not durable. Firm, straight grain, distinct rays. Requires careful seasoning. Easily wrought, polishes well. Small sizes. Used for cabinet work, interior joinery, turnery, brush stock. S.G. ·7 ; C.W. 3. Also see *New Zealand Beech*.

Silky Oak. *Cardwellia sublimis, Grevillea robusta, Orites excelsa* (*H.*). Queensland. Also called Lacewood. Pale pink to light brown, lustrous. Moderately hard and heavy ; tough, elastic, fairly strong and durable. Usually straight grain, but some beautiful wavy figure with striking silver grain. Rather coarse but even texture. Somewhat like American red oak. Excellent wood. Not difficult to work. Polishes well but does not fume readily. Important for veneers, furniture, superior joinery, bentwork, turnery, etc. S.G. ·6 ; C.W. 3.

Silky Wood. Sometimes applied to highly figured Canadian yellow birch, *q.v.*

Sill or **Cill.** The bottom horizontal member of framing, structural work, or of a lock or dock entrance. Also see *Dead Shore* and *Gantry*.

Silo. A tall tower or a battery of towers for the storing of grain or other granular materials.

Silvan. Applied to things associated with woods and forests.

Silver Ash. *Flindersia pubescens* and *F. schottiana* (*H.*). Queensland. White, lustrous, like Maple Silkwood in texture and weight. Also called Cudgerie. Moderately hard and heavy, tough, strong, not very durable. Pleasing figure when rotary peeled. Easily wrought and seasoned, stains readily, polishes well. Used chiefly for veneers and plywood, flooring, masts, shafts, bentwork, cabinet work, etc. S.G. ·65 ; C.W. 3.

Silver Aspen. *Pleiococca wilcoxiana* (*H.*). Queensland. White. Soft, moderately heavy, not strong or durable. Easily wrought, polishes well. Used for carving, turnery, inlays. S.G. ·6 ; C.W. 2.

Silverballi. Species of Laurel : *Nectandra, Ocotea,* and *Aniba*. See *Cirouaballi*.

Silver Banksia. *Banksia marginata* (*H.*). Victoria and South Australia. Pale yellow or buff colour. Beautiful grain and medullary rays make it a valuable cabinet wood. Used for veneers, cabinet work, boat knees, etc. S.G. ·56 ; C.W. 3·5.

Silver Basswood. *Panax elegans* (*H.*). Queensland. Also called

Celery Tree. Resembles *Tilia cordata.* Ivory, lustrous. Soft, light, tough, fairly strong and fissile. Straight, fine, close grain, some wavy, and rays ; even texture. Requires care in seasoning, easily wrought. Used for oars, sculls, turnery, bobbins, etc. S.G. ·48 ; C.W. 2.

Silver Beech. *Nothofagus menziesii.* New Zealand Beech, *q.v.*

Silver Birch. European birch, *q.v.*

Silver Fir. Applied to numerous species of Abies : *Abies alba*, (syn. *A. pectinata*), etc. Also called Swiss Pine. Pinkish white, lustrous. Soft, light, elastic, not durable. Easily wrought. Strasburg turpentine is an important extract. Used for musical instruments, toys, etc. S.G. ·35 ; C.W. 1·5. HIMALAYAN SILVER FIR. *Abies pindrow.* N. India. Similar to above. Good qualities and supplies. WESTERN SILVER FIR. *Amabilis Fir*, q.v. See *Grand Fir* and *Noble Fir.* Also see *Fir, Sitka*, and *Whitewood.*

Silver Fish. A thysanurous insect, *Lepisma saccharina.* It is very small, but a pest in kitchen floors, dressers, etc.

Silver Grain. The beautiful figure produced by conspicuous rays in quarter-sawn wood, as in oak. See *Annual Rings, Rays*, and *Grain.*

Silver Greywood. India, Andaman Isles. See *Chuglam.*

Silver Maple. *Acer saccharinum.* See *Maple.*

Silver Pine. *Dacrydium colensoi.* New Zealand Pine. Whitish to yellowish brown. Light, soft, durable. Straight grain and mottle. Similar to Rimu, *q.v.*, and used for similar purposes, also for sleepers, poles, etc. EUROPEAN SILVER PINE. *Abies alba.* See *Silver Fir.*

Silver Quandong. See *Quandong.*

Silver Spruce. *Picea sitchensis.* See *Sitka.*

Silvertop. *Eucalyptus sieberiana* (*H.*). Victoria. Light brown, black spots. Hard, heavy, strong, fissile, not durable. Little exported. Used for structural work, wagon building, flooring, decking, furniture. S.G. ·8.

Silver Wattle. See *Wattle.*

Silvia. (Syn. *Mezilaurus.*) See *Tapinhoan.*

Silvics. The science dealing with the growth of trees.

Simaruba. *Simaruba amara* and other *spp.* (*H.*). C. America. Also called Bitterwood, Bois-blanc, Marupa, Negrito, and Brazilian White Pine. Straw colour. Light, soft, resonant, and not durable, bitter taste. Straight coarse grain, firm texture. Easily wrought. Used for pattern making, organ pipes, interior joinery, etc. S.G. ·4 ; C.W. 1·5.

Simarubaceæ. The Bitterwood family. Includes *Ailanthus, Picræna, Picrasma, Quassia*, and *Simaruba.*

Simiri. See *Locust.*

Simpoh. *Dillenia* and *Wormia spp.* (*H.*). Malay. Reddish brown, white deposit. Hard, heavy, very variable in properties, fairly durable. Coarse grain, distinct rays. Difficult to season and work. Used for construction, furniture, planking, etc. S.G. ·68 ; C.W. 4.

Simul. *Bombax insigne* (*H.*). Andaman I. Also called Didu. Similar to Semul. See *Cottonwood* (*Indian*).

Sindora. See *Sepetir* and *Supa.*

Singga. *Balanocarpus penangianus* (*H.*). Malay. Also called Damar Hitam. Yellow. Fairly hard, heavy, and durable. Similar properties to Merewan, *q.v.* Beautiful wavy grain for decorative work, but small

sizes because of faulty hollow trees. Important for resin, which is nearly black and nearly transparent.
Single Iron. A plane iron without back iron.
Single Lath. A plasterer's lath ³⁄₁₆ in. thick. See *Laths*.
Single Floor. One consisting of common joists and floorboards.
Single-stick. A strong stick used in combat.
Single-tree. The cross-bar of a plough, carriage, etc., for holding the traces. A swingle- or whipple-tree.
Sink. 1. A trap-door in a theatre stage for raising or lowering scenery. 2. A shallow trough or basin under a water tap with drainage outlet. It is required for washing dishes, etc., in a scullery, pantry, etc.

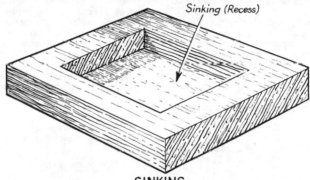

Sinking (Recess)

SINKING

Sinkers. 1. Roots of climbers or parasitic plants. In many cases they produce figured wood that increases its value. See *Figure*.
2. Logs too heavy for floating in the ponds at the docks. Also called dead-heads.
Sinking. A shallow recess. A part sunk below the surrounding surface. The term implies a recess cut from the solid, and is tested for depth by a *sinking square*. See *Sunk Panel* and *Lantern Light*.
Sipalis. A beetle destructive to hardwoods.
Siphonodon. See *Ivory Wood*.
Sipo. *Entandrophragma sp.* Similar to Sapele, *q.v.*
Siris, Black. *Albizzia odoratissima* (*H.*). India. Dark brown. Variegated streaks, lustrous, smooth. Moderately hard and heavy, strong, elastic, and fairly durable. Interlocked grain, mottle, coarse texture. Requires careful seasoning and difficult to work. Wax polish best. Esteemed for cabinet work and superior joinery, also used for wheelwright's and carriage work, propeller blades, etc. S.G. ·67 ; C.W. 4·5. Also see *Kokko*. WHITE SIRIS. *Albizzia procera*. Brown, variegated bands, lustrous, smooth. Variable properties. Similar to Kokko, *q.v.*, but lighter and softer, and used for similar purposes. S.G. ·45 to ·65. YELLOW SIRIS. *Albizzia xanthoxylon*. Australia. Similar to White Siris.
Siruaballi. Cirouaballi, *q.v.*

s.1.s.1.e. Abbreviation for *surfaced one side and one edge*. Several variations.

Sissoo. *Dalbergia sissoo.* A species of Indian Rosewood, *q.v.* Golden to dark brown with pleasing mottle and stripe.

Sister. Applied to a piece placed alongside a similar but more important or larger piece, as a sister or side keelson.

Sister Blocks. Like two single pulley blocks one above the other with the shell formed out of a single piece.

Site. The ground covered or to be covered by one or more buildings.

Sitka Spruce. *Picea sitchensis* (*S.*). Sitka Isle, W. Canada. The largest and most esteemed spruce. Pinkish white to pale brown, clean appearance, lustrous, nearly non-resinous. Moderately soft and light. It is very elastic, tough, and strong for its weight, resonant, smooth, and stable, not very durable. Seasons readily and easily wrought. Straight even grain, medium texture. Used for veneer cores, aircraft construction, hollow masts, sounding boards, food containers, construction, and general purposes. S.G. ·4 ; C.W. 1·2. See *Spruce* and *Picea*.

Sixes and Sevens. A particular type of crate used in the pottery trade.

Size. A powdered animal glue, often used to seal a surface to prevent absorption, raising of grain, etc.

Sizer. A machine for surfacing timber. A dimension planer.

SIZING TOOL AND GROOVES

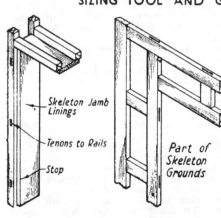

Skeleton Jamb Linings

Tenons to Rails

Stop

Part of Skeleton Grounds

SKELETON FRAMING

Sizing Grooves. Preparatory grooves in turnery to form a guide for the depth of cutting to the required outline.

Sizing Tool. A woodturner's tool for cutting grooves.

Skeed. See *Skid* (6).

Skeleton. 1. A light triangulated wood frame used as a templet for framing in shipbuilding. 2. The same as carcase, *q.v.*

Skeleton Floor. The structural parts of a floor, before the floorboards and ceiling are applied.

Skeleton Framing. Framed linings for openings. Light framing to be covered by boards, plywood, etc. Framing without panels, or framework with coverings.

Skeleton Grounds. Framed grounds for wide built-up architraves.

Skeleton Jamb Linings. Framed grounds to form the door frame with a planted lining to form the stops.

Skeleton Patterns. Used for large plane surfaces. The pattern consists of a frame which is bedded in the sand, and the sand between the members is swept out by a strickle. Large pipe bends are also made as skeleton patterns.

Skeleton Steps. Treads without risers in a flight of stairs.

Skep. A box or basket for carrying or hoisting granular materials or a number of small articles, garden produce, etc.

Skew. Implies out of square, twisted, or in an oblique position.

Skewer. A long, thin, pointed wood or metal pin.

Skewer Machine. A specially designed machine for making wood skewers.

Skew Fillet. A tilting fillet down the gable or verge of a roof.

Skew Nailing. Driving nails obliquely to the surface of the wood to give a more secure hold.

Skid. 1. A sticker. See *Air Seasoning* and *Piling*. 2. Devices under sail-planes, gliders, etc., to protect the machine when landing. Tail and landing skids to aircraft. 3. Pieces used for packing to a plane surface. 4. Logs or poles laid transversely for supporting logs, or to form a skidway for loading for transport. 5. See *Bummer*. 6. Strong pieces to protect the side planking of a ship. They extend from the main-wales to the top of the ship. Also called skeeds.

Skidway. An arrangement on which logs are rolled or slid for storage.

Skiff. A yacht, *q.v.* A small light boat.

Skilting. Same as skirting, *q.v.*

Skilvings. The rails of a cart.

Skin. 1. The surface of a piece of wood. 2. The planking of a vessel.

Skin Stress. The maximum or outer stress acting in a structural member. See *Stressed Skin.*

Skip. 1. A skep, *q.v.* 2. An area on the face of the wood missed by the knives in machine dressing.

Skirting or **Skirting Board.** A moulded base board or plinth to an inside wall. Also called a washboard. See *Finishings, Grounds, Double Skirting,* and *Pilaster.*

Skirts. Projecting eaves.

Skis. Snow shoes or runners. Specially shaped pieces cut from 7 ft. by 4 in. by 1 in. ash or hickory for ski-ing ; they are strapped to the shoes for sports on snow-clad slopes.

Skittles. Ninepins. A game in which turned wood pins are thrown down by a ball or disc.

Skived Joint. A spliced joint. The term is specially applied to the bevelled splayed joint used for jointing the endless belt of sandpaper for a belt sander.

Skylight. A glazed frame parallel to the surface of the roof. See

Roof Lights. Any form of light in a ship's deck : flat, pitched, dwarf, circular-headed, domical, or deck lights.

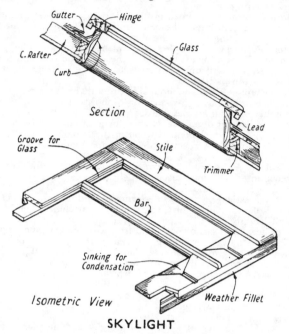

Section

Isometric View

SKYLIGHT

Skylines. Cables for "high lead" lumbering.
Skylux. A patent multiple gear for fanlights and continuous windows.
S.L. Abbreviation for *short lengths*.
Slabbing. Squaring a log.
Slab Cutting. See *Slash Sawn*.
Slab Edging. Making the outside cuts when converting or squaring a log.
Slabs. 1. The outer irregular pieces left from squaring a log, or the pieces cut away when preparing a handrail wreath, etc. 2. A thin piece cut from a board in boxmaking.
Slab Saw. A resaw, *q.v.*
Slab Slasher. A slasher, *q.v.*
Slack Blocks. Large wedges for supporting centres in bridge construction.
Slack Cooperage. Applied to barrels, casks, etc., for dry goods only.
Slamming Strips. 1. A planted door stop forming the rebate. 2. The edge strip to the shutting stile of a flush door.

Slash. 1. Debris left after felling trees. 2. Rough or random cutting with tools.

Slasher. 1. Several circular saws mounted on one shaft at intervals apart to saw wood into lengths, for firewood, etc. 2. A trimmer or edging saw.

Slash Figure or **Grain.** The figure or grain on a flat or slash sawn surface. See *Grain* and *Conversion*.

Slash Figure Check. The separation of the fibres along the growth rings on a flat-sawn surface. A shell shake.

Slash Knot. A spike knot, *q.v.*

Slash Pine. *Pinus caribæa.* C. America. Also called Cuban or Caribbean pine or Honduras pitch pine. It is coarser grained and usually more resinous than Longleaf pine, *q.v.* Also see *Pitch Pine*.

Slash Sawn. Applied to wood that is cut tangentially to the growth rings. See *Conversion*, *Flat Sawn*, and *Boxed Heart*.

Slate Battens. See *Battens*.

Slate Boarding. Sarking, *q.v.*

Slats. 1. Converted cedar for pencil manufacture. A standard slat is $7\frac{1}{4}$ in. by $2\frac{1}{2}$ in. by $\frac{1}{4}$ in. 2. Laths. Long, narrow, and thin slips of wood, as used for venetian blinds, crates, etc. 3. An auxiliary aerofoil, in aircraft. 4. Horizontal rails in the back of a chair.

Slat Saw. A small circular saw for cutting small stuff for special purposes, such as pencil slats.

Slavonian. A prefix applied to European beech, oak, etc., to denote source of origin.

Sled. A low small wood-framed vehicle, mounted on runners instead of wheels, for travelling on snow or ice. A sleigh.

Sledge. An elaborate sled in the form of a carriage drawn by horses, but also a variant of sled.

Sleeper. 1. A strong horizontal timber resting on the ground to support other timbers, etc. See *Derrick Crane* and *Gantry*. 2. The transverse timbers supporting railway lines. Sleepers must be elastic and resilient or the iron chairs to which the rails are fixed cut into the wood. The greatest difficulty in tropical countries is to resist the white ant. There are important gradings for railway sleepers in this country, controlling sizes, wane, and other defects. Treated redwood is generally used. 3. Valley boards resting on the common rafters to carry the feet of the jack rafters. They are used instead of valley rafters. Also see *Trestle Bridge*.

Sleeper Blocks. Pieces 8 ft. 11 in. by 10 in. by 10 in. to be sawn down the middle for sleepers.

Sleeper Plate. A wall plate on a sleeper wall.

Sleeper Walls. Honeycombed dwarf walls forming intermediate supports for ground-floor joists.

Sleeping Car. A railway carriage fitted with berths.

Sleepy Wood. A term applied to wood over-dried through imperfect kiln-seasoning.

Sleeve. A long *nut* with right- and left-hand threads to adjust the tension in a metal rod connection. See *Turnbuckle*.

Sleigh. See *Sledge*.

Slenderness Ratio. A term used in the design of columns and struts.

It is the ratio of the *effective length* to the *radius of gyration,* $l \div K$. It is the ratio *length to thickness* for a wood post or strut.

Slicer. A machine for cutting or slicing veneers and thin boards designed to eliminate the waste due to sawing. It operates like a huge plane and cuts slices from $\frac{1}{100}$ in. thick, usually $\frac{1}{28}$ in., with knives up to 16 ft. long. There are various types and they are constantly being improved. Logs are converted to flitches which are boiled or steamed. The flitch is dogged to a table which rises the thickness of the veneer after each operation. The knives are driven forward on a carriage and cut the wood with a shearing action to avoid crushing the fibres. See *Veneers* and *Conversion.*

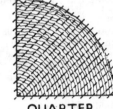

PLAIN SLICED   **QUARTER SLICED**

SLICING

Slide. An inclined plane for transporting goods. A log slide.

Slide Matched. The faces of two sheets of veneer matched together. It is common for striped quartered veneers. See *Veneer Matching* and *Matched.*

Sliders. 1. Lopers, *q.v.* **2.** Shelves or flaps that may be pulled or slid out of a carcase. **3.** Part of the mechanism that operates the pipes of an organ by means of the stops. They lead from one part of the wind-chest to the other and regulate the admission of the wind to the pipes. See *Pallet* and *Stop Action.*

Sliding Door. One that is not hinged or pivoted but slides on runners to open. Also see *Double Partition.*

Sliding Keel. A rectangular frame dropped vertically through the bottom to control a vessel against a side wind.

Sliding Sash. One that slides horizontally. See *Yorkshire Light.* Vertically sliding sashes are usually called balanced sashes. See *Sash and Frame Window.*

Sliding Seat. A seat in a rowing boat that moves with the action of the rower, on runners, to increase the length of the stroke.

Sliding Shutters. Those that slide either horizontally or vertically. See *Balanced* and *Lifting Shutters.* Also see *Shutters.*

SLIDING DOOR LOCK

Sliding Ways. The heavy timbers that slide down the standing-ways, *q.v.*, when launching a ship.

Sling Cleat. See *Ship's Cleats.*

Slings. 1. Hangers, *q.v.* **2.** Ropes or chains used to grip an object whilst hoisting. See *Knots.*

Slip. 1. See *Log Haul-up.* **2.** See *Parting Slip* and *Feather.* **3.** See *Slip-way.*

Slip Dovetail. A dovetail key.

Slip Feather. 1. A thin slip used to strengthen a mitred joint. **2.** A loose tongue cut diagonally across the grain.

Slip Key. A slip dovetail, *q.v.*

Slip Mortise. An open or a chase mortise, *q.v.*

Slipper. An arrangement on the sole of anything to facilitate sliding. See *Horsed Mould.*

Slippery Elm. *Ulmus fulva* (*H.*). N. America. Called red elm and marketed with American elm as elm or soft elm. See *Elm.*

Slipping. Same as lipping, *q.v.*

Slip Spline. A slip feather, *q.v.*

Slip Tenon. A tenon for an open or slot mortice, so that it can be driven in sideways.

SLIP FEATHER AND KEY

Slip-way. A long sloping way or gangway on which the cradle runs when launching a ship. An inclined plane on which a ship is built or drawn up for repairs.

Sliver. A splinter. A small piece of wood torn away or riven.

Sloanea. A genus with several species of hard, heavy, woods, in Central America, but they are not imported. See *Ironwood.*

Sloats. Narrow pieces holding framework together, as to a gate or the body of a cart. See *Slat.*

Sloe. Blackthorn, *q.v.*

Sloetia. See *Tempinis.*

Sloop. A sailing ship with one mast and rigged fore and aft. It has a standing bowsprit.

Sloot, v. Abbreviation for the name of the botanist van Slooten.

Slope of Grain. The angle between the direction of the grain and the axis of the stuff. Diagonal grain. It is generally stated as the tangent of the angle the slope makes with the edge. If the slope is less than 1 in 40 there is no loss of strength, but there is a loss of about 10 per cent. for 1 in 20, and about 30 per cent. for a slope of 1 in 10. See *Short Grain, Strength,* and *Stress Grade.*

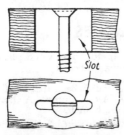

SLOT SCREWED JOINT

Slote. 1. A trap door to a stage. **2.** The main pieces for the shorter sides of a crate.

Slot Mortise. 1. A machine mortise formed by a revolving cutter that leaves the ends of the mortise rounded. **2.** See *Chase Mortise.* **3.** An open mortise. See *Slip Tenon.*

Slot Screw. A method of screwing which leaves the shank free to move in a slot to allow for expansion and contraction. See *Batten, Drawer,* and *Double Skirting.*

Slotted Collars. The collars holding the cutters on a French spindle machine.

Slotting Machine. A machine for mortising spade heads, etc.

Sluice. **1.** A flood-gate. See *Flume*. **2.** An inclined trough for washing out metals from ores.

Sluice Gate. A sliding partition or gate used as a dam for controlling the flow of water.

Slype. See *Tresaunte*.

Sm. Abbreviation for *surface measure*, and for the name of the botanist Smith.

S.M. Abbreviation for *standard matched* and for *single moulded*.

Smack. A small type of fishing vessel or coaster.

Small-boy. A low dressing chest. See *Tall-boy*.

Small Clear Specimen. A selected piece, free from defects, to test the properties of a particular species of wood.

Small Dimension Stock. Timber converted to special sizes for specified industries. It may be in the form of billets, blanks, bolts, flat, round, square, or built-up.

Small Fry. Planchettes, *q.v.* Thin pieces of wood up to 6 in. wide, slate battens, etc.

Small Knot. One between $\frac{1}{2}$ and $\frac{3}{4}$ in. diameter. See *Knots*.

Smoke Tree. *Cotinus americanus* (*H.*). U.S.A. Also called Chittam Wood. Bright orange. Light, soft, durable, coarse grain. Not imported.

Smooth Grain. A loose term applied to a smooth or silky surface, to fine grain, etc. See *Grain*.

Smoothing Plane. The plane used last when finishing the surface of wood. It may be wood or metal, and there are several varieties of

WOOD   METAL

SMOOTHING PLANES

the latter : Stanley, Norris, Sargent, Record, etc. The wood plane is about 8 in. long and shaped out of 3 in. by 3 in. beech.

S.N. Abbreviation for *shipping note*.

Snag. **1.** Any form of obstacle, obstruction, or drawback that occurs during working processes. See *Snagging*. **2.** A tree stump. A short branch. **3.** A rough part left by breaking off a branch.

Snag Boat. A steam boat with lifting appliances, etc., for removing obstructions.

Snagging. Removing the rough places where twigs have sprouted on the staves and rods for crate making.

Snail. Short and curly decorative figure ; the term is especially applied to figured walnut.

Snail Countersink. The ordinary type of countersink for softwoods.

Snake. A wavy saw-cut due to bad saw-fitting.

Snake Fence. One built of horizontal tree trunks with ends overlapping.

Snakewood. *Brosimum aubletti* (syn. *Piratinera guianensis*) (*H.*). B. Guiana. Also called leopard wood, speckled wood, tortoise-shell, and letterwood, *q.v.* Also *Colubrina spp.* C. America. Buckthorn family. Also called W. Indies greenheart, ironwood, sobragy, soldier wood, etc. Light reddish, darker patches. Very hard, heavy, strong, and durable. Fine texture. Difficult to work. Used for cabinet work, shipbuilding, structural work, sleepers, etc. S.G. 1·1 ; C.W. 4·5. INDIAN SNAKEWOOD or Strychnine Tree. *Strychnos nux vomica.* Creamy white to light brown, darker lines, dull, feels smooth, bitter taste. Hard, heavy, with close grain and indistinct growth rings. Interlocked, straight, and irregular grain, uneven texture. Difficult to season and work, warps freely. Used for shafts, tools, wheelwright's and cabinet work. Seeds of more value than wood. S.G. ·85. SURINAM SNAKEWOOD. Hoobooballi, *q.v.*

Snaky. Applied to a wavy saw-cut, due to poor saw-fitting.

Snape. An oblique cut to fit a splayed surface.

Snaped. The tapered end of a log to facilitate removal from the forest.

Snatch Block. A single sheave with a notch to put the rope into the block instead of threading it through.

Snath, Sneath, or **Sned.** The handle, or shaft, of a scythe.

snd. Abbreviation for *sound*.

s.n.d. Abbreviation for *sap no defect*.

Sneck. A local term for a door latch or fall bar.

Sneck Head. The catch for a latch.

Sneezewood. *Pteroxylon utile* (*H.*). S. Africa. Called Neishout. Yellowish. Hard, heavy, tough, strong. Very durable and resists insects. Close grain, attractive figure. Difficult to season and work. Irritant dust. Used for structural work, piling, cabinet making, fencing, etc. Local demands prohibit export. S.G. 1 ; C.W. 4.

Snipe. See *Nose* and *Snaped.*

Snipe's Bill. A small wood plane for working sunk mouldings under projecting members.

Snotter. The support for the lower end of the sprit, on a sailing ship.

Snow. A sailing ship, like a brig.

Snowball Tree. The English Guelder rose.

Snowboards. Long strips of wood, usually 3 in. by 1 in., nailed to bearers and placed

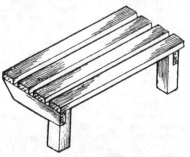

SNOW BOARDS

over roof gutters. They are often placed between north light roofs, and over V and parapet gutters. Spaces between the lags allow the snow to melt and trickle through to the gutter. Also called snow slats.

Snowdrop Tree. *Halesia carolina* (*H.*). Southern U.S.A. Salmon-pink colour. Moderately hard and heavy. Close firm grain, like birch, but more marked. Suitable for cabinet work, but little exported. S.G. ·62.

Snow Guards. Boards placed on edge, about $1\frac{1}{2}$ in. above an eaves gutter, to check the snow from falling below.

Snow-wood. White aspen, *q.v.*

Snug. A term applied to anything that is a good fit.

Soapberry. See *Sapindaceæ*.

Soapnut Tree. *Sapindus emarginatus* (*H.*). S. India. Bright yellow, variegated streaks, smooth. Very hard and heavy, with close grain and distinct rays. Difficult to work. Suitable for turnery, fancy articles, etc. S.G. 1.

Soaresia. Same as *Clarisia*. See *Oiti*.

Sobragy. See *Snakewood*.

Socket. 1. See *Pivot*. 2. The eye of a dovetail joint. See *Dovetail*. 3. An opening or cavity into which something is fitted endwise.

SOCKET CHISEL

Socket Chisel. A strong chisel with a socket to receive the handle. A bruzze.

Socle. A flat, plain, square member forming a base to a column, etc., instead of a pedestal. A plain plinth.

Sodium Fluoride. Used as a preservative against dry rot.

Sofa. A long seat or couch with upholstered bottom, back, and ends.

Sofa Table. A Sheraton table with flaps at the ends.

Soffit. The underside of stairs, arches, openings, eaves, etc. The horizontal or curved lining at the head of an opening. A larmier or drip. A narrow ceiling. See *Segmental Arch*, *Canopy*, and *Eaves*.

Soft Elm. *Ulmus americana*. See *Elm*.

Soft Maple. *Acer rubrum* and *A. saccharinum*. See *Maple*.

Soft Pine. Applied to several N. American soft pines, but it is suggested the name should be confined to B.C. ponderosa pine.

Soft Rot. A term applied to decayed wood when the residue is chiefly cellulose. See *White Rot* and *Hard Rot*.

Soft Tissue. Parenchyma, *q.v.*

Softwoods. Woods from coniferous or needle-leaved trees. Non-porous woods. Gymnosperms, *q.v.* The cellular structure is of a lower form of plant life than that of hardwoods, and consists of tracheids, tracheid ray, and parenchyma. Some softwoods are harder and heavier than many hardwoods, and the softest known woods are hardwoods. The softwoods used in this country are : cedar, chil, cypress, Douglas fir, numerous species of fir, hemlock, kauri, larch, matai, numerous species of pine, pitch pine, podo, rimu, sequoia, numerous species of spruce, tamarack, totara, yellow-wood, and yew. They are described alphabetically in the Glossary. Also see *Redwood* and *Whitewood*.

Solah. *Æschynomene sp.* (*H.*). Called Indian Pith. Very light and soft, but of little value. Used for sun helmets, floats, etc.

Soland. Abbreviation for the name of the botanist *Solander*.

Solarium or **Sollar.** A room or open gallery at the top of a building arranged to obtain the maximum amount of sunshine.

Soldier. **1.** A short vertical ground, as used for skirtings, etc. See *Double Skirting.* **2.** Heavy vertical timbers placed across several walings and strutted, for deep excavations. **3.** Profiles, *q.v.* **4.** Upright members in formwork, *q.v.*

Soldier Arch. A brick lintel with level soffit and vertical voussoirs.

Soldier Wood. See *Snakewood*.

Sole. **1.** A sill, *q.v.* **2.** The bottom surface of a plane, and of many

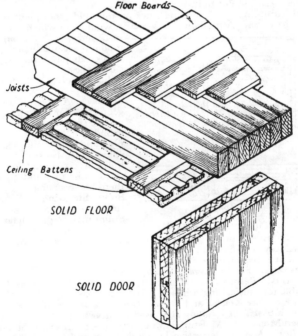

SOLID FLOOR

SOLID DOOR

SOLID CONSTRUCTION
(See p. 526.)

other objects : clog, plough, wood golf club, etc.   See *Compass Plane*.

Sole Bar. A longitudinal piece in the framing of a railway wagon.

Sole Piece or **Plate.** A sleeper carrying the feet of raking shores, *q.v.* A short piece buried in the ground to receive the struts to a post. See *Gates*.

Solera. Cyp, *q.v.*

Solid Bridging. Short pieces of board fixed vertically between joists to stiffen them. See *Strutting* and *Bridging*.

Solid Doors. **1.** Fire-resisting doors. They usually consist of three thicknesses of t. and g. boarding with the inner ply of boards horizontal. The whole is then covered with tin plate or sheet iron. **2.** The term is sometimes applied to flush doors, *q.v.*, especially when with solid cores. (See illus., p. 525.)

Solid Floor. **1.** A wood block floor on concrete. **2.** Wood floors in which the joists are in contact side by side, to make it fire-resisting.

Solid Frame. A door or window frame in which the stiles, etc., are of one piece and not built-up or cased.

Solid Moulding. A moulding stuck on the solid ; not planted.

Solid Mullion. A mullion to a Venetian window in which the centre sash only is hung and the mullions are of one piece. The sash cords pass over the fixed side sashes.

ELEVATION

PLAN

SOLID MULLION

Solid Panel. A flush panel, *q.v.*

Solids. A term used in geometry for certain recognised shapes : cone, cube, cylinder, ellipsoid, prism, prismoid, pyramid, sphere, spheroid. Also see *Polyhedra*.

Solid Steps. Laminated wood steps built up solid to make them fire-resisting.

Solid Strutting. Same as solid bridging, *q.v.*

Solignum. A proprietary preservative for wood.

Sollar. **1.** An upper chamber or garret. **2.** See *Solarium*. **3.** A platform in a shaft to a mine.

Sommer. See *Summer Beam*.

Sonneratia. See *Pagatpat, Perepat, Possi*.

Sonorous. Resonant. A quality of certain woods that makes them valuable for musical instruments. See *Resonance*.

Sophora. See *Kowhai* and *Pagoda Tree*.

Sopwith Staff. A levelling staff used in surveying.

Sorbed Moisture. Moisture in wood due to its hygroscopic properties.

Sorbitol. A sugar derivative in solution, for treating wood to make it more stable. The movement of wood, due to change of moisture content, is halved by the treatment.

Sorrenta Wood. Italian lime or sycamore immersed in a river at Sorrenta which turns the wood grey. It is used for decorative work, woodware, etc.

Sorting. Same as grading, *q.v.*

Sorting Boom. A boom to guide logs into a sorting jack.

Sorting Jack. A raft with an opening through which logs are passed to be sorted.

Sorting Table. A platform in a saw-mill on which lumber is assorted into sizes or grades.

Soss Hinge. A patent hinge that is mortised into the edge of the stuff so that it is not seen when the hinged piece is in its normal position. It is used for superior doors, falling leaves to tables, etc.

Souari. *Caryocar nuciferum* (*H.*). B. Guiana. See *Piquia*.

Soufflet. A form of quatrefoil in flamboyant tracery.

SOSS HINGE

Sound Board. See *Pallet* and *Sounding Board*.

Sound Boarding. Boards between joists to carry pugging for sound resistance.

Sound Chest. The main chest of an organ on which the pipes are erected. It may be 9 ft. by 5 ft. 6 in. or more in size. It consists of a main top piece, 2 in. thick, with a series of grooves to form wind channels and a cover board, ½ in. thick, over the grooves.

Sounding Board. 1. A structure over a pulpit or rostrum in a large building to reflect the speaker's voice towards the audience. It is usually a canopy of parabolic form and used where the acoustic properties are poor. 2. Wood plates in musical instruments to increase and propagate the sound. The sound board rests on the wind, or sound, chest of an organ and carries the pipes. (See illus., p. 258.)

Sounding Rod. A rod to measure the depth of water in the hold of a ship.

Sound Knot. See *Knots*.

Sound Merchantable. A loose term not signifying any established grading but suggesting that the wood is suitable for certain purposes. See *Merchantable*.

Sound Post. A support under the bridge of a violin, etc., to propagate the sounds to the body of the instrument.

Sound Proofing. Insulating walls, floors, etc., to prevent the passage of sound. Success depends upon scientific arrangement of the materials used. Two important factors for buildings are : mass, and the trapping of still air in small spaces or pores. Hence cellular or porous materials are effective in varying degrees. Many materials incorporate other qualities : fire-resisting, vermin-proof, etc. Some of the materials used are slag wool, Cabot's quilt, certain wall-boards, eel grass, granulated cork, pulp, hair felt, cellular plaster, seaweed, etc. There are numerous proprietary sound-resisting materials.

527

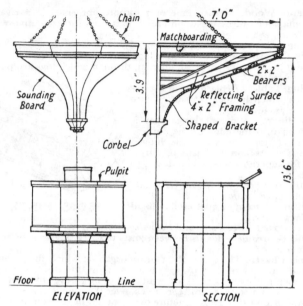

SOUNDING BOARD
TO PULPIT

Sound Wormy. Sound wood but with small worm holes, such as pin-holes. It is often used as a core for veneers.

Sour Gourd. See *Baobab.*

Sourwood. *Oxydendrum arboreum* (*H.*). U.S.A. Brown, tinged with red. Hard, moderately heavy, strong, and stiff. Shrinks greatly. Shock resisting. Uniform grain and texture. Not exported. Used for sled runners, tool handles, bearings, etc. S.G. ·54.

South Africa. There is a large variety of woods grown in S. Africa but little is exported at present except boxwood. The principal woods are : Assegai, Beukenhout, Camphor, Cedars, Chemnen, Els (red and white), Essenhout, Kamassi, Knobthorne, Knobwood, Ironwoods, Ivorywood, Matume, Pear (white and hard), Real Salie, Sneezewood, Stinkwoods, Tambooti, Terblans, Yellow-woods. Many others are described in the glossary. Government control and afforestation will probably allow for export in the near future. Also see *Cape* and *Rhodesian.*

South America. The most important woods exported to this country include : Balsa, Boxwood, Cedar, Chilian Pine, Kingwood, Lancewood, Lignum-vitæ, Logwood, Mahogany, Parana Pine, Partridgewood, Quassia, Quebracho, Roble, Rosewood, Tulipwood, Zebra-wood. Many other woods are described in the glossary.

South American Cedar. The suggested standard name for Cedrela spp. from tropical S. America. See *Cedar*.

Southern Balsam Fir. *Abies fraseri* (*S*.). U.S.A. Little variation from Balsam fir, *q.v.*

Southern Blue Gum. *Eucalyptus globulus.* See *Blue Gum.*

Southern Cypress. *Louisiana cypress.* See *Cypress.*

Southern Pine. Pitch pine, *q.v.*

Southern Red Cedar. *Juniperus silicicola.* See *Red Cedar.*

Southern Red Oak. *Quercus rubra.* American red oak. See *Oak.*

Southern White Cedar. *Chamæcyparis thyoides.* Similar to *Thuja occidentalis.* See *Cedar.*

Southern Whitewood. Magnolia, *q.v.*

Southern Yellow Pine. Pitch pine, *q.v.*

Southland Beech. *Nothofagus menziesii.* See *New Zealand Beech.*

Southwood. A patent reversible window for easy cleaning and glazing.

Sowdal. A saddle bar, *q.v.*

Soymida. See *Red Eyne.*

Space Frame. A term used in mechanics when the members of a frame are not all in one plane.

Spade Toe. A tapered foot with four equal sides resembling spades, characteristic of eighteenth-century furniture. See *Queen Anne.*

Spade Tree. The handle of a spade or shovel.

Spall. 1. A stay or tie, in shipbuilding. 2. To split off in chips or splinters.

Spalt. The waste wood left after conversion of shingle bolts to shingles.

Span. 1. The distance between the supports of an arch, beam, bridge, or roof truss. See *Clear* and *Effective Span.* 2. The distance between the wing tips of an aeroplane.

Spandrel or **Spandril.** Any triangular filling to bring to a level line, as under a stair string or between two adjoining arches, etc.

Spandrel Steps. Solid steps with the soffits shaped to the pitch of the stairs.

Spanes. The thin intermediate rails of a five-barred gate. Also called spleats.

Spanish. A prefix to distinguish woods imported from Spain, especially ash. Also see *Cedar, Chestnut,* and *Mahogany.*

Spanish Blind. An exterior blind with a sliding hood working on spring rollers.

Spanish Oak. *Quercus rubra.* American red oak. See *Oak.*

Spanker. See *Boom.*

Spanner. An adjustable tool for turning the nut of a bolt. Also see *Key* and *Wrench.*

Span-piece. A collar beam, *q.v.*

Span Rail. A shaped rail above eye level, in cabinet work. See *Curtain Rail.*

Span Roof. A roof with equal single pitch on each side. It is the most common type of roof. See *Roofs.*

Spar Deck. The upper deck of a ship, including forecastle and quarter-deck.

Spare Toe. A projection on the foot of a chair leg.

Spars. 1. Long structural units supporting auxiliary members of a main plane, or control surface, etc., of aircraft. Aeroplane struts. See *Fuselage* and *Bentwood*. **2.** Common rafters. **3.** A general term for the masts of a ship. The timbers hung to the masts for the sails, which are distinguished as booms, gaffs, and yards.

Spats. Fairings on the wheels of aircraft, for streamlining.

S.p.d. Abbreviation for *shipowner pays dues*.

Speaking. Same as squeaking. Applied to a hinged joint that is fitted too tightly and that rubs on the stop or rebate.

Spearfast. An improved saw tooth for hand cross-cut saws. It is very fast cutting for softwoods. Every fifth tooth has a specially deep gullet.

Spearwood. A native name for numerous woods that are suitable for making spears. Several Queensland species of Acacia are called Spearwood with a distinguishing prefix : Brigalow, Brown, Gidgee, Mulga, and Myall Spearwoods, *q.v.* They are very similar, but vary in colour. Very hard, heavy, strong, and durable. Small sizes, and used chiefly for cabinet work, fishing rods, fancy articles, etc. S.G. 1 ; C.W. 4·5. Also see *Wattle*.

Species. A subdivision of a genus, in the classification of trees. Species are distinguished by the fruit, flowers, leaves, bark, or wood. The Forest Products Research Laboratory has specimens of over 5,000 species of wood, suitable for woodworking in this country, and the number is continually increasing. See *Classification*.

Specification. 1. A statement defining the quality of materials and workmanship for a job. **2.** Agreed rules for the grading of wood.

Specific Gravity (S.G.). The relative weight of a substance compared with that of an equal volume of water. The weight per cubic foot of a substance is S.G. $\times 62\frac{1}{2}$ lb. The S.G. of wood varies considerably even in the same species, and depends upon the amount of wood substance per unit volume, hence upon the rate of growth in the same species. The S.G. of air-dry hardwoods varies between ·045 and 1·4, but the S.G. of softwoods is more uniform and varies between ·25 and ·8. A wood is considered *light* if less than ·5, and *heavy* if over ·7. The moisture content effects the density, and the values given in the glossary are average values for 15 per cent. m.c. Wood substance has a S.G. of about 1·5, or about 94 lb. per cub. ft.

Speckled Wood. Letterwood, *q.v.*

Spell. A form of trap or small spring board for throwing a ball into the air and used in the game of *knur and spell*.

Spendon. A stretcher piece connecting two sole bars, *q.v.*

Sper. Same as spar.

Spere. A shelter or screen to an entrance door.

Sperver. The frame at the top of a canopy or bed.

Sphæroma. Marine insects, *q.v.*, similar to *limnoria*, but larger and not so destructive.

Sphere. A globe or ball. A solid generated by a semicircle rotating about its diameter. Surface area $= 4\pi r^2$. Volume $= \frac{4}{3}\pi r^3$, where $r =$ radius.

Spheroid. A solid formed by an ellipse rotating about one of its axes. Nearly, but not quite, a sphere.

Spicey Cedar. *Tylostemon spp.* Nigeria. Also called Pogo. Yellowish to bright red, fairly lustrous. Moderately hard, heavy, and strong. Used for light construction, joinery, furniture, etc.

Spider. The assembled spokes for a wheel or pulley pattern.

Spider Bevel. A coachmaker's tool for multiple angles.

Spider Gauge. A grasshopper gauge, *q.v.*

Spigot. A plug for a cask.

Spike. A strong nail over 4 in. long. A long, pointed iron bar used as a fastening.

Spike Knot. A splay knot. One cut lengthwise of the branch. Also called horn, mule-ear, or slash knot. See *Knots*.

Spike Push Stick. See *Push Stick*.

Spile. **1.** A spigot. **2.** A splinter. **3.** A wood pin or wedge. **4.** A tapping gouge or spout used for tapping the sugar maple. **5.** A wood pile used for a foundation.

Spile Boards. Boards used for shoring up passage-ways in mines.

Spiling. **1.** Same as *piling* when applied to the foundations of buildings. **2.** The edge curve of a strake or plank in the hull of a vessel. **3.** Scribing equal distances from an uneven surface by means of a light rod or rule, instead of by compasses.

Spill. **1.** A wood shaving in spiral form. **2.** Same as spile (2).

Spindle. **1.** A steel bar to carry the knobs for a lock. Any long, slender rod, shaft, or pin used as an axis of rotation. **2.** A moulding machine in which the cutters are carried by a vertical spindle above the table. The types are French, square-head, rising and falling table. These machines may have a speed of 7,500 revs. per min. See *Head*. **3.** A small turned pillar in cabinet work. **4.** A rod carrying a bobbin, or one on which the cotton is wound, in cotton spinning. **5.** See *Plough*.

Spindle Block. See *Head*.

Spindle Blocking Machine. A machine for cutting holes in divisions of packing boxes for bottles, shells for guns, etc. Numerous holes can be cut in one operation by means of tubular saws for thick stuff or by drills for thin stuff, such as plywood.

Spindle Guards. The various appliances attached to spindle moulding

SPINDLE GUARDS

machines to conform with Home Office regulations for the protection of the machinist. There are numerous registered types.

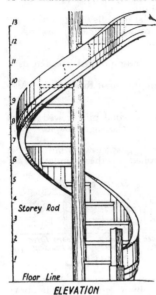

ELEVATION

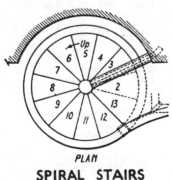

PLAN

SPIRAL STAIRS

Spindle Tree. *Euonymus europaeus* (*H*.), Europe. Also called Dogwood. White. Hard, heavy, tough, and difficult to split. Very even and smooth texture and grain. Fairly easy to work. Used for turnery, woodware, spindles, charcoal, gunpowder, etc. There are several other varieties, including: Evergreen Spindle, *E. japonicus*, Asia, and Broad-leaved Spindle, *E. latifolius*, Europe.

Spine. A woodworker's term for the heartwood of oak.

Spinet. An early keyboard musical instrument, the forerunner of the pianoforte.

Spinner. An extension of the propeller boss, for streamlining aircraft.

Spinney. A small thicket.

Spinning Gate. See *Gate* (2).

Spinning-wheel. An appliance for hand spinning. A spindle is rotated by a fly-wheel operated by a treadle.

Spiral. 1. A curve that continuously recedes from the centre about which it revolves. 2. Applied to anything winding about an axis, like a screw thread. See *Scroll* and *Helix*.

Spiral Grain. Grain in which the fibres are spirally aligned in the tree with regard to the axis. It is due to the sap taking a spiral course during growth. See *Interwoven Grain*.

Spiral Stairs. A stair, consisting of winders only, round a central newel. A circular staircase.

Spiral Turning. A twisted or screw-like turning of table and chair legs, spindles, columns, etc., characteristic of late seventeenth and eighteenth century architecture and furniture. There are numerous variations, distinguished as double, fiddle-head, fluted, latcher, open (double and triple), point, single bine, etc., and the turning is often tapered. See *Jacobean*.

Spire. 1. A tall pyramidal roof to a tower. 2. A slender stalk.

Spirelet. A small spire. One springing from a roof instead of from a tower.

Spirit Stain. A wood stain made of methylated spirits and the required colouring matter. It does not raise the grain like a water stain.

Spirket. The space between the floor timbers of a ship.

Spirketing. The inside planking of a ship. The strake wrought on the ends of the beams, or the strakes fitted to the port sills.

Spirostachys. See *Sandaleen* and *Tambooti.*

Spitter. A short pipe or outlet from eaves gutter to downpipe. See *Dished.*

Splads. Narrow pieces with spaces between. They are used instead of panels at the backs of certain furniture, ends of beds, etc., and called splad-backs. Also see *Splats.*

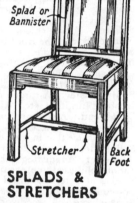

Splad or Bannister

Stretcher   *Back Foot*

SPLADS & STRETCHERS

Splash Block. A turned piece fixed on the dasher pole of a butter churn to prevent the milk from escaping from the churn.

Splash Board. A board to protect anything from mud splashes, as to a vehicle, or to a wall in course of erection.

Splats. 1. Thin narrow pieces filling the spaces in framing in the form of broken panels. 2. The upright pieces forming the filling to the back of a chair, etc. When narrow they are sometimes called banisters; when wide, as in Queen Anne chairs, they are called splats or splads. 3. Wood palings. 4. Cover strips for the joints of wallboards.

Splay. A bevel or slope. A large chamfer or one not at 45° to the face.

Splayed Edge. A bevel across the full thickness of the stuff.

Splayed Grounds. Bevelled on the edge to provide a key for plaster. See *Finishings.*

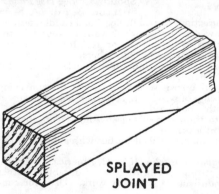

SPLAYED JOINT

Splayed Heading. A heading joint for floorboards in which the ends are splayed instead of butted.

Splayed Jambs and Linings. Jambs and linings not at right angles to the face of the wall.

Splayed Joint. A lengthening joint in which the two pieces are bevelled and overlap without increasing the cross-sectional area. A spliced joint.

Splay Knot. One cut lengthwise of the branch. A spike knot. See *Knots*.

Spleat. See *Spanes*.

Splice. 1. A splayed joint. 2. The joint between the Malacca cane handle and willow blade of a cricket bat. 3. The jointing of ropes by interweaving the strands.

Spline. 1. A long flexible strip of wood for setting out curves with inaccessable centre of radius. 2. A loose tongue or thin feather. A thin slip dividing two adjacent plane surfaces. See *Slip Feather*.

Splint. 1. A thin piece of wood, used to keep a fractured limb in position. 2. A flexible strip of wood used in wickerwork, basket-making, etc.

Splinter. A thin chip split off a piece of wood.

Splinter Bar. The front bar of the steering framework of a vehicle.

Split. 1. See *Fissile*. 2. A separation of the fibres of a piece of wood from face to face. 3. A pit prop cut lengthwise, with one flat and one round side. They are laid against the roof and supported by props. Also called *split bars*.

Split Baluster. A turned ornament sawn lengthwise and planted on Jacobean furniture, etc. Also called a mace or cannon. See *Jacobean*.

Split Head. A support for a light low scaffold for interior work. It consists of two pieces splayed outwards with a notch at the top to carry a bearer that in turn carries the scaffold boards.

Split Pattern. A divided pattern to facilitate moulding. See *Pattern*.

Spoke. 1. A radiating bar in a wheel. 2. A rung of a ladder. 3. A bar of wood used to prevent a wheel from turning. 4. The projecting handles of a steering-wheel.

Spoke Billet or **Bolt.** A sawn piece suitable for a spoke.

Spokeshave. A woodwork tool for planing small curved surfaces. It consists of a narrow blade fixed in a wood or metal stock, and requires both hands in use.

SPOKESHAVE

Spondias. See *Hog Plum*.

Sponson Beam. A beam or a curved projection protecting the paddle floats of a paddle steamer.

Spool. 1. A small hollow cylinder on which yarn is wound in the cotton and woollen industries. A bobbin or reel for cotton, etc. 2. Cast-iron separators, *q.v.*, between bolted timbers.

Spool Wood. Hardwood converted into small blocks suitable for spools or bobbins. Also called *spool bars*.

Spoon. 1. A bowl-shaped utensil with long handle for conveying liquids. 2. A shaving or chip of wood. 3. A form of oar with curved blade. 4. A golf club with wood head.

Spoon Back. Applied to the backs of Queen Anne chairs that are contoured to fit the back of a person.

Spoon Bit. A brace bit similar to a shell bit but with a more conical point.

Spoon, Golden. *Byrsonima crassifolia* (*H.*). C. America. Reddish. Rather hard, heavy, and strong, but brittle; fairly durable. Rather coarse grain, roey. Not difficult to season or work. Used for con-

struction, furniture, etc. *B. spicata* is similar but with much finer
texture and lustrous. Used for cabinet work, veneers, turnery, etc.
Spore. The reproductive body in plants that do not produce flowers.
A germ or seed.
Spot Board. A plasterer's mixing board. It is usually about 3 ft.
square.
Spot Market. Local market.
Spotted Blue Gum. *Eucalyptus maidenii.* See *Gum.*
Spotted Gum. *Eucalyptus maculata.* One of the hardest, most
pliable, and toughest of woods. Used for shafts of vehicles, tools,
levering poles, aircraft, etc. S.G. ·9. See *Gum.*
Spout or **Spouting.** 1. Pipes or troughs for conveying liquids. An
eaves gutter or a down-pipe. See *Down Spout, Boxing,* and *Lapped
Scarfed Joints.* 2. A lift or shoot
in a pawnbroker's shop for conveying pledged goods to place of
storage. 3. Ducts or shafts in
flour mills. They are distinguished as breeches, diamond-
shaped, and dog-legged.

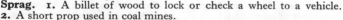

Triangular Fillet or Screed

SPOUT SCREED

Spout Screeds. Triangular
pieces sawn from the inside
of wood eaves gutters before
machining the gutters. They
are used as rails for fences, etc.
Spr. Abbreviation for the name
of the botanist Sprengel.
Sprag. 1. A billet of wood to lock or check a wheel to a vehicle.
2. A short prop used in coal mines.
Spray Board. A board on the gunwale of a ship to check spray.
Spraying. Applying paint, cellulose, etc., by spraying appliances.
Spreaders. Horizontal pieces of small section between layers of
shuttering to keep them in the correct position whilst placing the
concrete. See *Shuttering.*
Sprig. 1. A small nail with very small head. See *Nails.* 2. A small
shoot of a plant.
Sprig Bit. A bradawl.
Sprigging. Temporary nailing or fixing with very small nails.
Spring. 1. An apophyge, *q.v.* 2. A thin piece of tempered steel or
helical wire used to operate mechanisms. 3. A bend or turn through
an angle less than 90 degrees. 4. A plank or board curved in its own
plane. Also called *edge bend.* See *Warp.*
Spring Beam. 1. A fore and aft beam connecting the paddle beams
of a ship. 2. A beam with considerable deflection. 3. A tie beam.
4. A flexible bar at the top of a jig saw, etc.
Spring Board. 1. A board from which a tree-feller works when
cutting at some height from the ground. One end of the board is
fixed in a notch in the tree. 2. A flexible board used in gymnastics
to assist in jumping, diving, etc.
Spring Box. The frame of a sofa, etc.
Spring Cart. A light vehicle with the body on springs.

Spring Catch. A simple type of fastening for furniture doors, such as the ball or bullet catch.

Springer. 1. A triangular fillet nailed to the trimmer joist to form an abutment for a trimmer arch. 2. The lowest voussoir of an arch.

Spring Floor. See *Dance Floor*.

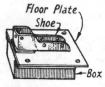

Floor Plate

Shoe

Box

SPRING HINGE

Spring Hinge. A boxed spring housed in the floor and controlling the movement of a swing door.

Springing or **Springing Line.** 1. The level from which an arch turns or springs. 2. A line on the face mould for a handrail wreath joining the centre of the cylinder to the end of the diagonal. 3. Springing means bowing when the curvature of a board or plank is in its own plane.

Spring Latch. See *Night* and *Rim Latch*.

Spring Mattress. A strong wood frame carrying some arrangement of springs for a bed.

Spring Sail. A shuttered sail for a windmill.

Spring Snib. A sash fastener.

Spring Wood. The early growth or inner part of an annual ring, *q.v.* Early wood.

Sprit. A small pole, especially applied to those suitable for rustic work.

Spritmast. A diagonal spar to extend a fore and aft sail. Also see *Tops* and *Top Mast*.

Sprit-sailyard. The yard immediately under the bowsprit.

Sprockets or **Sprocket Pieces.** 1. Short rafters fixed at a less pitch at the feet of common rafters to form projecting eaves. They form a break in the roof surface and are used for æsthetic reasons and to shorten the common rafters. See *Eaves*. 2. Wheel teeth engaging with the links of a chain.

Sprout. A root sucker or a stool shoot. A tree growing from the stump or root of another tree.

Sprout Forest. A coppice, *q.v.*

Spruce. *Picea spp.* (*S.*). Europe and N. America. Yellowish to reddish white. Light and soft to moderately light and soft. Usually little contrast between spring and summer wood. Very free from resin but often in pockets. Lustrous and often dappled on slash sawn wood. Strong and elastic for its weight, resonant, not durable, and fairly resistant to treatment. Easily seasoned but liable to warp. Easily wrought except for small, hard, and often loose knots. Used for interior construction, carcassing, cheap joinery, temporary timbering and scaffolding, boxes, wood pulp. Best qualities used as a substitute for white pine, veneer cores, ladders, flooring, shelving, slack cooperage, sounding boards, oars, etc. S.G. ·4 to ·55 ; C.W. 1·25. Black Spruce. *P. mariana.* Canadian spruce, superior to white spruce. Campbellton Spruce. Canadian spruce. Canadian Spruce. *P. alba, P. glauca, P. canadensis*, and other *spp.* More uniformly white than European. Chief supplies are white spruce. Usually imported in deal sizes or manufactured, and often named

after the port of shipment. CAT SPRUCE. Canadian white spruce.
COMMON SPRUCE. European spruce. EASTERN SPRUCE. *P. rubra* and
P. glauca. Eastern N. America. ENGELMANN SPRUCE. *P. Engel-
mannii.* British Columbia. EUROPEAN SPRUCE. *P. abies* or *P. excelsa.*
Whitewood or White deal, *q.v.* HEMLOCK SPRUCE. See *Hemlock*
(Eastern). HIMALAYAN SPRUCE. *P. Smithiana* or *P. morinda.* Ex-
cellent wood except for occasional dull red core which is of little value.
Good supplies. JAPANESE SPRUCE. *Abies mariesii* and *P. ajanensis.*
Similar to European. Called Todo matsu and Yezo matsu. MENZIES
SPRUCE. Sitka, *q.v.* MOUNTAIN SPRUCE. Engelmann spruce. NEW
BRUNSWICK SPRUCE. Canadian spruce. NORWAY SPRUCE. European
spruce. PRINCE ALBERT SPRUCE. Western Hemlock, *q.v.* QUEBEC
SPRUCE. Canadian spruce. RED SPRUCE. *P. rubra.* Canadian spruce,
superior to white spruce. SERBIAN SPRUCE. *P. omorika.* Jugo-Slavia.
See *Whitewood.* SILVER SPRUCE. Sitka, *q.v.* SITKA SPRUCE. Sitka,
q.v. SPRUCE DEAL. Red spruce, *q.v.* SPRUCE FIR. European spruce,
q.v. SPRUCE PINE. Eastern Hemlock, *q.v.* ST JOHN'S SPRUCE.
Canadian spruce. SWAMP SPRUCE. Canadian black spruce. TIDE-
LAND SPRUCE. Sitka, *q.v.* WHITE SPRUCE. *P. alba* and *P. glauca.*
Canadian spruce. WESTERN SPRUCE. Sitka, *q.v.* WEST VIRGINIAN
SPRUCE. *P. rubens.* Also see *Abies, Balsam, Hemlock, White Spruce,
Whitewood.*

Sprung. A term applied to a curved member that is distorted. Also
applied to a flexible member that is forced into position by bending.
See *Grating, Folded Floor,* and *Spring.*

Sprung Moulding. 1. A curved moulding in cabinet work. 2.
A thin moulding used for a cornice and attached to brackets or blocks
fixed to the frame.

Spud. 1. A dowel at the foot of a doorpost. 2. A tool for removing
the bark of trees. 3. A pointed piece of wood to make a hole in the
ground to receive young plants. 4. A small spade for removing weeds,
etc.

Spunk. Applied to partly decayed wood. Touchwood.

Spur. 1. A sharp projection. The marking point to a gauge. 2.
A strut. 3. A wheel with projecting teeth on the rim. 4. Machine
scribing cutters or lancets, *q.v.*

Spur Beams. Short timbers in turrets, steeples, domes, etc., for
carrying the feet of angle ribs, rafters, etc.

Spurge. See *Euphorbiaceæ.* A family of trees more important for
other products than wood : latex, tung, castor oil, etc. It has few
species of timber size.

Spur Mahogany. See *Miva Mahogany.*

Spurn. See *Root Spurn.*

Spurn Water. An angle fillet in ship joinery to throw off water.

Spur Tenon. See *Plug Tenon.*

Spurtle. Same as theavel, *q.v.*

Spy-hole. A peep-hole through which a person can see without being
seen.

Sq.E.Sd. An abbreviation for *square-edged and sound.* A grading
term for pitch pine.

Squab. A loose cushion seat for a couch or chair.

Square. **1.** A measurement of area of 100 sq. ft., used for all types of manufactured boards : flooring, matching, etc. The area is the nominal size before manufacture for t. and g. boards. **2.** The number of shingles required to cover 100 sq. ft. **3.** A piece square in section throughout its length, up to 6 in. by 6 in. square. **4.** A blank, *q.v.* **5.** A tool for testing right angles. See *Steel Square*. **6.** A pane of glass. **7.** See *Squares*.

Square Billet. A moulding characteristic of Norman architecture. See *Billet*.

Square Cut. One of the methods of preparing handrail wreaths, as distinct from *bevel cut*. The joints are square to the surface of the plank.

Square Dovetails. See *Lock Corners*.

Squared Log. A balk, *q.v.*

Square-edged. **1.** Stuff with plain edges at right angles to the face. Rectangular cross-section throughout the length. **2.** Lumber without wane.

Square Framed. A term applied to framing without mouldings.

Square Head. See *Head*.

Square Joint. A butt joint, *q.v.*

Square Mesh. See *Mesh*.

Square Panel. An ordinary plain panel without mouldings.

Square Piended Roof. See *Piend*.

Squares. Lever mechanisms in the tracker action of a pipe organ. See *Pedal*.

Square Shoot. **1.** A wood rain-water down-pipe. **2.** See *Shoot*.

Square Staff. A square fillet used instead of a bead at a salient corner of a plastered wall. See *Angle Bead*.

Square Templet. See *Templet*.

Square Timbers. The ribs, from the keel upwards, in the body of a ship. See *Cant* and *Knuckle Timbers*. Also see *Balk*.

Square-to Roof. A North-light roof, *q.v.* A roof with the sloping sides forming a right angle at the ridge.

Square Turning. Shaping balusters, newels, etc., continuously round the four sides. They may be prepared by machines.

Squaring. **1.** Correcting a frame so that it is rectangular. **2.** A term used when preparing a bill of quantities for working out the sizes of the different items on the dimension sheet into superficial areas, etc.

Squaring Bench. An appliance to slide on a saw bench and to carry the stock so that the ends can be squared or bevelled on the circular saw.

Squaring-off Machine. A special circular saw for cutting small stuff to a uniform length. It has a sliding table.

Squash Racket. See *Racket*.

Squinch Arch. A pendentive arch across an angle to carry the wall above, as when a square plan is brought to an octagonal plan.

Squint. Applied to things not at right angles. An oblique opening in a wall. A squint quoin is one in which the return wall is not at 90 degrees to the face wall. It may be obtuse or acute.

Stabiliser. A fixed horizontal surface in front of the elevator of an aeroplane to assist stability.

538

Stable. 1. Applied to wood that changes little in volume with variation of moisture content, or that resists absorption when seasoned. It is an important property of wood for interior joinery, cabinet-making, pattern-making, etc. **2.** A building for horses and cattle.

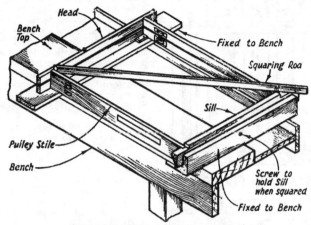

SQUARING SASH FRAME

Stabling. A set of stables for horses. Same as stables.

Stack. 1. A pile of timbers arranged for seasoning or storing. See *Air Seasoning*, *Kiln*, and *Piling*. **2.** See *Chockwood*. **3.** A rainwater pipe, from eaves to ground. **4.** A measurement of firewood of 108 cub. ft. **5.** The portion of a chimney above the roof.

Stacker. 1. A machine for loading timber on stacks and trucks. **2.** A timber piler.

Stack Stand. A strong wood frame on which grain is stacked to keep it off the ground.

Stadium. A place for public exhibitions of games and sports.

Staff. 1. A vertical stiffener to the bodywork of a vehicle. **2.** A stick used for defence or support. **3.** A pole or long handle. A flag-staff. On board ship they are distinguished as ensign, flag, and jack staffs. **4.** A graduated stick used in surveying and levelling. See *Sopwith*. **5.** See *Mace* (2).

Staff Bead. An angle or guard or return bead.

Staff Tree. See *Celastraceæ*.

Stage. 1. An elevated platform, especially for exhibitions or performances. A storey, step, or floor. See *Staging*. **2.** A slung platform on which men stand when unloading a ship.

Stage Door. An entrance to the back of a theatre, reserved for actors and staff.

Stagger. A term applied to multiplane aeroplanes in which one main plane is in advance of another main plane.

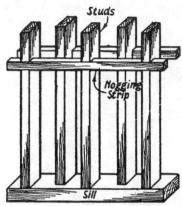

Studs

Nogging Strip

Sill

STAGGERED STUDS

Staggered. Applied to anything placed irregularly. Not in a straight line, or zigzag. Adjacent studs placed in a different plane so that they form two independent faces to the wall to make it more non-conducting. See *Hollow Partition*. Also applied to superposed planes of aircraft when not vertically over each other.

Staggered Butts. Shingles of irregular widths.

Stag-headed. Dry topped, *q.v.*

Staging. Scaffolding, *q.v.* A temporary platform. Also see *Stand*.

Stain. 1. A variation from the natural colour of wood due to any cause, such as fungi, incipient decay, chemical action. In most cases it may be avoided by quick conversion, well-ventilated covered stacks and clean softwood sticks. Woods that do not respond to this treatment and still develop stain should be dipped in antiseptic as soon as converted. See *Blue* and *Sap Stain*. 2. A dye to change the colour of a material. It is usually applied to wood to imitate a superior wood. The stain may have water, oil, or spirits as the vehicle. See *Dyes*.

Staircase. The compartment containing the stairs of a building. The complete stairs and finishings.

Stair-horse. Same as carriage piece, *q.v.*

Stair-rod. A wood or metal rod to secure a carpet to a stairs.

Stairs. A succession of steps connecting two different levels. The names given to the different types are: bracketed, close newel, continuous, dog-leg, double return, geometrical, half-turn, helical, newel, open newel, quarter-turn, solid newel, spiral, straight, winding. There is a large number of terms applied to stairs, explained alphabetically in the glossary. Stairs may be of wood, concrete, stone, or metal. Also see *Step, Tread, Landing, Strings, Flight*.

Stairway. The opening in a floor, or space, for the stairs.

Staith. A wharf, *q.v.*, especially the end of the wharf where loading takes place. A waterside depot for loading vessels.

Stake-boat. One moored at the start or finish of a boat race.

Stakes. Pointed sticks or posts for driving into the ground.

Staking or **Staking Out.** Driving stakes at the required positions and levels when setting out a site for building purposes.

Stalk. 1. Often used in the same sense as shank, *q.v.* 2. A chimney stack. 3. The vertical part of a reinforced concrete retaining wall.

4. The central support or stem of a small plant. A secondary stem or petiole supporting a leaf or flower.

Stall. **1.** A seat in a favoured or reserved position in a theatre, or for the clergy and choir in a church. See *Miserere*. **2.** A bench or table for the display of goods for sale. **3.** A division in a stable. **4.** A bay, between the book presses in a library, furnished with a bench.

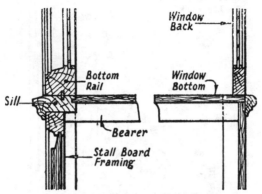

STALL BOARD

Stall Board. The sill or framing supporting the window frame for a shop front. Also see *Show Board*.

Stall Board Lights. A glazed frame under the stall board of a shop front, to light a basement.

Stall Riser. The riser between pavement and stall board of a shop front.

Stanchion. **1.** A mild steel pillar for supporting a beam, etc. A post supporting a deck beam. **2.** Sometimes applied to the mullions of a window. **3.** Removable bars for securing cattle in a stall. **4.** Upright pieces in the breastwork, bulkheads, etc., of a ship.

Stand. **1.** An erection or stage for spectators or performers. **2.** A frame on which anything is displayed or supported. **3.** An area of commercial trees suitable for timber. (See illus., p. 542.)

Standard. **1.** A measure of timber for softwoods. There are several different standards but only the Petrograd and Quebec standards are in general use. The Petrograd standard consists of 120 pieces (the long hundred or 10 dozen) 12 ft. by 11 in. by 1½ in., or its equivalent of 165 cub. ft. Other standards are : Christiania, 103⅓ cub. ft. ; Drammen, 121⅞ cub. ft. ; Dublin, 270 cub. ft. ; Gothenburg, 180 cub. ft. ; Quebec, 229⅓ cub. ft. The London Standard, 270 cub. ft., is now in disuse. **2.** A tree 1 to 2 ft. diam. at breast height. **3.** A temporary wood post as in scaffolding. See *Derrick Tower* and *Bricklayer's Scaffold*. **4.** A bench end. **5.** Sometimes applied to door posts and quarterings. **6.** Uprights in ship framing to receive panelling or sheeting for deck structures. **7.** The uprights in the side framing of a van or other vehicle. **8.** An upright post used as a support, as for a lamp-stand, hat-stand, etc. **9.** The upright supports for the mirror of a dressing table. **10.** Vertical inverted knees above the deck of a ship.

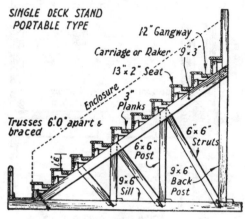

SINGLE DECK STAND PORTABLE TYPE

12" Gangway

Carriage or Raker 9"×3"

13"×2" Seat

Enclosure

3" Planks

Trusses 6'0" apart & braced

6"×6" Struts

6"×6" Post

9"×6" Sill

9"×6" Back Post

STAND

Standard Dozen. Twelve pieces 12 ft. by 11 in. by 1½ in.
Standard Hundred. See *Standard* (1).
Standard Knot. A sound knot between ¾ and 1½ in. diam., or not more than 1¼ in. diam. in hardwoods and cypress. See *Knots*.
Standard Length. An American term for the length to which timber is cut for general use. It varies for different woods.
Standing Timber. Trees of commercial size for timber.
Standing Ways. The wood sliding rails or guides used when launching a ship.
Standl. Abbreviation for the name of the botanist Standley.
Stand-rest. A sloping seat for resting in a half-standing position.
Stand Sheet. A glazed sash without frame. A fixed sheet.
Staple. A metal loop to receive a hook, pin, bolt, or padlock. See *Hasp*, *Locking Bar*, and *Box Staple*.
Starboard. The right-hand side of the hull of a ship, or of aircraft, looking forward.
Starleaved Gum. American red gum, *q.v.*
Starlings. Piles used to protect the piers of bridges. A cutwater, *q.v.*
Star Shakes. Several heart shakes together, shaped somewhat like a star. See *Shakes*.
Starter Frame. A shallow 2 in. high box used in shuttering to allow for the base of a column or wall to be cast with the floor or mass concrete.

Startex. Proprietary wood-fibre wall-boards.

Starting Gate. A movable barrier at the starting post for a horse race.

State Room. A superior type of cabin on board a passenger ship.

Statics. A branch of mechanics dealing with the equilibrium of forces and bodies at rest. See *Graphics*.

Station. A cross section through the hull of a ship.

Stationery Cabinet. A small cabinet to hold writing-paper, envelopes, etc. The pigeon-holes and small drawers fitted in the enclosed or top part of a writing desk.

Station Poles. Ranging poles, *q.v.*

Station Roof. A roof with central support only so that the trusses are in the form of cantilevers on each side of the support. It is often used on island platforms at railway stations. See *Umbrella Roof*.

Staudtia. See *Umaza*.

Stave. **1.** A staff. **2.** See *Staves*. **3.** The main pieces used in crate-making.

Staves. **1.** The separate parts of a cask. The staves are imported in three grades : *tight*, for liquids ; *slack*, for powders ; and inferior stuff for boxes rather than barrels. These grades are *Crown, Brack* (1), and *Brack* (2), and are in sizes to suit particular sizes of casks, *q.v.* **2.** Narrow boards to build up a curved surface. See *Cylinder, Jacket*, and *Wreathed String*. **3.** The rungs of a ladder. **4.** Cylindrical bars in a rack.

Stave, To. To caulk a joint.

Stavewood. **1.** Prepared wood for casks. **2.** Cudgerie, *q.v.*

Staving. Blocking the back of veneers in circular work.

Stay. **1.** A tie or brace to keep a post or frame rigid. See *Derrick Crane*. **2.** See *Casement Stay*. **3.** A prick post in fencing. **4.** A construction member of aircraft, usually of wire, to resist a tensile strain.

Stay-set Cap. A patent back iron, for a plane, that allows for occasional sharpening of the cutting iron without removing the cap.

Std. Abbreviation for *standard*.

Steam Chest. Any container of steam for steaming wood for bending or for heating cauls, etc.

Steaming. **1.** Placing wood in a steam bath to make it more plastic for bending, peeling, or slicing. **2.** Using saturated air at the required temperature to sterilise wood against fungi or insects ; or in kiln seasoning to prevent case-hardening and unequal seasoning. **3.** Steam treatment is applied to certain woods to darken them for decorative purposes.

Steam Nigger. A log canter. See *Nigger*.

Steam Treatment. See *Steaming*.

Steam Vat. For boiling veneer flitches in the vat preparatory to slicing or peeling.

Steeler. The aftmost or foremost plank in a strake of a wooden ship.

Steel Square. A large graduated square used by carpenters for obtaining lengths and bevels in timber framing, such as roofs.

Steeple. A pyramidal roof to a tower. It is not so slender as a spire.

Steer. A stack of wood, or rack, in which the pieces are arranged vertically. See *Piling* and *Air Seasoning*.

Steerage. **1.** The cheaper parts of the accommodation of a passenger ship. **2.** The steering end of a vessel.

Steering Wheel. A wheel to turn the rudder of a ship, or to steer a vehicle. See *Wheel*.

Steer's Bit. A patent expansion brace-bit, *q.v.*

Steeve. See *Steve*.

Steinberger Pine. White pine from the Rhine area.

Stele. **1.** The central core of a vascular plant. It consists of cambium, phloem, pericycle, pith, and xylem. **2.** A long handle to a brush, rake, etc.

Stem. **1.** The trunk of a tree. In deciduous trees it is less than the height of the tree. In some conifers it extends to the complete height. The following descriptive trade terms are applied : long or short,

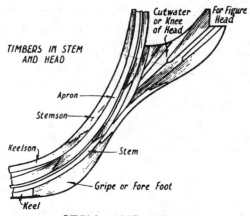

STEM AND HEAD

cylindrical or tapering, smooth or knotty, straight or crooked. **2.** The foremost structural member of a vessel. It is connected to the keel by the deadwood and to the upper strakes by the apron and breast-hook. It is the upright continuation of the keel on which the sides meet. Also see *Keel*.

Stemmatodaphne. A genus producing Malayan Medang, *q.v.*

Stemming Piece. A short piece of joist between trimmer and breast. It is used instead of a furring piece to carry the ends of the floor boards.

Stemonurus. A genus producing Malayan Dedaru.

Stemple. **1.** A supporting timber to a platform. **2.** Cross pieces in a mine that may be used as steps.

Stemson. A curved timber behind the apron of a vessel to support the scarfs.

Stenocarpus. See *White Oak*.

Step. **1.** A unit in a stairs usually consisting of tread and riser, but it may be solid. The proportions are : rise × going = about 66 in., or *twice rise* + one *going* = 23 to 24 in. The varieties of steps are : balanced, bull-nose, commode, curtail, dancing, drum end, flier, return, solid, winder. See *Stairs*. **2.** A break in the under-surface of the hull or float of a seaplane. It may be the main step or rear step. **3.** A block in which the foot of a post or ship's mast or capstan is secured.

Stephegyne. See *Binga* and *Kaim*.

Step Joint. A structural notched joint. When it has more than one notch it is a *double-step* joint, etc. See *Double Step*.

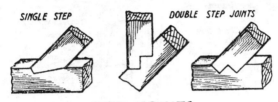

SINGLE STEP   DOUBLE STEP JOINTS

STEP JOINTS

Step Ladder. A self-supporting short ladder with flat steps. It is usually fitted with a framed stay hinged to the back.

Stepped String. A cut string, *q.v.*

Stepping. A grading term used in the pitch pine and Douglas fir trades.

Stepping Out. Setting out equal distances with compasses or dividers.

Steppings. The rectangular notchings in a cut string, for the treads and risers.

Sterculia. See *Imbira*, *Kurrajong*, *Maho*, *Ogugu*, *Okoko*, *Orodo*, *Papita*, and *Sawbya*.

Sterculiaceæ. The Cocoa family, and including the following genera : Brachychiton, Cola, Commersonia, Eriolæna, Fremontodendron, Guazuma, Heritiera, Kleinhovia, Melochia, Pterocymbium, Pterospermum, Sloanea, Sterculia, Tarrietia, Theobroma, Triplochiton.

Stere. A metric cubic measure of timber = 35·3 cub. ft.

Stereospermum. See *Pader* and *Paral*.

Stereum. A fungus, *S. sanguinolentum*, that attacks conifers and produces decay known as *red* or *stringy rot*, *mottled bark*, or *sapin rouge*.

Sterling. See *Starling*.

Stern. The rear, or aft, end of a vessel, plane, etc.

Stern Frame. The stern post, transoms, and fashion pieces at the stern of a ship.

Stern Post. The rearmost vertical member of the fuselage of an aeroplane. See *Fuselage*. The post or stanchion at the stern of a vessel. It is fixed to the keel and held in position by the transom and carries the rudder.

Sternsheets. The part of a boat between the rowers and the stern.

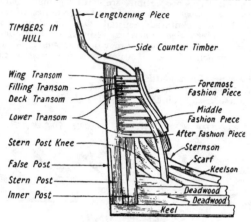

TIMBERS IN HULL

Lengthening Piece

Side Counter Timber

Wing Transom
Filling Transom
Deck Transom

Lower Transom

Foremost Fashion Piece

Middle Fashion Piece

After Fashion Piece

Stern Post Knee

Sternson

False Post

Scarf

Keelson

Stern Post

Inner Post

Deadwood

Deadwood

Keel

STERN TIMBERS

Sternson. The extremity of a keelson to which the stern post is fixed.

Sterospermum. See *Stereospermum.*

Stettin Fir. Redwood, *q.v.*

Steud. Abbreviation for the name of the botanist Steudel.

Steve. A pole used by stevedores for stowing cargo.

Stick. **1.** A small branch or shoot. Often applied to a straight trunk of a tree, or to a log. **2.** A cudgel, baton, staff, or walking stick. **3.** An implement used in printing. A composing stick. **4.** See *Sticker* (2). **5.** See *Winding Sticks.* **6.** Round timbers used as posts in trestle bridge construction. See *Trestle.*

Stick Bridge. A temporary bridge of undressed timbers. Also called a bush bridge.

Sticker. **1.** A machine used specially for shaping sash stuff, etc., for mass production. **2.** Thin strips of softwood placed between the layers of timbers when piling for seasoning to allow free circulation of air. See *Air Seasoning* and *Piling.* **3.** A wood rod connecting reciprocating levers in an organ, between key and pipe.

Sticking. **1.** See *Piling.* **2.** Shaping mouldings with plane or machine.

Sticking Board. A framed board for steadying small pieces whilst being stuck, or moulded, by hand.

Stiff or **Stiffness.** The property of wood that resists change of shape when acted upon by external forces, as a beam resisting bending when loaded. Rigid, or the reverse of flexible. See *Tests.*

Stiffeners. Parts of the framework of the fuselage of an aeroplane to strengthen the longerons, etc.

Stile. **1.** An arrangement of steps for surmounting a fence. A break in a fence for easy climbing. **2.** See *Stiles*.

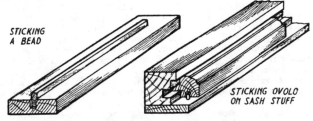

STICKING
A BEAD

STICKING OVOLO
ON SASH STUFF

STICKING BOARDS

Stiles. The outer vertical members of a piece of framing. See *Door* and *Sash*.

Still. Equipment for distillation. Used in the manufacture of spirits, as a whisky still, etc.

Stillage. A rack or frame on which things are laid, especially barrels, to keep them off the floor.

Stillengia. See *Chiu* and *Tallow Tree*.

Stilts. **1.** A pair of props attached to the legs of a man to lengthen the stride and enable him to reach to a considerable height without a ladder. Props to elevate anything. **2.** See *Starlings*.

Stinking Cedar. *Torreya taxifolia* or *Tumion taxifolium* (*S.*). Western U.S.A. Yellow, unpleasant odour. Small sizes and of little timber value. S.G. ·5.

Stinking Toes. Cana Fistola, *q.v.*

Stinkwood. *Ocotea bullata* (*H.*). South Africa. Called Cape olive or laurel. Two varieties : golden and black. Variegated golden brown to black, very lustrous, malodorous before seasoning. Hard, heavy, strong, tough, durable, and elastic. Beautiful wood but often with decayed heart. Resembles selected black walnut. Very difficult to season, shrinks greatly. Difficult to work, polishes well. Expensive. Highly esteemed for cabinet work and turnery. S.G. ·8 ; C.W. 4. RED STINKWOOD. S. African wild almond. WHITE STINKWOOD. *Celtis rhamnifolia.* S. Africa. Also called Camdeboo. A strong, elastic, wood valued for wheelwright's work, etc., but not exported. The name stinkwood is applied to several species of the Laurel family that produce malodorous woods.

Stirrup. **1.** Pieces of wood or iron by means of which a thing is suspended. A hanger. **2.** A tie.

Stirrup Strap. A strap fixed to a post and carrying a horizontal member. See *Cottered Joint* and *Framed Floor*.

St John's. See *Spruce*.

Stk. Abbreviation for *stock*.

Stnd. Abbreviation for *stained* and *standard*.

Stobs. Fixings for the feet of fence posts, or rough uprights for fencing.

Stock. **1.** The hub, or nave, of a wagon wheel. **2.** The timber held by a particular firm. **3.** Timber converted to the special requirements of a particular plant or industry : furniture stock, small dimension stock, etc. See *Dimension Timber*. **4.** The stem of a tree, especially the lower part. **5.** The wood part of a firearm, or the main part or body of a tool. See *Jack Plane*. **6.** The wood cross piece to receive the shank of an anchor. **7.** The main part of anything, especially if of wood, to which subsidiary parts are attached or held. **8.** Sometimes applied to the handles of certain tools. **9.** See *Stocks*. **10.** See *Horsed Mould*. **11.** A dowelling stock is a cooper's brace. **12.** The machinist's name for the material he is machining. Same as *stuff*.

Stockade. A defence of stakes to protect the enclosure. A palisade.
Stock Board. A board placed at the bottom of a brick mould. It has a planted *kick* to form the frog.
Stock Gang. A set of saws converting several similar sections simultaneously from the log.

Stockholm Tar. A wood preservative obtained by destructive distillation of waste wood from pines and firs.
Stock Lock. A dead lock. It consists of a heavy wood casing containing a bolt operated by a key.
Stocks. **1.** A framework of wood in which offenders were confined and exposed to public ridicule. **2.** The framework supporting a ship during construction.
Stock Sizes. Common sizes usually kept in stock by a firm, such as sash stuff, etc.
Stoep. A veranda to a S. African house. See *Stoop* (1).

STOCK LOCK

Stone Pine. Cembran Pine, *q.v.*
Stonewood. Turkish cornel. See *Cornus*.
Stool. **1.** The stump of a tree that will throw out new shoots which produces highly decorative wood. **2.** A framed support. A base for a chest of drawers, wardrobe, or cupboard. **3.** A seat without a support for one's back. **4.** A low bench for kneeling on, or for a foot rest. **5.** A portable piece of wood to which a bird is attached as a decoy. **6.** A commode, or night stool. **7.** A level projection on a weathered surface to form a seating for a post. **8.** Small channels fixed to the side of a ship for the dead-eyes for the back stays. **9.** A concrete block used as a foundation for a timber building.
Stooling. **1.** The act of throwing out shoots from a tree stump. Similar to pollarding. Second growth from the original roots. **2.** A flat seating on a weathered sill.
Stoop. **1.** The landing to a flight of steps leading to an external door. **2.** A mounting block, *q.v.* **3.** A receptacle, or basin, for holy water. A stoup. **4.** A tapered piece of oak spiked to the side of a post to strengthen the post.

Stoothing. Battens on walls.

Stop. **1.** A projection on the top of a bench to steady the stuff whilst being wrought. **2.** The edge of a rebate against which a door closes. See *Finishings*. **3.** An ornamental termination to a moulding or chamfer stuck on the solid. **4.** Anything that serves as a check, obstruction, or barrier. See *Floor Stop, Gates, Housing Joints*, and *Guide for Drawer*. **5.** Mechanism on musical instruments to control the sound. See *Stops*. **6.** A sudden change in the girth measurement of a log.

Stop Adzing. Squaring a log to different sectional areas.

Stop Bead. Same as inner bead, *q.v.*

Stop Chamfer. A chamfer with a stop, instead of running through. See *Chamfer*.

Stopped End. The end of a wood gutter. See *Lapped Scarfed Joints*.

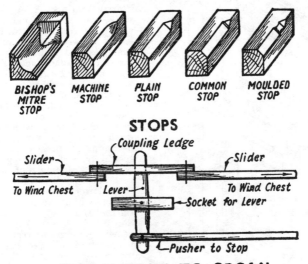

BISHOP'S MITRE STOP   MACHINE STOP   PLAIN STOP   COMMON STOP   MOULDED STOP

STOPS

Coupling Ledge

Slider   Slider

To Wind Chest   Lever   To Wind Chest

Socket for Lever

Pusher to Stop

STOP ACTION TO ORGAN

Stopped Lap Dovetail. See *Secret Lapped Dovetail*.

Stopping. **1.** Plastic materials for filling pores and cracks in wood. Glazier's putty or whiting and glue are used for painted surfaces, beeswax or beaumontage for polished surfaces, and litharge and glue for varnished pitchpine. There are also many proprietary materials. See *Fillers* and *Plastic Wood*.

Stops. **1.** The registers that bring the pipes or reeds of an organ into action as required. **2.** A piece to close the top of an organ pipe so that the pitch of the note is altered. The tone of a closed pipe is an octave lower than that of the same pipe open.

Storage Tissue. See *Parenchyma*.

Storey or **Story.** A part of a building between two consecutive floors. A horizontal division of a building. See *Floors*.

Storey Post. 1. A wood post supporting a beam to a floor. **2.** A newel extending from floor to floor.

Storey Rod. A board or strip for setting out the heights of the steps in a flight of stairs between two floor levels. See *Spiral Stairs*.

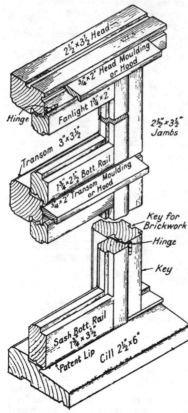

Storied. Applied to woods that have tiered arrangements of their elements, and that cause ripple marks in the converted wood.

Stormproof. See *Storm Window* (2).

Storm Window. 1. A frame with two sets of balanced sashes, or casements, with a space between to insulate against sound and to provide increased resistance against the weather. See *Double Windows*. **2.** The name applied to registered designs of mass-produced casement windows. They have specially designed throatings and rebates to withstand driving rain. In some cases overlapping rebates require a special type of hinge. **3.** A recess in a sloping roof for a vertical sash. It is the reverse of a dormer window and is entirely within the roof. See *Internal Dormer*.

Stoup. See *Stoop* (3).

Stove. A room with regulated temperature for correcting the moisture content of wood. See *Second Seasoning*.

Stoving. The act of drying wood, etc., in specially heated chambers. See *Seasoning* and *Kiln*.

Stowage. The cargo of a ship.

Stow Wood. Blocks for steadying stowage, especially barrels. Low-grade wood used for filling vacant places in a ship's hold.

STORMPROOF WINDOWS

St Petersburg. See *Petrograd*.

Straddle Poles. Scaffold poles astride a roof for a saddle scaffold.

The poles cross each other in pairs and are lashed together at the ridge and to the tops of the standards.

Straddle Scaffold. A saddle scaffold, *q.v.*

Straight Arch. One with a level soffit, or a camber arch. A lintel. See *Soldier Arch*.

Straightedge. A parallel piece of seasoned stable wood, such as yellow pine. It has straight edges and is used for testing levels and plane surfaces.

Straight Flight. A flight of stairs containing fliers only.

Straight Grain. Applied to a piece of wood when the principal wood cells run parallel to the length of the piece. It is applied to a log or tree when the cells are parallel to the axis of growth. See *Grain*.

Straight Joint. A butt joint, *q.v.*

Straight Lock. One screwed on the face of the door with the minimum cutting of the wood. Reverse of cut lock, *q.v.*

Strain. Alteration of shape due to stress. In structural design *strain* = *extension* ÷ *length*, or $e \div l$, where e represents contraction or extension. See *Stress* and *Modulus of Elasticity*. Also see *Permanent Set*.

Strainer. See *Turning Piece*.

Straining Beam. A horizontal timber between the heads of queen posts, *q.v.* A horizontal strut in structural framing. See *Partition*.

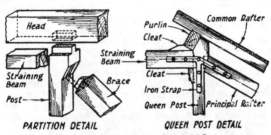

PARTITION DETAIL   QUEEN POST DETAIL

STRAINING BEAMS

Straining Piece. **1.** A horizontal piece receiving the thrust from the struts in a flying shore system. **2.** A compression member, or strut, in structural framing.

Straining Sill. A horizontal timber between the feet of queen posts, *q.v.*

Strakes. The pieces building up the covering of the hull of a ship. The shell of a ship. A line of planking extending the full length of the side of a vessel.

Strap. **1.** A batten on a wall for lath and plaster. **2.** A metal plate bolted or screwed across a joint in timber framing. A strap bolt is part plate and part bolt. See *Centering, Stirrup, Lap Joints*, and *Heel Strap*. **3.** See *Girthing Strap*.

Strap Hinge. A hinge with long plate for bolting or screwing to the face of gates or heavy doors. A band and gudgeon, *q.v.*

Strapped Wall. A wall battened for lath and plaster.

Strapping. Battens on a wall.

Strapwork. Elizabethan ornamentation imitating interlacing bands, or straps, and often with imitation rivets.

STRAPWORK

Strawberry Tree. *Arbutus unedo* (*H.*). S. Europe. Cultivated for beauty of tree rather than timber. Small sizes, used for fancy articles.

Straw Board. A proprietary insulating material, built up of layers of straw.

Streak. 1. Localised incipient decay. 2. Sometimes applied to stripe. Also see *Pitch Seam* and *Mineral Streaks*.

Streaked Birch. See *Threaded Box*.

Streamlined. Applied to moving objects that are designed to avoid eddies, or drag, behind the object, when exposed to the air stream due to moving at high speeds.

Strength. The strength of wood, of the same species, depends upon the conditions and rate of growth, position in tree, season of felling, and conversion. It varies with the moisture content, percentage of summer wood and heartwood, and specific gravity. Other important factors are : size and position of knots, and grain disturbance around knots, cross grain and slope of grain, *q.v.*, shakes, wane, and other defects, and decay. Knots in compression area give a slope of grain that is weak in sheer. See *Tests, Selection, Compression, Modulus of Rupture* and *Elasticity, Moisture Content*, and *Stress Grade*.

Stress. An internal resistance to an external force : tension, compression, shear, torque. The internal force, per square inch of section, resisting the action of the external forces. In structural design $stress = load \div sectional\ area$, or $f \div a$. See *Strain* and *Modulus of Elasticity*.

Stress Diagram. 1. A force, vector, or reciprocal diagram. A scale drawing from which the forces acting in the members of a framed structure can be measured. The lines of the scale drawing are parallel to the members of the frame and the external forces. See *Graphics*. 2. A diagram showing the distribution of the stresses acting in a structural member.

Stressed Skin Construction. A form of structural plywood construction. It consists of inner and outer skins, to resist the external forces, connected together by an inside core of wood ribs. The whole is rigidly resin-bonded to form a prefabricated unit. The principle is similar to that of a rolled steel joist where the resistance is chiefly in the flanges of the beam. This form of construction is only suitable for mass production and requires expensive equipment. It is extensively used in aircraft construction and applied to walls, roofs, and floors of prefabricated houses.

Stress Grade. A grading of timber, according to its value of f, for structural work. The grading applies to softwoods for joists, beams, columns, etc., and is applied to Douglas fir, larch, red pine, redwood or Scots pine, silver fir, spruce, and whitewood. The safe working stresses of *f* are 1,000, 1,200, and, for Douglas fir, 1,300 lb. per sq. in., and in each grade the size and position of permissible defects are stated. This method of grading ensures the economical use of wood for constructional work. It is of recent origin, in this country, and probably experience will suggest modifications and extensions of the grading rules. See *Strength*.

Stretcher. **1.** A frame on which fabrics are stretched whilst drying, or for convenience of working. **2.** A frame for carrying a helpless person in a horizontal position. **3.** A footboard in a rowing boat. **4.** A tie resisting tension only. **5.** Rails to tie the feet of furniture together. See *Splats* and *Splad*.

Stretchout. A development, *q.v.*, especially of the cylindrical string for a geometrical stairs.

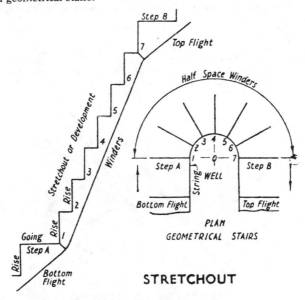

STRETCHOUT

Stria. A flute to a column, hence *striated*.

Strickle. A straight-edge to *strike* granular materials level with the top of the container, etc. A templet used in moulding for shaping the sand.

Striking. Removing temporary timbering, such as centering, when the work is completed. See *Easing*.

Striking Knife. A marking knife, *q.v.*

Striking Over. Continuing the lines on the edges and back of the stuff when setting-out.

Striking Plate. The plate into which the bolt and latch of a mortise lock engage. See *Mortise Lock*.

String-board. See *Strings* (1).

Stringcourse. A horizontal projecting band or moulding on the face of a building. It is a decorative feature and usually continues the level of the window sills or denotes the floor levels.

Stringer or **Stringer Beam.** 1. A long horizontal member in a trussed bridge. 2. Timbers tying together the heads of posts or standards, or of trestles in trestle bridges. 3. Walings.

Stringers. 1. The beams strung across the tops, or caps, of trestles in trestle-bridge construction. 2. Long pieces of wood holding a series of shorter pieces in position, as the sides of a ladder. 3. Longitudinal beams in ship and seaplane construction along the sides and deck. Auxiliary members, parallel to the main structural members, for building up the contour of frames, etc., especially in shipbuilding, aircraft, etc. See *Wriggle*.

Stringings. Narrow inlays up to $\frac{1}{8}$ in. wide and of one colour. See *Bandings*.

String Measure. See *Hoppus* and *Liverpool String Measure*.

String Piece. The horizontal member of a bow-string roof truss.

Strings. 1. The inclined supports for the ends of the steps of a flight of stairs. The various types are : bracketed, close, cut, outside, wall, and wreathed strings. See *Bracketed Stairs* and *Handrail*. 2. The

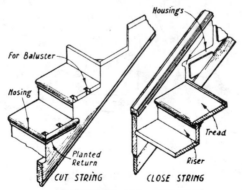

STRINGS TO STAIRS

sides of a ladder. 3. A slender member tying together the ends of a curved member. See *Belfast Truss* and *Stringer*. 4. The top range of planks on the inside of a ship.

Stringybark. *Eucalyptus obliqua* and *E. gigantea* (*H.*). Called Tasmanian Oak, *q.v.*, and Messmate. Pale straw to light brown. Some resemblance to oak. Hard, heavy, strong, and durable. Elastic, resilient. Straight, wavy, and interlocked grain. Moderately close texture. Subject to gum veins. Fairly difficult to work and requires care in seasoning. Used for all purposes : structural, piling, cabinet work, superior joinery, paving, sleepers, flooring, etc. S.G. ·9 ; C.W. 3·5.. Several other species of Eucalyptus are known as yellow, red, white, brown, and gum top stringybark. RED. *E. macrorrhyncha*. Victoria. Reddish brown. Used chiefly for fencing, poles, and sleepers. WHITE. *E. eugenoides*. N.S. Wales and Queensland. Brown. Hard, heavy, very durable. Large gum pockets. Resists insects. Used for exterior work. YELLOW. *E. muelleriana*. Victoria. Similar to white but smaller sizes. Difficult to work. Used for piling, wagon construction, sleepers, structural work. Other species are *E. acmenioides* and *E. carnea*. GUM TOP. Tasmania. Similar wood to *E. obliqua*. The name is applied to trees on which the bark is stringy below and smooth above. Other species : *E. Blaxlandi* and *E. capitellata*. New South Wales. Medium sizes and little exported. Also see *Messmate, Red Mountain Ash, Tingle-tingle, Swamp Gum,* and *White Stringybark* and *Yellow Stringybark*.

Strip. 1. A narrow piece of wood. A grading term for narrow lumber under 4 in. wide and under 2 in. thick. **2.** See *Sticker*.

Stripe Figure. Ribbon grain, in quarter-sawn wood. Stripe figure may also be produced by colour variation in the annual rings, but the trade term implies interlocked grain.

Stripping. 1. Removing the shuttering to concrete. **2.** Destroying the thread of a bolt or screw. **3.** Removing old paint before repainting.

Strix. The fillet between two flutes of a column.

Strombosia. See *Ataka, Dedali, Kamap, Obelu*.

Structural Veneers. Plain strong veneers suitable for plywood. They are so named to distinguish them from decorative or figured veneers.

Structural Work. Applied to large timber structures that are designed to carry heavy loads, such as stagings, centering, shoring, wharfs, bridges, piling, timbering to deep excavations, etc., which is often described as engineering work. See *Constructional Work*.

Structure of Wood. See *Hardwoods* and *Softwoods*.

Strut. An inclined compression member of a frame, as in a roof truss, centre, etc. Short members connecting the longerons of a fuselage. Upright braces connecting top and bottom wings of a biplane. See *Flying Shore* and *Fuselage*.

Strutting. Short timbers between joists to prevent the joists from canting and to stiffen them. See *Floor, Herring Bone, Keyed,* and *Solid Strutting*.

Strutting Piece. Same as straining piece, *q.v.*

Strychnos. See *Cape " Teak," Chitonga, Threaded Box,* and *Snakewood*.

Stub. The stump of a tree.

Stub Plane. A projecting part of a hull or fuselage to which is attached the main part of the plane, in aircraft.

Stub Tenon. A short tenon not passing through the material in which the mortise is formed. (See illus., p. 556.)

Stuck Moulding. One formed on the material. Not planted. See *Sticking*.

Stud. **1.** A nail with large ornamental head. **2.** A screw or bolt threaded at both ends. **3.** The rough vertical timbers, or quarterings, forming a partition, usually to carry lath and plaster or wall-boards. See *Hollow Partition*, *Grounds*, *Partition*, and *Shuttering*. **4.** See *Shelf Fittings*.

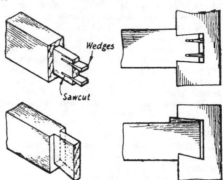

Wedges

Sawcut

STUB TENONS WITH SECRET WEDGING

Studding. **1.** A number of studs forming a partition. **2.** See *Boom*.

Studdles. Distance pieces, or struts, to keep the sets apart in mine shafts.

Stuff. A craftsman's name for converted wood that he is preparing for a job.

Stull. A large round prop in a mine.

Stump. **1.** The base of a tree after felling. **2.** See *Wickets*. **3.** The vertical piece supporting the outer end of the arm of a chair.

Stump Figure. Decorative wood from the base of a tree. It is highly decorative in many species, with ripple, burr, and wave.

Stump Tenon. A short tenon increased in thickness near the shoulder for strength.

Stump Wood. From the butt of a tree. It is usually half-round rotary cut for veneers.

Sty. An enclosure for pigs.

Style. Applied to a distinctive type of architecture or design.

STUMP TENON

Stylobate. A continuous unbroken pedestal for a row, or range, of columns.

Subaha. *Mitragyna stipulosa* or *M. macrophylla*. Abura, *q.v.*

Sub-plinth. A second, lower, plinth to a pedestal, etc.

Subsellia. See *Miserere* and *Sedilia*.

Sub-sill. A stall-board, *q.v.*

Sucupira. *Bowdichia brasiliensis* or *B. virgilioides* (*H.*). C. America. Also called Acapu, Amoteak, Brownheart, and Hudoke. Tan brown, darker stripes, lustrous. Very hard, heavy, strong, durable, tough. Straight, wavy, and ribbon grain, roey. Coarse texture. Excellent wood where strength and durability are required. Difficult to work. Used for structural work, cabinet work, sleepers, shipbuilding, wheelwright's work. S.G. ·9 ; C.W. 4·5. SUCUPIRA AMARELLA. *B. nitida* or *Ferrairea spectabilis*. Similar properties to above. Brownish yellow, darkens with exposure, streaked ; waxy appearance. Interwoven and roey grain. Used for cabinet work. S.G. ·9 ; C.W. 4·5.

Sudan Mahogany. *Khaya spp.* See *Nigerian Mahogany* and *Mahogany*.

Sudd. A floating barrier of logs.

Sudw. Abbreviation for the name of the botanist Sudworth.

Suffolk Latch. See *Norfolk Latch.*

Sugarberry. *Celtis lævigata.* Marketed as Hackberry, *q.v.*

Sugar Maple. See *Maple.*

Sugar Pine. *Pinus lambertiana* (*S.*). Western N. America. Included in Western white pine. California soft pine. Like mild Baltic redwood. One of the best soft pines, large sizes free from defects ; and good supplies. Yellowish white. Fairly durable and strong, resonant. Straight grain, very stable, coarser texture than white pine. Used extensively in organ building, and as white pine. S.G. ·4 ; C.W. 1·25. See *Pine* and *Western White Pine.*

Sugi. *Cryptomeria japonica* (*S.*). Called Japanese cedar. Brown, fragrant. Soft, light, weak, not durable. Straight and wavy grain, coarse texture. Distinct annual rings and fine rays. Easily wrought. Used for cabinet work and cheap lacquered furniture. S.G. ·28 ; C.W. 2.

Suite. A connected series, or set, of things. Applied to the furniture appropriate for a particular room, as bedroom suite, etc. A set of rooms.

Sumach. *Rhus typhina* (*H.*). Canada. Olive green, lustrous. Suitable for good cabinet work. CHINESE SUMACH. Tree of Heaven, *q.v.* A substitute for ash, but distinguished by wider rings and distinct rays.

Summer Beam. A large beam, or a beam serving as a lintel, for dead load only.

Summer-house. A small ornamental building, usually of wood, in a garden.

Summers. The rails, or joists, carrying the floorboards of a vehicle.

Summer Tree. Same as summer beam, *q.v.*

Summer Wood. The later growth of an annual ring, *q.v.* The cells have smaller cavities, thicker walls, and produce harder, denser, and darker wood than the spring wood in most species.

Sun Blind. See *Blind* and *Rolling Shutter.*

Sundeala. Proprietary wood-fibre wall-boards.

Sunderwood or **Sundri.** *Heritiera fomes, H. minor* (*H.*). India. Dark red brown, faint black streaks, dull but smooth. Very hard, heavy strong, durable, tough, and elastic, but not stable. Slightly interlocked grain, fine texture, fine ripple. Seasons well but slowly. Difficult to work due to hardness. Polishes well. Rather small sizes.

Used for all purposes. Structural work, boat-building, furniture, carriage work, implements, etc. S.G. ·95 ; C.W. 4.

Sunk. Recessed. A part lying below the main surface.

Sunk Bead. One not at the edge of the material, and slightly below the surface. See *Reeds*.

Sunk Face. See *Sunk Panel*.

Sunk Gutter. A secret gutter, *q.v.*

Sunk Mouldings. Planted mouldings lying below the face of the framing.

Sunk Panel. One formed by sinking a recess in the solid material, as in newels, bench ends, pilasters, etc. Also see *Fielded Panel* and *Pew*.

Sunk Shelf. A plate shelf. A recess is formed near the back of the shelf to receive a piece of crockery, such as a plate, vertically, so that it rests against the wall.

Sunlock. Hollock, *q.v.*

Sun Ray. A trade name for dark-coloured Prima Vera.

Sup. Abbreviation for *superficial*.

Supa. *Sindora supa* (*H.*). Philippines. Yellowish bronze to very dark brown with age. Oily, but dries out with seasoning. Moderately hard and heavy to hard and heavy. Not strong, fairly durable and resists insects. Interlocked grain, ribbon and handsome tangential figure, fine texture. Requires careful seasoning to prevent warp. Difficult to work because of grain and hardness. Finishes and polishes well. Used for furniture, superior joinery, musical intruments, flooring, trinkets, etc. S.G. ·65 ; C.W. 4.

Super. Proprietary wood-fibre wall-boards.

Super or **Superficial.** Square measure. *Foot super* means *square foot*.

Superimposed Carving. A separate carved piece planted in a large hollow of a moulding and common to Gothic architecture.

Supplies. The supplies of timber to this country are constantly changing due to political and economic reasons. In normal times woods are imported from nearly a hundred different countries. There are over 4,000 woods in common use throughout the world, and new varieties are continually being introduced into this country. Most of these woods are described in the glossary but some of them are still under investigation and reliable data is not yet available. See *Market Forms*.

Suradanni. *Hieronyma alchorneoides* (*H.*). B. Guiana. Dark red, black streaks. Hard, heavy, strong, durable. Difficult to work due to secretion. Used for shipbuilding, furniture, wheelwrights' work, etc. S.G. ·82 ; C.W. 4. The species *Melia cedrela* is also called Suradanni.

Surbase. The mouldings immediately above the base of a pedestal. The crowning moulding to a dado. See *Pilaster*.

Surface Check. A check of little depth, and chiefly confined to the surface. See *Checks*.

Surfaced. Applied to stuff that has been planed on one or more sides.

Surface Measure. A measure of area of surface only. When the stuff is 1 in. thick, *surface* and *board* measure are the same.

Surfacer. A planing machine. See *Machines*.

Surf-board. A board used for surf-riding.

Surin. Betis, *q.v.*
Surinam Mahogany. Cirouaballi, *q.v.*
Surinam "Teak." *Hymenæa courbaril.* Also called Guapino. See *Locust* and *Teak.*
Surround. 1. A wood frame to a metal casement window. 2.

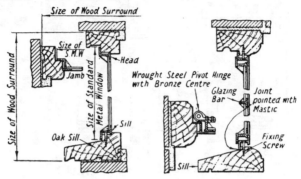

Size of Wood Surround
Size of Wood Surround
Size of S.M.W
Jamb
Head
Size of Standard Metal Window
Sill
Oak Sill
Wrought Steel Pivot Hinge with Bronze Centre
Glazing Bar
Joint pointed with Mastic
Fixing Screw
Sill

SURROUNDS FOR METAL WINDOWS

A decorative or exposed part round a plane surface, such as a floor. A border.
Suryan. Family *Meliaceæ* (*H.*). Genus not yet classified. Malay. Brown. Variable in weight and hardness, not durable. Fine grain and rays, with white deposit in pores. Supplies scarce. Used for furniture, planking, coffins, etc. S.G. ·46 to ·7.
Suspended Ceiling. One with independent ceiling joists suspended from the floor joists or roof timbers or from the bottom member of a steel truss.
Suspended Scaffold. A self-contained scaffold suspended by steel ropes from outrigger beams. Winches are fixed to the scaffold for winding up the scaffold in some cases.
Suspended Shuttering. Form work, for concrete floors and beams, suspended from the steelwork. (See illus., p. 560.)
S.W. Abbreviation for *sound wormy.*
Swag. A decorative carved loop representing leaves, husks, etc., and characteristic of Adam and several other styles of architecture. See *Festoon.*

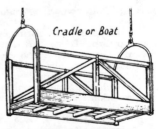

Cradle or Boat

SUSPENDED SCAFFOLD

Swage. A blacksmith's tool for shaping metal. A tool for spreading the points of saw teeth.
Swager. A machine for setting bandsaws.

Swage Saw. A thick, stiff, circular saw ground on one side to form a thin circumference. It is used for cutting thin stuff that will curl away from the saw.

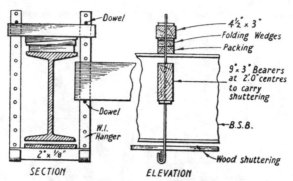

SUSPENDED SHUTTERING

Swallow Tail. A dovetail, *q.v.*

Swamp Ash. See *Ash.*

Swamp Box. *Tristania suaveolens* (*H.*). Queensland. Also called Swamp Mahogany. Red. Hard, heavy, tough, but brittle and not strong. Very durable and resistant to marine insects and white ant. Interwoven grain and fine even texture. Difficult to season but not difficult to work. Used for fender piles, flooring, mallets, etc. S.G. ·82.

Swamp Chestnut Oak. *Quercus prinus.* American white oak. See *Oak.*

Swamp Cottonwood. *Populus heterophylla.* See *Cottonwood* and *Populus.*

Swamp Cypress. Louisiana cypress, *q.v.*

Swamp Elm. *Ulmus americana.* White elm. See *Elm.*

Swamp Gum. *Eucalyptus regnans* and *E. ovata* (*H.*). Tasmania. Also called stringy gum and Tasmanian ash or oak. Similar to Victoria mountain ash. Cream colour. Moderately hard and heavy ; strong, fairly tough, stiff, and durable. Straight, coarse, even grain, uniform texture. Some fiddle-back figure. Not difficult to work and fumes well. Used for furniture, bentwork, construction, wheelwrights' work, cooperage, etc. S.G. ·8 ; C.W. 3. Also *Eucalyptus ptychocarpa.* W. Australia. Red, rather softer and more fibrous than other gums. Good timber but not yet exported. See *Eucalyptus* and *Gums.*

Swamp Mahogany. *Eucalyptus robusta* (*H.*). Australia. Excellent wood for most purposes. See *Eucalyptus* and *Swamp Box.*

Swamp Red Oak. *Quercus rubra.* American red oak. See *Oak.*

Swamp Tea Tree. *Melaleuca ericifolia* (*H.*). Tasmania. Brown. Small sizes. Used for decorative articles and inlays.

Swamp White Oak. *Quercus bicolor.* See *Oak.*
Swamp Yate. *Eucalyptus occidentalis* (*H.*). W. Australia. Similar, but inferior, to *Euc. cornuta*, and used for similar purposes. See *Yate.*
Swani. Salmon wood, *q.v.*
Swan Neck. A compound curve, or curve of contraflexure. A combination of ramp and knee to give a quick rise to a handrail. An S-shaped bend for a down-pipe for an overhanging eaves or for a parapet gutter. See *Knee to Handrail.*
Swan Neck Chisel. A specially shaped long chisel for levering out the core of a mortise for a mortise lock.
Swan River Mahogany. Jarrah, *q.v.*
Swartzia. See *Wamara.*
Sway Rods. Diagonal braces to resist wind pressure.

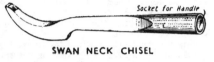

SWAN NECK CHISEL

Socket for Handle

Swedish. A prefix to several N. European woods, ash, redwood, etc., to denote source of origin.
Sweep. 1. A curve, especially one of large radius. 2. A long oar used for steering or propelling barges and sailing boats.
Sweetbark. *Sideroxylon Richardii* (*H.*). Queensland. Also called Blush Condoo and Rose Marara. Pinkish rose, faint flecks, lustrous, smooth. Moderately hard and heavy, not very durable. Straight grain good for bentwork, figured wood weak and brittle. Firm grain, close texture, distinct rays, fiddle back. Sandy nature is hard on cutting tools. Fumes and polishes well. Used for cabinet work, musical and other instruments, flooring, etc. S.G. ·68 ; C.W. 3·5.
Sweet Bay. Mediterranean Laurel, and U.S.A. Magnolia.
Sweet Birch. *Betula lenta.* Canada. See *Birch.*
Sweet Chestnut. See *Chestnut.*
Sweetening. A timber trades term for grading above the accepted quality.
Sweet Gum. See *Red Gum.*
Sweetia. See *Chichipate* or *Billywebb.*
Sweetwood. *Pradosia lactescens* (*H.*). Central America. Also called Buranhem. Dull grey, streaked with yellow and brown, smooth. Very hard, heavy, tough and strong, fairly durable. Straight grain, medium texture, visible rays. The wood is distinguished by its sweet taste. Difficult to season, but works readily. Used for vehicles, implements, oars, heavy construction not exposed to weather, spokes, felloes, bentwork, etc. S.G. ·9 ; C.W. 3·5. Species of *Nectandra, Ocotea,* and *Licaria* are also called Sweetwood.
Swell. 1. Entasis, *q.v.* 2. A mechanical device on an organ to control the volume of sound.
Swell Butted. Applied to a tree with specially large base.
Swell Chamber or **Box.** A chamber containing the pipes of an organ and one that is provided with movable shutters to control the volume of sound at the discretion of the player.
Swelling. See *Entasis.*
Swietenia. Mahogany, *q.v. Swietenia candollei,* Venezuelan

mahogany. *S. cirrhata*, Mexican. *S. macrophylla*, Brazilian, Central American, and probably Peruvian mahogany. *S. mahogoni*, Cuban mahogany. *S. humilis*, Central American mahogany. *S. krukovii*, Brazilian ᵊmahogany. *S. tessmannii*, a doubtful classification of Peruvian mahogany. Also see *Red Eyne* and *Satinwood* (E. Indies).

Swift. A cooper's two-handled trimmer for cleaning off irregularities.

Swill. Same as trug, *q.v.*

Swing Bar. A pivoted bar for securing a pair of gates. See *Fall Bar*.

Swing Boat. A boat-shaped car, or swing, used on fair grounds.

Swing Bridge. One that can be turned aside to allow ships to pass. It is counterbalanced and rotates on a pivot.

Swing Door. A door without stops to the frame so that the door opens in either direction. Special helical hinges or floor springs are used. See *Panic Bolts*.

Swinging Arches. Easing the centering to arches. The term is applied more to bridges consisting of a series of arches and when the easing is done by means of sand pistons.

Swinging Boom. See *Boom*.

Swinging Cross-cut Saw. See *Pendulum Cross-cut Saw*.

Swingle Tree. See *Single Tree*.

Swing Post. The hanging post for a gate.

Swing Sash. A pivoted sash or casement. See *Pivoted*.

Swing Saw. A machine crosscut or pendulum saw, *q.v.*

Swing Shelf. A hanging, or suspended, shelf.

Swintonia. One of the species producing Rengas, *q.v.*

Swirl. Irregular grain on the surface of wood as round a knot or at the intersection of branch and trunk.

Swirl Crotch. Figure obtained from that part of a tree where the crotch figure fades into the figure from the normal stem.

Swiss Pine. See *Silver Fir* and *Cembran Pine*. It is used for the tables of violins because of its resonance.

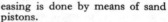

Pulley
Hangers
Balance Weight
Cast Iron Frame
Handle
Guard
Crosscut Saw
Pulley

SWINGING CROSSCUT

Switch. 1. A flexible twig. 2. A device to make or break a circuit or to change the course of an electric current.

Switchback. A steeply undulating track over which a car runs by its own impetus on fair grounds.

Switchboard. A board on which a number of electric switches are collected, and usually protected by a hinged case. A set of switches at a telephone exchange.

Swivel Hook. A rotating hook used in wardrobes.

Sword. 1. A metal support at the free corner of a Pullman berth on board ship. 2. See *Timber Sword*.

Sycamore. *Acer pseudoplatanus* and *Platanus occidentalis* (*H.*). Europe and N. America. The great maple or Scotch plane. Yellowish white to brownish, smooth, lustrous. Fairly hard and heavy. Strong and stable but not durable. Uniform grain and close texture, some wavy and roey, distinct rays. Slightly interlocked. Requires care in seasoning. Not difficult to work. Stains easily. Large sizes. Used for laundry, dairy, and butcher's appliances, food containers, beetles, furniture, musical instruments, turnery, plywood, veneers, cooperage, woodware, etc. Chemical treatment produces greywood or harewood, and it is steamed for *weathered sycamore*. American Sycamore, *P. occidentalis*, is also called buttonwood, buttonball, and American plane.

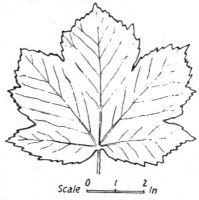

Scale 0 1 2 In

SYCAMORE

Quarter-sawn wood is often called lace wood. S.G. ·6 ; C.W. 3. ORIENTAL SYCAMORE. See *Mulberry*. SATIN SYCAMORE. *Geissois spp.* Queensland. Pinkish red. Fairly hard and heavy, but tough and strong, not durable. Fine even grain and texture, smooth and very lustrous. Stains and polishes well. Used for cabinet work, superior flooring, etc. S.G. ·65 ; C.W. 3. SILVER SYCAMORE. *Evodia micrococca.* Queensland. A useful wood with the characteristics of English sycamore but too small and scarce to have any commercial value except locally. S.G. ·6. Also see *Maple* and *Queensland Sycamore*.

Sydney Blue Gum. *Eucalyptus saligna.* See *Gum*.

Sylvan. See *Silvan*.

Sym. Abbreviation for the name of the botanist Symington.

Symphonia. See *Hog Gum, Osol,* and *Waika*.

Symplocos. See *White Hazlewood*.

Synara. Movingue or Lemonwood. A species of Peroba. See *Piqua Peroba*.

Syncarpia. See *Ironwood Box, Satinay,* and *Turpentine*.

Syringa. Mock Orange, *q.v.* See *Lilac*.

T

t. **1.** Abbreviation for *tensile force*. **2.** A symbol denoting a tangential surface or section of wood.

Tab. **1.** To keep account of, or tally, *q.v.* To check numbers or quantities. **2.** A small hinged inset in a control surface of an aeroplane.

Taban. *Palaquium spp.* Marketed as Nyatoh, *q.v.*

Tabasara. See *Prioria.*

Tabasco. A prefix to several woods denoting the source of origin, especially Central American Cedar and Mahogany.

Tabebuia. See *Prima vera* and *White Cedar.* The wood Fygrin may be of this genus. Also see *Encino, Peroba, Roble,* and *Tecoma.*

Tabernacle. A receptacle for anything holy or consecrated. A niche for a statue of a saint. A seat or niche with ornamental canopy.

Tabernacle Work. Ornamental open tracery work in churches, especially to the tops of niches, stalls, etc.

Table. **1.** Any wide flat surface. A piece of furniture consisting of a flat level surface, supported by legs, pillars, or pedestals. There is a great variety of tables used for domestic and other purposes : bed, billiard, card, chair, chess, coffee, commode, communion, console, curio, dining, dressing, draw, extension, fall, flap, folding, gaming garden, gate-leg, hall, kitchen, leaf, nested, needlework, occasional, pedestal, Pembroke, pier, pillar, pouch-work, refectory, Rudd's, serving, side, tea, telescope, tray-frame, trestle, trestle-end, tripod, wall, work, writing, etc. Tables are also named according to the shape of the top, as kidney, oval, round, etc., or according to the period or style, or by the name of the original designer. They are described alphabetically in the Glossary. See *Extension, Gate-leg,* and *Writing*

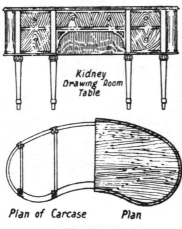

Kidney
Drawing Room
Table

Plan of Carcase   Plan

TABLE

Tables. Also see *Sideboard* and *Knuckle Joint.* **2.** A horizontal band of mouldings. **3.** The scarfings for a table joint, *q.v.* **4.** The surface of a machine on which the stuff runs. See *Cutter Block.* **5.** The top or face of a violin, usually of Swiss pine. **6.** A flat inscribed surface. **7.** The boards extending over the grooves in the wind-chest of an organ.

Table Chair. An arm-chair with hinged back that falls flat on to the arms so that it forms a table. Also called a chair table.

Table Flap. A hinged leaf to a table.

Table Guard. A raised rim or lip round a table on board ship, to prevent articles from sliding off the table. The guard can be lowered by sliding vertically on metal dovetail slides when not required, or for cleaning purposes. They are also called fiddle racks for long tables.

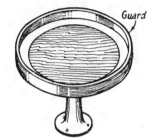

TABLE WITH GUARD

Table Haunch. A haunch that tapers to the outer edge so that it is not seen when assembled. A tapered haunch. See *Skylight.*

Table Joint or **Scarf.** A scarfed joint in which the fitted surfaces are parallel to the edges of the timbers with a vertical break at the middle, sometimes called the table. The joint is often strengthened by hardwood keys or folding wedges and indented ends. It is used for lengthening structural timbers. An oblique tabled scarf has the fitted surfaces inclined to the edges. See *Scarfed Joints* and *Fish Plates.*

Tablet. A small slab or table. A small flat piece of wood, etc., for a device or inscription.

Tabling. Indenting the ends of the pieces for a scarfed joint.

Tabouret. **1.** A wood frame used when embroidering. **2.** A small stool.

Tabular. Like a table. The flat plane part of anything.

Tacamahac. *Populus balsamifera.* See *Poplar.*

Tace. Same as sally, *q.v.*

Tachuelilla. *Zanthoxylum microcarpum.* See *Satinwood.*

Tacking. **1.** Fixing anything temporarily or not securely. **2.** Fixing thin material by means of tacks.

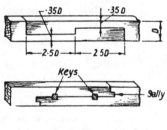

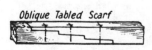

TABLE JOINTS OR SCARFS

Tackle. **1.** Lifting appliances. See *Pulley Blocks*. **2.** The tools and appliances, or gear, required for a specific occupation in work, sport, or games.

Tacks. Very short sharp nails with flat heads.

Tacky. Applied to paint, varnish, polish, glue, etc., when it is not quite hard. Sticky.

Tænia. See *Tenia*.

Taffrail. **1.** A guard rail to the deck of a ship, especially at the stern, not less than 3 ft. 6 in. above the surface of the deck. **2.** The upper curved part of the stern of a ship.

Tagger Wood. Eastern Wild Cinnamon. The species from Madagascar is highly esteemed in the East for coffins.

t.a.h. Abbreviation for *tape after hewn*.

Tail. The rear end of anything. See *Fuselage* and *Tail Unit*. The lower part of a shingle or slate.

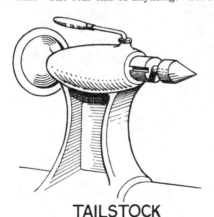

TAILSTOCK

Tail Bay. The end bay of a roof or floor.

Tail Board. An adjustable back to the body of a vehicle.

Tail Booms. Main longitudinal spars providing attachment to the tail unit of an aeroplane without fuselage.

Tail Float. A watertight compartment below the tail unit of a seaplane.

Tail Gate. The lower gate of a canal lock.

Tailing. Fixing the end of a projecting member securely in the wall. *Tailing in* or *tailing down*.

Tail Joist. A trimmed joist. See *Trimmer*.

Tail Piece. A piece of ebony at the bottom of a stringed instrument, such as a violin, etc., to which the strings are fastened.

Tail Plane. The aerofoils at the rear of the fuselage of an aeroplane, to give stability.

Tail Print. A projection on a pattern for a casting where an opening has to be cored below or above a parting.

Tail Stock. The movable stock, or head, to a lathe, *q.v.*

Tail Skid. See *Skids*.

Tail Trimmer. A trimmer against a wall to avoid building the ends of the joists in the wall.

Tail Unit. The rear of an aeroplane, including tail plane, fin, elevator, and rudder. See *Fuselage*.

Tainted. A term applied to wood that is out of condition through close piling : decayed, spotted, attacked by fungi.

Taking Off. Obtaining dimensions from drawings, etc., to estimate the amount of materials and the labour required for a job.

Takoradi Mahogany. *Khaya ivorensis.* African Mahogany, *q.v.*

Tala. *Celtis tala (H.).* C. America. Elm family. Yellow. Very variable in weight and hardness. Tough and strong but not durable. Rather coarse irregular grain. Fairly easy to work. Used for constructional work where strength and toughness are required. S.G. ·6 to ·84.

Talane. *Acacia speciosa (H.).* C. America. Dark brown. Hard, heavy, with beautiful grain. Little imported. Used for cabinet work and superior joinery.

Talauma. See *Magnolia* and *Safan.*

Talipot Palm. *Corypha umbraculifera.* Burma and Ceylon. Black and white stripes. Extremely hard and used for fancy articles.

Tall Boy. A high, comparatively narrow chest of drawers. A double chest of drawers with one carcase above another.

Tallow Tree. *Gleditschia* and *Gymnocladus chinensis (H.).* China. The latter species is similar to the Kentucky Coffee Tree, *q.v.* Both species are more important for the pods from which soap is made than for the wood. *Stillingia sebifera (H.).* China. White, fine, close grain. Used for furniture, engraving blocks, etc. Important for the tallow from the seeds.

Tallow Wood. *Eucalyptus microrys (H.).* N.S.W. and Queensland. Yellowish brown, darkening with age. Very hard, heavy, durable, and strong. Straight close grain, often interlocked. Fairly difficult to work and requires care in seasoning. Difficult to polish because of greasy nature. Considered the best all-round hardwood in New South Wales. Used for structural work, flooring, paving, sleepers, wagon and wheelwrights' work, turnery, carving, etc. S.G. 1. C.W. 3·5. TASMANIAN TALLOW WOOD. *Pittosporum bicolor.* Also called Cheesewood. Yellowish white, odorous until seasoned. Decorative, tough, flexible. Small sizes. Used for cabinet work, inlays, fishing rods, etc.

Tally. 1. A notched stick to record numbers. A record of the number of pieces and the grading of a quantity of lumber. 2. See *Chain.*

Tally Board. A piece of clean board about 2 ft. by 10 in. by 1 in. on which a tally-man records the sizes, etc., of the pieces of wood being measured.

Talon. An ogee moulding, *q.v.*

Talus. A slope or inclination.

Tamago. See *Lilac.*

Tamalan or **Tamalin.** *Dalbergia oliveri (H.).* Burma Tulipwood. Pinkish to reddish brown, darker lines, darkens with exposure, smooth, fragrant. Very hard, heavy, strong, and durable, fairly elastic. Straight, slightly interlocked grain, ripple, medium coarse texture. Seasons fairly well, difficult to work, polishes well. Used for cabinet work, superior joinery, bentwood, parquetry, wheelwrights' work, turnery, etc. S.G. 1 ; C.W. 4.

Tamarack. *Larix laricina.* One of the strongest softwoods of E. Canada, but not so good as Western Larch, and little used in building. S.G. ·5. See *Larch.*

Tamaricaceæ. A family of little timber value. Includes *Tamarix*. See *Ukan*.

Tamarind. *Tamarindus indica* (*H*.). E. Indies. Yellowish white, darker patches. Hard, heavy, and durable. Beautiful figure with fine rays. Close crooked grain. Difficult to work. Used for joinery, cabinet work, wheelwrights' work, etc. S.G. ·84; C.W. 4. The name is also applied to Moruro, *q.v.* HILL-TAMARIND. See *Jutahy*. Also see *White Tamarind* and *Sabicu*.

Tamarix. See *Ukan*.

Tambaiba or **Timbauba.** Timbo, *q.v.*

Tambooti. *Spirostachys africanus* (*H*.). S. Africa. Dark brown, variegated, fragrant, feels oily. Very hard, heavy and strong, stable. Compact close grain, beautiful figure. Not difficult to work. Scarce, but used for high-class furniture, decorative articles, turnery, etc., and suggested for gunstocks. S.G. 1; C.W. 3·5.

Tambour. Applied to anything drum-shaped. A circular frame on which material is stretched to be embroidered. A cylindrical stone for a column. A short stumpy column. See *Pedestal Sideboard*. **2.** A wood screen or vestibule to a church porch. **3.** The bell, campana, or vase of the Corinthian and Composite capitals. **4.** A defensive palisade to a gate, etc.

Tambour Fall. See *Roll Top*.

Tambours. Narrow, specially shaped, strips mounted on canvas, or steel strips, to form a flexible shutter or surface, as for a roll-top desk, rolling shutters, etc.

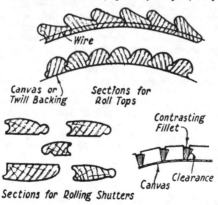

Canvas or Twill Backing

Wire

Sections for Roll Tops

Contrasting Fillet

Clearance

Canvas

Sections for Rolling Shutters

TAMBOURS

Tamo. Japanese ash, *q.v.*

Tampang or **Tapang.** *Artocarpus spp.* (*H*.). Malay. Similar but inferior to Keledang, *q.v.*, with which it is sometimes marketed.

Tampenis. Tempenis, *q.v.*

Tampin. A cone-shaped piece of box-wood to open the end of a lead pipe, in plumbing.

Tampion. A plug or bung or disc. A stopper in the end of an organ pipe.

Tan. Chinese birch.

T'an. *Dalbergia sp.* (*H*.). Chinese rosewood. Reddish black, veined. Hard, heavy, and durable. Beautiful wood for cabinet work and interior fittings. See *Rosewood*.

Tana. The lightest known wood. From Siam and Malay. It is chiefly pith and used for sun helmets and insulation.

Tanalith. A grade of Wolman Salts, *q.v.*, applied under vacuum and pressure or by soaking. It provides protection against fungi and insects with slight fire-resistance. It is intended for wood exposed to the weather.

Tanbark Oak. *Lithocarpus densiflora (H.).* U.S.A. Light to dark brown. Hard, heavy, strong. Chiefly sapwood which is not durable. Distinct rays. Only of local importance except for tannin. S.G. ·58.

Tan Barks. See *Tannic.*

T. and G. Abbreviation for *tongued and grooved.*

Tanekaha. *Phyllocladus trichomanoides (S.).* N. Zealand. Yellowish white. Moderately hard and heavy, very strong and stable, fairly durable. Very free from knots. Dense close grain. Easily seasoned and wrought. Used for sleepers, interiors of greenhouses, structural work, railway carriage work, bark valuable for tannin. S.G. ·68 ; C.W. 2.

Tang. The pointed end of a tool on which the handle is fixed.

Tangent. 1. A line touching a curve, and which does not cut the curve when produced. 2. A sticker, *q.v.*, in a pipe organ.

Tangential Section. The surface of wood that is cut tangential to the annual rings. See *Converting.*

Tangent System. One of the methods of preparing wreaths for handrails, *q.v.*

Tangile or **Tanguile.** *Shorea polysperma (H.).* Red Lauan. Also called Philippine, or Bataan, or Tanguile Mahogany. Light red to dark brownish red, fairly lustrous. Fairly light, moderately hard and strong, not durable. Resembles mahogany in grain and texture. Interlocked and cross grain, ribbon figure, fairly fine texture. Not difficult to season and work. Polishes well. Large sizes. Used for cabinet work, superior joinery, etc. S.G. ·6 ; C.W. 3·5.

Tanging. A term used by wheelwrights for preparing the ends of the spokes for the felloe. *Speeching* is fitting the spokes to the hub.

Tank. See *Vat.*

Tanker. A vessel used for carrying oil, or for distributing oil to other ships, for fuel.

Tannic. An acid derived from the bark, wood, and excrescences of many species of trees, and used in the manufacture of leather. Also called tannin. It changes its characteristics in the heartwood, becomes more insoluble, and influences the colour and increases the durability of the wood. See *Pliable.*

Tanwood. See *Quebracho.*

Tap, To. 1. Perforating the bark of growing trees to extract the sap, etc. 2. Forming the reverse thread in a drilled hole to engage with a screwed thread of a bolt or screw.

Tapana. *Hieronyma alchorneoides (H.).* W. Indies, Dark reddish brown. Hard, heavy, durable. Straight but beautiful grain. Small sizes and little exported. See *Urucurana.*

Tapang. See *Tualong* and *Tampang.*

Taper. Applied to anything that gradually decreases in thickness or width or diameter, as a wedge, plug, diminishing pipe, or brace bit,

A leg, in furniture, gradually diminishing in section and usually finishing with a spade toe.

Tapered Haunch. See *Table Haunch.*

Tapering Gutters. Parapet and vee gutters that vary in width to give the required fall. See *Gutters.*

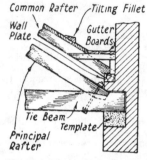

Common Rafter — Tilting Fillet — Wall Plate — Gutter Boards — Tie Beam — Template — Principal Rafter

TAPERING GUTTER

Taping. Securing a banding or veneer on a shaped edge until the glue sets. Two scrap pieces with nails at small intervals apart are used to carry the tapes.

Taping Machine. Used for taping the joints of veneers. The machines may be operated mechanically or by hand.

Tapinhoan. *Mezilaurus navalium* (*H.*). Brazil. Straw colour, fragrant. Hard, heavy, strong, durable. Even grain and texture. Used for structural work, piling, boat-building, cooperage, etc. S.G. ·88.

Tapioca. *Manihot utilissima* (*H.*). Argentine. Light yellow. Light, fairly soft, not durable. Used for interior work, cigar boxes, cheap furniture.

Tapiriri. *Tapirira guianensis* (*H.*). Central America. Also called Duka and Paó Pombo. Light straw colour or pinkish grey, darkening with exposure, lustrous, smooth. Light, moderately soft, not strong or durable. Often speckled with gum exudations. Straight grain, fine rays. Similar to Simaruba, *q.v.* Easily wrought and polishes well. S.G. ·4. *T. Marchandii*, also called Duka, is very similar but slightly harder and heavier. S.G. ·43.

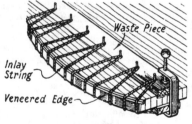

Waste Piece — Inlay String — Veneered Edge

TAPING

Tapir Wood. *Maytenus obtusifolia* (*H.*). Central America. Also called Carne d'Anta. Pale rose or brownish with purplish hue, dull but smooth. Very hard, heavy, tough and strong. Not durable. Straight and irregular grain, medium texture, visible rays. Not difficult to work but requires careful seasoning. Used for interior construction, turnery, etc. S.G. ·82 ; C.W. 3.

Tapped Trees. Trees that have been *bled*, or *boxed*, while standing, for resin, etc.

Tapsava. *Virola koschnyi* (*H.*). C. America. Usually called Banak and Sangre palo. Appearance of C. American cedar. Pale brown, darkening with exposure, some striped, lustrous. Fairly light and soft. Firm coarse texture. Straight grain, fissile. Easily wrought, polishes. Used chiefly for linings to furniture, drawer bottoms and sides, etc. Little imported. S.G. ·6 ; C.W. 2·75.

Tar. A dark viscous fluid obtained by destructive distillation of coniferous woods or from coal. It is used as a protective coating for wood and iron, and for walls.

Taraire or **Tarairi.** See *Tawa.*

Tarco or **Talco.** *Thoninia weinmannifolia* (*H.*). Argentine. Called White Ebony. Ivory white, beautiful grain. Small sizes. Used for decorative work, inlays, etc.

Tare. The allowance for container when assessing custom's duties on contents. The weight of a vehicle as distinct from the weight of the load.

Target. A circular plate of wood or iron at which one aims in shooting practice. It is painted in concentric circles with a central bull's eye.

Tarran. The registered name for several prefabricated building units.

Tarrietia. See *Red Tulip Oak, Lumbayau,* and *Mengkulang.*

Tar Spot. A disease that attacks sycamore trees. It is due to the fungus *Rhytisma acerinum.*

Tartan. 1. A single-masted vessel used in the Mediterranean. 2. Royal Tartan is the American name for Red Tulip Oak, *q.v.*

Tarus. The curb roll used at the intersection of the two surfaces in a mansard roof.

Tasimeter. Electrical apparatus for determining changes in pressure due to change of temperature or moisture content.

Tasmania. The forest trees include several species of Eucalypti, Blackwood, Beech or Myrtle, Leatherwood, Musk, Myrtle, Sassafras, Wattles (*Acacia spp.*), etc., and several softwoods : King William, Huon, and Celery Top pines, etc. There are also many secondary trees steadily increasing in importance on the market. They are described alphabetically.

Tasmanian Beech or **Myrtle.** See *Myrtle.*

Tasmanian Oak. *Eucalyptus gigantea, E. obliqua,* and *E. regnans* (*H.*). This is the suggested name for these woods, which are also called Alpine, Mountain, and Red Ash. Pale straw to light brown, darkening with exposure. Hard, heavy, strong, elastic, resilient. Very durable except in contact with ground. Resembles ash when converted but later it is like oak without ray figure. Straight grain, fairly coarse texture. Fairly difficult to season without collapse and warp, requires reconditioning. Fairly easy to work. Polishes well. Used for nearly all purposes : structural and constructional work, piling, paving, flooring, furniture, superior joinery, carving, etc. S.G. ·8 ; C.W. 3·5. Also see *Stringybark* and *Messmate.*

Tasmanian Pine. King William Pine, *q.v.*

Tassels. Torsels, *q.v.*

Tatajuba. *Bagassa guianensis* (*H.*). C. America. Also called Bagasse. Yellow, lustrous, russet brown on exposure. Hard, heavy, strong and durable, splits readily. Straight and irregular grain, rather coarse texture, distinct rays. Seasons readily, fairly easy to work, and polishes well. Used for joinery, cabinet work, and construction. S.G. ·8; C.W. 3.

Tatané or **Tataré.** See *Vinhatico*.

Tatri. Dillenia, *q.v.*

Taukkyan. *Terminalia arjuna* (*H.*). India and Burma. Also called Anjan, Arjun, Indian Laurelwood, etc. The wood is very attractive and may be compared with black walnut, streaked and figured with dark bands, but the tree is more valued as a decorative feature. The wood is very refractory and unstable. Satisfactory methods of seasoning will increase the demand. S.G. ·75. See *Laurel*.

Tauroniro. *Humiria sp.* (*H.*). B. Guiana. Dark brown. Very hard, heavy, and durable. Fine grain and texture. Used for furniture, turnery, construction, spokes. S.G. 1.

Tavoy Wood. See *Thingadu*.

Tawa. *Beilschmiedia tawa* (*H.*). New Zealand Chestnut. Also called Taraire. Brownish white. Hard, heavy, rather brittle, not durable. Straight grain. Difficult to season. Little appreciated and not exported. Suggested that if correctly seasoned it is suitable for flooring, joinery, etc. Used for packing cases, clothes pegs, pulp, etc.

Tawhai. *Nothofagus fusca*. New Zealand red beech, *q.v.*

Taxaceæ. Yew family. A sub-family of Coniferæ : Austrotaxus, Cephalotaxus, Dacrydium, Macrocachrys, Podocarpus, Saxegothæa, Taxus, Torreya. Also see *Podocarpaceæ*.

Taxodiaceæ. A sub-family of Coniferæ, including : Athrotaxis, Cryptomeria, Cunninghamia, Glyptostrobus. Sciadopitys, Sequoia, Taiwania, Taxodium.

Taxodium. See *Cypress* and *Faux Satine*.

Taxus. See *Yew*.

T.B. and S. Abbreviation for *top, bottom, and sides* in box shook manufacture.

Teak. *Tectona grandis* (*H.*). India, Burma. The only true teak. Best qualities were distinguished as Moulmein and Rangoon, but now marketed as Burma teak. Also exported from Siam, Saigon, and Java. Light golden brown, darkening with exposure to nearly black. White glistening deposit. Odorous, dull, feels oily. Moderately hard and heavy, to heavy, and very hard with age. Strong, stable, elastic, fire and acid resisting. Very durable and resists insects. Fissile, rather brittle. Sometimes with beautiful figure, but usually plain. Rather coarse grain and texture, but smooth. Not difficult to season, but girdled three years before felling. Easily wrought before seasoning and machines well. Difficulty of working increases with seasoning, as a secretion hardens which dulls the tools. Used for all purposes where strength, durability, appearance, and fire resistance are important : shipbuilding, railway carriage work, interior and exterior superior joinery, stairs, furniture. S.G. ·65 ; C.W. 3·5 to 5.

Because of its superior qualities many other woods are miscalled teak, but with a distinguishing prefix. See Cape, Johore, Mahobon, Philippine, Rhodesian, Surinam, Transvaal, and Uganda teak. Also see Bagac, Eng, Iroko, Merabau, Molave, Selangan batu, Turtosa, Yacal, and Yang. They are all excellent woods but no relation to teak. NATIVE TEAK. *Flindersia bennettiana*. New South Wales. No relation to teak. Yellowish. Hard, heavy, durable. Handsome grain and figure. Difficult to work. Large sizes. Used for structural work, joinery, etc. S.G. 1 ; C.W. 4.

Teapoy. A small tea-table.

Tea Tray. A small tray for carrying utensils when serving tea. See *Tray*.

Tea Tree. *Melaleuca ericifolia* and *M. laucadendron* (*H.*). Tasmania and Queensland. Called Swamp Tea, Bottlebrush, and Paper-bark. Also *Leptospermum ericoides* or *L. scoparium*. Called Mountain Tea or Manuka. (*H.*) Tasmania and New Zealand. The woods are pale brown, hard, heavy, dense, tough, durable, elastic, but not strong. Compact, fine, even, but short grain. Small sizes. Used for wheel-wrights' work, boat knees, small cabinet work, fancy articles, inlays, etc. S.G. ·75 ; C.W. 4. The name is also applied to Amyris wood and Cogwood.

Tea Trolley or **Wagon.** See *Wagon*.

Tea Wood. An alternative name for W. Indian Sandalwood and Brazilian Holly.

Teazel or **Teasel Post.** An angle post for a timber framed building.

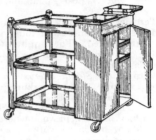

TEA WAGON

Teaze Tenons. Tenons reduced in width so that they may cross each other at right angles for rails on the same level in a corner post.

TEASE TENON

Angle Post

Sectional Plan

Teco. A patent timber connector, *q.v.*

Tecoma. See *Washiba*.

Teconia. See *W. Indies Ebony, Bethabara, Ipe Peroba, Lapacho, Peroba. Poui, Prima vera, Roble, Washiba.* Also see *Tabebuia*.

Tectona. See *Teak* and *Dahat*.

Tee. An umbrella-shaped finial, as on a pagoda.

Tee Gauge. A grasshopper gauge, *q.v.*

Tee Halving. See *Halving Joints*.

Tee Hinge. See *T-hinge*.

Teeth. 1. The serrated edges of saws. See *Saw Teeth*. 2. The pieces forming the faces of the sails in certain types of windmills.

Teijsm. Abbreviation for the name of the botanist Teijsmann.

Telamon. A carved male figure used as a column. See *Atlantes*.

Telescope Frame. Used for an extension table. Runners are fixed between slides ; or keyed slides, or lopers, slide in each other, like a telescope.

TEE HINGE

Telescopic Berth. A berth on board ship that can be extended so that it may be used as either a single or double bed.

Tellar or **Tellow.** A sapling left to stand for a timber tree when clearing the underwood.

Telopia. See *Waratah.*
Tembusa. *Fagræa gigantea* (*H.*). Malay. Also called Temesu. Yellow, pinkish tinge, odorous. Very hard, heavy, durable, and strong ; tough and elastic. Subject to insect attack. Difficult to distinguish sapwood from heartwood. Coarse grain, fine rays. Very difficult to season except as thin boards, and very difficult to work, especially when dry. Used for sea-water piling, boat keels, sleepers, structural work, wheels, posts, planking, furniture. S.G. ·9 ; C.W. 5. Also see *Anan.*
Temper or **Tempering.** 1. Preparing tool steel to the correct degree of hardness and toughness. The steel is first hardened and then re-heated. The required temper is decided by the colour of the film of oxide on the surface. 2. The term is also used for correcting a plastic mix.
Tempinis. *Slœtia sideroxylon* (*H.*). Malay. Yellow, turning red on exposure, darker markings, odorous. Very hard, heavy, strong, durable, and elastic. Close, straight grain, some irregular growth rings, fine rays. White deposit in pores. Difficult to work due to hardness. Excellent wood where strength and durability are important. Used for structural work, spokes, axe handles. Good substitute for Cuban mahogany. S.G. ·95 ; C.W. 4·5.
Template. 1. A stone slab built in a wall to distribute the pressure from a concentrated load, as a beam. 2. A temporary wedge used as a support in shipbuilding. Also see *Templet.*
Templet. A section mould or pattern made of thin material and used to set out the stuff to the required shape. A *square* templet is

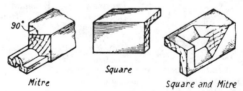

Mitre Square Square and Mitre

TEMPLETS

used for squaring over moulded corners. A *mitre* templet is used for mitering mouldings, as on sash stuff, preparatory to scribing.
Tempunai. *Artocarpus rigida* (*H.*). E. Indies. Usually marketed as Keledang, but inferior.
Temru. Indian ebony, *q.v.*
Ten. Abbreviation for the name of the botanist Tenore.
Tengar. *Ceriops candolleana* (*H.*). Malay. A mangrove. Yellow to orange. Little difference between sapwood and heartwood. Like Bakau and usually marketed under that name. Small sizes. S.G. ·95.
Tengkawang. Sengkwang, *q.v.*
Teng Mang. See *Camphor Tree.*
Tenia. The fillet separating the Doric frieze from the architrave.
Tenio or **Teneo.** *Weinmannia trichosperma* (*H.*). Central America.

Rose colour or brownish with darker stripes, lustrous, smooth. Moderately hard, heavy, strong, and durable. Stable. Straight grain, fine uniform texture, visible rays. Easily wrought and polishes well. Used chiefly for cabinet work. S.G. ·6 ; C.W. 2·5.

Tenon. The end of a piece reduced in section to fit in a recess or cavity of the same size, called a mortise. A tenon is usually in the form of a rectangular prism, with the width three to five times the thickness. The mortise is usually one third the thickness of the stuff. The various types are : barefaced, box, chase, double, dovetailed, frank, franked, forked, hammerhead, haunched, housed, loose, plug,

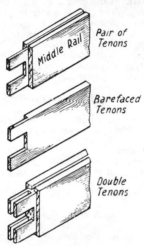

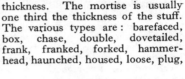

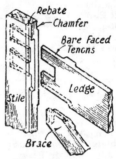

Pair of Tenons

Middle Rail

Barefaced Tenons

Double Tenons

TENONS

Rebate

Chamfer

Bare Faced Tenons

Ledge

Stile

Brace

TENONED BRACE

shouldered, stub, stump, teaze, tusk, twin. They are described alphabetically. Also see *Mortise*.

Tenon-bar Splice. A lengthening joint in heavy structural bearing timbers. Two steel bars are mortised through the timbers at a designed distance from the joint. Bolts are placed through the ends of the bars on each side of the timbers and tightened to pull the ends of the timbers together.

Tenoned Brace. A brace in a ledged and framed door, with short barefaced tenons to keep it in position. It is usually assembled with the rest of the framing.

Tenoner. A machine for forming tenons, scribing, etc. It may be double-ended to prepare both ends of a rail in one operation. See *Machines*.

Tenon Saw. A small saw with thin blade stiffened by a metal back, and used for cutting the shoulders of tenons, etc. It is about 14 in. long, with 10 or 12 teeth or points per inch, which are shaped like equilateral triangles. See *Mitre Box* and *Saws*.

Tension or **Tensile Force.** A force acting on a member and tending to lengthen the member, or pull the fibres apart lengthwise. A stretching force. See *Stress, Strain*, and *f*.

Tension Sleeve. A long cylindrical nut with left- and right-hand threads. It is used to join two rods together so that the total length can be adjusted by means of the sleeve or *screw shackle* to regulate the tension. It is used on wide gates to lift the nose and in long tie rods. See *Turnbuckle*.

Tension Wood. Applied to wood in which the cells are longer than normal wood elements and thin-walled. They may occur in parts of deciduous trees that are under tension, that is, on the convex side of leaning trees, the top side of a branch, etc. It differs from normal wood in the degree of lignification. See *Compression Wood*.

Tensovic. A proprietary name for specially hardened wood, chiefly beech and ash. Impregnation, probably, by synthetic resin, increases the hardness, stability, and tensile strength considerably.

Tent Bed. A bed with a canopy.

Tenter. A frame or machine with tenter hooks on which cloth is stretched.

Tenter Hook. 1. A square hook sharpened at both ends. They are partly driven, like a nail, in the top of a fence to serve as a protection or chevaux de frize. 2. The hooks on a tenter.

Tentest. Proprietary wood-fibre wallboard.

Tent Pegs. Wood stakes to which the stay ropes of a tent are secured. They are 13 in. to 4 ft. long and usually split from beech. See *Pegs*.

Tepesuchil. *Cordia alliodora.* See *Cyp*.

Terap. *Artocarpus incisa (H.).* E. Indies. Usually marketed as Keledang, but inferior.

Terblanz. *Faurea macnaughtonii (H.).* S. Africa. Dark brown, fragrant. Hard, heavy, close texture. Durable but brittle. A decorative wood used for furniture and general cabinet work. Little exported. S.G. ·95 ; C.W. 3.

Teredo navalis. The shipworm. A mollusc very destructive to wood in salt water. It cannot thrive in fresh water. It resembles a worm, and is up to 3 ft. long by ¾ in. diameter. They honeycomb the wood with holes up to more than 1 in. diameter, but the holes do not break into one another. They may continue parallel to each other with only the thickness of paper between the holes. Very few timbers can resist the teredo which can completely destroy a 12-in. pile in less than six months. Teak and greenheart are good. See *Marine Borers*.

Terentang. *Campnosperma spp. (H.).* Malay. Whitish to silvery grey, glistening deposit. Little difference between sapwood and heartwood. Soft, light, not durable. Subject to insect attack. Fine rays. Held in little esteem and not imported. S.G. ·32.

Term. A carved piece under each end of the taffrail of a ship. It is at the side timber of the stern and sometimes extends to the foot-rail of the balcony.

Termes. See *White Ant*.

Terminal. A finial, *q.v.* The shaped ends of newels, standards, etc.

Terminalia. The White Mangrove family. See *Afara, Araca, Aromilla, Badam, Bahera, Bingas, Chuglam, Eghoin, Foocadie, Framire, Guarabu, Harra, Hollock, Idigbo, Indian Almond, Indian Silver Grey-*

wood, *Kalumpit*, *Kindal*, *Kumbuk*, *Laurelwood*, *Limba*, *Mangrove*, *Mutti*, *Myrobolan*, *Naranjo*, *Araca*, *Nargusta*, *Panisaj*, *Taukkyan*, *Than*, *White Bombway*, and *White Chuglum*.

Terminaliaceæ. The White Mangrove family. Same as Combretaceæ, *q.v.*

Terminus. A pillar with the upper part carved as a statue or bust.

Termites. Similar to ants. Some species are very destructive to wood.

Tern Feet. Feet of furniture carved into a three scroll ornament, common in Louis XV. and Chippendale work.

Ternstrœmiaceæ. See *Theaceæ*.

Terraced Roof. A flat roof, *q.v.*

Teruntum. *Lumnitzera littorea* (*H.*). Malay. Dark grey, lighter grey on exposure, fragrant, glistening deposit. Little difference between sapwood and heartwood. Hard, moderately heavy, stable, very durable, rather knotty. Fine grain and rays. Rather small sizes. Seasons well. Used for structural work, flooring, tool handles, wheelwright's work, whole trunks used for piles. S.G. ·64.

Tessellated. Checkered. Inlaid with differently coloured squares like a chessboard.

Tesseræ. Small cubes of wood or other materials, especially glass and marble.

Tester. A flat canopy over a bed, pulpit, or tomb. A half-tester is a covering at the head of a bed only. An ornamental cover or decorated roof over the high altar of a church.

Tests. There are many tests applied to wood to prove its suitability for the many purposes for which it is required. Most of them must be conducted in a properly equipped laboratory under trained observation. The following properties and qualities are important : strength in compression, tension, bending, crushing, torsion, and shear ; deflection ; elasticity ; colour (fast or variable) ; figure and grain ; stability ; durability ; resilience ; resonance ; resistance to wear, etc. ; weight, or specific gravity ; moisture content ; plasticity ; workability. For internal examination X-ray apparatus is used. Also see *Identification, Photomicrograph, Resistance, Selection,* and *Texture*.

Tetraclinis. See *Thuya*.

Tetrahedron. A solid bounded by four equilateral triangles. It is one of the five regular solids. Surface area = (linear side)$^2 \times 1 \cdot 732$. Volume = (linear side)$^3 \times 0 \cdot 118$ or $\frac{1}{3} \times$ Base $\times$ Height.

Tetrameles. See *Thitpok*.

Tetramerista. See *Punah*.

Tewart Wood. See *Tuart*.

Texture. Determined by the relative size and distribution of the wood elements. It is variously described as coarse (large elements), fine (small elements), even (uniform size of elements

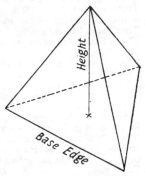

TETRAHEDRON

in spring and summer wood), hard, harsh, medium, mild, silky, smooth, soft, uneven (contrast in size in spring and summer wood), uniform, etc. The texture varies with different woods and rate of growth.

t. g. and b. Abbreviation for *tongued, grooved, and beaded*.

t. g. and v. Abbreviation for *tongued, grooved, and vee-jointed*.

Thalé. Indian Elm, *q.v.*

T-halving. See *Halving Joints*.

Than. *Terminalia oliveri* (*H.*). India. Yellowish grey to dark purple. Irregular heartwood which is rather dull but smooth. Very hard, heavy, strong, tough, and moderately durable. Fairly straight grain, very fine texture. Attractive figure. Difficult to season, must be converted green and slowly seasoned. Difficult to work, liable to pick-up, but finishes well. Suitable for superior joinery and cabinet work. S.G. ·84 ; C.W. 4·5.

Thatchboard. A proprietary insulating strawboard.

Thatching. Covering a roof with bundles of straw or reeds laced together with withes.

Theaceæ. The tea-plant family. Includes Caryocar. See *Piquiá*.

Theatre. 1. An operating room in a hospital. 2. A building for dramatic performances ; a playhouse.

Theavel. A Scottish name for a stick or spoon for stirring liquids.

Theobroma. The cocoa or cacáo tree. See *Cocoa*.

Therming. Square turning, *q.v.*

Thermoplastic Resin. Vinyl resin, which can be made plastic by reheating.

Thermostat. Apparatus for automatically regulating temperature.

Thermotex. A proprietary insulating board.

Thespesia. See *Maga, Parsipu*.

Thicknesser. 1. A woodworking machine for planing stuff to a uniform thickness. The stuff is fed by rollers under the cutters. 2. An under-framework of inferior wood to give an appearance of greater thickness to the tops of furniture.

Thicknessing Up. Gluing a strip under the edge of thin material to make it appear thicker and more solid.

Thickness Moulding. A bed moulding, *q.v.*

Thick Sap. Applied to *Guaicum sanctum*. See *Lignum-vitæ* and *Thin Sap*.

Thick Shagbark. See *Hickory Shellbark*.

Thick Stuff. Converted timber from 4½ in. in thickness.

Thimble. Same as ferrule, *q.v.*

Thingadu. *Parashorea stellata* (*H.*). India, Burma. Also called Tavoy Wood. Light yellowish to reddish brown, rather lustrous, feels rough. Moderately hard, heavy and strong, and fairly durable. Interlocked grain, even but very coarse texture, fine rays, resin canals. Not difficult to work but requires careful seasoning. Kiln seasoning emphasises contrast in colour, darker streaks becoming a rich golden brown. Suitable for cabinet work, superior joinery, etc. S.G. ·76 ; C.W. 3.

Thingan. *Hopea odorata* (*H.*). India, Burma, Andaman Isles. Also called Mai and Rimda. Yellowish brown, purplish cast on exposure, lustrous but dulls later, fairly smooth. White tangential lines. Heavy, very hard, strong, durable, elastic, resists insects. Interlocked, even, close grain. Some beautiful figure, with mottle and ray ; medium texture. Difficult to season and work, but finishes well. Large sizes. Used for general purposes, carriage building, shipbuilding. May be used for all purposes where a good non-ornamental wood is required. S.G. ·7 ; C.W. 4.

T-hinge. A long strap hinge shaped like the letter T and used for batten doors. The long arm is tapered and from 9 to 24 in. long. The cross-piece is from 4 to 6 in. long. It is screwed on the face of door and frame. See *Hatch* and *Tee Hinge*.

Thin Sap. Applied to *Guaicum officinale*. See *Lignum-vitæ* and *Thick Sap*.

Thin Stuff. Converted wood not more than 4 in. thick.

Thinwin. *Millettia pendula* (*H.*). India, Burma. Purplish brown to chocolate colour, variegated streaks, dull but smooth, slightly odorous. Similar to Partridge Wood. Very hard, heavy, durable, tough, and dense. Interlocked grain, mottle and ripple, medium coarse texture. Seasons fairly well but develops fine surface cracks. Beautiful wood for superior joinery and fittings, or for purposes where strength and durability are of primary importance. S.G. ·95 ; C.W. 5.

Third Fixings. A term sometimes applied to the final joinery finishings of buildings, and a continuation of *second* fixings. It usually implies the fittings following the initial decorating : locks, furniture, etc., but sometimes includes hanging of doors. See *Fixings*.

Thitka. *Pentace burmanica* (*H.*). Also called Burma mahogany. Yellow, darkens to deep honey colour, variegated markings, lustrous. Fairly hard, heavy, strong, and durable. Interlocked grain, roe and ribbon figure, ripple. Seasons fairly well but slowly, and liable to warp. Fairly difficult to work smooth. Polishes well. Appearance of mahogany and an excellent substitute. Large sizes. Used for cabinet work, superior joinery, furniture, boat-building, oars, instruments, etc. Also see *Marblewood*. S.G. ·62 ; C.W. 4.

Thitkado. Toon, *q.v.*

Thitmin. *Podocarpus neriifolia* (*S.*). India, Burma. Also called Princewood. Yellowish grey to brownish on exposure. Fairly light, soft, and strong, moderately durable. Straight grain, even medium coarse texture. Easily wrought. Not difficult to season but logs should be quickly converted. Large sizes. Used for most purposes where a good quality softwood is required, carpentry, joinery, furniture, tea boxes, spars, instruments, etc. See *Podocarpus*. S.G. ·55 ; C.W. 1·75.

Thitni. See *Amoora*.

Thitpok. *Tetrameles nudiflora* (*H.*). India and E. Indies. Also called Baing. Greyish white, to light golden brown on exposure, lustrous, feels rough. Light, soft, fairly strong for its weight, not durable. Subject to stain and insect attack. Fine close grain, interlocked in wide bands. Handsome fleck when quarter sawn, due to rays. Coarse texture. Easily seasoned and wrought. Logs should be quickly converted. Substitute for Balsa. Used for packing cases, tea boxes, rotary cut plywood, matching. S.G. ·32 ; C.W. 2.

Thitsein. Bahera, *q.v.*

Thitsho. *Pentace Griffithii.* Burma. Very similar to Thitka, *q.v.*

Thitsi. *Melanorrhœa usitata* (*H.*). India. Dark red, darkening with exposure, nearly black stripes, dull but smooth. Very hard, fairly heavy, strong, elastic, durable. Straight uneven grain, coarse texture, gum canals. Requires care in seasoning and quick conversion. Difficult to work due to hardness. Polishes well. Used for structural work, tools, implements. Lighter coloured wood used for furniture and superior joinery. Tree valued for its yield of black varnish. S.G. ·77 ; C.W. 4.

Thitya. *Shorea obtusa* (*H.*). Burma. Also called Itchwood ; Sal. Mahogany colour. Hard, heavy, durable. Close even grain. Secretion irritates the skin. Large sizes. Difficult to season without warp. See *Sal* and *Shorea.* S.G. ·9.

Tholas or **Thowl** or **Thole.** 1. A wood peg. A pin in the gunwale of a boat to serve as a fulcrum for the oar. 2. An obsolete term for a dome or cupola or a circular building.

Thom. Abbreviation for the name of the botanist Thomas.

Thomas' Rack. Perforated strips of steel in the carcase of a bookcase in which studs are inserted to carry the shelves. See *Shelf Fittings.*

Thoms. Abbreviation for the name of the botanist Thomson.

Thoninia. See *Tarco.*

Thorn. *Acacia nigrescens* (*H.*). S. Africa. Called Knobly Thorn. Hard, heavy, tough, and strong. Durable and resists insects. Not exported.

Thornapple. American hawthorn.

Thousand. See *Mille.*

Thrawn. Same as warped, *q.v.*

Thread. See *Screw Thread.*

Threaded Box. *Strychnos arborea* (*H.*). Queensland. Pale oatmeal colour, lustrous. Very hard, heavy, and tough. Straight close grain, fine texture. Also called Sagowood and Streaked Birch. Similar to Boxwood and used for similar purposes. S.G. 1 ; C.W. 4.

Three-centred. Applied to an approximate elliptical arch in which the outline is built-up of circular arcs struck from three centres of curvature. Also called a false ellipse. It simplifies the shaping of the voussoirs. Another form is a modified Tudor arch.

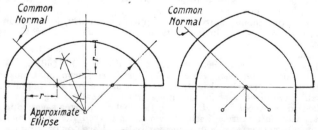

THREE-CENTRED ARCHES

Three-decker. Three stages or decks one above another. An old type of wood ship. An obsolete form of pulpit.

Three-four-five Rule. The application of trigonometry (Pythagoras' Theorem) for setting out a right angle. If the sides of a triangle are 3, 4, and 5 units long it is a right-angled triangle.

Three-light Window. A Venetian window, *q.v.* A window with two mullions forming three main divisions.

Three-men Boards. Applied to boards with transverse shakes or upsets because they require support at the middle during conveyance.

Three-way Strap. A wrought iron strap with three arms to tie together three members of a frame as at the apex of a king-post roof truss.

Threshold. The sill of an external doorway.

Throat. 1. The opening in a plane where the shavings escape. 2. The bottom of the space, or gullet, between two saw-teeth, *q.v.* 3. The end of a gaff next to the mast. The other end is called the peak.

Throating. A small groove on the underside of a projecting member, or between two surfaces nearly in contact, to prevent capillarity. They are formed on the under outer edge of window sills and transoms, and on the stiles and mullions of casement windows when the casements open.

Through and Through. The term used when converting logs by parallel cuts the full depth of the log.

Through Bill. See *Bill of Lading*.

Throwan or **Thrown Chair.** A chair built up chiefly of turned parts.

Throwing. 1. Felling trees. 2. Turning parts for chairs.

Thrust. The push exerted by an inclined compressive member in structural framing. The horizontal pressure of an arch at the crown and abutments, or of a strut.

Thuja. See *Thuya*.

Thujoideæ. A sub-family of Cupressaceæ. It includes the following genera : Actinostrobus, Callitris, Callitropsis, Diselma, Fokienia, Fitzroya, Librocedrus, Tetraclinis, Thujopsis, Thuja, Widdringtonia.

Thumb Cleat. See *Ship's Cleats*.

THUMB PLANES

(See p. 582.)

Thumb Latch. See *Norfolk Latch.*
Thumb Moulding. One consisting of a quarter ellipse and a small fillet and used on the edge of a projecting board as a table top. The outline resembles a thumb. See *Listel* and *Lining Up.*
Thumb Planes. Very small planes used for shaping mouldings by hand on circular work. (See illus., p. 581.)
Thumb Screw. A screw for securing sliding sashes. It is operated by finger and thumb. See *Plough* and *Pocket Screwing.*

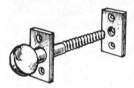

THUMB SCREW

Thumb Slot. See *Pocket Screwing.*
Thunder Shake. See *Shake* and *Upset.*
Thurming. Square turning, *q.v.*
Thuya. *Tetraclinis articulata* or *Callitris quadrivalvis.* N. Africa. Also called Thyine wood and Citron burl. Similar to *Thuja occidentalis.* Light chestnut brown. Hard and full of small knots, like Amboyna. Only obtainable as root burr, as the ordinary wood is not marketed in this country. The burr is highly decorative and only used as veneer. It is formed by stooling and by repeated destruction of coppice growth which produces increased growth below ground. It is costly and only used for decorative work. Also see *Western Red Cedar, White Cedar,* and *Arbor Vitæ.*
Thwacking Frame. An appliance or horse for bending pantiles.
Thwart. 1. A seat placed athwart a boat on which the rowers sit. A transverse seat. A sailing thwart is one in which the mast is stepped. **2.** The horizontal supports for the staging used when shipbuilding. They serve the same purpose as ledgers in a builder's scaffold.
Thwart Knee. A knee or bracket fixing the thwart to the side of a boat.
Thwart Marks. Sticker marks on wood after piling.
Thwartship. Across the breadth or beam of a ship.
Thyine Wood. See *Thuya.*
Thymelæceæ. The Leatherwood family. Includes Aquilaria, Brachythalamus, Daphnopsis, Direa, Funifera, Gyrinops, Lagetta, Linostoma, Lophostoma, Ovidia, Synaptolepis.
Tiamo. *Entandrophragma angolense, E. macrophyllum,* etc. Now called Gedu nohor. See *Sapele.*
Tiaong. *Shorea teysmanniana* (*H.*). Philippines. Red Lauan. Very similar to Tangile and usually marketed as such. Reddish, lustrous. Fairly hard, heavy, and strong, not durable. Cross grain, medium coarse texture. Seasons and works well. S.G. ·56. See *Tangile* and *Red Lauan.*
Tideland Spruce. Sitka, *q.v.*
Tie. 1. A tension member in a frame or truss. A piece used to prevent two other members from spreading or separating. **2.** A railway sleeper.
Tie Beam. The horizontal timber tying together the feet of the principal rafters of a roof truss. See *King Post, Roof Truss,* and *Pole Plate.*
Tie Bolt. Same as tie rod, especially when provided with a nut for increasing the tension.

Tied. Applied to a run for workmen to which cleats are fixed to prevent unequal sagging. See *Bridging Run.*

Tier. One of several stages built one above the other.

Tiercerons. Intermediate ribs in vaulting.

Tie Rod. An iron rod in tension. An iron bar used in place of a tie beam, or for tying together the parts of a structure.

Tier Pile. A pile of wide boards, single width only, arranged with sticks for air seasoning, *q.v.,* or a pile of logs.

Ties. Railway sleepers, *q.v.* See *Tie.*

Tige. 1. The shaft of a column. **2.** The stem or stalk of a plant.

Tiger Wood. *Machærium spp.* B. Guiana. Bright chestnut red, black spots. Hard, heavy, close grain. Used for decorative work. See *Negrillo.* Sometimes wrongly applied to *Nigerian Walnut* and *Zebrawood.*

Tight Cooperage. Casks used for liquids. See *Cooperage.*

Tight Knot. A sound firm knot. See *Knots.*

Tight Sheathing. See *Sheathing* (2).

Tight Side. A term applied to the concave side of knife-cut veneers which is in compression. The back or convex side often has slight ruptures and is called the loose side. The face of rotary cut veneers.

Tile Battens. Laths from $\frac{3}{4}$ to $1\frac{1}{4}$ in. thick and from $1\frac{1}{2}$ to $3\frac{1}{2}$ in. wide to which roof tiles are hung. See *Battens* and *Laths.*

Tile Pins. Oak pegs used instead of nails for fixing tiles, in chemical works, etc., where the chemicals would attack metal.

Tilia. See *Basswood, Estribeira, Lime,* and *Linden.*

Tiliaceæ. The Linden family. Includes : Apeiba, Belotia, Berrya, Brownlowia, Carpodiptera, Columbia, Diplodiscus, Echinocarpus, Elæocarpus, Goethalsea, Grewia, Heliocarpus, Luchea, Mollia, Pentace, Pityranthe, Schoutenia, Tilia.

Till. A drawer or receptacle to hold money in an office, shop, warehouse, etc. A drawer with bowl-shaped compartments.

Tillar. Same as tellar, *q.v.*

Tiller. 1. The fixing end of a frame saw blade. The other end is called the buckle. **2.** The lever fixed to the head of a rudder of a boat. It is used to turn the helm. **3.** A shoot springing from the root of a tree. A sucker.

Till Lock. A drawer lock.

Tilting Fillet. A strip of wood to tilt the bottom course of shingles or tiles at the eaves, etc., so that successive courses bed closely at the bottom. The fillet is usually triangular in section. See *Counter Battens, Eaves,* and *Parapet Gutter.*

Tilt, To. To incline. To set in a sloping position ; not horizontal or vertical.

Timah. Meranti, *q.v.*

Timber. Wood suitable for building and structural purposes, whether as standing trees, logs, or converted. The term is generally used in a wider sense and includes all kinds and forms of wood, especially when in bulk. There are numerous accepted meanings in this country according to the locality. In some countries the word has a statutory definition. Usually the term implies stuff of large section. Small stuff, especially when in the hands of the craftsman or machinist, or of a decorative character, is usually termed wood.   See *Lumber* and *Timbers*.

Timber Bob.   A two-wheeled vehicle for conveying timber.

Timber Bricks.   Wood blocks the size of a building brick built in a wall to provide a fixing for joinery, etc.   See *Nog*.

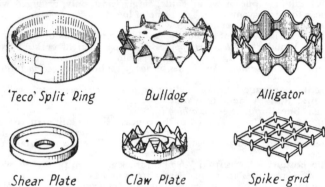

'Teco' Split Ring   Bulldog   Alligator

Shear Plate   Claw Plate   Spike-grid

TIMBER CONNECTORS

Timber Connectors.   Patented types of joint fasteners for heavy constructional work to prevent lateral movement and to simplify the arrangement of bolts and straps.   There are several types.   The

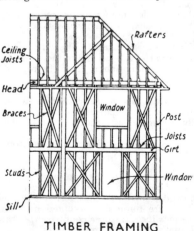

TIMBER FRAMING

Alligator is a corrugated ring with sharpened serrated edges. It is placed round the bolt and between the two timbers. When the bolt is tightened the toothed edges of the ring bed into the timbers and form a secure joint against lateral movement due to shear, compression, or tension. The rings are up to $4\frac{7}{8}$ in. diameter by $1\frac{1}{8}$ in. deep, for a 1-in. bolt. *Bulldog* joint plates are similar in action but they are rect-angular or circular plates pierced for the bolt, and bite into the wood easily. The edges are cut and turned up and down alternately at right angles to the plate in the form of sharp prongs. The *Teco*

584

connector is in the form of a split ring and is used on heavy constructional work. It requires a prepared circular groove. Other types are shear and claw plates and spike-grids.

Timber Dog. See *Dog.*

Timber Drier. See *Kilns.*

Timber Exchange. A place where those engaged in the buying and selling of timber gather together, and where fortnightly auctions take place of the stocks lying at the docks.

Timber Framed. Applied to a building in which timber is the only structural part of the building, except the foundations. The panels of lath and plaster, or other materials, only serve to keep out the weather ; or the outside may be boarded or shingled. See *Half-timbered.*

Timber Hitch. A hitch knot used in hoisting. See *Knots.*

Timbering. Temporary timbers used for supporting the sides of excavations. The requirements vary according to the depth, the firmness of the earth, and the pressure on surrounding areas.

Timber Measure. See *Standard, Board Measure, Hoppus* and *String Measure,* and *Market Forms.*

Timbers. 1. The variety of woods used in this country is constantly changing and is governed by political and economic conditions. The woods obtainable are described alphabetically in the Glossary, and many others are included that will probably appear on the market in the near future ; about 300 varieties are in common use. Over 5,000 kinds have been received at the Forest Products Research Laboratory for examination, and the number is increasing. Many woods, especially Asiatic, have a large number of vernacular names but only the better known names have been included or those likely to be used in this country. See *British Trees.* **2.** The framework ribs of a boat to which the planking

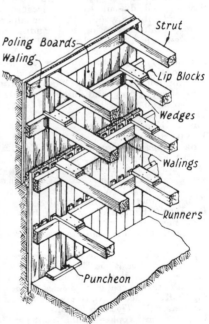

TIMBERING FOR
EXCAVATION

or skin is attached. They may be carried round from gunwale to gunwale, or they may spring from the keel on both sides. They are distinguished as square, cant, and knuckle timbers, and fashion-pieces. Other timbers are called floor, filling, top, and short-top timbers. See *Keel, Stem, Stern*.

Timber Slips. 1. Slides for conveying timber from forest to saw-mill. 2. See *Pallet*.

Timber Sword. A flexible steel implement with a slot at the point to insert a tape. It is used to pass the tape under a log when measuring the log.

Timbo. *Enterolobium spp.* (*H*.). Tropical America. A cheap sub-stitute for cedar. Not imported.

Time Charter. A charter party, *q.v.*, under which a vessel is supplied for a stated period.

Tindalo. *Pahudia rhomboidea* (*H*.). Philippines. Also called Oro. Pale orange to dark red with age, slightly odorous. Hard, heavy, strong, and durable. Straight grain, some crossed, fairly fine texture. Seasons well, difficult to work because of hardness. Polishes well. Excellent cabinet wood. Used for superior joinery, musical instru-ments, furniture, flooring, tool handles, etc. S.G. ·85 ; C.W. 4.

Tingle-tingle. Stringy bark, *q.v.* RED TINGLE-TINGLE. *Eucalyptus Jacksoni* (*H*.). W. Australia. Red. Very similar in characteristics and quality to Jarrah. An excellent wood for structural and con-structional work, and in increasing demand for cabinet work and superior joinery. YELLOW TINGLE-TINGLE. *Euc. Guilfoylei*. Very hard, heavy, strong, and durable. Dense but fissile. Straight grain. Similar in quality to Tuart, *q.v.* S.G. 1 ; C.W. 4.

Tinsa. Sandan, *q.v.*

Tinya. Khasia pine, *q.v.*

Tinyu. Beefwood, *q.v.*

Tipa. *Tipuana speciosa* (*H*.). Tropical America. Yellowish to reddish brown, rosy hue, smooth, lustrous, fine stripes. Moderately hard, heavy, and strong, but not durable. Fine grain and texture, roey. Fairly difficult to work. Used for cabinet work, carriage work, etc. S.G. ·65 ; C.W. 4. TIPA COLARADO. *Machærium pseudo tipa* (*H*.). S. America. Similar to above.

Tip Bins. Hinged or swivelled bins to kitchen cabinets, to hold flour, sugar, etc.

Tip-cart. One that can tip over backwards or sideways to deposit loose materials. The body is swivelled to the shafts or chassis

Tipped Saw. Circular saws with teeth of tungsten-carbide alloy or carbaloy. See *Widia Saw*.

Tipple. A wood or steel structure near a mine to which the tubs are hauled to tip the coal into vehicles, barges, or railway trucks.

Tip Staff. A decorative rod used as a badge of office.

Tipuana. See *Tipa* and *Picho*.

Tire. A term applied to the front part of a band saw blade, when referring to the tension in the saw.

Titoki. *Alectryon excelsum* (*H*.). New Zealand. Light red. Hard, heavy, strong, tough, and elastic. Straight grain. Used for wheel-wright's work, carriage building, shafts, tools, etc.

T.L.O. Abbreviation for *total loss only*.

Toads Back. Applied to a handrail with a flat curve on the upper surface.

Toat. Same as tote (1), *q.v.*

t.o.b. Abbreviation for *tape over bark*.

Toboggan. A small sledge for pleasure coasting down snow-covered slopes.

Tobros. Guanacaste, *q.v.*

Tochi. *Æsculus turbinata* (*H.*). Japanese Horse Chestnut. Golden brown. Very similar to the English variety. Only figured and mottled wood is imported.

Toddalia. See *Ironwood* (*White*).

Toe. 1. The bottom of the shutting edge of a door. 2. The front end of a plane. See *Heel*. 3. The foot of a strut. See *Housing Joints*.

Toe Board. A scaffold board standing on edge, as a protection. See *Saddle scaffold*.

Toe Nailing. Skew nailing the foot of a post or strut.

To-fall. Same as lean-to, *q.v.*

Toggle. 1. A short piece of hardwood with two holes for fixing the return end of a rope in any required position. Also called a bowser. 2. A short bar or pin placed through the eye or bight at the end of a rope, or through an end-to-end joint.

Toggle Chain. A short chain with ring and toggle hook for regulating the binding chain round logs.

Toggle Joint. A knee or elbow joint. An end-to-end joint.

Toilet Glass. A mirror for a dressing-table.

Toilet Seat. A W.C. or closet seat.

Toilet Table. A dressing-table.

Tolerance. 1. An amount less than the quoted dimensions allowed in certain specifications, *i.e.*, ⅛ in. scant on occasional pieces in thickness, ¼ in. in width, and 2 in. in length. If the scant exceeds the above in width and thickness the pieces are tallied in the next inch class, and for length in the next foot class. 2. See *Limits*. 3. See *Clearance*.

Toll Gate. A gate closing a road, bridge, etc., over which a person or authority has right-of-way and the privilege of making a charge for the use of the road or bridge. Also called toll-bar or toll-house.

Tolu. Saqui-saqui, *q.v.*

Toluifera. See *Myroxylon* and *Oleo Vermelho*.

Tomahawk. 1. A worn plane iron, with handle attached, for scraping off glue and paper from veneering. 2. An American pattern of axe.

Tompian. Same as tampion. The plug for a flute.

Toms. Horizontal grounds for panelling on a ship.

Ton. Usually implies 42 cub. ft. in shipping, but varies between 40 and 60 cub. ft. for timber according to the kind of wood and whether round or converted.

Tonawanda Pine. *Pinus strobus*. See *White Pine*.

Tonca Bean. *Dipteryx odorata* (*H.*). C. America. Brownish, variegated streaks, feels waxy. Very hard, heavy, strong, and durable. Interwoven grain, roey, fine texture. Very difficult to work. Polishes well. Used for cogs, fancy articles, decorative work. A substitute for Lignum-vitæ. S.G. 1·2 ; C.W. 5. Also includes *Coumarouna spp.*

Tondino. A round moulding ; like a ring. A hollowed platter with a wide flat rim.

Tongs. A pair of pivoted bent levers for gripping a log when hoisting with lifting tackle.

Tongue. A reduction of the thickness of the edge of a board by rebating opposite faces. The tongue fits into a groove in an adjacent board to strengthen the joint, keep the faces flush, and to avoid an open joint in the event of shrinkage. A *loose tongue* serves the same purpose but is a long thin strip of wood or metal to fit in grooves in adjacent boards. *Double tongues* are used in cabinet work, for airtight work and canted corners. The joint is similar to the hook joint, *q.v.* Also see *Angle Joints* and *Floor Boards*.

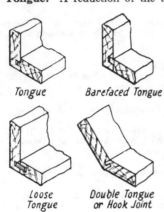

Tongue   *Barefaced Tongue*

Loose Tongue   *Double Tongue or Hook Joint*

TONGUES

Tongue and Groove. A term used to describe flooring, matching, etc., in which one edge is worked with a tongue and the other edge with a groove. See *Flooring*.

Tonk's Fittings. Adjustable metal bearers for book shelves, etc. See *Shelf Fittings*.

Tonneau. The body or rear part of certain types of vehicles.

Tonquin. Plain wood of *Chloroxylon swietenia* or E. Indies Satinwood.

Tooart. See *Tuart.*

Tool Chest. A specially designed box in which tools are kept.

Tooling. Simple carving with gouge or vee-tool.

Tool Marks. A defect graded as imperfect manufacture, *q.v.*

Tool Pad. A combination tool consisting of interchangeable small bits fitted into the handle by means of chuck and jaws.

Tools. The numerous hand implements used by the craftsman to facilitate mechanical operations. The tools used in woodworking are described alphabetically. A woodworker's kit is expensive, but good quality tools are necessary and only the best should be purchased. Protection and care will increase their life and produce better work. Saws should be

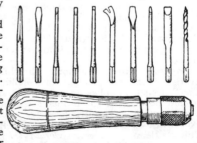

TOOL PAD

sheathed and brace bits kept in a baize, moleskin, or leather roll. Planes should be oiled and the irons knocked back when not in use. The edges of chisels and gouges should be protected, and the oilstone kept covered. A box or tool chest provides better protection than a bass or bag. All metal tools should be protected against corrosion.

Toon. *Cedrela toona (H.).* Indian or Moulmein Cedar. Mahogany family, and resembles a soft mahogany in colour, texture, and general characteristics. Light brick red to rich reddish brown, darker streaks, beautifully veined, smooth, and fairly lustrous. Moderately light, soft, and durable. Fragrant and stable, but rather brittle and not strong. Resists insects and easily treated. Straight grain, rays. Fairly close, rather uneven texture. Trees should be girdled and logs quickly converted to avoid warp and shakes. Should be kiln seasoned and allowed to condition. Easily wrought and polishes well. Used for furniture, superior joinery, cabinet work, carriage work, musical instruments, cigar boxes, toys, etc. S.G. ·56 ; C.W. 2·5. *C. serrata* is similar to above but stronger. See *Mahogany*.

Tooth. See *Saw Teeth*.

Toothing Plane. One with a cutting iron grooved on the back to give a serrated cutting edge. It is used to prepare the groundwork for veneers to provide a key for the glue.

Tooth Ornament. An early Gothic moulding consisting of a series of small pyramids on a member of a large moulding. See *Dog Tooth*.

Top. 1. A child's toy, usually pear-shaped, that rotates on a metal point and is propelled by a whip. 2. A form of platform round the lower mast head, like a scaffold.

Top Beam. A collar beam, *q.v.*

Top Carriage. The upper part of the fore carriage and wheel plate of a vehicle.

Top Cutting. 1. Cutting up the tops of felled trees into cordwood, etc. 2. Cutting the top end of a log.

Topgallant. The third mast above the deck. It is above the top-mast and below the royal-mast. See *Masts*.

Top Hung. Applied to casements that are hinged on the top edge. They open outwards by means of a quadrant or stay.

Top Lighting. Any form of lighting from overhead : lantern, sky-light, etc. Also see *Clerestory*.

Top Log. The uppermost log from the bole of a tree.

Topman. A top sawyer, *q.v.*

Top Masts. The top masts of a ship include fore, main, mizzen, and sprit. See *Tops*.

Topping. Sharpening the tops of the teeth of a circular saw by hand filing.

Topping Plane. A cooper's plane. It is smaller than a jack plane, with a curved body but flat sole.

Top Rail. The top horizontal member of a piece of framing.

Tops. 1. The arrangement of the top of the main mast to receive the foot of the top mast, *q.v.* 2. The head of a felled tree useless for timber.

Top Sawyer. The one who takes the upper position in a saw-pit, *q.v* .

Top Sides. Upper part of a ship's sides above water level.

Torch. Bastard torch, *Ocotea spp*. Black torch, *Ocotea* and *Erithalis spp*. White torch is Amyris wood. Yellow torch, *Exostema spp*.
Torchére. A stand for a candlestick. A large candelabrum.
Torchwood. Applied to the Burseraceæ family and to several tropical American woods, from trees from which oil and resin are extracted. Especially applied to *Amyris balsamifera*. See *Sandalwood*.
Torn Grain. A defect due to the tearing away of the fibres, by plane or cutters, below the dressed surface. See *Chipped Grain*.
Torque. A twisting or turning force.
Torr. Abbreviation for the name of the botanist Torrey.
Torrak. Wood sawn from dead or over-mature trees.
Torresea. See *Umburana*.
Torreya. See *Kaya* and *Stinking Cedar*.
Torsels. 1. The pieces of wood that tie under the ends of a mantle-tree. 2. A twisted scroll. 3. A template.
Tortoise Shell Wood. See *Bocote* and *Letterwood*.
Torus. 1. A large semicircular moulding or bead, usually with a fillet, and used for the moulding on a skirting board. See *Attic Base*. 2. The central thickened part of the pit membrane of a bordered pit.

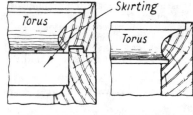

Tosh Nailed. Same as skew-nailed, *q.v.*
Totara. *Podocarpus totara* and *P. Hallii* (*S.*). New Zealand. Reddish brown, lustrous, smooth. Resembles pencil cedar. Light, fairly soft, strong, very durable, stable, rather brittle. Straight, close grain. Easily seasoned and wrought. Figured wood and

TORUS MOULDING

burrs used for veneers. Used for all purposes : structural work, joinery, sleepers, paving, flooring, furniture, pattern making, poles, etc. S.G. ·5 ; C.W. 2.
Tote. 1. The handle of a bench plane : jack- and try-plane. 2. Abbreviation for totalisator, an automatic betting machine on race-courses, etc.
Touchwood. Soft white decayed wood, due to the action of certain fungi. It is easily fired and used as tinder.
Toughness. A property of wood that enables it to bend considerably without breaking.
Toupie. A solid cutter for a spindle machine for shallow mouldings. The cutter is circular with the edge turned to the contour of the required moulding, and slots made to form cutting edges.
Touriga. *Calophyllum touriga* (*H.*). Queensland. Brown Touriga. Indian red. Very hard, heavy, strong and durable. Interwoven grain, dense fine texture, visible darker rays. Too hard for ordinary purposes. Used where strength and durability are of primary importance. S.G. ·95. Also see *Red Touriga*.
Tow. A number of rafts of wood towed together.

Towanero or **Toweroenierou.** Bastard Bullet Wood, *q.v.*
Towel-horse. A wood frame on which towels, etc., are hung.
Tower. 1. The tall supports for a derrick crane. See *Derrick Tower*.
2. A tall building in the form of a shaft. The plan may be circular, square, or polygonal.
Tower Bolt. A fastening in the form of a sliding cylindrical bolt. The bolt is controlled by keepers riveted to a back plate.

Tower Gantry. A derrick tower, *q.v.*
Towhai. *Weinmannia racemosa* (*H.*). New Zealand. Deep red. Hard, heavy, and strong. An ornamental wood used chiefly for cabinet work. S.G. ·72. Also see *Rose Marara*.
Toxylon. See *Osage Orange*.
Trab. A beam or wall plate.

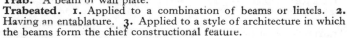

TOWER BOLT

Trabeated. 1. Applied to a combination of beams or lintels. **2.** Having an entablature. **3.** Applied to a style of architecture in which the beams form the chief constructional feature.
Trabeation. Same as entablature, *q.v.*
Trace. 1. The line of intersection between two planes in geometry. **2.** The point where a line intersects a plane. See *Planes and Traces*.
Tracery. Mullions and bars arranged in ornamental geometrical

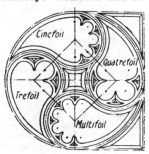

TRACERY PANEL

formation as in Gothic windows, panels, etc. The ornamentation is obtained by flowing lines and foliation. See *Cusps, Foils, Bar Tracery, Cincfoil, Plate Tracery*, and *Chancel Screen*.
Tracheides. Wood cells with bordered pits, commonly called wood fibres. The narrow vertically elongated single type of cells forming the main portion, about 90 per cent., of the wood of conifers. They correspond to the fibres in hardwoods and occur in the wood of certain deciduous trees. Tracheides are long tubular cells, usually about ⅛ in. long, with rounded ends and with pits in their side walls forming means of communication with adjacent tracheides. Thin-walled tracheides are chiefly concerned with the conduction of sap and thick-walled tracheides with strengthening the structure of the wood. Thin- and thick-walled tracheides occur in alternating zones : spring and summer wood. See *Softwoods*.
Trachelium. The neck of a column.
Tracing. A copy of a drawing obtained by tracing or inking over the original on transparent paper or linen so that prints may be obtained from the tracing. See *Printing* and *Blue Print*.
Trackers. Part of the mechanism of a pipe organ between key and pipe. See *Pedal* and *Pallet*.
Track Spike. Heavy, specially made spikes for securing railway lines for temporary work instead of using chairs.

Tractor. Applied to the propellor of an aircraft in front of the engine.
Trade Names. See *Names of Woods*.
Trail. A continuous ornamental band of leaves, tendrils, etc., in carving and plaster work. See *Vignette*.
Trail Board. A carved piece between the cheeks to the knee of the head of a ship.
Trailer. **1.** A vehicle without motive power for attaching to another vehicle. **2.** A plant that requires vertical support.
Trailing Edge. A light piece of wood forming the back edge of the wing of an aeroplane.
Tram. **1.** A piece of wood forming a track or roadway, especially in mines. A short sleeper. A beam or log. **2.** The shaft of a cart.
Tram Car. **1.** A four-wheeled wagon used in coal mines and running on rails. **2.** A large public conveyance running on a track or tramway.
Trametes. Fungi, especially *Trametes pini*, very destructive to the heartwood of conifers. It produces decay known as red ring or pocket rot.
Trammel. **1.** A beam compass, *q.v.* **2.** An appliance for drawing elliptical curves. It consists of two intersecting grooved pieces controlling a beam compass that describes the ellipse, *q.v.*

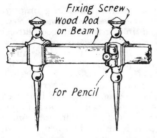

TRAMMEL

Transept. The transverse part of a church built in the shape of a cross. The short arm of the cross forming a right angle with the nave.
Transom. **1.** A strengthening intermediate horizontal member of a frame or opening between head and sill. See *Window* and *Door Frame*. **2.** The horizontal members of the projecting stern of a boat. The stern piece of a vessel with a square or rounded aft end. They are distinguished as *main* or *half* transom and transom knee. Boards forming the end of the boat. They are secured to the sternpost and give shape to the stern. Other terms used are deck, wing, and helmport transom. See *Stern Timbers*. **3.** The top bed carrying the perch bolt of a vehicle. **4.** A beam across a saw-pit.
Transom Knees. The knees at the stern of a ship springing from the transom.
Transom Window. A window above the transom of a door. A fanlight.
Transporter Mast. A mooring mast fixed on a movable platform.
Transverse Section. A cross section. A section at right angles to the grain of the wood.
Transverse Shake. See *Rupture* or *Upset*.
Trap. **1.** A scaffold projecting over a putlog sufficiently to be a source of danger. **2.** An appliance used in the game of ·trap-ball for throwing the ball into the air. **3.** A horse-carriage with two wheels.
Trap Door. A door in a horizontal surface, as in a ceiling or floor.

Trapezoid. A quadrilateral with unequal sides but with two opposite sides parallel. *Trapezium.* A kite-shaped figure. A quadrilateral with no parallel sides. These definitions are reversed in mathematics.

Trap Tree. A tree deadened or felled to attract bark beetles. The bark is then destroyed. The object is to protect growing trees.

Traumatic Canals. Resin canals in long or short rows parallel to the growth rings and due to injury to the growing tree.

Travel. A term by craftsmen to indicate horizontal distances. The horizontal distance between the ends of an inclined member.

Traveller. The trussed beams carrying a travelling crane or movable winch. The beams are supported on strong rails and posts and used for transporting heavy goods from one part of a works or goods yard to another.

Traveller Gantry. A strong staging carrying a traveller, which in turn carries the lifting gear.

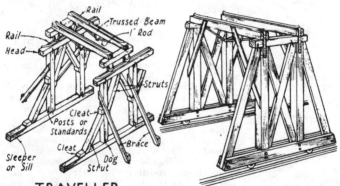

**TRAVELLER
GANTRY**   **TRAVELLING GANTRY**

Travelling Cradle. A swinging scaffold or cradle suspended from a track projecting over the top of a building. It can be moved along the track, and raised or lowered as required. See *Suspended Scaffold.*

Travelling Crane. See *Traveller.*

Travelling Gantry. A gantry mounted on wheels so that it can move as a whole along rails.

Traverse. 1. A curtained screen, for privacy. 2. See *Traversing.*

Traverser. A movable platform in large garages for manipulating the cars.

Traversing. Planing a wide board across the grain at an angle of about 45 degrees. Anything laid, or built, or worked crosswise.

Trawl Board. A wood *kite* used with certain types of trawl nets to keep the mouth of the net open.

Trawl Bobbin. Large wood bobbins attached to trawl nets so that the net passes easily over the bed of the sea.

Trawler. A ship using a trawl net for fishing.

Tray. 1. A flat surface with projecting rims and handles for conveying food, crockery, etc. A salver. See *Waiter*. A shallow

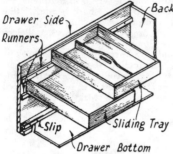

Drawer Side

Runners

Slip   Sliding Tray

Drawer Bottom

TRAY FITTINGS TO DRAWER

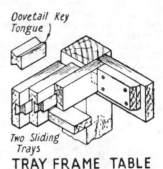

Dovetail Key Tongue

Two Sliding Trays

TRAY FRAME TABLE

drawer for displaying goods, such as jewellery, confectionery, etc. A shallow lidless box. 2. A temporary fleak, *q.v.*

Tray Frame Table. An extension dining-table in which each extending section is a framed tray. The rails are tied together by means of dovetail key tongues adjusted for easy sliding.

Treacle Moulding. A projecting thumb moulding or nosing to a hinged cover with a deep hollow in the under surface like a continuous finger hole.

Tread. The horizontal or top surface of a step. See *Handrail*, *Step*, and *Going*.

Treadle or **Treddle.** The part of a machine operated by the foot to provide the motive power, as to a lathe, loom, sewing machine, etc.

Treatment. See *Preservation*.

Tredgold Joint. See *Halving Joints*.

TREACLE MOULDING

Tree. 1. The name applied to numerous wood parts of machines, especially if in the form of a cross or support. 2. A specially shaped adjustable wood block for keeping footwear in correct shape. Also see *Axle*, *Boot*, *Cross*, *Ridge*, *Roof*, *Saddle*, *Spade*, *Trestle* and *Whipple Tree*, and *Trees*.

Tree Class. A classification of trees according to size : seedling, shoot, small and large sapling, small and large pole, standard, veteran.

Tree Felling Machine. A portable cross-cut saw for cutting through the trunk of a tree by power.

Treen. Applied to a collection of objects made of wood.

Tree-nail. A hardwood pin, usually of cleft oak, to secure a tenon or to fix laminations together, or to secure the planks of a ship's side

Tree of Heaven. *Ailanthus glandulosa* (*H.*). Bitterwood family. N. China and E. Indies. The leaf is too valuable for silk-worm cultivation for the wood to have any commercial value. It resembles ash but yellowish and coarser.

Trees. Perennial plants with self-supporting trunk and branches and from which wood is obtained. Trees are divided into two big classes : Angiosperms and Gymnosperms. The former is subdivided into Monocotyledons (palms, bamboos, etc.) and Dicotyledons which provide the *hardwoods* of commerce. Gymnosperms are subdivided into Cycadaceæ, Gnetaceæ, and Coniferæ. The last-named family provides the *softwoods* of commerce. There are over 30,000 woody species, providing every variety of wood, of which about 5,000 are in common use, and thousands of square miles of unexplored forests. See *Classification, British Trees, Hardwoods, Selection, Softwoods, Timber*.

Treetex. Proprietary wood-fibre wallboards.

Trefoil. A tracery foliation consisting of three foils and cusps. See *Tracery*.

Trefoil Arch. One consisting of four segments arranged on an equilateral triangle.

Treilage. The framework of an espalier, *q.v.*

Trellis. An espalier, *q.v.* Wood framing with open panels formed of thin narrow laths or wire. The laths cross each other, usually diagonally, to form lattice work, *q.v.* See *Interwoven Fence* and *Mesh*.

Trembling Aspen. *Populus tremuloides*. See *Aspen* and *Poplar*.

Tremie. A long box-shaped funnel for depositing plastic concrete under water. It is traversed over the required site by a crane so that the concrete is deposited in the required position.

TREFOIL ARCH

Trenail. See *Tree-nail*.

Trench. 1. A long narrow excavation for drains, foundations, etc. 2. A long narrow rectangular or dovetailed groove across the grain of a piece of wood. See *Housing Joints*.

Trencher. A wood platter. A bread board. See *Woodware*.

Trenching. Forming trenches.

Trench Timbering. The temporary timbering used to support the sides of an excavation. See *Timbering*.

Tresaunte. A passage in a large house between hall and offices. A part of the cloisters of a monastery.

Trestle. 1. A bearer with four splayed legs braced together and used for scaffolding, etc. A scaffold trestle is usually about 5 ft. high for working up to ceiling height. 2. An open braced frame of heavy timbers for supporting a bridge, viaduct, etc. The frame often consists of round timbers called sticks.

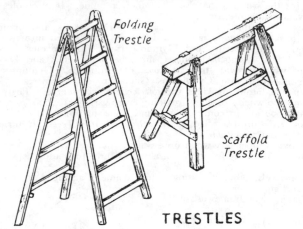

Folding Trestle

Scaffold Trestle

TRESTLES

Trestle Bridge. One supported by trestles.

Trestle Table. One consisting of loose planks or ledged boards supported by trestles.

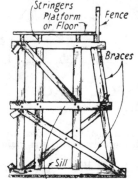

Stringers Platform or Floor

Fence

Braces

Sill

Intertie Sleepers Half Elevation

TRESTLE OR BENT FOR BRIDGE

Trestle Trees. Horizontal pieces on the lower mast of a ship to support the top mast.

Trewia. *Trewia nudiflora* (H.). India. Also called Gamari, *q.v.* Pale brownish grey, smooth. Light, soft, not durable. Subject to fungal stain and insect attack. Straight grain, medium texture. Easily wrought but requires careful seasoning. Used for packing cases, matching, carving, matches. S.G. ·35 ; C.W. 2.

Triangle. A three-sided plane figure. Triangles are named according to the sides : *equilateral* (all sides equal), *isosceles* (two sides equal), *scalene* (sides unequal) ; or according to the angles : *right angled* (90 degrees), *acute angled* (all angles less than 90 degrees), or *obtuse angled* one angle greater than 90 degrees). Area$=\frac{1}{2}$ base $\times$ perpendicular height.

Triangle of Forces. A law in mechanics which enables the designer to obtain the stresses acting in the members of a structural frame, graphically. " If three forces acting at a point are in equilibrium, then the forces can be represented in magnitude, direction, and sense, as the sides of a triangle, taken in proper order." See *Graphic Statics*.

Triangulation. **1.** Arranging the members of a frame or truss in triangles to build up a structurally *perfect* frame, which allows for accurate and rigid design. **2.** A method of surveying by dividing the site into a number of triangles for the application of trigonometry.

Tribune. A pulpit, rostrum, or platform for speech-making. A raised dais for certain officials.

Trichia. See *Bejuco*.

Trichilia. See *Cape Mahogany* and *Pimenteira*.

Triforium. A gallery or arcade above the arches of the nave of a church. When it is without windows it is also called a blind-storey.

Triglyph. A block repeated at regular intervals in the Doric frieze. It is ornamented with two full and two half grooves or flutes. See *Doric*.

Trigonometry. Mathematics dealing with the relations between the sides and angles of triangles.

Trim or **Trimming.** **1.** Straightening or squaring the end of a board or moulding. Correcting the mitred end of a moulding. **2.** Framing an opening in roof, floor, or ceiling timbers.

Trimmed Joist. A floor joist supported at one or both ends by a trimmer. See *Hearth*.

Trimmer. **1.** The joist carried by two trimming joists and in turn carrying the trimmed joists. The arrangement is to trim an opening in a floor for a staircase, fireplace, etc. See *Hearth* and *Tusk Tenon*. **2.** See *Trimming Machine*.

Trimmer Arch. An arch between trimmer and chimney breast to support the hearth.

Trimming Joist. The larger joists in a floor carrying the trimmer. see *Hearth*.

Trimming Machine. One for preparing the mitred end of a moulding or the squared end of a board. It may be operated by treadle or by lever handle. Also

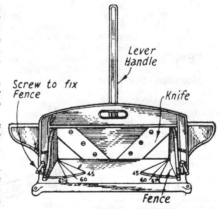

TRIMMER

called a trimmer, mitring machine, or guillotine, *q.v.*

Trims. Prepared woodwork for the finishings of buildings, such as door and window linings or casings, architraves, etc. See *Buck*.

Trincomalee. *Berrya amomilla* (*H.*). India. Also called Trincowood and Halmalille, etc. Dark red or brown, darker lines. Dull, feels greasy. Very hard, heavy, tough, and durable ; strong. Inter-locked grain, fine even texture. Requires careful seasoning to avoid surface cracks. Log seasoning perhaps the best. Rather difficult to work, especially to saw ; polishes well. Used for carriage construction, bent work, spokes, implements, furniture, superior joinery, etc. S.G. ·95 ; C.W. 4.

Tring. The name applied to two varieties of mangroves ; *Bruguiera* and *Rhizophora*, from Malay. They are both hard, durable woods but not exported to this country. See *Mangroves*.

Tringle. **1.** A curtain rod. **2.** A small fillet over the Doric triglyph. **3.** Any small square ornament or member, as a reglet, listel, platband, etc.

Trinidad Cedar. Central American cedar. See *Cedar*.

Triolith. A grade of Wolman Salts, *q.v.*, applied under vacuum and pressure or by soaking. It gives protection against fungi and slight fire-resistance. It is intended for interior timbers.

Trip. **1.** A mechanism to produce instantaneous release of a machine part. **2.** Any mechanical device for transferring the stuff from the live rollers behind the saw, after conversion, towards the machines.

Tripartite. Divided into three parts.

Triplane. An aeroplane with three supporting planes or wings one above the other.

Triplaris. The Ant tree, *q.v.* The wood is of little commercial importance.

Triple or **Tripod Mast.** A very tall mast built in the form of a slender tripod. They are usually built of scarfed timbers and may be over 100 ft. high, as used for electric power distribution before the introduction of the Grid Scheme pylons.

**TRIPLE
TENON**

Triple Tenon. A three-tenon joint sometimes used on thick framing.

Triplochiton. See *Ayous* and *Obeche*.

Triplochitonaceæ. A family including Mansonia and Triplochiton.

Tripod. A support consisting of three legs secured together at the top and the feet spread apart. A three-legged stool. See *Triple Mast*.

Tripod Table. One with a central pillar and three branched feet. See *Pillar Table*.

Triptych. Same as diptych, *q.v.*, but with three leaves.

Tristania. See *Brush Box, Kanooka, Swamp Box*, and *Pelawan*.

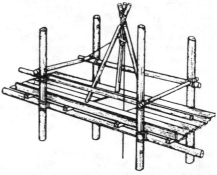

TRIPOD

Trochilus. A scotia, *q.v.* A concave moulding the outline of which is built up of two circular arcs of different radii.

Trolley. A small truck or narrow cart with four small wheels often flanged to run on rails. Similar to a bogey, but the body is fixed to the ends of the axles. Also see *Tea Trolley*.

Trompillo. *Guarea trichilioides (H.).* C. America. Resembles mahogany and Spanish cedar, and used for similar purposes. Pinkish grey to mahogany colour, occasionally streaked, lustrous, fragrant. Fairly hard, heavy, strong, and durable. Straight grain, roey, medium texture, faint rays. Used for cabinet work, superior joinery, etc. Also see *Marinheiro*. S.G. ·54 ; C.W. 3. The name is applied to other genera.

Tropical Cypress. Cypress pine, *q.v.*

Trough. A long narrow channel for conveying liquids. A spout for conveying water. A large vessel for holding water or foodstuffs for animals.

Trough Gutter. A parallel or box gutter, *q.v.*

Trow. A vessel with flat stem, usually ketch-rigged, used on the River Severn.

Truck. 1. A wood ball or disc at the top of a mast or flagstaff, pierced for the halyards. Lignum-vitæ beads strung on a rope for a parrel. *Shroud* trucks are short cylindrical pieces bored lengthways, grooved down the sides and round the middle. *Flag-staff* trucks are flat, round pieces with a small sheave on each side. 2. A small wood wheel or cylinder. 3. A type of barrow with two low wheels for conveying heavy packages. 4. An open wagon. A railway wagon.

Truckle Bed. A small, very low bed that can be wheeled under a larger bed.

True Cedar. Lebanon cedar. See *Cedar*.

Trueing Up. Straightening the surfaces of wood by hand. Preparing a plane surface.

True Service Tree. *Pyrus sorbus.* See *Service Tree*.

True Shape Section. A templet or section mould for a radial section of a circular moulding, etc., or for a right-angled section of a raking moulding.

True Wood. Heartwood, *q.v.*

Trug. 1. A wood pail. 2. A basket formed of very thin strips of wood.

Trumeau. A pier glass ; or a pier.

Trumpet Arch. An arch in a thick wall with conical-shaped head. A semicircular arch with splayed jambs and head.

Truncated. Applied to a solid when the top is cut away. See *Frustum*.

Truncheon. A wood club or cudgel. A short staff or baton of authority. Also applied to a tree when the branches have been lopped off, which produces rapid growth.

Trundle. 1. A small carriage or truck with low wheels. 2. A pinion, or small wheel, or castor. 3. See *Truckle Bed*.

Trunk. 1. The stem of a tree. 2. A large rectangular wood pipe or tube. 3. A trough to convey water. A flume. 4. A box or chest. 5. The shaft of a column.

Trunk Hatch. A shelter to the saloon stairs on board ship.

Trunnel. Same as tree-nail, *q.v.*

Trunnions. The projecting supports about which anything rotates.

Truss. 1. A rigid triangulated framework of wood or iron, designed and arranged to transfer the loads acting on the frame to the supports. See *Roof Truss*. 2. An ornamental projecting head to a pilaster. They often form part of a shop front to cover the ends of the fascia and blind. See *Rolling Shutter*. 3. A bracket or console. A large corbel or modillion. See *Ancone* and *Console*.

Trussed Beam. A compound wood beam usually consisting of a cast-iron saddle under the middle with tension rods fixed at the saddle and at the top of the ends to resist bending. See *Traveller Gantry*.

Trussed Centre. A framed centre.

Trussed Partition. A framed partition. It is usually supported at the ends only and often carries a floor and ceiling. See *Partition*.

Trussed Rafter Roof. A wood roof with the collar of each couple stiffened by braces. A roof truss consisting of two trussed rafters tied together by a horizontal bar. A scissors truss, *q.v.*

Trusses. 1. See *Truss*. 2. Fittings to keep the middle of the lower yards to the mast of a ship.

Truss Hoops. Iron or ash hoops used in cooperage.

Trussing. 1. The timbers that form a truss. 2. Triangulating a frame to make a self-contained truss.

Try-plane. The largest of the woodworker's bench planes. It is used for trueing-up or straightening plane surfaces, jointing edges, etc. The body is about 23 in. long by 3¼ in. square, and it has a closed handle. Otherwise it is like a jack plane.

Tso. Chinese oak.

T-square. A tee-shaped instrument used for drawing straight horizontal lines in mechanical drawing and as a base for the set square.

Tsuga. See *Hemlock*.

Tualang. *Koompassia parvifolia* and *K. excelsa* (*H.*). Philippines, Malay. Red, darker streaks, darkening with age. Variable in weight, very hard and fissile. Not durable and subject to insect attack and dry rot. Irregular coarse and crooked grain, fine rays, ripple, very unequal texture. Not appreciated because of uneven quality and difficulty of working. Selected figured wood used for furniture, etc. *K. excelsa* is much heavier, more uniform, and better wood. S.G. ·7 to 1·1 ; C.W. 4.

Tuan. Chinese lime.

Tuart. *Eucalyptus gomphocephala* (*H.*). W. Australia. Cream colour. Very hard, heavy, dense, strong, and durable. Elastic and stable, tough and difficult to split. Close, interlocked, and some curly grain, distinct rays. Large sizes. Used for wheelwright's and wagon work, bobbins, turnery, etc. Home demands prevent export. S.G. 1·1 ; C.W. 4·5.

Tub. 1. An open wood vessel or cask for liquids. 2. A clumsy or badly designed boat.

Tubbing. A continuous watertight lining to a mine shaft.

Tuck. The position where the ends of the bottom planks of a ship are collected together, just under the counter or stern.

Tudor. Applied to buildings, furniture, etc., characteristic of the

transitional style of architecture prevailing about 1500 to 1603 A.D.
The chief characteristics were the depressed or four-centred pointed
arches, carved embattled mould-
ings, pendants, panelled walls,
fan tracery vaulting. It includes
the Elizabethan style and is
sometimes called late Perpendicular.

Tudor Flower. A trefoil ornament characteristic of the Tudor
style of architecture. It was
used in continuous formation as
a crest on cornices, etc.

Tudor Rose. A decorative
feature of the Tudor style.

Tufboard. Proprietary wood-
fibre wall-boards.

Tug. A metal hoop from which
lifting tackle is suspended.

Tug-boat. A steam boat used
for towing other vessels.

Tulip Cedar. *Melia azedarach*
(*H*.). Queensland. Also Indian
Persian lilac. Buff, variegated
streaks. Light, soft, not durable, or strong, or stable.
Difficult to season. Fairly difficult to work and polish due to
woolly texture. Used for decorative cabinet work, fancy articles,
superior joinery, etc. S.G. ·5 ;
C.W. 3·5. See *Lilac*.

Tulip Lancewood. See *Tulipwood*.

Tulip Oak. See *Red Tulip Oak*.

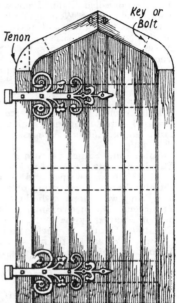

TUDOR DOOR

Tulip Plum. *Pleiogynium solandri* (*H*.). Queensland. Variegated
browns. Hard, heavy, strong, tough, durable. Close straight grain,
fine texture. Difficult to season and work,
polishes well. Scarce. Decorative wood
used for fancy articles, turnery, walking
sticks, etc. S.G. 1 ; C.W. 4.

Tulip Poplar. *Liriodendron tulipifera.*
Canary Wood, *q.v.*

Tulip Satinwood. *Rhodosphæra rhodan-
thema* (*H*.). Queensland. Variegated
browns, lustrous, speckled. Fairly heavy
and soft. Not strong, or stable, or durable ;
fissile. Straight grain, bird's eye, silky ray

TUDOR ROSE

grain. Not difficult to season or work, polishes well, small sizes.
Decorative wood used for cabinet work, turnery, inlays, panelling, etc.
S.G. ·65 ; C.W. 3.

Tulip Tree. See *Canary Wood* and *Blue Mahoe*.

Tulipwood. *Harpullia pendula (H.).* New South Wales and Queensland. Also called Tulip Lancewood. Variegated, from yellow to black. Like Olive Wood. Very hard, heavy, and tough. Fairly durable. Close grain and texture. Difficult to season and work, polishes well, small sizes. Valuable cabinet wood, used for fancy articles, small cabinet work, etc. S.G. 1 ; C.W. 4. BURMA TULIPWOOD. See *Tamalan*.

Tulip Wood. Not classified but probably *Dalbergia sp. (H.).* Brazil. Also called Rosewood, Bois de Rose, and Pinkwood. Yellowish to rose red, variegated stripes, fades with exposure, fragrant, smooth. Very hard and heavy. Fissile and splintery, subject to defects. Valuable cabinet wood, but scarce except in small sizes. Sometimes confused with Bubinga and Satiné. S.G. ·95 ; C.W. 3·5.

Tult. Implies *dozen*. A measure of timber consisting of twelve logs 18 ft. long.

Tumbled-in. Same as trimmed, *q.v.*

Tumbled Joints. Applied to joints of framing in which the accuracy is modified to suit warped and twisted members.

Tumble-home. Narrowing of the hull of a ship towards the deck line. An inward slope of the hull of a flying boat. See *Flare*.

Tumbler. The mechanism in a lock that holds the bolt in position until it is turned by the key.

Tumbling Drum. A machine for testing the strength and efficiency of boxes as containers for transport.

Tumbling Home. The curving inwards of the side of a ship from the lower deck upwards.

Tumion. See *Stinking Cedar (Torreya taxifolium)* and *Nutmeg (T. californicum)*.

Tumu. *Bruguiera gymnorhiza (H.).* Malay. A mangrove, difficult to distinguish from Bakau. Very variable in weight, hard, strong, fairly durable but easily attacked by dry rot. Used for piling, posts, etc. Not imported. S.G. ·74 to ·94.

Tun. A large cask or vat. Once a measure of 252 gallons, or four hogsheads.

Tun-dish. A wood funnel.

Tuning Horn. A wood or brass cone used in tuning organ pipes.

Tupelo. *Nyssa aquatica (H.).* U.S.A. Called Tupelo Gum, Nyssa, and Bay Poplar. Usually included with American whitewood. Similar to red gum but tougher. Grey to light brown or ivory. Soft, light, stiff, strong, tough, difficult to split, hard wearing, not durable. Interlocked grain, very uniform texture, smooth, ribbon stripe. Easily wrought but difficult to season without warp. Used for fixed interior work, panels, electric casings, turnery, inferior cabinet work, flooring, plywood. S.G. ·5 ; C.W. 2. Other species of *Nyssa* are called Black, Swamp Black, and Sour Tupelo Gums.

Tuque. *Nectandra concinna (H.).* Venezuela. Also called Angelino. Greenish yellow to olive, fairly lustrous, smooth. Moderately hard and heavy. Strong and durable. Mostly straight grain. Fairly easily wrought and seasoned. Used chiefly as a substitute for Greenheart. S.G. ·53 ; C.W. 3.

Turkey Box. *Buxus sempervirens.* See *Box.*
Turkey Oak. *Quercus cerris.* See *Oak.*
Turkish Cornel. *Cornus sp.* See *Cornus.*

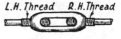

L.H. Thread   R.H. Thread

Turnbuckle. 1. A small metal plate turning on a central screw and fixed to a door frame to keep the door closed. A turn button. 2. A catch fastening on the inside

TURNBUCKLE

of a door operated by a knob outside ; a cupboard turn. 3. A button, *q.v.* 4. A screw shackle, *q.v.* See *Tension Sleeve.*
Turnery. Articles turned in a lathe or turning machine, *q.v.* The art of turning, or shaping, wood in a lathe, *q.v.*
Turn-in. A small horizontal curve to a handrail to fix into a wall or newel.

TURNING BOX

Turning Box. A specially made box or tray for scratching and inlaying turnery.
Turning Machine. There is a great variety of automatic machines for the mass production of balusters, table and chair legs, lamp standards, candlesticks, bobbins, knobs, handles, shafts, wheels, spokes, cues, toys, pill boxes, beads, etc. Nearly any size and shape may be obtained by means of cams and formers : round, oval, polygonal, or square, with twist or spiral. The stuff may be straight, bent, or irregular, for such as cabriole legs, axe handles, etc. The machines may produce anything from small pins or beads up to stuff 8 ft. long by 15 in. diameter. Also see *Lathe.*

Turning Piece. A solid *centre* for a flat segmental or camber arch. It consists of a single piece with the top edge shaped to the required curvature.

Turning Plane. A plane used in woodcraft for shaping rake handles, etc. The plane is revolved round the long handle, which is held in a vice.

Turning Saw. A compass saw, *q.v.*
Turnip Wood. Red Bean, *q.v.*
Turn-over Board. One on which the bottom part of a mould is rammed.

Turnpike. A gate to stop traffic until toll is paid. See *Toll-gate.* A turnstile, *q.v.*
Turn-screw. A screw-driver.
Turnstile. A short post with horizontal arms that rotate to allow anyone to enter or leave a building or enclosure.
Turnstile Gate. See *Kissing Gate.*

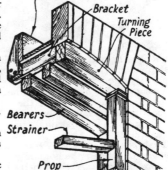

Shuttering for Lintel
Bracket
Turning Piece
Bearers
Strainer
Prop

TURNING PIECE

Turn Tread. A winder, *q.v.*
Turpentine. *Syncarpia laurifolia* (*H.*). New South Wales. Also called Luster. Pinkish brown to brownish red, lustrous. Hard, heavy, strong, very durable, rather brittle. Resists insects, teredo, and fire. Firm short grain, some wavy, close texture. Shrinks and warps badly. Not difficult to work, but gritty. Polishes well. Used for marine piling, jetties, flooring, paving, sleepers, ship-building, etc. S.G. 1 ; C.W. 4. Brush Turpentine, *S. leptopetala*, is a similar wood.

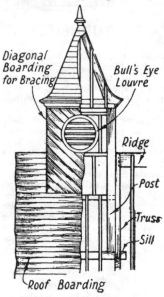

Diagonal Boarding for Bracing / Bull's Eye Louvre / Ridge / Post / Truss / Sill / Roof Boarding

TURRET

Turræanthus. See *Olon* and *Avodire*.
Turret. 1. A small tower on a building. It may be for ventilation, or a bell, or for observation purposes, or only for ornament. See *Ogee Turret*. 2. A cockpit in an aeroplane for a gunner. See *Fuselage*.
Turret Steps. Winding stairs. Solid newel stairs in a circular chamber.
Turtosa. *Oldfieldia africana* (*H.*). Called African oak and African " teak," Fou, etc. Very hard, heavy, strong, tough, and durable. It does not resemble either oak or teak, and is no relation to these woods. It has many of their qualities and was used as a substitute, but it is harder, and transport difficulties make it scarce. Used for shipbuilding, stairs, etc., and where strength and durability are of primary importance. See *Oak*.
Tuscan. One of the five orders of architecture, *q.v.* It is usually regarded as a variation of the Doric and is characterised by lack of superfluous ornamentation.
Tushing. Transporting felled trees by dragging along the ground.
Tusk Nailing. Same as skew-nailing. Two nails driven at opposite angles to increase the holding power.
Tusk Tenon. A combined tenon and housing joint for *bearing* timbers at right angles to each other, as in floor joists. It is used between trimmer and trimming joist. The tenon is usually of extra length and mortised for a key to tighten the joint. See *Hearth* and *Framed Floor*.
Twig. A thin slender branch.
Twig Burr. Decorative wood obtained from irregular grain due to growth round lopped twigs. See *Burr* and *Pollarding*.

Twigging. Osier twigs used for binding the ends of ash hoops in cooperage.

Twin Tenon. A divided tenon. The name is also often applied to a double tenon, *q.v.* See *Diminished Stile.*

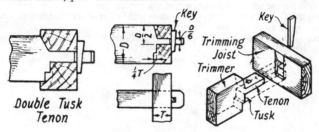

Double Tusk Tenon

TUSK TENON

Twist Bit. A brace bit that allows for clean accurate boring. There are several registered types. See *Bits.*

Twist or **Twisting.** Spiral warping of a board. Winding. See *Warp.*

Twisted Fibres. See *Interlocked* and *Spiral Grain.*

Twist Turning. Spiral turning, characteristic of Jacobean furniture, *q.v.* See *Spiral Turning.*

Two-bolt Lock. One combining latch and lock. The former is operated by a knob and spindle or by a slide, and the latter by a key.

Two-handed Saw. A large saw with handle at each end and operated by two men for cross-cutting stuff of large section.

Two-light Frame. A window frame with two main divisions as produced by one mullion. See *Windows.*

Two-to-one. A term used in stair-building for a particular type of pitch chart.

Two-way Column. An angle column with a re-entrant angle to face in two directions.

TWIST BIT

Tyloses. A disarrangement in the normal cell formation due to the protoplast of a ray or wood parenchyma cell extending into adjacent inactive elements. Bubble-like, or sac-like, ingrowths that practically fill the vessels and that form in many trees when changing from sapwood to heartwood. They are sometimes visible on the face of converted hardwoods and add to the decorative effect of the wood because of their high lustre. They reduce the permeability of the wood.

Tylostemon. See *Spicey Cedar.*

Tymp. A short horizontal timber for supporting the roof of a mine or underground excavation.

Tympanum. The segmental or triangular panel of a pediment, *q.v.*

The filling between the head of an opening, or lintel, and an arch above.

Type.   **1.** A cupola roof to a turret.   **2.** A canopy over a pulpit. **3.** A rectangular piece of metal or hardwood with a raised letter on the end for printing the characters on paper, etc.

Type Sash.   A window made of sash stuff of uniform thickness with a part opening on pivots.   See *Fixed Sash*.

Tyre.   The hoop of a wheel.   It is pulled in position by *tyre dogs* while hot.

U

U/a. Abbreviation for *unassorted*.
Uapaca. *Uapaca staudtii (H.).* Nigeria. Also called Akun. Reddish brown, darker zones. Hard, heavy, strong, durable. Straight and interlocked grain, compact, fibrous. Little figure. Substitute for teak. Not difficult to work but liable to pick-up. S.G. ·8 ; C.W. 3.
Ubilesson. Sapele, *q.v.*
Ubula. Opepe, *q.v.*
Uganda Mahogany. *Khaya anthötheca,* African mahogany. See *Mahogany*.
Uganda Walnut. *Lovoa brownii.* Also called Nkoba. See *Nigerian Walnut*.
Uqueopokin. *Hanoa klaineana (H.).* Nigeria. Creamy yellow. A soft, light wood, with coarse, fibrous, open grain. Easily wrought, liable to pick-up. Used for the same purposes as Obeche. S.G. ·4 ; C.W. 3.
Uhu. *Diospyrus insculpta (H.).* Benin ebony. Light in colour with very little black heart. Characteristics of ebony, but very resilient, tough, and long-fibred. Small sizes. Used for fancy articles, cabinet work, bentwork, etc. See *Ebony*.
Uira-pepe. See *Uria-pepe*.
Ukan. *Tamarix aticulata (H.).* India. White, yellow tinge, lustrous, feels rough but smooth. Fairly hard, heavy, and strong, not durable. Fibrous spiral grain, coarse texture. Rays give lustrous fleck on radial surface. Requires careful seasoning. Not appreciated in this country. Used for implements, brushes, small ornaments. S.G. ·7 ; C.W. 4.
Ukay. A marketing term used in the grading of Western American woods, such as Douglas Fir. *Ukay No.* 1 *merchantable* is the best quality for carcassing, and *Ukay Select merchantable* is for exterior work. See *Merchantable*.
Ukpenekwï. *Phialodiscus unijugatus (H.).* Nigeria. Variegated yellowish brown, lustrous, fine pore markings. Fairly hard, heavy, strong, and durable. Straight, compact, fibrous grain. Not difficult to work but liable to pick-up. Under investigation. Suitable for superior joinery, plywood, etc. S.G. ·8 ; C.W. 3.
Ulmaceæ. Elm family. Includes Aphananthe, Celtis, Holoptilea, Phyllostylon, Planera, Trema, Ulmus.
Ulmo. *Eucryphia cordifolia (H.).* C. America. Also the leatherwood of Tasmania. Light chocolate brown, variegated, lustrous, smooth. Moderately hard, heavy, strong, and tough ; fairly durable. Straight grain, fine and uniform texture. Easily wrought. Used for furniture, flooring, vehicles, piling, etc. S.G. ·63 ; C.W. 2·5.
Ulmus. See *Elm. Ulmus alata,* winged elm. *U. americana,* white elm. *U. crassifolia,* cedar elm. *U. campestris,* English, or red elm.

U. fulva, red, or slippery elm. *U. glabra*, smooth-leaved wych elm. *U. hollandica*, Dutch elm. *U. integrifolia* or *Holoptea integrifolia*, Indian elm, *q.v. U. lancifolia*, Indian elm. *U. major*, Dutch elm. *U. montana*, wych, or mountain, or white elm. *U. nitens*, feather-leaved elm. *U. procera*, English or red elm. *U. racemosa*, rock elm. *U. scraba*, rough-leaved wych elm. *U. spp.*, Japanese elm. *U. stricta*, Cornish and Jersey elm. *U. thomasi*, rock elm. *U. vegeta*, Huntingdon elm. *U. wallichiana*, Indian elm.

Ulono. Iroko, *q.v.*

Ultimate Strength. Breaking strength. The load that causes rupture when testing materials to destruction. See *Safe Load, Factor of Safety*, and *f*.

Umaza. *Staudtia stipitata* (*H.*). Nigeria. Chestnut brown. Hard and heavy. Close fibrous grain, even texture. Easily wrought, some liable to pick-up. Under investigation. S.G. ·8.

Umbellularia. See *Pacific Myrtle*.

Umbo. 1. The boss of a shield. **2.** A protruberance on a tree.

Umbrella Roof. A station, or cantilever, roof. It is supported by a central post only.

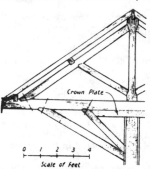

Umbrella Wood. *Musanga smithii.* See *Aga.*

Umburana. *Amburana* (syn. *Torresea*) *cearensis* (*H.*). Tropical America. Reddish yellow, variegated markings, fragrant, feels waxy. Moderately light and soft, fairly durable. Roey grain, coarse texture, distinct rays. Easily wrought. Used for furniture, joinery, etc. S.G. ·6 ; C.W. 2.

Um, Doni. Waterwood, *q.v.*

Umgesisa. *Cussonia umbellifera* (*H.*). S. Africa. Not exported.

Umgusi. Rhodesian " teak," *q.v.*

Umiry. Bastard bullet wood, *q.v.*

Crown Plate

0 1 2 3 4
Scale of Feet

UMBRELLA ROOF

Umokæze. *Cœlocaryon oxycarpum* (*H.*). Nigeria. Greyish to yellowish brown, lustrous. Fairly hard but light in weight. Straight open grain. Easily wrought. Under investigation. Suitable for joinery, plywood, etc. S.G. ·45.

Umzimbiti. *Millettia caffra* (*H.*). S. Africa. Not exported.

Undecagon. A polygon with eleven sides. Area =(side)$^2 \times 9 \cdot 366$.

Under-bark. A term used in the measurement of logs when the thickness of the bark is deducted. See *Over-bark*.

Underbench Guard. A guard, or apron, to a circular saw under the bench top. See *Saw Bench*.

Underbrush. Large woody forest plants that do not make trees.

Undercarriage. The lower part of aircraft, including skids, floats, or wheels.

Under Cover. Applied to wood stored in covered sheds.

Undercroft. An apartment or chapel below ground.

Undercut. 1. Not square. The shoulders of a tenon cut slightly out of square to ensure a tight joint on the face. **2.** Projecting members of a moulding with a hollow underneath to form a drip. **3.** Applied to a carving cut away so that some of it is separate from the ground. **4.** Applied to mouldings of Louis XIV and XV. **5.** The notch cut in a tree to assist the tree in falling in the required direction.

Underfoot. Same as underpin, *q.v.*

Underlining. Felt, building paper, etc., used for insulating or weather-resisting purposes.

Underpin. To strengthen the foundations of a wall. The work usually entails the use of props, or dead shores, and needles and sometimes raking shores.

Underside. Opposite side to *face side*, q.v.

Underwood. **1.** Bushes and small trees growing under timber trees. **2.** Quick growing shoots from one stool. The stump of a tree that regularly throws out new shoots.

Undulating. Applied to an edge or surface that is shaped with alternating convex and concave curves.

Uneven Grain. Applied to wood in which the growth rings are irregular in width.

Uneven Texture. **1.** Applied to wood in which there is a contrast between the spring and summer wood, as in pitch pine, larch, etc. **2.** A description of wood with considerable difference in the size of the elements.

Uniform. See *Contract Forms*.

Uniform Load. A load evenly distributed along a beam or structural frame producing equal reactions. A load may also be uniformly increasing along a structural member.

Uniseriate. Applied to rays one cell wide, as in the sparsely developed rays of softwoods, or in poplar and willow. A ray consisting of one row of cells as shown in the tangential section.

Unit Stress. The stress per unit of sectional area. See *Stress*.

Universal Cramp. A cramp for holding a mitre in position until the glue sets.

Universal Plane. A Stanley adjustable plane for rebating, ploughing, sticking mouldings, etc. It was originally called the " 55 " plane because it was supplied with that number of cutters, but many more have been added

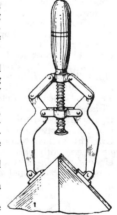

UNIVERSAL CRAMP

and blanks are provided to be cut to any required contour. It is extremely useful when machines are not available or where the quantity does not justify the use of machinery. (See illus., p. 610.)

Universal Riser Chart. A term used in stair-building for a particular type of pitch chart.

Universal Woodworker. A multiple machine that performs a number of operations : planing, sawing, moulding, tenoning, etc. Also called a general joiner.

Unmerchantable. Not up to the required quality for marketing. See *Merchantable*.

Unsorted. Applied to merchantable wood that is not graded. All the product of a log above a certain quality.

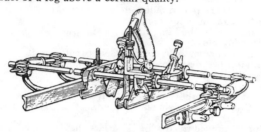

UNIVERSAL PLANE

Unstable. Liable to vary greatly in volume with change of moisture content. See *Stable*.

Untrimmed Floor. One with common, or bridging, joists only. There is no trimmed opening for fireplace or staircase, etc.

Unwrought. Off the saw. Not planed or wrought, *q.v.*

Up. The nosing to the landing at the top of a flight of stairs, or the last riser at the top of the flight.

Upas Tree. *Antiaris toxicaria* (*H.*). E. Indies, Java. Whitish. Light, soft, fibrous. Easily wrought. Of little timber value. Poisonous sap. Used for shelving, carpentry, etc.

Upland Hickory. *Carya alba.* See *Hickory*.

Upper Breast Hook. See *Keel*.

Uppers. Better grades of timber in American lumbering.

Upright Yellow-wood. *Podocarpus thunbergii*. Like yellow-wood, *q.v.*, but slightly harder.

Upsets. 1. Cross fractures in timber. The fibres are broken across the grain, or crippled by compression. Also called thunder shakes, rupture, felling, and transverse shakes. Various causes are suggested, such as felling across obstructions, jamming whilst floating down a river, thunder, internal strain in growing tree due to soft spongy inner wood and more dense and compact later growth. Upsets are common in mahogany and are a serious defect. Very often they are difficult to detect until the wood is wrought, and the stuff may break apart. **2.** A swage.

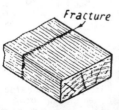

UPSET

Urandra. See *Daru* or *Dedaru*.

Urb. Abbreviation for the name of the botanist Urban.

Urea. Formula $NH_2.CO.NH_2$. A white crystalline solid. It is made by decomposing ammonium carbamate, which is produced by the combination of dry carbon dioxide and dry ammonia. Urea is readily soluble in water. The plasticity of green wood is greatly increased by soaking in a saturated solution of 1 lb. of urea to 1 pint of water for several days, at ordinary temperatures. The wood is bent and heated in an oven. It sets quickly, when cooled, to the required shape. Urea is also used in synthetic resins (see *Glues* and *Plastics*), and is an important wood preservative, making the wood almost immune from fungal and insect attack. It has no ill effects on the wood, glue, paint, or iron, and it increases the stability.

Uria-pepé. *Holocalyx balansæ* (*H.*). C. America. Also called Alecrin and Ibirapepe. Purplish brown, dark streaks. Very hard, heavy, tough, strong, and durable. Roey grain, fine texture. Fairly difficult to work, polishes well. Used for structural work, construction, turnery. A substitute for hickory. S.G. 1 ; C.W. 3·5.

Uricury. *Attalea speciosa.* A South American palm.

Urn. A turned ornament shaped like a vase and common to certain styles of furniture and decoration.

Urticaceæ. A family that includes Bœhmeria, Gyrotænia, Laportea, Myriocarpa, Urera.

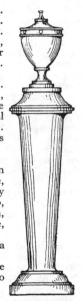

Urucurana. *Hieronymia* and *Alchornea spp.* (*H.*). C. American. Reddish brown, fairly lustrous, smooth. Moderately hard, heavy, and durable ; warps freely. Irregular grain, some roey, coarse uneven texture, visible rays. Fairly difficult to work. Used for construction, vehicles, cabinet work, structures. S.G. ·72 ; C.W. 3·5.

Urunday. *Astronium spp.* (*H.*). Tropical America. Reddish brown, claret red on exposure. Very hard, heavy, strong, and durable. Straight grain, fine texture. Difficult to work. Used for structural work, piling, sleepers, cabinet work, turnery. S.G. 1·1 ; C.W. 4. See *Guayacan.* Urunday blanco is *Diplokeleba sp.*

U/S. Abbreviation for *unsorted softwoods.*

U.S.A. Large quantities of timber are imported from the United States of America. *Eastern :* spruce, pine, birch, chestnut, cypress, hickory, loblolly pine, maple, oak, poplar, pitch pine, red gum, tupelo, walnut. *Western :* Douglas fir, hemlock, larch, pencil cedar, ponderosa pine, sequoia, sugar pine, Western red cedar, white pine, etc.

Used Length. The actual merchantable length of a tree. The sum of the several logs cut from a tree.

Utility Furniture. The name applied to war-time furniture produced under Government licence and to Government specifications.

Utskott. Same as wrack, *q.v.*

Uva or **Uvero.** See *Sea Grape.*

Uwen. See *Eben.*

Uwowengu. Nigerian Kokko, *q.v.*

URN AND PEDESTAL

V

V. Abbreviation for *Vee*, q.v.

Vaalbosch. Real Salie, *q.v.*

Vacciniaceæ. A sub-family of Ericaceæ. Not important for timber.

Vaiya. Vayila, *q.v.*

Valet. Abbreviation for the name of the botanist Valeton.

Valley. A recessed or re-entrant angle in a roof. The intersection of two roof surfaces forming an external angle less than 180 degrees. The reverse of a hip, *q.v.* A V-shaped channel or gutter. Valleys may be open, secret, laced, or swept, *q.v.* See *Open Valley* and *Roofs*.

Valley Board. A board running down a valley to carry the zinc or lead.

Valley Rafter. The timber down the valley, usually from ridge to eaves, and on which the jack rafters intersect. See *Roofs*.

Van. 1. A covered vehicle or truck for the conveyance of goods by road or rail. They are distinguished as furniture, goods, guards, luggage, mail van, etc. 2. A caravan.

Vane. 1. A blade of an aeroplane's or ship's propeller. The sail of a windmill. 2. A weathercock.

Vaquetero. *Lucuma spp. (H.).* Venezuela. Pale brown, dull but smooth. Very hard and heavy. Strong and durable. Straight grain. Not very difficult to work. S.G. ·95 ; C.W. 3·5.

Vargord. A bar or spar used instead of a bowline on some luggers.

Variegated. Marked with irregular streaks or patches of different colours.

Varnish. Resin dissolved in oil or spirits and used as a finishing coat for painted or stained wood.

Varnish Tree. *Rhus vernicifera.* Chinese lacquer tree. The black varnish tree is *Melanorrhœa sp.* See *Rengas*.

Vascular System. A term applied, in wood growth, to the strands of elongated cells near the centre of the stem in young shoots, and under the bark in later growth when it is called the cambium. It serves to conduct the moisture from the roots to nourish new growth.

Vascular Tissue. Specialised conducting tissue. It consists of phlœm and xylem.

Vase. 1. An ornamental vessel of glass, pottery, wood, etc. See *Urn*. 2. The body of the Corinthian capital.

Vasicentre. Applied to parenchyma that completely borders the vessels, as in ash.

Vateria. See *White Dhup*.

Vatica. See *Mascal Wood*, *Narig*, and *Resak*.

Vats. Circular tanks, used in various industries, for holding liquids. The woods commonly used are Douglas fir, kauri pine, larch, oak, pitch pine, red cypress, redwood, according to the particular requirements. The sizes vary from 50 to 100,000 gals. capacity. The names *tank* and *back* are usually confined to rectangular vessels. Vats may be varied in shape to suit the position and may be lined with glass, metal, rubber, etc.

Vault.   **1.** A chamber with vaulted roof.   **2.** An underground chamber.   **3.** An arched roof.

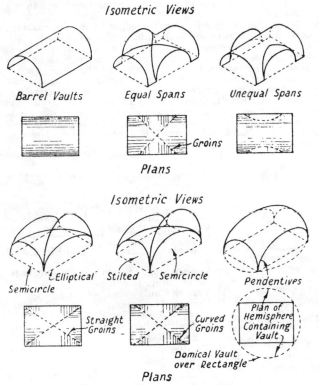

Isometric Views

Barrel Vaults   *Equal Spans*   *Unequal Spans*

Plans   Groins

Isometric Views

Semicircle   *Elliptical*   *Stilted*   *Semicircle*   *Pendentives*

Straight Groins   *Curved Groins*   Plan of Hemisphere Containing Vault

Domical Vault over Rectangle

Plans

VAULTS AND GROINS

Vaulting.   **1.** A number of vaults collectively.   **2.** The covering of chambers in arched form, usually with brick, stone, or concrete. The three types are barrel or wagon, groined, and ribbed ; and they may be elaborated into conoidal, fan, lierne, or stellar. Other descriptions and terms used are compound, cross, diagonal, double, rampant, simple, single, stilted, surbased, and surmounted. The outline may be conical, cylindrical, elliptical, Gothic or pointed, or spherical.

Vayila. *Pœciloneuron indicum* (*H.*). India. Also called Vaiya, Kirballi, etc. Dark red. Hard, heavy, strong, and durable. Straight grain, medium coarse texture. Used for structural work, sleepers, etc. Not imported.

Vedi. Indian babul, *q.v.*

Vee Gutter. A triangular gutter at the intersection of two sloping roofs, as to a double lean-to roof, or between two North-light roofs. See *Gutters*.

Vee Joint. Small chamfers on the edges of match-boarding, etc., to break the joints between the boards, to hide shrinkage.

Vee Roof. A double lean-to roof, forming a vee gutter.

Vehicle. **1.** A carriage, cart, or wagon, for the conveyance of people or goods. The various types are barouche, brougham, buggy, cab, car, cart, char-a-banc, coach, coupé, dog-cart, handcart, hansom, landau, landaulette, lorry, omnibus, pantechnicon, phaeton, sociable, tramcar, van, victoria, wagon, etc. There are numerous types of vehicles for transporting timber described in the Glossary. **2.** Oil, spirits, water, gold size, or varnish in which the base and pigments of paint are distributed and suspended. The oil may be boiled or raw linseed, China wood, poppy, or nut oil.

Vein. A strip, or streak, of different colour than the surrounding surface.

Veiner. A carver's parting tool.

Vell. Abbreviation for the name of the botanist Velloza.

Vellapiney. White Dhup, *q.v.*

Vencola. Quira, *q.v.*

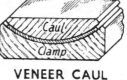

VENEER CAUL

Veneer Bag. Steam and vacuum bags used in veneering. See *Glue*.

Veneer Caul. See *Caul*.

Veneer Hammer. A piece of sheet brass about 6 in. long in a wood handle. It is worked in a zig-zag manner from

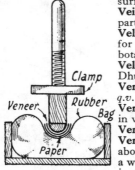

VENEER BAG

the centre outwards to squeeze out surplus glue when veneering.

Veneering. The art of facing a groundwork of suitable material with a very thin layer of wood. The facing, or veneer, is usually of superior and more decorative wood that could not be used for the groundwork because of weakness or expense. The ground-

VENEER HAMMER

work, core, or carcase, is usually of cheaper wood, but straight in grain and stronger, and suitable for glue. Veneers are laid by means of press, veneer hammer, caul, or saddle, to squeeze out the surplus glue and to keep the veneer in position until the glue is set. For shaped work, sand boxes and bags are used, and, for intricate work, steam and vacuum bags. The art has been practised for thousands of years and it is becoming of increasing importance because of the scarcity and appreciation of rare woods. See *Caul, Cylinder, Curtail Step, Glue, Plastics,* and *Veneers.*

Veneer Knife. The large knife, or cutter, used to produce veneers, *q.v.*

Veneer Matching. Applying veneer to obtain some particular effect. Also see *Matched.*

Veneer Pins. Used for tacking the veneer to the base, or core, until the glue has set.

Veneers. Nearly any kind of wood may be used for veneers, according to the requirements. See *Plywood.* Woods with close firm grain are best, but most valuable veneers are from roots and burrs. For furniture and superior joinery decorative hardwoods are used. There are three methods of cutting the veneers according to the wood and the kind of grain required. Highly decorative woods, such as burrs, very hard woods, and woods that do not lend themselves to more economical methods, are

'V' Match Herringbone

Diamond Match

Reversed Diamond

Inverted 'V' Match

Basket Weave

Block Design

VENEER MATCHING

sawn with a special type of circular saw, up to 18 ft. diameter. Although eleven veneers per inch can be sawn, the method is wasteful, and if the wood is suitable slicing is preferred. Knife cutting, or *slicing,* is similar to taking off a large shaving with a plane. The knives are 7 to 10 ft. long and slice the wood with a shearing action. Even with the greatest care some woods are *torn, or distressed,* by this method. The third method, *peeling* or rotary cutting, is used when the appearance is of secondary importance or large dimension stock is required. This method may be compared to unrolling linoleum. The log is peeled round the circumference and it may be peeled to any width according to the quality of the log. The circumferential grain obtained by peeling is usually uninteresting and the veneers are used chiefly for plywood. There are exceptions, however, such as bird's-eye maple. *Half-rotary slicing* is a compromise between slicing and peeling to obtain ribbon and silver grain. The segments, or flitches, are quarter sawn and rotate against the knife by means of a specially arranged stay log.

The logs are cut to length into flitches, which are steamed or boiled before slicing or peeling. · They are quartered for slicing and half-rotary slicing. The thickness of peeled veneers may be from $\frac{1}{80}$ to $\frac{3}{8}$ in. The most satisfactory thickness for slicing is suggested as $\frac{1}{28}$ in., but sliced sheets may be from 4 to 140 per inch. Sawn veneers may be of any thickness according to the wood and the purpose for which it is required. See *Plywood*, *Slicing*, and *Rotary Cutting*.

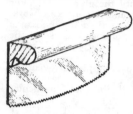

VENEER SAW

Veneer Saws. 1. Circular saws for cutting logs, etc., into veneers. They may be up to 18 ft. diameter and built up of 25 segments. See *Veneers*. **2.** A specially shaped saw used by the cabinetmaker for cutting thick sawn veneers to the required shape. The blade is about 5 in. long with the cutting edge slightly curved. It is screwed to the side of a specially shaped handle for easy gripping, and so that the handle is clear of the straight-edge. The points, or teeth, are similar to those of a dovetail saw.

Venesta. A proprietary plywood.

Venetian Blind. A window blind consisting of slats, or thin laths, like louvres, held together by tapes. They open and close, and rise and fall, by operating cords. A jalousie.

Venetian Door. A door frame with side, or wing, lights in the same frame.

Venetian Window. A window with three pairs of sashes and two mullions. The centre pair of sliding sashes are wider than the outer pairs, which are usually fixed. See *Sliding Sashes*, *Boxed* and *Solid Mullion*, and *Window*.

Venezuelan Boxwood. *Gossypiospermum*, or *Casearia, præcox*. Called Maracaibo boxwood and Zapatero, *q.v.*

Venezuelan Mahogany. *Swietenia candollei*. Like Honduras mahogany. See *Mahogany*.

Vengai. *Pterocarpus marsupium* (*H.*). E. Indies. See *Bijasal*.

Head Rail
Ladder Tape
Pull Cord
Lath
Tilting Cord
Bottom Lath

VENETIAN BLIND

Vent. 1. A small opening or flue to allow liquids or gases to escape. An aperture in a cask. **2.** Abbreviation for the name of the botanist Ventenat.

Ventil. A valve in the wind trunk of a pipe organ for stopping the wind from certain pipes, or silencing them.

Ventilating Bead. A wide inner bead at the bottom of the frame for sliding sashes. It enables the bottom sash to be open a little without causing a draught. See *Mullion* and *Sash and Frame.*

Ventilating Duct and **Flue.** A conduit for transmitting air.

Ventilating Frets. Small frames between the deck plating of a ship and the top runner, with open panels for ventilation. Any decorative fret pierced for ventilation.

Ventilating Shutter. A shutter that allows for ventilation. See *Venwood.*

Ventilating Trunk. Same as *ventilating duct* but larger. A ventilating shaft.

Ventilator. 1. Any arrangement for keeping the air fresh in a room, such as grids, fans, jalousies, louvres ; or any of the registered or patented types : Archimedian, Bax, Sheringham, Tobin, etc. In modern large buildings the Plenum or vacuum systems and special air-conditioning plants are used. 2. Hardwood planks, usually teak, used to form air passages in cargoes of grain from Burma.

Ventiloc. A patent sliding sash window.

Vent Peg. A plug to a barrel.

Venwood. A registered type of shutter to overcome the difficulties of the blackout as a precaution against air raids. It may be removable when not required, hung, or sliding.

Vera. Lignum-vitæ, *q.v.* Also *Phlebotaenia cuneata*, Cuba, which is a similar wood. S.G. 1·1. Also *Antirrhœa spp.* See *Vera-wood.*

Verandah. An external gallery or platform to a building. It is supported by slender columns or cantilevers, with a lean-to or flat roof, and enclosed by a balustrade. A covered path against the wall of a house often partially enclosed by glazed frames.

Vera-wood. *Bulnesia arborea* (*H.*). Tropical America. Also marketed as Congo cypress, Guayacan, Lignum-vitæ, Maracaibo, Venesia, Vera amarillo, etc. Olive green to brown, variegated stripes, feels waxy, fragrant. Very hard, heavy, strong, durable. Interwoven grain, fine uniform texture. Substitute for Lignum-vitæ, *q.v.* S.G. 1·1 ; C.W. 4·5.

Verbenaceæ. Teak family. Includes Avicennia, Callicarpa, Citharexylum, Gmelina, Petitia, Pseudocarpidium, Tectona, and Vitex.

Verda. Indian silver greywood, *q.v.*

Verge. 1. A mace, rod, or staff of office. 2. The extreme side or edge of anything. A border or margin. 3. See *Verges.*

Verge Board. A barge board, *q.v.*

Verge Moulding. A moulding under the projecting roof slates at a gable.

Verges. The overhanging edges of roof coverings at a gable. The intersection between gable wall and roof.

Vermelho. *Centrolobium spp.* (*H.*). Brazil. Not definitely classified. Also an adjective applied to several woods : Amarella, Arariba, Balaustre, Branca, Canary Wood, etc. Olive brown, darker stripes. The wood is hard and heavy, and used chiefly as veneers in this country. Also see *Branca* and *Kartang.*

Vermilion Wood. Padauk (Andaman), *q.v.*

Veroity. *Quararibea sp.* (uncertain) (*H.*). Tropical America. White to greyish white. Tends to stain dark after cutting. Rather hard and heavy, tough and strong, not durable and subject to worm attack. Straight grain, smooth. Requires careful seasoning. Resembles holly in colour and working qualities. S.G. ·72 ; C.W. 3.

Vertical Grain. The grain on quarter-sawn wood. See *Conversion*.

Vertical Plane. A vertical surface, in geometry, on which the elevations, in orthographic projection, are projected. See *Planes and Traces*.

Vertical Sliding Sashes. See *Sash and Frame Window*.

Vertical Trace. The point where a line intersects the vertical plane or a line where an auxiliary plane intersects the vertical plane. See *Planes and Traces*.

Vervona. Sequoia burr. It is somewhat like thuya burr, and esteemed as a decorative veneer.

Vesica. A panel or decoration in the form of a pointed oval. The outline consists of two circular arcs.

Vessel. 1. A vase. A utensil for holding liquids or solids. 2. A ship. The name is generally applied to any craft larger than a boat. 2. See *Vessels*.

Vessel Lines. The vessels of wood seen on a longitudinal surface.

Vessels. Cellular ducts containing and circulating the sap in plants. Elongated tube-like cells in wood tissue due to cell fusion, that assist in conducting water. They are only present in hardwoods, and called pores. They are placed one over the other with the ends missing, or well perforated, so that they form a continuous pipe-like tube.

VESICA PISCIS

Vestibule. A small room at the entrance to a building forming a screen or protection to a larger room or hall. A small antechamber to an apartment.

Vestry. A room near to the choir of a church in which vestments and sacred utensils are kept.

Vestured Pit. A bordered pit with its cavity partly or wholly lined with projections due to the overhanging secondary wall. See *Pits*.

Veteran. A tree over 2 ft. diameter at breast height.

V.G. Abbreviation for *vertical grain*.

Viaduct. A long bridge supported by a series of piers or trestles over a valley or low level. It may serve as a road, railway, or waterway.

Viburnum. See *Guelder Rose, Wayfaring Tree*, and *Laurustrinus*.

Vice or **Vise.** An adjustable appliance to a bench, for metal or woodworking, to grip and hold the material firmly whilst it is being wrought.

Victorian Oak. See *Tasmanian Oak* and *Mountain Ash*.
Vidal. The patents controlling the construction of stressed skin aircraft, and double-curvature work generally, that is built up of resin-bonded veneers.

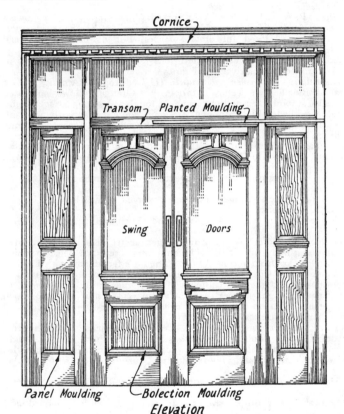

Cornice

Transom *Planted Moulding*

Swing *Doors*

Panel Moulding *Bolection Moulding*
Elevation

VESTIBULE FRAMING

Vidpani. *Pityrantha verrucosa* (*H.*). Ceylon. Bright yellow. Hard, heavy, tough, and very strong. A good substitute for boxwood, but difficult to work smooth. S.G. ·82.
Viewly. A term sometimes applied to imported wood implying " in good condition."

Vignette. Carved ornamentation imitating vine leaves and grapes. A small ornamental design without border.

Villa. 1. A detached suburban dwelling house. 2. A country house of some pretension.

Villaresia. See *Silky Beech*.

Vine. *Vitis vinifera* (*H.*). Widespread. The grape vine. Yellowish grey. Hard and heavy. Hard bony grain, rather rough texture, distinct rays. Used for small articles, but of little commercial value.

Vinery. A specially constructed glass-house, usually in the form of a lean-to, for the cultivation of vines.

Vinhatico. *Plathymenia reticulata* (*H.*). Tropical America. Also called Brazilian or Yellow Mahogany. Orange yellow, variegated streaks, smooth. Fairly light and soft, tough and strong for its weight, stable, fairly durable. Straight firm grain, some roey. Easily wrought, polishes well. Used for cabinet work, superior joinery, flooring, shipbuilding, etc. S.G. ·6 ; C.W. 2. Also *Enterolobium ellipticum* or *Pithecolobium vinhatico*. Called Gold Wood, Tataré, and Yellow-wood. Yellow to bright orange, variegated, lustrous, smooth, darkens with exposure. Moderately light and soft, durable and strong. Some roey grain, rather fine texture. Uses as above. S.G. ·6 ; C.W. 2.5.

Viol. A musical stringed instrument played with a bow. The size and the number of strings vary.

Violet Wood or **Violetta.** Kingwood, Purpleheart, and Raspberry Jam Wood, *q.v.* Also see *Amaranth*.

Violin. A fiddle. A musical stringed instrument consisting of a hollow flat body with long neck containing screws, or pegs, to which the strings are fixed and held taut. The neck terminates with a scroll-shaped head. Over seventy different pieces of wood are required in the making. Other instruments of the same family are : *viola*, or tenor violin, slightly larger than the violin ; *violincello* (baritone), much larger and rests on the floor ; *violone*, or double-bass, the largest of this family of musical instruments.

Viraru. *Ruprechtia spp.* (*H.*). Central America. Reddish grey, reddish streaks, dull but smooth. Moderately hard, heavy, tough, strong, fairly durable. Straight grain, fine texture. Easily wrought. Used for construction, joinery, flooring, furniture. S.G. ·67 ; C.W. 2·5.

Virgin. Applied to forests not touched by felling, hence no second growth.

Virginal. A form of spinet. It preceded the harpsichord and clavecin.

Virginia. A prefix applied to loblolly pitch pine, *Pinus tæda* ; pencil cedar, *Juniperus virginiana* ; red spruce, *Picea rubra*.

Virola. (Myristica.) See *Banak, Bicuiba, Chuglam, Pindi, Tapsava*.

Vitex. See *Jocote, Leban, Milla, Molave, Puriri*.

Vitis. See *Vine*.

Vitrine. A display cabinet with daylight background. The back is usually of frosted glass.

Vitrite. A proprietary synthetic resin glue.

Vitruvian Scroll. A scroll ornament named after Vitruvius.

Vitta. A fillet ornament in a frieze, capital, etc.

Vochysia. See *San Juan, Quaruba,* and *Yemeri.*

Vochysiaceæ. A family including Callisthene, Erisma, Erismadelphus, Qualea, Salvertia, Vochysia.

Volet. 1. A parrel. **2.** One wing or panel of a triptych.

Volhynian Oak. Similar to Austrian oak, from Southern Poland and the Ukraine. See *Oak.*

Volute. A scroll in spiral form, characteristic of several architectural orders. See *Ionic Volute.*

VOLUTE

Vomitory. An inclined passage or staircase leading from the entrance hall or corridor of a theatre to the middle of the balcony, for convenience of exit or entrance.

Vouacapoua. See *Acapú.*

W

w. A symbol for *load per foot run* or *weight per cubic foot.*

W. A symbol for *breaking* or *total load,* and abbreviation for *white.*

Wabble Saw. A drunken saw, *q.v.*

Wadaduri or **Wadodorie.** Monkey pot, *q.v.*

Wafer. A pellet, *q.v.*

Wagon or **Waggon.** 1. A four-, or more, wheeled vehicle for transporting heavy goods by road or rail. 2. A small domestic four-wheeled conveyance for use between kitchen and dining-room. A dinner wagon or trolley. It may be fitted with drawer for cutlery, etc., and have falling or extension leaves. There are numerous registered types with mechanical devices for raising or lowering the trays. See *Tea Trolley.*

Wagon-bevelling. Simple bevelled ornamentation on the edges and corners of furniture.

Wagonette. A four-wheeled pleasure vehicle with seats running lengthwise, now superseded by the charabanc or motor coach.

Wagon-lit. A sleeping car on continental railways.

Wagon Roof. A trussed rafter roof, *q.v.*, with ashler pieces at the feet of the couples and foot-plates from the feet of the rafters to the feet of the ashlering.

Wagon Vault. A barrel vault, *q.v.*

Wagtail. The parting slip to a sash and frame window.

Waibaima. *Licaria spp.* B. Guiana. A species of greenheart, *q.v.*

Waika Chewstick. *Symphonia globulifera.* British Honduras. Hog gum, *q.v.* Also called Amarilla and Leche.

Wainscot Plank

Wainscot Billet

WAINSCOT BILLET

Waikey. *Inga spp.* (*H.*). B. Guiana. Several varieties of soft light woods of little timber value.

Wainscot Billet. An oak log converted economically to give silver grain.

Wainscoting. Wall panelling of wood. Applied to boards or panelled framing covering the walls of a room, especially when of oak. Panelling or boarding up to dado height. See *Dado Framing.*

Wainscot Oak. Selected oak cut radially to obtain silver grain.

Wainscot Plank. The middle part of the log left after converting into billets.

Waist. The part between the quarter-deck and forecastle of a ship.

Waist Rail. The intermediate rail in the framework of the driver's cabin to the body of a vehicle.

Wales. Strong planks spaced at intervals along a ship's side to strengthen the decks. They resemble hoops round the sides and bows, and are distinguished as main and channel wales. Also see *Walings.*

Walile. Nigerian boxboard, *q.v.*

Walings. Horizontal timbers forming guides for sheet piling, or tying together the poling boards in timbering for excavations. A strap across the face of small timbers. See *Guide Piles* and *Timbering*.

Walking Line. An imaginary line 18 in. from the centre line of the handrail and used when setting out the winders for a stairs.

Walkway. A gangway from the fuselage to the tail in a large aeroplane.

Wall. **1.** The membrane in wood structure that encloses the cell contents. It is compound, consisting of several layers, in a mature cell. **2.** Abbreviation for the name of the botanist Wallich.

Wallaba. *Eperua falcata* and *E. caudata* (*H.*). Tropical America. Reddish brown to purplish, dark gum streaks, odorous, smooth. Resembles Rhodesian teak. Very hard and heavy. Strong, durable, fissile. Straight, coarse, even, grain and texture. Distinct rays. Difficult to work and polish because of gum exudation. Used for structural work, shingles, poles, etc. S.G. ·9 ; C.W. 4.

Wallaceodendron. See *Banuyo*.

Wall-boards. Manufactured sheets for covering large areas. There is a great variety, consisting of asbestos-cement, asbestos-wood, plaster, wood fibre, and plywood. Wood fibre boards are made from all kinds of wood fibre, from cane to any kind of waste softwood, and

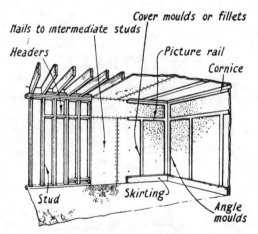

WALL BOARDS

may have a finished surface or be intended chiefly for insulating. The three main types are insulating, medium hard, and hardboard. They are prepared from similar emulsions and the degree of hardness depends upon heat and pressure. Other variations depend upon the processes of preparing the fibrous pulp. Chipping machines prepare the waste wood which is then reduced by grinding and de-fibrating. The emulsion is fed through rollers to produce a wide ribbon-like board which is cut to size and dried in a kiln. The hardness is produced by hot-plate presses with a pressure of up to 2,500 tons, and variation to the finished surface is produced during this process. Hardboards are sufficiently dense to be waterproof, and they may be steamed and bent to nearly any shape.

The sizes of wall-boards vary, but they are usually up to 16 ft. by 4 ft. If the width is increased up to 6 ft., the length is decreased. Hardboards are usually $\frac{1}{8}$ or $\frac{3}{16}$ in. thick, but insulating boards are up to $\frac{3}{4}$ in. thick. There is a large number of well-known proprietary makes of wood-fibre wall-board : Ankar, Beaver, Bordex, B.P., Canec, Celotex, Donnacona, Ensonit, Ensowal, Essex, Fiberlic, Fyburstone, Gliksten, Huntonit, Insulite, Insulwood, Kenmore, Lloyd, L.W., Maf, Metco, Masonite, Maftex, Plastergon, Presdwood, Startex, Sundeala, Super, Tentest, Treetex, Tufboard, Wellinlith, etc. Wall-boards are used for all purposes : ceilings, walls, floors, furniture, formwork, motor cars, boats, aeroplanes, panels, fittings, joinery, etc. See *Building Boards* and *Plywood*.

Wall Box. A cast-iron box built in a wall to carry the end of a timber such as a beam or binder. It allows for ventilating the end of the timber and distributes the pressure.

Wall Hangers. Pressed steel or cast-iron stirrups for carrying the ends of constructional timbers to avoid building the end of the timber in the wall. See *Hanger*.

Wallhold. The length, or bearing, of a constructional timber resting on the supporting wall.

Plate Shelf

WALL PANELLING

Wall-hook. A strong nail with right-angled head for attaching anything to a wall.

Wall Panelling. See *Panelling*.

Wall Piece. A vertical timber fixed to a wall by holdfasts or wall-hooks, to receive the thrust, and distribute the pressure, from a strut or shore. See *Shores*.

Wall Plate. A horizontal timber on a wall to provide a fixing and level bearing for, and to distribute the pressure from, spars and joists. See also *Wall Piece*. See *Hearth, Collar-tie Roof, Eaves*.

Wall-pocket. A small recess in a wall to receive the end of a constructional timber. The pocket should be coated with bitumen to keep the wood dry, and should be sufficiently large to provide ventilation.

Wall Post. A post or stile of a frame attached to a wall. See *Angel Beam*.

Wall String. The string of a flight of stairs against a wall.
Wall Table. A drop table, *q.v.*
Wallum Oak. *Banksia æmula* (*H.*). Queensland. Indian red, lustrous. Light, soft, tough, not strong or durable. Interwoven grain, ray and mottle. Difficult to season and work, polishes well, small sizes. Used for decorative cabinet work, fancy articles, etc. S.G. ·6; C.W. 4.
Walnut. *Juglans regia* (*H.*). S. Europe, Asia Minor. Usually distinguished by source of origin. Light brown to nearly

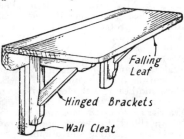

Falling Leaf

Hinged Brackets

Wall Cleat

WALL TABLE

black, darker markings, smooth, lustrous. Hard, moderately heavy, strong, tough, fairly durable, but sapwood is subject to attack by worms. Very stable when seasoned. Shock resisting. Fine even grain and texture, beautiful figure due to annual rings. Requires care in seasoning, not difficult to work and polishes well. Esteemed for cabinet and carved work. Burrs, stumps, crotches and butts are highly valued for veneers. Also used for gun stocks, rifle butts, bent work, superior joinery, propeller blades, etc. S.G. ·65; C.W. 3·5. AMERICAN BLACK, *Juglans nigra.* U.S.A. Similar to *J. regia,* but darker and not so decorative, and used for similar purposes. S.G. ·58. ANCONA.

0 1 2
Scale of Ins

WALNUT

J. regia. Figured Italian walnut. The name is generally applied to figured wood from Mediterranean countries. AFRICAN. See *Nigerian Walnut* and *Lovoa.* ARGENTINE. *J. australis.* AUSTRALIAN. See *Queensland Walnut.* BENIN. See *Nigerian Walnut* and *Lovoa.* BLACK SEA. *J. regia.* BRAZILIAN. See *Imbuia* and *Laurel.* CALIFORNIAN. *J. Lindsii.* Also called Claro walnut. Tan brown, variegated stripes, black spots. Highly decorative with wavy growth lines. CAUCASIAN. *J. regia.* Russia and Asia Minor. CIRCASSIAN. *J. regia.* Selected S. European and Asia Minor walnut. CLARO. See *Californian.* EAST INDIES. See *Siris* and *Kokko.* ENGLISH. *J. regia.* FRENCH.

J. regia. GOLD COAST. See *Nigerian Walnut.* INDIAN. Indian silver greywood, *q.v.* ITALIAN. *J. regia.* See *Ancona.* JAMAICAN. See *Nogal.* JAPANESE. *J. sieboldiana.* KURUMI. Japanese walnut. MANCHURIAN. *J. mandschurica.* Resembles *J. regia,* but lighter, milder, not so variegated, and with straighter grain. MEXICAN. See *Guanacaste.* NEW GUINEA. *Dracontomelum spp.* Also called Pacific or Papuan walnut, and Loup or Lup. See *Dao.* NIGERIAN. See *Nigerian Walnut.* ORIENTAL. See *Queensland Walnut.* PACIFIC or

PAPUAN. Same as *New Guinea*. PERSIAN. *J. regia*. The common name for Mediterranean walnut. PERUVIAN. *J. neotropica*. QUEENSLAND. See *Queensland Walnut*. RUSSIAN. *J. regia*. SATIN. See *Red Gum*. SHAGBARK. Hickory shagbark, *q.v.* SOUTH AMERICAN. See *Guanacaste*. WEST AFRICAN. See *Nigerian Walnut*. WEST INDIES. See *Nogal*. WHITE. See *Butternut* and *Hickory Shagbark*. Also see *Blush Walnut* and *Canary Ash*.

Walnut Bean. Queensland Walnut, *q.v.*

Walsura. *Walsura piscidia* (*H.*). S. India. Purplish plum colour. Hard and heavy, with close uniform grain. Fine distinct rays. Suitable for cabinet work, turnery, etc. S.G. ·95.

Wamara. *Swartzia tomentosa* (*H.*). Tropical America. Also called brown ebony, ironwood, and clubwood. Purplish to black. Very hard, heavy, strong and durable. Straight, fine, uniform grain and texture. Difficult to work, polishes well. Too hard for ordinary purposes. Used for fancy articles, cabinet work, turnery, etc. S.G. 1·1 ; C.W. 5. Also called Womara.

Wana. Cirouaballi, *q.v.*

Wand. A slender rod used as a symbol of office ; a conductor's baton.

W. and A. Abbreviation for the names of the botanists Wright and Arnott.

Wandering Heart. Wood in which the pith is crooked so that the heart outcrops on the converted wood. It is not so strong as straight-grained wood.

Wandoo. *Eucalyptus redunca, E. elata*, or *E. wandoo* (*H.*). W. Australia. Also called Spotted gum, White gum, and Powder bark. Similar to Tuart, *q.v.* Light brown. Very hard, heavy, strong, and durable. Too hard and heavy for ordinary purposes. Little exported. Used for sleepers, piling, structural work, cogs, wheelwright's work, and wagon construction. S.G. 1·1.

Wane or **Waney Edge.** Applied to converted wood in which the corner is missing at the circumference of the log, due to too economical conversion. The defect denotes the presence of sapwood.

Wapa. Wallaba, *q.v.*

Waras. *Heterophragma roxburghii* (*H.*). India. Olive grey to greyish brown, lustrous but dulls later, feels rough. Moderately hard, heavy, and strong. Straight and interlocked grain, distinct rays. Not difficult to work except to a smooth surface. Suitable where a highly finished surface is not required. S.G. ·6 ; C.W. 3. See *Petthan*.

Waratah. *Telopea truncata* (*H.*). Tasmania. Hard and compact, handsome grain, small sizes. Used for inlays, etc.

Waratah Oak. *Orites excelsa*. See *Silky Oak*.

Warb. Abbreviation for the name of the botanist Warburg.

Warburgia. See *Musiga*.

Ward. Abbreviation for the name of the botanist Warder. Also see *Wards*.

Warding File. A flat thin file for cutting keys for locks.

Wardrobe. A portable or fixed closet, or cupboard, or small room, for the protection of clothes.

Wardrobe Lock. Like a cupboard lock, but with a spring latch in addition to the bolt operated by the key.

Wards. 1. The metal rings in a lock intended to prevent any but the correct key from turning ; or the slots in the key to fit the rings. 2. See *Bailey.* 3. Rooms for patients in hospitals and similar institutions.

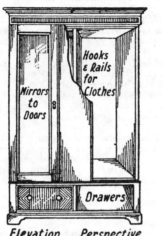

Elevation   Perspective   Section

WARDROBE

Warikuri or **Warakuri.** Brazil white cedar. See *White Cedar.*
Warping. Casting, twisting, or *in wind.* Not a plane surface. The different kinds of distortion, or warp, are described as bowing, cupping, springing, and twisting, *q.v.* Warping can often be avoided by careful piling and seasoning, and by the method of conversion. See *Shrinkage, Seasoning,* and *Conversion.*

Warri. A species of African mahogany.

Warri Wood. Maple Silkwood, *q.v.*

Wart. A protuberance on a tree. Abnormal growth.

Washboard. 1. A skirting board. 2. A corrugated board on which clothes are scrubbed. 3. Applied to imperfect machining that

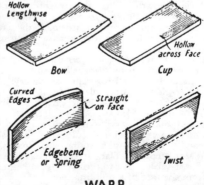

WARP

shows an undulating surface. Ridges due to faulty sawing or machining.

Washboarding. See *Ribbing.*

Washboard Tree. See *Oak Walnut.*

Washer. A metal plate or disc to distribute the pressure from the nut or head of a bolt. See *Timber Connectors.*

Washiba. *Tabebuia spp.* (*H.*). Tropical America. Also called Bowwood. Olive brown, lustrous. Very hard, heavy, strong, durable, and elastic. Resembles greenheart and used for similar purposes. S.G. ·95.

Wash Stand. A piece of bedroom furniture, usually with marble top, for holding the basin, etc., for personal ablutions.

Wash Strake. A plank above the gunwale of a ship.

Wash Tub. A large wood tube in which clothes are washed.

Waste. The unmerchantable stuff left after converting a round log ; for hardwoods it may be as high as 30 per cent.

Water Ash. See *Ash.*

Water Bar. **1.** An oak, galvanised iron, or copper strip fitted into a window sill and frame sill to prevent water from penetrating. It is usually about 1 in. by ⅜ in., if of oak, and it should be bedded in thick white lead paint. See *Hopper Window.* **2.** Patent water excluders for the bottom of doors and french casements, usually for doors opening inwards. See *Weather Bars.*

Water Butt. A cask used to collect rain-water.

Water Checks. Throatings to casements, etc.

Water Elm. *Ulmus americana.* See *Elm.*

Water-gate. **1.** A gate to a house or grounds giving access to a river or lake. **2.** A gate to release or check the flow of water.

Water Hickory. *Hickoria,* or *Carya, aquatica.* See *Hickory.*

Water Leaf. A carved detail like an elongated laurel leaf, characteristic of Sheraton and Hepplewhite furniture.

Water Maple. *Acer saccharinum.* See *Maple.*

Watermark. A stain in certain woods due to attack by fungus during growth. The stain in some cases increases the value of the wood for decorative purposes, as in horse-chestnut, and has no serious effects on the wood. Watermark disease in cricket-bat willow is a very serious defect. It makes the wood of little value, and eventually kills the tree.

Water Pear. *Eugenia gerrardi* (*H.*). S. Africa. Greyish-brown, variegated, with reddish-yellow tinges. Fine even grain and texture, some curly. Used for superior joinery, cabinet work, etc. S.G. ·5.

Water-plane. A sea-plane, *q.v.*

Water Shoot. · A pipe or gutter for discharging water, especially from a roof. Also see *Chute* and *Shoot.*

Water Stains. **1.** Pigments mixed with water and used for staining woods to any required colour. Water stains raise the grain of the wood more than spirit stains. **2.** Brown discoloration in hemlock.

Water Streaks. Dark streaks in oak due to injuries to the trees.

Water-table. A projecting ledge or moulding to throw off the water from the face of a building, etc.

Water Tower. A structure, often of wood, to carry a high tank to give the necessary head, or pressure, to a water-supply system, or for cooling the water.

Water-ways. Long pieces connecting the decks to the side of a ship and forming a channel for water, which is carried away by the scuppers. The deck plank nearest to the bulwark.

Water-wheel. A wheel fitted with vanes so that it turns by a flow of water.

Waterwood. *Chimarrhis cymosa* (*H.*). W. Indies. Also called Bois Riviere. Excellent wood for cabinet work and superior joinery. Not imported. Also *Syzygium cordata.* S. Africa. Natal swamps. Not durable except under water. Easily wrought. Used for interior joinery, flooring, etc. Not exported. S.G. ·7. C.W. 2·5.

Wattle. **1.** *Acacia spp.* Australia, Tasmania. Several species: *A. dealbata, A. decurrens, A. mollis.* Distinguished as silver and black wattle. Sometimes called mimosa. The wood is yellowish-red, lustrous, hard, heavy, and tough. It is used chiefly for tannin. S.G. ·75. See *Blackboy, Grass Tree,* and *Spearwood.* **2.** Wickerwork hurdles.

Wattle and Dab. Rough timber framework covered with interlaced wickerwork to receive plaster.

Wave Moulding. **1.** An undulating face on a moulding formed by cam action on the rotary cutters during manu- facture. **2.** A Gothic moulding consisting of a convex and two concave curves.

Wavy Grain. Due to the regular undulation of the elements with the layers crossing each other. It is prevalent in birch, mahogany, maple, sycamore, etc. The fibres are arranged in short, fairly uniform ripples or waves. A constant changing of direction of the fibres due to wrink- ling of the fibres, as under large branches or where the roots merge into the stem. If wavy-grained wood is split radially the ex-

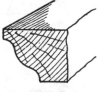

WAVE MOULDING

posed surfaces are wavy or undulating. See *Compression Wood.*

Wawa. Obeche, *q.v.*

Wax Polish. Beeswax, with a little turpentine to make it work freely, used for polishing wood, especially hardwood floors and fumed oak.

Way. A piece of wood laid on loose ground or sand to serve as a runner.

Wayfaring Tree. *Viburnum lantana* (*H.*). Britain. Many local names : Cottoner, mealy tree, whipcrop, etc. A bush-like tree of little timber value.

Wayleave. An arrangement, when purchasing standing trees, to provide convenient passage for the transport of the felled trees.

Way-post. A sign-post, *q.v.*

W.b. Abbreviation for *weatherboards.*

Wear. To remove the surface of wood by abrasion, attrition, or friction, as on floors. Hardness is generally a measure of the resis- tance to wear, but the resistance of the grain structure against dis- integration is a more important factor. Specific gravity is also a measure, to some extent, of the resistance. Conversion is important, but differs with the species of wood. In most cases rift-sawn wood wears better, but in other cases slash-sawn has greater resistance. Laboratory testing with an abrasive machine is important for deciding the wearing properties of untried woods. Recent research shows that periodic oiling increases the resistance considerably ; hence a naturally oily wood is more resistant, other qualities being equal.

Weather Bar. Metal fittings for the bottom of doors and french casements to exclude water, especially when they open inwards. There are several registered types. See *Water Bar*. Also see *Hopper Window*.

Weather Board. See *Weather Moulding*.

Weather Boarding. Horizontal boards covering exposed surfaces and overlapping to keep out rain-water. They are called *clap-boards*,

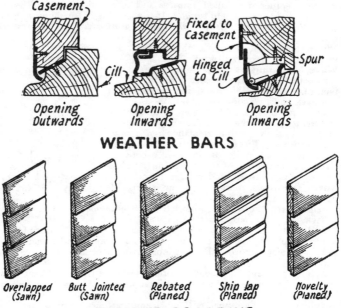

WEATHER BARS

Overlapped (Sawn)   Butt Jointed (Sawn)   Rebated (Planed)   Ship lap (Planed)   Novelty (Planed)

WEATHER BOARDING

q.v., when bevelled and in short lengths, as used vertically or horizontally for fencing, and *sidings* when in long lengths. They are distinguished as bevelled, rebated, and novelty sidings, *q.v.*

Weather Check. A throating, *q.v.*

Weather Deck. An exposed deck on board ship.

Weathered Oak. See *Limed Oak*.

Weathered Sycamore. Figured sycamore treated with steam, to darken the colour to light brown for cabinet work.
Weather Fillet. A protection against driving rain. See *Skylight*.
Weathering. 1. Sloping or bevelling a surface to throw off water. 2. Resisting alternate wet and dry conditions. 3. The chemical and mechanical discoloration and disintegration of wood surfaces due to exposure to weather, dust, light ; and also fibre change due to varied moisture content. See *Decay* and *Durability*.
Weather Moulding. A board or moulding fixed to the bottom rail of an exterior door to keep out draughts and rain.
Webbing. A strong hemp fabric, about 2 in. wide, used as a foundation for the seats of chairs, etc., for connecting the laths of a venetian blind, to form the hinges of a clothes horse, etc.
Wedge. A tapered piece of wood or metal to give a gradually increasing pressure or to be used as a fixing. It is used for cleaving, riving, splitting, etc. The efficiency depends upon the taper, which depends upon the coefficient of friction of the materials in contact. The wedge is one of the mechanical powers, or simple machines. See *Folding* and *Fox Wedging*.
Weed Tree. Applied to a tree that is of little timber value.
Weep Hole. A hole to allow collected water to escape, as condensation on the inside of a window. See *Lantern Light* and *Casement Window*.
Weeping Pine. *Pinus excelsa.* Himalayas. So called because of long drooping needles or leaves. See *Pine*.
Weevil. A general term for various species of beetles destructive to wood.
Weight. 1. The weight of dry wood depends upon the cellular space, that is, the proportion of wood substance to air space. If the cell walls are thick and the cavities small, then the wood is heavy. The actual wood substance has a specific gravity of about 1·55, or 97 lb. per cubic foot. It varies between 1·4 and 1·62. Balsa has about 7 per cent. wood content, Lignum-vitæ has about 85 per cent., and Palo diablo about 90 per cent. In the same species the amount of cellular space depends upon the rate of growth. See *Specific Gravity*. 2. See *Sash Weight* and *Semicircular Head*.
Weinmannia. See *Rose Marara, Tenio, Towhai*.
Weir. A dam or fence of stakes across a river.
Welanga. *Pterospermum suberifolium* (*H.*). India. Dark red. Fairly hard, heavy, and tough. Fine uniform grain, ray flecks. Handsome figure. Used for cabinet work, turnery, etc. S.G. ·72.
Well. 1. The space between the outer strings of a stairs with continuous or return flights. See *Geometrical* and *Open Newel Stairs*. 2. See *Bench*.
Wellingtonia. See *Sequoia*.
Wellinlith. Proprietary wood-fibre wall-boards.
Welsh Dresser. A kitchen dresser with drawers, and framed open shelves and cupboards above.
Welsh Groin. One formed at the intersection of vaults of unequal heights.
Wenge. *Millentia laurentii.* A highly decorative cabinet wood from the Belgian Congo.

West Africa.   There is a large number of woods imported from W. Africa, and the number is continually increasing. The well-known woods include Abura, Afara, Bubinga, Cedar, Cherry Mahogany, Ekki, Iroko, Ironwood, Mahogany, Obeche, Odum, Ofum, Opepe, Sapele, Walnut, etc. See *Nigerian Woods*.

Western.   A prefix applied to the woods from N.W. America such as red cedar, white pine, balsam, hemlock, larch, cottonwood, etc.

Western Arbor-vitæ.   Western red cedar, *q.v.*

Western Balm Cottonwood.   *Populus trichocarpus.*   Black cotton-wood, *q.v.*   Also see *Poplar*.

Western Balsam.   *Abies grandis.*   Grand or Lowland fir, *q.v.*

Western Firs.   The several species of *Abies* from Western North America.

Western Hemlock.   *Tsuga heterophylla.*   See *Hemlock*.

Western Jack Pine.   *Pinus contorta.*   Lodge-pole pine, *q.v.*

Western Larch.   *Larix occidentalis.*   See *Larch*.

Western Pine.   Applied to several species from N.W. America, but chiefly to ponderosa, sugar, and Idaho white pine, *q.v.*   They are excellent soft pines ; large sizes free from defects, plentiful supplies, and used for nearly all purposes.

Western Plane.   Buttonwood, *q.v.*

Western Red Cedar.   *Thuja plicata (S.).*   N.W. America. Called British Columbia and Pacific red cedar. Reddish brown, weathers to grey. Soft, light, durable. Aromatic oil resists insects. Strong for weight, rather brittle, fissile, stable. Straight grain, medium to rather coarse texture. Distinct growth rings. Seasons and works easily. Used for shingles, sidings, joinery, cabinet work, greenhouses, shelters, etc. S.G. ·36. C.W. 1·8.

Western Silver Fir.   *Abies amabilis.*   Amabilis fir. See *Fir*.

Western Spruce.   *Picea sitchensis.*   Sitka spruce, *q.v.*

Western Tamarack.   *Larix occidentalis.*   See *Larch*.

Western White Pine.   *Pinus monticola.*   Western Canada and U.S.A. See *White Pine*.

Western Yellow Pine.   *Pinus ponderosa.*   See *Ponderosa Pine*.

West Indian Boxwood.   *Casearia*, or *Gossypiospermum, præcox*. Mara-caibo boxwood. A good substitute for ordinary box, *q.v.*   S.G. ·82.

West India Satinwood.   *Fagara flava* or *Zanthoxylum flavum*. See *Satinwood*.

West Indies.   The chief imports from the West Indies include balsa, boxwood (Maracaibo), cedar, Cuban mahogany, locust or coubaril, sandalwood or amyris wood, sapodilla, satinwood. They are usually marketed as Central American. Numerous other woods are also described in the Glossary.

Westland Pine.   *Dacrydium colensoi.*   New Zealand silver pine, *q.v.*

Wet-cemented.   Applied to plywoods in which the veneers are not dried, or seasoned, before the adhesive is applied. The drying of the green wood depends upon the heat of the plattens during manufacture. See *Dry-cemented*.

Wet Cooperage.   Casks, etc., for liquids. See *Cooper*. They are usually of oak or chestnut.

Wet Dock.   See *Docks*.

Wet Mill. A saw-mill where logs are stored in water.
Wet Process. See *Wet-cemented*.
Wet Rot. Decomposition of wood due to the weather. The term is usually applied to fungal decay of exterior woodwork exposed to alternate wet and dry. It is also applied to species of dry rot that require a lot of moisture. See *Rot* and *Dry Rot*.
Weymouth Pine. *Pinus strobus*. White or yellow pine, *q.v.*
Wharf. A structure for receiving and discharging goods from and to ships.
Wharf Borer. *Nacerda melanura*. An insect, resembling the Longhorn beetle, that attacks damp and decayed wood.
What-not. A decorative set of shelves for displaying small ornaments, etc. It is also called an etagère.
Wheatsheaf Back. A Chippendale chair in which the splad somewhat resembles a sheaf of wheat.
Wheel Arch. A bentwood or inclined bottom rail at the front of the body of a vehicle.
Wheelback. Applied to Hepplewhite chairs in which the back has decorative splads, radiating from a carved boss, like the spokes of a wheel.
Wheel House. **1.** The rising at the front of the body of a vehicle to provide clearance for the wheels. **2.** The shelter for the steering gear on board ship.
Wheeling Step. A winder, *q.v.*
Wheels. A circular framework, consisting of hub or nave, spokes, and felloe, that rotates to facilitate the movement of a vehicle, etc. The parts of a wheel for a machine are named boss or hub, spokes, and rim. Wheels are distinguished according to their special purpose : carriage, cart, spinning, steering, etc. Also see *Catherine Wheel*.

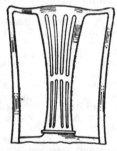

WHEATSHEAF BACK (CHIPPENDALE)

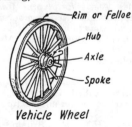

Vehicle Wheel

Steering Wheel

WHEELS

Wheel Window. See *Rose Window*.
Wheelwright. A woodworker engaged in the making of wheels and the heavier types of horse-drawn vehicles.

Wheelwright's Tenon. One with squared shoulders and the tenon narrower than the width of the rail. It is tapered in width so that it wedges tightly into the mortise, and is secured by pinning.

Whelm. A hollow trunk of a tree used as a watercourse. They are usually of elm.

Wherry. 1. A barge used on the Norfolk Broads and Tyne. 2. A rowing boat for conveying passengers on rivers, lakes, etc.

Whetting. Sharpening, *q.v.*

Whim or **Whin.** A pole and two wheels used for hauling timber. There are numerous other names applied to this type of vehicle : timber bob, gin, tug, drug, etc.

Whimble. A shell gimlet.

Whip-and-derry. Rope and pulley used for hoisting.

Whipcrop. See *Wayfaring Tree*.

Whipple-tree. See *Single-tree*.

Whippy. 1. A term applied to a log, or smaller stuff, that is bent in the peeling machine or lathe. 2. Springy or flexible.

Whip Saw. 1. A strong two-handed saw used in pit sawing. 2. A fret or band saw.

Whip Sawing. Same as pit sawing, *q.v.*

Whipstaff. Steering gear of a ship used instead of a steering wheel. A long wood lever used in old ships for moving the tiller.

White. A prefix applied to numerous woods to distinguish the species, and also to the sapwood of certain woods.

White Afara. Eghoin, *q.v.*

White Alder. White beam, *q.v.* Also *Platylophus trifoliaties*. S. Africa. A hard, tough, durable wood, but not exported. Also see *Alder*.

White Ant. *Termes spp.* The most destructive insect to both growing tree and seasoned wood, and the most difficult to eradicate. Few timbers are immune from attack in tropical and sub-tropical countries. Cypress pine, several Eucalypts, and a few tropical woods are resistant. Arsenic and creosote are effective insecticides. See *Preservation*.

White Ash. *Fraxinus americana.* Also called cane ash. See *Canadian Ash*. The name is usually applied to all species of American ash except *Fraxinus nigra*. AUSTRALIAN WHITE ASH. *Eucalyptus fraxinoides*. Similar to European ash in some respects and used as a substitute in Australia and in aircraft. See *Ash* and *Mountain Ash*.

White Aspen. *Pleiococca wilcoxiana* (*H.*). Queensland. Also called snow-wood. Only exported as plywood.

White Balsam. *Abies concolor.* Colorado white fir, *q.v.*

Whitebark. See *Queensland Greenheart*.

White Basswood. *Tilia heterophylla.* See *Basswood*. Also *Duboisia myoporoides* (*H.*). Queensland. Also called Corkwood. Cream colour. Light, soft, with straight firm even grain and texture. Used chiefly for carving, turnery, toys, etc. S.G. ·36.

Whitebeam. *Pyrus*, or *Sorbus*, *aria* (*H.*). Gt. Britain, Europe. Similar to Rowan. Pinkish grey to brownish, with darker lines due to summer wood. Moderately hard and heavy, tough and strong. Straight grain, fine uniform texture. Requires care in seasoning to avoid warp. Fairly difficult to work but finishes well. Used for turnery, carving, handles, mallet heads, etc. S.G. ·72. C.W. 3.

White Bean. See *White Siris.*
White Beech. *Gmelina leichhardtii (H.).* N.S. Wales. Queensland grey " teak," whitish brown. Moderately hard and heavy, brittle and not strong, stable, fairly durable. Close even grain. Used for wheelwright's work, cooper's work, carving, etc. Scarce and not exported.
White Birch. *Betula papyrifera.* Paper birch. See *Birch.* The name is also applied to European birch, *q.v.*, and to the sapwood of *Betula lutea,* or Canadian yellow birch.
White Bombway or **Bombwe.** *Terminalia procera (H.).* India. Like black chuglam. Light brown, darker streaks, lustrous but feels rather rough. Moderately hard and heavy, stable, strong, not durable, liable to insect attack. Straight and interlocked even grain, coarse texture. Requires careful seasoning and fairly difficult to work, polishes well. Large sizes. Used for furniture, better class joinery, carriage work, etc. S.G. ·62 ; C.W. 3·5.
White Borer. *Xylotrichus quadrupes.* An insect very destructive to growing trees in eastern countries.
White Cedar. The following are usually marketed as white cedar : Canoe cedar, Arbor-vitæ, Alaska or Yellow cedar, Port Orford cedar, Incense cedar and Idaho cedar. They are similar to red cedar except for colour and fragrance. BORNEO WHITE CEDAR. White seraya, *q.v.* BRAZIL WHITE CEDAR. *Tabebuia longipes (H.).* British Guiana. Called Warakuri. Straw colour. Hard, heavy, strong, durable, close grain but not smooth, fine rays. Little exported. S.G. ·72. INDIAN WHITE CEDAR. *Dysoxylum binéctariferum* and *D. malabaricum (H.).* Reddish grey to reddish brown, purplish cast, faint stripes, lustrous, smooth. Moderately hard and heavy, strong, durable. Straight, interlocked grain, even fine texture. Easily wrought but requires careful seasoning. Limited supplies. Used for construction, superior joinery, casks, tanks, etc. S.G. ·7 ; C.W. 2·5. NORTHERN WHITE CEDAR. *Thuja occidentalis (S.).* U.S.A. Also called Arbor-vitæ. Pale brown, variegated, slightly fragrant, soft and brashy. Used for posts and poles. S.G. ·32. SOUTHERN WHITE CEDAR. *Chamæcyparis thyoides.* U.S.A. Reddish brown, fragrant. Superior wood to Northern. Used for posts, poles, shingles, planking, woodware, etc. S.G. ·32. See *Cedar.*
Whitechapel Cart. A tradesman's light two-wheeled cart.
White Cheesewood. *Alstonia scholaris (H.).* Queensland. Creamy white. Light, soft, not strong or durable. Straight grain, uniform texture. Easily wrought. Requires careful seasoning. Excellent wood for carving. Used for cheap furniture, crates, etc. S.G. ·5 ; C.W. 2.
White Chuglam. *Terminalia bialata.* See *Chuglam.*
White Cooperage. Applied to such articles as pails, churns, tubs, buckets, etc. Sycamore is commonly used.
White Cornelwood. *Citrus australis (H.).* Queensland. Also called Native Lime. Cream colour, lustrous, smooth. Very hard, heavy, strong, tough, durable. Straight grain, fine even texture. Used for handles, shafts, bentwork, carving, inlays, etc. S.G. ·8.
White Cypress. *Callitris glauca.* See *Cypress.* Also *Taxodium*

distichum. Louisiana, or Bald, cypress, *q.v.* Also a general term for light-coloured cypress.

White Deal. *Picea abies* (*P. excelsa*) and *Abies alba.* European spruce. See *Whitewood* (Baltic). The name white deal should be discontinued.

White Dhup. *Vateria indica* (*H.*). India. Light yellow to light brown, lustrous, feels coarse and fibrous. Moderately hard, heavy, and strong, not durable. Interlocked grain, medium coarse texture, rays and flecks. Requires careful seasoning, easily wrought. Often sold as Malabar white pine. Used for cases, planking, flooring, matchboarding, etc. S.G. ·6 ; C.W. 2·5. See *Dhup* and *Villapiney.*

White Elm. *Ulmus americana.* American elm or Canadian orhamwood. See *Elm.*

White Els. See *Els.*

White Fir. *Abies concolor.* Colorado white fir. Also applied to Alpine, Amabilis, Grand, and Noble fir, and European spruce. See *Fir* and *Whitewood (Baltic).*

White Guarea. *Guarea cedrata.* See *Obobonufua.*

White Gum. *Eucalyptus viminalis* or *E. hæmastoma.* See *Gum.*

White Hazlewood. *Symplocos spicata* (*H.*). Queensland. Not exported except as plywood.

White Hemlock. *Tsuga canadensis.* Eastern hemlock, *q.v.*

White Hickory. Refers to the sapwood of hickory, *q.v.*

White Ironbark. *Eucalyptus paniculata* (*H.*). Australia. Also called Grey ironbark. Pale brown, darkening with exposure. Very hard, heavy, strong, and durable. Gum secretion. Used for structural work, poles, sleepers. S.G. 1·1 ; C.W. 4. Also see *Gum* (*yellow*) and *Ironbark.*

White Ironwood. *Toddalia lanceolata* (*H.*). S. Africa. White. Hard, heavy, strong, tough, elastic, not durable. Close even grain. In demand for spokes, shafts, handles of striking tools, etc. Not exported.

White Lauan. *Shorea spp., Parashorea spp.,* and *Pentacme spp.* The group includes numerous woods : white lauan proper, almon, bagtikan, kalunti, mayapis and mindanao. *Pentacme contorta* and *P. mindanensis* (*H.*). Philippines. Greyish, indistinct sapwood. Moderately hard, heavy, and strong, stable, not durable. Interlocked grain, fairly coarse texture. Requires careful seasoning, fairly difficult to work. Stains and polishes well. Large sizes. Used as a substitute for mahogany, cabinet work, superior joinery, etc. S.G. ·55 ; C.W. 3. See *Lauan* and *Red Lauan.*

White Mahogany. *Eucalyptus acmenoides.* S. and E. Australia. Also called Yellow Stringybark. No resemblance to mahogany and more like Queensland walnut but harder and heavier. Yellowish brown. Very hard, heavy, strong and durable. Close straight grain. Medium sizes. Used for plain cabinet work, superior joinery, sleepers, structural work, etc. S.G. ·85 ; C.W. 4. The name white mahogany is also applied to *Khaya anthotheca* (African mahogany), *Canarium euphyllum* (Dhup), White Lauan, White Seraya, and Prima Vera. It is suggested the name should be discontinued.

White Mangrove. See *Mangrove.*

White Maple. The sapwood of rock maple. *Acer saccharum.* See *Maple.*

White Meranti. Yellow meranti. *Shorea bracteolata* and other *spp.* See *Meranti.*

White Mountain Ash. *Eucalyptus regnans.* Used for the same purposes as European ash. See *Mountain Ash.*

White Oak. Applied to several species of American oak. See *Oak.*

White Olivier. *Terminalia amazonia.* Nargusta, *q.v.*

White Pear. *Apodytes dimidiata* (*H.*). S. Africa, Greyish brown, purplish streaks. Hard, fairly heavy, strong, close fine grain. Valued for felloes. S.G. ·75. Also see *Pear.*

White Peroba. *Paratecoma peroba.* See *Peroba.*

White Pine. (Northern.) *Pinus strobus.* See *Yellow Pine.* WESTERN WHITE PINE, *P. lambertiana* (Sugar Pine) and *P. monticola* (Idaho White Pine). Both are similar in many respects to *Pinus strobus.* NEW ZEALAND WHITE PINE. *Podocarpus dacrydioides.* See *Kahikatea.*

White Pocket Rot. Decay caused by the fungus *Fomes annosus.* Also called red rot. Very destructive to the heartwood of softwoods

White Poplar. *Populus alba.* See *Poplar.* Also applied to Canadian aspen, *Populus tremuloides.*

White Rot. Fungal decay in which the coloured constituents are removed leaving a white substance, chiefly cellulose. The lignin is destroyed or modified. Also called soft rot. See *Rot.*

White Seriah or **Seraya.** *Shorea plagata.* See *Seraya.*

White Silkwood. *Flindersia acuminata* (*H.*). Queensland. Similar to Maple Silkwood, *q.v.*, except for colour.

White Silver Pine. New Zealand silver pine or Westland pine, *q.v.*

White Siris. *Albizzia procera.* See *Siris.* Also *Ailanthus inerbiflora* (*H.*). Queensland. Also called White Bean and Sassafras. Silvery white. Very soft and light, but rather tough. Not strong or durable. Straight grain, slightly woolly texture. Easily wrought, smooth. Requires careful seasoning. Used chiefly for toys, brush stock, boxes, small cabinet work. S.G. ·4 ; C.W. 2.

White Spruce. *Picea glauca* and *P. alba.* Canada. Very similar to Whitewood (Baltic), *q.v.* See *Picea, Spruce,* and *Fir.*

White Stinkwood. Camdeboo. See *Stinkwood.*

White Stringybark. *Eucalyptus eugenioides.* Victoria, Australia. Pale brown. Hard, heavy, fissile, very durable. Excellent wood and not difficult to work. Used for general purposes, shipbuilding, sleepers, carving, etc. S.G. ·85 ; C.W. 3. See *Stringybark.*

White Tamarind. *Cupania xylocarpa* (*H.*). Queensland. Creamy white, lustrous. Hard, heavy, tough, strong, elastic. Straight fine grain. Seasons and works satisfactorily, polishes well. Esteemed for billiard cues, also used for floors and superior joinery and tool handles. S.G. ·8 ; C.W. 3. A similar Queensland wood, *Cupania pseudorhus,* is called Pink Tamarind. See *Tamarind.*

White Thingan. See *Thingan.*

Whitethorn. Same as hawthorn, *q.v.*

White Walnut. *Juglans cinerea.* Butternut, *q.v.*

White Willow. *Salix alba.* See *Willow.*

Whitewood. Applied to several different woods. AFRICAN WHITE-WOOD, Obeche, *q.v.* AMERICAN, Canary whitewood, see *Canary Wood*. BALTIC WHITEWOOD, *Picea abies* (*S.*). N. Europe. Yellowish white to white. Soft, light, fairly strong, not durable. Easily seasoned and wrought, except for small hard knots. See *White Deal*. Shipments from Jugoslavia also include *Picea omorika*, and from Central Europe *Abies alba*, or silver fir. Used for general purposes, cheap constructional work, and joinery, flooring, matching, food containers, shuttering, temporary timbering, etc. S.G. ·5 ; C.W. 1·5. See *Fir*. BORNEO WHITEWOOD, White Seraya, *q.v.* SOUTHERN WHITEWOOD, Magnolia, *q.v.* TASMANIAN WHITEWOOD, Tallowwood, *q.v.*

Whittle. 1. To pare, shape, or carve, by removing shavings with a knife. 2. Indiscriminate or aimless cutting of wood or trees. 3. A sheath knife.

Whole Timbers. Unsawn balks.

Whorl. 1. A disc used to steady or balance a spindle. 2. A single turn of a spiral.

Whorled Foot. A carved foot to furniture carved into the shape of a whorl. It is a characteristic of the Queen Anne period.

Wicker or **Wickerwork.** Pliant stems, such as osier, willow, etc., interwoven into baskets, light furniture, etc.

Gate Post   Grill

Wicket Gate

WICKET GATE

Wicket. 1. A small door framed in a large gate, for foot traffic. 2. The lower half of a stable door. 3. The stumps, or sticks, used in the game of cricket.

Widdringtonia. Boom, or Clanwilliam, cedar (*W. juniperoides*), *q.v.* Mlanji cedar (*W. whytei*), *q.v.*

Wide Ringed. Applied to coarse-grained woods ; due to too rapid radial growth.

Widia Saws. Circular saws with specially hardened inserted teeth. The teeth are inserted in grooves and pinned. The saws are specially useful for very hard woods and may be used for months without sharpening.

Wild Cherry. *Prunus avium.* See *Cherry*.

Wild Cinnamon. *Cinnamomum zeylanicum* (*H.*). India, Burma. Light brownish grey to yellowish brown, rather lustrous, smooth, and fragrant. Variable weight and hardness, fairly durable, liable to insect attack. Straight grain, some wavy, medium even texture. S.G. ·5 to ·72. Also *C. cecicodaphne* and *C. inunctum*, India and Burma. These two species have no market names but they provide superior wood to the above, with qualities of *C. tavoyanum*. See *Cinnamon*.

Wild Durian. *Cullenia excelsa.* See *Durian*.

Wilde Sering. *Burkea africana* (*H.*). S. Africa. Yellowish brown to dark red, lustrous. Moderately hard and heavy, tough, strong, and

stable. Handsome grain. Esteemed for cabinet work. Not exported
S.G. ·72.

Wild Lemon. *Xymalos monospora (H.).* S. Africa. An ornamental wood
appreciated for cabinet work and superior joinery, but not exported.

Wild Olive. Mastic, *q.v.*

Wild Pear. *Pirus piraster.* See *Service Tree.*

Wild Pine. *Pinus sylvestris.* See *Redwood.*

Wild Service Tree. *Pyrus torminalis.* See *Service Tree.*

Wilga. *Geijera spp. (H.).* Queensland. Also called Green Satinheart, Flintwood, Glasswood, etc. Cream, variegated, lustrous.
Extremely hard, heavy, strong, resilient. Close, dense, short grain,
even texture. Used for tools, hand-screws, skate rollers, fishing rods,
etc. S.G. 1·1 ; C.W. 5.

Willd. Abbreviation for the name of the botanist Willdenow.

Willesden Paper. A proprietary waterproof paper used as a lining
on roofing boards and sheathing. It is tough and rot-proof, and may
be obtained up to three-ply. The preparation consists of impregnating
with cupramonium, obtained from ammonia solution and copper
turnings.

William and Mary. Applied to furniture
characteristic of the style prevailing during
the years 1690-1702. It was marked by
Dutch influence and the increased use of
walnut. Cabriole legs, both smooth and
with carved knee, and club, claw, or ball
feet were characteristics. Marquetry and
scrolled Flemish legs were common.

Williott Diagrams Deflection diagrams
used in the design of framed structures.

Willow. *Salix spp. (H.).* England, Europe,
N. America. About 160 species. The willow
family includes *Populus* and *Salix.* Yellowish

**WILLIAM
& MARY**

brown or fawn colour. Light, soft, stable, resilient, tough and
flexible, difficult to fracture, dents easily. Very easily wrought.
Better qualities used for cricket bats and artificial limbs. Also used for
slack cooperage, veneer cores, boxes, baskets, linings to carts, brakes, etc.

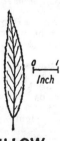

WILLOW

S.G. ·4; C.W. 1·2. ALMOND LEAVED W. *S. triandra.*
Used for willow rods for basket work. AMERICAN
W. See *Black W.* and *Sauce colorado.* BAY LEAVED
W. *S. pentandra.* BEDFORD W. *S. viridis.* BLACK
W. *S. nigra.* Called Southern willow and sold as
Swamp walnut. Harder than English willow.
Used for veneers, furniture, etc. CLOSE BARK W.
S. cœrulea. Cricket bat willow. Grows rapidly ; annual
rings sometimes ¾ in. apart. Whitish, straight grain.
The best wood is used for cricket bats and artificial
limbs. COMMON W. White willow, *q.v.* CRACK
W. *S. fragilis.* One of the largest willows. In-
ferior to close bark willow. Frequently pollarded.
Reddish colour. Used for sieve rims, etc. DARK
LEAVED W. *S. nigricans.* GOAT W. *S. caprea.*

GOLDEN W. *S. vitellina.* HYBRID W. *S. viridis.* Allied to close bark willow. Used for cheaper quality bats, etc. INDIAN W. *S. tetrasperma.* Pale red. Light, soft, even grain, porous. Inferior to English. JAPANESE W. *S. urbaniana.* A substitute for close bark willow, but inferior, and brownish in colour. LAPLAND W. *S. lapponicum.* A shrub. OPEN BARK W. Crack willow, *q.v.* PURPLE W. *S. purpurea, S. hippophaëfolia,* and *S. rubra.* Provides willow rods for basket weaving. SADLER'S W. *S. sadleri.* A shrub. SAUCE COLORADO W. *S. humboldtiania.* C. America willow. Not imported. SMALL TREE W. *S. arbuscula.* TAN LEAVED W. *S. phylicifolia.* TWIG W. *S. viminalis.* WEEPING W. *S. babylonica.* A native of Asia. Of little timber value. WHITE W. *S. alba.* One of the largest willows. Wood of similar quality to crack willow. WHORTLE LEAVED W. *S. myrsinites.* A shrub. WOOLLY W. *S. lanata.* A shrub. Also see *Aspen, Cottonwood, Osier, Sallow,* and *Salix.*

Willow Myrtle. See *Peppermint.*

Willow Oak. *Quercus phellos.* A species of American red oak. See *Oak.*

Willow Rods. Flexible twigs for basket weaving obtained from several species of willow.

Wils. Abbreviation for the names of the botanists E. H. Wilson and P. Wilson.

Wimble. A type of auger, *q.v.*

Winch. A machine for hoisting heavy loads. It is arranged on the principle of the *wheel and axle,* or *windlass.* Also called a crab.

Wind. See *Winding.*

Wind Beam. A collar beam, *q.v.*

Wind Bent. A braced frame to resist wind stresses.

Wind Brace. A brace in a framed structure to resist distortion by the wind.

Wind Break. 1. A tree broken by the wind. **2.** A fence or hedge to break the force of the wind.

Wind Chest. A long wood box in an organ for storing the wind, which is admitted to the pipes as required through small holes in the top of the chest. There is a chest to each manual and one to the pedals. Sometimes the chest is divided. See *Slider, Pallet,* and *Sound Chest.*

Wind Cutter. The upper lip to the mouth of an organ pipe, *q.v.*

Winders. Radiating steps to form a change of direction from that of the flyers in a flight of stairs. See *Geometrical Stairs, Balanced Kite Winders,* and *Stretchout.*

Winding. Warped, or twisted, or cast. Not a plane surface. See *Warping.*

Winding Gear. Appliances for lifting or moving heavy loads : capstan, crab, winch, windlass, etc., with the necessary blocks and ropes.

Winding Stairs. Spiral and solid newel stairs, *q.v.*

Winding Sticks. Two parallel strips of wood used to test a plane surface. By sighting the top edges of the sticks one can tell whether the surface is twisted or not.

WINDING STICKS

Wind Jammer. A large sailing ship for long voyages.
Windlass. A lifting appliance in which the rope winds round a horizontal drum.
Window. A glazed frame in an opening to provide light, ventilation, or ornamentation. The area of glass for habitable rooms should equal at least one-tenth of the floor area, and at least half of the window should open for ventilation purposes. Windows may be of wood or metal. They are classified as fast sheets, casements, pivoted sashes, sliding sashes, and Yorkshire lights. Wood frames may be cased or boxed, or solid. Windows are also distinguished as bay, borrowed, bow, bull's eye, double, dormer, french, hopper or hospital, lay lights, lantern lights, mullion, sash and frame, skylight, storm, and venetian.

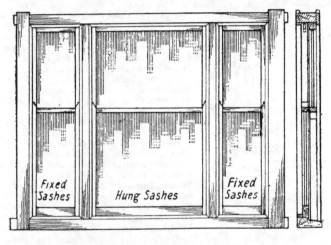

WINDOW (VENETIAN)

They are also distinguished according to their architectural features, or by the number of divisions, or lights, in the frame. There are numerous patented and registered types, and combinations of the above types. The size of a window is usually denoted by the size of the wall opening.
Window Back. 1. The inside boarding or framing between the floor and the window bottom of an ordinary window. See *Lifting Shutters.* 2. The framing forming the back of the enclosure to a shop window. See *Stall Board.*
Window Blind. See *Blind* and *Venetian Blind.*

Window Board. **1.** The show-board or window bottom on the inside of a shop window. **2.** A shelf fixed to the inside of the sill of a window frame. See *Jamb*, *Hopper Window*, and *Lifting Shutters*.

Window Bottom. See *Stall Board*.

Window Box. **1.** A long trough on the sill of a window opening containing plants. **2.** See *Sash and Frame*.

Window Finishings. Architraves, linings, shutters, etc., to a window.

Window Fittings. **1.** Blinds, curtains, pelmets, etc., and the necessary fixings. **2.** The fittings required in a shop window for the display of goods.

Window Frame. A cased or a solid frame in which the sashes are fixed, hinged, pivoted, or sliding. See *Window*.

Window Linings. **1.** Thin boards to cover the reveal between the window frame and face of wall. They usually serve as a screed for plastering and the edge is covered by an architrave. See *Hospital Window*. **2.** Any thin material to cover the joint between frame and brickwork. See *Fixed Sash*. **3.** The separate parts to build up a boxed frame, *q.v.* Also see *Jamb*.

Window Sash. See *Sash*.

Window Seat. A fixed seat to a window back under a window to a living-room, especially to a bay window.

Wind Shakes. Applied to cup or ring shakes and upsets, *q.v.*

Windsor Chair. A wood chair with curved sides and high hoop-shaped back filled with turned balusters. The legs are usually turned, and braced by stretcher rails.

Wind Stick. The propeller of an aeroplane.

Wind Tie. A diagonal tie to brace a structure against wind pressure.

Wind Trunk. A square wood tube conveying the wind from the bellows to the several parts of an organ.

WINDSOR CHAIR

Wine Press. An appliance for squeezing the juice from grapes.

Winet Saws. Widia saws, *q.v.*

Wing. **1.** A projecting and smaller part at the side or end of a larger building. **2.** The wide projecting surfaces that help to lift and support an aeroplane. The surface and parts of the frame are of plywood. The box spars run lengthwise and are stiffened by shaped ribs and diagonal ties. The shaped ribs terminate with a shaped block at the front and taper to the trailing edge at the back. **3.** See *Wings*.

Wing Compass. A type of compass in which a set screw fixes the position of the movable leg on a wing or quadrant stay.

Wing Nut. A nut to a bolt, with projections so that it may be turned by hand and without spanner. See *Banjo* and *Twist Bit*.

Wing Rails. Side rails.

Wings. The sides of a theatre stage. See *Wing*.

Wing Skid. A small skid or runner under the wing tip of an aeroplane to avoid damage when landing.
Wing Transom. See *Stern Timbers.*
Wing Window. A smaller window at the side of a larger window or doorway. See *Flanking Window* and *French Door.*

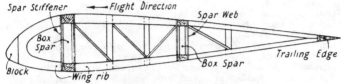

WING SECTION

Winning Post. A post or other object to denote the end of a race in sporting events.
Winteranaceæ. Same family as Cancllaceæ, *q.v.*
Winters' Bark Tree. *Drimys winteri* (*H.*). S. and N. America. Light brown. Fairly light and soft. Tough and strong for its weight, not durable. Straight uniform grain, fine texture. Easily wrought. Somewhat like basswood. Used for construction, interior joinery, box making. Bark important for medicine.
W.I. Puncheon. Abbreviation for *West Indies puncheon staves.*
Wira. *Hemicyclia sepiaria* (*H.*). Ceylon. Whitish sapwood, small dark heartwood. Very hard and heavy. Close texture. Small sizes. Used for turnery and small fancy articles. S.G. 1.
Wire Edge. **1.** A very thin unsupported edge left after grinding or sharpening a cutting tool. After sharpening the bevelled side of the tool it is turned over and rubbed flat on the oil-stone to remove the wire edge. See *Sharpening* and *Oil Stone.* **2.** Slash figure check on the edge of a piece of wood.
Wistaria, Wild. *Bolusanthes speciosus* (*H.*). S. Africa. Of little timber value and not exported. It is very durable and resists insects, and used chiefly for fencing.
Witches' Broom. A disease due to the fungus *Exoascus.* It often attacks hornbeam and birch trees.
Witch Hazel. Red gum, *q.v.* In this country the name is also applied to hornbeam and wych elm. See *Wych Hazel.*
Witch Knots. A tangled clump of twigs on oak branches, etc., caused by fungus.
Withes. **1.** Reeds for interlacing thatching to roofs. **2.** The divisions between the flues of a chimney.
Withy. **1.** A flexible band of osier twigs or other pliant shoots. **2.** The crack willow, *q.v.*
Wit Sering. *Kirkia acuminata* (*H.*). S. Africa. Greyish brown. Difficult to season. A secretion makes the wood gritty and difficult to work. Wavy and irregular grain produces a burr walnut effect. Esteemed for cabinet work. S.G. ·6.
Wobble Saw. See *Drunken Saw.*

Wodier. *Odina wodier* (*H.*). India. Resembles cigar-box cedar. Numerous vernacular names. Light yellowish brown to brownish red, lustrous, smooth. Fairly hard and light. Straight and interlocked grain, medium even texture. Seasons well, easily wrought but liable to pick up if slash sawn. Gum content clogs saw teeth. Used as a substitute for plain mahogany and for nearly all purposes in India. S.G. ·6 ; C.W. 3.

Wold. A woodland or small forest.

Wolder. A fishing boat rigged as a lugger or as a dandy, and used in the Wold area.

Wolf. A term used in forestry for a tree that has grown faster than the surrounding trees and attained a greater height.

Wolmanol. A grade of Wolman Salts, *q.v.*, for application by brush or spray. It is supplied in liquid form ready for use. The *A* grade is intended for use on exposed timber.

Wolman Salts. A group of aqueous wood preservatives of fluoride-phenol-chrome composition, with or without the addition of insecticidal and fire-resistant chemicals. They are applied by vacuum and pressure impregnation or by open tank soaking. Special grades may be obtained for brushing or spraying. Plants for the treatment of timber are operating throughout the country and Wolmanised timber is obtainable through any timber merchant. The treated wood is clean, odourless, and may be painted. See *Minolith, Tanalith, Triolith,* and *Wolmanol.* Also see *Preservation.*

Womara. See *Wamara.*

Wood. 1. The lignified part of a tree within the cambium ; an aggregate of plant cells ; the xylem part of fibro-vascular tissue. See *Wood Elements.* Owing to its many qualities such as appearance, elasticity, non-conduction of heat, resonance, etc., together with its combination of lightness and strength and ease of working, wood is used for nearly all purposes. See *Annual Rings, Cells, Growth, Parenchyma, Timber, Timbers, Vessels, Wood Substance.* 2. A tract of tree-covered land. A small forest.

Wood-agate. Petrified wood.

Wood Apple. *Feronia elephantum* (*H.*). India. Yellowish grey, darker streaks, rather dull. Hard, heavy, durable, and fairly strong. Straight and narrowly interlocked grain, distinct rays, even medium texture. An inferior substitute for boxwood. Not imported. S.G. ·82.

Wood Beetles. See *Beetle* and *Death Watch Beetle.*

Wood-block. A block of fine-grained wood engraved for printing.

Wood Blocks. Blocks used as floor coverings over concrete. They are usually 9 in. by 3 in. by 1¼ in. See *Block Floor.*

Wood Carving. 1. The art of cutting figures or designs in wood. 2. A piece of carving in wood.

Woodcraft. Skill and knowledge appertaining to forestry and forest conditions.

Wood-cut. 1. An engraving or design cut on the face of wood for printing purposes. 2. The design or picture so produced. See *Wood Engraving.*

Woodcutter. 1. One who makes wood-cuts. A wood engraver. 2. One engaged in felling trees.

Wood Engraving. A design cut on the end grain of wood for multiple reproduction. See *Woodcut*.

Wood Elements. The cellular units composing wood : ray cells, parenchyma, vessels or pores, fibres and tracheids, *q.v.* Also see *Wood*.

Wood Fibre. 1. Wood pulp, *q.v.* **2.** See *Fibres*. **3.** Wood wool, *q.v.*

Wood Finishings. Protective and decorative coatings : french polish, cellulose lacquer, liming, and wax. The term is also applied to painting, varnishing, etc. Also see *Finishings*.

Wood Flour. Waste wood specially prepared in the form of meal and used as a filler for numerous synthetic materials, jointless flooring, wall boards, linoleum, explosives, plastics, plastic wood, etc.

Woodlock. 1. A patent friction stay for casements. **2.** A thick piece fitted on the rudder of a ship to keep it in position.

Wood Meal. Wood flour, *q.v.*

Wood Nogs. See *Nogs*.

Wood Oil. Tung oil ; used for varnish, putty, soap, etc.

Wood Oil Tree. *Aleurites cordata* (P.). China. The Tung Tree. White. Hard, heavy, durable, stable. Resists insects and heat. Used for musical instruments, joinery, etc.

Wood Pulp. Wood reduced to pulp for making paper, fibre-boards, etc. It may be produced mechanically or chemically. The operations include barking, sawing, chopping, crushing, digesting or boiling, bleaching, treatment for sale as *half-stuff*, and soda recovery. It is produced chemically by means of caustic soda or bisulphate of lime.

Wood Rays. See *Rays*.

Wood Screw. 1. A metal screw for use in wood. See *Screws*. **2.** A large screw or helical thread formed on wood, as for hand-screws, vice-screws, etc.

Woodshed. A place for cutting and storing firewood.

Wood Slip. See *Pallet*.

Wood Substance. The chief chemical constituent in cellulose, together with lignum, hemicellulose, and small amounts of other chemicals. It is like a stiff gel of a fibrous or crystalline structure.

Wood Tar. Tar obtained from wood by distillation, as distinguished from coal-tar.

Wood-ware. Dairy and domestic kitchen and table utensils made of wood : bread platters, bowls, ladles, pastry-boards, trenchers, pails, wash-boards, etc. Also sometimes applied to fitments, cupboards, etc., of a loose or semi-loose type, especially when of mass production.

Wood Wool. Shavings used for packing and for surgical dressings after treatment. It is made from light, straight-grained wood, free from odour, resin and gum, and brittleness when dry : basswood, cottonwood, etc. A wood-wool machine is used for the purpose. Wood-wool is also used as a substitute for hair in plaster and many other purposes. Also see *Excelsior*.

Woodwork. Applied to the wood used in buildings, especially to the joinery.

Woollybutt. *Eucalyptus longifolia*. New South Wales and Victoria. Called Alpine ash. Similar to Jarrah in properties and colour.

Reddish. Hard, heavy, strong, and durable. Interlocked grain. Used for sleepers, structural work, paving, carriage work, all kinds of exterior work. Little imported. S.G. ·95. *Euc. botryoides*, Queensland, and *Euc. miniata*, W. Australia, are also called woollybutts and are similar to *E. longifolia*.

Workability. See *Comparative Workability* and *Working Qualities*.

Work-box. A receptacle for sewing materials.

Working. 1. Applied to the movement or dimensional changes of wood due to variation in moisture content, *q.v.* 2. Shaping and preparing wood.

Working Factor. Same as *factor of safety*, q.v.

Working Qualities. The properties of wood with which the labour costs vary. One wood may entail six times as much labour as another wood for the same preparation. The difference is due to density, hardness, conversion, variation in grain structure and direction, seasoning, resin, mineral content, knots, stability, etc. Most woods work easier when seasoned, but some are more difficult due to hardening of secretions. Interlocked grain invariably adds to the labour costs. See *Comparative Workability* and *Selection*.

WORK-TABLE

Work-table. A small table with receptacle for sewing materials, etc. Also called a pouch table.

Worm-holes. The holes, over $\frac{1}{8}$ in. diameter, formed by wood-boring larvæ. See *Pin-hole* and *Shot-hole*.

Wormia. See *Simpoh*.

Worm Roller. A screw roller, *q.v.*

Worms. Applied to insects that attack either wood or growing trees by boring. See *Teredo, Limnoria,* and *Worm-holes*.

Wormwood. *Eucalyptus rudies.* W. Australia. Also called Flooded Gum. See *Gum*.

Wot. Abbreviation for *wrought*.

Wovenboard. Woven fencing. Wovenoak. See *Interwoven Fencing*.

Wrack. Applied to inferior qualities of wood from the Baltic. Used chiefly for packing cases. It is the sixth grade and usually the lowest grade exported. Double wrack is the seventh quality in grading.

Wrack Wainscot. Billets or logs not included in the usual merchantable wood. The degrading may be due to sizes or to quality.

Wrapping. Building-up veneered surfaces for double-curvature work. The veneers are in narrow widths to bend round the curved surface in a spiral manner.

Wreath. A part of a continuous handrail going round a curved plan.

Wreathed Column. One shaped or carved in helical form. A twisted or spiral column.

Wreathed Stairs. A geometrical stairs, *q.v.*
Wreathed String or **Piece.** The curved part of a continuous string

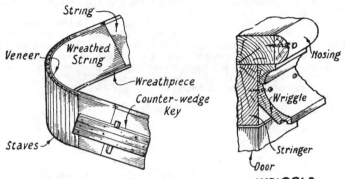

WREATHED STRING   **WRIGGLE**

for a geometrical stairs, *q.v.* That part of the string immediately
under the wreath.
Wrench. A spanner, *q.v.*
Wrest Block or **Plank.** The part of a piano frame to which the wrest
pins for the strings are fixed. It is usually of quarter-sawn beech.
Wriggle. A protection over an exposed door or window on board
ship to throw the water away from the opening.
Wrightia. See *Lanete* and *Pala*.
Wringer. A machine with rollers for squeezing the water out of
clothes. See *Rollers*.
Wrinkled Barks. Applied to several species of Eucalyptus because
of the condition of the bark. See *Blackbutt, Bloodwood, Mahogany
Gum*, and *Woollybutt*.
Writhed. Twisted. A term used by
the cratemaker when twisting the rods
whilst hot.
Writing Bureau. An elaborate type
of writing desk well equipped with
drawers and pigeon-holes for papers
and documents.
Writing Desk. A piece of furniture
specially shaped for writing on, and
with a receptacle for holding papers.
The types vary from a box with hinged
sloping lid to elaborate pieces of furniture. See *Kneehole, Cylinder Fall,
Escritoire, Pedestal*, and *Roll Top Desk*.

BUREAU

Writing Table. A flat writing desk. See *Kneehole Desk*.
Wrongs. The largest branches of an oak tree suitable for conversion.
Wrot. Abbreviation for *wrought*.

Wrought Stuff. Planed-up wood, as distinct from sawn stuff.
Wu-chang. *Cinnamomum sp.* Chinese cinnamon tree.
Wu-tien. *Castanopsis sp.* China. Like Indian chestnut. See *Chestnut*.
Wych Elm. *Ulmus glabra* (*H.*). Gt. Britain. Golden brown to brown with greenish streaks or tinge. Moderately hard and heavy ; strong and tough. Straight grain, distinct rings give handsome figure on slash-sawn wood, coarse texture. Not difficult to work. Requires care in seasoning, but less liable to shakes than common elm. Used for wheelwright's work, agricultural implements, bent work, boat-building, selected wood for fixed joinery. See *Elm* and *Ulmus*.
Wych Hazel. Wych elm, *q.v.*
Wythe. See *Withe*.

X

X. A symbol denoting a section of wood at right angles to the grain. A transverse or cross section.

Xanthostemon. See Penda.

Xanthoxylum. (*Fagara.*) See *Cohucho* and *West Indies Satinwoods*. Also see *Zanthoxylon*.

Xebec. A three-masted sailing vessel common to the Mediterranean Sea.

Xestobium. See *Death Watch Beetle*.

Ximenia. See *Hog Plum* and *Olive* (*Wild*).

X tgd. Abbreviation for *cross-tongued*.

Xylem. The woody elements of vascular tissue. Woody tissue as contrasted with phlœm. The inner new cells in the cambium layer. The principle water-conducting and strengthening tissue of the stem, etc., of vascular plants.

Xylia. See *Irul, Jamba, Pyinkado*.

Xylocarpus. (*Carapa.*) See *Nyireh, Pussur Wood, Puzzle Fruit Tree*.

Xylomelum. See *Pear* (*Native*).

Xylophone. A musical percussion instrument.

Xylopia. See *Berberine* and *Pindahyba*. Also see *Zylopia*.

Xylotrichus. See *White Borer*.

Xylotrupes gideon. A beetle very destructive to growing trees.

Xymalos. See *Wild Lemon*.

Y

y. A symbol, used in structural design, for the distance from the neutral axis to the extreme fibres of the section.

Y. Abbreviation for *yellow*.

Yaba. *Andira inermis*. Cuba. See *Andira* and *Moca*.

Yaca. *Santalum yasi* (*H*.). Fijian sandalwood. Resembles Rimu. Yellow brown, variegated, fragrant, fairly lustrous. See *Sandalwood*.

Yacal. *Shorea* and *Hopea spp.*, and *Isoptera sp.* (*H*.). Philippines. It is the name applied to the harder and heavier species of the Lauan family. Like Thingan but harder and heavier. Light yellow brown. Very hard, heavy, and durable; resists insects. Close interwoven ribbon grain. Firm fine texture. Requires careful seasoning. Difficult to work but finishes well. Good substitute for teak for many purposes. Used for structural work, piling, shipbuilding, flooring, etc. S.G. 1; C.W. 4. Also see *Damar Laut*.

Yacca. *Podocarpus coriaceus* (*S*.). Jamaica. Cypress or Yellowwood. See *Podo* and *Podocarpus*.

Yachidamo. Japanese ash, or Tamo, *q.v.*

Yacht. 1. A light sailing racing vessel. 2. A vessel for private use and designed for pleasure cruising, whether propelled mechanically or by sails only.

Yagrume. *Didymopanax morototsni* (*H*.). C. America. Yellowish grey. Fairly light and soft, but firm, rather brittle, not durable. Straight grain, rather coarse texture. Easily wrought. Suggests alder. Used for carpentry, boxboards, etc. S.G. ·45.

Yale Lock. A cylindrical lock with a corrugated key that operates a series of pins in the barrel of the lock to move the latch. It is popular because of its neat appearance and the difficulty of duplicating the key.

Yamane. Gumhar, *q.v.*

Yamao. *Guarea trichilioides*. Cuba. See *Marinheiro*.

Yana. *Ximenia americana*. Cuba. See *Hog Plum*.

Yang. *Dipterocarpus tuberculatus*. Siam. Same as Gurjun and Eng, *q.v.* The name yang is also applied to Chinese poplar and the banyan tree.

Yapunyah. *Eucalyptus ochrophloia*. Queensland. Compares favourably with the best of the Eucalypts, *q.v.*

Yard. 1. A long spar, tapering towards the ends, fastened to the mast of a ship to support the sail. They are distinguished as main, mizzen, square, lateen, cross-jack, sprit-sail, fore-topsail, driver, royal, studding sail, top gallant, etc. 2. An enclosed area, usually paved, belonging to a building or business. 3. A measure of length of 36 in. 4. Nine square feet, or a square yard, of flooring, matching, etc. In the timber trade it often implies 9 sq. ft. before machining.

Yardarm. The end of a yard, especially if for a square sail.

Yarder. The engine that collects the felled logs on to the drag.

Yard Keeper. A timber importer who stores timber. See *Brass Plater*.

Yard Stain. Stain that develops during air-seasoning. See *Stain*.

Yard Stick. A graduated rod, 3 ft. long, used for measuring.

Yarn Beam. A roller in a loom on which the warp threads are wound in weaving.

Yarn Winder. A wood stand carrying revolving arms for winding wool or yarn.

Yarran. A species of myall, *q.v.*

Yarrow Guard. A patent guard for a spindle moulding machine.

Yarúa. *Peltophorum brasilense.* Cuba. See *Moruro* or *Cana fistula* and *Sea Grape.* Species of *Coccoloba* and *Brasilettia* are also called Yarua.

Yaruru. *Aspidosperma sp.* Paddlewood, *q.v.*

Yate. *Eucalyptus cornuta* (*H.*). Australia. Pale yellow brown. Extremely hard, heavy, strong, tough, and very dense. Perhaps the strongest and hardest wood known. Tensile strength nearly equal to wrought iron. Very difficult to work. Scarce and little exported. Used for shafts, spokes, wheelwright's work, etc. S.G. 1·2; C.W. 6. BLACK YATE. See *Yorrell*.

Yawl. 1. A ship's small boat. 2. A small cutter-rigged sailing vessel. The smallest type of two-masted sailing ship. The after mast is very small compared with the main mast.

Yaya. *Oxandra lanceolata.* Cuban lancewood. See *Lancewood*.

Yd. Abbreviation for *yard*.

Yealm. A bundle of thatch for a thatched roof.

Yellow. A prefix applied to numerous woods to distinguish the species.

Yellow Birch. *Betula lutea.* Canada. See *Birch*.

Yellow Box. *Eucalyptus melliodora* (*H.*). Victoria, Australia. Yellowish brown. Very hard, heavy, strong, tough, durable. Wavy interlocked grain. Difficult to work, polishes well. Used for structural work, sleepers, piling, planking. Little exported. S.G. 1; C.W. 4·5. See *Yellow Wood*.

Yellow Boxwood. *Sideroxylon pohlmanianum* and *S. myrsinoides* (*H.*). Queensland. Another species is called bulletwood. Very similar to European box and used for similar purposes. S.G. ·9.

Yellow Buckeye. *Æsculus octandra* (*H.*). U.S.A. Pale yellow, lustrous. Ripple marks. S.G. ·46. See *Horse Chestnut*.

Yellow Cedar. *Cupressus nootkatensis* or *Chamæcyparis n.* See *Cedar*.

Yellow Cheesewood. 1. *Eucalyptus raveretiana.* See *Ironbox*. 2. *Sarcocephalus cordatus* (*H.*). Queensland. The Leichardt tree. Only exported as veneers.

Yellow Cypress. *Thuja excelsa.* Western Canada, and *Taxodium distichum*, U.S.A. See *Cypress.* Also applied to yellow cedar, *q.v.*

Yellow Deal. *Pinus sylvestris.* Baltic redwood. See *Redwood*.

Yellow Fir. Douglas fir, *q.v.*

Yellow Gum. *Liquidambar styraciflua.* American red gum, *q.v.*

Yellowheart. *Fagræa fragrans* (*H.*). India, Burma. Called Anan or Burma yellowheart. Yellowish brown to old gold, dull to fairly lustrous, smooth, feels oily, odorous. Hard, heavy, strong, durable, resists insects. Straight and interlocked grain, even medium coarse texture. Requires care in seasoning, stains easily. Fairly difficult to work smooth. A valuable wood where strength and durability are important. Little imported. S.G. ·8 ; C.W. 4.

Yellow Meranti. *Shorea bracteolata* and other *spp.* See *Meranti*.

Yellow Pine. **1.** *Pinus strobus (S.).* Canada. Called white, Weymouth, and Quebec pine ; also cork pine, soft white pine, and Northern white pine. Pale yellow. Light, soft, very stable, not strong or durable. Fine resin ducts, that show as short hair-like markings on planed surfaces due to the resin catching dust. Very uniform texture. Easily wrought and seasoned. Large sizes free from knots or defects. Esteemed for interior joinery, pattern making, carving, cores for veneer. Expensive owing to demands. S.G. ·45 ; C.W. 1. Very similar woods are Western white pine (*P. monticola*), sugar pine (*P. lambertiana*), Siberian pine (*P. cembra*), Korean pine (*P. koraiensis*), Chinese pine (*P. armandi*), Japanese pine (*P. parviflora*), North Indian pine (*P. excelsa*), Western yellow pine (*P. ponderosa*). **2.** The American name for the various species of pitch pine, especially when used for flooring. **3.** New Zealand silver pine, *q.v.*

Yellow Poplar. *Liriodendron tulipifera.* Canary wood, *q.v.*

Yellow Sanders. Applied to Wild Olive or Hog Plum, Granadillo, and W. Indian Satinwood.

Yellow Satinwood. *Distemonanthus benthamianus (H.).* Nigeria. Called Anyaran, Ayan, and Movingui. Lemon yellow, lustrous, smooth. Hard, heavy, durable. Fine even grain and texture. Very decorative due to darker stripes and roe figure. Requires careful seasoning to avoid discoloration. Not difficult to work but liable to pick up ; polishes well. Used for cabinet work, superior joinery, turnery, etc. S.G. ·8 ; C.W. 3.

Yellow Siris. *Albizzia xanthoxylon.* Australia. See *Siris*.

Yellow Stain. A stain in hardwoods, especially oak, suggested due to the fungus *Penicillium divaricatum*.

Yellow Stringybark. *Eucalyptus acmenioides* and *E. muelleriana.* Australia. Pale yellowish brown. Hard, heavy, fissile, very durable, fire-resisting. Characteristics of the good class Eucalypts, *q.v.* Used for structural and constructional work, piling, sleepers, paving, carriage- and ship-building. S.G. 1 ; C.W. 4. See *Stringybark*.

Yellow Tulipwood. See *Grey Boxwood*.

Yellow-wood. *Podocarpus spp. (S.).* S. and E. Africa. Yellow. Resembles yellow pine. Fairly light, hard for its weight, smooth, lustrous. Fine grain and texture. Easily wrought but requires care in seasoning; should be kiln-conditioned. Stains and polishes well. Large sizes free from knots and other defects. The strength and mechanical properties are equal to those of Baltic redwood. Used for all interior purposes, food containers, etc. S.G. ·5 ; C.W. 2. S. AFRICAN YELLOW-WOOD. *P. elongata*, common yellow-wood, *P. falcatus*, falcate yellow-wood, and *P. heckelii*. KENYA YELLOW-WOOD. *P. gracilior* and *P. milanjianus*. There are other species in the Southern hemisphere but they are all very similar woods. See *Podo, Podocarpus, Musengera, Yacca*, and *Geelhout*. AMERICAN YELLOW-WOOD. *Clad-*

rastis lutea (*H.*).   Yellow to light brown or with brown streaks. Moderately hard and heavy, fairly stable. Gum deposits, visible rays. Used for cabinet work, veneers, gunstocks. S.G. ·54; C.W. 3.

Yellow Wood. The name often applied to W. Indies satinwood, Florida boxwood, Fustic, and Brazilian Vinhatico, *q.v.*

Yellow-wood Ash. *Flindersia oxleyana* (*H.*). Queensland and N.S. Wales. Called Long Jack. Resembles W. Indies satinwood. Yellow, lustrous, smooth. Hard, heavy, strong, durable, tough, elastic, rather fissile. Close firm grain and texture. Polishes well. Used for cabinet work, bent-work, vehicle shafts, implements, tool handles, turnery, carving. S.G. ·78; C.W. 3.

Yemane. Gumhar, *q.v.*

Yemeri. *Vochysia hondurensis* (*H.*). British Honduras. Similar to San Juan, *q.v.* Of little commercial value.

Yerool. Pyinkado, *q.v.*

Yew. *Taxus bacata* (*S.*). Europe, Asia, widespread. Orange to rich brown, purplish tint, stripes and dark spots, smooth, lustrous. Hard, heavy, strong, durable, tough, elastic. Not resinous. Handsome grain and figure. Close even grain, fine texture. Finishes well and acquires a natural polish. Little appreciated in this country except as burrs. Used for brushes, woodware, gate posts, parquet flooring, bows, small furniture, burrs for veneers. Stained black and called German ebony. S.G. ·8 IRISH YEW. *T. fastigiata.* Similar to *T. bacata*, but preferred for bows. AMERICAN or PACIFIC YEW. *T. brevifolia.* Similar to above but rather darker in colour, lighter in weight, and with more uniform grain. S.G. ·65.

Yii. Chinese elm.

Ying mu. The butt of the Chinese camphor tree. Valued for cabinet work.

Yinma. Chickrassy, *q.v.*

Yoke. *Piptadenia peregrina* (*H.*). Trinidad. See *Piptadenia* and *Yokes.*

Yoke-elm. An obsolete name for hornbeam.

Yokes. 1. Strong pieces of wood bolted round the forms or shuttering for concrete columns. They keep the panels of the formwork in position whilst the concrete is poured. **2.** A wood bar or frame to fit a person's shoulders, for carrying pails, etc. **3.** A cross piece forming part of the harness for animals drawing heavy loads. **4.** A cross-bar carrying the steering lines of a boat.

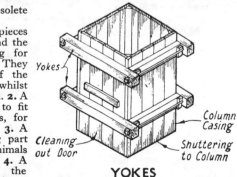

Column Casing

Shuttering to Column

Cleaning out Door

YOKES

5. A cross-bar from which a bell is hung. **6.** Sometimes applied to the head of a window frame.

Yokewood. *Catalpa longissima* (*H.*). W. Indies. Brownish grey. Hard, heavy, durable, and fairly strong. Also called Roble.
Yole. A small cargo sailing boat, used in the Orkneys.
Yoma Wood. Padauk, *q.v.*
Yon. *Anogeissus acuminata* (*H.*). Burma. Yellowish grey, small chocolate brown heartwood, lustrous, smooth. Hard, heavy, strong, especially in shear, elastic, tough, not very stable or durable. Straight, irregular, and interlocked grain. Fine texture, pith flecks. Good substitute for ash and hickory for handles, shafts, etc. Difficult to season and work. S.G. ·8 ; C.W. 4.

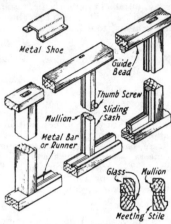

YORKSHIRE LIGHT

York Gum. *Eucalyptus fœcunda.* Australia. Pale brown, white markings. Very hard, heavy, dense, and difficult to split. Interlocked grain, beautiful figure. Used for wheelwright's work, wagons, etc. S.G. 1·1.
Yorkshire Light. A solid window frame with a sash sliding horizontally.
Yorrell. *Eucalyptus gracilis* (*H.*). W. Australia. Deep brown. Very hard, heavy, strong. Subject to termite attack. Used for vehicles, agricultural implements, poles, etc.
Young's Modulus. See *Modulus of Elasticity*.
Yuba. *Eucalyptus obliqua.* Tasmanian oak, *q.v.*
Yule Logs. Short lengths of round wood used as firewood as part of Christmas festivities.

Z

Z. A symbol for *modulus of section*, q.v.

Zambesi Redwood. Rhodesian teak, *q.v.*

Zanthoxylon. See *Harewood, Ruda, Satinwood, Tachuelilla*, and *Xanthoxylon*.

Zapatero. W. Indian or Venezuelan or Maracaibo boxwood, *q.v.*

Zapota. *Achras zapota.* Sapodilla, *q.v.*

Zapote, Mamey. *Calocarpum mammosum.* See *Sapote*.

Zareba. A palisaded enclosure.

Zebrano. *Cynometra sp.* or *Brachystegia sp.* Classification uncertain. (*H.*) W. Africa. Also called zingana and African zebrawood. Golden yellow, brown to nearly black streaks, lustrous. Highly decorative, due to contrasting colours. Hard, heavy, stable. Difficult to work. Used for fancy articles, cabinet work, etc. S.G ·7 ; C.W. 4·5. See *Ekpaghoi.*

Zebra Wood. *Astronium fraxiniofolium* (*H.*). S. America. Also called Bois Serpent, Goncalo alves, Kingwood, Locust wood, Tiger wood. Orange brown to dark brown, wide variegated bands. Colours more uniform with age. Very hard, heavy, strong, durable. Subject to gum specks making the surface feel oily. Straight fine grain, roey, close texture, faint rays. Rather small sizes. Difficult to work. Polishes well. Used for furniture, veneers, inlays, superior joinery. S.G. ·92 ; C.W. 4·5. *Pithecolobium racemiflorum.* C. America. Also called Bois serpent, snakewood, etc. Pale brown, purplish to black markings. Similar to above in properties and characteristics, but rather coarser texture. Difficult to work smooth. Used for furniture, decorative work, spokes, etc. S.G. 1·1 ; C.W. 5. Several highly decorative variegated woods are known by this name : Arang, Buey, Rengas, Snakewood, Zebrano, etc. Also see *Marblewood* and *Kingwood*. Red Zebrawood is the varnish tree of Burma. The wood is dark red with yellow markings and has hard close grain.

Zelkowa. *Zelkowa ulmoides* (*H.*). S. Russia, Persia. Like elm, but harder. Not imported. Also Keaki, *q.v.*

Zeta. A room over a church porch.

Zigzag Moulding. A dancette or chevron moulding, *q.v.*

Zigzag Partition. A partition with staggered studs to form two independent faces, for sound insulation. It is not very effective, however, unless some form of insulating material is placed between the two sets of studs. See *Staggered*.

Zingana. Zebrawood, *q.v.*

Ziricote. *Cordia dodecandra* (*H.*). C. America, W. Indies. Brown, with nearly black lines, lustrous, smooth. Hard and heavy with fine close grain. Used for decorative cabinet work and turnery. S.G. ·95 ; C.W. 4·5. Also see *Cyp*.

Zizyphus. See *Jujube*, *Cogwood*, and *Mistol*. *Zizyphus jujuba* and *Z. xylopyrus* are Indian species but they are not imported and have no market names.

Zoccola. A platform or step to a pedestal.

Zocle. See *Socle*.

Zollernia. See *Santa Wood* or *Páo santo*.

Zonate. Applied to banded pores or parenchyma. Arranged concentrically, as shown in a transverse section.

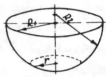

ZONE OF SPHERE

Zone. 1. A belt or band running round an object. A horizontal strip of the surface of a dome or sphere. See *Development*. 2. A portion of a sphere between two parallel planes. Area of curved surface $=2\pi R_2 H$ and volume $=\frac{1}{6}\pi H(H^2+3R_1{}^2+3r^2)$, where $H=$ distance between parallel surfaces. 3. Applied to the two parts of a growth ring, as spring and summer zones.

Zoophorus. A frieze, *q.v.*, with carvings of men and animals.

Zorra. See *Quamwood* and *Rain Tree*.

Z-pile. A separate pile or stack for timbers on which the marked measurements do not agree with the official measurements.

Zucc. Abbreviation for the name of the botanist Zuccarini.

Zulus. A type of Scottish fishing craft.

Zygophyllaceæ. The Lignum-vitæ family. Includes Bulnesia, Guaiacum, Larrea, and Porlieria.

Zygophyllum. Same as Bulnesia, *q.v.*

Zylopia. *Zylopia quintassii* (*H.*). Nigeria. Also called Aghako and Opalufon. Light grey or light yellow to cream. Hard, heavy, strong, durable. Close straight grain. Fairly easy to work but some liable to pick-up. Substitute for lancewood. Used for paddles, boatbuilding, carving, etc. S.G. ·8 ; C.W. 2·5. Same as *Xylopia*.

BOOKS

OF

RELATED INTEREST

FROM

STOBART DAVIES

WORLD WOODS IN COLOUR
William Lincoln

275 commercial world timbers in full colour, describing general characteristics, properties and uses table. 300 pages.

WHAT WOOD IS THAT?
A Manual of Wood Identification
Herbert L Edlin

40 actual wood samples are analysed in terms of 14 key characteristics to enable the reader to identify woods quickly and accurately. 160 pages.

THE COMMERCIAL WOODS OF AFRICA
A Descriptive Full-Colour Guide
Peter Phongphaew

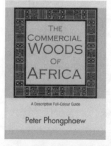

Profiles 90 commercial African woods in full colour, with descriptive text and map indicating tree's habitat in Africa. 216 pages.

PURPOSE MADE JOINERY
2nd Edition
Edward Foad, MCIOB

High quality purpose-made joinery skills employs many of the traditional craft principles and skills and this book reflects this in its coverage of modern materials, equipment and techniques. Fully illustrated with detailed drawings. 314 pages.

CIRCULAR SAWS
Their Manufacture, Maintenance & Application in the Woodworking Industries
Eric Stephenson

Every conceivable aspect of manufacture, use, maintenance, grinding, and sharpening of circular saws of all types. 288 pages, 1,000 illustrations.

THE CONVERSION & SEASONING OF WOOD
Wm. H Brown

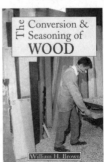

A guide to principles and practice covering all aspects of timber conversion from the log and dealing with proven methods of seasoning. 222 pages.

MODERN PRACTICAL JOINERY
George Ellis

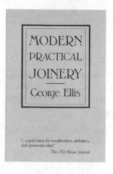

This vast coverage of internal joinery includes windows, doors, stairs, handrails, mouldings, shopfitting and showcase work, all clearly detailed and illustrated with hundreds of line drawings. Nearly 500 pages and 27 chapters.

CIRCULAR WORK IN CARPENTRY AND JOINERY
George Collings

A practical guide on circular work of single and double curvature. Fully illustrated with detailed drawings. 120 pages.

MODERN PRACTICAL STAIRBUILDING AND HANDRAILING
George Ellis

Reprinted 2001 from the original classic 1932 edition. It has 51 chapters, 108 plates and numberous detailed drawings and black and white photos. An excellent reference for the stairbuilder.

TREATISE ON STAIRBUILDING AND HANDRAILING
W & A Mowat

Originally published 1900, this remains one of the best illustrated and authoritative works to date. Invaluable for all stairbuilding and restoration. 390 pages.

BUILDING CABINET DOORS & DRAWERS
Danny Proulx

A practical book that includes all styles and a thorough discussion of suitable joinery techniques. 112 pages including 200 illustrations.

SHEDS THE DO-IT-YOURSELF GUIDE
David Stiles

Guide to building your own purpose-built shed, including several examples from the simple to the more ambitious. 182 pages illustrated in colour throughout.

JAPANESE WOODWORKING TOOLS
Their Tradition, Spirit and Use
Toshio Odate

A complete guide to Japanese tools, showing how each tool works, how it is sharpened and cared for and its use. 192 pages.

THE BOOK OF BOXES
The Complete Practical Guide to Design and Construction
Andrew Crawford

This all-colour illustrated book provides detailed instructions on planning and making a wide variety of attractive boxes including fitting locks and hinges, and decorative techniques such as marquetry, parquetry, inlaying, tunbridgeware, etc. 144 pages.

MAKING WOODEN CLOCK CASES
Tim & Peter Ashby

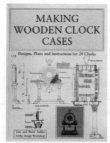

This book presents complete measured drawings and detailed plans of 23 clocks for the craftsman to make. 222 pages.

WOODTURNING DESIGN
Mike Darlow

Covers the concepts and process of woodturning design using more than 500 full-colour photos and drawings sourced from around the world. 280 pages.

HOW TO BUILD CLASSIC GARDEN FURNITURE
Danny Proulx

A step-by-step guide to making elegant garden furniture, including 20 projects. 128 pages illustrated in colour throughout.

ARTS AND CRAFT FURNITURE
Projects You Can Build for the Home
Blair Howard

15 super step-by-step projects to make in the Arts & Crafts style of Stickley, Limbert and other exponents of this popular period. 126 pages illustrated in colour throughout.

FURNITURE FOR THE 21ST CENTURY
Betty Norbury

A visual feast of today's furniture designers' best work. 192 pages including 420 fabulous colour photographs.

MAKING SHOJI
Toshio Odate

Complete step-by-step instructions on the construction of shoji – Japanese sliding doors and screens. 120 pages, including 200 illustrations.

CELEBRATING BOXES
Peter Lloyd and Andrew Crawford

68 of the world's finest designer-makers share their design ideas and aspirations with the reader. 152 pages including over 250 colour photos.